ANNUAL REPORTS IN MEDICINAL CHEMISTRY
Volume 41

ANNUAL REPORTS IN MEDICINAL CHEMISTRY
Volume 41

Sponsored by the Division of Medicinal Chemistry of the American Chemical Society

EDITOR-IN-CHIEF:
ANTHONY WOOD
PFIZER GLOBAL RESEARCH & DEVELOPMENT
SANDWICH LABORATORIES
UNITED KINGDOM

SECTION EDITORS
MARK G. BOCK · DALIA COHEN · MANOJ C. DESAI · DAVID MYLES · ALBERT J. ROBICHAUD · ANDREW STAMFORD

EDITORIAL ASSISTANT
HANNAH YOUNG

AMSTERDAM • BOSTON • HEIDELBERG • LONDON • NEW YORK • OXFORD
PARIS • SAN DIEGO • SAN FRANCISCO • SINGAPORE • SYDNEY • TOKYO
Academic Press is an imprint of Elsevier

Academic Press is an imprint of Elsevier
84 Theobald's Road, London WC1X 8RR, UK
Radarweg 29, PO Box 211, 1000 AE Amsterdam, The Netherlands
The Boulevard, Langford Lane, Kidlington, Oxford OX5 1GB, UK
30 Corporate Drive, Suite 400, Burlington, MA 01803, USA
525 B Street, Suite 1900, San Diego, CA 92101-4495, USA

First edition 2006

Copyright © 2006 Elsevier Inc. All rights reserved

No part of this publication may be reproduced, stored in a retrieval system
or transmitted in any form or by any means electronic, mechanical, photocopying,
recording or otherwise without the prior written permission of the publisher

Permissions may be sought directly from Elsevier's Science & Technology Rights
Department in Oxford, UK: phone (+44) (0) 1865 843830; fax (+44) (0) 1865 853333;
email: permissions@elsevier.com. Alternatively you can submit your request online by
visiting the Elsevier web site at http://elsevier.com/locate/permissions, and selecting
Obtaining permission to use Elsevier material

Notice
No responsibility is assumed by the publisher for any injury and/or damage to persons
or property as a matter of products liability, negligence or otherwise, or from any use
or operation of any methods, products, instructions or ideas contained in the material
herein. Because of rapid advances in the medical sciences, in particular, independent
verification of diagnoses and drug dosages should be made

ISBN-13: 978-0-12-040541-1
ISBN-10: 0-12-040541-5
ISSN: 0065-7743

For information on all Academic Press publications
visit our website at books.elsevier.com

Printed and bound in USA
06 07 08 09 10 10 9 8 7 6 5 4 3 2 1

Working together to grow
libraries in developing countries

www.elsevier.com | www.bookaid.org | www.sabre.org

ELSEVIER   BOOK AID International   Sabre Foundation

CONTENTS

CONTRIBUTORS   xi

PREFACE   xiii

I. CENTRAL NERVOUS SYSTEM DISEASES

Section Editor: Albert J. Robichaud, Wyeth Research,
                  Princeton, New Jersey

1. Novel Approaches for the Treatment of Schizophrenia   3
   Bruce N. Rogers and Christopher J. Schmidt,
   Pfizer Global Research & Development, Pfizer Inc.,
   Eastern Point Rd, Groton, CT, USA

2. Recent Strategies for the Development of New Antidepressant   23
   Drugs
   Deborah A. Evrard,
   Chemical & Screening Sciences, Wyeth Research, CN 8000,
   Princeton, NJ, USA

3. Neuroprotective Agents for the Treatment of Ischemic and   39
   Hemorrhagic Stroke
   Simon N. Haydar[1] and Warren D. Hirst[2],
   [1]Chemical & Screening Sciences and [2]Neuroscience,
   Wyeth Research, CN 8000, Princeton, NJ 08543, USA

4. Novel Sodium Channel Blockers for the Treatment of Neuropathic Pain   59
   Brian Marron
   Icagen Inc., 4222 Emperor Blvd., Durham, NC 27703, USA

II. CARDIOVASCULAR AND METABOLIC DISEASES

Section Editor: Andrew Stamford, Schering-Plough Research Institute,
                  2015 Galloping Hill Road, Kenilworth, New Jersey

5. Centrally Acting Anti-Obesity Agents   77
   David Hepworth[1], Philip A. Carpino[2] and Shawn C. Black[2],

¹*Pfizer Global Research & Development – Sandwich Laboratories, Sandwich, CT 13 9NJ, UK and* ²*Pfizer Global Research and Development – Groton Laboratories, Groton, CT 06340, USA*

6. Nuclear Hormone Receptor Modulators for the Treatment of Diabetes and Dyslipidemia   99
   Peter T. Meinke, Harold B. Wood and Jason W. Szewczyk, Merck Research Laboratories, P.O. Box 2000, Rahway, NJ 07065, USA

7. 11 β-Hydroxysteroid Dehydrogenase Type 1 Inhibitors   127
   Craig D. Boyle, Timothy J. Kowalski and Lili Zhang, Schering-Plough Research Institute, 2015 Galloping Hill Road, Kenilworth, NJ 07033, USA

8. Glucokinase Activators for the Treatment of Type 2 Diabetes   141
   Theodore O. Johnson and Paul S. Humphries, Pfizer Global Research and Development, San Diego, CA 92121, USA

9. Renin Inhibitors   155
   Colin M. Tice, Vitae Pharmaceuticals, 502 West Office Center Drive, Fort Washington, PA 19034, USA

10. Antiarrhythmic Agents   169
    Mark J. Suto and Douglas S. Krafte, Icagen Inc., 4222 Emperor Blvd, Suite 390, Durham, NC, USA

III. INFLAMMATORY, PULMONARY, AND GASTROINTESTINAL DISEASES

Section Editor: Mark G. Bock, Novartis Institutes for BioMedical Research Inc., Cambridge, Massachusetts

11. Progress in Anti-SARS Coronavirus Chemistry, Biology and Chemotherapy   183
    Arun K. Ghosh[1], Kai Xi[1], Michael E. Johnson[2], Susan C. Baker[3] and Andrew D. Mesecar[2],
    ¹*Departments of Chemistry and Medicinal Chemistry, Purdue University, West Lafayette, IN, USA,* ²*Center for Pharmaceutical Biotechnology and Department of Medicinal Chemistry and Pharmacognosy,*

Contents vii

   University of Illinois at Chicago, IL, USA and [3]Department of Microbiology and Immunology, Loyola University Medical Center, IL, USA

12. Inhibitors of the Expression of Vascular Cell Adhesion   197
   Molecule-1
   Charles Q. Meng,
   AtheroGenics, Inc., 8995 Westside Parkway, Alpharetta, GA 30004, USA

13. Recent Advances in Gastrointestinal Prokinetic Agents   211
   David A. Sandham[1], and Hans-Jürgen Pfannkuche[2]
   [1]Global Discovery Chemistry, Horsham Research Centre, Wimblehurst Road, Horsham RH12 5AB, UK and [2]Novartis Institutes of Biomedical Research, Gastrointestinal Diseases Area, Postfach, CH-4002 Basel, Switzerland

14. PGD_2 Antagonists   221
   Julio C. Medina and Jiwen Liu,
   Amgen Inc., South San Francisco, CA 94080, USA

15. Progress in the Development of Inhaled, Long-Acting β_2-Adrenoceptor   237
   Agonists
   Paul A. Glossop and David A. Price,
   Pfizer Global Research and Development, Sandwich Laboratories, Ramsgate Road, Sandwich, Kent, CT13 9NJ, UK

IV. CANCER AND INFECTIOUS DISEASES

Section Editor: David Myles, Kosan Biosciences, Inc.,
       Hayward California

16. The Acyl Sulfonamide Antiproliferatives and Other Novel Antitumor   251
   Agents
   Mary M. Mader,
   Lilly Research Laboratories, Eli Lilly and Company, Lilly Corporate Center, Indianapolis, IN 46285, USA

17. Progress on Mitotic Kinesin Inhibitors as Anti-cancer   263
   Therapeutics
   Gustave Bergnes, Xiangping Qian and Andrew A. Wolff,
   Cytokinetics, Inc., 280 E Grand Ave, South San Francisco, CA 94080, USA

18. Developing Infectious Disease Strategies for the Developing World   275
   Paul J. Lee and Leonard R. Krilov,
   Division of Pediatric Infectious Diseases, Winthrop University
   Hospital, Mineola, NY 11501, USA

19. Influenza Neuraminidase Inhibitors as Antiviral Agents   287
   Y. Sudhakara Babu, Pooran Chand and Pravin L. Kotian,
   BioCryst Pharmaceuticals, Inc., 2190 Parkway Lake Dr., Birmingham,
   AL 35244, USA

20. Recent Developments in Antifungal Drug Discovery   299
   Roberto Di Santo,
   Dipartimento di Studi Farmaceutici, Istituto Pasteur – Fondazione
   Cenci Bolognetti, Università di Roma "La Sapienza",
   P.le Aldo Moro 5, 00185 Rome, Italy

V. TOPICS IN BIOLOGY

Section Editor: Dalia Cohen, Rosetta Genomics Inc., Technology Center of New Jersey, New Jersey

21. Genomic Data Mining and Its Impact on Drug Discovery   319
   N. R. Nirmala,
   Genome and Proteome Sciences Department, Novartis Institutes for
   Biomedical Research, Cambridge, MA 02139, USA

22. HTS of cDNA and RNAi for Target Identification   331
   Mark A. Labow,
   Genome and Proteome Sciences Department, Novartis Institutes for
   Biomedical Research Inc., Cambridge, MA 02139, USA

23. Genomic Approaches to Identify Molecular Basis of Multi-Factorial Diseases   337
   Chandrika Kumar,
   Novartis Institute for Biomedical Research, Cambridge,
   MA 02139, USA

VI. TOPICS IN DRUG DESIGN AND DISCOVERY

Section Editor: Manoj C. Desai, Gilead Sciences, Inc., Foster City, California

24.	Structure–Activity Relationships for *In vitro* and *In vivo* Toxicity Julian Blagg, *Pfizer Global Research and Development, Sandwich,* *Kent CT13 9NJ, UK*	353
25.	Importance of Early Assessment of Bioactivation in Drug Discovery Cornelis E.C.A. Hop, Amit S. Kalgutkar and John R. Soglia, *Pharmacokinetics, Dynamics and Metabolism, Pfizer Global* *Research and Development, Eastern Point Road, MS8118D-2026,* *Groton, CT, USA*	369
26.	The Application of Pulmonary Inhalation Technology to Drug Discovery A.R. Clark, R.K. Wolff, M.A. Eldon, S.K. Dwivedi, *Nektar Therapeutics, San Carlos, CA, USA*	383
27.	Recent Advances in Oral Prodrug Discovery Aesop Cho, *Gilead Sciences, Inc., 333 Lakeside Drive, Foster City, CA, USA*	395
28.	Oxytocin Antagonists and Agonists Alan D. Borthwick, *GlaxoSmithKline Research and Development, Medicines Research* *Centre, Gunnels Wood Road, Stevenage, Herts SG1 2NY, UK*	409

VII. TRENDS AND PERSPECTIVES

Section Editor: Anthony Wood, Pfizer Global Research & Development, Sandwich Laboratories, Sandwich, Kent, UK

29.	Knowledge and Intelligence in Drug Design Andrew L. Hopkins [1] and Alex Polinsky [2], [1]*Pfizer Global Research & Development, Ramsgate Road,* *Sandwich, Kent CT 13 9NJ, UK and* [2]*Pfizer Global Research* *and Development, 10770 Science Center Drive, San Diego,* *CA 92121, USA*	425
30.	To Market, To Market – 2005 Shridhar Hegde and Michelle Schmidt, *Pfizer Global Research & Development, St. Louis, MO 63017, USA*	439

	Contents
COMPOUND NAME, CODE NUMBER AND SUBJECT INDEX, VOLUME 41	479
CUMULATIVE CHAPTER TITLES KEYWORD INDEX, VOLUME 1–41	485
CUMULATIVE NCE INTRODUCTION INDEX, 1983–2005	503
CUMULATIVE NCE INTRODUCTION INDEX, 1983–2005 (BY INDICATION)	523

Color Plate Section between pages 322 and 323

CONTRIBUTORS

Babu Y. Sudhakara	287	Krilov Leonard R	275
Baker Susan C.	183	Kumar Chandrika	337
Bergnes Gustave	263	Labow Mark A.	331
Black Shawn C.	77	Lee Paul J.	275
Blagg Julian	353	Liu Jiwen	221
Borthwick Alan D.	409	Mader Mary M.	251
Boyle Craig D.	127	Marron Brian	59
Chand Pooran	287	Medina Julio C.	221
Cho Aesop	395	Meinke Peter T.	99
Clark A.R.	383	Meng Charles Q.	197
Carpino Philip A.	77	Mesecar Andrew D.	183
Di Santo Roberto	299	Nirmala Nanguneri R.	319
Dwivedi S.K.	383	Pfannkuche Hans-Jürgen	211
Eldon M.A.	383	Polinsky Alex	425
Evrard Deborah A.	23	Price David A.	237
Ghosh Arun K	183	Qian Xiangping	263
Glossop Paul A.	237	Rogers Bruce N.	3
Haydar Simon N.	39	Sandham David A.	211
Hegde Shridhar	439	Schmidt Christopher J.	3
Hepworth David	77	Schmidt Michelle	439
Hirst Warren D.	39	Soglia John R.	369
Hop Cornelis E.C.A.	369	Suto Mark J.	169
Hopkins Andrew L.	425	Szewczyk Jason W.	99
Humphries Paul S.	141	Tice Colin M.	155
Johnson, Michael E.	183	Wolff Andrew A.	263
Johnson Theodore O.	141	Wolff R.K.	383
Kalgutkar Amit S.	369	Wood Harold B.	99
Kotian Pravin L.	287	Xi Kai	183
Kowalski Timothy J.	127	Zhang Lili	127
Krafte Douglas S.	169		

PREFACE

Annual Reports in Medicinal Chemistry continues to focus on providing timely and critical reviews of important topics in medicinal chemistry together with an emphasis on emerging topics in the biological sciences, which are expected to provide the basis for future drug discovery.

Volume 41 mostly retains the familiar format of previous volumes, this year with 30 chapters. Sections I–IV are disease-oriented and generally provide updates on specific medicinal agents and approaches considered to have developed significantly since previous review in this series. As in past volumes, annual updates have been limited only to the most active or topical areas of research in favor of specifically focused and mechanistically oriented chapters, where the objective is to provide the reader with the most important new results in a particular field.

Sections V and VI continue to emphasize important enabling topics in biology, medicinal chemistry, and drug design illustrating critical interfaces among these disciplines. Section V, Topics in Biology, is focused on the application of genomic approaches to target identification exemplifying the power of emerging biological technologies like RNAi. Chapters in Section VI, Topics in Drug Design and Discovery address the highly prevalent problem of drug toxicity as well as progress in pulmonary drug delivery and oral pro-drug design.

Volume 41 concludes with a thought-provoking chapter on Knowledge Management and last but not the least our regular chapter "To Market, To Market" covering NCE and NBE introductions worldwide in 2005.

In addition to the chapter reviews, a comprehensive set of indices has been included to enable the reader to easily locate topics in Volumes 1–41 of this series.

Volume 41 *of Annual Reports in Medicinal Chemistry* was assembled with the indispensable editorial assistance of Hannah Young and I would like to thank her for her hard work and enduring support. I would also like to recognize the innovative and enthusiastic section editors who have worked hard to deliver the content of this book and my sincere thanks go to them again this year. Finally, I hope that you the reader will enjoy and profit from the contents of this volume.

Anthony Wood
Sandwich, UK
May 2006

Section 1:
Central Nervous System Diseases

Editor: Albert J. Robichaud
Wyeth Research
Princeton
New Jersey

Novel Approaches for the Treatment of Schizophrenia

Bruce N. Rogers and Christopher J. Schmidt

Pfizer Global Research & Development, Pfizer Inc., Eastern Point Road, Groton, CT, USA

Contents

1. Introduction	3
2. Approaches targeting altered NMDA receptor function	4
2.1. AMPA potentiators	4
2.2. GlyT1 inhibitors	6
2.3. Metabotropic glutamate receptors	7
3. Approaches primarily targeting cognition	10
3.1. Alpha7 Nicotine acetylcholine receptor agonists and positive allosteric modulators	10
3.2. Dopamine D1/D5 receptor agonists	11
3.3. Serotonin 5-HT_6 antagonists	12
4. Approaches primarily targeting psychosis	13
4.1. Dopamine D3 receptor antagonist	13
4.2. Serotonin 5-HT_{2C} receptor agonists	14
5. Emerging mechanisms and conclusions	15
References	16

1. INTRODUCTION

Schizophrenia is a complex psychiatric disorder characterized by the variable expression of three major categories of symptoms: positive, negative and cognitive symptoms. Psychotic or positive symptoms include hallucinations and delusions, while negative symptoms include apathy, anhedonia and social and emotional withdrawal. Cognitive symptoms, although recognized in the earliest descriptions of schizophrenia, have recently come to center stage in schizophrenia research. The first pharmacotherapies developed for the treatment of schizophrenia were aptly termed antipsychotic agents as they primarily affected positive symptoms. These initial therapies, discovered serendipitously in the mid-1950s, were ultimately determined to be D2 dopamine receptor antagonists and their efficacy gave rise to the proposal that schizophrenia was a disorder of excessive dopaminergic activity – the dopamine hypothesis of schizophrenia. The dopamine hypothesis has driven drug discovery in this field for decades. The most recently marketed agents, referred to as atypical antipsychotics due to their reduced side-effect profile, are active at several additional transmitter receptors, but still owe their efficacy to D2 receptor blockade. Despite recent imaging studies confirming a dopaminergic hyperreactivity in patients [1], schizophrenia is now viewed as the result of a neurochemical imbalance across multiple transmitter systems, with both genetic and developmental contributions [2]. Reduced glutamatergic neurotransmission, particularly at the NMDA receptor subtype, is now believed to underlie even the dopaminergic abnormalities [3]. The hypoglutamatergic hypothesis is supported by the observation that NMDA receptor antagonists can reproduce the positive, negative and cognitive symptoms of schizophrenia in healthy volunteers and exacerbate these symptoms in patients [4].

Although antidopaminergic therapies still dominate the treatment of schizophrenia, our growing understanding of the neurochemical deficits underlying this disorder has led to a corresponding broadening in the search for novel treatment approaches. Very recently, this search has been further invigorated by the realization that although current therapies are effective treatments for psychosis, it is their cognitive symptoms that ultimately prevent schizophrenia patients from achieving functional recovery [5].

In this brief review, we begin with a description of several new approaches targeting the proposed deficit in NMDA receptor-mediated neurotransmission. As this is believed to be the primary neurochemical deficit in schizophrenia, it is possible that such therapies, if successful, will ultimately produce improvements in all aspects of the disorder. The recent focus on the cognitive deficits of schizophrenia has led to a number of approaches specifically designed to target this symptom domain. The most promising of these mechanisms will be described next. Finally, despite their marketing success, upwards of 30% of patients do not respond adequately to the atypical antipsychotic agents and many patients have residual positive symptoms leading to problems of compliance and relapse. Thus, there yet remains a need for improved antipsychotic agents and several promising approaches to this challenge will be described. The goal of this review is not to be exhaustive, but rather to describe approaches where recent examples of progress in the area of schizophrenia research have occurred.

2. APPROACHES TARGETING ALTERED NMDA RECEPTOR FUNCTION

2.1. AMPA potentiators

Excitatory glutamatergic neurotransmission involves the activation of both ligand-gated ion channels and ligand-activated second messenger-coupled receptors. The ionotropic glutamate receptors include the N-methyl-D-aspartate (NMDA), α-amino-3-hydroxy-5-methyl-4-isoxazolepropanoic acid (AMPA) and the less well-studied kainate subtypes. There are also eight classes of metabotropic glutamate receptors some of which will be described later in this review. NMDA and AMPA receptors interact in a dependent, almost circular fashion and play a complementary role in controlling excitatory neurotransmission. AMPA receptors mediate the majority of fast excitatory transmission in the brain. These are hetero- or homotetrameric receptors composed of GluR1 – GluR4 subunits with each subunit also existing as one of the two splice variant termed "flip" and "flop" [6,7]. RNA editing provides the opportunity for even greater structural and functional diversity [8]. While AMPA receptors' gate primarily Na^+, the NMDA receptor is a high conductance, slowly activating Ca^{2+} channel. At normal membrane potentials, the

NMDA receptor channel is subject to voltage-dependent Mg^{2+} blockade and its opening requires membrane depolarization by AMPA receptors. Thus, manipulations increasing AMPA receptor activity have the potential to augment NMDA receptor function as well. At the same time, NMDA receptor activation is involved in the membrane trafficking of AMPA receptors, a process believed to underlie basic forms of neuroplasticity such as long-term potentiation (LTP) [9], and long-term depression (LTD) [10]. The enhanced synaptic efficacy of LTP is due in part to the rapid insertion of "silent" AMPA receptors into the neuronal membrane following NMDA receptor activation.

Direct activation of AMPA receptors carries the risk of producing seizures, excitotoxicity and a loss of efficacy due to desensitization. However, the discovery of positive allosteric modulators (PAMs) offers a mechanism for enhancing receptor activity while avoiding these issues [11]. As indicated by their name, AMPA PAMs do not interact with the agonist/glutamate site on the receptor but rather at an allosteric site where they affect the kinetics of receptor desensitization and/or deactivation. Crystallographic studies of the AMPA receptor have confirmed the location of an allosteric site on the extracellular domain of the receptor. Binding of a PAM at this site stabilizes the receptor in its' active, agonist-bound conformation [12]. The presence of multiple allosteric sites is suggested by experiments demonstrating that some potentiators affect receptor deactivation to prolong signal duration, while others slow desensitization and enhance signal amplitude [13]. Regardless of the molecular mechanism, these allosteric sites provide a target for amplifying activity at AMPA receptors without eliminating either signal content or the homeostatic processes that maintain excitatory neurotransmission in the physiological range. It is important to emphasize that despite the virtually ubiquitous distribution of AMPA receptors throughout the brain, AMPA potentiators have been observed to have specific rather than global effects on CNS function. This is presumed to be due to both subtype selectivity among compounds and regional differences in the expression pattern of the subunits and splice variants. Pyrrolidinone nootropics such as aniracetam (**1**) and piracetam were among the first agents shown to enhance the rate of AMPA receptor signaling. Modifications of these structures led to the benzamide potentiators or AMPAkines including CX-516 (**2**) and CX-717 (**3**). Compounds in this group tend to increase AMPA receptor signaling by slowing dissociation of the agonist from the receptor rather than by affecting peak current amplitude. Despite significant pharmacokinetic issues, preliminary clinical studies with CX-516 have yielded encouraging results in studies of human memory and as an adjunctive therapy in schizophrenia [6]. A second generation AMPAkine, ORG 624448, was selected for evaluation in the first NIMH sponsored MATRICS study of cognitive therapy for schizophrenia [14–16].

Aniracetam (**1**)   CX-516 (**2**)   CX-717 (**3**)

The benzothiadiazine potentiators include the diuretic cyclothiazide (**4**) and more brain penetrant molecules such as IDRA-21 (**5**) and its analogs. Like the benzamides, agents from this class have been shown to facilitate LTP *in vitro* and *in vivo*, improve cognitive performance in both rodent and primate models and to enhance the behavioral activity of antipsychotic agents [17]. Only preclinical information is available on the activity of a series of biarylsulfonamides disclosed by Eli Lilly and Co. Exemplified by LY404187 (**6**), these agents share the therapeutic potential of the earlier potentiators. They are active in animal models of cognitive function, but also have neurotrophic/neuroprotective effects [18]. Their ability to increase levels of brain-derived neurotrophic factor (BDNF) may underlie this activity and contribute to their efficacy in models of behavioral despair. The latter effect is consistent with suggestions that AMPA potentiators may also be useful in the treatment of depression.

Cyclothiazide (**4**)   IDRA-21 (**5**)   LY404187 (**6**)

2.2. GlyT1 inhibitors

The NMDA/glutamate receptor is unique in requiring the binding of a second co-agonist for activation. Both D-serine and glycine are believed to act as endogenous ligands at the NMDA co-agonist site. The affinity of glycine at the NMDA receptor varies somewhat among configurations of the NMDA receptor but is in the range of 0.1 μM. Extracellular glycine concentrations in the CNS are estimated to be between 6 and 10 μM indicating that the glycine site of the NMDA receptor should be saturated. However, preclinical studies suggest that this is not the case and that NMDA receptor activity, particularly receptors containing the NR2A subunit [19], can be enhanced by the addition of agonists at the glycine site [20]. The concept of increasing NMDA receptor function *via* the activation of the co-agonist site is supported by the results of several double blind trials in which high-dose glycine (up to 60 g/day), or glycine site agonists such as D-serine, have been administered to schizophrenic patients, stabilized on either typical or atypical antipsychotic agents. While these studies have generally not demonstrated improvements in positive symptoms, presumably due to the presence of the antipsychotic agent, the majority have demonstrated improvements in negative or cognitive symptoms [3].

Synaptic glycine concentrations are maintained at subsaturating levels by two high-affinity glycine transporters, GlyT1 and GlyT2 [21]. Both are members of the Na$^+$, Cl$^-$ dependent transporter family that includes monoaminergic transporters such as SERT and NET. GlyT1 is widely expressed in glia and some neurons throughout the CNS and is believed to regulate extracellular glycine concentration in the vicinity of NMDA receptors. In contrast, the GlyT2 transporter controls glycine concentrations at the inhibitory glycine receptor in the spinal cord.

Preclinical evidence confirms that NMDA receptor function can be augmented *via* GlyT1 inhibition indicating that this transporter may be targeted for the treatment of schizophrenia much in the same way the serotonin transporter is targeted for depression. Early *in vitro* studies of the GlyT1 transporter used inhibitors such as the glycine ester, glycyldodecylamide and the competitive inhibitor, sarcosine (*N*-methylglycine) [22]. A breakthrough was achieved with the discovery of ALX-5407 (NFPS) (**7**), a sarcosine derivative of fluoxetine. NFPS is a single digit nanomolar inhibitor of glycine transport by GlyT1 with no activity at GlyT2. A number of related compounds have been described, although the majority resemble **7** in suffering from behavioral toxicity, apparent irreversibility at the transporter [23] and the need for high systemic exposures to gain access to the CNS [24]. One such analog, NPTS (**8**), is completely inactive following peripheral administration, but has proven to be a useful GlyT1 ligand [25]. Although the patent literature indicates that many companies have moved beyond these amino acid analogs to achieve better drug-like properties, very little information is available on the pharmacological characterization of these agents [26]. One exception is Sanofi's SSR-504734 (**9**), which has been profiled in a series of neurophysiological and behavioral assays in rodents [27]. Relevant to its' use in schizophrenia, **9** ($IC_{50} = 18$ nM for uptake at human GlyT1) increased extracellular glycine in the rodent brain as measured by microdialysis, enhanced NMDA-mediated currents in rat hippocampal slices, reversed MK-801-induced hyperactivity and normalized a spontaneous prepulse inhibition deficit in mice. The improved pharmaceutical properties and behavioral activity of SSR-504734 (**9**) confirms the potential of the GlyT1 transporter as a drug target for augmenting glutamatergic function.

ALX-5407/ NFPS (**7**)   NPTS (**8**)   SSR-504734 (**9**)

2.3. Metabotropic glutamate receptors

The metabotropic glutamate receptors are members of the G-protein coupled receptor family. There are eight metabotropic receptors grouped into three families. Group I receptors (mGluR1 and 5) are linked to Gαq and increase phosphotidylinositol turnover to elevate intracellular Ca^{2+}. Both groups II (mGluR2 and 3) and III (mGluR4, 6, 7 and 8) are negatively linked to adenylyl cyclase *via* Gαi. Group II mGluRs are located presynaptically on glutamate terminals where they may act as autoreceptors regulating glutamate release *in vivo*. Unlike the group II and III receptors, group I (mGluR1 and mGluR5) are primarily postsynaptic and their effect on intracellular Ca^{2+} allows them to modulate the activity of other signaling pathways. Each of these receptor subtypes has potential utility for the treatment of schizophrenia.

2.3.1. mGluR5 agonists and potentiators

The mGluR5 subtype of receptors have been shown to augment NMDA receptor activity. Activation of mGluR5 receptors therefore has the same therapeutic potential described for the AMPA potentiators or GlyT1 inhibitors. Not surprisingly, as in the case of AMPA receptor agonists, there are significant drawbacks to the development of direct mGluR receptor agonists including the added difficulty of achieving subtype selectivity within a group (i.e. mGluR5 vs. mGluR1). Once again, however, these receptors possess subtype-specific allosteric sites; negative and positive allosteric modulators as well as neutral ligands have now been described for the mGluR5 receptor [28]. Like AMPA receptors, mGluR receptors possess a large extracellular domain. In this case, however, the extracellular domain contains the agonist binding site, while the allosteric sites are found within the transmembrane domain of the receptor [29].

Merck has described three series of mGluR5 potentiators. The benzaldazine series, exemplified by DFB (10), has been reported to exhibit the full spectrum of negative, neutral and positive allosteric modulators of the mGluR5 receptor. Micromolar concentrations of DFB, although without affect alone, produce a two- to threefold leftward shift in the concentration response curve for glutamate-induced increases in intracellular Ca^{2+} without altering the maximal response to glutamate. Exploration of the SAR around a benzamide screening lead yielded CPPHA (11), which produced a six- to ninefold shift in the glutamate response at submicromolar concentrations [30]. Unlike DFB (10), CPPHA (11) did not displace the binding of the negative allosteric modulator, [3]MPEP, suggesting the two potentiators have different binding sites [31]. A pyrazole series exemplified by 3-cyano-N-(1,3-diphenyl-1H-pyrazol-5-yl)benzamide (CDPPB (12)) yielded the first compound deemed to have sufficient potency and selectivity for *in vivo* characterization. In cells transfected with the human mGluR5 receptor, 12 produced up to fourfold shifts in the concentration response curve for glutamate-induced increases in intracellular Ca^{2+} with an EC_{50} value of 20 nM. Consistent with the potential for mGluR5 PAMs in the treatment of schizophrenia, 12 reduces amphetamine-induced locomotor activity and normalizes amphetamine-induced disruption of prepulse inhibition [32].

Addex Pharmaceuticals has characterized a series of 3-oxadiazolyl piperidines as mGluR5 potentiators. Although less potent than the compounds described by Merck, ADX47273 (13) (EC_{50} 351 nM) produces a >10-fold leftward shift in the glutamate-induced increase in intracellular Ca^{2+} but is inactive alone. *In vivo* 13 antagonized amphetamine-induced hyperactivity as well as apomorphine disruption of prepulse inhibition in rats [33].

2.3.2. mGluR2 agonists and potentiators

Group II metabotropic glutamate receptors, mGluR2 and mGluR3, are located presynaptically on glutamate terminals where they may act as autoreceptors regulating glutamate release [34]. The potential for these receptors in the treatment of schizophrenia has been established, and is reviewed in several recent accounts [35,36]. The largest dataset for this mechanism has been generated using the mGluR2/3 agonist LY354740 (**14**) and antagonist LY341495 (**15**) [37,38]. Recent reports from the Lilly group on the extremely potent mGluR2/3 agonist LY404039 (**16**) [39], and peptide pro-drug LY2140023 (**17**) have augmented the arsenal of agonists available for clinical evaluation [40]. Compound **16** is reported to be orally active and ~30× more potent than **14** in PCP and amphetamine-induced locomotor activity models of psychosis. In addition, **16** has been reported to increase cortical dopamine turnover in rats, an effect predictive of potential procognitive activity (see below). Despite the availability of these excellent compounds, it has been very difficult to achieve group II subtype selectivity in an agonist. In the past year, the first example of a selective agonist was reported with **18**, which demonstrates good agonist potency ($EC_{50} = 160\,nM$) at the mGluR2 receptor and is an antagonist ($IC_{50} = 1050\,nM$) at the mGluR3 receptor [41].

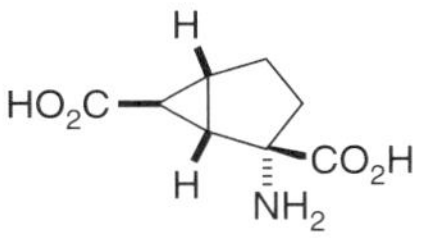

LY354740 (**14**) 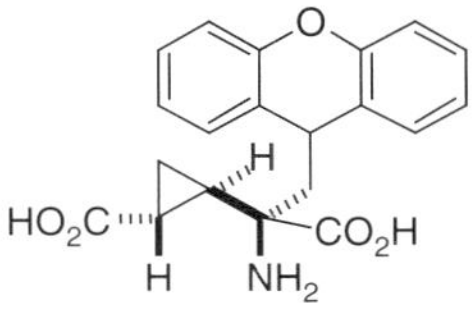   LY341495 (**15**) 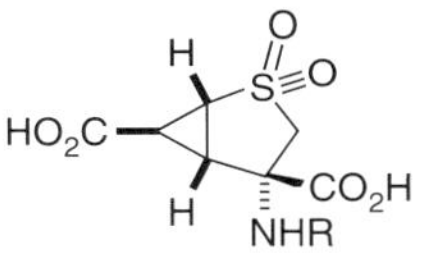   R = H, LY404039 (**16**)   **18**
                                       R = L-Ala, LY2140023 (**17**)

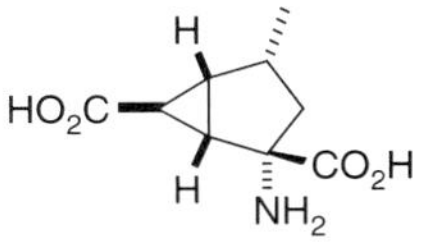

Several noteworthy reports in the mGluR2 area have appeared in the past 4 years with regard to selective mGluR2 potentiators [42]. The first reported selective potentiator of mGluR2 was LY181837 (**19**) [43]. Extensive SAR studies at Lilly led to the discovery of LY487379 (**20**), a potent potentiator ($EC_{50} = 270\,nM$) with *vivo* activity but only modest oral bioavailability [44]. For example, **20** antagonized PCP-induced hyperlocomotor activity, suggestive of potential antipsychotic efficacy [45]. Merck has also disclosed a series of potent mGluR2 PAMs exemplified by **21**, with *in vivo* activity in a ketamine-induced hyperlocomotor assay (40 mg/kg, i.p.) [46].

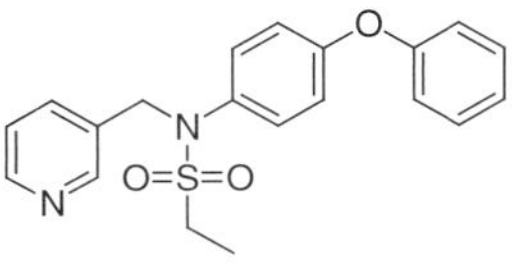

LY181837 (**19**) 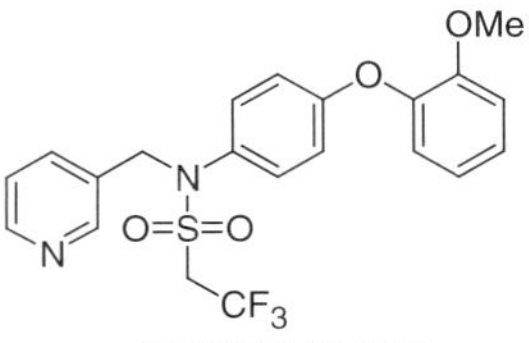   LY487379 (**20**) 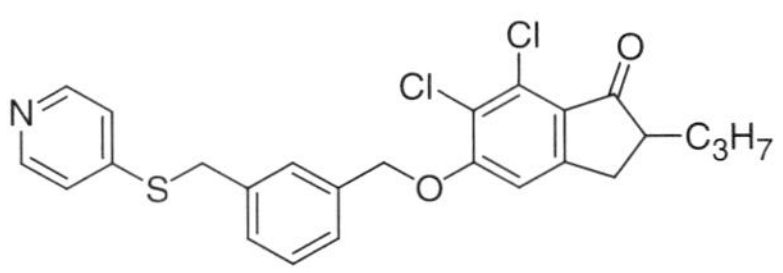   **21**

3. APPROACHES PRIMARILY TARGETING COGNITION

3.1. Alpha7 Nicotine acetylcholine receptor agonists and positive allosteric modulators

It is believed that neuronal nicotinic acetylcholine receptors (nAChRs) are involved in a variety of attention and cognitive processes [47]. These Ca^{2+} permeable, ligand-gated ion channels modulate synaptic transmission in key regions of the CNS involved in learning and memory, including the hippocampus, thalamus and cerebral cortex [48–50]. Among the nAChRs, physiological, pharmacological and human genetic data suggest a link between the loss of α7 nAChR and sensory gating deficits in schizophrenia [51]. Conversely, improvements in sensory processing are thought to correlate with enhanced cognitive performance in animal models and in patients with schizophrenia [52], suggesting a role for selective α7 nAChR agonists in treatment of cognitive dysfunction in schizophrenia.

Nicotinic ligand diversity has expanded greatly over the past decade, and several recent key reviews focus on the identification of selective agents [53–56]. Recent disclosures describe the *in vivo* activity of agonists based on the azabicyclic diamine template. The selective partial agonist SSR180711A (**22**) [57–59], is a diazabicyclononane carbamate derivative ($K_i = 50$ nM, $EC_{50} = 800$ nM) active in object recognition, Morris water maze and an MK-801-induced memory deficit model. The selective α7 nAChR agonist PNU–282,987 (**23**) ($K_i = 24$ nM, $EC_{50} = 128$ nM) [60,61], was recently reported to restore P50 gating deficits in rodent models. Researchers at Mitsubishi recently reported that the potent spiro-oxazolidinone (**24**) ($K_i = 9$ nM) significantly improves MK-801-induced auditory gating deficits [62]. PHA-543613 (**25**), an agonist of the α7 nAChR ($K_i = 9$ nM, $EC_{50} = 65$ nM) also demonstrated *in vivo* activity in both an amphetamine-induced P50 gating deficit model and object recognition [63].

SSR180711A (**22**)   PNU-282,987 (**23**)   **24**   PHA-543613 (**25**)

Interest continues to grow for the use of positive allosteric modulators of the alpha 7 nAChRs, and recent reviews cite their therapeutic potential in the treatment of cognition [64,65]. Positive allosteric modulators increase the probability of channel opening, while decreasing the inherent agonist potential for receptor desensitization. Much of the early work by Gurley and coworkers was accomplished utilizing 5-hydroxy indole (**26**) as a modulator of the alpha 7 receptor [66]. The recent disclosure of the selective alpha 7 positive allosteric modulator, PNU-120596 (**27**) provides a selective tool for further exploration of this mechanism [67,68]. Compound **27** increases agonist-evoked Ca^{2+} flux mediated by an engineered variant of the human alpha 7 nAChR, and enhances the acetylcholine-evoked inward currents in hippocampal interneurons. The compound also demonstrates robust *in vivo* activity in an amphetamine-induced P50 gating deficit model (0.1–3 mg/kg, i.v.). This compound represents a new class of ligands that activate the alpha 7 receptor and thus provide a potential mechanism to treat the cognitive deficits in schizophrenia.

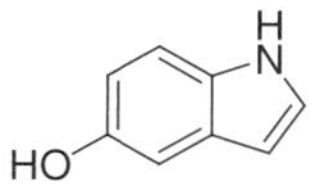

3.2. Dopamine D1/D5 receptor agonists

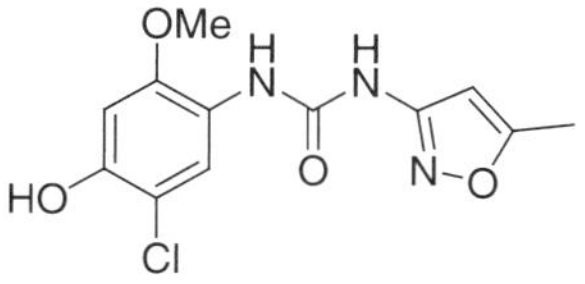

Many of the cognitive abnormalities in schizophrenia are similar to those resulting from damage to the prefrontal cortex (PFC) including attentional abnormalities, problems in reasoning and judgment and working memory deficits [69]. Together with evidence of reduced prefrontal dopaminergic function in schizophrenia, these observations have lead to the view that positive symptoms are due to subcortical hyperdopaminergia, while cognitive deficits are the results of hypodopaminergia in the PFC [70]. There are in fact data indicating that reduced dopaminergic activity in the PFC may cause over activation of the subcortical dopamine system [71,72]. Clinical reports that PFC D1 receptors are upregulated in schizophrenia due to a localized decrease in dopaminergic activity and that D1 receptor antagonists aggravate psychotic symptoms, are consistent with this hypothesis [73,74].

The first D1 selective compounds were benzazepines such as the antagonist, SCH23390 (**28**) and the partial agonist, SKF-38393 (**29**). Although **29** was a valuable tool in early studies of D1 receptor function, the development of full agonists such as the phenanthridine, dihydrexidine and the isochroman, A-68930 (**31**) was an important step in exploring the full therapeutic potential of D1 agonists. For example, although active in rodent models of Parkinson's disease (PD), the observation that **29** was ineffective in both primates and human studies initially dampened interest in this area [75–77]. However, full agonists were later shown to be effective in primate PD models and, in the case of dihydrexidine, have shown promising activity in limited human trials [78]. Although poor oral bioavailability, dose-limiting side effects and the development of tolerance has precluded the clinical development of these agents for PD, they have been useful preclinical compounds [77]. Dihydrexidine (**30**) has been shown to improve delay and scopolamine-induced deficits in passive avoidance in rodents as well as improve a delayed response task in primates [78]. Similarly SKF-812197 (**32**), a commercially available benzazepine D1 agonist, has been shown to have procognitive activity in a number of preclinical studies [79–81].

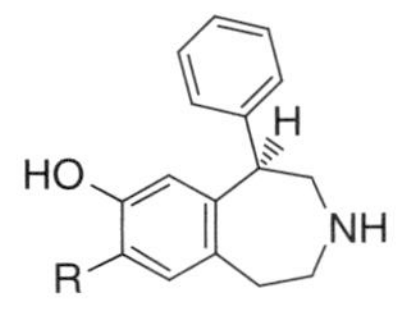

R = Cl, SCH23390 (**28**)
R = OH, SKF-38393 (**29**)

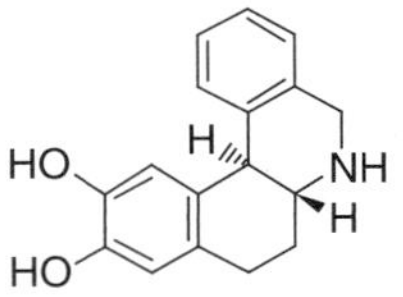

Dihydrexidine (**30**)

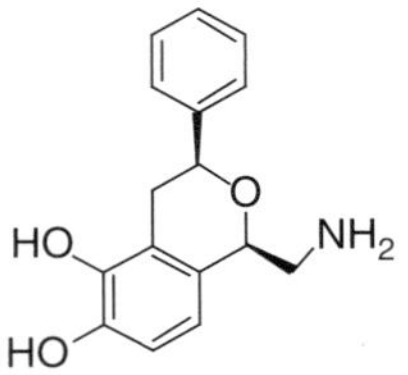

A-68930 (**31**)

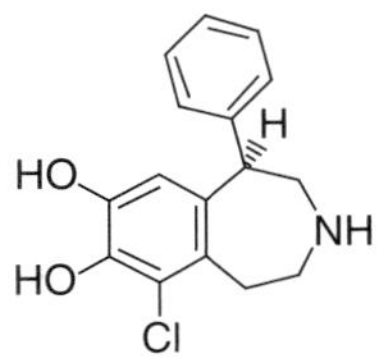

SKF-812197 (**32**)

Dihydrexidine (**30**) and related compounds such as dinapsoline (**33**), A-77636 (**34**) and A-86929 (**35**) are all rigid dopamine analogs defined by a *trans*-β-aryldopamine pharmacophore [82]. SAR around the core of **33** has been further explored [83], based on preclinical results suggesting a reduced rate of D1 receptor desensitization with this compound [84]. Although this pharmacophore predicts full D1 agonist activity, modest structural alterations dramatically affect selectivity. For example, tethering the β-phenyl ring back to the catechol ring *via* an ether linkage produces dinoxyline (**36**), a potent agonist at all five dopamine receptor subtypes [85].

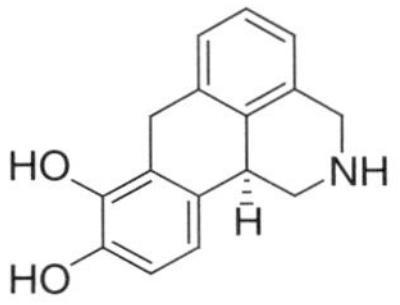

Dinapsoline (**33**)

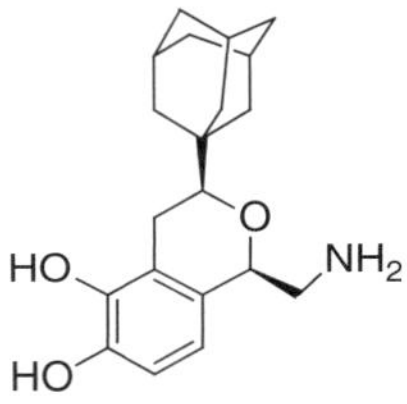

A-7763 (**34**)

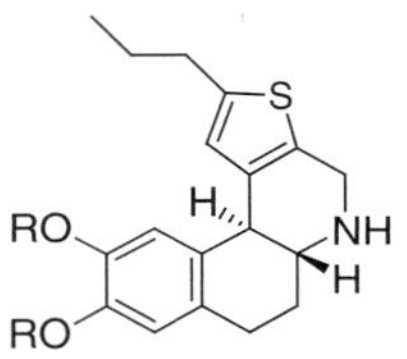

R = H, A-86929 (**35**)
R = Ac, Adrogolide (**37**)

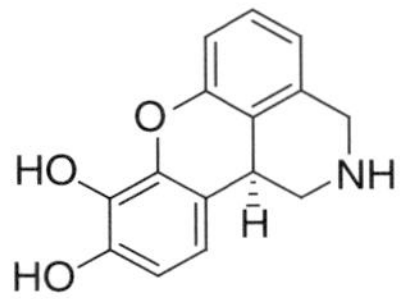

Dinoxyline (**36**)

Loss of efficacy with repeated drug administration has hampered the development of D1 agonists for the treatment of PD and may pose a similar challenge to the use of these agents for the treatment of schizophrenia [86]. Pharmacokinetics may play a significant role in D1 desensitization [87]. Compound **34**, with a half-life of 37 h, is reported to produce significant receptor desensitization following a single dose, while t.i.d dosing with **35** maintained activity for up to 30 days of treatment [77]. An effort to improve the pharmaceutical properties of **35** led to the study of its' diacetyl analog as a pro-drug [88]. Adrogolide (**37**) (also known as ABT-431 or DAS-431) is rapidly converted to A-86929 in plasma but suffers from poor oral bioavailability (<4%). Nonetheless, i.v. administration of ABT-431 has been reported to produce an antiparkinsonian effect in patients [89]. The pro-drug of A-86929 (**37**) has also been shown to reverse D2 antagonist-induced cognitive deficits in primates [90].

3.3. Serotonin 5-HT$_6$ antagonists

5-HT$_6$ receptors are positively coupled to adenylyl cyclase [90] and are found in key brain regions associated with learning and memory including the olfactory tubercles, cerebral cortex, nucleus accumbens, striatum and hippocampus [91]. In addition, the atypical antipsychotics clozapine and olanzapine possess high affinity for the 5-HT$_6$ receptor [92]. The potential of 5-HT$_6$ antagonists in the treatment of cognitive disorders has been discussed in several recent reviews [93,94], and the availability of selective ligands for this receptor continues to expand [95].

Ligands containing the indole template continue to be of high interest in the 5-HT$_6$ area with several additional reports in the past year. A series of conformationally constrained N-arylsulfonyltryptamine derivatives characterized by **38** were reported to be very potent 5-HT$_6$ antagonists ($K_i = 4.0$ nM) [96]. Although no selectivity data were reported, **38** demonstrated full antagonist activity (IC$_{50} = 1.0$ nM) in a cAMP accumulation assay. An independent report on bicyclic heteroarylpiperazines [97] recently described an aryl sulfonamide indole template exemplified by **39** ($pK_i = 8.6$). This compound also provides >100-fold selectivity in a 50-receptor panel, and demonstrated potent *ex vivo* binding (ED$_{50} = 3$ mg/kg, p.o.) relative to a standard 5-HT$_6$ antagonist, SB-271046 (ED$_{50} = 11$ mg/kg). Finally, this compound shows a brain–blood ratio of 3:1 and has good oral bioavailability in two species (49% in rat and 90% in dog). A three-dimensional pharmacophore model for 5-HT$_6$ receptor antagonists also appeared in the past year, which may assist in the design of future aryl sulfonamide ligands for the receptor [97].

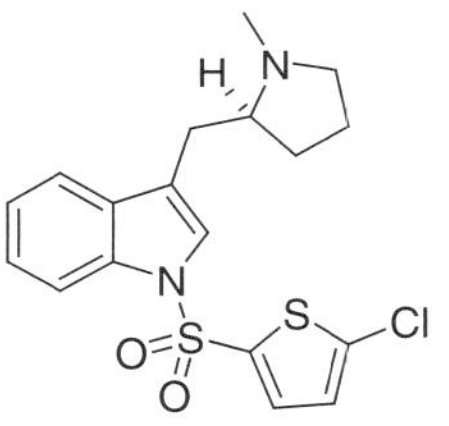

38 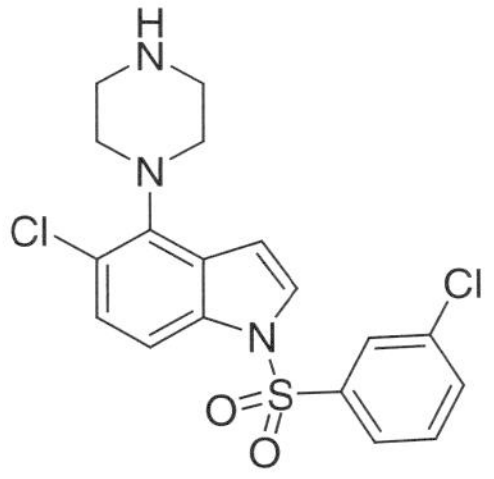   **39**

4. APPROACHES PRIMARILY TARGETING PSYCHOSIS

4.1. Dopamine D3 receptor antagonist

Schizophrenia is recognized as a multifactorial disease, but the dopamine hypothesis remains a prevailing theory to describe the disorder [98]. Most of the available evidence suggests that dopamine D3 receptor antagonists may be effective antipsychotic agents, although some studies have suggested potential efficacy against the negative and cognitive symptoms of schizophrenia as well [99]. It is also believed that a selective compound for D3 vs. D2 should afford an agent with lower EPS potential [100,101]. The D3 receptor is primarily located in the limbic brain areas (islands of Calleja, ventral striatum/nucleus accumbens, dentate gyrus and striate cortex), regions implicated in schizophrenia. The medicinal chemistry of selective D3 antagonists continues to evolve [102], with several recent reports of selective agents.

Additional research around the tetrahydroisoquinoline series (SB-277011) of selective D3 antagonists continues to appear in the literature. The tetrahydrobenzazepine, SB-414796 (**40**) represents a potent and selective D3 antagonist (D3 pK_i = 8.4, D3/D2 selectivity = 100×) with high oral bioavailability, CNS penetration, and demonstration of a reduction in firing of dopaminergic cells in ventral tegmental area (VTA), a prediction of antipsychotic activity [103]. The arylalkylpiperazine ST-280 (**41**) demonstrates potent and selective D3 binding activity (K_i = 0.5 nM, D3/D2 selectivity = 153x) and represents a potential binding ligand for the receptor. The arylalkylpiperazine, **42**, is among the most potent (D3 K_i = 0.18 nM) and selective D3 full antagonist reported to date (>10,000x over D2) [104]. A recent series of *trans*-olefin phenyl piperizines resulting in **43** was shown to have excellent D3 potency (K_i = 0.7 nM) and selectivity (133x vs. D2) [105]. Finally, in a recent series of disclosures from Abbott, the benzazepinone series, exemplified by A-706149 (**44**) demonstrates potent and selective antagonist activity (D3 K_i = 0.8 nM, 296x D2 selective) and oral activity (dosed at 2.15 mg/kg, p.o.) in a rat huddling deficit model of social interaction [106]. These recent disclosures of potent and selective D3 antagonist may afford the compounds necessary to examine their potential in the treatment of schizophrenia.

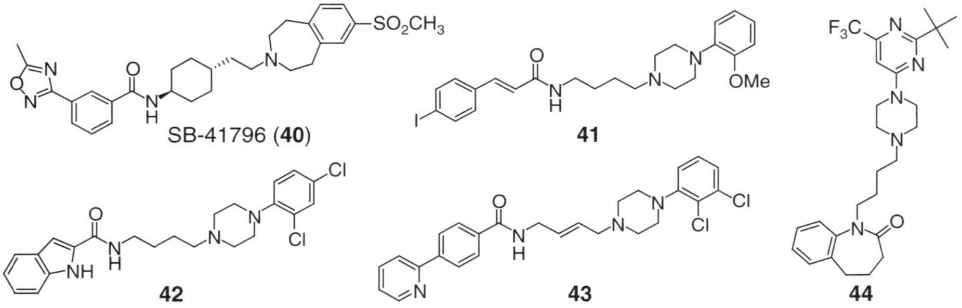

4.2. Serotonin 5-HT$_{2C}$ receptor agonists

The serotonin (5-HT) receptor family are among the most studied of the neurotransmitters. The 5-HT$_{2C}$ receptor subclass has seen a great deal of research during the past 10 years, and evidence has established the potential as a target for treating anxiety, depression [107], schizophrenia [108,109] and most notably obesity [110,111]. Until recently, attaining selectivity in the 5-HT$_{2C}$ receptor subclass over the closely related 5-HT$_{2A}$ and 5-HT$_{2B}$ receptors had been elusive, but selective agents are now emerging [112]. The rational for the use of selective 5-HT$_{2C}$ agonists in the treatment of schizophrenia is based upon the ability of 5-HT$_{2C}$ receptor activation to reduce dopamine neurotransmission in the mesolimbic system, and the potential to reduce side effects mediated by current antipsychotics such as weight gain [113].

Several noteworthy reports in the area of schizophrenia research have recently appeared. The tetracyclic indole derivative WAY-163909 (**45**) has been established in the literature as a potent and selective agonist (5-HT$_{2C}$ K_i = 10.5 nM, EC$_{50}$ = 8 nM; 5-HT$_{2B}$ K_i = 485 nM, EC$_{50}$ = 185 nM; 5-HT$_{2A}$ K_i = 212 nM, EC$_{50}$ = no effect) [114]. Recent reports from Wyeth on the activity of **45** in antipsychotic models include activity in rat conditioned avoidance assay (ID$_{50}$ = 1.3 mg/kg), mouse PCP-induced hyperactivity (0.1 mg/kg) and MK-801-induced disruption of prepulse inhibition (MED = 5.4 mg/kg). Additionally, at doses up to 30 mg/kg **45** showed no cataleptic potential in mouse. Taken together these data suggest the potential utility of a 5-HT$_{2C}$ receptor agonist as an atypical antipsychotic. Several additional reports of 5-HT$_{2C}$ ligands have been reported in the literature in the past year. Researchers at Vernalis reported indole **46** to be a full 5-HT$_{2C}$ agonist with good potency (K_i = 1.6 nM, EC$_{50}$ = 2.9 nM (99% efficacy)) with modest selectivity over the 5-HT$_{2A}$ and 5-HT$_{2B}$ receptors (5-HT$_{2A}$ K_i = 31 nM, EC$_{50}$ = 32 nM (88% efficacy) and 5-HT$_{2B}$ K_i = 12 nM, EC$_{50}$ = 1.1 nM (65% efficacy)) [115]. The pyrroloisoquinoline VER-2692 (**47**) is reported to be a potent 5-HT$_{2C}$ ligand (K_i = 1.6 nM) with modest selectivity over the 5-HT$_{2A}$ and 5-HT$_{2B}$ receptors (K_i = 31 and 12 nM, respectively), and *in vivo* activity at 3 mg/kg in a feeding model [116].

WAY-163909 (**45**)   **46**   VER-2692 (**47**)

5. EMERGING MECHANISMS AND CONCLUSIONS

Schizophrenia is the focus of intense preclinical and clinical research and is an area in which discoveries in the basic sciences are rapidly reviewed in the context of their potential therapeutic implications. Although it is not feasible to cover either the depth or breadth of the available subject matter in a review of this size, we will conclude with a brief mention of potential targets awaiting the development of the necessary biology and chemistry to establish their importance in this disease.

Phosphodiesterases are a small gene family now under active evaluation for their potential as CNS drug targets. PDE4 inhibitors such as rolipram are known to enhance LTP and to improve performance in animal models of cognitive function [117]. Bayer has recently reported that a selective PDE2 inhibitor, Bay 60-7550, increases cGMP in hippocampal and cortical cultures, enhances LTP and improves performance in a novel object recognition task [118]. PDE10A is a dual substrate phosphodiesterase expressed at very high levels in the medium spinney neurons of the striatal complex [119]. Recent rodent studies with a nonselective PDE inhibitor, papaverine, indicate that this enzyme may be a novel target for the treatment of psychosis [120].

Although the importance of cholinergic function in cognition is well established, studies using cholinesterase inhibitors in schizophrenia have yielded inconclusive results [121]. However, the M1/M4 agonist, xanomeline is reported to be active in models predictive of antipsychotic activity and intriguing results in Alzheimer's patients are supportive of this conclusion [122]. Finally, there is a growing literature suggesting that CB-1 receptor antagonists could have utility in the treatment of the cognitive deficits associated with schizophrenia [123,124]. Experimentation with selective agents such as rimonabant [125,126] is providing biochemical and pharmacological support for the eventual clinical evaluation of this hypothesis.

As a complex, multidimensional disorder, schizophrenia represents a major challenge to psychopharmacology and medicinal chemistry. Significant progress has been made with the introduction of atypical antipsychotics, but new agents are necessary to fully control this devastating disease. Our growing understanding of the underlying genetic and biochemical alterations in schizophrenia continue to provide a rich substrate for the development of novel medicines. The ultimate validity of these targets as viable drug therapies can of course only be established in the clinic.

REFERENCES

[1] M. Laruelle, A. Ai-Dargham, R. Gil, L. Kegeles and R. Innis, *Biol. Psychiat.*, 1999, **46**, 56.
[2] S. Marenco and D. R. Weinberger, *Dev. Psychopathol.*, 2000, **12**, 501.
[3] D. C. Javitt, *Mol. Psychiat.*, 2004, **9**, 984.
[4] B. J. Morris, S. M. Cochran and J. A. Pratt, *Curr. Opin. Pharmacol.*, 2005, **5**, 101.
[5] M. A. Geyer and C. A. Tamminga, *Psychopharmacology*, 2004, **174**, 1.
[6] G. Lynch, *Curr. Opin. Pharmacol.*, 2006, **6**, 82.
[7] M. D. Black, *Psychopharmacology*, 2005, **179**, 154.
[8] M. O'Neill, D. Bleaman, D. M. Zimmerman and E. S. Nisenbaum, *Curr. Drug Targets – CNS Neuro. Disord.*, 2004, **3**, 181.
[9] W. Lu, H. Man, W. Ju, W. S. Trimble, J. F. MacDonald and Y. T. Wang, *Neuron*, 2001, **29**, 243.
[10] E. C. Beattie, R. C. Carroll1, X. Yu, W. Morishita, H. Yasuda, M. von Zastrow and R. C. Malenka, *Nat. Neurosci.*, 2000, **3**, 1291.
[11] U. Staubli, G. Rogers and G. Lynch, *Proc. Natl. Acad. Sci. USA*, 1994, **91**, 777.
[12] R. Jin, S. Clarke, A. Weeks, J. Dudman, E. Gouaux and K. Partin, *J. Neurosci.*, 2005, **25**, 9027.
[13] J. N. C. Kew and J. A. Kemp, *Psychopharmacology*, 2005, **179**, 4.
[14] MATRICS is Measurement and Treatment Research to Improve Cognition in Schizophrenia. For a full description of this NIMH sponsored initiative see: http://www.matrics.ucla.edu/index.shtml.
[15] M. F. Green, K. H. Nuechterlein, J. M. Gold, D. M. Barch, J. Cohen, S. Essock, W. S. Fenton, F. Freseg, T. E. Goldberg, R. K. Heaton, R. S. E. Keefe, R. S. Kern, H. Kraemer, E. Stover, D. R. Weinberger, S. Zalcman and S. R. Marder, *Bio. Psychiat.*, 2004, **56**, 301.
[16] The actual structure of this compound is not known at this time.
[17] M. D. Black, *Psychopharmacology*, 2005, **179**, 154.
[18] J. C. Quirk and E. S. Nisenbaum, *CNS Drug Rev*, 2002, **8**, 255.
[19] D. E. Chapman, K. A. Keefe and K. S. Wilcox, *J Neurophysiol*, 2003, **89**, 69.
[20] R. Bergeron, T. M. Meyer, J. T. Coyle and R. W. Greene, Proc. Natl. Acad. Sci., *USA*, 1998, **95**, 15730.
[21] J. Gomeza, K. Ohno and H. Betz, *Curr. Opin. Drug Discov. Dev.*, 2004, **6**, 675.

[22] D. C. Javitt and M. Fruscianter, *Psychopharmacology*, 1997, **129**, 96.
[23] K. R. Aubrey and R. J. Vandenberg, *Br. J Pharmacol.*, 2001, **134**, 1429.
[24] S. M. Lechner, *Curr. Opini. Pharmacol.*, 2006, **6**, 75.
[25] J. A. Lowe, S. E. Drozda, K. Fisher, C. Strick, L. Lebel, C. J. Schmidt, D. Hiller and K. S. Zandi, *Bioorganic & Med. Chem. Lett.*, 2003, **13**, 1291.
[26] K. Hashimoto, *Recent Patents on CNS Discovery*, 2006, **1**, 43.
[27] R. Depoortere, G. Dargazanli, G. Estenne-Bouhtou, A. Coste, C. Lanneau, C. Desvignes, M. Poncelet, M. Heaulme, V. Santucci, M. Decobert, A. Cudennec, C. Voltz, D. Boulay, J. P. Terranova, J. Stemmelin, P. Roger, B. Marabout, M. Sevrin, X. Vige, B. Biton, R. Steinberg, D. Francon, R. Alonso, P. Avenet, F. Oury-Donat, G. Perrault, G. Griebel, P. George, P. Soubrie and B. Scatton, *Neuropsychopharmacology*, 2005, **30**, 1963.
[28] J. A. O'Brien, W. Lemaire, T.-B. Chen, R. S. L. Chang, M. A. Jacobson, S. N. Ha, C. W. Lindsley, H. J. Schaffhauser, C. Aur, D. J. Pettibone, P. J. Conn and D. L. Williams, *Mol.Pharmacol.*, 2003, **64**, 731.
[29] A. Ritzen, M. Mathiesen and C. Thomsen, *Basic Clin. Pharmacol. Toxicol.*, 2005, **97**, 2002.
[30] D. L. Williams and C. W. Lindsley, *Curr. Top. Med. Chem.*, 2005, **5**, 825.
[31] J. A. O'Brien, W. Lemaire, M. Wittmann, M. A. Jacobson, S. N. Ha, D. D. Wisnoski, C. W. Lindsley, H. J. Schaffhauser, B. Rowe, C. Sur, M. E. Duggan, D. J. Pettibone, P. J. Conn and D. L. Williams, Jr, *J. Pharmacol. Exp. Ther.*, 2004, **309**, 568.
[32] G. G. Kinney, J. A. O'Brien, W. Lemaire, M. Burno, D. J. Bickel, M. K. Clements, T.-B. Chen, D. D. Wisnoski, C. W. Lindsley, P. R. Tiller, S. Smith, M. A. Jacobson, C. Sur, M. E. Duggan, D. J. Pettibone, P. J. Conn and D. L. Williams, *J. Pharmacol. Exp. Ther.*, 2005, **313**, 199.
[33] (a) V. Mutel, A. S. Bessis, M. Epping-Jordan, E. Le Poul, B. Ludwig, S. M. Poli and J. P. Rocher, 5th Annu. Meet. on Metabotropic Glutamate Receptors, Taromina, Italy, 2005; (b) *Neuropharmacology*, 2005, (Supp. 1), **41**, Abstract 94.
[34] G. Guillerm, M. Muzard, C. Glapski, S. Pilard and E. De Clercq, *J. Med. Chem.*, 2006, **49**, 1223.
[35] D. D. Schoepp and G. J. Marek, *Curr. Drug Targets – CNS*, 2002, **1**, 215.
[36] G. J. Marek, *Curr. Opin. Pharmacol.*, 2004, **4**, 18.
[37] J. A. Monn, M. J. Valli, S. M. Massey, M. M. Hansen, T. J. Kress, J. P. Wepsiec, A. R. Harkness, J. L. Grutsch, Jr., R. A. Wright, B. G. Johnson, S. L. Andis, A. Kingston, R. Tomlinson, R. Lewis, K. R. Griffey, J. P. Tizzano and D. D. Schoepp, *J. Med. Chem.*, 1999, **42**, 1027.
[38] J. Cartmell, J. A. Monn and D. D. Schoepp, *J. Pharmacol. Exp. Ther.*, 1999, **291**, 161.
[39] (a) D. D. Schoepp, B. G. Johnson, R. A. Wright, M. Valli, S. Massey, K. Perry, J. Witkin, D. McKinzie, J. Vandergriff, K. Rasmussen, G. Marek, E. Nisenbaum, K. Griffey, J. P. Tizzano and J. A. Monn , 5th Annu. Meet. on Metabotropic Glutamate Receptors, Taromina, Italy, 2005; (b) *Neuropharmacology*, 2005, **41**, (Supp. 1), Abstract 133.
[40] (a) J. A. Monn, T. C. Britton, M. Valli, S. Massey, S. S. Henry, A. DeDios, B. G. Johnson, R. A. Wright, A. H. Dantzig, L. B. Tabas, T. Lindstrom, M. P. Clay, S. A. Sweetana, E. D. Moher, J. L. Gno, S. T. Patil, S. Glatt, S. Lowe, S. Shee, M. Skinner, S. Anderson, J. M. Hitchcock and D. D. Schoepp, 5th Annu. Meet. on Metabotropic Glutamate Receptors, Taromina, Italy, 2005; (b) *Neuropharmacology*, 2005, **41**, (Supp. 1), Abstract 89.
[41] C. Dominguez, L. Prieto, M. J. Valli, S. M. Massey, M. Bures, R. A. Wright, B. G. Johnson, S. L. Andis, A. Kingston, D. D. Schoepp and J. A. Monn, *J. Med. Chem.*, 2005, **48**, 3605.

[42] M. T. Rudd and J. A. McCauley, *Curr. Top. Med. Chem.*, 2005, **5**, 869.
[43] M. P. Johnson, M. Baez, G. E. Jagdmann, T. C. Britton, T. H. Large, D. O. Callagaro, J. P. Tizzano, J. A. Monn and D. D. Schoepp, *J. Med. Chem.*, 2003, **46**, 3189.
[44] D. A. Barda, Z. Q. Wang, T. C. Britton, S. S. Henry, G. E. Jagdmann, D. S. Coleman, M. P. Johnson, S. L. Andis and D. D. Schoepp, *Bioorg. Med. Chem. Lett.*, 2004, **14**, 3099.
[45] M. P. Johnson, D. Barda, T. C. Britton, R. Emkey, W. J. Hornback, G. E. Jagdmann, D. L. McKinzie, E. S. Nisenbaum, J. P. Tizzano and D. D. Schoepp, *Psychopharmacology*, 2005, **179**, 271.
[46] A. B. Pinkerton, R. V. Cube, J. H. Hutchinson, J. K. James, M. F. Gardner, B. A. Rowe, H. Schaffhauser, D. E. Rodriguez, U. C. Campbell, L. P. Daggett and J.-M. Vernier, *Bioorg. Med. Chem. Lett.*, 2005, **14**, 1565.
[47] E. D. Levin and B. B. Simon, *Psychopharmacology*, 1998, **138**, 217.
[48] J. A. Dani, *Biol. Psychiat.*, 2001, **49**, 166.
[49] M. R. Picciotto, B. J. Caldarone, S. L. King and V. Zachariou, *Neuropsychopharmacology*, 2000, **22**, 451.
[50] Neuronal Nicotinic Receptors: Pharmacology and Therapeutic Opportunities (eds S. P. Arneric and J. D. Brioni) Wiley-Liss, New York, 1999.
[51] R. Freedman, H. Coon, M. Myles-Worsley, A. Orr-Urtreger, A. Olincy, A. Davis, M. Polymeropoulos, J. Holik, J. Hopkins, M. Hoff, J. Rosenthal, M. C. Waldo, F. Reimherr, P. Wender, J. Yaw, D. A. Young, C. R. Breese, C. Adams, D. Patterson, L. E. Adler, L. Kruglyak, S. Leonard and W. Byerley, *Proc. Natl. Acad. Sci. USA*, 1997, **94**, 587.
[52] H. T. Nagamoto, L. E. Adler, K. A. McRae, P. Huettl, E. Cawthra, G. Gerhardt, R. Hea and J. Griffith, *Neuropsychobiology*, 1999, **39**, 10.
[53] S. R. Breining, A. A. Mazurov and C. H. Miller, *Ann. Rep. Med. Chem.*, 2005, **40**, 3.
[54] A. A. Jensen, B. Frolund, T. Liljefors and P. Krogsgaard-Larsen, *J. Med. Chem.*, 2005, **48**, 4705.
[55] K. Hashimoto, K. Koike, E. Shimizu and M. Iyo, *Curr. Med. Chem.*, 2005, **5**, 171, and references therein.
[56] W. H. Bunnelle, M .J. Dart and M. R. Schrimpf, *Curr. Med. Chem.*, 2004, **4**, 299, and references therein.
[57] P. Pichat, O. E. Bergis, J. Terranova, V. Vantucci, C. Gueudet, D. Franon, V. Voltz, R. Steinberg, G. Griebel, B. Scatton, F. Avenet, F. Oury-Donat and P. Soubrie, *Soc. Neurosci.*, 2004, **34**, San Diego, Abstract 583.3.
[58] O. E. Bergis, P. Pichat, R. Santamaria, R. Biton, L. Rouquier, R. Steinberg, G. Giebel, F. Oury-Donat, P. Avenet, P. Soubri and B. Scatton, *Soc. Neurosci.* 2004, **34**, San Diego, Abstract 583.2.
[59] B. Biton, O. E. Bergis, F. Galli, A. Nedelec, A. Lochead, S. Jegham, C. Lanneau, P. Granger, J. Leonardon, P. Avenet, D. Godet, A. Coste, X. Vig, F. Oury-Donat, P. George, P. Soubri, G. Griebel and B. Scatton, *Soc. Neurosci.*, 2004, **34**, San Diego, Abstract 583.1.
[60] M. Hajós, R. S. Hurst, W. E. Hoffmann, M. Krause, T. M. Wall, N. R. Higdon and V. E. Groppi, *J. Pharmacol. Exp. Ther.*, 2005, **312**, 1213.
[61] A. L. Bodnar, L. A. Cortes-Burgos, K. K. Cook, D. M. Dinh, V. E. Groppi, M. Hajos, N. R. Higdon, W. E. Hoffmann, R. S. Hurst, J. K. Myers, B. N. Rogers, T. M. Wall, M. L. Wolfe and E. Wong, *J. Med. Chem.*, 2005, **48**, 905.
[62] R. Tatsumi, M. Fujio, H. Satoh, J. Katayama, S. Takanashi, K. Hashimoto and H. Tanaka, *J. Med. Chem.*, 2005, **48**, 2678.
[63] D. G. Wishka, D. P. Walker, K. M. Yates, S. C. Reitz, S. Jia, J. K. Myers, K. L. Olson, E. J. Jacobsen, M. L. Wolfe, V. E. Groppi, A. J. Hanchar, B. A. Thornburgh,

L. A. Cortes-Burgos, E. H. F. Wong, B. A. Staton, T. J. Raub, N. R. Higdon, T. M. Wall, R. S. Hurst, R. R. Walters, W. E. Hoffmann, M. Hajos, S. Franklin, G. Carey, L. H. Gold, K. K. Cook, S. B. Sands, S. X. Zhao, J. R. Soglia, A. S. Kalgutkar, S. P. Arneric and B. N. Rogers, *J. Med. Chem.*, 2006, **49**, 4425–4436.

[64] M. N. Romanelli and F. Gualtieri, *Med. Res. Rev.*, 2003, **23**, 393.
[65] J.-P. Changeux and S. J. Edelstein, *Curr. Opinion Neurobiol.*, 2001, **11**, 369.
[66] R. Zwart, G. De Filippi, L. M. Broad, G. I. McPhie, K. H. Pearson, T. Baldwinson and E. Sher, *Neuropharmacology*, 2002, **43**, 374.
[67] R. S. Hurst, M. Hajos, M. Raggenbass, T. M. Wall, N. R. Higdon, J. A. Lawson, K. L. Rutherford-Root, M. B. Berkenpas, W. E. Hoffmann, D. W. Piotrowski, V. E. Groppi, G. Allaman, R. Ogier, S. Bertrand, D. Bertrand and S. P. Arneric, *J. Neurosci.*, 2005, **25**, 4396.
[68] D. W. Piotrowski, B. N. Rogers, D. M. Wilhite, B. J. Margolis, T. N. Vetman, D. Sarapa, D. G. Wishka, K. M. Yates, B. A. Acker, E. J. Jacobsen, W. W. McWhorter, V. O. Badescu, W. Xu, D. P. Walker, J. K. Myers, M. B. Berkenpas, D. M. Dinh, R. S. Hurst, T. M. Wall, N. R. Higdon, M. Hajós, W. E. Hoffmann, K. Lee and V. E. Groppi, Thirteenth RSC-SCI Medicinal Chemistry Symposium, Cambridge, UK, September 14–17, 2005.
[69] C. Pantelis, F. Z. Barber, T. R. Barnes, H. E Nelson, A. M. Owens and T. W. Robbins, *Schizophrenia Res*, 1999, **37**, 251.
[70] A. Abi-Darham and M. Laruelle, *Eur. Psychia.*, 2005, **20**, 15.
[71] P. S. Goldman-Rakic, S. A. Castner, T. H. Svensson, L. J. Siever and G. V. Williams, *Psychopharmacology*, 2004, **174**, 3.
[72] A. Abi-Dargham and H. Moore, *Neuroscientist*, 2003, **9**, 404.
[73] G. C. Sedvall and P. Karlsson, *Neuropsychopharmacology*, 1999, **21**, S181.
[74] A. Abi-Dargham, O. Mawlawi, I. Lombardo, R. Gil, D. Martinez, Y. Huang, D. R. Hwang, J. Keilp, L. Kochan, R. Van Heertum, J. M. Gorman and M. Laruelle, *J. Neurosci.*, 2002, **22**, 370.
[75] M. Nomoto, P. Jenner and C. D. Marsden, *Neurosci. Lett.*, 1985, **57**, 37.
[76] A. Braun, G. Fabbrini, M. M. Mouradian, C. Serrati, P. Barone and T. N. Chase, *J. Neural Trans.*, 1987, **68**, 41.
[77] M. William, S. Wright and G. K. Lloyd, *Trend Pharmacol. Sci.*, 1997, **18**, 307.
[78] P. Salami, R. Isacson and B. Kull, *CNS Drug Rev.*, 2004, **10**, 230.
[79] S. Bandyopadhyay, C. Gonzalez-Islas and J. J. Hablitz, *J. Neurophysio.*, 2005, **93**, 864.
[80] S. B. Floresco and A. G. Phillips, *Behav. Neurosci.*, 2001, **115**, 934.
[81] A. I. Hersi, W. Rowe, P. Gaudreau and R. Quirion, *Neuroscience*, 1995, **69**, 1067.
[82] D. M. Mottola, *J. Med. Chem.*, 1996, **39**, 393.
[83] S. Y. Sit, K. Xie, S. Jacutin-Porte, K. M. Boy, J. Seanz, M. T. Taber, A. G. Gulwadi, C. D. Korpinen, K. D. Burris, T. F. Molski, E. Ryan, C. Xu, T. Verdoorn, G. Johnson, D. E. Nichols and R. B. Mailman, *Bioorg. Med. Chem.*, 2004, **12**, 715.
[84] A. G. Gulwadi, C. D. Korpinen, R. B. Mailman, D. E. Nichols, S. Y. Sit and M. T. Taber, *J. Pharmacol. Exp. Ther.*, 2001, **296**, 338.
[85] R. A. Grubbs, M. M. Lewis, C. Owens-Vance, E. A. Gay, A. K. Jassen, R. B. Mailman and D. E. Nichols, *Bioorg. Med. Chem.*, 2004, **12**, 1403.
[86] M. Wade and G. G. Nomikos, *Psychopharmacology*, 2005, **182**, 393.
[87] J. Ryman-Rasmussen, D. E. Nichols and R. B. Mailman, *Mol. Pharmacol.*, 2005, **68**, 1039.
[88] W. J. Giardina and M. Williams, *CNS Drug Rev*, 2001, **7**, 305.
[89] O. Rascol, J. G. Nutt, O. Blin, C. G. Gopetz, J. M. Trugman, C. Soubrouillard, J. H. Carter, L. J. Currie, N. Fabre, C. Thalamus, W. J. Giardina and S. Wright, *Arch. Neurol.*, 2001, **58**, 249.

[90] S. A. Castner, G. V. Williams and P. S. Goldman-Rakic, *Science*, 2000, **287**, 2020.
[91] R. P. Ward, *Neuroscience*, 1995, **64**, 1105.
[92] B. L. Roth, S. M. Hanizaavarch and A. E. Blum, *Psychopharmacology*, 2004, **174**, 17.
[93] E. S. Mitchell and J. F. Neumaier, *Pharmacol. Ther.*, 2005, **108**, 320.
[94] S. L. Davies, J. S. Silvestre and X. Guitart, *Drug of the Future*, 2005, **30**, 479.
[95] W. E. Childers and A. J. Robichaud, *Ann. Rep. Med. Chem.*, 2005, **40**, 17.
[96] D. C. Cole, W. J. Lennox, J. R. Stock, J. W. Ellingboe, H. Mazandarani, D. L. Smith, G. Zhang, G. J. Tawa and L. E. Schechter, *Bioorg. Med. Chem. Lett.*, 2005, **15**, 4780.
[97] M. L. Lopez-Rodriguez, B. Benhamu, T. de la Fuente, A. Sanz, L. Pardo and M. Campillo, *J. Med. Chem.*, 2005, **48**, 4216.
[98] R. R. Luedtke and R. H. Mach, *Curr. Pharm. Design*, 2003, **9**, 643.
[99] J. N. Joyce and M. J. Millan, *Drug Discov. Today*, 2005, **10**, 917.
[100] M. Millan, L. Seguin, A. Gobert, D. Cussac and M. Brocco, *Psychopharmacology*, 2004, **174**, 341.
[101] J. A. Siuciak and R. A. Fujiwara, *Psychopharmacology*, 2004, **175**, 163.
[102] A. E. Hackling and H. Stark, *ChemBioChem*, 2002, **3**, 946.
[103] G. J. Macdonald, C. L. Branch, M. S. Hadley, C. N. Johnson, D. J. Nash, A. B. Smith, G. Stemp, K. M. Thewlis, A. K. K. Vong, N. E. Austin, P. Jeffrey, K. Y. Winborn, I. Boyfield, J. J. Hagan, D. N. Middlemiss, C. Reavill, G. J. Riley, J. M. Watson, M. Wood, S. G. Parker and C. R. Ashby, *J. Med. Chem.*, 2003, **46**, 4952.
[104] G. Campiani, S. Butini, F. Trotta, C. Fattorusso, B. Catalanotti, F. Aiello, S. Gemma, V. Nacci, E. Novellino, J. A. Stark, A. Cagnotto, E. Fumagalli, F. Carnovali, L. Cervo and T. Mennini, *J. Med. Chem.*, 2003, **46**, 3822.
[105] P. Grundt, E. E. Carlson, J. Cao, C. J. Bennett, E. McElveen, M. Taylor, R. R. Luedtke and A. H. Newman, *J. Med. Chem.*, 2005, **48**, 839.
[106] H. Geneste, G. Backfisch, W. Braje, J. Delzer, A. Haupt, C. W. Hutchins, L. L. King, W. Lubisch, G. Steiner, H.-J. Teschendorf, L. Unger and W. Wernet, *Bioorg. Med. Chem. Lett.*, 2006, **16**, 658.
[107] M. J. Millan, *Therapie*, 2005, **60**, 441.
[108] G. P. Reynolds, L. A. Templeman and Z. J. Zhang, *Prog. Neuropsychopharmacol. Biol. Psychiat.*, 2005, **29**, 1021.
[109] V. Di Matteo, M. Cacchio, C. Di Giulio and E. Esposito, *Pharmacol. Biochem. Behav.*, 2002, **71**, 727.
[110] B. M Smith, W. J. Thomsen and A. J Grottick, *Expert Opin. Invest. Drugs*, 2006, **15**, 257.
[111] M. J. Bickerdike, *Curr. Top. Med. Chem.*, 2003, **3**, 885.
[112] M. J. Bishop and B. M. Nilsson, *Expert Opin. Ther. Patents*, 2003, **13**, 1691.
[113] M. Giorgetti and L. H. Tecott, *Eur. J. Pharmacol.*, 2004, **488**, 1–9.
[114] J. Dunlop, A. L. Sabb, H. Mazandarani, J. Zhang, S. Kalgaonker, E. Shukhina, S. Sukoff, R. L. Vogel, G. Stack, L. Schechter, B. L. Harrison and S. Rosenzweig-Lipson, *J. Pharmacol. Exp. Ther.*, 2005, **313**, 862.
[115] S. Rover, D. R. Adams, A. Benardeau, J. M. Bentley, M. J. Bickerdike, A. Bourson, I. A. Cliffe, P. Coassolo, J. E. Davidson, C. T. Dourish, P. Hebeisen, G. A. Kennett, A. R. Knight, C. S. Malcolm, P. Mattei, A. Misra, K. J. Mizrahi, M. Muller, R. H. Porter, H. Richter, S. Taylor and S. P. Vickers, *Bioorg. Med. Chem. Lett.*, 2005, **15**, 3604.
[116] D. R. Adams, J. M. Bentley, K. R. Benwell, M. J. Bickerdike, C. D. Bodkin, I. A. Cliffe, C. T. Dourish, A. R. George, G. A. Kennett, A. R. Knight, C. S. Malcolm, H. L. Mansell, A. Misra, K. Quirk, J. R. A. Roffey and S. P. Vickers, *BioOrg. Med. Chem. Lett.*, 2006, **16**, 677–680.

[117] G. M. Rose, A. Hopper, M. De Vivo and A. M. Tehim, *Curr. Pharm. Design*, 2005, **11**, 3329.
[118] F. G. Boess, M. Hendrix, F. J. van der Staay, C. Erb, R. Schreiber, W. van Staveren, J. de Vente, J. Prickaerts, A. Blokland and G. Koenig, *Neuropharmacology*, 2004, **47**, 1081.
[119] J. A. Siuciak, S. A. McCarthy, D. S. Chapin, R. A. Fujiwara, L. C. James, R. D. Williams, J. L. Stock, J. D. McNeish, C. A. Strick, F. S. Menniti and C. J. Schmidt, *Neuropharmacology*, 2006, **51**, 374–385.
[120] J. A. Siuciak, D. S. Chapin, J. F. Harms, L. A. Lebel, L. C. James, S. A. McCarthy, L. K. Chambers, A. Shrikehande, S. K. Wong, F. S. Menniti and C. J. Schmidt, *Neuropharmacology*, 2006, **51**, 386–396.
[121] J. I. Friedman, *Psychopharmacology*, 2004, **174**, 45.
[122] N. R. Mirza, D. Peters and R. G. Sparkes, *CNS Drug Rev*, 2003, **9**, 159.
[123] B. Dean, S. Sundram, R. Brandbury, E. Scarr and D. Copolov, *Neuroscience*, 2001, **103**, 9–15.
[124] F. M. Leweke, T. Detre, J. Koral and P. Fajans, *Am. J. Psychiat.*, 1973, **130**, 1319.
[125] J. H. Lange and C. G. Kruse, *Drug Discov. Today*, 2005, **10**, 693.
[126] F. Barth, *Ann. Rep. Med. Chem.*, 2005, **40**, 103.

Recent Strategies for the Development of New Antidepressant Drugs

Deborah A. Evrard

Chemical & Screening Sciences, Wyeth Research, CN 8000, Princeton, NJ, USA

Contents

1. Introduction	23
2. Monoamine reuptake inhibitors	24
2.1. SNRIs	24
2.2. Triple reuptake inhibitors	25
3. SSRIs with monoaminergic receptor modulation	26
3.1. SSRI + 5-HT1A antagonism	26
3.2. SSRI + 5-HT1D antagonism	28
3.3. SSRI + α2 antagonism	28
4. Serotonergic receptor modulation	29
4.1. 5-HT1A/1B/1D antagonists	29
4.2. 5-HT2C receptor modulators	30
4.3. 5-HT7 receptor antagonists	30
5. Neuropeptide receptor modulation	31
5.1. Neurokinins	31
5.2. Corticotropin-releasing factor	31
5.3. Melanocortin MC4 antagonists	33
5.4. MCH1 antagonists	33
5.5. Galanin	35
5.6. Vasopressin	35
6. Conclusion	36
References	36

1. INTRODUCTION

Depression is one of the most common neuropsychiatric conditions and has a tremendous socioeconomic impact; it is the leading cause of disability in the U.S. and the fourth leading cause of disability globally [1]. Although depression is viewed as a highly treatable condition, and there are a number of pharmacologic and non-pharmacologic treatments currently available, there are still clear opportunities for improvement of existing therapies. Much research has been focused at reducing the lag time (typically 2–6 weeks) that occurs before clinical efficacy is observed with currently available antidepressants; strategies for developing faster-acting antidepressants have been recently reviewed [2]. There has also been increased interest in identifying new antidepressants with less sexual dysfunction, which reportedly occurs in 40–70% of patients treated with selective serotonin reuptake inhibitors (SSRIs) and results in significant patient non-compliance [3].

An area where there is arguably the most opportunity for improvement is that of antidepressant efficacy. It is estimated that almost half of depressed patients continue to have residual symptoms even after adequate treatment, and that roughly 20–30% of patients show minimal or no response to the currently available drug therapies. It is the hope that drugs acting by newer mechanisms will meet at least some, if not all, of these unmet needs. This review will highlight some recent developments in monoamine-based strategies and then discuss relatively newer approaches targeting neuropeptide receptors. These and other strategies have been discussed in several recent reviews [2,4–9].

2. MONOAMINE REUPTAKE INHIBITORS

Nearly all antidepressants currently in use act by increasing the concentrations of the monoamine neurotransmitters serotonin (5-HT), norepinephrine (NE), and/or dopamine (DA), primarily by inhibition of reuptake mechanisms (for a review, see Ref. [10]). Selective serotonin reuptake inhibitors (SSRIs) were introduced in the 1980s and include fluoxetine, paroxetine, sertraline, and escitalopram. Serotonin and NE reuptake inhibitors (SNRIs) include venlafaxine and duloxetine. The "atypical" antidepressant buprorion has been reported to inhibit the NE transporter (NET) and the DA transporter (DAT). Although new SSRIs continue to be identified, only those possessing additional receptor modulating properties will be discussed in this review (see Section 3).

2.1. SNRIs

Venlafaxine has become mainline therapy for major depressive disorder (MDD) and has been shown to have a higher remission rate than the SSRIs [11]. Duloxetine was launched in the U.S. in 2004 and is reported to have more balanced affinities at the serotonin transporter (SERT) and NET [12]. The next SNRI to be launched is likely to be desvenlafaxine (**1**) succinate, which is the succinate salt of an active metabolite of venlafaxine, and is currently undergoing FDA review.

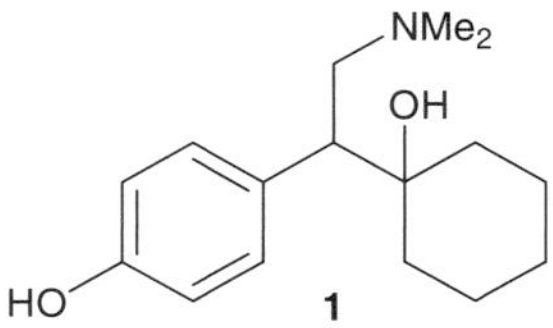

The SNRI F-98214-TA, **2**, is a potent inhibitor of the uptake of both 5-HT and NE in rat synaptosomes (IC50 = 1.9 and 11.2 nM, respectively) and was active in the tail suspension test (MED = 10 mg/kg), the forced swim test (30 mg/kg, p.o.), and the olfactory bulbectomy model (30 mg/kg/day, p.o., 14 days). In most assays F-98214-TA was more potent than fluoxetine, venlafaxine, or desipramine [13].

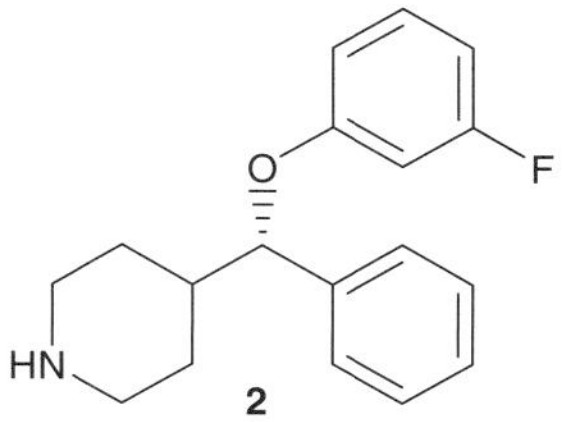

2.2. Triple reuptake inhibitors

Triple reuptake inhibitors (TRIs), which increase DA levels in addition to serotonin and NE, are expected to be as efficacious as monoamine oxidase inhibitors (MAOIs) without being limited by the same side effects and dietary restrictions that accompany MAOI use. The rationale for including DAT inhibition is partially based on the well-established role of dopaminergic systems in motivation and reward. Anhedonia and lack of interest, which are core symptoms of MDD, result from dopaminergic impairment in corticolimbic areas, and depressed patients have been shown to have decreased DA release by nerve terminals in the mesolimbic system [14]. TRIs may also have a faster onset of action in addition to increased efficacy [15]. Animal models demonstrate sensitization of the dopaminergic response after chronic treatment with desipramine, and the time interval required to effect this increased sensitivity may contribute to the lag time in achieving therapeutic antidepressant response.

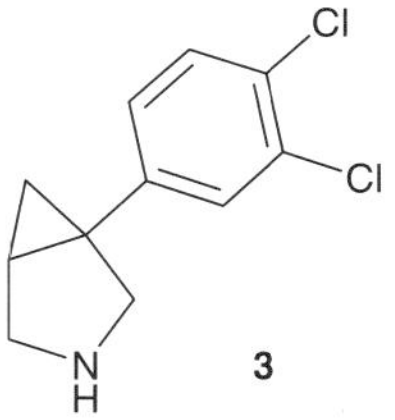

The TRIs DOV-216303, (rac)-3, and DOV-21947, (+)-3, are undergoing clinical development. Inhibition of tritiated amine uptake by human recombinant transporters expressed in HEK cells by DOV-216303 was reported to be IC50 = 13.8, 20.3, and 78 nM for SERT, NET, and DAT, respectively [15]. The corresponding IC50 values for DOV-21947 are 12.3 nM (SERT), 22.8 nM (NET), and 96 nM (DAT) [15]. Both compounds were orally active in the forced swim test. DOV-216,303 has been shown to be safe and tolerated in normal human volunteers [16], and was reported to show efficacy in Phase II clinical trials [17]. DOV-21947 has also been reported to be well tolerated with no dose-limiting side effects in initial Phase I trials.

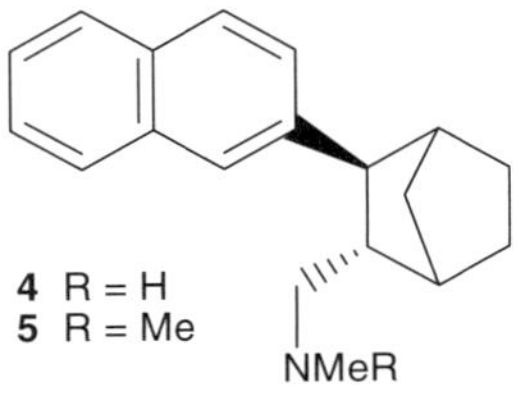

4 R = H
5 R = Me

Similar inhibition of the three transporters was reported for bicyclic compound **4**, which blocked the uptake of tritiated amines in synaptosomes (IC50 = 9, 28, and 86 nM at SERT, NET, and DAT, respectively) [18]. The methylated analog **5** showed comparable activity (IC50 = 9 nM for SERT, 25 nM for NET, and 76 nM for DAT). Both compounds were active in assays confirming *in vivo* TRI, but no information on antidepressant activity in animal models was reported.

3. SSRIS WITH MONOAMINERGIC RECEPTOR MODULATION

3.1. SSRI + 5-HT1A antagonism

An approach to a faster-acting antidepressant that received much attention over the past decade is the combination of SERT inhibition with 5-HT1A receptor antagonism. Activation of somatodendritic 5-HT1A autoreceptors by increased 5-HT levels (produced from an SSRI) causes a reduction in neuronal firing until the autoreceptors desensitize. The length of time required for this process of desensitization has been correlated with the time required to reach clinically therapeutic effects (typically 2–4 weeks) with SSRIs. Antagonism of the 5-HT1A autoreceptor would block the negative feedback mechanism on neuronal firing, so that the effects of the SSRI would be seen more rapidly.

Efforts to identify a single molecule possessing both SSRI and 5-HT1A antagonist activity have been revealed in a number of publications over the last few years. Compounds utilizing pindolol as a starting point are exemplified by compound **6** (5-HT1A K_i = 14.35 nM, SERT K_i = 0.86 nM), which elevated rat hypothalamic 5-HT levels to 713% of baseline (measured by microdialysis) after acute dosing at 10 mg/kg, p.o. [19].

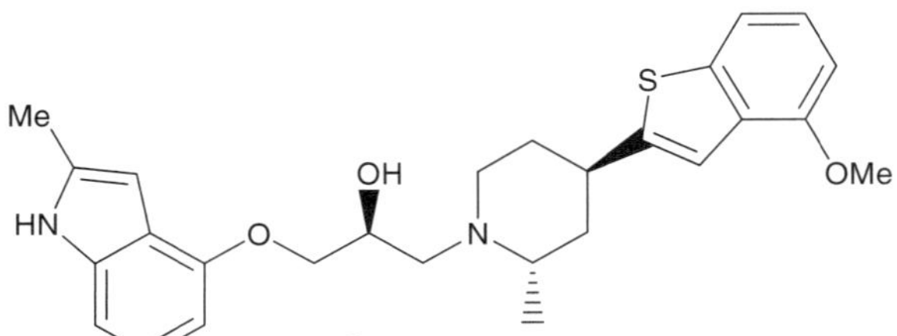

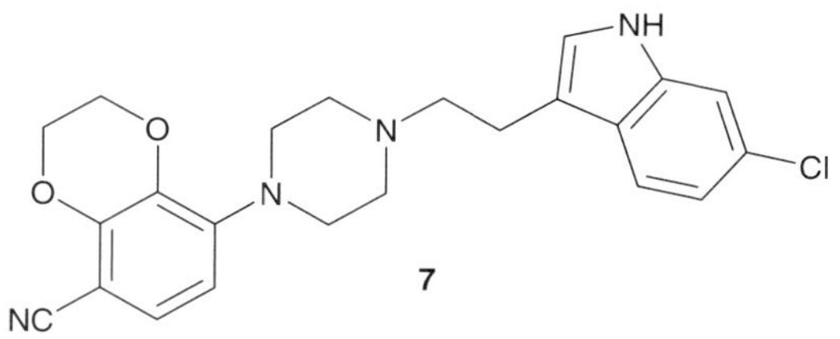

Recent Strategies for the Development of New Antidepressant Drugs 27

The dual SSRI-5-HT1A antagonist Lu 36-274 (**7**, 5-HT1A $K_i = 4.5$ nM, SERT IC50 = 6 nM) combines the 5-HT1A binding properties of a phenylpiperazine moiety with the SERT inhibitory ability of a 3-indolyl-alkylamine moiety [20]. Lu 36-275 is selective versus D2 dopaminergic receptors ($K_i = 430$ nM) and to lesser degree versus α1 adrenoreceptors ($K_i = 110$ nM). Rat cortical 5-HT levels were increased to 500% of control after an acute dose of 16 mg/kg, sc. In the schedule-induced polydipsia (SIP) model Lu 36-274 reduced water intake following a single dose of 32 mg/kg, p.o., suggesting a fast-acting antidepressant profile [21].

Other compounds utilizing the alkylindole SERT ligand include **8** and **9**. The benzoxazine derivative **8** had potent affinities for both 5-HT1A ($K_i = 9.31$ nM) and SERT ($K_i = 3.13$ nM), but showed substantial 5-HT1A agonism (66%) and no selectivity versus α1 [22]. The aryloxyalkylamine derivative **9** (5-HT1A $K_i = 26.6$ nM, SERT $K_i = 0.47$ nM) demonstrated full 5-HT1A antagonism, but was also not selective over α1 [23].

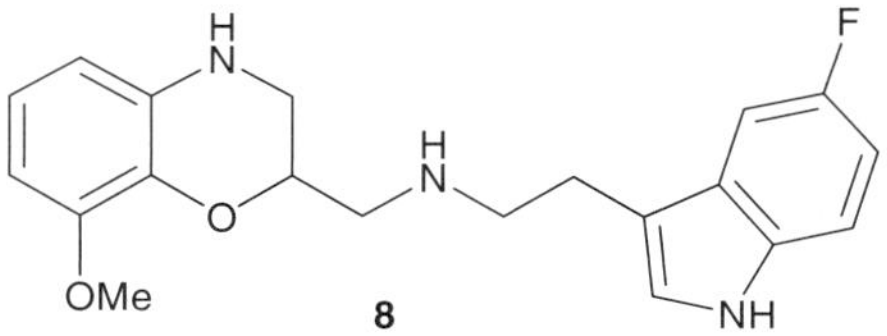

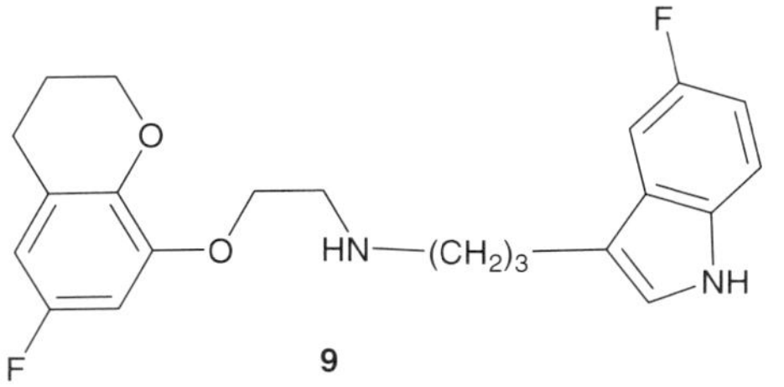

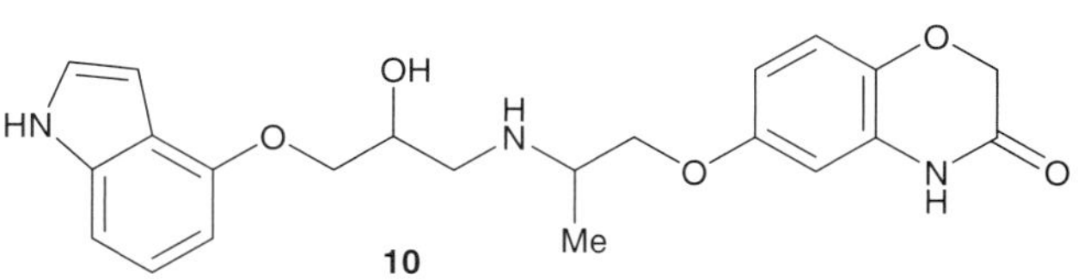

The two series of molecules exemplified by **10** and **11** utilize a 3,4-dihydro-2H-benzoxazinone derivative as the SERT component [24]. The pindolol derivative **10**

combines high affinty for the 5-HT1A receptor ($pK_i = 9.1$) with moderate affinity for SERT ($pK_i = 7.3$), but also posseses high β2 adrenergic receptor affinity ($pK_i = 9.2$). Further SAR efforts resulted in SB-649915 (**11**) that retains the high affinity for 5-HT1A ($pK_i = 9.5$) plus has excellent SERT affinity ($pK_i = 8.2$). Although SB-649915 was selective over β2 and a number of other biogenic amine receptors, it was found to have significant affinity for 5-HT1B and 5-HT1D receptors (pK_i 8.1 and 8.7, respectively). Activity in animal models predictive of antidepressant response has not yet been reported.

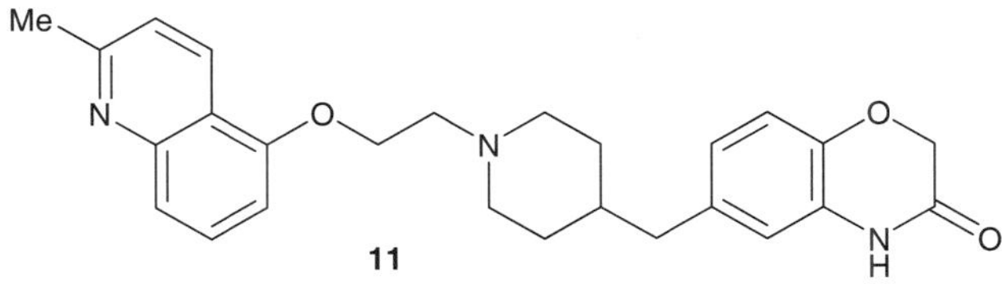

11

3.2. SSRI + 5-HT1D antagonism

The 5-HT1D receptor functions as an autoreceptor controlling the release of 5-HT from nerve terminals. Therefore a drug with a combination of SERT inhibition and 5-HT1D antagonism may also be of interest as a more rapid-acting antidepressant. Compound **12** was identified as a potent SSRI ($K_i = 0.11$ nM) that also exhibits 5-HT1D affinity ($K_i = 56$ nM), with less affinity for the 5-HT1B receptor ($K_i = 281$ nM) [25]. In microdialysis studies in guinea pigs **12** increased hypothalamic 5-HT levels as high as 300% of baseline. *In vivo* behavioral studies in animal models predictive of antidepressant response have not been reported.

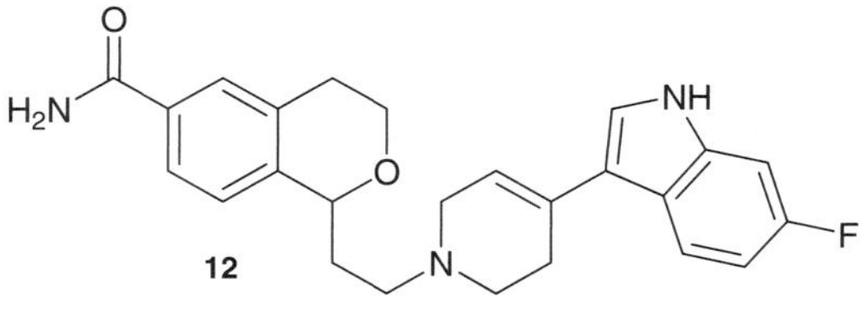

12

3.3. SSRI + α2 antagonism

α2 adrenergic receptors act as autoreceptors for NE neurons in much the same way that 5-HT1 receptors regulate serotonergic neurons. Thus, antagonism of α2 receptors would prevent the negative feedback NE exerts on its own synthesis, neuronal firing, and release [26]. α2 antagonism also results in increased extracellular dopamine, acetylcholine, and 5-HT levels *in vivo* in rats and humans [27]. It has therefore been proposed that a combination of SERT inhibition and α2 antagonism could result in rapid-onset antidepressant action, and possibly increased efficacy. Clinical studies using yohimbine and other α2 antagonists in combination with SSRIs support these hypotheses [28].

SAR studies that began with a lead found by random screening led to compounds such as **13** (α2A $K_i = 0.3$, SERT $K_i = 4.5$ nM) and **14** (α2A $K_i = 2.4$, SERT $K_i = 8.9$ nM) that are good candidates for further *in vivo* characterization [27].

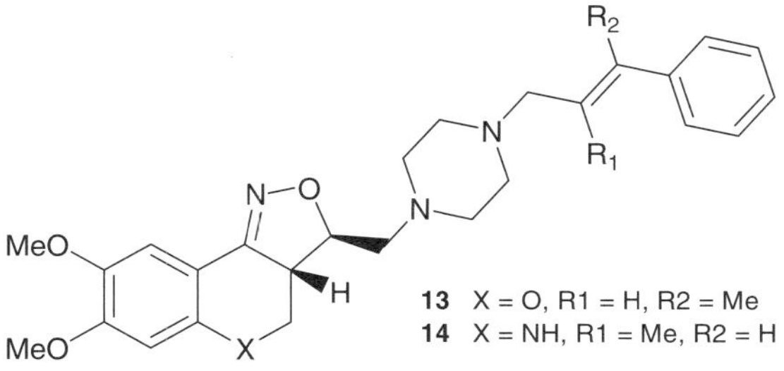

13 X = O, R1 = H, R2 = Me
14 X = NH, R1 = Me, R2 = H

4. SEROTONERGIC RECEPTOR MODULATION

Advances in the development of selective serotonin receptor modulators was recently reviewed [29]. Although several subtypes of 5-HT receptors may have roles in depression and anxiety, only the 5-HT1 family, 5-HT2C and 5-HT7 receptors will be highlighted here.

4.1. 5-HT1A/1B/1D antagonists

Serotonergic neurotransmission is controlled by three autoreceptors, the 5-HT1A and 5-HT1D receptors (mentioned above) and the 5-HT1B receptor, which limits the synthesis and release of 5-HT from the nerve terminal. It is hypothesized that a drug that antagonizes all three autoreceptors simultaneously would *acutely* mimic their chronic desensitization. The resulting immediate and sustained increase in levels of synaptic 5-HT should translate into a rapidly acting antidepressant [30]. Compound **15** (pK_i for 5-HT1A, 1B, 1D = 8.9, 9.0, and 9.2, respectively) showed *in vivo* effects in models indicative of 5-HT1B antagonism and is undergoing testing in preclinical depression models [30].

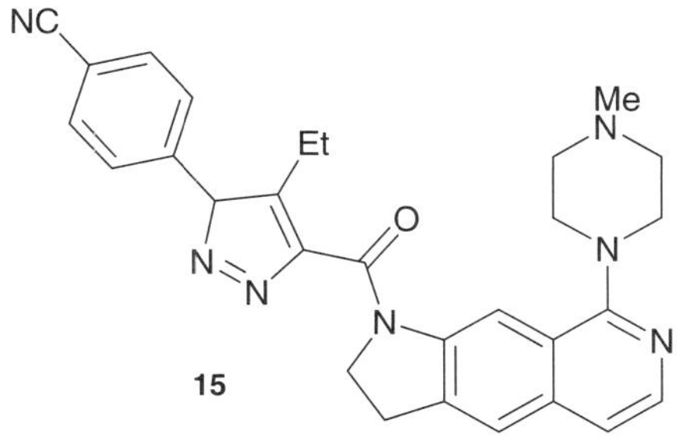

4.2. 5-HT2C receptor modulators

Selective 5-HT2C agonists have shown antidepressant-like effects in several animal models of depression [5]. Of the recently identified 5-HT2C agonists, Ro 60-0175, WAY-161503, and WAY-163909 have been the most extensively characterized in depression models. Results in chronic mild stress and olfactory bulbectomy models of depression suggest a more rapid onset of antidepressant action: typical antidepressants require 2–3 weeks of dosing to show effectiveness, while 5-HT2C agonists are active in less than one week in both models.

5-HT2C antagonists may also have a role in the treatment of depression. It has been shown that the 5-HT2C antagonists SB-242084 and RS 102221 potentiate the action of the SSRI citalopram on rat cortical and hippocampal 5-HT levels, although the 5-HT2C antagonists alone had no effect [31]. It has also been claimed that the clinical antidepressant efficacy of agomelatine, a melatonin agonist, may be in part due to 5-HT2C antagonism [32].

4.3. 5-HT7 receptor antagonists

The 5-HT7 receptor has been postulated to play a role in depression based partly on its distribution in the brain and also on the observation that chronic treatment with antidepressants results in downregulation of this receptor. Additional support for this hypothesis has been gained by a recent study showing that 5-HT7 receptor knockout mice demonstrate antidepressant-like profiles in forced swim and tail suspension tests [33]. The identification of selective antagonist tool compounds has helped add further evidence in support of the hypothesis. For example, SB-269970, **16**, exhibits antidepressant-like activity, even though this compound has a rather poor PK profile. SB-656104-A, **17**, has been identified as a 5-HT7 antagonist with better PK properties and has been shown to modulate REM sleep in rats in a manner consistent with potential antidepressant-like activity [34].

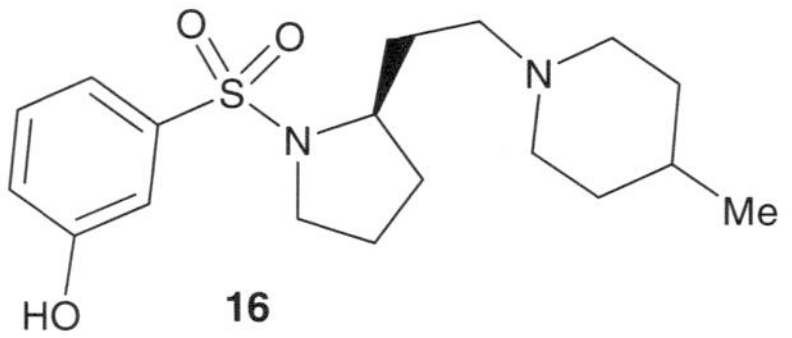

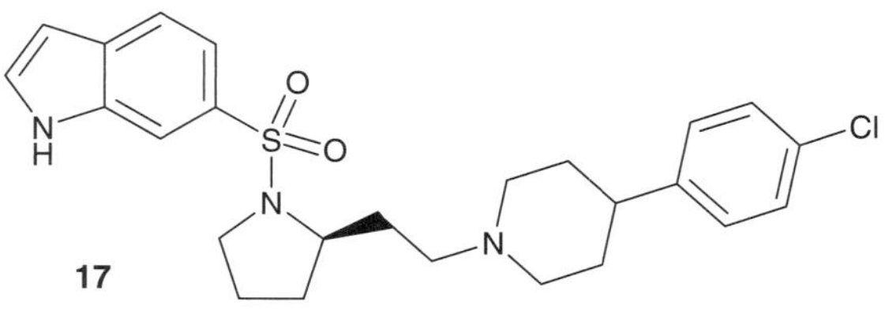

5. NEUROPEPTIDE RECEPTOR MODULATION

5.1. Neurokinins

The potential for NK1 antagonists as antidepressants has been the subject of a tremendous amount of research effort over the last decade [35]. At least three compounds (aprepitant, L759274, and CP-122721) showed early evidence of antidepressant effects in clinical studies. The failure of aprepitant to separate from placebo in Phase III studies was a great disappointment [6]. Nonetheless, NK1 antagonists continue to be of interest for depression therapy since they have been shown to potentiate the activity of SSRIs. The NK1 antagonist vestipitant is currently being evaluated in combination with paroxetine for anxiety and depression. Interestingly, both of these activities have been attained in a single compound, **18**, which is both a potent SERT inhibitor (pIC50 = 8.0) and an NK1 antagonist (pIC50 = 8.5) [36]. This compound reportedly increased rat 5-HT levels up to 250% of baseline (measured by microdialysis), and was orally active in the isolation-induced guinea-pig pup vocalization test of anxiety.

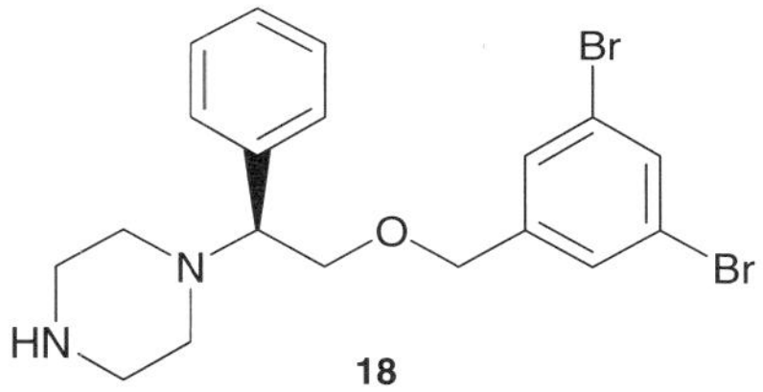

NK2 antagonists have primarily been investigated for the treatment of inflammatory conditions such as obstructive airways disease [5]. However, studies in preclinical animal models also suggest a potential role for NK2 antagonists as novel antidepressants. Notably, the NK2 antagonist saredutant displayed antidepressant-like properties in the forced swim test and in maternal separation of guinea pig pups [4]. Clinical trials for the treatment of depression are now in progress.

5.2. Corticotropin-releasing factor

Corticotropin-releasing factor (CRF) is a well-known regulator of the hypothalamic-pituitary-adrenal (HPA) axis, which is activated in response to stress. Hyperactivity of the HPA axis has been linked to depression in humans, and both the elevation of CRF concentrations in the cerebral spinal fluid (CSF) and an increase in the number of CRF-containing neurons in the paraventricular nucleus have been observed in depressed patients. In addition, some antidepressants (e.g., desipramine and fluoxetine) have been shown to decrease CRF levels in the CSF. Furthermore, a number of CRF1 antagonists exhibits anxiolytic and antidepressant activity in certain animal models [37,38]. For example, antalarmin (CP-154,526) produces antidepressant-like responses in the rat learned helplessness and mouse chronic mild stress models of depression. Several CRF1 antagonists have reached human clinical trials, but to date only one has published data: R121919/NBI-30775, **19**, demonstrated antidepressant efficacy in a small open-label Phase II study. Unfortunately, this compound was terminated due to increased liver enzymes [39].

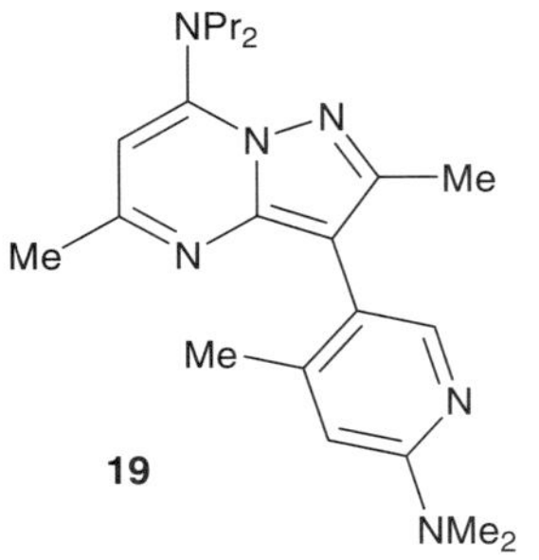

Further SAR efforts on **19** included reduction of the conformational freedom by incorporation into a tricyclic core in an attempt to alter PK properties. A resulting compound, NBI-35965, **20**, (pK$_i$ = 8.5) reduced CRF-induced ACTH release in normal rats (MED = 10 mg/kg, p.o.) and was active in the restraint stress model of ACTH release in mice at 20 mg/kg, po [40]. Further modification to a more polar core generated **21** (K$_i$ = 2 nM), which significantly attenuated CRF-induced ACTH release at 10 mg/kg, p.o. [40]. Activity in depression models was not reported.

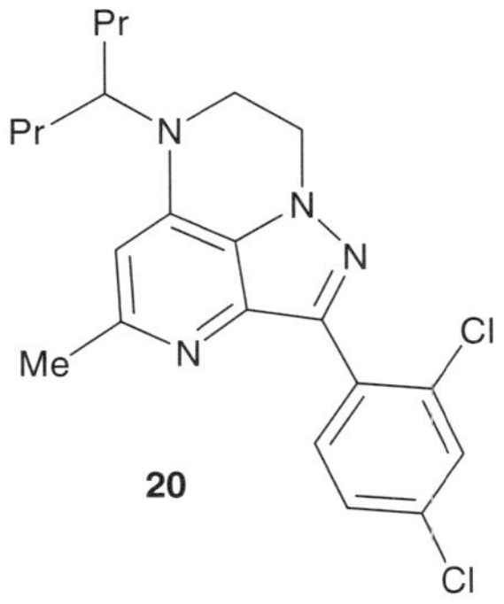

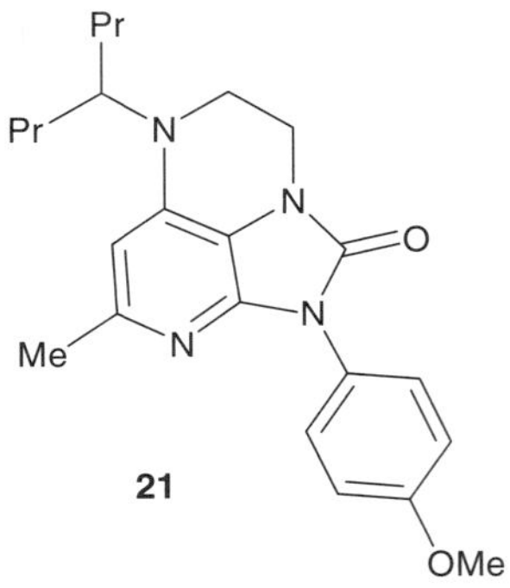

Like the majority of CRF-1 antagonists, R278995/CRA0450, **22**, failed to show any effect in rodent behavioral despair models such as rat forced swim and mouse tail suspension tests. However, **22** was active after acute administration in the learned helplessness paradigm. It was also active acutely and chronically (10-day dosing) in the rat olfactory bulbectomy model of depression [41].

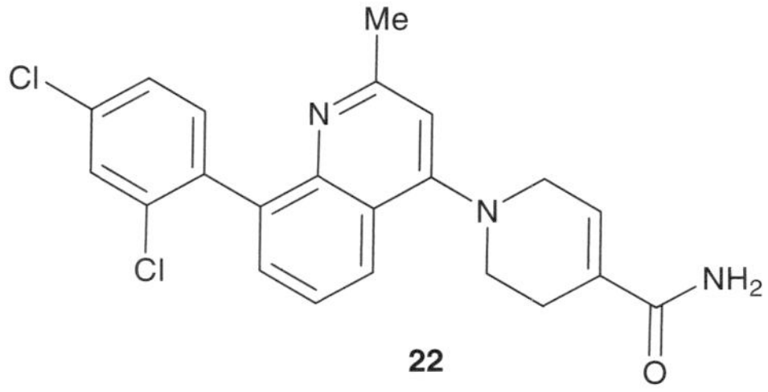

SSR125543, **23**, was orally active in the rat forced swim test (30 mg/kg, p.o.), and showed positive effects in the mouse chronic mild stress paradigm [42]. Another particulary interesting feature of **23** was the demonstration of neurogenesis in mice after chronic (28-day) dosing [43]. This observation is noteworthy since increased neurogenesis has been shown to result from chronic treatment with clinically effective antidepressants and may be important for an efficacious response [44].

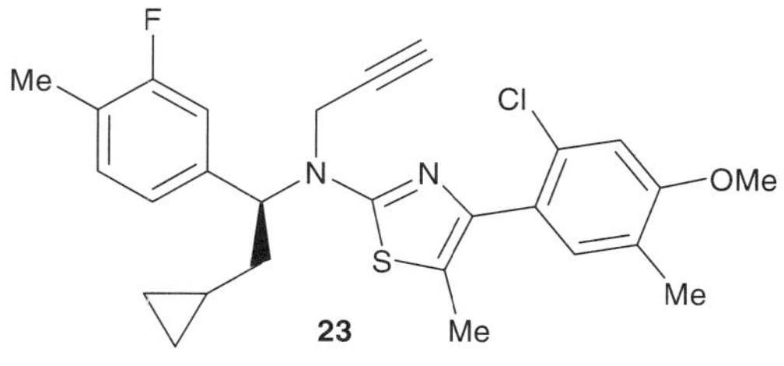

5.3. Melanocortin MC4 antagonists

Melanocortins participate in the HPA axis response to stress, at least partly through the release of CRF. The MC4 receptor has been identified as the receptor of particular interest in relationship to stress-related disorders including anxiety and depression [45]. The non-peptide MC4 antagonist MCL0129, **24**, ($K_i = 7.9$ nM) demonstrated antidepressant effects in the forced swim and learned helplessness models, and also showed anxiolytic activity after oral administration [46]. Interestingly, the same series of compounds led to the serendipitous discovery of MCL0042 (**25**), a dual SSRI-MC4 antagonist (MC4: $IC_{50} = 124$ nM; SERT: $IC_{50} = 42$) that was active in the olfactory bulbectomized rat model of depression [45].

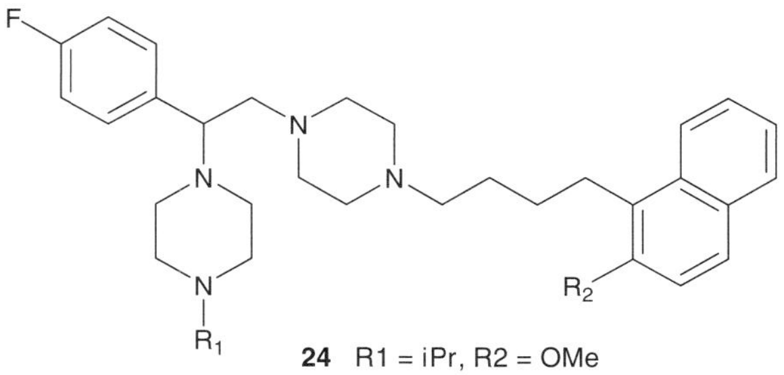

24 R1 = iPr, R2 = OMe
25 R1 = Me, R2 = H

5.4. MCH1 antagonists

Melanin concentrating hormone (MCH) is involved in a number of physiological processes including feeding, energy homeostasis, sexual behavior, mood regulation, and stress responses. Interest in this neuropeptide has largely been focused on the discovery of MCH1 receptor antagonists for the treatment of obesity [47]. However, interest is rising in the role of MCH in mood regulation and stress responses, with the possibility of developing MCH1 antagonists as antidepressant drugs [45,48].

SNAP-7941, **26**, was the first MCH1R antagonist ($K_b = 0.5$ nM) to demonstrate an antidepressant-like effect in the rat forced swim test [49]. It also showed anxiolytic activity in the rat social interaction test and the guinea-pig pup vocalization test. Since that initial report, a number of other MCH1R antagonists have also been reported to have anxiolytic and/or antidepressant-like effects in preclinical animal models. For example, in addition to demonstrating anxiolytic activity, SNAP-94847, **27**, showed antidepressant-like activity in the mouse forced swim test [48]. In addition, the potential for rapid onset of action was demonstrated in the chronic mild stress model [45]. Chronic dosing (28 days) with **27** in mice also resulted in increased neurogenesis.

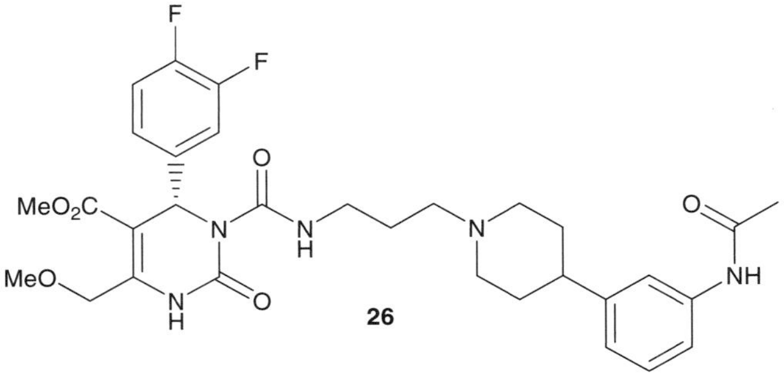

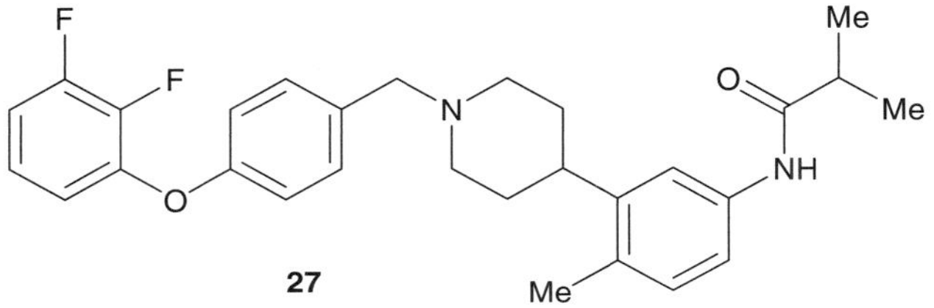

The quinazoline derivatives ATC0065, **28** (IC50 = 16 nM) and ATC0175, **29**, IC50 = 7.2 nM) were also orally active in animal models of anxiety and in the rat forced swim antidepressant test [45].

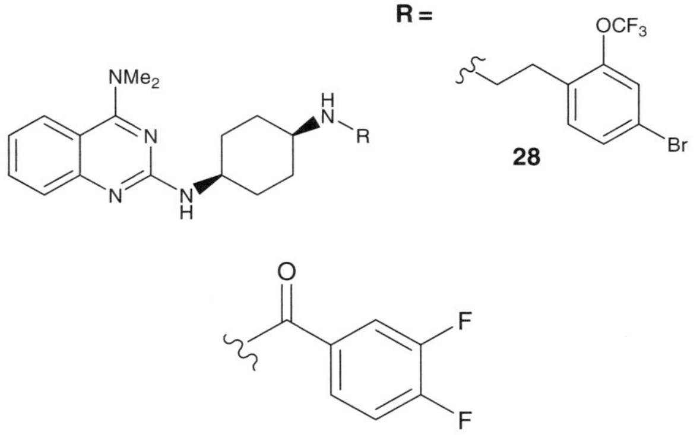

5.5. Galanin

The neuropeptide galanin has been shown to have mood-regulating effects, possibly through modification of serotonergic neurotransmission [2]. Galanin reduces the gene expression of the enzyme tryptophan hydroxylase, which is the rate-limiting enzyme in 5-HT biosynthesis. The selective galanin-3 receptor (Gal3-R) antagonist SNAP-37889, **30**, ($K_i = 17.4$ nM) has shown anxiolytic and antidepressant-like activity in animal models: in the rat forced swim test it was active at 3 and 10 mg/kg, p.o., producing effects comparable to fluoxetine [50]. SNAP-398299, **31**, ($K_i = 5.33$ nM) is a highly water-soluble analog of **30** that produced an anxiolytic-like effect (MED = 1 mg/kg, p.o.) in the rat social interaction test.

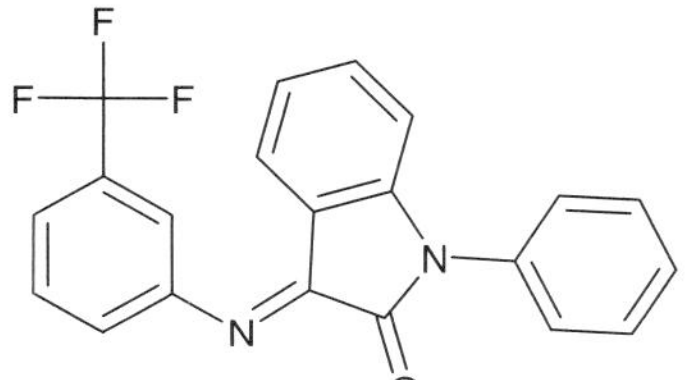

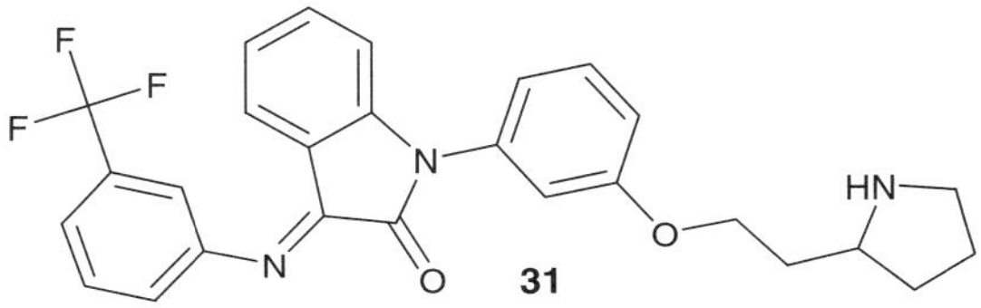

5.6. Vasopressin

Arginine vasopressin (AVP) has been implicated in emotional behavior including depression and anxiety; evidence was obtained initally through studies with non-selective V1a/b receptor antagonists [51]. SSR149415, **32**, was the first non-peptide selective V1b antagonist ($K_i = 1.54$ nM) to be described [42]. Anxiolytic and antidepressant effects were demonstrated in animal models, including the chronic mild stress paradigm in mice [51,52]. In that model, increased neurogenesis was also demonstrated [43]

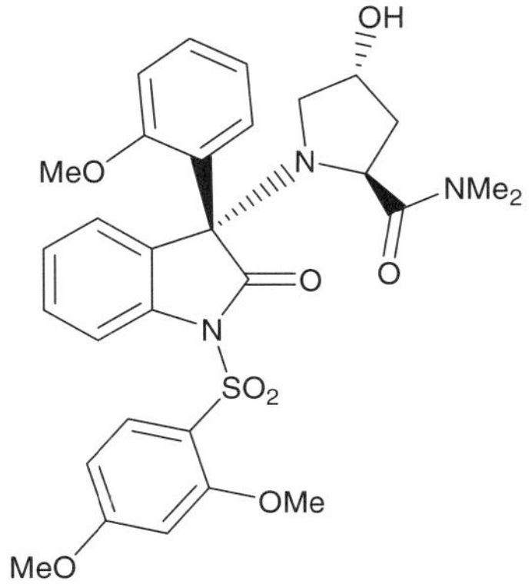

6. CONCLUSION

Clearly there is a large amount of research being done in the search for new antidepressants. Due to space limitations, this review has highlighted only some of the strategies being pursued. Other emerging areas of research include the modulation of glutamatergic neurotransmission [53], GABA [54], cytokines [55,56], and neurotrophins [57]. While much research remains focused on direct modulation of monoaminergic systems, other strategies involve manipulation of separate biological systems that are linked either downstream or upstream to the serotonergic system. Although animal models provide evidence to support many of these strategies, clinical studies in humans will be needed not only to demonstrate efficacy, but also to delineate any advantages of drugs acting through these mechanisms compared to existing therapies.

REFERENCES

[1] M. H. Trivedi, B. Kleiber and T. L. Greer, *Drug Dev. Res.*, 2005, **65**, 335.
[2] A. Adell, E. Castro, P. Celada, A. Bortolozzi, A. Pazos and F. Artigas, *Drug Discov. Today*, 2005, **10**, 578.
[3] H. G. Nurnberg, *J. Psychiat. Pract.*, 2001, **7**, 92.
[4] D. A. Slattery, A. L. Hudson and D. J. Nutt, *Fundam. Clin. Pharmacol.*, 2004, **18**, 1.
[5] L. E. Schechter, R. H. Ring, C. E. Beyer, Z. A. Hughes, X. Khawaja, J. E. Malberg and S. Rosenzweig-Lipson, *NeuroRx*, 2005, **2**, 590.
[6] P. E. Holtzheimer and C. B. Nemeroff, *NeuroRx*, 2006, **3**, 42.
[7] O. Berton and E. J. Nestler, *Nat. Rev. Neurosci.*, 2006, **7**, 137.
[8] I. A. Antonijevic, *Psychoneuroendocrinology*, 2006, **31**, 1.
[9] T. C. Baghai, H.-J. Moller and R. Rupprecht, *Curr. Pharm. Design.*, 2006, **12**, 503.
[10] M. W. Walter, *Drug. Dev. Res.*, 2005, **65**, 97.
[11] C. Shelton, R. Entsuah, S. K. Padmanabhan and P. E. Vinall, *Int. Clin. Psychopharmacol.*, 2005, **20**, 233.
[12] F. P. Bymaster, T. C. Lee, M. P. Knadler, M. J. Detke and S. Iyengar, *Curr. Pharm. Design.*, 2005, **11**, 1475.
[13] I. Artaiz, A. Zazpe, A. Innerarity, E. Del Olmo, A. Diaz, J. A. Ruiz-Ortega, E. Castro, R. Pena, L. Labeaga, A. Pazos and A. Orjales, *Psychopharmacology*, 2005, **182**, 400.
[14] E. Esposito, *Curr. Drug Targets*, 2006, **7**, 177.
[15] P. Skolnick, P. Popik, A. Janowsky, B. Beer and A. S. Lippa, *Life Sci*, 2003, **73**, 3175.
[16] B. Beer, J. Stark, P. Krieter, P. Czobor, G. Beer, A. Lippa and P. Skolnick, *J. Clin. Pharmacol.*, 2004, **44**, 1360.
[17] PR Newswire 12/19/05.
[18] L. Axford, J. R. Boot, T. M. Hotten, M. Keenan, F. M. Martin, S. Milutinovic, N. A. Moore, M. F. O'Neill, I. A. Pullar, D. E. Tupper, K. R. Van Belle and V. Vivien, *Bioorg. Med. Chem. Lett.*, 2003, **13**, 3277.
[19] K. Takeuchi, T. J. Kohn, N. A. Honigschmidt, V. P. Rocco, P. G. Spinazze, S. K. Hemrick-Luecke, L. K. Thompson, D. C. Evans, K. Rasmussen, D. Koger, D. Lodge, L. J. Martin, J. Shaw, P. G. Threlkeld and D. T. Wong, *Bioorg. Med. Chem. Lett.*, 2006, **16**, 2347.
[20] E. K. Moltzen, I. Mikkelsen, B. Bjornholm, L. T. Brennum, S. Hogg, and C. Sanchez, Poster 958.16, Society for Neuroscience, 33rd Annu. Meet., November 8–12, 2003, New Orleans, USA.

[21] A. Mork and S. Hogg, Poster 958.17, Society for Neurosciences, 33rd Annu. Meet., November 8–12, 2003, New Orleans, LA.
[22] D. Zhou, B. L. Harrison, U. Shah, T. H. Andree, G. A. Hornby, R. Scerni, L. E. Schechter, D. L. Smith, K. M. Sullivan and R. E. Mewshaw, *Bioorg. Med. Chem. Lett.*, 2006, **16**, 1338.
[23] R. E. Mewshaw, D. Zhou, P. Zhou, X. Shi, G. Hornby, T. Spangler, R. Scerni, D. Smith, L. E. Schechter and T. H. Andree, *J. Med. Chem.*, 2004, **47**, 3823.
[24] P. J. Atkinson, S. M. Bromidge, M. S. Duxon, L. M. Gaster, M. S. Hadley, B. Hammond, C. N. Johnson, D. N. Middlemiss, S. E. North, G. W. Price, H. K. Rami, G. J. Riley, C. M. Scott, T. E. Shaw, K. R. Starr, G. Stemp, K. M. Thewlis, D. R. Thomas, M. Thompson, A. K. Vong and J. M. Watson, *Bioorg. Med. Chem. Lett.*, 2005, **15**, 737.
[25] G. H. Timms, J. R. Boot, R. J. Broadmore, S. L. Carney, J. Cooper, J. D. Findlay, J. Gilmore, S. Mitchell, N. A. Moore, I. Pullar, G. J. Sanger, R. Tomlinson, B. B. Tree and S. Wedley, *Bioorg. Med. Chem. Lett.*, 2004, **14**, 2469.
[26] R. W. Invernizzi and S. Garattini, *Prog. Neuro-Psychopharmacol. Biol. Pyschiat.*, 2004, **28**, 819.
[27] J. I. Andres, J. Alcazar, J. M. Alonso, R. M. Alvarez, M. H. Bakker, I. Biesmans, J. M. Cid, A. I. DeLucas, J. Fernandez, L. M. Font, K. A. Hens, L. Iturrino, I. Lenaerts, S. Martinez, A. A. Megens, J. Pastor, P. C. M. Vermote and T. Steckler, *J. Med. Chem.*, 2005, **48**, 2054.
[28] G. Sanacora, R. M. Berman, A. Cappiello, D. A. Oren, A. Kugaya, N. J. Liu, R. Gueorguieva, D. Fasula and D. S. Charney, *Neuropsychopharmacology*, 2004, **29**, 1166.
[29] W. E. Childers and A. J. Robichaud, *Annu. Rep. Med. Chem.*, 2005, **40**, 17.
[30] T. D. Heightman, L. M. Gaster, S. L. Pardoe, J. P. Pilleux, M. S. Hadley, D. N. Middlemiss, G. W. Price, C. Roberts, C. M. Scott, J. M. Watson, L. J. Gordon, V. A. Holland, J. Powles, G. J. Riley, T. O. Stean, B. K. Trail, N. Upton, N. E. Austin, A. D. Ayrton, T. Coleman and L. Cutler, *Bioorg. Med. Chem. Lett.*, 2005, **15**, 4370.
[31] T. I. Cremers, M. Giorgetti, F. J. Bosker, S. Hogg, J. Arnt, A. Mork, G. Honig, K. P. Bogeso, B. H. Westerink, H. den Boer, H. V. Wikstrom and L. H. Tecott, *Neuropsychopharmacology*, 2004, **29**, 1782.
[32] F. Rouillon, *Int. Clin. Psychopharmacol.*, 2006, **21** (Suppl. 1), S31.
[33] P. B. Hedlund, S. Huitron-Resendiz, S. J. Henriksen and J. G. Sutcliffe, *Biol. Psychiat.*, 2005, **58**, 831.
[34] D. R. Thomas, S. Melotto, M. Massagrande, A. D. Gribble, P. Jeffrey, A. J. Stevens, N. J. Deeks, P. J. Eddershaw, S. H. Fenwick, G. Riley, T. Stean, C. M. Scott, M. J. Hill, D. N. Middlemiss, J. J. Hagan, G. W. Price and I. T. Forbes, *Br. J. Pharmacol.*, 2003, **139**, 705.
[35] S. McLean, *Curr. Pharm. Design.*, 2005, **11**, 1529.
[36] T. Ryckmans, O. Berton, R. Grimee, T. Kogej, Y. Lamberty, P. Pasau, P. Talaga and C. Genicot, *Bioorg. Med. Chem. Lett.*, 2002, **12**, 3195.
[37] D. M. Nielsen, *Life Sci*, 2006, **78**, 909.
[38] D. H. Overstreet, D. J. Knapp and G. R. Breese, *Drug Dev. Res.*, 2005, **65**, 191.
[39] S. Jordan, R. Chen, V. Koprivica, R. Hamilton, R. E. Whitehead, K. Tottori and T. Kikuchi, *Eur. J. Pharmacol.*, 2005, **517**, 165.
[40] R. S. Gross, Z. Guo, B. Dyck, T. Coon, C. Q. Huang, R. F. Lowe, D. Marinkovic, M. Moorjani, J. Nelson, S. Zamani-Kord, D. E. Grigoriadis, S. R. J. Hoare, P. D. Crowe, J. H. Bu, M. Haddach, J. McCarthy, J. Saunders, R. Sullivan, T. K. Chen and J. P. Williams, *J. Med. Chem.*, 2005, **48**, 5780.
[41] S. Chaki, A. Nakazato, L. Kennis, M. Nakamura, C. Mackie, M. Sugiura, P. Vinken, D. Ashton, X. Langlois and T. Steckler, *Eur. J. Pharmacol.*, 2004, **485**, 145.

[42] C. S.-L. Gal, J. Wagnon, J. Simiand, G. Griebel, C. Lacour, G. Guillon, C. Barberis, G. Brossard, P. Soubrie, D. Nisato, M. Pascal, R. Pruss, B. Scatton, J.-P. Maffrand and G. Le Fur, *J. Pharmacol. Exp. Ther.*, 2002, **300**, 1122.
[43] R. Alonso, G. Griebel, G. Pavone, J. Stemmelin, G. Le Fur and P. Soubrie, *Mol. Psychiat.*, 2004, **9**, 278.
[44] L. Santarelli, M. Saxe, C. Gross, A. Surget, F. Battaglia, S. Dulawa, N. Weisstaub, J. Lee, R. Duman, O. Arancio, C. Belzung and R. Hen, *Science*, 2003, **301**, 805.
[45] S. Chaki, Y. Oshida, S.-I. Ogawa, T. Funakoshi, T. Shimazaki, T. Okubo, A. Nakazato and S. Okuyama, *Pharmacol. Biochem. Behav.*, 2005, **82**, 621.
[46] S. Chaki, S. Hirota, T. Funakoshi, Y. Suzuki, S. Suetake, T. Okubo, T. Ishii, A. Nakazato and S. Okuyama, *J. Pharmacol. Exp. Ther.*, 2003, **304**, 818.
[47] M. D. McBriar and T. J. Kowalski, *Annu. Rep. Med. Chem.*, 2005, **40**, 119.
[48] B. Dyck, *Drug Dev. Res.*, 2005, **65**, 291.
[49] B. Borowsky, M. M. Durkin, K. Ogozalek, M. R. Marzabadi, J. DeLeon, R. Heurich, H. Lichtblau, Z. Shaposhnik, I. Daniewska, T. P. Blackburn, T. A. Branchek, C. Gerald, P. J. Vaysse and C. Forray, *Nat. Med.*, 2002, **8**, 825.
[50] C. J. Swanson, T. P. Blackburn, X. Zhang, K. Zheng, Z. Q. Xu, T. Hokfelt, T. D. Wolinsky, M. J. Konkel, H. Chen, H. Zhong, M. W. Walker, D. A. Craig, C. P. Gerald and T. A. Branchek, *Proc. Natl. Acad. Sci. USA*, 2005, **102**, 17489.
[51] D. H. Overstreet and G. Griebel, *Pharmacol. Biochem. Behav.*, 2005, **82**, 223.
[52] G. Griebel, J. Simiand, C. Serradeil-Le Gal, J. Wagnon, M. Pascal, B. Scatton, J.-P. Maffrand and P. Soubrie, *Proc. Natl. Acad. Sci. USA*, 2002, **99**, 6370.
[53] A. Kugaya and G. Sanacora, *CNS Spectros*, 2005, **10**, 808.
[54] D. A. Slattery, S. Desrayaud and J. F. Cryan, *J. Pharmacol. Exp. Ther.*, 2005, **312**, 290.
[55] O. J. G. Schiepers, M. C. Wichers and M. Maes, *Prog. Neuro-Psychopharmacol. Biol. Pyschiat.*, 2005, **29**, 201.
[56] H. Anisman, Z. Merali, M. O. Poulter and S. Hayley, *Curr. Pharm. Design.*, 2005, **11**, 963.
[57] A. A. Russo-Neustadt and M. J. Chen, *Curr. Pharm. Design.*, 2005, **11**, 1495.

Neuroprotective Agents for the Treatment of Ischemic and Hemorrhagic Stroke

Simon N. Haydar[1] and Warren D. Hirst[2]

[1]*Chemical & Screening Sciences, Wyeth Research, CN 8000, Princeton, NJ 08543, USA*
[2]*Neuroscience, Wyeth Research, CN 8000, Princeton, NJ 08543, USA*

Contents

1. Introduction	39
1.1. Neuroprotection by caspase inhibitors	40
1.2. Neuroprotection by poly(ADP-ribose)polymerase inhibitors	42
1.3. Peroxisome proliferator-activator receptor gamma (PPARγ) ligands	45
1.4. C-Jun-N-terminal kinase inhibitors	47
1.5. Src kinase inhibitors	50
1.6. Estrogens	51
2. Conclusion	53
References	53

1. INTRODUCTION

Ischemic stroke is a common life-threatening neurological disorder with severely limited therapeutic options. In the United States of America, stroke is the third (in Japan the first) leading cause of death and the major cause of disability. Approximately 700,000 stroke cases (new and recurrent) occur each year in the USA alone [1]. Care of the more than 5.5 million stroke survivors in the USA is estimated to cost over $58 billion [2]. Over 80% of stroke cases are ischemic, resulting from an obstruction of blood flow in a major cerebral artery by thrombi or emboli. Hemorrhagic stroke accounts for 15–20% of stroke cases. At present, the only approved therapy for acute ischemic stroke is intravascular thrombolysis of the obstructing blood clot using recombinant tissue plasminogen activator (rTPA): Genentech's Altepase™ (Activase). Given the associated risk of life-threatening hemorrhage and a restrictive 3h treatment window, only about 1–2% of patients receive Altepase™ treatment [3]. Clearly there is an urgent need for a safe, efficacious and widely applicable neuroprotective agent.

Multiple cellular processes are activated in response to ischemic stroke. The ischemic core, near the occluded vessel, undergoes rapid anoxic cell death. However, in the ischemic penumbra, tissue may be spared by restored perfusion, albeit limited, from collateral blood supply, and/or as a result of spontaneous or therapeutic reperfusion and cellular mechanisms of neuroprotection [4]. This tissue is, however, at risk and is affected by multiple stresses, which manifest themselves from minutes to hours and days following ischemia. Evidence indicates that, left untreated, the penumbral region becomes part of the core [5]. In minutes, excitotoxicity results from elevated glutamate levels leading to Ca^{2+} influx and production of reactive oxygen species from the inability of impaired mitochondria to use oxygen in electron transport [6]. In hours, there is expression and activation of proapoptotic signaling intermediates, and initiation of inflammatory processes. Extending from days to weeks following the stroke, a glial scar develops expressing extracellular matrix molecules, which inhibit axon extension [7]. Angiogenesis [8] and neurogenesis [9] suggest a previously under-recognized capacity for self-repair after ischemic stroke.

Despite progress in understanding the mechanisms of ischemic damage, and identification of agents effective in animal models of stroke, clinical efficacy has not been achieved. Several drugs that have advanced into phase III clinical trials have been discontinued due to a lack of efficacy or safety concerns, with many more failures at earlier stages [10,11]. A series of Stroke Therapy Academic Industry Roundtable (STAIR) conferences have provided recommendations and guidance for pre-clinical and clinical development of acute stroke therapies [12,13]. One molecule, whose preclinical development has closely adhered to the STAIR recommendations, is NXY-059, a free radical trapping agent [14,15]. Recent data from a phase III clinical trial (SAINT 1 trial) have shown that administration of NXY-059 within 6 h of onset of acute ischemic stroke significantly improved the primary outcome of reduced disability at 90 days. However, it did not significantly improve other outcome measures, including neurologic function. This suggests that despite the strong pre-clinical data, additional clinical studies are required to confirm whether this drug is beneficial for treatment of ischemic stroke [16].

The present review focuses on molecules and mechanisms that are at considerably earlier stages of development but are novel approaches to neuroprotection that may become the next generation of drugs for the treatment of acute ischemic and hemorrhagic stroke.

1.1. Neuroprotection by caspase inhibitors

Caspases are a family of cysteine proteases that are expressed as inactive zymogens and undergo proteolytic maturation in a sequential manner in which initiator caspases cleave and activate the effector caspases 3, 6 and 7. Caspase-3, among effector caspases, has been implicated in neuronal apoptosis during normal brain development and in delayed neuronal cell death after brain injury [17,18]. The latter observations have suggested that this is a potential therapeutic target which has led to the development of selective and potent caspase-3 inhibitors. Inhibitors of caspase-3 have been shown to reduce the amount of cellular and tissue damage in cell culture and animal models of disease. To date, most inhibitors of caspase-3 have been peptide-based compounds that either reversibly or irreversibly inhibit the catalytic activity of this enzyme [19,20]. *In vivo* studies have shown that the peptide-based inhibitors of caspase-3, M-826 (**1**) and M-867 (**2**), reduce apoptosis [9] and tissue damage in animal models of sepsis and neonatal hypoxic-ischemic brain injury [21]. In a rat model of neonatal hypoxia-ischemia (H-I), an *i.c.v.* injection of

M-826 (**1**), a reversible inhibitor of caspase-3, blocks caspase-3 activation and is neuroprotective against neonatal H-I induced brain injury [22].

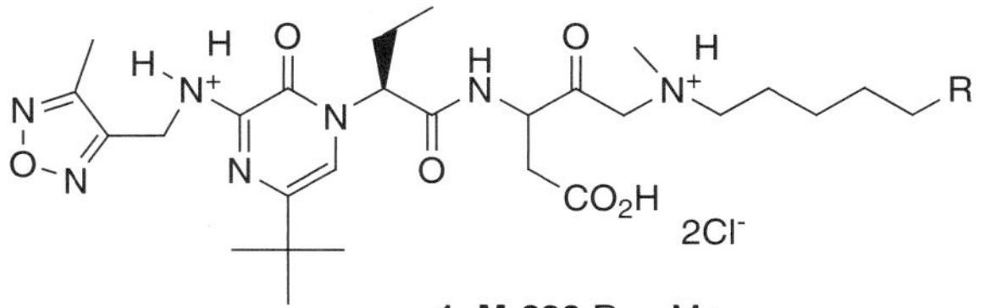

1, M-826 R = Me
2, M-867 R = H

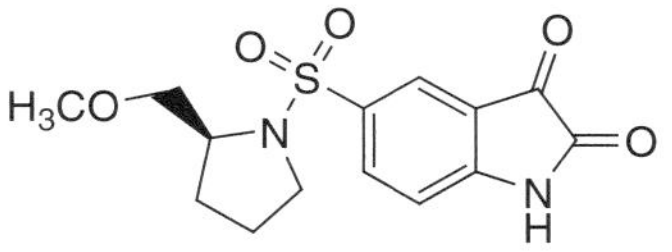

3
IC_{50}: 44 nM

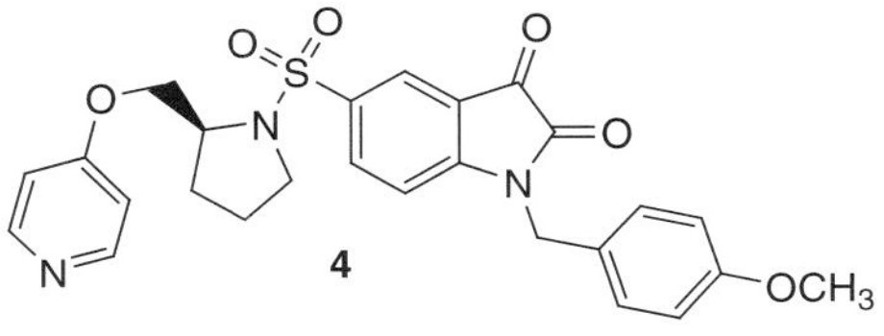

4
IC_{50}: 3.9 nM

One of the potential problems of peptide-based caspase inhibitors is their poor metabolic stability, limited cell penetration and unfavorable physico-chemical profile [23]. To overcome this problem, research efforts have focused on identifying nonpeptide small molecule inhibitors. Recently, a number of isatin-based inhibitors of caspase-3 and -7 have been reported. In particular, one compound, (S)-(+)-5-1[1-(2-methoxymethyl-pyrrolidine)sulfonyl]isatin, **3**, has been shown to reduce tissue damage in an isolated rabbit heart model of ischemic injury [24]. Additional structure-activity relationship studies have revealed that replacement of the 2-methoxymethyl group with a pyridine-3-oxymethyl moiety and the introduction of an alkyl group on the isatin nitrogen group resulted in an improvement in potency for inhibiting caspase-3. This is exemplified with compound **4** which was found to have an IC_{50} of 3.9 nM for caspase-3 and similar potency against caspase-7, but at least 100 fold less potent versus caspase-1, -6 and -8 [25]. Alternatively, 1,3-dioxo-2,3-dihydro-1H-pyrrolo[3,4-c]quinolines constitute an interesting group of physiologically active molecules. For example, compounds **5** and **6** within this series inhibited caspase-3 with $IC_{50} = 4$ and 6 nM, respectively [26–29]. This series of compounds were shown to have a noncompetitive and reversible character of caspase-3 inhibition, with moderate selectivity against caspase-3.

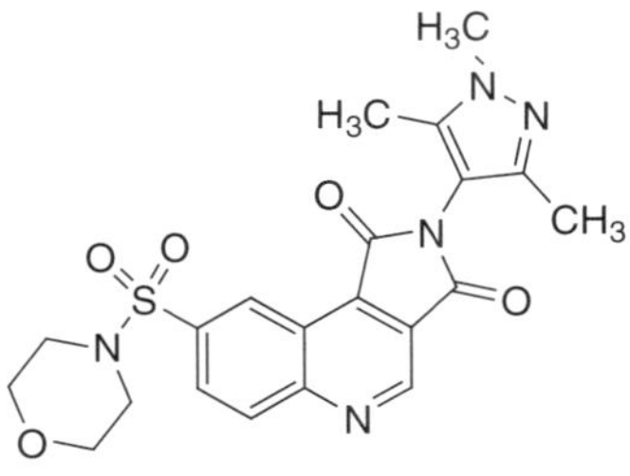

5
IC$_{50}$: 4 nM

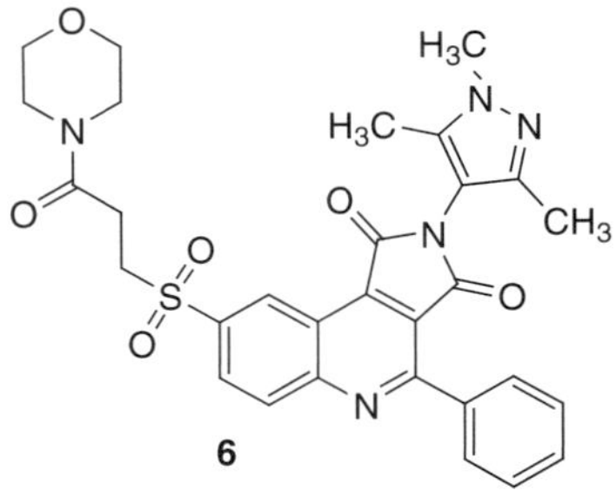

6
IC$_{50}$: 6 nM

1.2. Neuroprotection by poly(ADP-ribose)polymerase inhibitors

Poly(ADP-ribose) polymerase-1 (PARP-1) is a chromatin-bound nuclear enzyme involved in a variety of physiological functions related to genomic repair, including DNA replication and repair, cellular proliferation and differentiation, and apoptosis [30]. PARP-1 functions as a DNA damage sensor and signaling molecule. Upon binding to DNA breaks, activated PARP cleaves NAD$^+$, the nicotinamide, and ADP-ribose and polymerizes the latter into nuclear acceptor proteins including histones, transcription factors and PARP itself. A cellular suicide mechanism of necrosis and apoptosis by PARP activation has been implicated in the pathogenesis of brain injury and neurodegenerative disorders, and PARP inhibitors have been shown to be effective in animal models for stroke, traumatic brain injury and Parkinson's disease [31,32]. Therefore, inhibition of PARP by pharmacological agents may be useful for the therapy of neurodegenerative disorders and several other diseases. In stroke, many studies have demonstrated that PARP inhibitors such as 3-aminobenzamide (**7**) [33], PJ34 (**8**) [34] and 3,4-dihydro-5-[4-(1-piperidinyl)butoxy]1-(2*H*)-isoquinoline (**9**) lead to significant reduction in brain damage and improvement of neurological outcome in a focal cerebral ischemia model. Several PARP inhibitors were advanced in the clinic for the treatment of cerebrovascular ischemia. Currently, the most advanced compound is INO-1001 (**10**, IC$_{50}$ < 5 nM), which is being developed by Inotek [35,36]. Results from phase I have confirmed its safety and tolerance. A phase II study with INO-1001 was initiated in late 2003 but no results have been published to date.

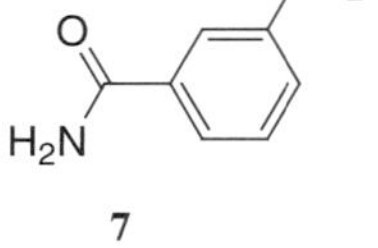

7

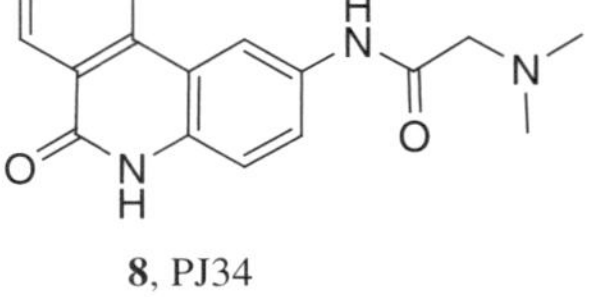

8, PJ34

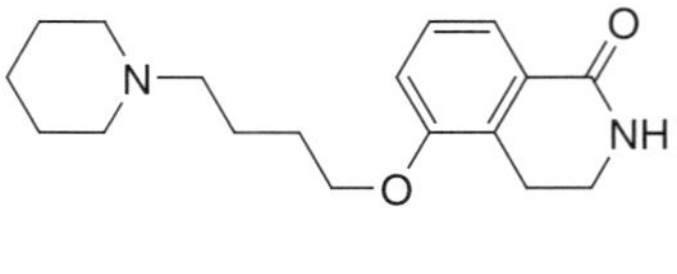

9, DPQ

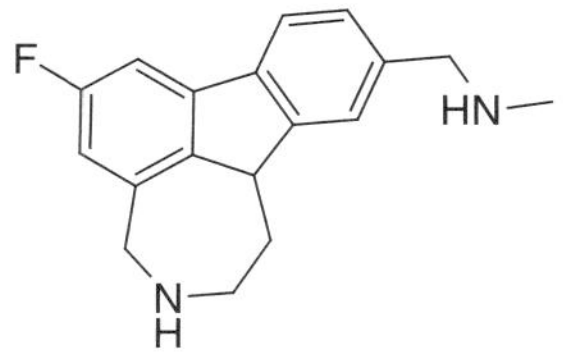

10, INO-1001

Ono pharmaceuticals has reported the development of ONO-2231, the lead from a series of PARP inhibitors (including ONO-1924H, **11**) [37] for the potential intravenous treatment of ischemic stroke. A UK phase I trial on ONO-2331 was underway by May 2005. In November 2004, the company published results showing that ONO-1924H, administered *i.v.*, led to a significant *in vivo* decrease in cerebral damage in rats with cytotoxicity induced by hydrogen peroxide.

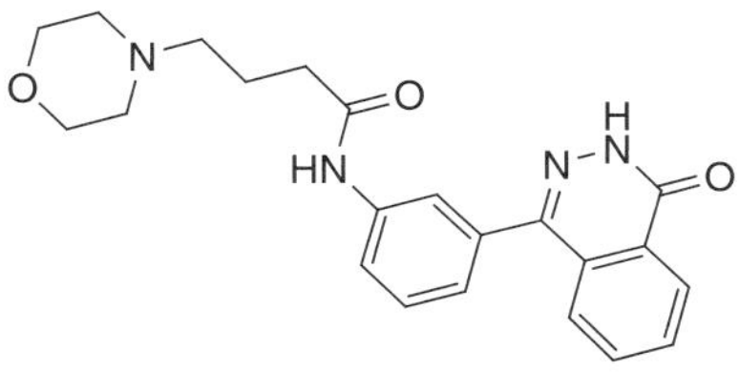

11, ONO-1924H

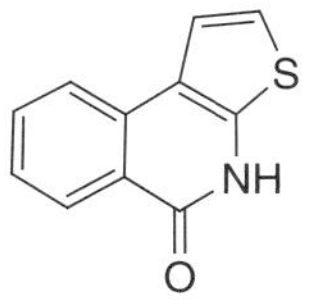

12, TIQ-A

GlaxoSmithKline, under license from the University of Florence, is investigating a series of poly (ADP-ribose) polymerase enzyme inhibitors, including TIQ-A (**12**) [38], for the potential treatment of cerebrovascular ischemia. *In vitro*, TIQ-A showed an IC_{50} value of 0.45 µM for PARP-1 inhibition and displayed significant neuroprotective activity when rat cortical cell cultures were exposed to oxygen and glucose deprivation [38]. In the transient rat model of focal ischemia, administration of TIQ-A (**12**) (3 and 10 mg/kg, *i.p.*) at 0 and 120 min after the occlusion reduced infarct volume by 40% and 70%, respectively. In a permanent occlusion model, administration of TIQ-A reduced the infarct volume by 35%, 24 h after occlusion [38,39]. Additional compounds that are being investigated for the treatment of cerebrovascular ischemia include GP-6150 (**13**), which constitutes a prototype of a series of tetracyclic compounds that inhibited PARP activity at IC_{50} values ranging from 0.046 to 5 µM. GPI 6150 is described as a powerful PARP inhibitor with an IC_{50} value of 0.15 µM (K_i value of 0.060 µM) [40]. GPI-6150, at 15 mg/kg *i.p.*, given 30 min before and 3 h after brain injury induced by fluid percussion, attenuated the size of the lesion without affecting TUNEL-positive apoptotic cells [32]. Several compounds from MGI pharma (formerly Guilford) have recently been identified as PARP inhibitors. These include compounds, such as **14**, having IC$_{50}$ values for PARP inhibition in the range of 0.131–40 µM, with one example shown to reduce infarct volume at 10 mg/kg *i.p.* in a rat model using transient middle cerebral artery occlusion [41,42].

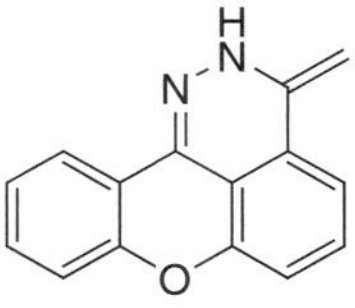

13, GP-6150

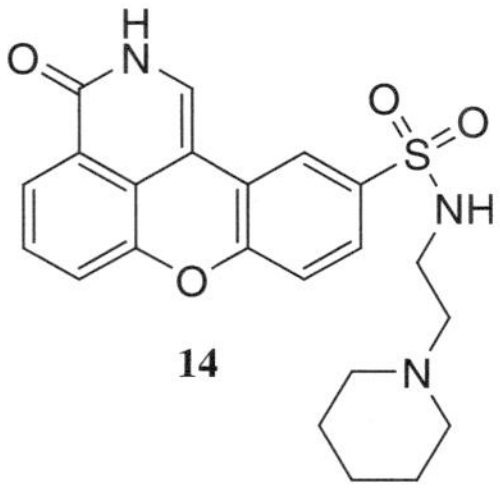

14

FR247304 (**15**) is a novel PARP-1 inhibitor that has recently been identified through structure-based drug design [43]. In an enzyme kinetic analysis, FR247304 exhibits potent and competitive inhibition of PARP-1 activity, with a K_i value of 35 nM. In cell death model, treatment with FR247304 (10 nM–10 µM) significantly reduced NAD depletion by PARP-1 inhibition and attenuated cell death after exposure to 100 µM hydrogen peroxide. After 90 min of middle cerebral artery occlusion in rats, poly(ADP-ribosyl)ation and NAD depletion were markedly increased in the cortex and striatum from 1 h after reperfusion. The increased poly(ADP-ribose) immunoreactivity and NAD depletion were attenuated by FR247304 (32 mg/kg *i.p.*) treatment and FR247304 significantly decreased ischemic brain damage measured at 24 h after reperfusion [43]. Other compounds in this class that demonstrated significant neuroprotective properties are FR257516 (**16**), FR197262 (**17**) and FR142057 (**18**) [44–48].

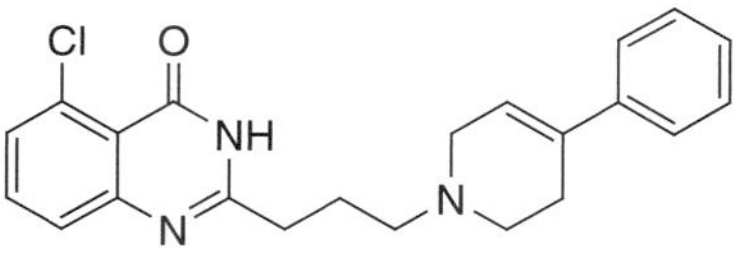

15, FR247304

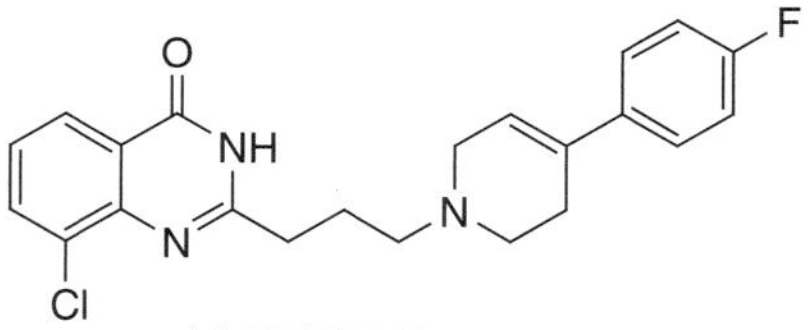

16, FR257516

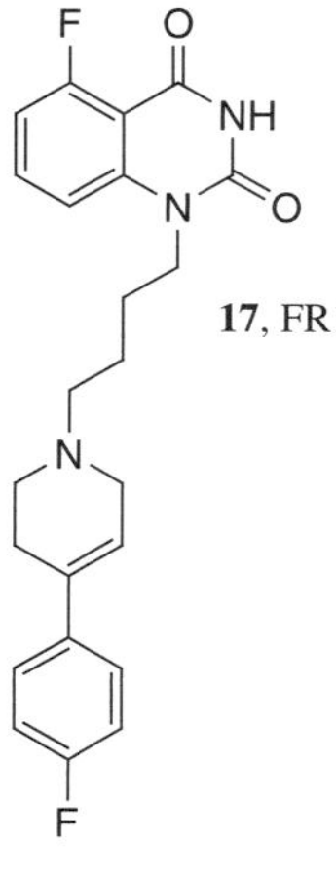

17, FR197262

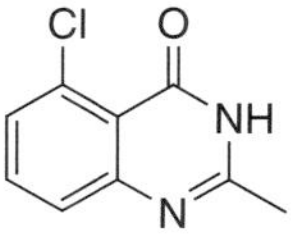

18, FR142057

Despite significant progress in the understanding of the medicinal chemistry and pharmacology of PARP, crucial issues remain to be resolved. In particular, selectivity for PARP-1 over PARP-2 is of utmost importance. Many of the reported compounds in this class are nonselective inhibitors that are potentially mutagenic due to inhibition of PARP-2 [49]. Furthermore, solubility, formulations and brain penetration are key issues that remain to be optimized.

1.3. Peroxisome proliferator-activator receptor gamma (PPARγ) ligands

Peroxisome proliferator-activated receptors (PPARs) are orphan receptors belonging to the steroids/thyroid/retinoid receptor super family of ligand-activated transcription factors. There are three PPAR isoforms (PPARα, -β, -γ), each of which is differentially expressed and displays a distinct pattern of ligand specificity [50]. PPARs are implicated in several physiological processes, such as the regulation of lipoprotein, lipid metabolism and glucose homeostasis. Recent observations indicate that PPAR activators could reduce the inflammation induced in different inflammatory pathologies including asthma, hypertensive heart disease, hepatic inflammation and cerebral ischemia [51]. *In vivo*, PPARγ agonists have been shown to modulate inflammatory responses in the brain and to reduce infract size following transient focal ischemia [52,53]. Cerebral ischemia is frequently accompanied by inflammation, which can worsen neuronal injury [54]. Activation of PPARγ reduces inflammation and the expression of nicotinamide adenine dinucleotide phosphate (NADPH) oxidase [55]. In addition, PPARγ activators increase levels of CuZn-superoxide dismutase (SOD) in cultured endothelium, suggesting an additional mechanism by which it may exert protective effects within the brain [56]. To date, a large number of structurally diverse synthetic ligands have been discovered which modulate PPARγ activity. The first compounds reported as high-affinity PPAR agonists are a class of compounds known as 'glitazones' or thiazolidinediones (TZDs). Initial compounds in this class, Troglitazone (**19**), Pioglitazone (**20**), Rosiglitazone (**21**), were subtype selective, high-affinity agonists for PPARγ [54,57,58]. Troglitazone was the first drug to be approved for treating type 2 diabetes, but was withdrawn from the market in 1999 due to hepatic toxicity. However, Pioglitazone and Rosiglitazone are now available and reportedly show no hepatic side effects [52,59,60]. The latter two drugs in this class were shown to exert protective effects against cerebral ischemia/reperfusion injury by reducing oxidative stress and inflammatory response. Both drugs were administered 30 min prior to ischemia and the rapid onset of their protective effects might suggest a potential role of PPARγ agonist in modulating the early events occurring during transitory ischemic attacks [56,61,62].

19, Troglitazone

20, Pioglitazone

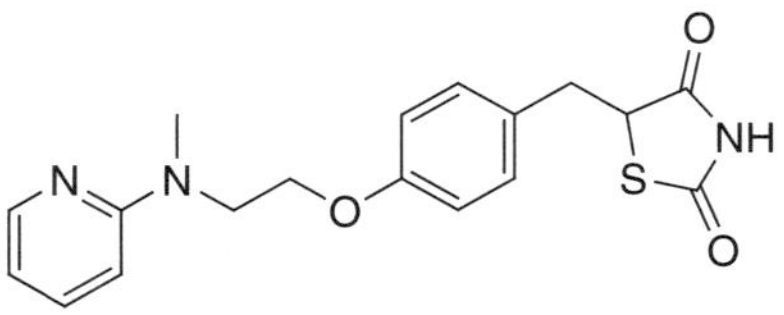

21, Rosiglitazone

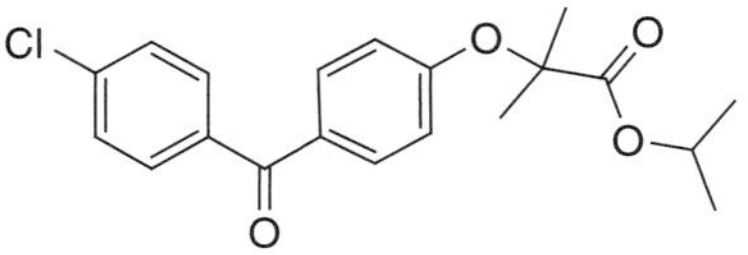

22, Fenofibrate

Recent observations indicate that PPARα activators could reduce the post-traumatic inflammation induced in acute ischemic stroke. Indeed treatment with fenofibrate (**22**), a known PPARα activator, reduces cerebral infarct size caused by cerebral ischemia [63]. Although the authors reported that 100 mg/kg fenofibrate exerted neurological recovery-promoting effect, this dosage did not reduce the brain lesion, suggesting that a high dose associated with a long-term treatment with fenofibrate may induce adverse effects on cell death. One possible interpretation of these results is that high dose of fenofibrate may exert massive anti-inflammatory effect by inhibiting cytokines (tumor necrosis factor α, interleukin-1) production. It is known that blockade of cytokines in the early phase is beneficial but their presence is also fundamental for the processes of tissue repair and regeneration. Thus a high dose of fenofibrate could counteract these beneficial effects [51,64].

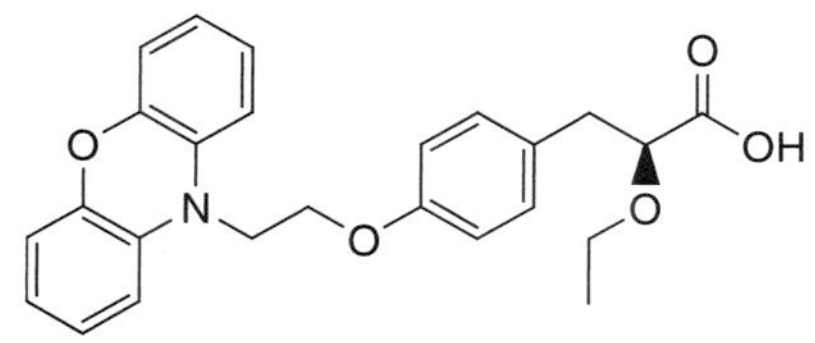

23, Ragaglitazar

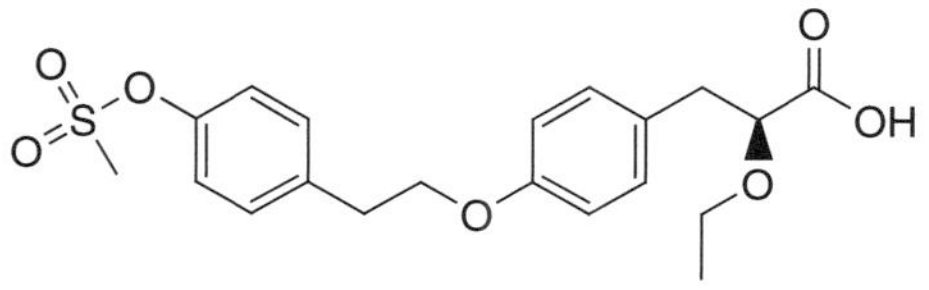

24, Tesaglitazar

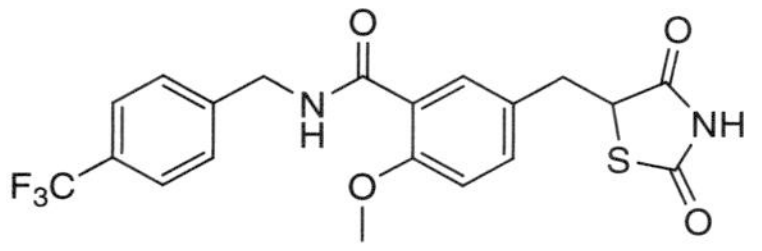

25, KRP297

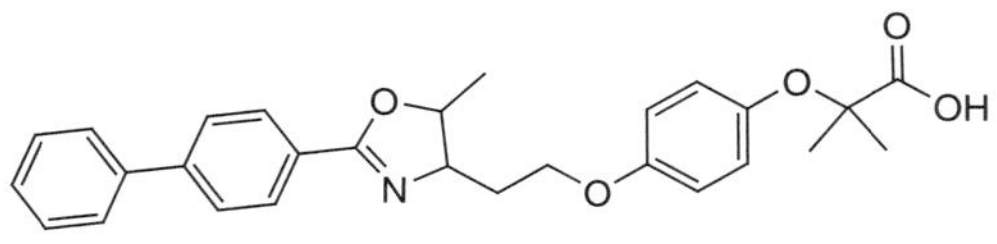

26, LY465608

Despite the clear benefits of the currently marketed PPARγ agonists, they have been implicated with increased weight gain, which is accompanied by an increase in subcutaneous fat mass. New dual-acting PPARα and -γ agonists, designed to combine the beneficial effects seen with insulin sensitizers and fibrates, have received increased attention. The dual agonists have been suggested to reduce the weight gain associated with adipogenesis resulting from the PPARγ activation. Recent reports of dual-acting agonists, such as ragaglitazar (**23**) [65], tesaglitazar (**24**) [66], KRP 297 (**25**) [67,68] and LY465608 (**26**) [68], may offer improvements to this class.

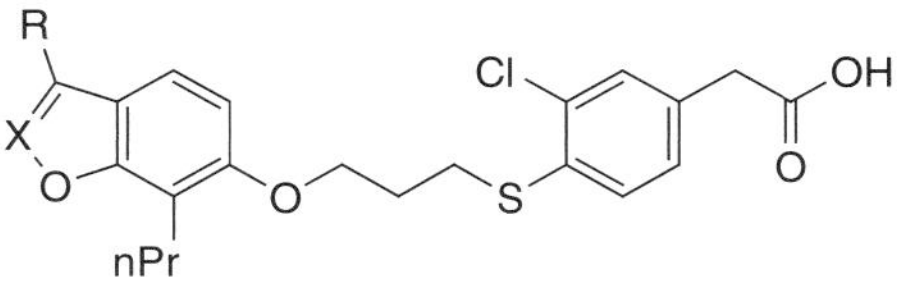

27, R=Ph, X=CH
28, R=Et, X=N

In addition to these novel reports of PPARα/γ dual agonists, a few reports have appeared highlighting ligands that are PPARα/γ/δ pan agonists. Exemplified by compounds **27** and **28**, they have reported EC_{50} values of 10–200 nM at all the human PPAR subtypes [69].

In the past few years, we have seen a substantial increase in our understanding of the effects of selective PPAR-γ ligands, particularly of the TZDs. We know today that the therapeutic effects of PPAR-γ ligands extends beyond their use as insulin-sensitizers, as a number of these compounds exert beneficial effects in conditions associated with both peripheral (not described here) and central ischemia and inflammation. The biggest hurdle currently associated with this mechanism is that all of the pre-clinical studies have administered the compounds before or concomitant with the occlusion, which, unless patients are already taking these drugs at the time of their stroke, may limit their acute use as a therapeutic agent due to a prohibitively small window of treatment opportunity.

1.4. C-Jun-N-terminal kinase inhibitors

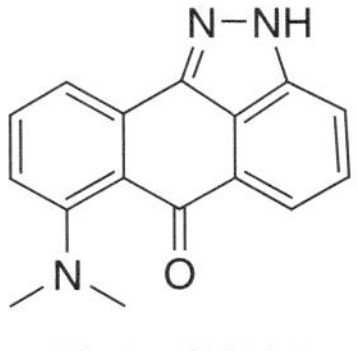

29, SP-600125

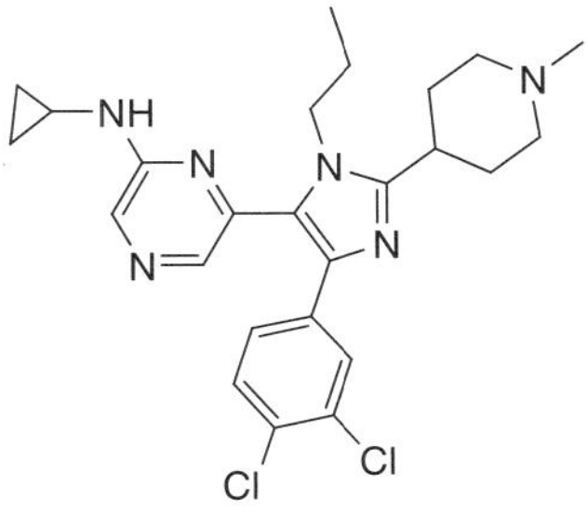

30
IC_{50} ~7 nM (JNK1, 2, 3)

JNK, a mitogen-activated kinase, has been shown to play a role in neuronal death. The JNK pathway is an important mediator of cell death. It is activated in focal cerebral ischemia in the mouse and mediates neuronal death. This is supported by the protection of JNK3 knockout mice from kainic-acid-induced seizures, apoptosis and ischemia [70,71]. Thus, JNK inhibition is a target to prevent neurotoxicity following excess excitatory amino acid release, as seen following acute ischemia [72]. Several studies have demonstrated the benefits of peptide inhibitors of JNK in experimental models of ischemic stroke, traumatic nerve injury, neurodegenerative diseases and hearing loss [72,73]. In recent published results the neuroprotective action of D-JNKI1, a cell penetrating and protease-resistance peptide selectively inhibiting the c-Jun-N-terminal kinase, was demonstrated [74]. D-JNKI1 was shown to provide neuroprotection in two models of mild focal cerebral ischemia: 30 min endoluminal ("suture") middle cerebral artery occlusion (MCAO) in the mouse and permanent distal MCAO in young rats, with extended therapeutic windows. Considerable effort is being directed to the development of inhibitors of the JNK pathway, either by inhibiting JNKs themselves or by inhibiting the activation of the JNKs [74,75]. A series of pyrazoloanthrone derivatives, exemplified by SP-600125 (**29**), are among the first reported JNK inhibitors by Celgene which demonstrate IC_{50} values of 110 nM for JNK1 and JNK2 and 150 nM for JNK3. In recent studies, SP-600125 was shown to have protected transient brain ischemia/reperfusion-induced neuronal death in rat hippocampal CA1 region by inhibiting the activation of nuclear substrate (c-Jun) and the inactivation of non-nuclear substrate (Bcl-2) induced by ischemic insult [76]. A common structural feature among small molecule JNK inhibitors is the 2-aminopyrazoleanthrone moiety found in several recent reports. Researchers at Merck described phenylimidazole derivative as JNK inhibitors and highlighted compound **30** as an example of this series with an IC_{50} value of 7 nM, but with no JNK isoform selectivity [77,78]. Eisai has claimed compound **31** featuring a fused ring reported to be potent pan-JNK inhibitors [78]. Takeda disclosed a series of compounds with a pyrido-thiazole moiety as JNK inhibitors, exemplified by **32**, which has *in vitro* IC_{50} values of 30–210 nM against JNK isoforms [79]. In a cellular assay using THP-1 cells, compounds of this class inhibited TNFα production with IC_{50} values of 2–100 nM. Serono disclosed a series of benzothiazole-substitued 2-amino-pyrimidine-based JNK inhibitors. A representative example, AS-601245 (**33**) exhibited 10–100 fold selectivity for JNK over a range of kinases, including P38, ERK1 and Src. This compound was neuroprotective in models of focal cerebral ischemia in rat and global ischemia in gerbil [80].

31
IC_{50} = 27 nM (JNK1)
IC_{50} = 13 nM (JNK2)
IC_{50} = 14 nM (JNK3)

32
IC_{50} = 97 nM (JNK1)
IC_{50} = 57 nM (TNF α production)

33, AS-601245
IC_{50} = 150 nM (JNK1)
IC_{50} = 220 nM (JNK2)
IC_{50} = 70 nM (JNK3)

Eisai has described a variety of 1*H*-indazole derivatives as JNK inhibitors, with IC$_{50}$ values of approximately 50–400 nM against JNK3. Some of these compounds have good selectivity against ERK2, such as compound **34** [81]. More recently, a series of compounds featuring a 7-azaindole core, such as compound **35**, was shown to inhibit JNK3 with and IC$_{50}$ value of 520 nM [82].

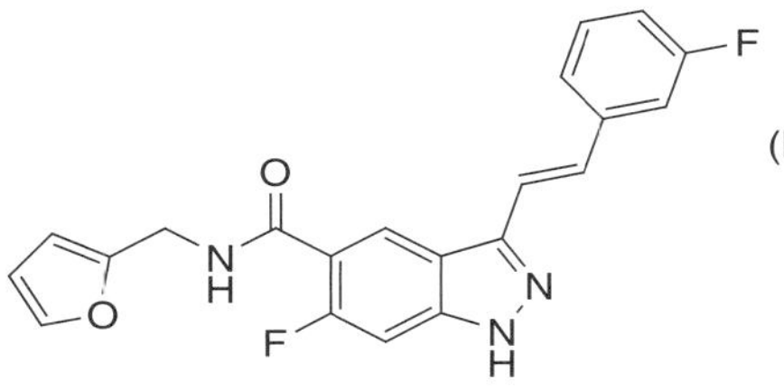

34

IC$_{50}$ = 55 nM (JNK3)
IC$_{50}$ = 22 µM (ERK2)

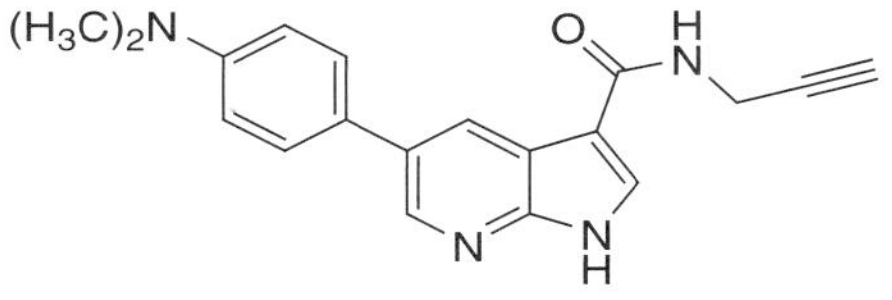

35

IC$_{50}$ = 520 nM (JNK1)

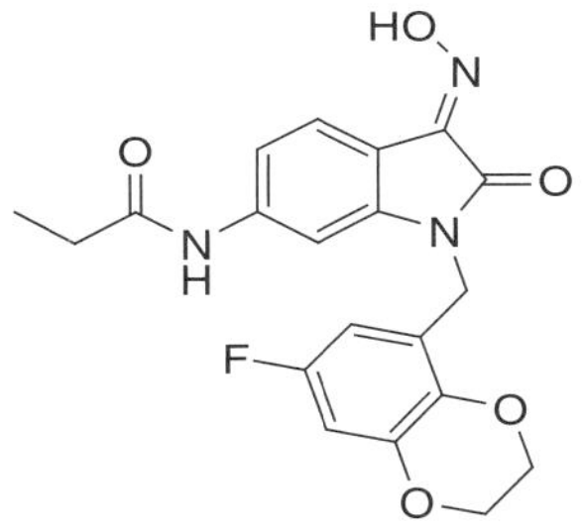

36

IC$_{50}$ < 100 nM (JNK3)

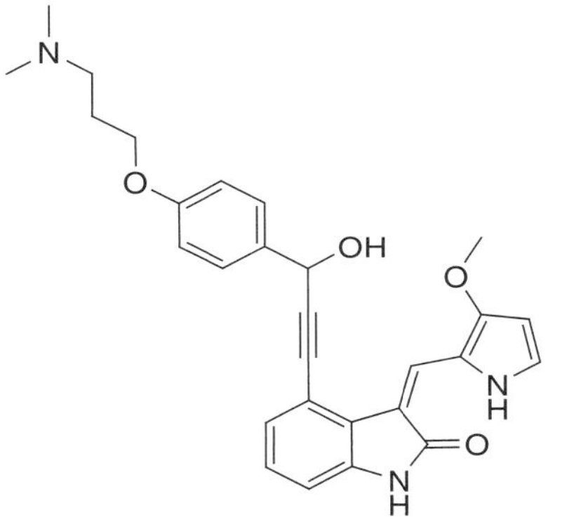

37

IC$_{50}$ = 34 nM (JNK1)

Oxindole-based compounds have also been shown to be potent JNK inhibitors. Vertex reported a series of 3-oximido-3-oxindole analogs for the treatment of stroke and neurodegenerative diseases. A number of compounds containing benzo-1,3-dioxolane groups, such as compound **36** inhibited JNK3 with IC$_{50}$ values of <1 µM. Alkynyl analog **37**, which is also representative of this class and reported by Hoffmann La Roche, inhibited JNK with an IC$_{50}$ value of 34 nM. In summary, there is mounting evidence that inhibition of JNK3 may represent a new and effective strategy to treat ischemic stroke [83,84]. However, one of the lingering issues is how much JNK isoform selectivity is required for drug discovery to alleviate any risk from adverse side effects, including a compromised cellular immune response and cellular hyperproliferation. It will be of great interest to see how many of these JNK inhibitors progress to clinical trials for the treatment of cerebral ischemia.

1.5. Src kinase inhibitors

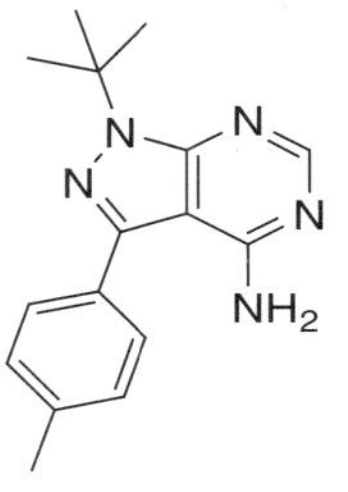

38, PP1

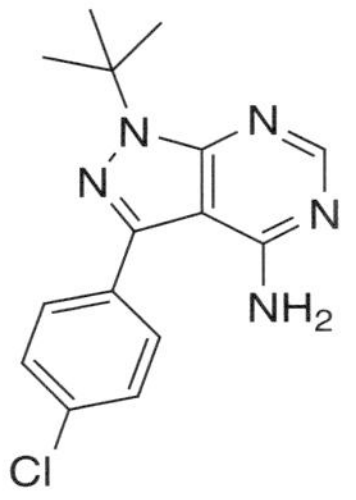

39, PP2

Src-family protein tyrosine kinases (SFKs) are important signaling enzymes controlling cell growth, proliferation and migration. In the brain, SFKs have miscellaneous physiological roles. Accumulating evidence suggests that Src controls the activity of the N-methyl-D-aspartate (NMDA) receptor, which is of relevance to neuronal long-term potentiation [85,86]. In cerebral ischemia, the role of SFKs is unclear [87]. It has been proposed that vasogenic edema induced by vascular endothelial growth factor (VEGF) can aggravate the ischemic brain injury [88]. Based on findings that Src is required for VEGF-induced vascular leakage, it is suggested that Src kinases mediate ischemic injury through this specific mechanism [89,90].

The pyrazolo-pyrimidinyl-amine compounds PP1 (**38**) and PP2 (**39**) were identified as Src kinase inhibitors, which can preferentially inhibit SFKs with nM IC_{50} values. They have been used extensively as inhibitors to study the cellular signaling of SFKs. Researchers at the Scripps Research Institute have shown that PP1- suppressed vascular permeability and protected wild-type mice from ischemia-induced brain damage without influencing VEGF expression, when given up to 6 h following stroke [91,92]. This suggests that Src might be a therapeutic target for cerebral ischemia. Several companies have described various classes of SFK inhibitors. It has been reported that certain 4-anilino-3-cyanoquinoline derivatives, such as compound **40** (SKI-606), are potent Src kinase inhibitors [93–95]. Most recently SKI-606, which inhibits c-Src with an IC_{50} value of 1.2 nM and is a dual Src/Abl kinase inhibitors, has reached phase I trials for solid tumors and, recently, SKI-606 was also shown to reduce edema and functional deficits in a rat intracerebral hemorrhage model [96].

Numerous reports in the literature show that Src kinases inhibitors reduce edema and necrosis of coronary tissue resulting from occlusion of coronary vasculature. Additional SFK inhibitors include macrocyclic dienone derivatives such as herbimycin A (**41**), geldanamycin (**42**), radicicol R3246 (**43**); the pyrido[2,3-d]pyrimidine derivatives such as PD-173955 (**44**).

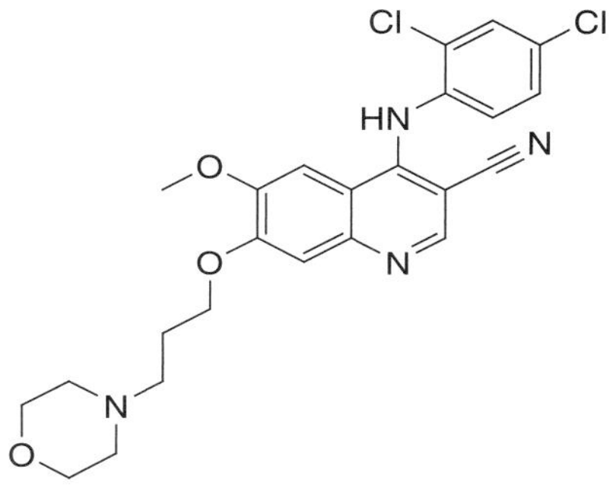

40, SKI-606

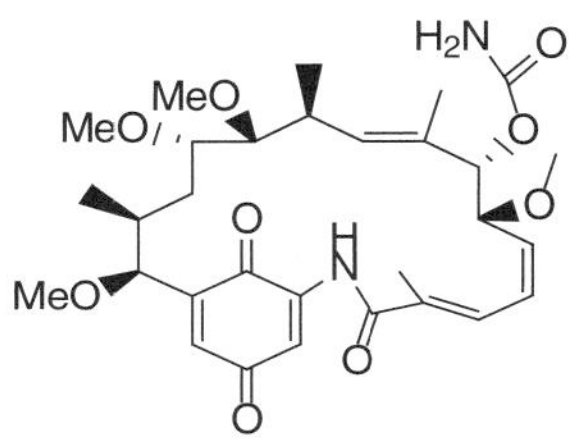

41

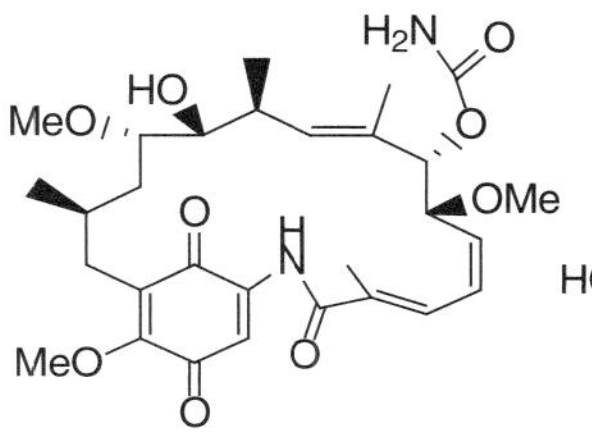

42

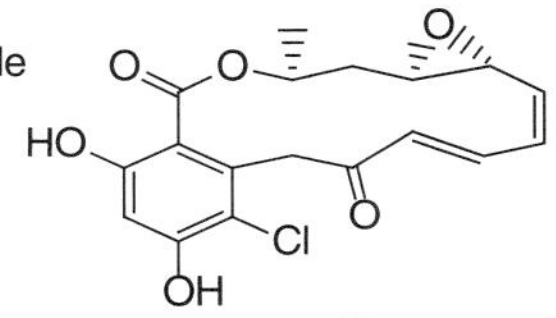

43

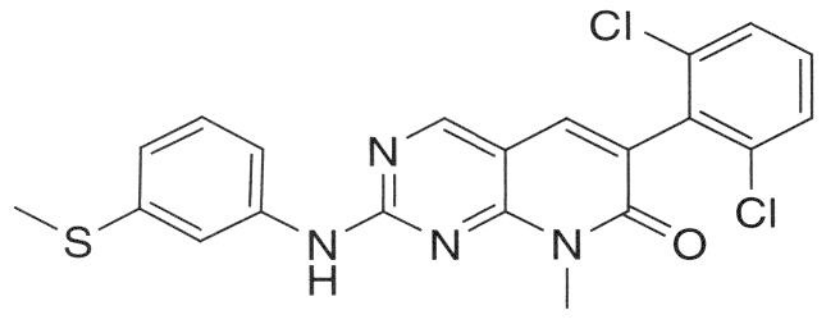

44

Given that Src kinase inhibitors may affect multiple pathways invoked following ischemic stroke resulting in reduced edema, improved cerebral perfusion, decreased infarct volume and improved neurological outcome in a therapeutically relevant time frame making this an exciting novel therapeutic target in the pathophysiology of cerebral ischemia [88,97].

1.6. Estrogens

There is a growing interest in the actions of estrogens as neuroprotectants against neurodegenerative diseases and acute brain damage caused by stroke. Estrogens exert their neuroprotective effects predominantly through antioxidant effects of steroids and attenuation of NMDA-receptor activation [98,99]. In *in vivo* studies, the neuroprotective effects of estrogens have been demonstrated in all acute cerebral ischemia scenarios and it has been shown that estrogens could have a longer therapeutic window that extends up to 6 hours after ischemic insult [100,101]. Evidence from both *in vitro* and *in vivo* studies suggested that the neuroprotective effects of estrogens do not require ER-dependent gene transcription. 17β-estradiol (**45**) can preserve mitochondrial function, cell viability and ATP levels in human lens cells during oxidative stress [102]. Several other synthesized estrogen analogs have also been reported to possess neuroprotective properties. Structure-activity relationship studies indicate that the most essential structural motif that elicits neuroprotective functions is the phenolic ring of the steroid. Furthermore, additional studies improving the potency at the receptor site did not afford improved neuroprotection. Although ERs have been found throughout the CNS, no phenotypic change in CNS has been found in both ERα and ERβ knockout mice, suggesting that rapid onset of neuroprotective action induced by estrogens are unlikely mediated through genomic mechanisms. The neuroprotective actions against stroke have also been seen with selective ER modulators such as tamoxifen (**46**) [103] and LY353381 (**47**) [104].

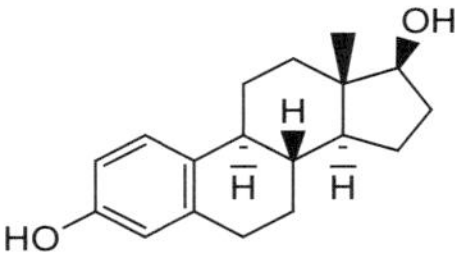

45, 17β-estradiol

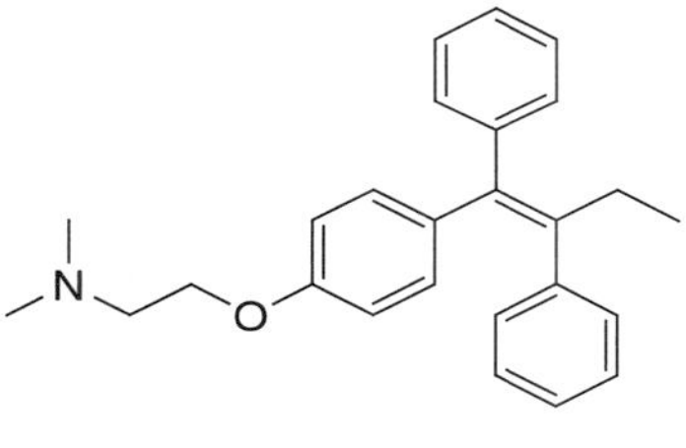

46, tamoxifen

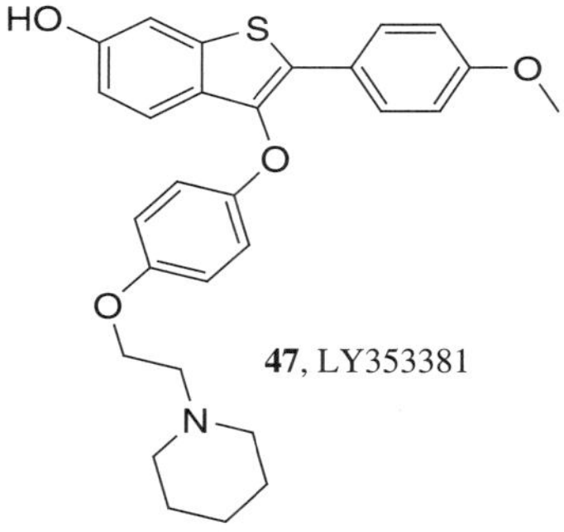

47, LY353381

Furthermore, neuroprotective actions of other nonfeminizing estrogen analogs such as 17α-estradiol (**48**) [105] and adamantyl estrogen analog (**49**) [106] have also been demonstrated in stroke models. The actions of these nonreceptor binding estrogen analogs indicated that ERs are not required for the neuroprotective effects of estrogens. Neuroprotection by estrogen holds great promise for improving the clinical management of neural trauma.

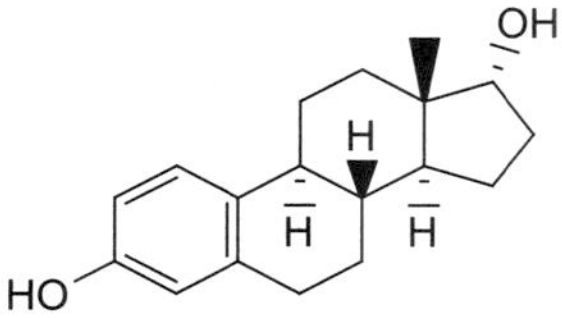

48, 17α-estradiol

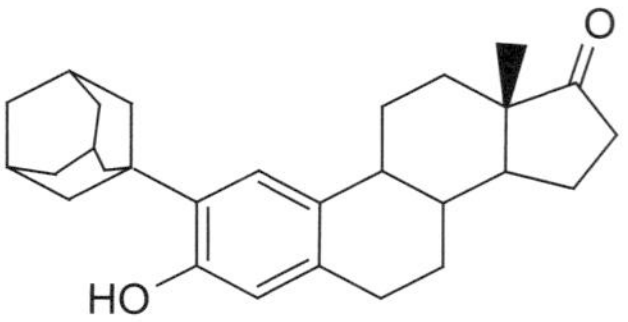

49

2. CONCLUSION

The failure of drugs as neuroprotective agents for acute stroke may be attributed to a number of factors including toxicity, poor tolerability, lack of efficacy and, perhaps, most significantly limitations in study design. The latter factor has been addressed by recent STAIR recommendations. Regarding efficacy, it could be noted that a number of the drugs tested to date, such as sodium channel blockers [107], potassium channel activators [108], glycine antagonists [109] or opioid antagonists [110] may target a single, specific component of the ischemic cascade. This suggests that any selective drug that targets a specific event in the ischemic cascade may not have a large clinical impact. Protection of neurons during ischemia/reperfusion is likely to require a polypharmacy approach or use of a compound that has pleiotropic actions affecting multiple neurotoxic processes, such as those reviewed above.

REFERENCES

[1] C. Leary Megan and L. Saver Jeffrey, *Cerebrovas. Diseases (Basel, Switzerland)*, 2003, **16**, 280.
[2] T. Thom, N. Haase, W. Rosamond, V. J. Howard, J. Rumsfeld, T. Manolio, Z.-J. Zheng, K. Flegal, C. O'Donnell, S. Kittner, D. Lloyd-Jones, D. C. Goff, Jr., Y. Hong, Members of the Statistics Committee and Stroke Statistics, R. Adams, G. Friday, K. Furie, P. Gorelick, B. Kissela, J. Marler, J. Meigs, V. Roger, S. Sidney, P. Sorlie, J. Steinberger, S. Wasserthiel-Smoller, M. Wilson and P. Wolf, *Circulation* **113** (2006) e85.
[3] S. D. Reed, S. C. Cramer, D. K. Blough, K. Meyer, J. G. Jarvik and D. Z. Wang, *Stroke*, 2001, **32**, 1832.
[4] L. Hinkle Janice and L. Bowman, *J. Neurosci. Nurs.: J Am. Asso. Neurosc. Nurses*, 2003, **35**, 114.
[5] M. D. Ginsberg, *Stroke*, 2003, **34**, 214.
[6] B. Ovbiagele, S. Kidwell Chelsea, S. Starkman and L. Saver Jeffrey, *Curr. Neurol. Neurosci. Rep.*, 2003, **3**, 9.
[7] M. D. N. Kiran and S. Panickar, *Glia*, 2005, **50**, 287.
[8] D. A. Greenberg and K. Jin, *Nature*, 2005, **438**, 954.
[9] R. J. Lichtenwalner and J. M. Parent **26** (2005) 1.
[10] K. Muir and P. Teal, *J. Neurol.*, 2005, **252**, 1011.
[11] C. S. Kidwell, D. S. Liebeskind, S. Starkman and J. L. Saver, *Stroke*, 2001, **32**, 1349.
[12] M. Fisher, *Stroke*, 2003, **34**, 1539.
[13] M. Fisher, for the Stroke Therapy Academic Industry Roundtable, *Stroke*, 2005, **36**, 1808.
[14] A. R. Green and T. Ashwood, *Curr. Drug Targets: CNS Neurol. Disord.*, 2005, **4**, 109.
[15] A. R. Green, T. Ashwood, T. Odergren and D. M. Jackson, *Pharmaco. Ther.*, 2003, **100**, 195.
[16] K. R. Lees, J. A. Zivin, T. Ashwood, A. Davalos, S. M. Davis, H.-C. Diener, J. Grotta, P. Lyden, A. Shuaib and H.-G. Hardemark, W. W. Wasiewski for the Stroke–Acute Ischemic NXYTTI, *N. Engl. J. Med.*, 2006, **354**, 588.
[17] J. J. Legos, D. Lee and J. A. Erhardt, *Expert Opin. Emerging Drugs*, 2001, **6**, 81.

[18] V. Y. Wong, P. M. Keller, M. E. Nuttall, K. Kikly, W. E. DeWolf, Jr., D. Lee, S. M. Ali, D. P. Nadeau, E. T. Grygielko, N. J. Laping and D. P. Brooks, *Eur. J. Pharmacol.*, 2001, **433**, 135.

[19] E. L. Grimm, R. Aspiotis, C. Bayly, M. Garcia-Calvo, A. Giroux, Y. Han, D. McKay, D. Nicholson, E. Peterson, D. Rasper, J. Renaud, S. Roy, J. Tam, P. Tawa, N. Thornberry, J. Vaillancourt, R. Zamboni and S. Xanthoudakis, *Abstracts of Papers, 220th ACS National Meeting, Washington, DC, United States, August 20–24, 2000*, MEDI.

[20] I. C. Choong, W. Lew, D. Lee, P. Pham, M. T. Burdett, J. W. Lam, C. Wiesmann, T. N. Luong, B. Fahr, W. L. DeLano, R. S. McDowell, D. A. Allen, D. A. Erlanson, E. M. Gordon and T. O'Brien, *J. Med. Chem.*, 2002, **45**, 5005.

[21] H. Han Byung, D. Xu, J. Choi, Y. Han, S. Xanthoudakis, S. Roy, J. Tam, J. Vaillancourt, J. Colucci, R. Siman, A. Giroux, S. Robertson George, R. Zamboni, W. Nicholson Donald and M. Holtzman David, *J. Biol. Chem.*, 2002, **277**, 30128.

[22] R. S. Hotchkiss, K. C. Chang, P. E. Swanson, K. W. Tinsley, J. J. Hui, P. Klender, S. Xanthoudakis, S. Roy, C. Black, E. Grimm, R. Aspiotis, Y. Han, D. W. Nicholson and I. E. Karl, *Nat. Immunol.*, 2000, **1**, 496.

[23] J. W. Becker, J. Rotonda, S. M. Soisson, R. Aspiotis, C. Bayly, S. Francoeur, M. Gallant, M. Garcia-Calvo, A. Giroux, E. Grimm, Y. Han, D. McKay, D. W. Nicholson, E. Peterson, J. Renaud, S. Roy, N. Thornberry and R. Zamboni, *J. Med. Chem.*, 2004, **47**, 2466.

[24] J. G. Chapman, W. P. Magee, H. A. Stukenbrok, G. E. Beckius, A. J. Milici and W. R. Tracey, *Eur. J. Pharmacol.*, 2002, **456**, 59.

[25] D. Lee, S. A. Long, J. H. Murray, J. L. Adams, M. E. Nuttall, D. P. Nadeau, K. Kikly, J. D. Winkler, C. M. Sung, M. D. Ryan, M. A. Levy, P. M. Keller and W. E. DeWolf, Jr., *J. Med. Chem.*, 2001, **44**, 2015.

[26] D. V. Kravchenko, V. M. Kysil, S. E. Tkachenko, S. Maliarchouk, I. M. Okun and A. V. Ivachtchenko, *Farmaco*, 2005, **60**, 804.

[27] D. V. Kravchenko, Y. A. Kuzovkova, V. M. Kysil, S. E. Tkachenko, S. Maliarchouk, I. M. Okun, K. V. Balakin and A. V. Ivachtchenko, *J. Med. Chem.*, 2005, **48**, 3680.

[28] D. V. Kravchenko, V. V. Kysil, A. P. Ilyn, S. E. Tkachenko, S. Maliarchouk, I. M. Okun and A. V. Ivachtchenko, *Bioorg. Med. Chem. Lett.*, 2005, **15**, 1841.

[29] D. Kravchenko, Y. A. Kuzovkova, V. Kysil, S. Tkachenko, S. Maliartchouk, I. Okun and A. Ivachtchenko, *Abstracts of Papers, 229th ACS National Meeting, San Diego, CA, United States, March 13–17, 2005*, MEDI.

[30] L. Virag and C. Szabo, *Pharmacol. Rev.*, 2002, **54**, 375.

[31] Y. Zhang, X. Zhang, T. S. Park and J. M. Gidday, *J. Cereb. Blood Flow Metab.*, 2005, **25**, 868.

[32] J. H. Greenberg, M. C. LaPlaca and T. K. McIntosh, *PARP as a Therapeutic Target*, 2002, **83**.

[33] J. Y. Couturier, D.-Z. Li, N. Croci, M. Plotkine and I. Margaill, *Exp. Neurol.*, 2003, **184**, 973.

[34] G. S. Scott, R. B. Kean, T. Mikheeva, M. J. Fabis, J. G. Mabley, C. Szabo and D. C. Hooper, *J. Pharmacol. Exp. Ther.*, 2004, **310**, 1053.

[35] V. C. Besson, Z. Zsengeller, M. Plotkine, C. Szabo and C. Marchand-Verrecchia, *Brain Res*, 2005, **1041**, 149.

[36] G. Szabo, P. Soos, S. Mandera, U. Heger, C. Flechtenmacher, S. Baehrle, L. Seres, A. Cziraki, A. Gries, Z. Zsengeller, C. F. Vahl, S. Hagl and C. Szabo, *Shock*, 2004, **21**, 426.

[37] Y. Kamanaka, K. Kondo, Y. Ikeda, W. Kamoshima, T. Kitajima, Y. Suzuki, Y. Nakamura and K. Umemura, *Life Sci*, 2004, **76**, 151.
[38] A. Chiarugi, E. Meli, M. Calvani, R. Picca, R. Baronti, E. Camaioni, G. Costantino, M. Marinozzi, D. E. Pellegrini-Giampietro, R. Pellicciari and F. Moroni, *J. Pharmacol. Exp. Ther.*, 2003, **305**, 943.
[39] A. Chiarugi, *Pharmacol. Res.*, 2005, **52**, 15.
[40] M. C. LaPlaca, J. Zhang, R. Raghupathi, J. H. Li, F. Smith, F. M. Bareyre, S. H. Snyder, D. I. Graham and T. K. McIntosh, *J. Neurotrauma*, 2001, **18**, 369.
[41] P. F. Jackson, J.-H. Li, K. M. Maclin and J. Zhang, US6635642, 2003.
[42] P. F. Jackson, J.-H. Li, K. M. Maclin and J. Zhang, WO991164, 1999.
[43] A. Iwashita, N. Tojo, S. Matsuura, S. Yamazaki, K. Kamijo, J. Ishida, H. Yamamoto, K. Hattori, N. Matsuoka and S. Mutoh, *J. Pharmacol. Exp. Ther.*, 2004, **310**, 425.
[44] A. Iwashita, K. Mihara, S. Yamazaki, S. Matsuura, J. Ishida, H. Yamamoto, K. Hattori, N. Matsuoka and S. Mutoh, *J. Pharmacol. Exp. Ther.*, 2004, **310**, 1114.
[45] K. Hattori, Y. Kido, H. Yamamoto, J. Ishida, K. Kamijo, K. Murano, M. Ohkubo, T. Kinoshita, A. Iwashita, K. Mihara, S. Yamazaki, N. Matsuoka, Y. Teramura and H. Miyake, *J. Med. Chem.*, 2004, **47**, 4151.
[46] A. Iwashita, S. Yamazaki, K. Mihara, K. Hattori, H. Yamamoto, J. Ishida, N. Matsuoka and S. Mutoh, *J. Pharmacol. Exp. Ther.*, 2004, **309**, 1067.
[47] H. Yamamoto, K. Mukoyoshi and K. Hattori, WO2003080581, 2003.
[48] J. Ishida, K. Hattori, Y. Kido, and H. Yamamoto, WO2003063874, 2003.
[49] W.-M. Tong, U. Cortes and Z.-Q. Wang, *Biochim. Biophys. Acta, Rev. Cancer*, 2001, **1552**, 27.
[50] H. E. Xu, M. H. Lambert, V. G. Montana, K. D. Plunket, L. B. Moore, J. L. Collins, J. A. Oplinger, S. A. Kliewer, R. T. Gampe, Jr., D. D. McKee, J. T. Moore and T. M. Willson, *Proc. Nat. Acad. Sci. USA*, 2001, **98**, 13919.
[51] D. Deplanque, P. Gele, O. Petrault, I. Six, C. Furman, M. Bouly, S. Nion, B. Dupuis, D. Leys, J.-C. Fruchart, R. Cecchelli, B. Staels, P. Duriez and R. Bordet, *J. Neurosci.*, 2003, **23**, 6264.
[52] M. Collino, M. Aragno, R. Mastrocola, M. Gallicchio, A. C. Rosa, C. Dianzani, O. Danni, C. Thiemermann and R. Fantozzi, *Eur. J. Pharmacol.*, 2006, **530**, 70.
[53] R. Darteil, B. K. Caumont and J. Najib, FR2850969, 2004.
[54] S. Sundararajan, J. L. Gamboa, N. A. Victor, E. W. Wanderi, W. D. Lust and G. E. Landreth, *Neuroscience (Oxford, UK)*, 2005, **130**, 685.
[55] I. Inoue, S.-I. Goto, T. Matsunaga, T. Nakajima, T. Awata, S. Hokari, T. Komoda and S. Katayama, *Metab. Clin. Exp.*, 2001, **50**, 3.
[56] T. Shimazu, I. Inoue, N. Araki, Y. Asano, M. Sawada, D. Furuya, H. Nagoya and J. H. Greenberg, *Stroke*, 2005, **36**, 353.
[57] S. Uryu, J. Harada, M. Hisamoto and T. Oda, *Brain Res*, 2002, **924**, 229.
[58] P. Aoun, D. G. Watson and J. W. Simpkins, *Eur. J. Pharmacol.*, 2003, **472**, 65.
[59] B. Schutz, J. Reimann, L. Dumitrescu-Ozimek, K. Kappes-Horn, E. Landreth Gary, B. Schurmann, A. Zimmer and T. Heneka Michael, *Neurosci.: Off. J. Soc. Neurosc.*, 2005, **25**, 7805.
[60] Y. Zhao, A. Patzer, P. Gohlke, T. Herdegen and J. Culman, *Eur. J. Neurosci.*, 2005, **22**, 278.
[61] B. Schuetz, J. Reimann, L. Dumitrescu-Ozimek, K. Kappes-Horn, G. E. Landreth, B. Schuermann, A. Zimmer and M. T. Heneka, *J. Neurosci.*, 2005, **25**, 7805.
[62] M. Kiaei, K. Kipiani, J. Chen, Y. Calingasan Noel and M. F. Beal, *Exp. Neurol.*, 2005, **191**, 331.

[63] H. Inoue, X.-F. Jiang, T. Katayama, S. Osada, K. Umesono and S. Namura, *Neurosci. Lett.*, 2003, **352**, 203.
[64] V. C. Besson, X. R. Chen, M. Plotkine and C. Marchand-Verrecchia, *Neurosci. Lett.*, 2005, **388**, 7.
[65] G. R. Madhavan, R. Chakrabarti, S. K. B. Kumar, P. Misra, R. N. V. S. Mamidi, V. Balraju, K. Kasiram, R. K. Babu, J. Suresh, B. B. Lohray, V. B. Lohrayb, J. Iqbal and R. Rajagopalan, *Eur. J. Med. Chem.*, 2001, **36**, 627.
[66] B. B. Lohray, V. B. Lohray, A. C. Bajji, S. Kalchar, R. R. Poondra, S. Padakanti, R. Chakrabarti, R. K. Vikramadithyan, P. Misra, S. Juluri, N. V. Mamidi and R. Rajagopalan, *J. Med. Chem.*, 2001, **44**, 2675.
[67] M. Nomura, S. Kinoshita, H. Satoh, T. Maeda, K. Murakami, M. Tsunoda, H. Miyachi and K. Awano, *Bioorg. Med. Chem. Lett.*, 1999, **9**, 533.
[68] D. A. Brooks, G. J. Etgen, C. J. Rito, A. J. Shuker, S. J. Dominianni, A. M. Warshawsky, R. Ardecky, J. R. Paterniti, J. Tyhonas, D. S. Karanewsky, R. F. Kauffman, C. L. Broderick, B. A. Oldham, C. Montrose-Rafizadeh, L. L. Winneroski, M. M. Faul and J. R. McCarthy, *J. Med. Chem.*, 2001, **44**, 2061.
[69] A. B. Jones, *Med. Res. Rev.*, 2001, **21**, 540.
[70] T. Borsello, P. G. H. Clarke, L. Hirt, A. Vercelli, M. Repici, D. F. Schorderet, J. Bogousslavsky and C. Bonny, *Nat. Med. (New York, NY)*, 2003, **9**, 1180.
[71] C.-y. Kuan, A. J. Whitmarsh, D. D. Yang, G. Liao, A. J. Schloemer, C. Dong, J. Bao, K. J. Banasiak, G. G. Haddad, R. A. Flavell, R. J. Davis and P. Rakic, *Proc. Nat Acad. Sci. USA*, 2003, **100**, 15184.
[72] C. Bonny, T. Borsello and A. Zine, *Rev. Neurosci.*, 2005, **16**, 57.
[73] D. C. Wu, W. Ye, X. M. Che and G. Y. Yang, *J. Cereb. Blood Flow Metab.: off. J. Int. Soc. Cereb. Blood Flow Metab.*, 2000, **20**, 1320.
[74] C.-Y. Kuan and R. E. Burke, *Curr. Drug Targets: CNS Neurol. Disord.*, 2005, **4**, 63.
[75] L. Hirt, J. Badaut, J. Thevenet, C. Granziera, L. Regli, F. Maurer, C. Bonny and J. Bogousslavsky, *Stroke*, 2004, **35**, 1738.
[76] G. Scapin, S. B. Patel, J. Lisnock, J. W. Becker and P. V. LoGrasso, *Chem. Biol.*, 2003, **10**, 705.
[77] P. Lograsso, J.-M. Lisnock-Geissler, S. Xanthoudakis, J. Tam, J. G. Bilsland, S. J. Harper and L. Young, WO2001091749, 2001.
[78] P. Graczyk, H. Numata, A. Khan, V. Palmer, D. P. Medland, H. Oinuma and G. Bhatia, WO2002081475, 2002.
[79] S. Ohkawa, K. Naruo, S. Miwatashi, H. Kimura and T. Kawamoto, WO2002062792, 2002.
[80] S. Carboni, A. Hiver, C. Szyndralewiez, P. Gaillard, J.-P. Gotteland and P.-A. Vitte, *J. Pharmacol. Exp. Ther.*, 2004, **310**, 25.
[81] N. Ohi, N. Sato, M. Soejima, T. Doko, T. Terauchi, Y. Naoe and T. Motoki, WO2003101968, 2003.
[82] P. Graczyk, A. Khan, G. Bhatia and Y. Iimura, WO2004078756, 2004.
[83] W. L. Corbett, K.-C. Luk and P. E. Mahaney, WO20000359909, 2000.
[84] K.-C. Luk, P. E. Mahaney and S. G. Mischke, WO2000035906, 2000.
[85] X. M. Yu and M. W. Salter, *Proc. Nat. Acad. Sci. USA*, 1999, **96**, 7697.
[86] X. M. Yu and M. W. Salter, *Nature*, 1998, **396**, 469.
[87] X. M. Yu, R. Askalan and G. J. Keil, II. and M. W. Salter,, *Science*, 1997, **275**, 674.
[88] R. Paul, Z. G. Zhang, B. P. Eliceiri, Q. Jiang, A. D. Boccia, R. L. Zhang, M. Chopp and D. A. Cheresh, *Nat. Med. (New York)*, 2001, **7**, 222.
[89] D. A. Cheresh, R. Paul and B. Eliceiri, US2003130209, 2004.
[90] D. A. Cheresh, R. Paul and B. Eliceiri, US2004214836, 2004.

[91] R. Karni, S. Mizrachi, E. Reiss-Sklan, A. Gazit, O. Livnah and A. Levitzki, *FEBS Lett.*, 2003, **537**, 47.
[92] M. Warmuth, N. Simon, O. Mitina, R. Mathes, D. Fabbro, P. W. Manley, E. Buchdunger, K. Forster, I. Moarefi and M. Hallek, *Blood*, 2003, **101**, 664.
[93] D. H. Boschelli, B. Wu, A. C. B. Sosa, H. Durutlic, J. J. Chen, Y. Wang, J. M. Golas, J. Lucas and F. Boschelli, *J. Med. Chem.*, 2005, **48**, 3891.
[94] D. H. Boschelli, *Med. Chem. Rev. – Online*, 2004, **1**, 457.
[95] D. W. Losordo, WO2004032709, 2004.
[96] D. M. Berger, M. Dutia, G. Birnberg, D. Powell, D. H. Boschelli, Y. D. Wang, M. Ravi, D. Yaczko, J. Golas, J. Lucas and F. Boschelli, *J. Med. Chem.*, 2005, **48**, 5909.
[97] L. S, C. Y, and Z. MM, 33rd *Annual Society for Neuroscience Conference*, 2003.
[98] P. S. Green, S. H. Yang and J. W. Simpkins, *Novartis Found. Symp.*, 2000, **230**, 202.
[99] Y. Wen, S. Yang, R. Liu, E. Perez, K. D. Yi, P. Koulen and J. W. Simpkins, *Brain Res.*, 2004, **1008**, 147.
[100] L. M. Garcia-Segura, I. Azcoitia and L. L. DonCarlos, *Prog. Neurobiol. (Oxford)*, 2000, **63**, 29.
[101] J. Wang, P. S. Green and J. W. Simpkins, *J. Neurochem.*, 2001, **77**, 804.
[102] R. Liu, Y. Wen, E. Perez, X. Wang, A. L. Day, J. W. Simpkins and S.-H. Yang, *Brain Res*, 2005, **1060**, 55.
[103] K. Osuka, P. J. Feustel, A. A. Mongin, B. I. Tranmer and H. K. Kimelberg, *J. Neurochem.*, 2001, **76**, 1842.
[104] M. I. Rossberg, S. J. Murphy, R. J. Traystman and P. D. Hurn, *Stroke*, 2000, **31**, 3041.
[105] P. S. Green, S. H. Yang, K. R. Nilsson, A. S. Kumar, D. F. Covey and J. W. Simpkins, *Endocrinology*, 2001, **142**, 400.
[106] R. Liu, S.-H. Yang, E. Perez, K. D. Yi, S. S. Wu, K. Eberst, L. Prokai, K. Prokai-Tatrai, Z. Y. Cai, D. F. Covey, A. L. Day and J. W. Simpkins, *Stroke*, 2002, **33**, 2485.
[107] H. C. Diener, M. Cortens, G. Ford, J. Grotta, W. Hacke, M. Kaste, P. J. Koudstaal and T. Wessel, *Stroke*, 2000, **31**, 2543.
[108] B. S. Jensen, *CNS Drug Rev*, 2002, **8**, 353.
[109] R. L. Sacco, J. T. DeRosa, E. C. Haley, Jr., B. Levin, P. Ordronneau, S. J. Phillips, T. Rundek, R. G. Snipes and J. L. P. Thompson, *JAMA*, 2001, **285**, 1719.
[110] W. M. Clark, E. C. Raps, D. C. Tong and R. E. Kelly, *Stroke*, 2000, **31**, 1234.

Novel Sodium Channel Blockers for the Treatment of Neuropathic Pain

Brian Marron

Icagen Inc., 4222 Emperor Blvd., Durham, NC 27703, USA

Contents

1. Introduction	59
2. Background	60
2.1. Channel structure	60
2.2. Biophysics of channel block	60
3. Biological target validation	61
4. Discovery of novel small molecule blockers	62
4.1. Screening platforms	62
4.2. Clinical compound	62
4.3. Subtype selective blockers	63
4.4. Diamides blockers	63
4.5. Phenyloxyphenyl blockers	65
4.6. Biaryl blockers	67
4.7. Miscellaneous blockers	67
5. Conclusions	70
References	70

1. INTRODUCTION

Neuropathic pain is described as pain resulting from injury to or chronic changes in peripheral and/or central sensory pathways where the painful state exists without apparent noxious input. Current treatment options do not provide adequate relief for many patients and many of the agents have dose-limiting side effects [1]. Sodium channel blockers in use for neuropathic pain were originally developed for other therapeutic indications and include lamotrogine (**1**), carbamazepine (**2**), amitriptyline (**3**) and mexiletine (**4**). This report will cover the last three years of the discovery of novel voltage-gated sodium channel blockers for the treatment of neuropathic pain. A review of sodium channels blockers that covered neuronal sodium channel blockers for all indications was published in 2001 [2] and a review describing small molecule blockers of sodium channels has recently been published [1].

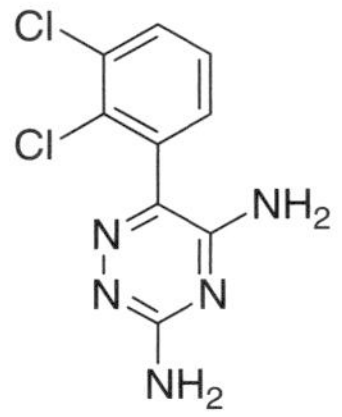

The number of reports of novel sodium channel blockers has increased in recent years, possibly due to the availability of new screening platforms as well as the increase in the level of biological validation of the role of the various sodium channels in neuropathic pain. In addition, non-selective sodium channel blockers have been used clinically for a number of years generating a measure of target validation. Biological studies have provided insight of the role of individual sodium channels in various pain models. Therefore, the discovery of selective compounds and compounds with limited CNS penetration would provide pharmacological means to evaluate the therapeutic potential of these approaches.

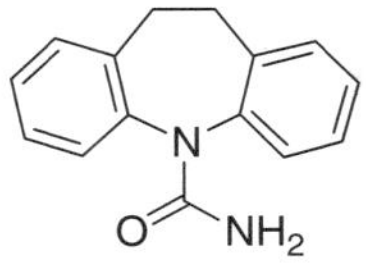

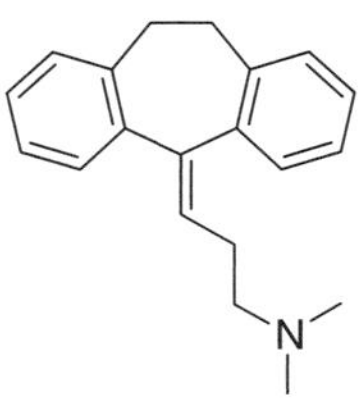

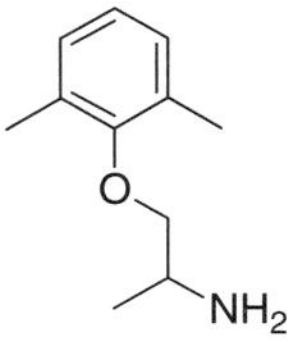

2. BACKGROUND

Sodium channels are a family of nine proteins that control the flow of sodium ions across cell membranes and are involved in all levels of nerve conduction and propagation, and in the determination of neuronal excitability.

2.1. Channel structure

Functional sodium channels consist of four homologous domains of six transmembrane helices as an α subunit that can associate with a β subunit [3]. They can be pharmacologically divided into two classes based on their sensitivity to tetrodotoxin (TTX) with six TTX-sensitive (TTXs) and three TTX-resistant (TTXr) channels. Table 1 provides a summary of the gene name, protein name, TTX sensitivity and distribution. The $Na_v1.X$ nomenclature will be used throughout this report. Expression of the channels is spatially and temporally regulated in mammals and the individual subtypes exhibit a range of biophysical properties [4,5]. Intellectual property on the proteins has been reviewed recently [6].

2.2. Biophysics of channel block

The Na channels are responsible for generating action potentials and propagating impulses in excitable cells. Cells that contain sodium channels typically exhibit negative resting membrane potentials primed for activation upon depolarization. Once depolarized, the Sodium channels activate (open state) quickly and then rapidly

Table 1. Voltage Gated Sodium Channels: Classification, Names and Distribution

Gene	Name	TTX	Localization
SCN1A	$Na_v1.1$	6 nM	CNS, DRG
SCN2A	$Na_v1.2$ (brain type II)	18 nM	CNS
SCN3A	$Na_v1.3$	15 nM	CNS, DRG
SCN4A	$Na_v1.3$ (SkM1)	5 nM	Skeletal muscle
SCN5A	$Na_v1.5$ (H1)	2 µM	Heart
SCN8A	$Na_v1.6$ (PN4)	6 nM	CNS, DRG
SCN9A	$Na_v1.7$ (PN1)	2 nM	CNS, DRG
SCN10A	$Na_v1.8$ (PN3)	60 µM	DRG
SCN11A	$Na_v1.9$ (NaN)	2 µM	DRG

transition to a closed or inactivated state. Closed channels are available for reopening but inactivated channels must recover from inactivation. Compounds can have different affinities for the various states and binding can be voltage- or state-dependant. In addition to the state-dependence, channels also exhibit a use or frequency-dependence (also called phasic block). A cell that is firing rapidly such as a damaged nerve will result in the sodium channels cycling more rapidly through the conformational states, open, closed and inactivated. Therefore, compounds that preferentially bind to the open or inactivated state "accumulate" and block will increase with higher frequency activity. Non-selective sodium channel blockers currently in use to treat neuropathic pain preferentially bind to either the open or, more often, the inactivited state resulting in both voltage and use dependence [7].

3. BIOLOGICAL TARGET VALIDATION

Over the past few years many reviews on the role of voltage-gated sodium channels in pain and overviews of biological target validation have been published [8–14]. In addition, human sodium channelopathies have been identified linking channels to pain states [15]. The roles of the various sodium channels in pain have been elucidated based on their biophysical properties, expression and localization or changes in expression in pain states, and the use of biological validation techniques such as antisense, conditional knock downs and knock-outs. Of the sodium channels, $Na_v1.3$ [16–19], $Na_v1.7$ [20], $Na_v1.8$ [21–24] and $Na_v1.9$ [25] present the best opportunities for pain therapeutics; however, little has been reported on the screening of $Na_v1.9$.

Non-selective sodium channel blockers developed as local anesthetics [26], anticonvulsants, antidepressants, or antiarrhythmics have found application in the treatment of various pain conditions [27]. All of these agents have CNS penetration and have undesired effects and off-target activities. In spite of the fact that these agents are non-selective against $Na_v1.5$, the heart sodium channel, they demonstrate few cardiovascular liabilities. The observed therapeutic index is thought to be a result of their use and voltage-dependent properties as neuronal firing frequencies are often 3–100 fold higher than that of cardiac myocytes.

4. DISCOVERY OF NOVEL SMALL MOLECULE BLOCKERS

The distribution of the channels and biological validation leads one to expect that a selective blocker of $Na_v1.3$, $Na_v1.7$ or $Na_v1.8$ could demonstrate efficacy in neuropathic or other serious pain states. Although the role of the central component has not been tested with selective agents, a non-CNS penetrating derivative of lidocaine is reported to have demonstrated efficacy in animal pain models [28]. One might expect that a non-CNS penetrating drug might enjoy a greater therapeutic index than currently used agents especially if peripherally distributed channels are targeted. Only non-selective agents are currently on the market, so the therapeutic potential of a selective agent is untested.

The compounds discussed in the report have been discovered both by modifying known compounds and from screening efforts.

4.1. Screening platforms

A number of screening platforms have been used to identify ion channel modulators [29]. These high-throughput assays coupled with the advent of higher throughput electrophysiology (EP) platforms have helped make higher throughput screening of Na channel blockers more accessible. The assays described used in the characterization and discovery of novel blockers include displacement binding assays with radiolabeled toxins or functional assays whereby the channel is opened with a chemical agent then the ability of a compound to block the modified channel is measured. Confirmation generally employs EP methodology using cloned or native channels. The pain assays and effects of sodium channel blockers in them are referred to in the sodium channel reviews and in the cited references and include formalin, CCI (Bennett), SNL (Chung) and CFA [30–32]. Generally, all compounds described below were found to be voltage- and state-dependent.

4.2. Clinical compound

Ralfinamide (**5**) is reported to be in clinical trial for neuropathic pain. Discovered as an anticonvulsant using a pharmacological model it has shown activity in pain models. It has been reported to block sodium channel currents in dorsal root ganglion neurons (DRGs) in a voltage- and use- dependent manner; TTXr $IC_{50} = 10\,\mu M$ with 3-fold selectivity over TTXs currents under depolarizing conditions [33–36].

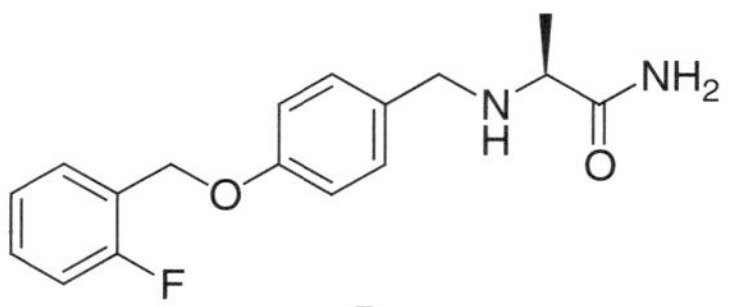

4.3. Subtype selective blockers

There have been a several reports of selective agents in the recent patent literature. In addition, in a press release Vertex has stated that they have entered into a deal with GSK to develop a selective Na channel blocker for neuropathic pain.

Biarylpyridine derivatives **6** have been claimed to be at least 5-fold selective for $Na_v1.8$ over the TTXs currents endogenously expressed in the SH-SY5Y neuroblastoma cell line [37].

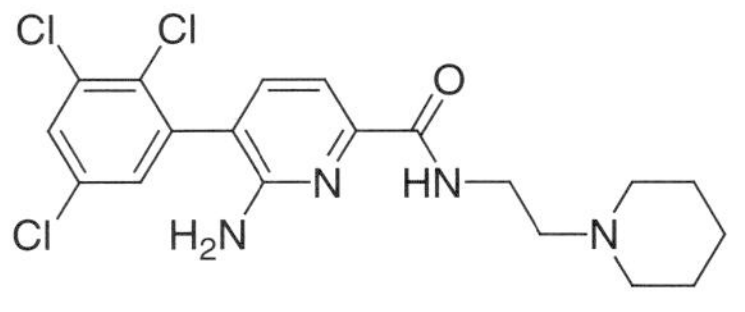

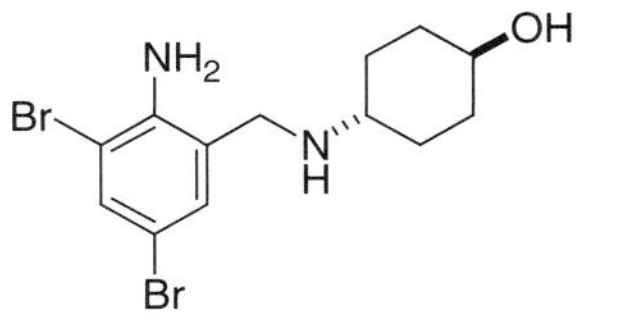

Ambroxol **7**, a secretolytic compound on the market for the treatment of respiratory diseases, has been reported to have about 3-fold selectivity for $Na_v1.8$ ($IC_{50} = 35.2\,\mu M$ tonic block and $22.5\,\mu M$ phasic block) over $Na_v1.2$ currents [38]. The inhibition is not frequency-dependant and the activity of **7** has been compared to several local analgesics in pain models [39]. It demonstrated activity in animal models of pain including formalin, partial nerve ligation and CCI when dosed at 1 g/kg to achieve plasma levels corresponding to levels reached in clinical use. In more recent studies its ability to block TTXr currents in sensory neurons and selectivity of block against recombinant $Na_v1.2$ channels has been described [40].

4.4. Diamides blockers

Compound **8**, $IC_{50} = 0.15\,\mu M$ against $Na_v1.7$, was recently identified via HTS screening in a functional assay and has been extensively profiled [41,42]. It was nonselective when tested against $Na_v1.2$, $Na_v1.5$ and $Na_v1.7$ in EP. This compound demonstrated both voltage- and use-dependence and blocked TTXr currents in dissociated mouse DRGs with an $IC_{50} = 0.025\,\mu M$. It demonstrated activity in the rat formalin assay with either intradermal administration or i.v. administration. Poor PK properties (oral bioavailability of 2%, mean residence time 0.23 h) limited the duration of the analgesic effect. This compound shared the properties of most currently used non-selective agents except for a 100-fold improvement in potency.

With **8** as a starting point, many succinamide analogues were synthesized varying the chain length, steric constraints, acyclic bridges as well as analogues with cyclic constraints including both aromatic and carbocyclic linkers [43]. Of the many analogues synthesized, the *trans*-cyclopentane dicarboxamide **9** (racemic) constraint was the most potent, $IC_{50} = 0.06\,\mu M$. The corresponding *cis*-cyclopentane dicarboxamide was 10-fold less potent and the *trans*-substituted cyclohexane **10** was even less active with an $IC_{50} = 2.1\,\mu M$. The malediamide was 3-fold less potent than **8**. Interestingly, the pthalamide **11** was equipotent to **8**.

The dicarboxamide **9** suffered from poor PK presumably due to amide bond hydrolysis. N-Methylation of the aminomethylbiphenylsulfonamide amide nitrogen was tolerated with marginal improvement in PK, but low bioavailability was still an issue. Replacing the biphenyl with alkyl-aryls resulted in a significant loss of activity. Replacement of the sulfonamide was tolerated but substituting other positions resulted in the loss of activity. The bithiophene could be replaced with bi-phenyls or *para*-substituted benzyl amines and amide nitrogen methylation was tolerated. As a result of these efforts **12**, referred to as CDA54, $Na_v1.7$ $IC_{50} = 0.21\,\mu M$, racemic, was identified. It had a 44% oral bioavailability and a 1h half-life. This compound reduced pain behavior in the rat formalin model by 29% at 10 mg/kg p.o. It was further profiled in EP, selectivity assays, PK and additional pain assays [44].

Further profiling showed that **12** exhibited no selectivity against $Na_v1.2$, $Na_v1.5$ and was both voltage- and use-dependent. It showed activity against native TTXr currents in dissociated DRGs with an $IC_{50} = 0.09\,\mu M$. It possessed in vivo efficacy at 10 mg/kg in two nerve injury models, CCI tactile allodynia and Spinal Nerve ligation (Chung), while not affecting acute nociception or motor coordination. This compound does not have CNS penetration, eliminating the potential for CNS side effects. Despite its lack of selectivity against the $Na_v1.5$ cardiac channel, no cardiovascular effects were demonstrated in anesthetized dogs with a plasma concentration of $6.7\,\mu M$.

Optimization of the PK parameters of **12** was conducted to reduce the instability in human liver microsomes [45]. Metabolism studies indicated that the instability could be attributed to the biphenylsulfonamide amide portion of the molecule. In contrast to earlier reports [44], the biphenyl could be replaced as demonstrated in **13**. Replacement of the N-methyl amide was explored with a variety of isosteres with the 1-hydroxy-3-phenylprop-1-yl substitution providing compounds exemplified by **14** that retained activity at $Na_v1.7$, $IC_{50} = 0.64\,\mu M$, as the racemate. A slight eudismic ratio was realized when the enantiomers were separated. The racemic mixture of **14** and the more potent enantiomer, **ent-14**, were found to be more stable in microsomes than **12**. This translated to greater *in vivo* exposure for **14** and **ent-14**. However, no improvement in anti-allodynic efficacy over **12** was realized in the CFA model with either of these compounds.

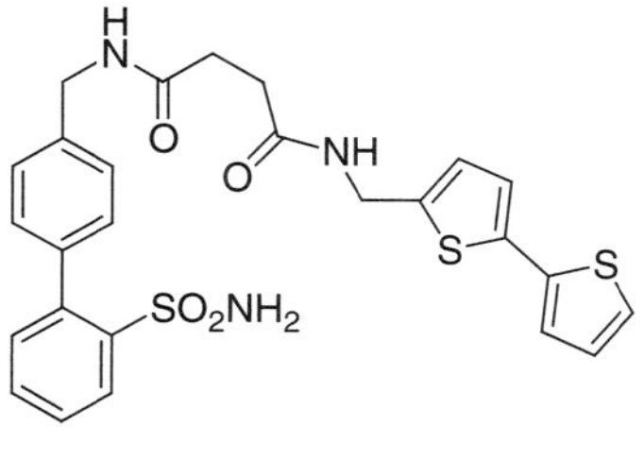

8

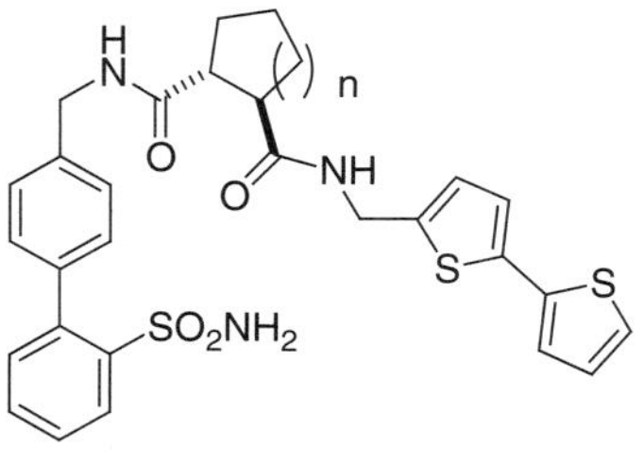

9 n = 1
10 n = 2

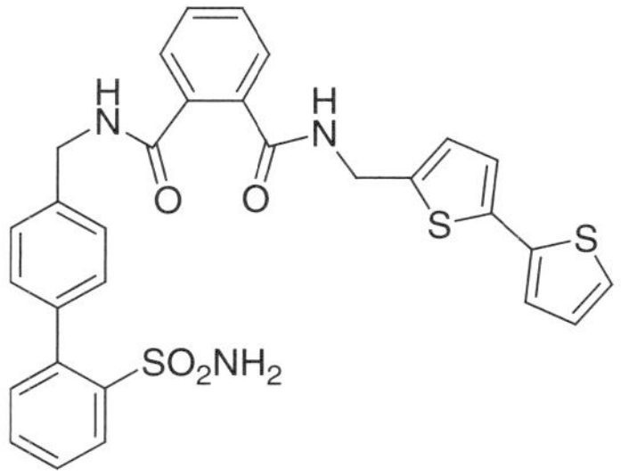

[Structures of compounds 11 and 12]

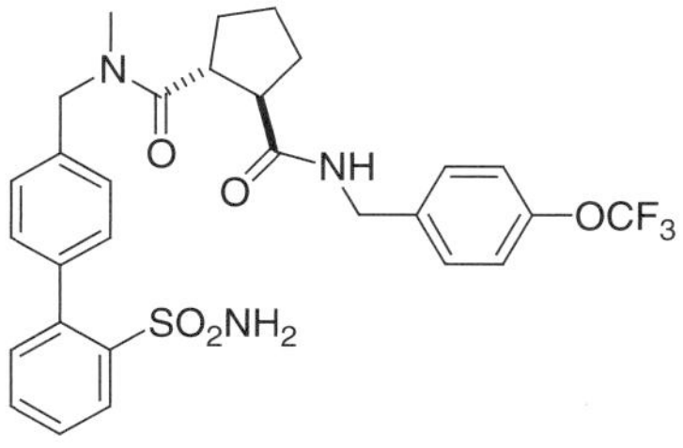

In addition, these compounds carried an hERG liability as measured by inhibition in a ^{35}S-labeled MK-0499 binding assay. Interestingly simultaneous substitution on both phenyls of **14** reduced potency. The remaining amide in **14** could be replaced with reverse amides, ethers or reverse sulfonamides. For example, 3,5-dichlorophenyl-substituted molecule **15** (IC$_{50}$ = >3 μM) was inactive, while the matched analogue, **16**, with the 4-trifluoromethoxy remained active (IC$_{50}$ = 0.33 μM) across all of the amide isosteres.

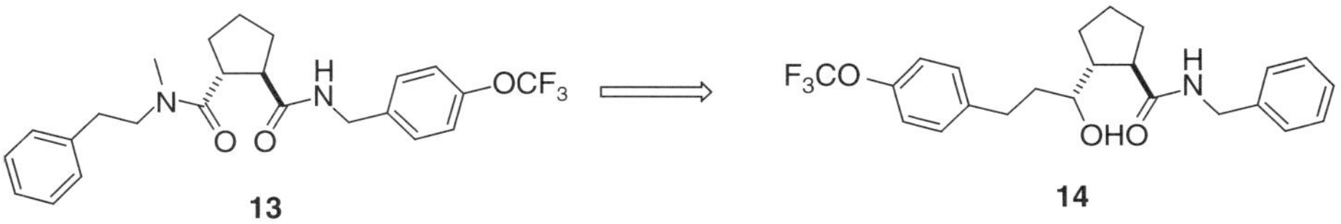

[Structures of compounds 13, 14, 15, 16]

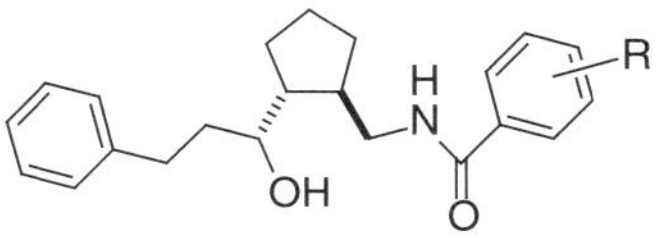

15 R = 3,4 di-Cl
16 R = 4-OCF$_3$

4.5. Phenyloxyphenyl blockers

V102862, **17**, a phenoxyphenyl benzaldehyde semicarbazone was originally described in the literature as an anticonvulsant and was found to be a non-selective sodium channel blocker (IC$_{50}$ = 610 nM, rat brain Na$_v$1.2) with activity in the Chung assay [46]. It has sub-optimal PK properties and has been used as a starting point to develop several series of sodium channel blockers that replace the labile semicarbazone with a variety of heterocyclic bioisosteres. Replacement with a pyridyl carboxamide, as in **18**, keeps the heteroatoms in the same orientation as in **17** with an improvement in potency, IC$_{50}$ = 96 nM, against rNa$_v$1.2, and activity in rat DRGs, IC$_{50}$ = 101 nM TTXs currents and IC$_{50}$ = 270 nM TTXr currents [47,48]. The other pyridyl regioisomers were >20-fold less potent.

Based on PK results, **18** and **19** (IC$_{50}$ = 123 nM) were evaluated in the Chung model. Dose-dependent reversersal of tactile allodynia was observed with **18** at 3 and 10 mg/kg p.o. with an ED$_{50}$ of 10 mg/kg. Anti-allodynic activity for **19** was also observed in the Chung model with an MED of 3 mg/kg p.o. and a therapeutic index (TI) relative to CNS side effects, of 10. This compares favorably to an MED of 100 mg/kg p.o. and a TI of 1 for carbamazepine. It is interesting to note that while most of the compounds in this series were non-selective, **20** (IC_{50} = 180 nM TTXs rDRG) was about 9-fold TTXs/TTXr selective in rDRGs. The role of the carboxamide was explored further and based on a previously reported compound, **21** was synthesized providing a potent (rNa$_v$1.2 IC_{50} = 75 nM) and state-dependant blocker that did not possess the carboxamide.

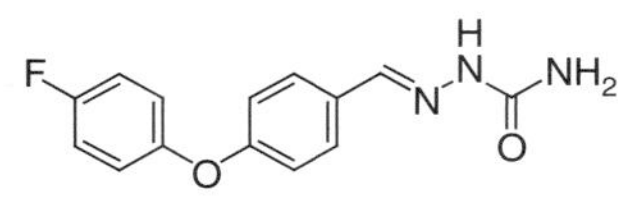

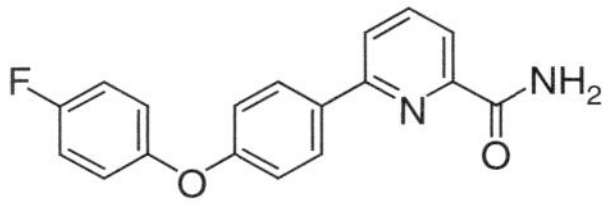

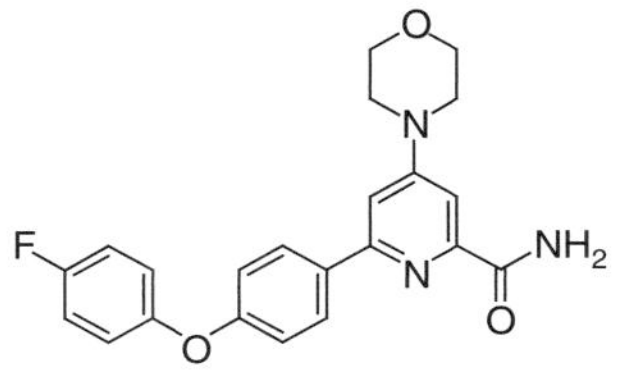

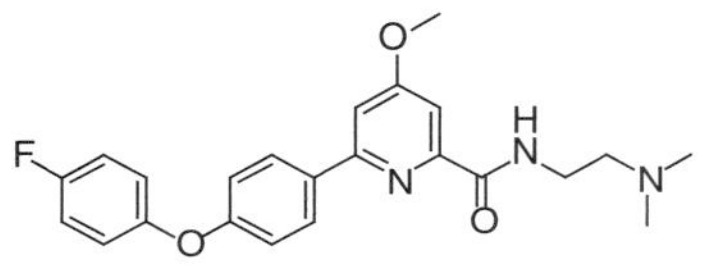

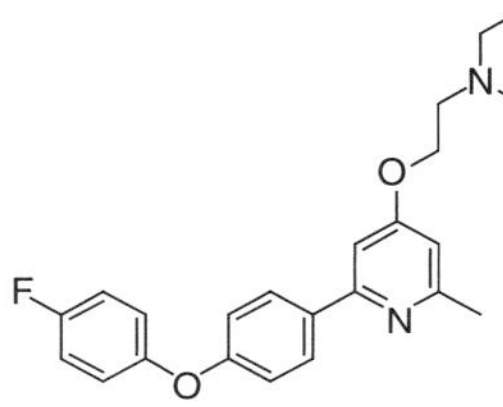

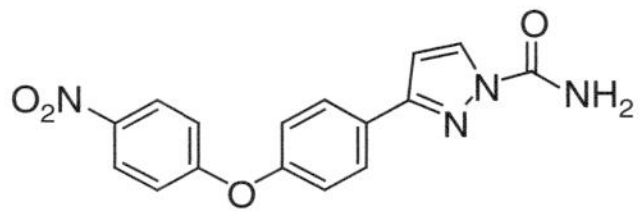

Other bioisosteres used to mimic the semicarbazone have included pyrazoles exemplified by **22** which demonstrated activity in the Chung model at 10 mg/kg p.o. [49]. The primary carboxamide was not required for potency, but the distal phenyl was. The nitro group could be replaced with 4-F, 2,4-di-F or hydrogen.

Phenoxyphenyl thiazolidine-2,4-dione **23** and oxazolidine-2,4-dione **24** have been described with activity against Na$_v$1.7 [50]. Additionally, the biaryl analogue, **25**, was shown to have similar activity.

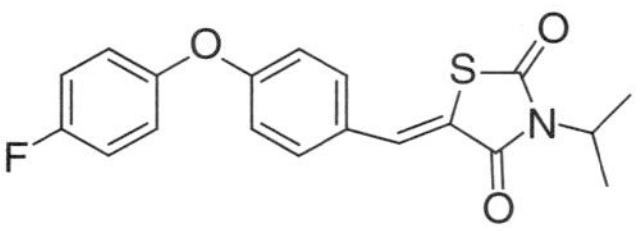

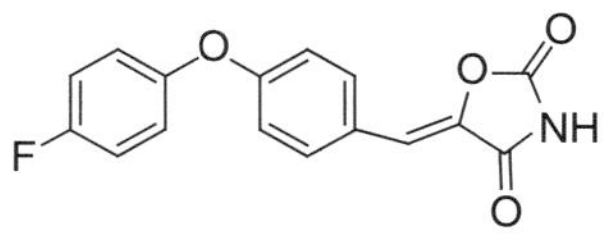

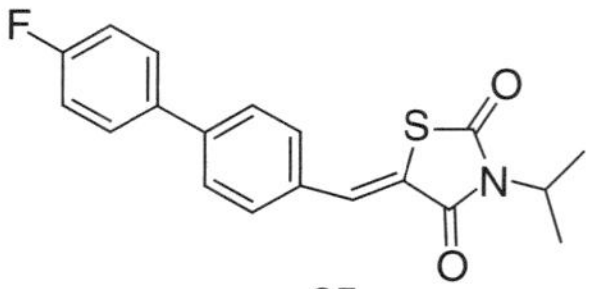

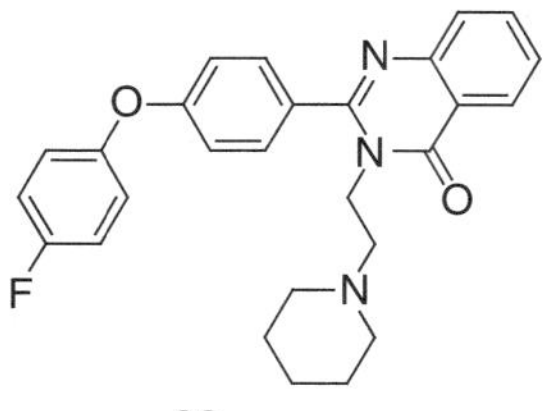

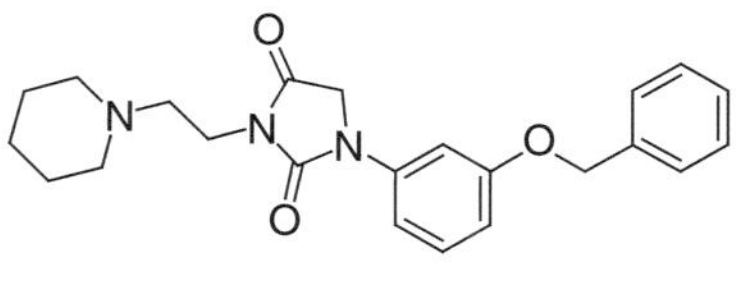

Additional isosteres include 1*H*-quinazolin-4-one, benzopyrimidin-4-one **26** and 1,3-thiazin-4-ones [51], and phenoxy- and benzyloxy-hydantoins **27** both with binding activity in rNa$_v$1.2 [52].

4.6. Biaryl blockers

A number of biaryl sodium channel blockers have been described in patents recently with claims for the treatment of chronic and neuropathic pain. Compounds were claimed as Na$_v$1.7 blockers with confirmation in EP, but no selectivity or efficacy data were provided. These series include biphenyl derivatives containing thiazoles **28**, imidazoles and oxazoles [53] and various substituted triazoles **29** [54,55]. In addition, pyridyl and pyrimidyl central rings **30** bound to a variety of six-member heterocycles including pyrimidines, pyridines and pyrazines [56], and binary-substituted pyrazinones **31** [57] have been reported. It is noteworthy that a carboxamide moiety is prominent in many of the claimed structures as in the phenoxyphenyl series, e.g. **18**.

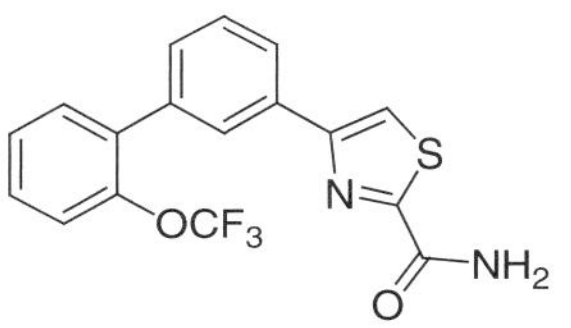

28

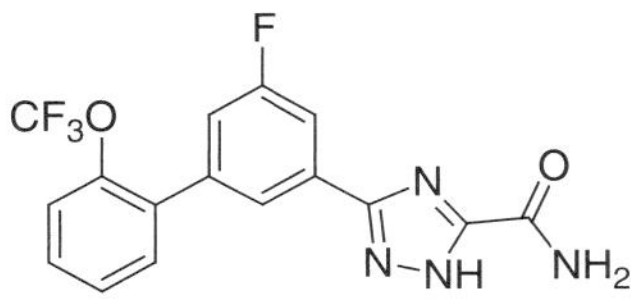

29

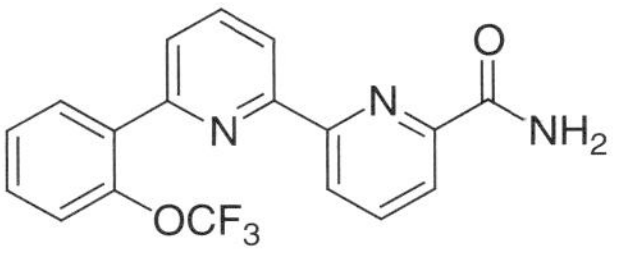

30

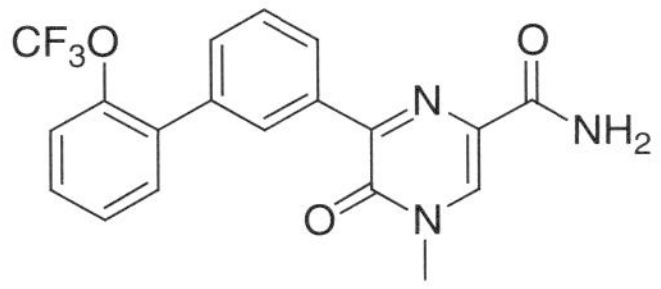

31

4.7. Miscellaneous blockers

The known [58] potent cardiac sodium channel blocker **32** was found to inhibit Na$_v$1.7 in a VIPR format (IC$_{50}$ = 4 µM) and was used to develop a new series of blockers [59]. When injected locally, **32** was active in a dose-dependant manner in the phase II response of the formalin model (MED 1.5 µg), but was inactive when dosed i.v. at 3 mg/kg (13% inhibition). In addition to PK liabilities, it also displayed considerable hERG activity (IC$_{50}$ = 180 nM, displacement of MK-499 binding). Modifying the basicity of the amine by incorporating it into an amide bond, **33**, resulted in a reduction of hERG activity (88% inhib at 10 µM) but with a loss in Na$_v$1.7 activity as well, IC$_{50}$ = 19.4 µM. Activity against Na$_v$1.7 was regained and a further reduction in hERG activity was realized (Na$_v$1.7 IC$_{50}$ = 1.3 µM; hERG

$IC_{50} = 7.7\,\mu M$) with the pyrrolidine amide, **34**. Oral bioavailability was low for this compound (9.4%) and the $t1/2$ was 0.48 h. It exhibited a 68% reduction in the phase II response in formalin assay at 3 mg/kg, i.v. Attempts to replace the p-chlorophenylthiazole side chain resulted mostly in inactive compounds except for replacement with p-chlorophenyl-1,2,4-oxadiazole and bithiophenemethyl as in **8**. Substitution of the methoxy group with larger moieties resulted in reduced potency, but replacement with Cl, CH_3 and H were tolerated with slight improvements in potency.

A conformationally constrained analogue **35** was prepared to determine its effects on potency and PK properties. Although potency was slightly diminished ($IC_{50} = 3.6\,\mu M$, EP $IC_{50} = 0.21\,\mu M$, $IC_{50} = 5.4\,\mu M$ hERG) the *in vitro* $t1/2$ doubled to 0.88 h.

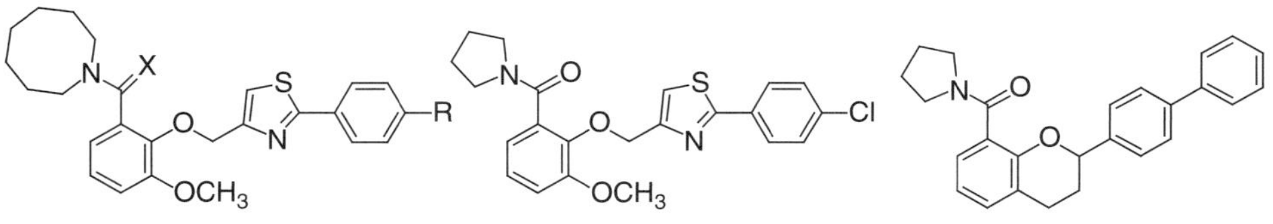

32 R - H, X = H,H
33 R - Cl, X = O

34

35

Arylimidazoles derived from β-carboline **36** have also been described with activity assessed in a binding assay to $rNa_v1.2$ [60]. Compounds were further evaluated in a functional assay in blocking the cytotoxic effects of veratridine, a sodium channel opener, in SH-SY5Y neuroblastoma cells. The SAR showed that the R^1 position required a lipohphilic group to maintain activity. For example, when the n-hexyl of **37** ($IC_{50} = 16$ nM) was shortened to a methyl, **38**, the potency decreased to an IC_{50} of 1460 nM in the binding assay.

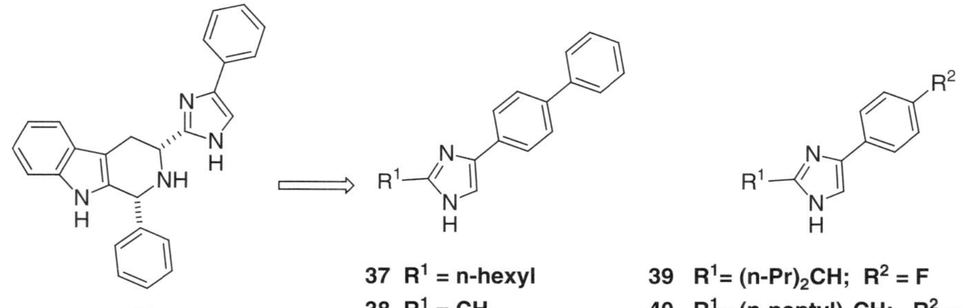

36

37 R^1 = n-hexyl
38 R^1 = CH_3

39 R^1= (n-Pr)$_2$CH; R^2 = F
40 R^1= (n-pentyl)$_2$CH; R^2 = F
41 R^1= (n-Pr)$_2$CH; R^2 = tBu

Furthermore, R^2 seemed to tolerate a variety of groups but the potency was dependant on the size of the lipophilic group in R^1. For example, **39**, $IC_{50} = 518$ nM, improves potency to 146 nM by increasing the size of the lipophilic group at R^1 as in **40**. A further increase was realized by substituting the fluorine at R^2 with a tBu group to provide **41** with an $IC_{50} = 30$ nM. No other sodium channel selectivity data or biophysical characterization were provided.

A series of non-selective benzimidazoles, exemplified by **42** has been described. In addition to the amide, ureas are exemplified as well as indole amides and they are claimed to be blockers of sodium and $Ca_V2.2$ channels [61].

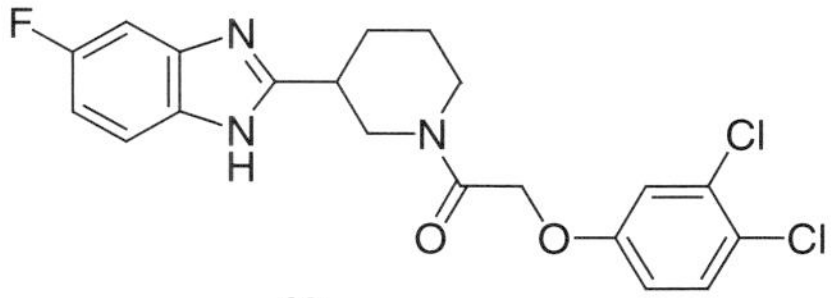

42

Recently, three related series of quinazolines [62] and pyrimidines [63,64], **43**, **44**, **45**, claimed as voltage-gated ion channel blockers for chronic and neuropathic pain with screening assays were described for $Na_V1.2$, $Na_V1.3$ and $Ca_V2.2$. No other biological data or selectivity data were presented, and therefore it is difficult to determine their potential as blockers.

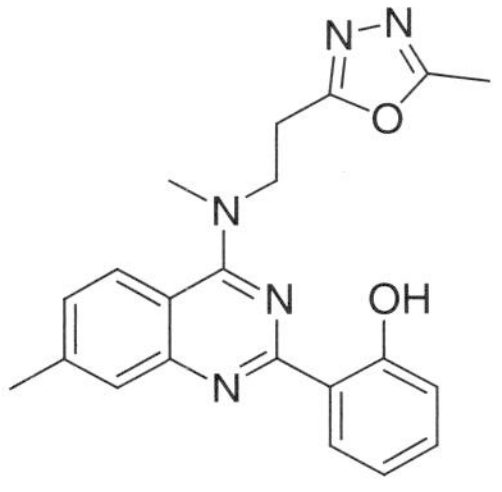

43 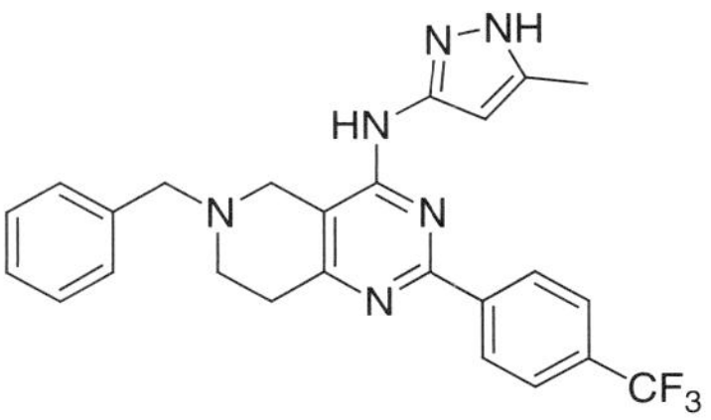 **44** 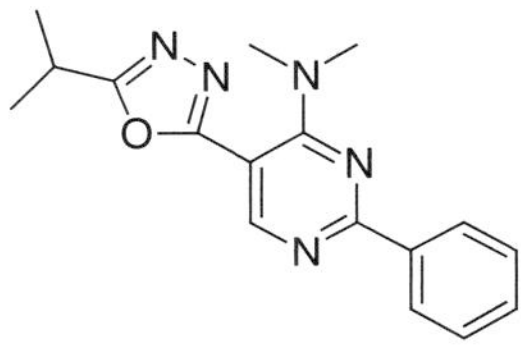 **45**

Aminothiazolesulfonamide phenyl amides, **46**, with a number of carbamates and ureas exemplified have been claimed as sodium and calcium channel blockers for chronic and neuropathic pain with screeing assays described for $Na_V1.2$, $Na_V1.3$ and $Ca_V2.2$ [65,66].

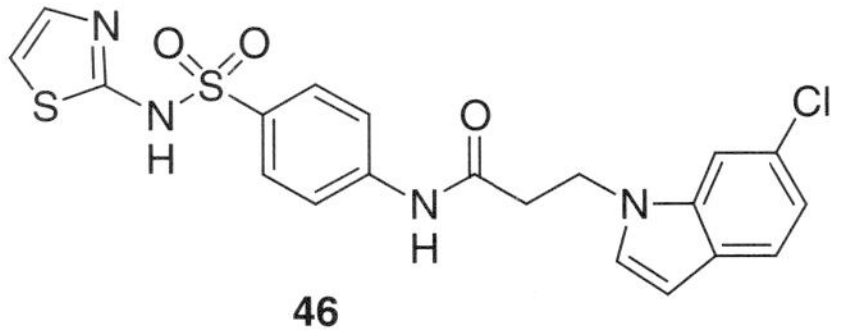

46

A number of novel series have been identified by screening against $Na_V1.8$. Amino amide (**47**) derivatives [67] and substituted benzyloxy phenyls **48** where the amide has been removed [68], and azacyclic amides **49** ($IC_{50} = 190$ nM) [69] were identified by screening in a SH-SY5Y neuroblastoma cell line stably expressing human $Na_V1.8$ on a FLIPR platform.

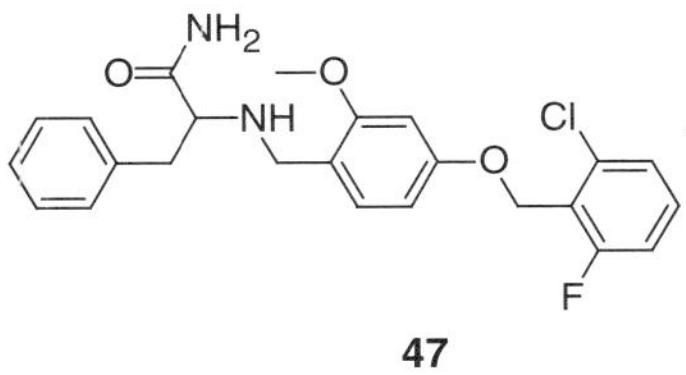

47 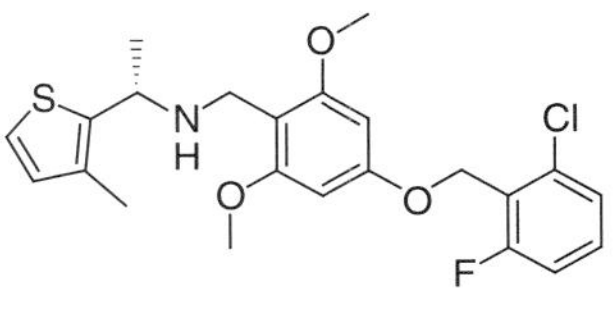 **48** 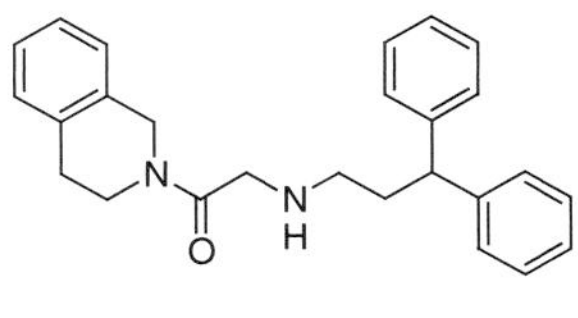 **49**

Other series that have been claimed as $Na_v1.8$ blockers include pyrazolopyrimidines **50**, pyrazoles **51** and piperidines **52**, all of which were identified in a flux assay [70–73]. Selectivity data were not provided.

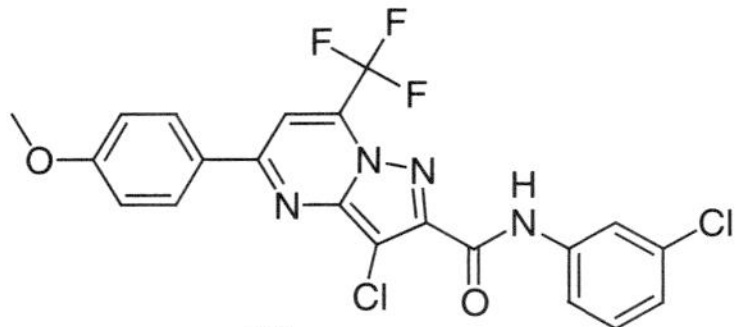

5. CONCLUSIONS

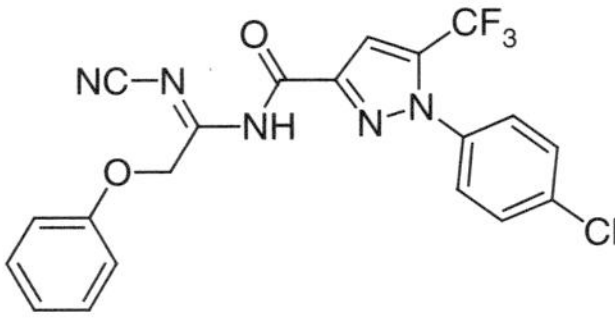

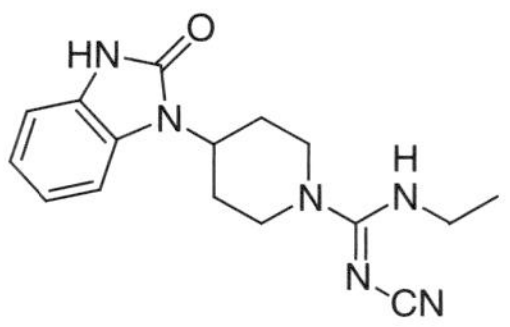

Many novel non-selective small molecule sodium channel blockers have been disclosed in the literature over the past three years. Although efficacy in animal models has been achieved with non-selective peripherally acting agents, there have only been a few reports of selective subtype blockers, and to date no efficacy data have been provided for them. The discovery of subtype selective blockers will provide a means to further elucidate the role of sodium channels in pain and potentially provide therapeutically useful compounds to address a large unmet medical need.

REFERENCES

[1] J. E. Gonzalez, A. P. Termin and D. M. Wilson, in *Voltage-Gated Ion Channels as Drug Targets (Methods and Principles in Medicinal Chemistry)* (eds D. J. Triggle, M. Gopalakrishnan, D. Rampe, W. Zheng, R. Mannhold, H. Kubinyi and G. Folkers), Wiley-VCH Verlag GmbH & Co. KGaA, Weinheim, 2006, p. 168.
[2] T. Anger, D. J. Madge, M. Mulla and D. Riddall, *J. Med. Chem.*, 2001, **44**, 115.
[3] W. A. Catterall, A. L. Goldin and S. G. Waxman, *Pharmacol. Rev.*, 2005, **55**, 397.
[4] D. S. Krafte, M. Chapman and K. McCormack, in *Voltage-Gated Ion Channels as Drug Targets methods and Principles in Medicinal Chemistry* (eds D. J. Triggle, M. Gopalakrishnan, D. Rampe, W. Zheng, R. Mannhold, H. Kubinyi, G. Folkers), Wiley-VCH Verlag GmbH & Co. KGaA, Weinheim, 2006, p. 151.
[5] J. K. J. Diss, S. P. Fraser and M. B. A. Djamgoz, *Eur. Biophys. J.*, 2004, **33**, 180.
[6] V. K. Gribkoff and R. J. Winquist, *Expert Opin. Ther. Patents*, 2005, **15**, 1751.
[7] J. C. Hunter and D. Loughhead, *Curr. Opin. Cent. Peripher Nerv. Syst. Investig. Drugs*, 1999, **1**, 72.
[8] R. A. Kinloch and P. J. Cox, *Expert Opin. Ther. Targets*, 2005, **9**, 685.
[9] M. Devor, *J. Pain*, 2006, **7**, (Suppl 1), S3.
[10] J. Lai, F. Porreca, J. C. Hunter and M. S. Gold, *Annu. Rev. Pharmacol. Toxicol.*, 2004, **44**, 371.
[11] J. N. Wood and J. Boorman, *Curr. Top. Med. Chem.*, 2005, **5**, 529.
[12] J. N. Wood, J. P. Boorman, K. Okuse and M. D. Baker, *J. Neurobiol.*, 2004, **61**, 55.
[13] S. Hong, T. J. Morrow, P. E. Paulson, L. L. Isom and J. W. Wiley, *J. Biol. Chem.*, 2004, **279**, 29341.
[14] S. Hong and J. W. Wiley, *Biochem. Biophys. Res. Com.*, 2006, **339**, 652.

[15] S. D. Dib-Hajj, A. M. Rush, T. R. Cummins, F. M. Hisama, S. Novella, L. Tyrrell, L. Marshall and S. G. Waxman, *Brain*, 2005, 128 (pt8) 1847.
[16] S. Dib-Hajj, J. AA. Black, P. Felts and S. G. Waxman, *Proc. Natl. Acad. Sci. USA*, 1996, **93**, 14950.
[17] B. C. Hains, J. P. Klein, C. Y. Saab, M. J. Craner, J. A. Black and S. G. Waxman, *J. Neurosci.*, 2003, **23**, 8881.
[18] E. M. Garry, A. Delaney, H. AA. Anderson, E. C. Sirinathsinghji, R. H. Clapp, W. J. Martin, P. R. Kinchington, D. L. Krah, C. Abbadie and S. M. Fleetwood-Walker, *Pain*, 2005, **118**, 97.
[19] J. A. Lindia, M. G. Kohler, W. J. Martin and C. Abbadie, C, Pain 117 (2005) 145.
[20] M. A. Nassar, C. L. Stirling, G. Forlani, M. D. Baker, E. A. Matthews, A. H. Dickenson and J. N. Wood, *Proc. Natl. Acad. Sci. USA*, 2004, **101**, 12706.
[21] C. Roza, J. M. A. Laird, V. Souslova, J. N. Wood and F. Cervero, *J. Physio.*, 2003, **550**, 921.
[22] J. M. A. Laird, V. Souslova, J. N. Wood and F. Cervero, *J. Neurosci.*, 2002, **22**, 8352.
[23] L. C. Stirling, G. Forlani, M. D. Baker, J. N. Wood, E. A. Matthews, A. H. Dickenson and M. A. Nassar, *Pain*, 2005, **113**, 27.
[24] M. S. Gold, D. Weinreich, C. S. Kim, R. Wang, J. Treanor, F. Porreca and J. Lai, *J. Neurosci.*, 2003, **23**, 158.
[25] B. T. Priest, B. A. Murphy, J. A. Lindia, C. Diaz, C. Abbadie, A. M. Ritter, P. Liberator, L. M. Iyer, S. F. Kash, M. G. Kohler, G. J. Kaczorowski, D. E. MacIntyre and W. J. Martin, *Proc. Natl. Acad. Sci. USA*, 2005, **28**, 9382.
[26] H. A. Fozzard, P. J. Lee and G. M. Lipkind, *Curr. Pharm. Design*, 2005, **11**, 2671.
[27] E. Kalso, *Curr. Pharm. Design*, 2005, **11**, 3005.
[28] Q. Chem, T. King, T. W. Vanderah, M. H. Ossipov, T. P. Malan, Jr., J. Lai and F. Porreca, *J. Pain*, 2004, **5**, 281.
[29] P. J. Birch, L. V. Dekker, I. F. James, A. Southan and D. Cronk, *DDT*, 2004, **9**, 410.
[30] G. Blackburn-Munro, N. Ibsen and H. K. Erichsen, *Eur. J. Pharm.*, 2002, **445**, 231.
[31] H. K. Erichsen, J. -X. Hao, X. -J. Xu and G. Blackburn-Munro, *Eur. J. Pharm.*, 2003, **458**, 275.
[32] N. Attal and D. Bouhassira, *J. Pain*, 2006, **7** (Suppl. 1), S31.
[33] P. Pevarello, A. Bonsignori, P. Dostert, F. Heidempergher, V. Pinciroli, M. Colombo, R. A. McArthur, P. Salvati, C. Post, R. G. Fariello and M. Varasi, *J. Med. Chem.*, 1998, **41**, 579.
[34] P. Salvati, R. Maj, C. Caccia, M. A. Cervini, M. G. Fornaretto, E. Lamberti, P. Pevarello, G. A. Skeen, H. S. White, H. H. Wolf, L. Faravelli, M. Mazzanti, E. Mancinelli, M. Varasi and R. G. Fariello, *J. Pharmacol. Exp. Ther.*, 1999, **288**, 1151.
[35] O. Veneroni, R. Maj, M. Calabresi, L. Faravelli R. G. Fariello and P. Salvati, *Pain*, 2003, **102**, 17.
[36] T. C. Stummann, P. Salvati, R. G. Fariello and L. Faravelli, *Eur. J. Pharm.*, 2005, **510**, 197.
[37] C. A. L. Lane, G. N. Maw, D. J. Rawson and L. R. Thompson, WO06011050, 2006.
[38] T. Weiser and N. Wilson, *Mol. Pharm.*, 2003, **62**, 443.
[39] T. Weiser, *Neurosci. Lett.*, 2006, **295**, 179.
[40] W. Gaida, K. Klinder, K. Arndt and T. Weiser, *Neuropharm.*, 2005, **49**, 1220.
[41] M. H. Fisher, C. Li, J. Liang, P. Y. Meinke, D. Ok, W. H. Parsons, P. P. Shao and S. Tyagarajan, WO03101381, 2003.
[42] B. T. Priest, M. L. Garcia, R. E. Middleton, R. M. Brochu, S. Clark, G. Dai, I. E. Dick, J. P. Felix, C. J. Liu, B. S. Reiseter, W. A. Schmalhofer, P. P. Shao, Y. S. Tang, M. Z. Chou, M. G. Kohler, M. M. Smith, V. A. Warren, B. S. Williams, C. J. Cohen, W. J.

Martin, P. T. Meinke, W. H. Parsons, K. A. Wafford and G. J. Kaczorowski, *Biochem. itself*, 2004, **43**, 9866.

[43] P. P. Shao, D. Ok, M. H. Fisher, M. L. Garcia, G. J. Kaczorowski, C. Li, K. A. Lyons, W. J. Martin, P. T. Meinke, B. T. Priest, M. M. Smith, M. J. Wyvratt, F. Yea and W. H. Parsons, *Bioorg. Med. Chem. Lett.*, 2005, **15**, 1901.

[44] R. M. Brochu, I. E. Dick, J. W. Tarpley, E. McGowan, D. Gunner, J. Herrington, P. P. Shao, D. Ok, C. Li, W. H. Parsons, G. L. Stump, C. P. Regan, J. J. Lynch Jr, K. A. Lyons, O. B. McManus, S. Clark, Z. Ali, G. J. Kaczorowski, W. J. Martin and B. T. Priest, *Mol. Pharm.*, 2006, **69**, 823.

[45] D. Ok, C. Li, C. Abbadie, J. P. Felix, H. H. Fisher, M. L. Garcia, G. J. Kaczorowski, K. A. Lyons, W. J. Martin, B. T. Priest, M. M. Smith, B. S. Williams, M. J. Wyvratt and W. H. Parsons, *Bioorg. Med. Chem. Lett.*, 2006, **16**, 1358.

[46] V. I. Ilyin, D. D. Hodges, E. R. Whittemore, R. B. Carter, S. X. Cai and R. M. Woodward, *Br. J. Pharm.*, 2005, **144**, 801.

[47] B. Shao, S. Victory, V. I. Ilyin, R. R. Goehring, Q. Sun, D. Hogenkamp, D. D. Hodges, K. Islam, D. Sha, C. Zhang, P. Nguyen, S. Robledo, G. Sakellaropoulos and R. B. Carter, *J. Med. Chem.*, 2004, **47**, 4277.

[48] B. Shao, R. R. Goehring, S. F. Victory and Q. Sun, WO03022276, 2003.

[49] J. Yang, P. Gharagozloo, J. Yao, V. I. Ilyin, R. B. Carter, P. Nguyen, S. Robledo, R. M. Woodward and D. J. Hogenkamp., *J. Med. Chem.*, 2004, **47**, 1547.

[50] M. B. Ayer, P. K. Chakravarty, P. T. Meinke, W. H. Parsons and S. Tyagarajan, US050165072, 2005

[51] Q. Sun, D. J. Kyle and S. F. Victory, US040152695, 2004.

[52] Q. Sun and D. K. Kyle, WO04010950, 2004.

[53] P. K. Chakravarty, M. H. Fisher, W. H. Parsons, S. Tyagarjan and B. Zhou, W04094395, 2004.

[54] M. K. Park, P. K. Chakravarty, B. Zhou, E. Gonzalez, H. Ok, B. Palucki, W. H. Parsons and R. Sisco, WO05047270, 2005.

[55] M. K. Park, P. K. Chakravarty, B. Zhou, E. Gonzalez, H. Ok, B. Palucki, W. H. Parsons, R. Sisco and M. H. Fisher, US060020006, 2006.

[56] P. K. Chakravarty, M. H. Fisher, W. H. Parsons, J. Liang and B. Zhou, WO04084824, 2004.

[57] J. Liang, P. K. Chakravarty, D. E. Pan, W. H. Parsons, P. P. Shao and F. Ye, WO05097136, 2005.

[58] J. R. McCulough, G. Trube and M. L. Conder, *Biophys. J.*, 1987, **51**, 260a.

[59] J. Liang, R. M. Brochu, C. J. Cohen, I. E. Dick, J. P. Felix, M. H. Fisher, M. L. Garcia, G. J. Kaczorowski, K. A. Lyons, P. T. Meinke, B. T. Priest, W. A. Schmalhofer, M. M. Smith, J. W. Tarpley, B. S. Williams, W. J. Martin and W. H. Parsons, *Bioorg. Med. Chem. Lett.*, 2005, **15**, 2943–2947.

[60] A.-M. Liberatore, J. Schulz, J. Pommier, M.-A. Barthelemy, M. Huchet, P.-E. Chabrier and D. Bigg, *Bioorg. Med. Chem. Lett.*, 2004, **14**, 3521.

[61] D. M. Wilson, A. P. Termin, J. E. Gonzalez III, N. Zimmermann, Y. Zhang and L. T. D. Fanning, US050209282, 2005.

[62] J. E. Gonzalez III, D. M. Wilson, A. Peter, D. J. Grootenhuis, Y. Zhang, B. J. Petzoldt, L. T. D. Fanning, T. D. Neubert, R. Tung, E. Martinborough and N. Zimmermann, US040248890, 2004.

[63] D. M. Wilson, A. P. Termin, T. D. Neubert, J. Wang, Y. Zhang, J. E. Gonzalez III, E. Martinborough and N. Zimmermann, US050187217, 2005.

[64] D. M. Wilson, E. Martinborough, T. D. Neubert, A. P. Termin, J. E. Gonzalez III and N. Zimmermann, WO05003099, 2005.

[65] J. E. Gonzalez III, A. P. Termin, E. Martinborough and N. Zimmermann, US060025415, 2006.
[66] J. E. Gonzalez III, A. P. Termin, E. Martinborough and N. Zimmermann, WO05013914, 2005.
[67] R. J. Hamlyn, D. C. Tickle, M. R. Huckstep and R. Lynch, WO4087125, 2004.
[68] N. S. Jennings, S. Stokes, R. J. Hamlyn, D. C. Tickle , M. R. Huckstep, R. Lynch and L. J. S. Knutsen, WO05000309, 2005.
[69] R. J. Hamlyn, M. R. Huckstep, R. Lynch, S. Stokes, D. C. Tickle and L. Patient, WO05005392, 2005.
[70] R. N. Atkinson, M. F. Gross and M. A. Van Rhee, WO03037900, 2003.
[71] R. N. Atkinson and M. F. Gross, WO03037274, 2003.
[72] R. N. Atkinson, I. Drizin, R. J. Gregg, M. F. Gross, M. E. Kort and L. Shi, US050020564, 2005.
[73] M. F. Gross, R. N. Atkinson and S. M. Johnson, WO03037890, 2003.

Section 2
Cardiovascular and Metabolic Diseases

Editor: Andrew Stamford
Schering-Plough Research Institute
2015 Galloping Hill Road
Kenilworth New Jersey

Centrally Acting Anti-Obesity Agents

David Hepworth[1], Philip A. Carpino[2] and Shawn C. Black[2]

[1]*Pfizer Global Research & Development – Sandwich Laboratories, Sandwich, CT13 9NJ, UK*
[2]*Pfizer Global Research & Development – Groton Laboratories, Groton, CT 06340, USA*

Contents

1. Introduction	77
2. CB1R Antagonists	78
3. Neuropeptide receptor modulators	81
3.1. MC4R agonists	81
3.2. MCH1R antagonists	82
3.3. GHSR antagonists	83
3.4. Neuropeptide Y family modulators	85
3.5. Opioid receptor antagonists	87
4. Aminergic receptor modulators	88
4.1. 5HT2CR agonists	88
4.2. 5HT6R antagonists	90
4.3. H3R antagonists	91
5. Monoamine re-uptake inhibitors	92
6. Conclusions	92
References	92

1. INTRODUCTION

The prevalence of obesity, a disease characterized by excess adiposity and a body mass index (BMI) greater than $30\,kg/m^2$ (or $27\,kg/m^2$ with at least one co-morbidity) has rapidly increased over the past decade, especially in Western Europe and the United States. This rise has been attributed to several factors including increased sedentary lifestyles and diets high in calories and saturated fats. In a small percentage of obese subjects, genetic factors may contribute to the etiology of the disease. Humans have evolved efficient mechanisms for storing, rather than catabolizing, excess energy primarily to protect against periods of starvation, and these mechanisms are likely contributors to the epidemic in societies which have an abundant supply of affordable and readily available high-energy foods. Obesity is not simply a cosmetic concern to societies throughout the world, but rather a serious health problem. It increases the risk of developing co-morbidities such as diabetes mellitus and hypertension; furthermore it elevates the relative risk of mortality due to cardiovascular disease. Obesity now ranks as the second leading cause of preventable death after smoking in the United States [1].

Treatment of obesity involves a combination of diet, exercise, pharmacotherapy and, in cases of morbid obesity, bariatric surgery. Dietary changes alone are often not successful in maintaining weight loss while bariatric surgery can be associated with superior (relative to diet) weight loss it also has associated risks of complication (including mortality). Only two drugs are currently approved for chronic weight loss treatment: a pancreatic lipase inhibitor called orlistat (Xenical®) and a monoamine re-uptake inhibitor sibutramine (Meridia®). Both have different adverse effect profiles that limit more widespread use. A cannabinoid receptor type 1 (CB1R) antagonist, rimonabant (**1**), is currently under review by the US Food and Drug Administration (FDA). A significant unmet medical need exists for safe and effective pharmacotherapies to reduce both body weight and the consequences of co-morbidities in the obese population.

The regulation of body weight is a complex process involving multiple, overlapping (and often redundant) pathways [2]. Under homeostatic conditions, energy expenditure is closely matched with energy intake. When energy intake exceeds energy expenditure for long periods of time, a state of positive energy balance occurs, resulting in the disposition of adipose tissue in different regions of the body. Energy intake is derived from eating behavior which is governed by opposing hormonal and neural processes that stimulate or inhibit food intake. There is also a hedonic component to the consumption of food. The principal homeostatic and reward systems that control appetite and body weight are located within the central nervous system (CNS). A small region at the base of the brain called the hypothalamus processes short- and long- term signals regarding the body's energy status from nutrients (glucose, triglycerides), adipose tissue (leptin), pancreas (insulin, glucagon-like peptide 1) and the gastrointestinal tract (ghrelin, cholecystokinin, peptide YY, pancreatic polypeptide (PP)) [3]. The hypothalamus integrates these diverse signals and responds to the body's energy needs by stimulating catabolic processes during periods of positive energy balance or anabolic pathways during periods of energy deficit.

This review will describe new agents in the literature (2005–2006) that target central pathways involved in controlling food intake, body weight, and reward and are under active investigation.

2. CB1R ANTAGONISTS

The endocannabinoid system consists of endogenous ligands (i.e., anandamide and 2-arachidonoyl glycerol) that bind to and activate either of two receptor subtypes, CB1R or CB2R, that are distributed throughout the periphery (gut, liver, spleen, immune cells) and the CNS [4,5]. Endocannabinoids and exogenous cannabinoids (notably Δ9-tetrahydrocannabinol, the principal active constituent in *Cannabis sativa L.*, or marijuana) have been shown to stimulate appetite [6]. CB1R was hypothesized to play a role in the regulation of appetite because of its location in the hypothalamus. This was confirmed using a combination of CB1R selective ligands and CB1 knockout (KO) mice [7–11]. Recently, 2-year clinical efficacy data with rimonabant, a selective CB1R antagonist, have been published, claiming rimonabant-treated patients had significantly reduced waist circumference and plasma triglycerides and an increase in HDL cholesterol and adiponectin concentrations in addition to significant and sustained reductions of body weight [12].

Significant interest in CB1R antagonism as a potential weight loss therapy followed publication of the rimonabant clinical data. Several reviews describe recent medicinal chemistry developments around pyrazolyl derivatives, such as rimonabant and other new chemotypes [13–16]. On the basis of literature and patent application reports from the past year, rimonabant appears to be still widely used as a lead for the design of new compounds, although structurally distinct series have now appeared.

The H-bond acceptor group in rimonabant is believed to stabilize an Asp366–Lys192 salt bridge in the uncoupled state of the receptor, leading to inverse agonist activity [17]. Analogs of rimonabant have been reported in which a hydrogen bond acceptor group is located at the C4 position of a pyrazolyl ring [18]. The functional consequence of moving the H-bond acceptor group from the C3 position as in rimonabant to the C4 position in the new analogs is not known. Compounds such as 2 exhibit CB1R binding affinities between 10 nM-10 μM.

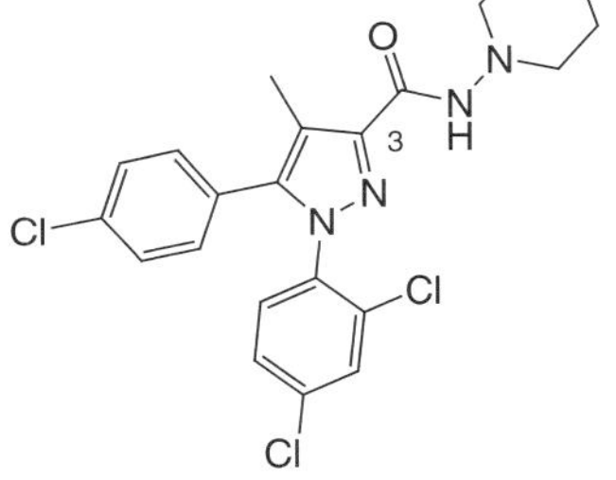

1 (Rimonabant)

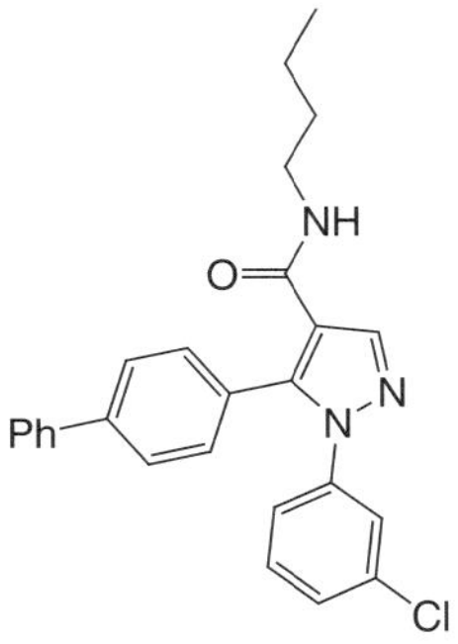

2

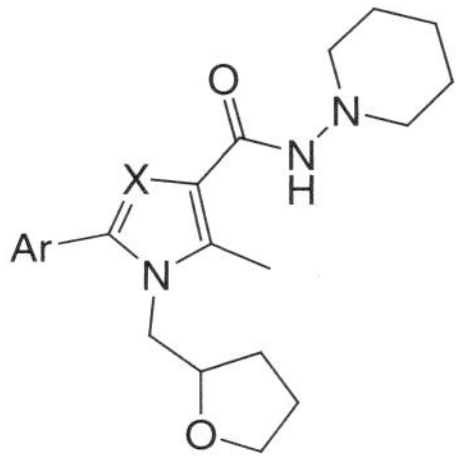

3 Ar = 2,5-di-CF$_3$-Ph, X = CH
4 Ar = 4-CF$_3$O-Ph, X = N

A series of pyrrolyl and imidazolyl derivatives was identified that contain N1 alkyl groups in place of N1 aryl moieties as found in rimonabant [19]. Similar compounds in the pyrazolyl series have exhibited higher affinities for CB2R over CB1R. The pyrrolyl derivative 3 (hCB1R IC$_{50}$ = 8 nM) and the imidazolyl derivative 4 (hCB1R IC$_{50}$ = 120 nM) are reportedly at least 10-fold more selective for CB1R over CB2R.

Bicyclic 2,6-dihydro-pyrazolo[4,3-d]pyrimidin-7-ones such as 5 were prepared as part of a program to explore different spatial orientations of the key pharmacophoric groups in rimonabant [20]. The N6 1-piperidinyl derivative 5 and rimonabant were equipotent in the binding assay despite poor overlap of the N6 group in the former compound with the 1-piperidinyl group in the proposed low energy conformer of the latter compound [21]. Optimization of the SAR in 5 led to the discovery of the N6 trifluoroethyl analog 6 (hCB1R K_i = 0.3 nM). Pharmacokinetic analysis of 6 revealed bioavailability of 62% in the rat and a brain-to-plasma (B/P) ratio of 7.2 at the 1.5 h time-point. Compound 6 decreased food intake in a dose-dependent manner following oral administration in a fasting-induced re-feeding model in rats. Replacement of the 2,6-dihydro-pyrazolo[4,3-d]pyrimidin-7-one core with other 5,6-bicyclic heterocyclic moieties as in **7** led to a small loss of binding affinity.

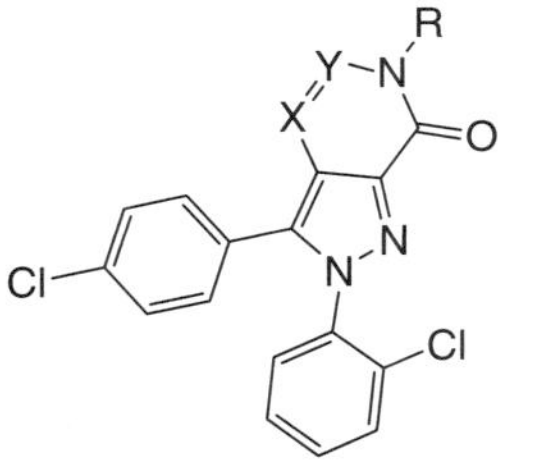

5 X = N, Y = CH, R = 1-piperidinyl
6 X = N, Y = CH, R = CH$_2$CF$_3$
7 X = CH, Y = N, R = CH$_2$CF$_3$

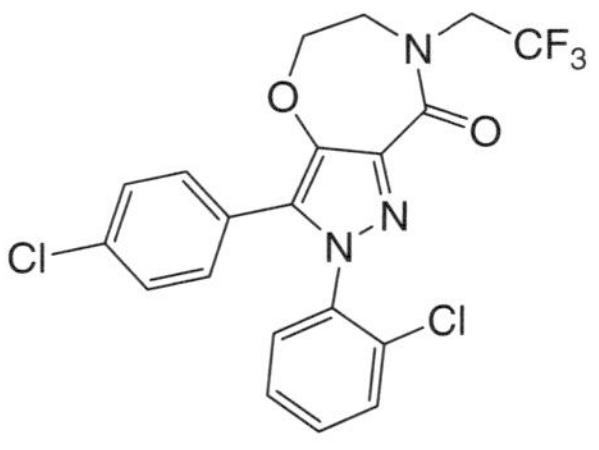

Another bicyclic series exemplified by **8** has been described, with binding affinities less than 200 nM [22]. These 6,7-dihydro-2H-pyrazolo[3,4-f][1,4]oxazepin-8(5H)-one derivatives are conformationally constrained analogs of rimonabant in which a C4 oxygen group has been used as an anchor group for connecting the core heterocycle with the C3 carbamoyl group. As with compounds such as **5**, the potential H-bond acceptor group in **8** projects into a different region of space compared to the corresponding group in the purported energetically favored conformer of rimonabant [21].

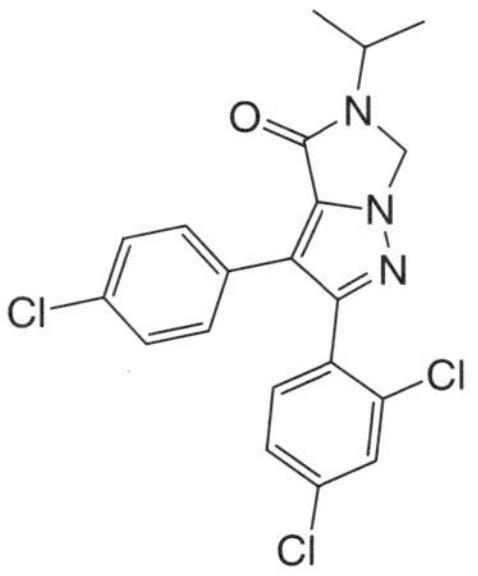

A class of potent 5,5-bicyclic derivatives exemplified by **9** (hCB1R K_i < 8 nM) was recently described [23]. Like the pyrazoyl derivative **2**, these 5,6-dihydroimidazo[1,5-b]pyrazol-4-one derivatives contain an H-bond acceptor group in a region of space that would be occupied by the C4 methyl group in rimonabant, assuming overlap of similar aromatic residues. The functional activities of these compounds have not been described.

New 6,6-bicyclic 1,8-naphthyridinones (e.g., **10**) were identified by appending a second ring between the C2 and C3 positions of pyridyl derivatives such as **11** which were reported to exhibit B/P ratios of 0.03–0.26 at 0.2–4 h post dosing [24]. The design of these bicyclic compounds was guided by the hypothesis that altering certain physicochemical properties in **11** such as lipophilicity (e.g., calculated logD or cLogD) would lead to improvements in pharmacokinetic properties. The cLogD value of **10** was significantly lower than that of **11** (i.e., 3.1 versus 7.9). Compound **10** was slightly less potent than the pyridyl analog **11** in the binding assay (hCB1R K_i = 7.5 nM versus hCB1R K_i = 1.3 nM). The oral bioavailability of **10** was 93% and its B/P ratio was 2.7 at 4 h after an intravenous dose. This compound decreased food intake and body weight following oral administration in a dose-dependent manner in a diet-induced obesity (DIO) rat model. Removal of the *N*-methyl group in the core heterocycle resulted in a 10-fold loss of binding affinity.

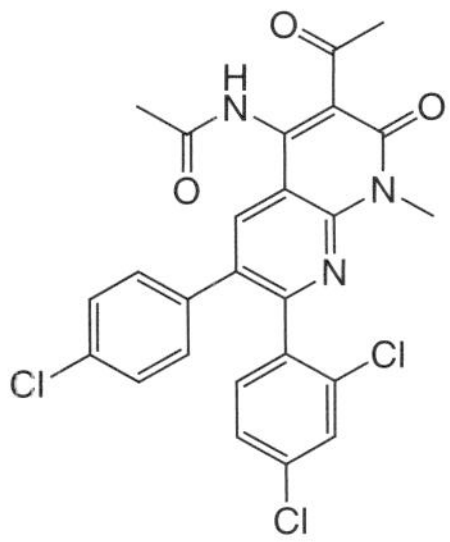

10 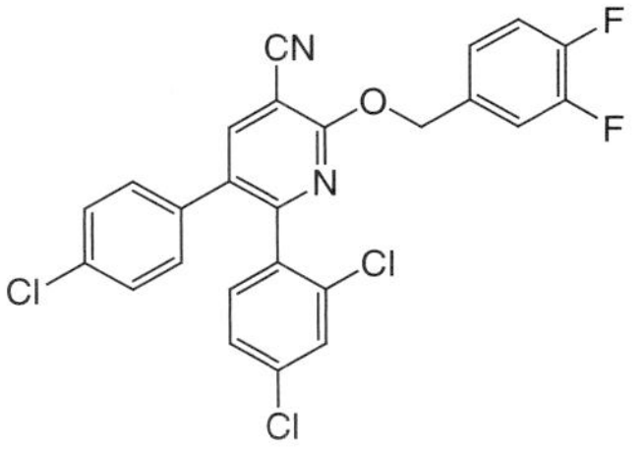 **11** 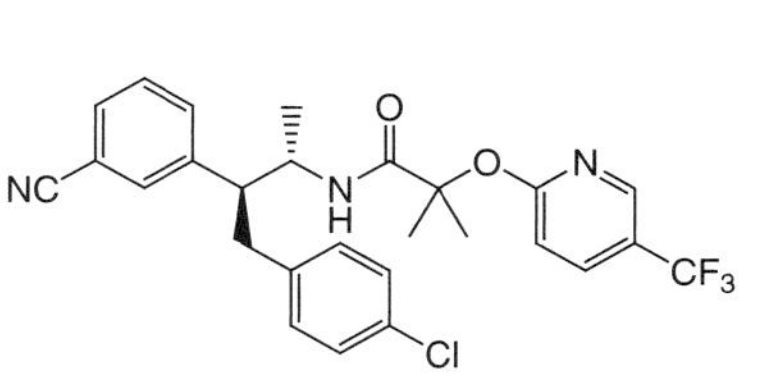 **12**

Several structurally distinct series compounds have been disclosed in the patent application literature, although minimal biological characterization has been provided. A patent application has described a stereoselective synthesis and new polymorphic forms of the carbamoyl derivative **12**, a compound previously claimed as a CB1R antagonist [25,26]. Two other patent applications have claimed novel 1,2,3,4-tetrahydroquinoline compounds such as **13** and indolyl derivatives exemplified by **14** as CB1R antagonists [27,28].

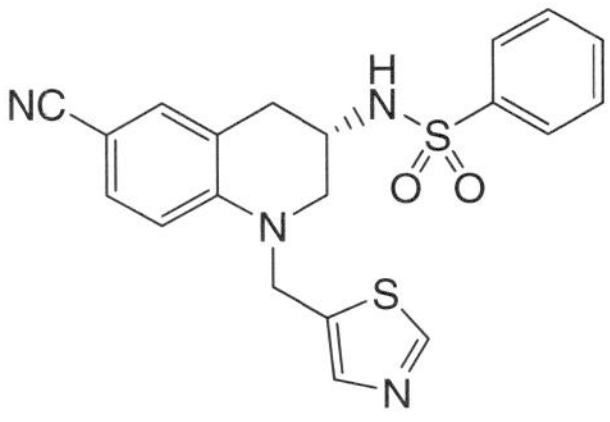

13 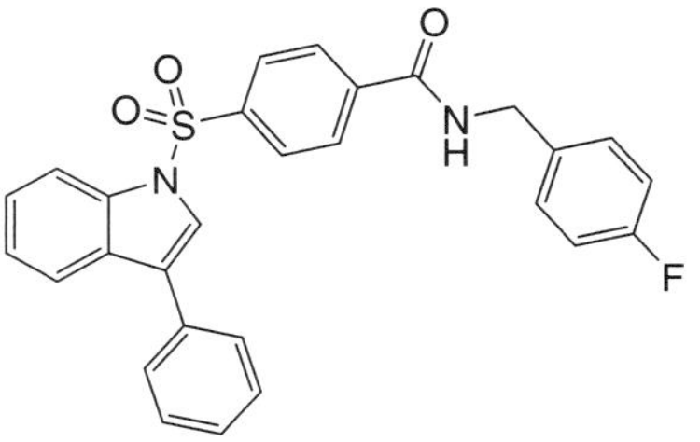 **14**

3. NEUROPEPTIDE RECEPTOR MODULATORS

3.1. MC4R agonists

The search for melanocortin-4 receptor (MC4R) agonist drugs continues to be a highly active area of research, with over 100 publications and patent applications appearing since the start of 2005. The biology of this area has recently been reviewed [29], and some of the rationale for this approach has been described in a recent journal issue dedicated to advances in melanocortin receptor pharmacology [30]. A review of small molecule melanocortin ligands (mainly focused on MC4R agonists) covering the literature to mid-2005 has also appeared recently [31]. Interesting biochemistry of the MC4R was reported in the past year suggesting that the receptor is constitutively active *in vivo* and that this is achieved by 'self-activation' from the N-terminus [32]. A model of the receptor-binding site derived using a combination of site-directed mutagenesis and molecular modeling has also appeared, which may provide guidance for agonist design [33].

The majority of MC4R patent applications and publications continue to be focused around two main structural classes – piperidyl amides derived from 4-halophenylalanine (class (1)), and piperidyl amides derived from 3,4-disubstituted pyrrolidines (class (2)). In class (1), compound **15** (RY764) appears to be the most advanced member to be publicly disclosed having been optimized for potency, selectivity for the MC4R, pharmacokinetic properties, and reduced formation of covalent protein adducts (a marker of toxicity) [34]. Discovery of this agent appears to have been prompted by problems encountered with the previously reported and structurally related agent **16** (MB243) [35,36]. Compound **15** was shown to be efficacious in rodent models of food intake.

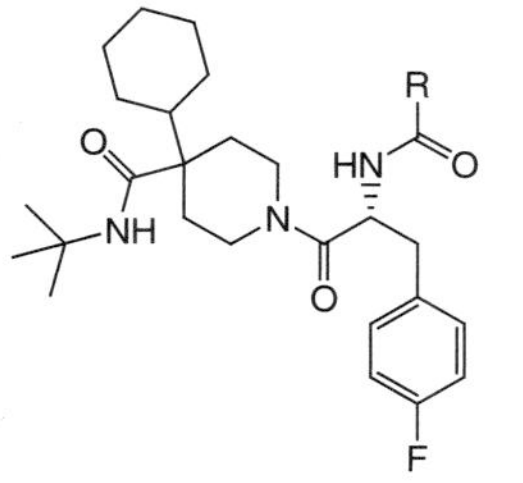

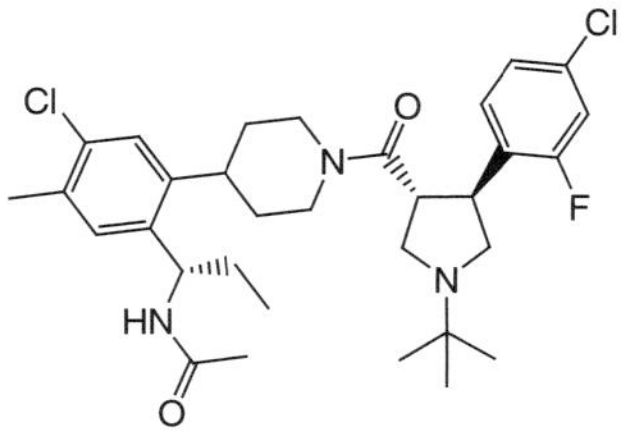

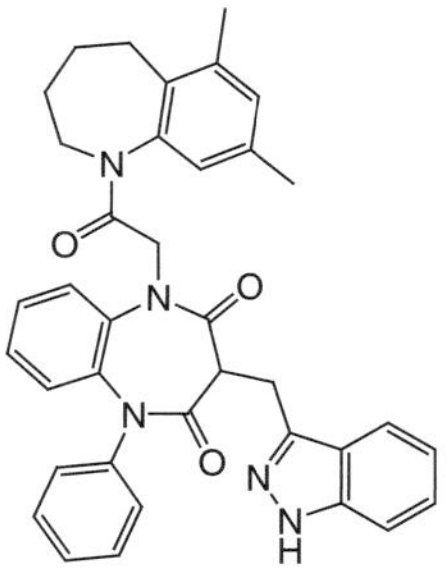

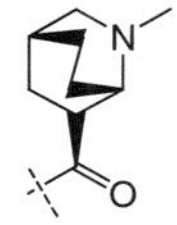

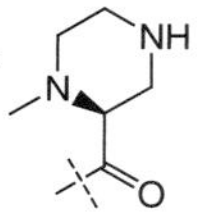

Most of the work around class (2) agents has appeared in the patent application literature with limited detailed SAR. The genesis of this series has been described in a recent presentation [37]. Compound **17** is typical of examples from a recent application in which MC4R activity is claimed to be $<10\,\mu M$ [38].

Alternative small molecule MC4R agonist templates are starting to emerge in the patent application literature, for example the benzodiazepine **18** was shown to have a pEC_{50} value of 7.22 in an MC4R functional screen [39].

Interestingly, while the majority of patent applications and publications appear to be focused toward small molecule MC4R agonists, peptidic agonists are still under active investigation. The cyclic derivative Ac-cyclo[hCys-His-(4Cl-D-Phe)-Arg-Trp-penicillamine]NH$_2$ (**19**) was recently reported to have an MC4R K_i value of 0.03 nM [40].

3.2. MCH1R antagonists

Activity remains intense in this area and, from information in the public domain, at least two agents have now entered clinical trials [41]. Medicinal chemistry efforts toward this target have yielded an extraordinary number and variety of differentiated

Centrally Acting Anti-Obesity Agents

chemical series, most of which have been described in several recent reviews [42–45]. The biology of this target has also been recently reviewed [46].

Several papers describe SAR around different classes of basic MCH1R antagonist series. It has been recently disclosed that several of these series carried a cardiovascular (CV) safety liability unconnected with the MCH1R or human-ether-a-go-go related gene (hERG) pharmacology [47]. A discovery effort to preserve MCH1R activity while removing any CV liabilities resulted in the identification of a series of coumarin derivatives (e.g., **20**) with a clean cardiovascular profile. A compound from this series (e.g., **21**) was shown to be efficacious in a DIO mouse model of obesity at 10 mg/kg *po*. [48].

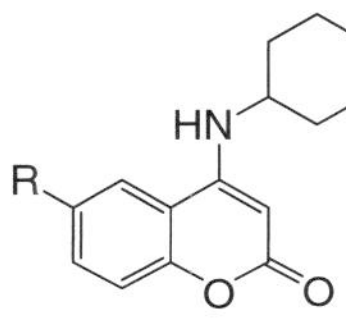

20 R = OMe
21 R = Cl
22 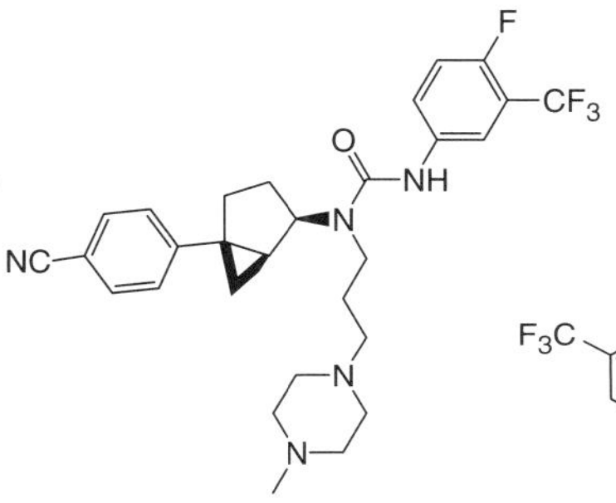
23 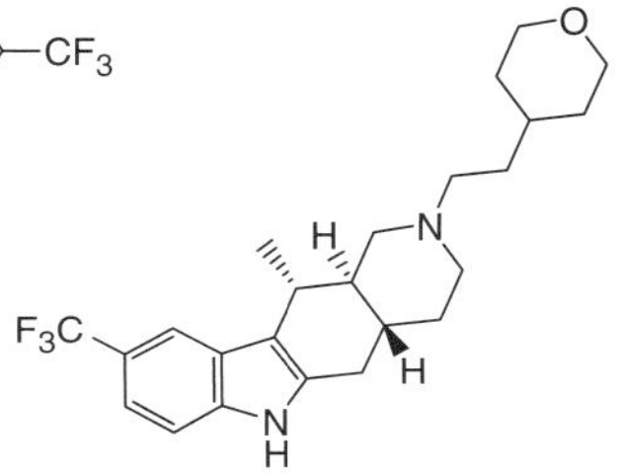

Detailed SAR and *in vivo* pharmacology has been presented for a bicyclo[3.1.0]hexyl- (e.g., **22**) and a closely related bicyclic[4.1.0]heptyl-urea series [49,50]. Interestingly, anorectic activities of these compounds in rodent models correlated with *ex vivo* receptor occupancy (RO) of MCH1R. A high (>70%) and sustained RO was reported to be required to drive significant efficacy in the models [51].

Studies in a carbazole series, which had previously only been disclosed in the patent application literature, have been described leading to **23** (T0910792) a potent and selective MCH1R antagonist (K_i = 0.3 nM) reportedly showing efficacy in numerous obesity models [52], including reduction of body weight and food consumption in cynomolgus monkeys at 10 mg/kg [53].

3.3. GHSR antagonists

Ghrelin is an octanoylated 28 amino acid peptide hormone that is synthesized principally in the oxyntic mucosal cells of the stomach. It binds to and activates the growth hormone secretagogue receptor (GHSR; ghrelin receptor) which is expressed primarily (but certainly not exclusively) in the hypothalamus and the pituitary gland [54]. Intravenous administration of ghrelin has been reported to increase food intake in rodents and humans [55,56]. Plasma ghrelin levels rise pre-prandially and fall post-prandially, indicating a possible role for the hormone as an endogenous gut-derived, orexigenic signal. In the past year, it has been shown that ghrelin and GHSR KO mice are resistant to weight gain when placed on a high fat diet and that the GHSR exhibits ligand-independent activity in transiently transfected cells [57]. While there remains much unknown about ghrelin and human obesity, the abundant animal data and human data demonstrating the orexigenic effect of ghrelin, as well as its cyclical concentrations in humans linked to periods of increased food intake, support discovery of GHSR antagonists or inverse agonists for obesity treatment.

A series of 2,4-diaminopyrimidine GHSR antagonists was identified through high-throughput screening (HTS) [58]. The initial hit **24** exhibited an IC_{50} value of 310 nM in a binding assay and an IC_{50} value of 1.4 µM in a FLIPR Ca^{2+} mobilization assay. Replacement of the C6 ethyl group with a benzyloxymethyl group led to the discovery of **25** (A-778193, GHSR IC_{50} = 19 nM; FLIPR IC_{50} = 180 nM). Selectivity screening revealed off-target activity at dihydrofolate reductase (IC_{50} = 300 nM). Compound **25** was reported to have a B/P ratio of 3 at 1 h and oral bioavailability of 5.5% and was effective in decreasing food intake and body weight in several acute rat models. However, it was also disclosed that **25** decreased food intake in GHSR KO mice, although not to the same extent as in wild-type animals. A second-generation GHSR tool compound **26** was identified with weak DHFR inhibitory activity by replacing the 4-chlorobenzyl group with a 4-methylsulfonylbenzyl group and the C6 benzyloxymethyl group with a tetrahydrofuranyl moiety [59].

24 R = Et; R′ = CH$_2$(4-Cl-Ph) **27** Ar = 2,6-di-Cl-Ph, R = Me **29** R = H; R′ = H

25 R = CH$_2$OCH$_2$Ph **28** Ar = 2,6-di-Cl-Ph, R = [1,3-dioxane] **30** R = OMe
R′ = CH$_2$(4-Cl-Ph)
R′ = N(Me)CO$_2$CH$_2$CH(CH$_3$)$_2$

26 R = [tetrahydrofuranyl]
R′ = CH$_2$(4-MeSO$_2$Ph)

Several other series of GHSR antagonists have also been reported. An isoxazole derivative **27** (GHSR binding IC_{50} = 130 nM; Fluorometric Imaging Plate Reader (FLIPR) IC_{50} = 180 nM) was identified from HTS [60]. Extension of the C5 methyl group resulted in the discovery of the 1,3-dioxanepropyl derivative **28** (IC_{50} = 6 nM, FLIPR IC_{50} = 14 nM) [61]. Compounds in this series were reported to be highly cleared in microsomal preparations. Scaffold manipulation of **28** led to the identification of a series of tetralin derivatives, such as **29** [62]. Introduction of a C2 isobutylcarbamate group geminal to the carbamoyl group and a C8 methoxy group on the tetralin ring in **29** afforded **30** (IC_{50} = 16 nM; FLIPR IC_{50} = 29 nM). The bioavailability of **30** was

19% in the rat, in contrast to <0.6% for an analog of **29**. The improved bioavailability of **30** was attributed to quaternization of the C2 position which rendered the amide group less resistant toward metabolic hydrolysis. Compound **30** showed only weak activity against a panel of receptors. Despite their *in vitro* binding activities, compounds in this tetralin series did not show any efficacy in animal models [21].

Recently, bicyclic and tricyclic GHSR antagonists have also been disclosed in two different patent applications. The 6,7-bicyclic compound, the 3,4-dihydro-1H-benzo[e][1,4]diazepine-2,5-dione **31**, showed an IC_{50} value <10 nM in a GHSR competitive binding assay [63]. The β-carboline derivative **32** was less potent, with a K_i value of 59 nM [64].

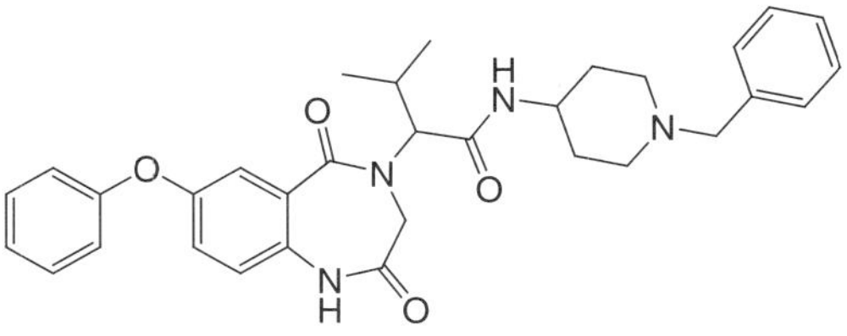

31 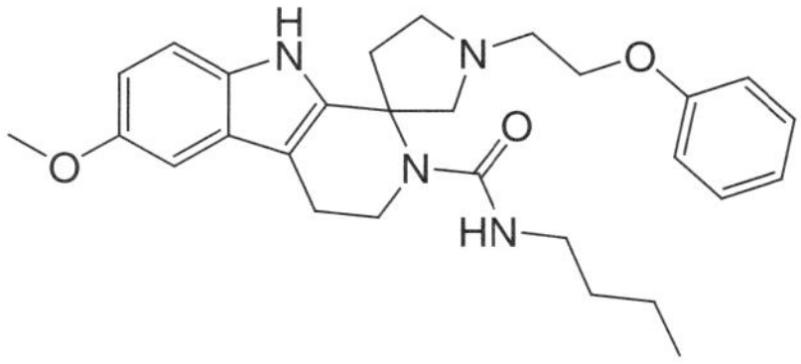 **32**

3.4. Neuropeptide Y family modulators

NPY, PYY, PYY_{3-36}, and PP are a class of polypeptide hormones that belong to the neuropeptide Y family. These peptides act through five receptor subtypes – NPY Y1–Y5 (Y6 subtype is found in mouse but not in primates or rats). Y1R is localized primarily in the brain and in the vasculature, while Y5R is found almost exclusively in the hypothalamus. Y2R is present in both the gastrointestinal tract and in the hypothalamus and is an autoreceptor, which negatively regulates NPY secretion. Y4R is located in both the brainstem and the NTS. NPY and PYY bind preferentially to Y1R, Y2R, and Y5R. PYY is cleaved by dipeptidylpeptidase-IV (DPP-IV) to give PYY_{3-36} which is selective for Y2R over the other receptors. PP is the only member of this peptide family that binds preferentially to Y4R. Roles for Y1R and Y5R in natural feeding have not been fully established. In contrast, Y2R and Y4R are believed to participate in the regulation of body weight since PYY_{3-36} and PP have been shown to reduce food intake in rodent and non-human primate models [65,66]. Modified versions of PYY_{3-36} are currently under clinical evaluation [67]. The mechanisms-of-action by which PYY_{3-36} and PP exert their anorectic activities are not known, but may involve vagal afferent pathways that proceed through the brain stem to the hypothalamus [68]. A review of the major developments in this area was last published in 2003 [69].

3.4.1. NPY Y2R and Y4R agonists

New Y2R and Y4R agonists are peptidic derivatives derived from NPY or PYY. N-terminal acylation of PYY fragments such as PYY_{26-36} and PYY_{27-36} has provided Y2R agonists such as **33** with good binding affinities and receptor subtype selectivities [70]. PEGylated analogs of these fragments have been prepared to extend to half-life. A series of 2,7-D/L-diaminosuberic acid derivatives containing two pentapeptide fragments derived from the C-terminal of NPY (i.e., NPY_{32-36}) have exhibited good Y4R binding and functional activity and selectivity against the other NPY receptor subtypes [71]. Compound **34** ($K_i = 0.95$ nM; $EC_{50} = 14.8$ nM) inhibited 2 and 4 h food intake in fasted mice following intraperitoneal administration (100 nmol/mouse), but had no effects on food intake in Y4R KO mice. Both compound **34** and PP exhibited similar binding affinities, but PP was >100-fold more active in a cAMP functional assay. The difference between the binding and functional activities of these compounds was attributed to the presence of low $[Na^+]$ concentrations in the membrane preparations used for binding.

3.4.2. NPY Y5R antagonists

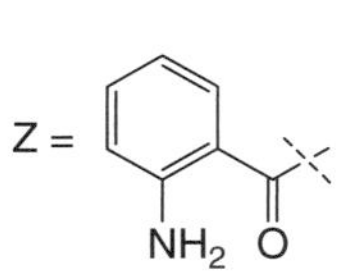

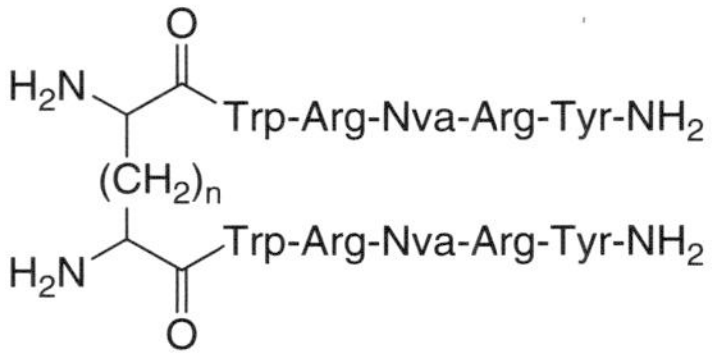

The published data on Y5R antagonists have provided a conflicting picture of the role of this receptor in body weight regulation. Several antagonists have been shown to decrease food intake in various rodent models, but were later found to be anorectic in Y5R KO mice [72]. Other antagonists with good central exposure were inactive in DIO rodent models despite high binding affinities and good functional potencies [73]. In the last year, however, several reports have appeared that suggest Y5R may be physiologically important in obesity. Most noteworthy, extensive effort has been devoted to proving specificity-of-action.

A tetrahydrocarbazole urea derivative **35** (FMS586, $IC_{50} = 4.3$ nM) was shown to inhibit fasting-induced re-feeding behavior and transiently suppress natural feeding behavior in Wistar rats after oral doses of 25–100 mg/kg [74]. Compound **35** did not induce *c-Fos* expression after oral administration as measured by immunohistochemical analysis, but blocked hPP-induced *c-Fos* expression in Y5R-positive cells, thus demonstrating *in vivo* Y5 antagonism.

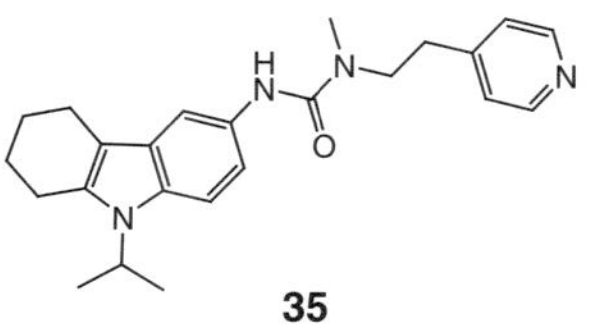

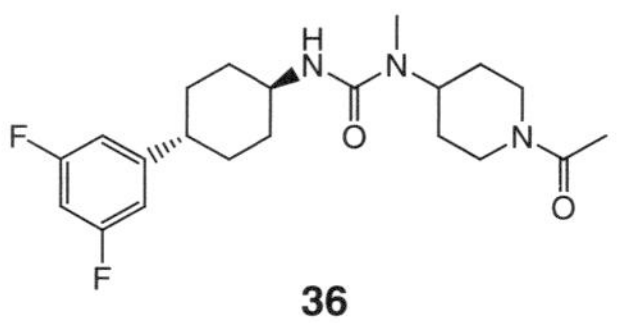

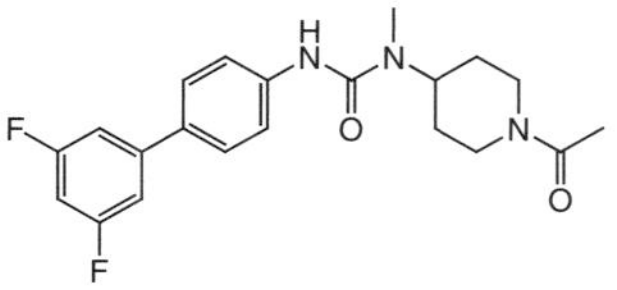

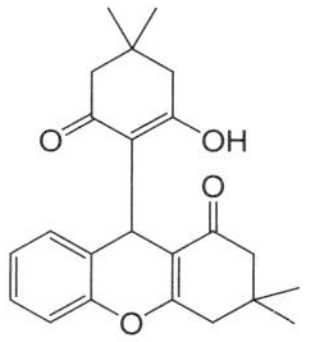

38 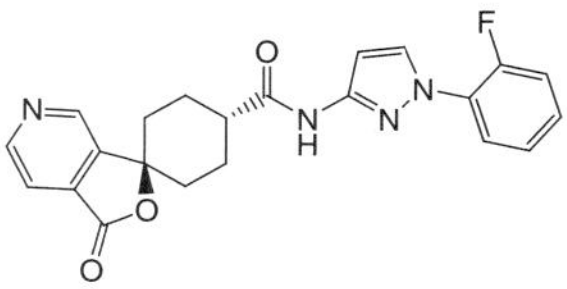 **39**

A phenylcyclohexylamine derivative **36** (rat and human $K_i = 1.8$ nM) was also found to decrease food intake and body weight in a 28-day study in DIO rats [75]. This cyclohexyl compound was derived from a biphenyl analog **37** ($K_i = 1.1$ nM) which had a mutagenicity liability. Interestingly, compounds **36** and **37** were equipotent in the binding assay despite different three-dimensional conformations.

A tricyclic derivative **38** (L-152804) has been studied in different rodent models of obesity [76,77]. This compound showed moderate Y5R binding affinity (mY5R = 44 nM, rY5 $K_i = 31$ nM, Ca^{2+} $IC_{50} = 210$ nM) and was claimed to have little off-target activity (no significant cross reactivity with 120 other binding assays and seven enzyme assays). Compound **38** decreased body-weight gain in DIO mice, reduced adipose tissue mass, and improved DIO-associated hyperinsulinemia. It was specifically reported that high and sustained ROs were required for activity. Compound **38** was inactive in Y5R KO mice. A patent application has claimed that **38** can prevent the decrease in metabolic rate and energy expenditure that can occur with food restriction and loss of body weight [78].

The carbamoyl derivative **39** has been previously disclosed in the patent application literature as a Y5R antagonist [79]. A new paper describes methods for quantifying **39** in human plasma and urine to support human pharmacokinetic studies [80]. This class of compounds is reported to decrease appetite while increasing metabolic rate [79,80].

3.5. Opioid receptor antagonists

The endogenous opioid system has been known to regulate appetitive behavior for over 30 years [81]. In preclinical species, both endogenous peptidic and non-peptidic opioid receptor agonists have exhibited orexigenic properties. Non-selective opioid receptor antagonists including naloxone have shown anorectic activity in rodents [82]. Opiates mediate their activity through three subtypes called the μ-, δ-, and κ-opioid receptors [83]. The μ-opioid receptor plays an important role in feeding behavior since administration of the selective μ-opioid agonist DAMGO within the nucleus accumbens selectively increases consumption of high fat food in rats [84]. A selective opioid receptor antagonist naltrexone has produced mixed results in humans, having either efficacy or no efficacy in promoting weight loss [85–88].

Despite mixed results observed with antagonists in the clinic, the discovery of new selective or pan-selective opioid receptor some antagonists for use as potential weight loss agents has continued. A series of pan-selective opioid antagonists has been identified that reduce food intake and body weight in rodent models [89]. The phenyl-piperidine antagonist **40** was not efficacious in rodent feeding models following oral administration, primarily because of extensive first pass metabolism [90]. Removal of the aromatic hydroxyl group resulted in loss of potency. Replacement of the hydroxyl group with carbamoyl and carbamyl groups led to discovery of **41**, a pan-selective antagonist with bioavailability of 32% in the rat. Compound **41** decreased food intake in fasted Long–Evans rats following a 3 mg/kg oral dose.

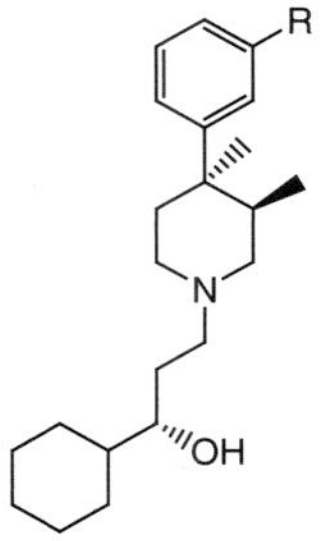

40 R = OH

41 R = CONH$_2$

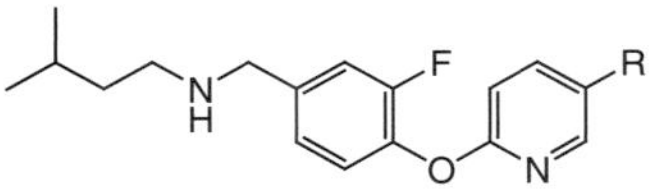

42 R = CONH$_2$

43 R = (4,5-dihydro-1H-imidazolyl)

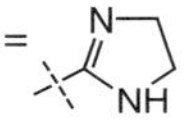

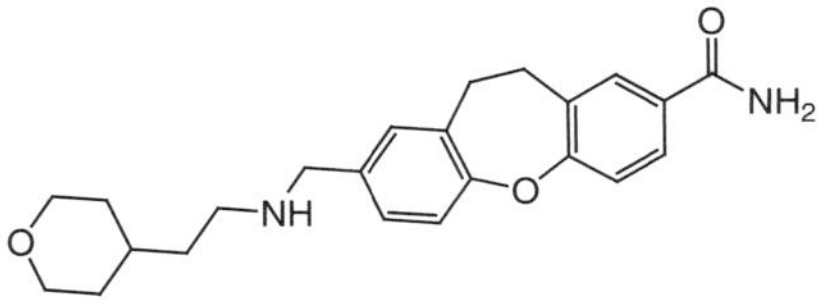

44

A series of structurally unrelated nicotinamide derivatives such as **42** has also been shown to exhibit μ-, κ-, and δ-opioid receptor antagonist activities and cause weight loss in rodent models of **41**. Analogs have been recently described in a series of patent applications. Replacement of the carboxamide group in **42** with heterocyclic groups led to the identification of the 4,5-dihydro-1H-imidazolyl derivative **43** (GTP-γ-^{35}S K_i = 10, 34, and 120 nM at μ-, κ-, and δ-opioid receptors, respectively) [91]. Introduction of an alkyl tether between the aryl rings in a diphenylether analog of **42** provided the tricyclic dibenzo[b,f]oxepine compound **44** which showed good functional activities at the different subtypes (GTP-γ-^{35}S K_b = 0.7, 2.2, and 2.1 nM at μ-, κ-, and δ-opioid receptors) [92].

4. AMINERGIC RECEPTOR MODULATORS

4.1. 5HT2CR agonists

The 5-HT2C receptor (5HT2CR) was identified in 1986 and is widely expressed in the CNS [93]. Several lines of evidence support a role for this receptor in body weight regulation. KO mice exhibit a phenotype characterized by increased body weight relative to wild-type littermates (average of 13% increase), increased food intake, and a significantly greater percentage of adipose tissue (48% increase) [94]. The non-selective 5HT2CR agonist mCPP was found to reduce food intake in wild-type mice, but not in 5HT2CR KO mice. Furthermore, the 5-HT-releasing compound d-fenfluramine, which inhibited food intake in both rodents and man, was shown to have weaker anorectic activity in 5HT2CR KO mice [95]. At least one 5HT2CR agonist has been reported to be in clinical development as described in a 2006 review of the medicinal chemistry in this area [93].

The anti-obesity effects of several different 5HT2CR agonists have been recently reported. The tetracyclic analog **45** (WAY163909) is a novel 5HT2CR full agonist that was discovered by using a pharmacophoric model for 5HT2CR binding activity. Compound **45** showed greater than 20-fold binding selectivity for 5HT2 CR over both 5HT2A receptor (5HT2AR) and the 5HT2B receptor (5HT2BR) [96]. Functionally, the compound stimulated intracellular calcium mobilization with an EC_{50} value of 8 nM. It showed no agonist activity at 5HT2AR, but partial agonist activity at 5HT2BR ($EC_{50} = 185$ nM, $E_{max} = 40\%$). It also produced dose-dependent reductions in food intake in normal SD rats, an effect blocked by a 5HT2CR antagonist. In a 10-day study, **45** caused a 56% decrease in body weight, with reductions also observed in triglyceride levels. No toleration of the anorectic effect was observed, in contrast to studies with other 5HT2CR agonists [97].

The non-selective tricyclic derivative hexahydro-pyrido[3′,2′:4,5]pyrrolo[1,2-a] pyrazine **46** was used as a starting point for the discovery of a selective 5HT2C receptor agonist in the same series [98]. Introduction of a methyl group on the piperazinyl ring in the core heterocycle and separation of the 4R,10aR di astereomer provided compounds with good 5HT2CR binding, but little subtype selectivity. Variation of the substituent arrangement on the aryl group improved the binding selectivity. An analog **47** was identified that showed greater than 10-fold selectivity for 5HT2CR ($K_i = 1.9$ nM) compared to either 5HT2AR ($K_i = 40$ nM) or 5HT2BR ($K_i = 19$ nM). This compound was shown to reduce food intake in a dose-dependent manner in a fasting-induced re-feeding model in Wistar rats (MED = 10 mg/kg po).

Several issues were identified with **47** including phospholipidosis and hERG activity ($IC_{50} = 2.5$ μM), attributed to the lipophilicity and amphiphilicity of the core heterocycle [98]. Replacement of the tricyclic ring in **47** with a hexahydro-pyrido[3′,2′:4,5]pyrrolo[1,2-a]pyrazine group resulted in compounds such as **48** with reduced hERG activity and 20- and 48-fold binding selectivity for 5HT2CR over 5HT2AR and 5HT2BR. Compound **48** was also shown to decrease 2 h food intake in a fasting-induced re-feeding model in Wistar rats in a dose-dependent manner, with a MED of 10 mg/kg po, an effect antagonized by the 5HT2CR antagonist SB-242084.

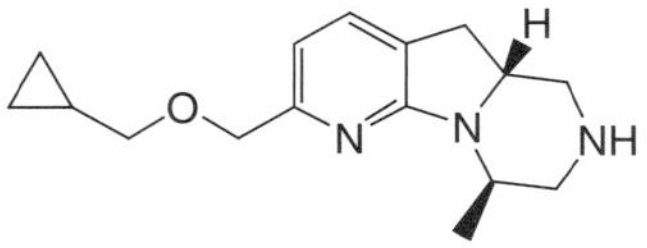

48 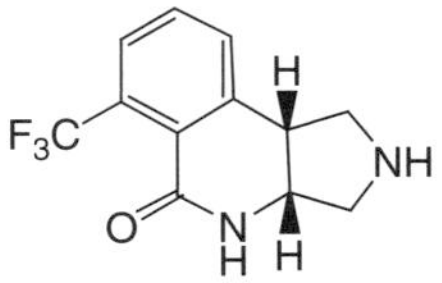 **49** 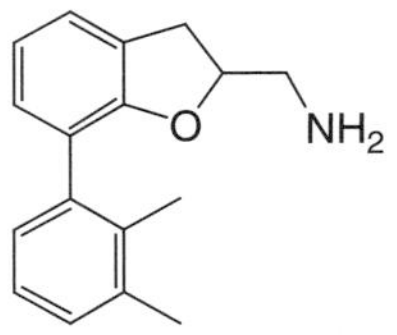 **50**

Several new 5HT2CR chemotypes have recently been disclosed in the patent application literature. A tricyclic compound **49** was reported to exhibit high selectivity for 5HT2CR over the other subtypes [99]. A bicyclic dihydrobenzofuranyl alkanamine **50** was also claimed to exhibit high binding affinity for 5HT2CR, with a K_i value of 1.1 nM [100].

4.2. 5HT6R antagonists

The 5-HT6 receptor (5HT6R) is expressed almost exclusively in the CNS. Several pharmacological studies and genetic models support a role for this receptor in obesity. 5HT6R KO mice were found to be resistant to body weight gain on a high fat diet [101]. Administration of 5HT6R antisense oligonucleotide complementary to bases 1–18 of the rat 5HT6R cDNA initiation sequence was shown to reduce both food intake and body weight of diet-induced obese rats over a 6-day period [102]. A brain penetrant small molecule antagonist **51** ($pK_i = 8.6$) was reported to inhibit fasting-induced re-feeding behavior in male Wistar rats at 24 h post dose [103,104]. Mechanistically, these 5HT6R antagonists are believed to decrease γ-aminobutyric acid-mediated signaling in the hypothalamus, resulting in release of α-MSH [104].

Numerous 5HT6R antagonists have been reported in the past year, especially in the patent application literature, but it has been difficult to tell which compounds may be useful as weight loss agents. A patent application claims a series of benzoxazinone-derived piperidinesulfonamides, such as **52** ($K_i = 152$ nM), for the treatment of food intake disorders and other diseases. A recent review describes progress in this area [104].

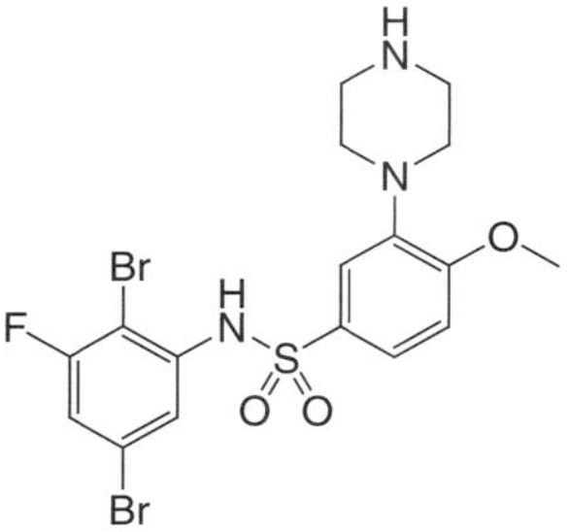

51 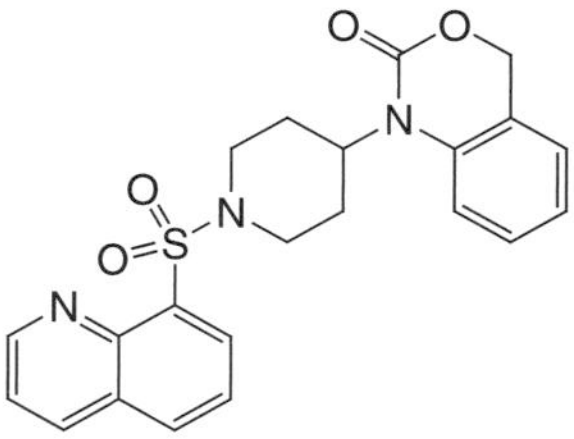 **52**

4.3. H3R antagonists

The histamine H3 receptor (H3R) is a G-protein coupled receptor localized principally in the CNS that plays a role in cognitive and homeostatic control, including appetitive behavior. Targeted disruption of H3R in mice leads to an obese phenotype, including increased weight, food intake, adiposity, and reduced energy expenditure [105]. Some potent H3R antagonists have been reported to decrease food intake, but show minimal activity in other CNS models. Other antagonists have shown efficacy in CNS models, but not obesity models. The reasons for this discrepancy are not known, but may be due to different splice variants in the hypothalamus versus the other areas of the brain. Several recent reviews describe in more detail some of the other reasons for the varying activities of H3 antagonists in obesity versus CNS models [106,107].

A piperazine derivative **53** (NNC 38-1049) was recently described as a potent H3R competitive antagonist [hK_b = 2.3 nM (cAMP assay), rK_i = 5 nM (GTP-γ-^{35}S assay)]. This compound was brain penetrant, with hypothalamic exposure of 3000 ng/g tissue following a 60 mg/kg oral dose [108]. It was also shown to increase histamine levels in the PVN, which has a large population of H1R and H3R and is the likely site of action for the anti-obesity effects. Oral administration of **53** to DIO rats resulted in a reduction of food intake and a significant decrease in body weight. No changes were observed in the behavioral satiety sequence or in pica consumption following intraperitoneal administration.

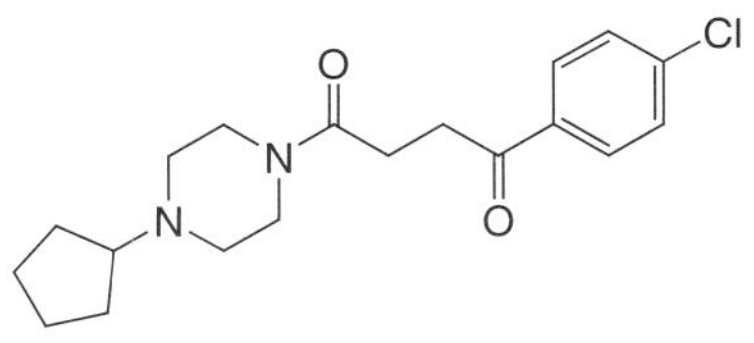

53

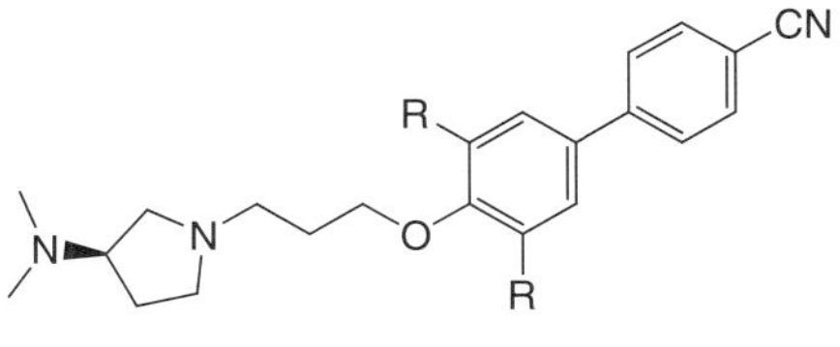

54 R = F

55 R = H

The biphenyl derivative **54** (A-423579, hH3R pK_i = 8.69, rH3R pK_i = 8.27) has been shown to cause weight loss in rodent models [109]. Compound **54** exhibited both antagonist and inverse agonist activities (GTP-γ-^{35}S inverse agonism pEC$_{50}$ = 7.71). This pyrrolidine derivative was efficacious in DIO rats following oral administration. Weight loss with **54** was gradual and cumulative, unlike that observed for the positive control in the study, sibutramine, which diminished over time. Compound **54** did not decrease food intake in H3R KO mice [110]. The des-fluoro analog **55** was equipotent to **54** in a binding assay, but was found to have genetic toxicology issues [109]. As previously noted for compounds that target H3R, **54** was efficacious only in anti-obesity models – it did not show any significant activity in an inhibitory avoidance model that was developed to characterize the attentional and cognitive effects of H3R antagonists [110].

Many new H3R chemotypes have been reported in the patent application literature in the past year. However it is not clear which compounds may be useful as weight-loss agents or as agents to treat CNS indications such as attention-deficit hyperactivity disorder (ADHD). Several reviews describe new H3R antagonists that have recently been disclosed [106,107].

5. MONOAMINE RE-UPTAKE INHIBITORS

Interest in monamine re-uptake inhibitors for obesity treatment is supported by clinical efficacy for sibutramine, bupropion, and more recently, reports of weight loss efficacy for 2,3-disubstituted tropane derivatives [111]. Owing to the multiple indications under investigation for this mechanism, a thorough review is beyond the scope of this article.

6. CONCLUSIONS

Obesity is a chronic disease associated with numerous co-morbidities and whose prevalence is increasing throughout the world. There is considerable unmet medical need for safe and effective treatments, providing a considerable impetus for the ongoing research in this field. Much attention has been focused on the discovery of centrally acting agents, with the most active areas of research being described in this report. While CB1R antagonists are most likely to comprise the next wave of weight loss drugs, the considerable efforts of researchers in this area make it likely that agents targeting other central pathways will advance through clinical trials and reach the patient.

REFERENCES

[1] A. H. Mokdad, J. S. Marks, D. F. Stroup and J. L. Gerberding, *J. Am. Med. Assoc.*, 2004, **291**, 1238.
[2] M. W. Schwartz and D. Porte, Jr., *Science*, 2005, **307**, 375.
[3] S. P. Kalra, M. G. Dube, S. Pu, B. Xu, T. L. Horvath and P. S. Kalra, *Endocr. Rev.*, 1999, **20**, 68.
[4] L. Hanus, L. S. Abu, E. Fride, A. Breuer, Z. Vogel, D. E. Shalev, I. Kustanovich and R. Mechoulam, *Proc. Natl. Acad. Sci. USA.*, 2001, **98**, 3662.
[5] A. C. Howlett, F. Barth, T. I. Bonner, G. Cabral, P. Casellas, W. A. Devane, C. C. Felder, M. Herkenham, K. Mackie, B. R. Martin, R. Mechoulam and R. G. Pertwee, *Pharmacol. Rev.*, 2002, **54**, 161.
[6] C. M. Williams and T. C. Kirkham, *Psychopharmacology*, 1999, **143**, 315.
[7] M. Rinaldi-Carmona, F. Barth, M. Heaulme, D. Shire, B. Calandra, C. Congy, S. Martinez, J. Maruani, G. Neliat and D. Caput, *FEBS Lett.*, 1994, **350**, 240.
[8] G. Colombo, R. Agabio, G. Diaz, C. Lobina, R. Reali and G. L. Gessa, *Life Sci.*, 1998, **63**, 113.

[9] T. C. Ravinet, M. Arnone, C. Delgorge, N. Gonalons, P. Keane, J.-P. Maffrand and P. Soubrie, *Am. J. Physiol. Regul. Integr. Comp. Physiol.*, 2003, **284**, R345.
[10] A. L. Hildebrandt, S. D. M. Kelly and S. C. Black, *Eur. J. Pharmacol.*, 2003, **462**, 125.
[11] V. Di Marzo, S. K. Goparaju, L. Wang, J. Liu, S. Bátkai, Z. Járai, F. Fezza, G. I. Miura, R. D. Palmiter, T. Sugiura and G. Kunos, *Nature*, 2001, **410**, 822.
[12] J.-P. Despres, A. Golay and L. Sjoestroem, *N. Engl. J. Med.*, 2005, **353**, 2121.
[13] J. H. M. Lange and C. G. Kruse, *Drug Discov. Today*, 2005, **10**, 693.
[14] G. G. Muccioli and D. M. Lambert, *Curr. Med. Chem.*, 2005, **12**, 1361.
[15] R. A. Smith and Z. Fathi, *IDrugs*, 2005, **8**, 53.
[16] F. Barth, *Annu. Rep. Med. Chem.*, 2005, **40**, 103.
[17] D. P. Hurst, D. L. Lynch, N. J. Barnett, S. M. Hyatt, H. H. Seltzman, M. Zhong, Z. H. Song, J. J. Nie, D. Lewis and P. H. Reggio, *Mol. Pharmacol.*, 2002, **62**, 1274.
[18] A. Pendri, S. Gerritz, D. S. Dodd and C. Sun, *US Patent Application* 2005080087, 2005.
[19] A. Mayweg, R. Narquizian, P. Pflieger and S. Roever, *US Patent Application* 2005250769, 2005.
[20] P. A. Carpino, D. A. Griffith, S. Sakya, R. L. Dow, S. C. Black, J. R. Hadcock, P. A. Iredale, D. O. Scott, M. W. Fichtner, C. R. Rose, R. Day, J. Dibrino, M. Butler, D. B. DeBartolo, D. Dutcher, D. Gautreau, J. S. Lizano, R. E. O'Connor, M. A. Sands, D. Kelly-Sullivan and K. M. Ward, *Bioorg. Med. Chem. Lett.*, 2006, **16**, 731.
[21] J. Y. Shim, W. J. Welsh, E. Cartier, J. L. Edwards and A. C. Howlett, *J. Med. Chem.*, 2002, **45**, 1447.
[22] P. A. Carpino, R. L. Dow and D. A. Griffith, *US Patent Application* 2005101592, 2005.
[23] D. A. Griffith and S. M. Sakya, *WO Patent Application* 2005061507, 2005.
[24] J. S. Debenham, C. B. Madsen-Duggan, T. F. Walsh, J. Wang, X. Tong, G. A. Doss, J. Lao, T. M. Fong, M.-T. Schaeffer, J. C. Xiao, C. R. R. C. Huang, C.-P. Shen, Y. Feng, D. J. Marsh, D. S. Stribling, L. P. Shearman, A. M. Strack, D. E. MacIntyre, L. H. T. Van der Ploeg and M. T. Goulet, *Bioorg. Med. Chem. Lett.*, 2006, **16**, 681.
[25] K. R. Campos, A. Klapars, J. C. McWilliams, C. S. Shultz, D. J. Wallace, A. M. Chen, L. F. Frey, A. V. Peresypkin, Y. Wang, R. M. Wenslow and C.-Y. Chen, *WO Patent Application* 2006017045, 2006.
[26] W. K. Hagmann, L. S. Lin, S. K. Shah, R. N. Guthikonda, H. Qi, L. L. Chang, P. Liu, H. M. Armstrong, J. P. Jewell and T. J. Lanza, Jr., *US Patent Application* 2003077847, 2003.
[27] P. M. Sher, C. Sun, R. B. Sulsky, G. Wu and W. R. Ewing, *US Patent Application* 2005009870, 2005.
[28] J. R. Allen, A. K. Amegadzie, K. M. Gardinier, G. S. Gregory, S. A. Hitchcock, P. J. Hoogestraat, W. D. Jones, Jr. and D. L. Smith, *WO Patent Application* 2005066126, 2005.
[29] A. P. Coll, B. G. Challis, G. S. H. Yeo, I. S. Farooqi and S. O'Rahilly, *Curr. Opin. Endocrinol. Diabetes*, 2005, **12**, 205.
[30] C. Haskell-Luevano, *Peptides*, 2006, **27**, 257.
[31] A. Todorovic and C. Haskell-Luevano, *Peptides*, 2005, **26**, 2026.
[32] R. A. H. Adan, *Trends Pharmacol. Sci.*, 2006, **27**, 183.
[33] K. Hogan, S. Peluso, S. Gould, I. Parsons, D. Ryan, L. Wu and I. Visiers, *J. Med. Chem.*, 2006, **49**, 911.
[34] Z. Ye, L. Guo, K. J. Barakat, P. G. Pollard, B. L. Palucki, I. K. Sebhat, R. K. Bakshi, R. Tang, R. N. Kalyani, A. Vongs, A. S. Chen, H. Y. Chen, C. I. Rosenblum, T. MacNeil, D. H. Weinberg, Q. Peng, C. Tamvakopoulos, R. R. Miller, R. A. Stearns, D. E. Cashen, W. J. Martin, J. M. Metzger, A. M. Strack, D. E. MacIntyre, L. H. T. Van der Ploeg, A. A. Patchett, M. J. Wyvratt and R. P. Nargund, *Bioorg. Med. Chem. Lett.*, 2005, **15**, 3501.

[35] B. L. Palucki, M. K. Park, R. P. Nargund, R. Tang, T. MacNeil, D. H. Weinberg, A. Vongs, C. I. Rosenblum, G. A. Doss, R. R. Miller, R. A. Stearns, Q. Peng, C. Tamvakopoulos, L. H. T. Van der Ploeg and A. A. Patchett, *Bioorg. Med. Chem. Lett.*, 2005, **15**, 1993.

[36] G. A. Doss, R. R. Miller, Z. Zhang, Y. Teffera, R. P. Nargund, B. Palucki, M. K. Park, Y. S. Tang, D. C. Evans, T. A. Baillie and R. A. Stearns, *Chem. Res. Toxicol.*, 2005, **18**, 271.

[37] F. Ujjainwalla, *Abstracts of Papers, 230th ACS National Meeting, Washington, DC, United States, Aug. 28–Sept. 1, 2005*, 2005, MEDI.

[38] K. J. Barakat, L. Guo, J. Liu, R. P. Nargund, I. K. Sebhat and Z. Ye, *WO Patent Application* 2006019787, 2006.

[39] J. R. Szewczyk and K. H. Donaldson, *US Patent Application* 2006003991, 2006.

[40] D. B. Flora, M. L. Heiman, J. L. Hertel, H. M. Hsiung, J. P. Mayer, D. L. Smiley, L. Z. Yan and L. Zhang, *WO Patent Application*, 2005000339, 2005.

[41] Pipeline information available at www.gsk.com and www.amgen.com; see also IDdb, Thomson Scientific.

[42] H. J. Dyke and N. C. Ray, *Expert Opin. Ther. Pat.*, 2005, **15**, 1303.

[43] T. J. Kowalski and M. D. McBriar, *Expert Opin. Invest. Drugs*, 2004, **13**, 1113.

[44] A. Browning, *Expert Opin. Ther. Pat.*, 2004, **14**, 313.

[45] T. J. Kowalski and M. D. McBriar, *Annu. Rep. Med. Chem.*, 2005, **40**, 119.

[46] G. J. Hervieu, *Expert Opin. Ther. Targets*, 2006, **10**, 211.

[47] P. R. Kym, A. J. Souers, T. J. Campbell, J. K. Lynch, A. S. Judd, R. Iyengar, A. Vasudevan, J. Gao, J. C. Freeman, D. Wodka, M. Mulhern, G. Zhao, S. H. Wagaw, J. J. Napier, S. Brodjian, B. D. Dayton, R. M. Reilly, J. A. Segreti, R. M. Fryer, L. C. Preusser, G. A. Reinhart, L. Hernandez, K. C. Marsh, H. L. Sham, C. A. Collins and J. S. Polakowski, *J. Med. Chem.*, 2006, **49**, 2339.

[48] P. R. Kym, R. Iyengar, A. J. Souers, J. K. Lynch, A. S. Judd, J. Gao, J. Freeman, M. Mulhern, G. Zhao, A. Vasudevan, D. Wodka, C. Blackburn, J. Brown, J. L. Che, C. Cullis, S. J. Lai, M. LaMarche, T. Marsilje, J. Roses, T. Sells, B. Geddes, E. Govek, M. Patane, D. Fry, B. D. Dayton, S. Brodjian, D. Falls, M. Brune, E. Bush, R. Shapiro, V. Knourek-Segel, T. Fey, C. McDowell, G. A. Reinhart, L. C. Preusser, K. Marsh, L. Hernandez, H. L. Sham and C. A. Collins, *J. Med. Chem.*, 2005, **48**, 5888.

[49] M. D. McBriar, H. Guzik, S. Shapiro, J. Paruchova, R. Xu, A. Palani, J. W. Clader, K. Cox, W. J. Greenlee, B. E. Hawes, T. J. Kowalski, K. O'Neill, B. D. Spar, B. Weig, D. J. Weston, C. Farley and J. Cook, *J. Med. Chem.*, 2006, **49**, 2294.

[50] R. Xu, S. Li, J. Paruchova, M. D. McBriar, H. Guzik, A. Palani, J. W. Clader, K. Cox, W. J. Greenlee, B. E. Hawes, T. J. Kowalski, K. O'Neill, B. D. Spar, B. Weig and D. J. Weston, *Bioorg. Med. Chem.*, 2006, **14**, 3285.

[51] T. J. Kowalski, B. D. Spar, B. Weig, C. Farley, J. Cook, L. Ghibaudi, S. Fried, K. O'Neill, R. A. Del Vecchio, M. McBriar, H. Guzik, J. Clader, B. E. Hawes and J. Hwa, *Eur. J. Pharmacol.*, 2006, **535**, 182.

[52] L. Li, P. Fan, X. Chen, J. Mihalic, Y. Fu, K. Dai, L. Liang, M. Reed, M. Wright, P. Timmermans, J.-L. Chen and J. Jaen, *Abstracts of Papers, 231st ACS National Meeting, Atlanta, GA, United States, March 26–30*, 2006, MEDI.

[53] See entry for T0910792 in Investigational Drugs database (IDdb), Thomson Scientific.

[54] M. Kojima, H. Hosoda, Y. Date, M. Nakazato, H. Matsuo and K. Kangawa, *Nature*, 1999, **402**, 656.

[55] A. M. Wren, C. J. Small, C. R. Abbott, W. S. Dhillo, L. J. Seal, M. A. Cohen, R. L. Batterham, S. Taheri, S. A. Stanley, M. A. Ghatei and S. R. Bloom, *Diabetes*, 2001, **50**, 2540.

[56] A. M. Wren, L. J. Seal, M. A. Cohen, A. E. Brynes, G. S. Frost, K. G. Murphy, W. S. Dhillo, M. A. Ghatei and S. R. Bloom, *J. Clin. Endocrinol. Metab.*, 2001, **86**, 5992.
[57] B. Holst and T. W. Schwartz, *Trends Pharmacol. Sci.*, 2004, **25**, 113.
[58] M. D. Serby, H. Zhao, B. G. Szczepankiewicz, C. Kosogof, Z. Xin, B. Liu, M. Liu, L. T. J. Nelson, W. Kaszubska, H. D. Falls, V. Schaefer, E. N. Bush, R. Shapiro, B. A. Droz, V. E. Knourek-Segel, T. A. Fey, M. E. Brune, D. W. A. Beno, T. M. Turner, C. A. Collins, P. B. Jacobson, H. L. Sham and G. Liu, *J. Med. Chem.*, 2006, **49**, 2568.
[59] B. Liu, M. Liu, Z. Xin, H. Zhao, M. D. Serby, C. Kosogof, L. T. J. Nelson, B. G. Szczepankiewicz, W. Kaszubska, V. G. Schaefer, H. D. Falls, C. W. Lin, C. A. Collins, H. L. Sham and G. Liu, *Bioorg. Med. Chem. Lett.*, 2006, **16**, 1864.
[60] Z. Xin, H. Zhao, M. D. Serby, B. Liu, V. G. Schaefer, D. H. Falls, W. Kaszubska, C. A. Colins, H. L. Sham and G. Liu, *Bioorg. Med. Chem. Lett.*, 2005, **15**, 1201.
[61] B. Liu, G. Liu, Z. Xin, M. D. Serby, H. Zhao, V. G. Schaefer, H. D. Falls, W. Kaszubska, C. A. Collins and H. L. Sham, *Bioorg. Med. Chem. Lett.*, 2004, **14**, 5223.
[62] H. Zhao, Z. Xin, J. R. Patel, L. T. J. Nelson, B. Liu, B. G. Szczepankiewicz, V. G. Schaefer, H. D. Falls, W. Kaszubska, C. A. Collins, H. L. Sham and G. Liu, *Bioorg. Med. Chem. Lett.*, 2005, **15**, 1825.
[63] X. Chen, R. V. Connors, K. Dai, Y. Fu, J. C. Jaen, Y.-J. Kim, L. Li, M. E. Lizarzaburu, J. T. Mihalic and S. J. Shuttleworth, *WO Patent Application* 2006020959, 2006.
[64] P. Brandt, U. Bremberg, R. Crossley, M. Graffner Nordberg, A. Jenmalm Jensen, E. Ringberg and T. Ward, *WO Patent Application* 2005048916, 2005.
[65] R. L. Batterham, C. W. Le Roux, M. A. Cohen, A. J. Park, S. M. Ellis, M. Patterson, G. S. Frost, M. A. Ghatei and S. R. Bloom, *J. Clin. Endocrinol. Metab.*, 2003, **88**, 3989.
[66] F. H. Koegler, P. J. Enriori, S. K. Billes, D. L. Takahashi, M. S. Martin, R. L. Clark, A. E. Evans, K. L. Grove, J. L. Cameron and M. A. Cowley, *Diabetes*, 2005, **54**, 3198.
[67] D. Renshaw and R. L. Batterham, *Curr. Drug Targets*, 2005, **6**, 171.
[68] S. Koda, Y. Date, N. Murakami, T. Shimbara, T. Hanada, K. Toshinai, A. Niijima, M. Furuya, N. Inomata, K. Osuye and M. Nakazato, *Endocrinology*, 2005, **146**, 2369.
[69] A. W. Stamford, J. Hwa and M. van Heek, *Annu. Rep. Med. Chem.*, 2003, **38**, 61.
[70] K. Lumb, L. Decarr, P. Coish and S. O'Connor, *WO Patent Application* 2005053726, 2005.
[71] A. Balasubramaniam, D. E. Mullins, S. Lin, W. Zhai, Z. Tao, V. C. Dhawan, M. Guzzi, J. J. Knittel, K. Slack, H. Herzog and E. M. Parker, *J. Med. Chem.*, 2006, **49**, 2661.
[72] O. Della-Zuana, L. Revereault, A. Beck-Sickinger, A. Monge, D. H. Caignard, J. L. Fauchere, J. M. Henlin, V. Audinot, J. A. Boutin, S. Chamorro, M. Feletou and N. Levens, *Int. J. Obes.*, 2004, **28**, 628.
[73] A. V. Turnbull, L. Ellershaw, D. J. Masters, S. Birtles, S. Boyer, D. Carroll, P. Clarkson, S. J. G. Loxham, P. McAulay, J. Teague, L. K. M. Foote, J. E. Pease and M. H. Block, *Diabetes*, 2002, **51**, 2441.
[74] N. Kakui, J. Tanaka, Y. Tabata, K. Asai, N. Masuda, T. Miyara, Y. Nakatani, F. Ohsawa, N. Nishikawa, M. Sugai, M. Suzuki, K. Aoki and H. Kitaguchi, *J. Pharmacol. Exp. Ther.*, 2006, **317**, 562.
[75] G. Li, Y. Huang, J. Kelly, A. Stamford, W. Greenlee, D. Mullins, M. Guzzi, X. Zhang, J. J. Hwa, J. Gao, L. Ghibaudi, S. Weihaus and K. C. Cheng, *Abstracts of Papers, 230th ACS National Meeting*, Washington, DC, United States, Aug. 28–Sept. 1, 2005, MEDI.
[76] A. Kanatani, A. Ishihara, H. Iwaasa, K. Nakamura, O. Okamoto, M. Hidaka, J. Ito, T. Fukuroda, D. J. MacNeil, L. H. T. Van der Ploeg, Y. Ishii, T. Okabe, T. Fukami and M. Ihara, *Biochem. Biophys. Res. Commun.*, 2000, **272**, 169.

[77] A. Ishihara, A. Kanatani, S. Mashiko, T. Tanaka, M. Hidaka, A. Gomori, H. Iwaasa, N. Murai, S.-i. Egashira, T. Murai, Y. Mitobe, H. Matsushita, O. Okamoto, N. Sato, M. Jitsuoka, T. Fukuroda, T. Ohe, X. Guan, D. J. MacNeil, L. H. T. Van der Ploeg, M. Nishikibe, Y. Ishii, M. Ihara and T. Fukami, *Proc. Natl. Acad. Sci. USA.*, 2006, **103**, 7154.
[78] D. J. MacNeil, J. H. McIntyre, L. H. T. Van Der Ploeg and A. Ishihara, *WO Patent Application* 2004009015, 2004.
[79] T. Fukami, A. Kanatani, A. Ishihara, T. Ishii, T. Takahashi, Y. Haga, T. Sakamoto and T. Itoh, *WO Patent Application* 2001014376, 2001.
[80] R. Simpson, A. Patti and B. Matuszewski, *J. Liq. Chromatogr. R. T.*, 2006, **29**, 273.
[81] R. J. Bodnar, *Peptides*, 2004, **25**, 697.
[82] S. G. Holtzman, *J. Pharmacol. Exp. Ther.*, 1974, **189**, 51.
[83] R. J. Bodnar, *Expert Opin. Invest. Drugs*, 1998, **7**, 485.
[84] M. Zhang, B. A. Gosnell and A. E. Kelley, *J. Pharmacol. Exp. Ther.*, 1998, **285**, 908.
[85] M. R. Yeomans and R. W. Gray, *Physiol. Behav.*, 1997, **62**, 15.
[86] R. Malcolm, P. M. O'Neil, J. D. Sexauer, F. E. Riddle, H. S. Currey and C. Counts, *Int. J. Obes.*, 1985, **9**, 347.
[87] R. L. Atkinson, L. K. Berke, C. R. Drake, M. L. Bibbs, F. L. Williams and D. L. Kaiser, *Clin. Pharmacol. Ther.*, 1985, **38**, 419.
[88] F. Fruzzetti, C. Bersi, D. Parrini, C. Ricci and R. Genazzani Andrea, *Fertil. Steril.*, 2002, **77**, 936.
[89] M. A. Statnick, F. C. Tinsley, B. J. Eastwood, T. M. Suter, C. H. Mitch and M. L. Heiman, *Am. J. Physiol. Regul. Integr. Comp. Physiol.*, 2003, **284**, R1399.
[90] N. Diaz, M. Benvenga, P. Emmerson, R. Favors, M. Mangold, J. McKinzie, N. Patel, S. Peters, S. Quimby, H. Shannon, M. Siegel, M. Statnick, E. Thomas, J. Woodland, P. Surface and C. Mitch, *Bioorg. Med. Chem. Lett.*, 2005, **15**, 3844.
[91] M. G. De la Torre and C. H. Mitch, *WO Patent Application* 2005090286, 2005
[92] H. B. Broughton, N. Diaz Buezo, C. H. Mitch and C. Pedregal-Tercero, *WO Patent Application* 2005090337, 2005.
[93] B. M. Smith, W. J. Thomsen and A. J. Grottick, *Expert Opin. Invest. Drugs*, 2006, **15**, 257.
[94] L. H. Tecott, L. M. Sun, S. F. Akana, A. M. Strack, D. H. Lowenstein, M. F. Dallman and D. Julius, *Nature*, 1995, **374**, 542.
[95] S. P. Vickers, P. G. Clifton, C. T. Dourish and L. H. Tecott, *Psychopharmacology*, 1999, **143**, 309.
[96] J. Dunlop, A. L. Sabb, H. Mazandarani, J. Zhang, S. Kalgaonker, E. Shukhina, S. Sukoff, R. L. Vogel, G. Stack, L. Schechter, B. L. Harrison and S. Rosenzweig-Lipson, *J. Pharmacol. Exp. Ther.*, 2005, **313**, 862.
[97] S. P. Vickers, N. Easton, L. J. Webster, A. Wyatt, M. J. Bickerdike, C. T. Dourish and G. A. Kennett, *Psychopharmacology*, 2003, **167**, 274.
[98] H. G. F. Richter, D. R. Adams, A. Benardeau, M. J. Bickerdike, J. M. Bentley, T. J. Blench, I. A. Cliffe, C. Dourish, P. Hebeisen, G. A. Kennett, A. R. Knight, C. S. Malcolm, P. Mattei, A. Misra, J. Mizrahi, N. J. T. Monck, J. M. Plancher, S. Roever, J. R. A. Roffey, S. Taylor and S. P. Vickers, *Bioorg. Med. Chem. Lett.*, 2006, **16**, 1207.
[99] J. M. Fevig, J. Feng and S. Ahmad, *US Patent Application* 2006014777, 2006.
[100] J. L. Gross, M. J. Williams, G. P. Stack, H. Gao and D. Zhou, *WO Patent Application* 2005044812, 2005.
[101] P. Caldirola, *SMI Conference on Obesity and Related Disorders (Feb 17-18, London)— see IDdb3 ref RF479308, Thomson Scientific*, 2003.
[102] M. L. Woolley, J. C. Bentley, A. J. Sleight, C. A. Marsden and K. C. F. Fone, *Neuropharmacology*, 2001, **41**, 210.

[103] G. Perez-Garcia and A. Meneses, *Pharmacol. Biochem. Behav.*, 2005, **81**, 673.
[104] J. Holenz, P. J. Pauwels, J. L. Diaz, R. Merce, X. Codony and H. Buschmann, *Drug Discov. Today*, 2006, **11**, 283.
[105] K. Takahashi, H. Suwa, T. Ishikawa and H. Kotani, *J. Clin. Invest.*, 2002, **110**, 1791.
[106] T. A. Esbenshade, G. B. Fox and M. D. Cowart, *Mol. Interventions*, 2006, **6**, 77.
[107] R. Aslanian and N.-Y. Shih, *Annu. Rep. Med. Chem.*, 2004, **39**, 57.
[108] K. Malmlof, F. Zaragoza, V. Golozoubova, H. H. F. Refsgaard, T. Cremers, K. Raun, B. S. Wulff, P. B. Johansen, B. Westerink and K. Rimvall, *Int. J. Obes.*, 2005, **29**, 1402.
[109] A. A. Hancock, M. S. Diehl, T. A. Fey, E. N. Bush, R. Faghih, T. R. Miller, K. M. Krueger, J. K. Pratt, M. D. Cowart, R. W. Dickinson, R. Shapiro, V. E. Knourek-Segel, B. A. Droz, C. A. McDowell, G. Krishna, M. E. Brune, T. A. Esbenshade and P. B. Jacobson, *Inflamm. Res.*, 2005, **54**, S27.
[110] A. A. Hancock, *Biochem. Pharmacol.*, 2006, **71**, 1103.
[111] J. Reess, A. Raschig, S. Pollentier, O. Graff, B. O. Mikkelsen and M. Priskorn, *WO Patent Application* 2005070427, 2005.

Nuclear Hormone Receptor Modulators for the Treatment of Diabetes and Dyslipidemia

Peter T. Meinke, Harold B. Wood and Jason W. Szewczyk

Merck Research Laboratories, PO Box 2000, Rahway, NJ 07065, USA

Contents

1. Introduction	99
2. PPARs	99
2.1. PPARγ	100
2.2. PPARα	102
2.3. PPARδ	104
2.4. PPARα/γ dual agonists	105
2.5. Other PPAR dual and pan agonists	107
2.6. Selective PPARγ modulators	108
3. RXR modulators	111
4. LXR	112
4.1. LXR subtype selective agonists	114
4.2. Current patent literature	115
5. FXR	117
6. Conclusion	118
References	118

1. INTRODUCTION

The burgeoning epidemic of diabetes and dyslipidemia currently afflicting much of the developed world is continuing unabated, according to recent CDC statistics. Metabolic syndrome, an undesirable cluster of disorders afflicting individuals that includes obesity, hyperglycemia, dyslipidemia and hypertension, contributes to the etiology of insulin insensitivity. Concerted endeavors to identify superior methods to treat multiple aspects of the metabolic syndrome have fueled a veritable renaissance in the field of nuclear hormone receptor (NHR) modulators. These biological targets, which include the peroxisome-proliferator-activated receptors (PPAR), retinoid X receptor (RXR), liver X receptor (LXR) and farnesoid X receptor (FXR) are recognized as pharmaceutically viable intervention points and each is described in detail below.

2. PPARS

The PPAR subtype family is comprised of three members (α, γ and δ) that function as lipid sensors and transcriptional regulators of nutrient homeostasis. Comprehensive reviews detailing multiple aspects of PPAR biology have recently appeared [1–4]. In a highly choreographed sequence of events, binding of an agonist to the PPAR ligand-binding domain (LBD) induces conformational changes leading to co-repressor release and co-activator recruitment and the formation of an obligate heterodimer with RXR. Following further conformational changes, this heterodimer binds to regulatory peroxisome-proliferator-response elements (PPREs), initiating chromatin remodeling and transcriptional activation of literally hundreds of genes. The PPAR receptor subfamily LBDs are remarkably flexible and capable of accommodating a diverse array of ligands. Despite this promiscuity, creative research has yielded isoform selective agents, various dual agonists and pan agonists (*vide infra*).

2.1. PPARγ

Abundant evidence exists indicating that adipose tissue indeed serves as the primary target for the insulin-sensitizing effects of PPARγ agonists whose action alters the regulation of genes involved in lipid uptake, metabolism and synthesis. PPARγ agonism promotes adipocyte differentiation and its activation causes adipocyte lipid accumulation (the "glitazone paradox") [5], decreases circulating lipids such as free fatty acids (FFAs) via decreased FFA synthesis and increased triglyceride (TG) formation, ultimately ameliorating hyperglycemia by reversing lipotoxicity in liver and muscle [6]. PPARγ agonism additionally modulates adipose endocrine activity by regulating adiponectin and resistin concentrations, whose functions include the modulation of liver and muscle insulin sensitivity [1]. Two genes, necdin and E2F4, thought to be integral in regulating cell differentiation processes, recently were identified as PPARγ target genes following their modulation by rosiglitazone treatment (8 weeks) in humans [7]. Anti-inflammatory benefits also have been ascribed to PPARγ activation, due to desirable reductions seen in pro-inflammatory cytokine and chemokines derived from adipose tissue. Indeed, in rodent models, anti-atheroslerotic effects have been observed, although in man the data currently remain ambiguous.

While three thiazolidinone (TZD)-derived PPARγ agonists (troglitazone, rosiglitazone, pioglitazone) advanced to the market, the subsequent withdrawal of troglitazone due to hepatotoxicity stimulated the search for non-TZD-containing structures, although no association of rosiglitazone or pioglitazone with this liability exists. Despite the unambiguous efficacy of PPARγ full agonists, their intimate association with numerous adverse effects (cardiomegaly and tumorgenicity in preclinical species, along with weight gain, fluid retention, edema and its sequela clinically) has precluded their wider application. In actuality, multiple development candidates failed due to either edema- or carcogenicity-related liabilities. In response to these safety concerns, the FDA mandated that all clinical investigations of all PPAR agents, irrespective of subtype, may not exceed 6 months duration without the prior completion of 2-year carcinogenicity studies. While weight gain effects are ascribed, at least in part, to adipogenic effects, the precise mechanism(s) by which PPARγ agonists cause fluid retention remains unclear. However, PPARγ is expressed at high levels in the kidney's tubular nephron and mice lacking this target are resistant to TZD-mediated increases in weight gain and fluid retention, suggesting that the epithelial Na^+ channel (ENaC) is a direct target of PPARγ ligands [8,9].

Direct comparison of pioglitazone and rosiglitazone in dyslipidemic diabetics indicated that while both TZDs were comparable with respect to glycemic control, weight gain, edema and congestive heart failure, they had distinct effects on certain lipid parameters [10]. Specifically, pioglitazone treatment not only lowered TG levels while rosiglitazone therapy raised them modestly, but it also had superior effects on HDL-C and LDL-C concentrations. Outcomes studies utilizing pioglitazone (PROACTIVE) and rosiglitazone (RECORD) were initiated in order to assess whether chronic TZD treatment can also ameliorate cardiovascular risks. While the RECORD trial remains ongoing [11], the PROACTIVE study, while failing to meet its primary endpoints, did reduce overall all-cause mortality and adverse cardiovascular events but was complicated by a fourfold increase in the incidence of edema that exacerbated heart failure risk [12]. Other outcomes studies provided clear clinical evidence for β-cell preservation as a consequence of chronic TZD treatment, significantly lowering the risk of progression to frank diabetes in patients treated with troglitazone (TRIPOD) or pioglitazone (PIPOD) [13].

Newly reported PPARγ agonists include nitrated analogs of prevalent fatty acids that function as putative high affinity endogenous ligands with potentially significant implications for the pathology of metabolic syndrome. Nitrolinoleic acid (**1**, LNO_2) [14] and nitrooleic acid (**2**, $OA-NO_2$) [15], detected at physiologically relevant levels in human plasma and urine, induced preadipocyte differentiation in 3T3-L1 adipocytes. TZD **3**, while potently inducing TG accumulation in 3T3-L1 cells and correcting hyperglycemia in KKA^y mice, was less cytotoxic in cultured rat hepatocytes than was rosiglitazone [16]. The potent benzoxazinone **4** normalized plasma glucose levels in *db/db* mice at doses as low as 1 mg/kg/day [17,18]. The structurally novel pyran **5** resulted from efforts to identify non-TZD containing PPARγ selective agonists [19]. Pyran **5** activated PPARγ at potencies comparable to rosiglitazone (no *in vivo* data provided). Perhaps the most unusual PPARγ agonist disclosed is the potent dimer **6** which was derived from a structurally related PPARα/γ dual agonist [20]. Molecular modeling studies were used to clarify the origin of its PPARγ binding selectivity. This dimeric structure retained modest PPARα and PPARδ activities (EC_{50} values of 1200 and 1800 nM, respectively)

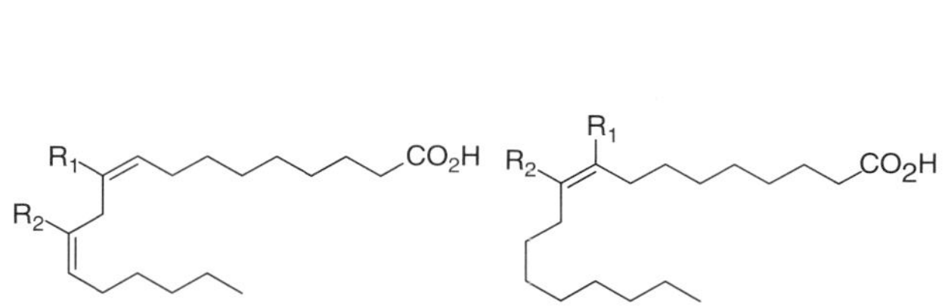

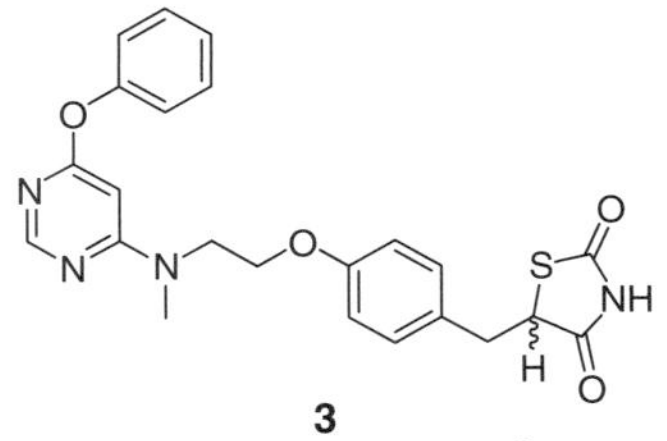

3
(hγ EC$_{50}$ not reported)

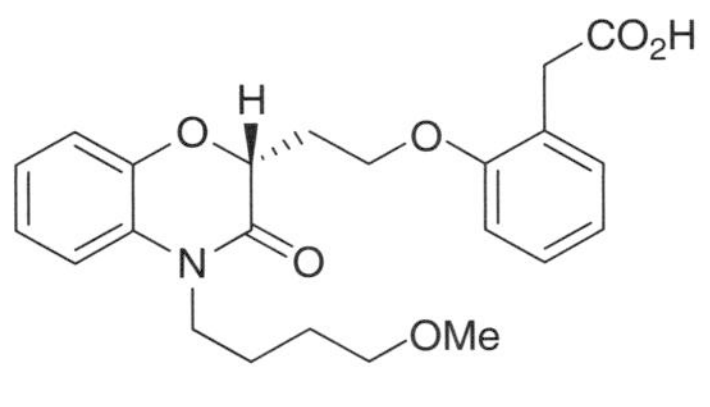

4
(hγ EC$_{50}$ = 110 nM)

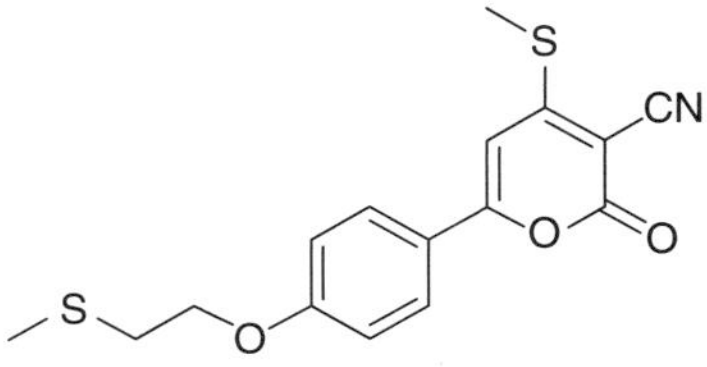

5
(hγ EC$_{50}$ not reported)

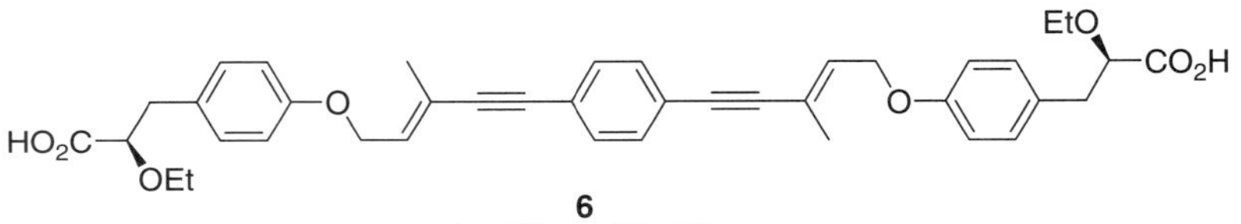

6
(hγ EC$_{50}$ = 80 nM)

New potential applications for PPARγ agonists include the pathogenesis of neurodegenerative diseases, such as multiple sclerosis (MS), Parkinson's (PD) and Alzheimer's (AD). While the data associated with PD is currently limited to beneficial effects in murine models [21], intriguing evidence suggestive of utility for PPARγ agonists for MS [22] and AD [23–25] has recently appeared. On the basis of data from PPARγ effects in experimental models that mimic MS, small clinical studies suggest that pioglitazone's benefits to MS patients may be ascribed to its anti-inflammatory effects [22] whereas a growing body of evidence has linked the pathogenesis of AD to hyperinsulinemia [25] in which clinical investigations of rosiglitazone treatment suggest a genetically defined subpopulation (APOE ε4 negative patients) showed evidence of improved cognition [24].

2.2. PPARα

While PPARα is predominantly expressed in liver, it is also present in tissues (e.g. heart, skeletal muscle) that extract their energy requirements from lipids. Circulating fatty acids migrate to the liver where they are metabolized to provide fuel for peripheral tissues, a role elucidated in part through seminal studies of PPARα−/− mice [26].

While in rodents, PPARα agonists induce hepatic peroxisome proliferation leading to significant hepatoxicities, these effects fortunately are not recapitulated in man in part due to the 10-fold lower levels of PPARα expression in humans. Weak PPARα agonists such as fenofibrate, gemfibrozil and bezafibrate are remarkably effective at lowering TG levels with lesser effects on raising HDL-C in dyslipidemic populations [27]. These desirable effects are mediated through transcriptional activation of multiple genes that control these functions [2]. PPARα agonists also are known to possess anti-inflammatory effects on vascular tissue and were proven to reduce cardiovascular risk in outcomes studies using gemfibrozil (VA-HIT) [28], fenofibrate (FIELD) [29] or bezafibrate (BIP) [30].

Comparatively, few new PPARα selective agonists have been reported (excluding patent disclosures). The 1,3-dicarbonyl **7** was shown to be a potent PPARα selective agonist and exhibited modest glucose-lowering activity in *db/db* mouse models [31]. Benzisoxazole **8** is a potent PPARα agonist, which was highly efficacious in hyperlipidemic and hyperglycemic models [32]. Cyclic fibrate **9** also showed desirable utility in preclinical dyslipidemic models [33], while the PPARα-weighted agonist **10** was highly efficacious in the *db/db* model [34].

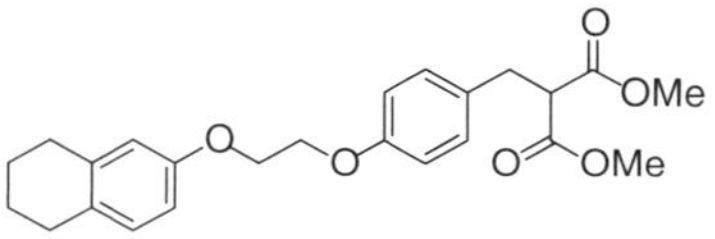

7
(hα / hγ / hδ EC$_{50}$ = 30 / 1500 / 3100 nM)

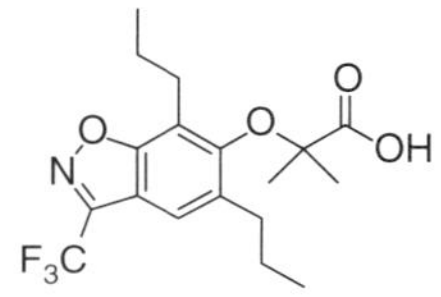

8
(hα EC$_{50}$ = 300 nM)

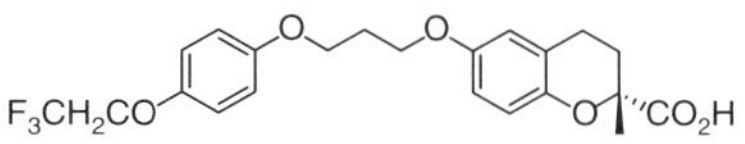

9
(hα / hγ EC$_{50}$ = 40 / 3000 nM)

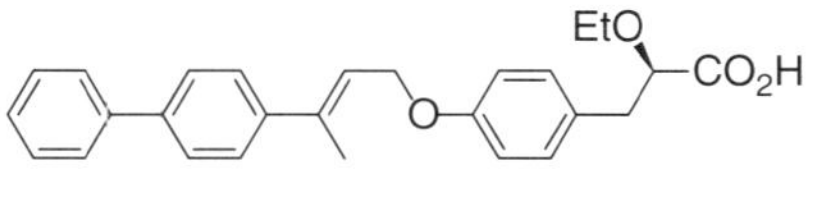

10
(hα / hγ / hδ EC$_{50}$ = 67 / 1130 / 6900 nM)

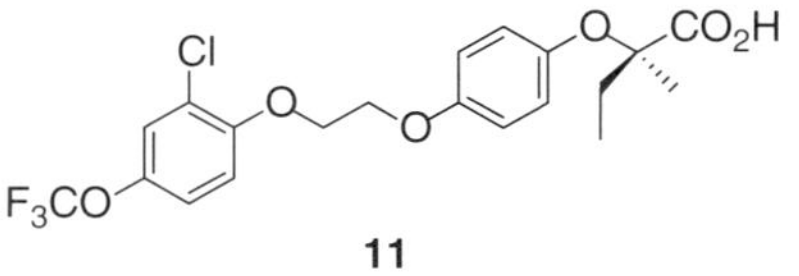

11
(hα EC$_{50}$ = 3 nM)

Myopathy and rhabdomyelosis are well recognized but comparatively rare adverse events associated with all fibrates, which are weak PPARα agonists. As the mechanism underlying these adverse effects is poorly understood (PPARα expression in muscle is very low), it remains an open question as to whether a potent PPARα hyperlipidemic agent would exacerbate the frequency and severity of clinical myotoxicity. An *in vitro* model using differentiated rat skeletal muscle cultures was developed to evaluate myotoxicity [35]. In this assay, high-affinity PPARα agonists induce myotoxicities at lower drug concentrations than do weaker PPARα agonists (1 nM for **11** versus micromolar levels for bezafibrate and gemfibrozil).

PPARα agonists show promise in treating individuals afflicted with hepatitis C virus (HCV), a major cause of liver disease. As HCV particles are associated with circulating LDL [36] and the expression of hepatic PPARα in HCV-positive individuals is dramatically impaired [37], a pilot study (8 week) treating HCV patients with bezafibrate was initiated that showed a statistically significant reduction in HCV mRNA titers [38].

2.3. PPARδ

Of the PPAR isoforms, the biological function of the PPARδ subtype, which is expressed ubiquitously, remains the most controversial in part because only recently were sufficiently potent and selective ligands identified and as mechanistic studies yielded conflicting results. For instance, PPARδ null mice are prone to epithelial cell proliferation [39], suggesting that a PPARδ agonist would inhibit proliferation whereas *in vitro*, high affinity PPARδ agonists stimulate proliferative changes in human breast and prostate cells [40]. Treatment of *Apc*^{min} mice, which are predisposed to colon polyp formation, with a PPARδ agonist significantly increased the number and size of polyps formed [41], a result consistent with related studies [42]. Other investigations using PPARδ−/− mice, however, yielded conflicting data, suggesting that PPARδ lacks a specific role in cancer but instead decreased adiposity remarkably in all depots examined [43]. The changes in adiposity are consistent with other observations that PPARδ agonism induces genes associated with fatty acid β-oxidation in muscle fibers and can ameliorate diabetes in *ob/ob* mice [44].

GW501516 (**12**), and its fluorinated analog GW0742 (**13**), were the first potent PPARδ agonists identified and their discovery was guided by crystallographic studies [45]. T0913659 (**14**) is a potent but modestly PPARδ selective agonist [46]. A PPARα selective agonist served as inspiration for **15**, which was another potent PPARδ-selective agonist [47].

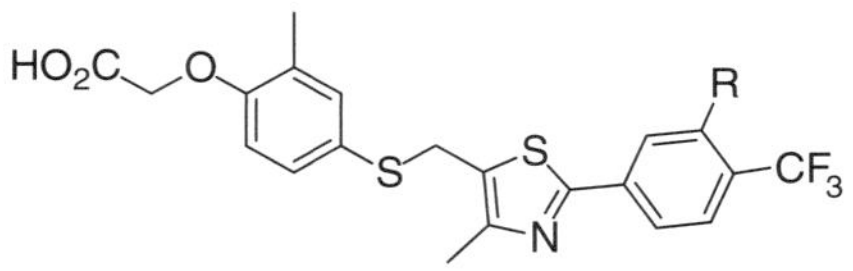

12: GW501516 R= H (hδ EC$_{50}$ = 1 nM)
13: GW0742 R= F (hδ EC$_{50}$ = 1 nM)

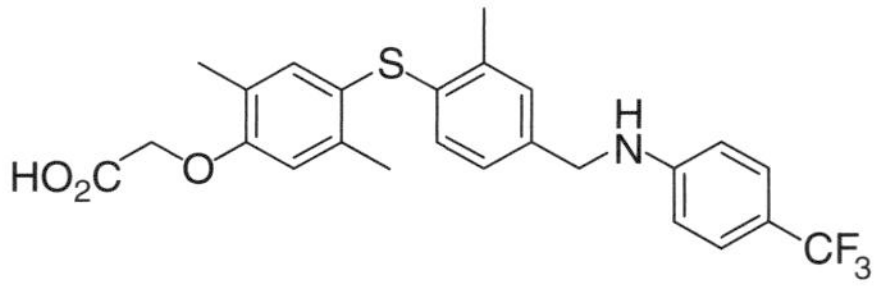

14: T0913659
(hγ / hγ / hδ EC$_{50}$ = 510 / 1500 / 9 nM)

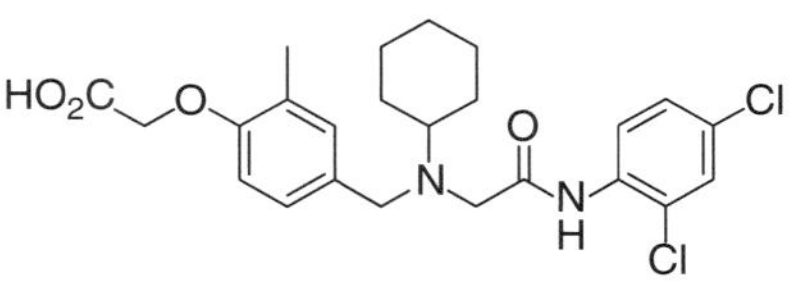

15
(hα / hγ / hδ EC$_{50}$ >100K / >100K / 3 nM)

These new selective agents proved invaluable for the characterization of PPARδ effects, suggesting that PPARδ agonism may attenuate metabolic syndrome-related abnormalities. For instance, GW0742 reduced atherosclerosis in murine models [48]. In primates, GW501516 increased HDL-C significantly (80%), while lowering LDL-C substantially (−29%) [49,50]. Similarly, T0913659 exhibited beneficial lipid effects in monkeys [46]. In LDLR−/− mice, PPARδ agonism was shown to increase CD36 expression and decrease inflammation, but had no beneficial effects on vascular lesion formation nor to lower macrophage lipid accumulation [51]. GW501516 has advanced to Phase II clinical study for dyslipidemia; results from preliminary trials (6 weeks) showed significant reductions in total cholesterol, TG and apolipoprotein B concentrations [52].

As noted previously, myopathy is an adverse event clinically associated with PPARα agonist monotherapy, although its etiology is poorly understood. To date, no reports of myopathy ascribed to PPARδ have appeared, despite its abundant expression levels in muscle fibers and PPARδ agonist effects on lipid accumulation. However, in the previously described rat-cultured myotube assay, GW501516 also had significant effects, although these could not be rigorously ascribed to PPARδ agonism [35]. Ultimately, given the limitations of preclinical tools, clinical evaluation of potent PPARδ selective agonists is necessitated to generate unambiguous evidence regarding potential myopathic liabilities of PPARδ agonism.

2.4. PPARα/γ dual agonists

As clinical evidence suggests both hyperglycemia and dyslipidemia (e.g. atherosclerosis) contribute to morbidity, single entity PPARα/γ dual agonists may confer broad benefits in the diabetic population. Indeed, concomitant therapy using rosiglitazone and fenofibrate yielded complementary effects on glycemic and lipid parameters with few adverse events [53]. While preclinical evaluation of PPARα/γ dual agonists is confounded by PPARα-mediated reductions of hyperglycemia or hyperinsulinemia in rodent efficacy models [32,54,55] or PPARγ contributions to lipid lowering in dyslipidemia models [56], engagement of isoform-specific gene changes is readily monitored. Also, PPARα agonists reduce body weight gain in rodents (in stark contrast to PPARγ agonist effects), so while in principle PPARα/γ dual agonists could alleviate weight gain effects, clinical data to date indicates no such benefit exists.

An array of structurally diverse insulin sensitizers with additional PPARα activity has been extensively characterized, showing robust effects preclinically; several also progressed to clinical evaluation. KRP-297 (**16**) [57,58], while exhibiting robust glycemic and lipid effects clinically [59], was discontinued due to the formation of rodent vascular tumors. Similarly, ragaglitazar (**17**) [60,61] reduced key glycemic and lipid parameters in man [62], but was discontinued due to the formation of rodent bladder tumors. Muraglitazar's (**18**) [63] development, while clinically manifesting useful glycemic and lipid changes [64], has been delayed due to an increased incidence of congestive heart failure. Naveglitazar (**19**) [65] was shown to significantly reduce HbA1c levels and fasting triglycerides while increasing HDL-C levels in Type 2 diabetes mellitus (T2DM) patients after 12 weeks [66]. Similarly, tesaglitazar (**20**) [67], in clinical trials showed beneficial effects on parameters of lipid and glucose metabolism in an insulin-resistant population [68]. Both have completed Phase II trials. Phase III studies for both tesaglitazar and naveglitazar have been delayed pending completion of 2-year rodent carcinogenicity study outcomes as per FDA guidelines.

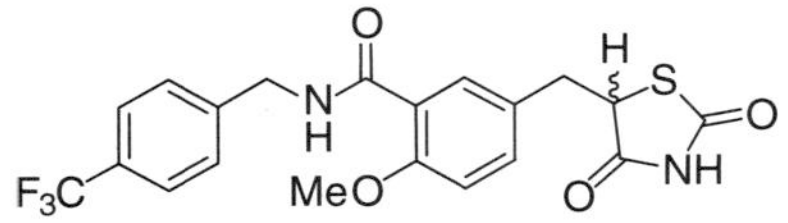

16: KRP-297 (MK-0767)
(hα / hγ EC_{50} = 228 / 326 nM)

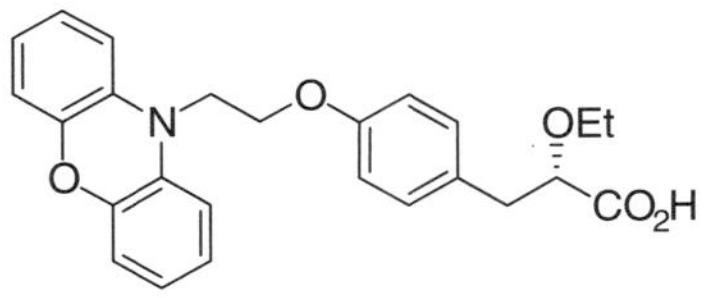

17: Ragaglitazar
(hα / hγ EC_{50} = 270 / 324 nM)

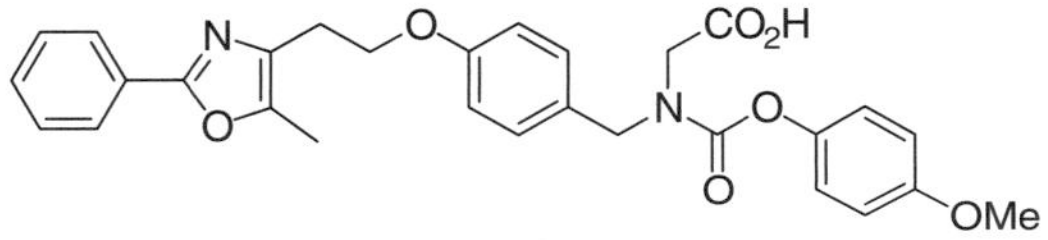

18: Muraglitazar
(hα / hγ EC_{50} = 320 / 190 nM)

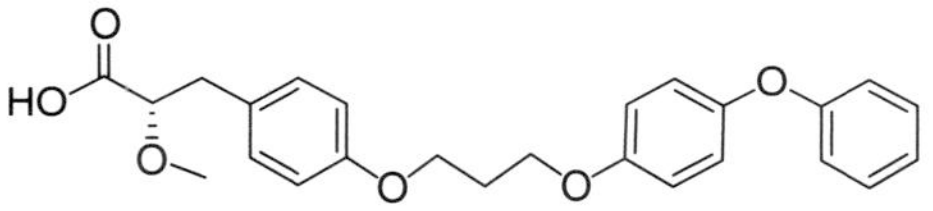

19: Naveglitazar (LY818)
(hα / hγ EC_{50} = 2816 / 361 nM)

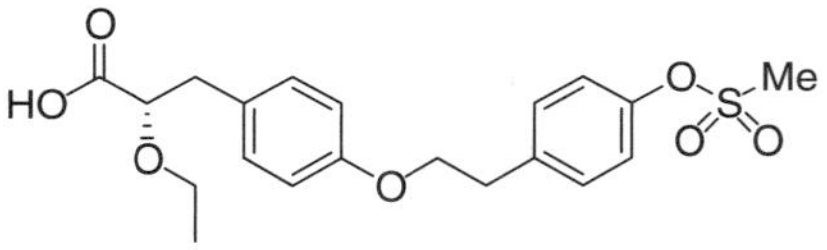

20: Tesaglitazar
(hα / hγ EC_{50} = 1200 / 1300 nM)

Additional ligands include TZD18 (**21**), which potently activated PPARα and γ genes *in vivo* and was highly efficacious in diabetes models as well as in hamster and dog models of dyslipidemia [69,70]. PPARα-selective 8-(*S*)-HETE served as a starting point for creative medicinal chemistry efforts yielding the dual agonist **22** [71,72]. The cyclic fibrate **23**, evolved from TZD18 and the *gem*-dimethyl moiety of fenofibrate, exhibited substantial antihyperglycemic and hypolipidemic activities that exceeded those of the positive controls (rosiglitazone and fenofibrate) [73]. The α-ethoxy acid **24** normalized plasma glucose in rodents [74], as did indane **25** [75] and isoxazolone **26** [76]. The PPARα-weighted phenyl acetic derivative **27** exhibited significant correction of hyperglycemia with reduced lipid proliferation and cardiac hypertrophy in safety studies [77].

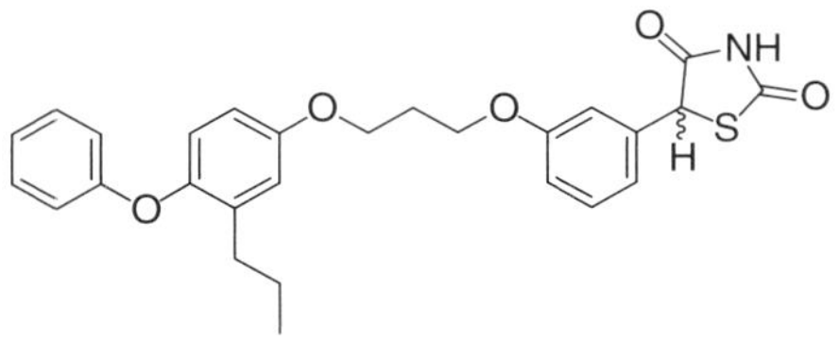

21: TZD18
(hα / hγ EC$_{50}$ = 26 / 14 nM)

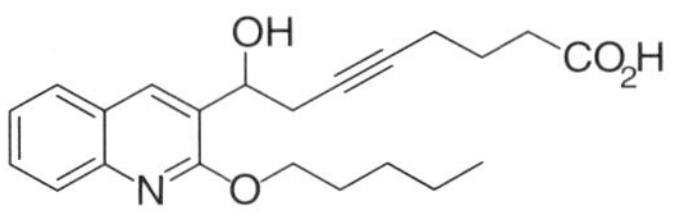

22
(hα / hγ EC$_{50}$ = 114 / 617 nM)

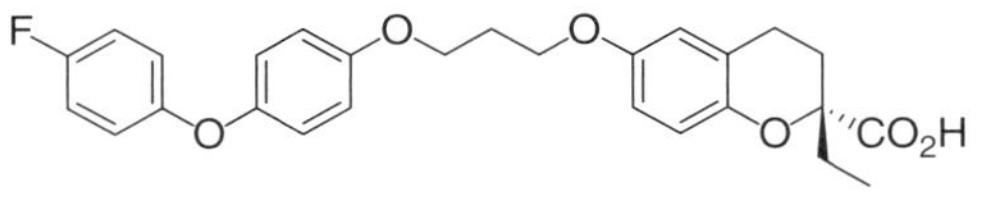

23
(hα / hγ EC$_{50}$ = 40 / 170 nM)

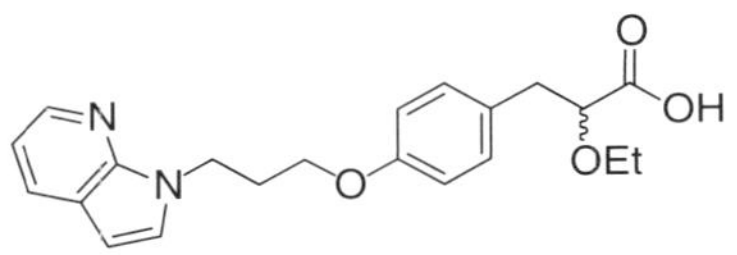

24
(hα / hγ EC$_{50}$ = 110 / 750 nM)

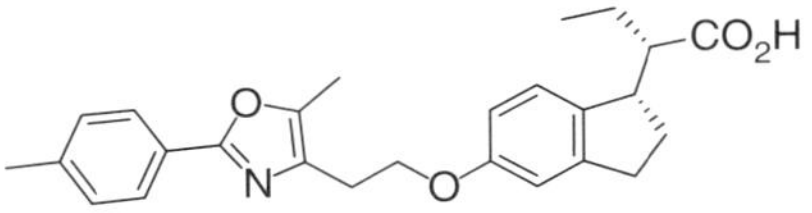

25
(hα / hγ EC$_{50}$ = 78 / 30 nM)

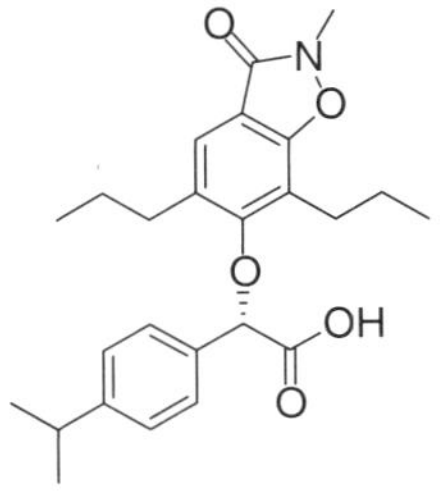

26
(hα / hγ EC$_{50}$ = 74 / 210 nM)

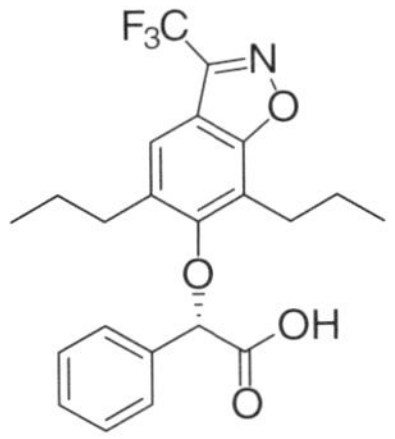

27
(hα / hγ EC$_{50}$ = 15 / 365 nM)

2.5. Other PPAR dual and pan agonists

Considerable effort has been expended to identify insulin sensitizers that retain useful activity on one or both of the other two PPAR isoforms. These include a series of PPARα/δ dual agonists (i.e. **28**), though no *in vivo* data were provided [78]. The indole acetic acid **29** was a potent PPARα/γ/δ pan agonist with excellent efficacy in a rodent diabetes model [79,80]. A series of pan agonists with comparable receptor potencies were reported; **30** showed significant activity in the *db/db* model [81]. Similarly, **31** was shown to be a potent PPAR pan agonist [82]. No *in vivo* data were provided that could clearly assess the PPARα and PPARδ effects for these compounds.

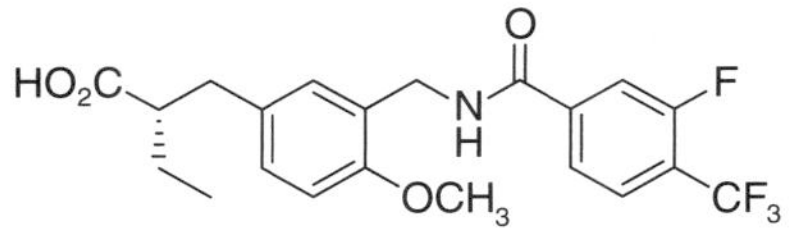

28
(hα / hγ / hδ EC$_{50}$ = 12 / 4900 / 23 nM)

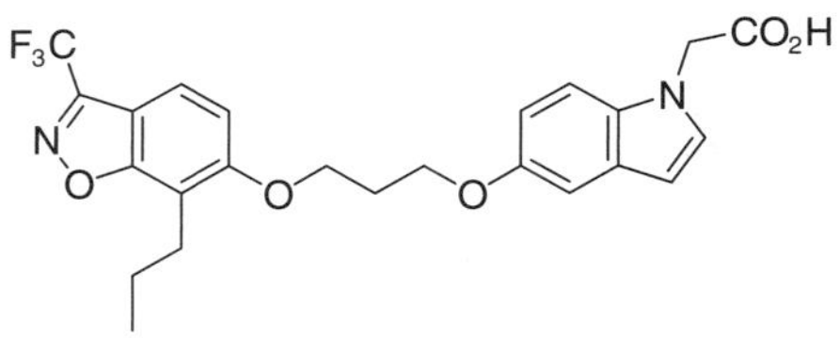

29
(hα / hγ / hδ EC$_{50}$ = 14 / 230 / 10 nM)

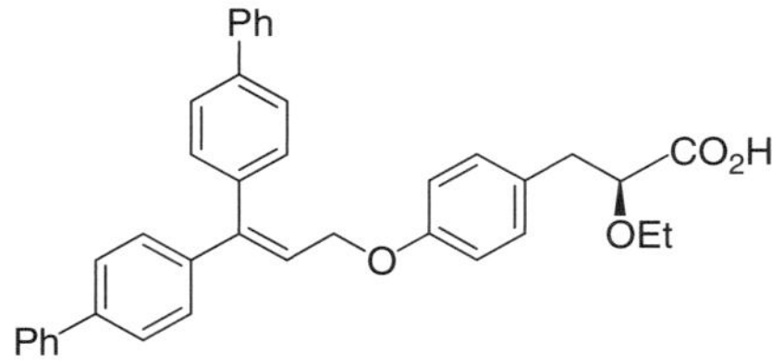

30
(hα / hγ / hδ EC$_{50}$ = 1100 / 300 / 500 nM)

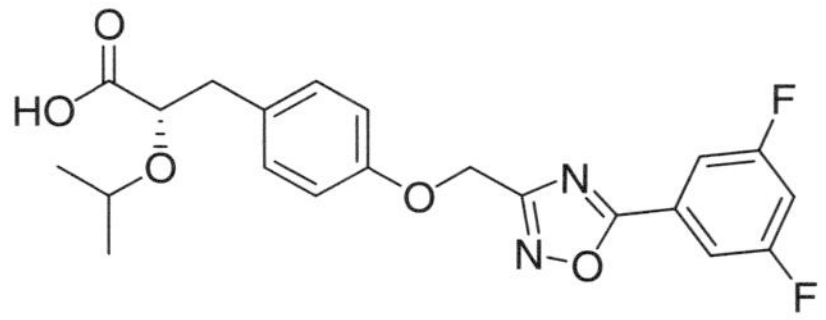

31
(hα / hγ / hδ EC$_{50}$ = 40 / 100 / 160 nM)

2.6. Selective PPARγ modulators

In attempts to circumvent the previously described limitations of PPARγ full agonists, zealous efforts are underway to identify and characterize selective PPARγ modulators (SPPARγMs or partial agonists) that retain desirable efficacy and exhibit superior tolerability, an endeavor further complicated by the lack of a consistent definition for SPPARγMs and a precise understanding of the molecular mechanisms by which PPARγ full agonists exert their deleterious effects. In principle, however, a SPPARγM could stabilize the PPAR LBD in a novel manner relative to PPARγ full agonists, thereby altering specific co-activator and/or co-repressor interactions, culminating in differential gene expression leading to an altered biological response. Compelling evidence recently has emerged suggesting that structurally diverse SPPARγMs indeed stabilize the PPARγ LBD in a myriad of unique ways, ultimately leading to a differential transcriptional response, while retaining the useful antihyperglycemic properties of PPARγ full agonists in diabetes models with ameliorated adverse effects in safety studies. Naturally, rigorous clinical trials will be required to establish whether the insulin-sensitizing effects of SPPARγMs can be fully dissociated from the aforementioned weight gain and edema/fluid retention effects in man.

Several SPPARγMs have advanced into the clinic, including MBX-102 (**32**), FK614 (**33**) and AMG-131 (formerly T-131) (**34**). MBX-102 (metaglidasen) is the (−) enantiomer of halofenate [83], a weak PPARγ ligand that exhibits both partial agonist and antagonist activity [84]. MBX-102-regulated PPARγ target genes in 3T3-L1 adipocytes and was efficacious in rodent diabetes models but did not increase body or heart weight in early safety studies. Currently in Phase II clinical trials, metaglidasen increased adiponectin concentrations and lowered HbA1c levels with less weight gain than noted in the placebo arm. FK614 was comparable in intrinsic potency to pioglitazone in PPARγ transactivation assays but exhibited lower maximal activation [85]. In comparison with full agonists, FK614 exhibited differential effects on known PPARγ co-activators such as SCR-1 and CBP [86] and appeared to bind PPARγ ligand binding domain in a manner distinct from full agonists [87], providing a physical basis for its unique biological profile. FK614 retained robust efficacy in *db/db* and *ob/ob* diabetes models, but ameliorated common PPARγ adverse effects relative to full agonists in SD rat safety studies [85]. While FK614 advanced to Phase II clinical trials, its development was recently discontinued when no advantage was demonstrated in clinical trials. While less information regarding AMG-131 (formerly T-131) has been disclosed, this SPPARγM was more efficacious than rosiglitazone in the *fa/fa* rat model and exhibited an improved safety profile [88,89]. Currently in Phase II clinical trials, AMG-131 increased adiponectin levels in a time- and dosage-dependent manner to an extent comparable to that seen with full agonists [89]. In an interesting development, the antihypertensive agents telmisartan (**35**) and irbesartan (**36**) were found to also function as weak partial agonists of PPARγ, regulating known PPARγ target genes (e.g. FABP, CD36, PEPCK-C) in 3T3-L1 preadipocytes, lowering serum glucose levels in rodents on high fat diet without significant body weight gain [90,91] and inducing the PPARγ biomarker adiponectin [90,92]. Protease protection studies suggest that telmisartan stabilizes PPARγ in a manner distinct from that seen with full agonists [90]. Telmisartan, in pilot studies (12 months) of hypertensive diabetics co-medicated with antihyperglycemic sulfonylureas, not only normalized hypertension, but also significantly lowered plasma insulin levels (decreases in fasting plasma glucose and HbA1c were not significant) and raised serum adiponectin concentrations without concomitant changes in body weight or fluid retention [93,94]. Comprehensive studies will be required to rigorously establish whether these new agents indeed confer advantages over traditional TZDs.

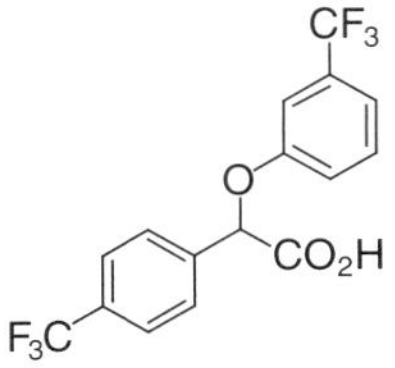

32: MBX-102
(hγ EC_{50} = 18 μM)

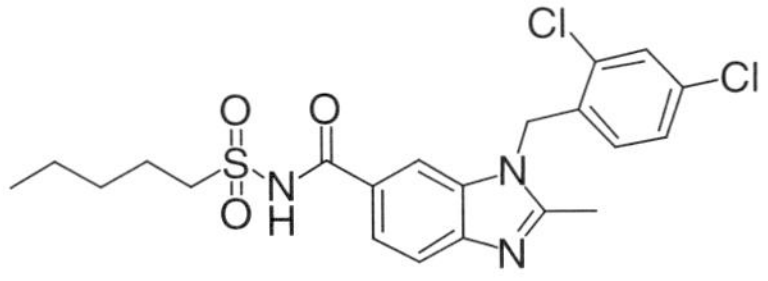

33: FK614
(hγ EC_{50} = 10 nM)

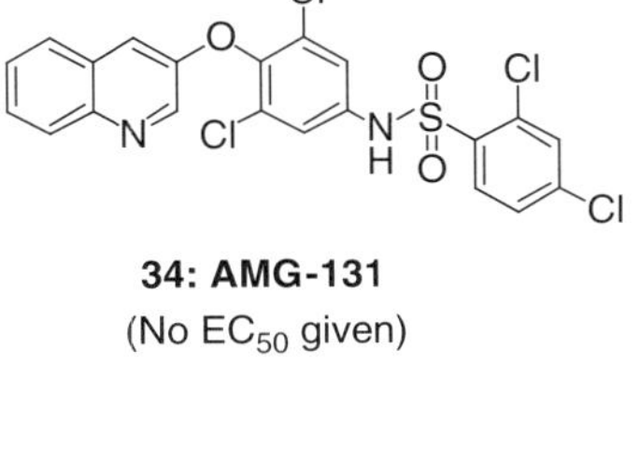

34: AMG-131
(No EC_{50} given)

35: Telmisartan
(hγ EC$_{50}$ = 4.5 uM, 25% - 30% max.)

36: Irbesartan
(No EC$_{50}$ given)

Considerable structural diversity exists among more recently disclosed SPPARγMs (**37–43**), most of which have comparable or superior intrinsic potency relative to rosiglitazone with submaximal responses in transactivation assays while retaining useful activity in rodent diabetes models. The indole-2-carboxylic acid **37** [95], one member of a more extensive series [96], bound to the PPARγ LBD in a unique manner, showed qualitative differences on PPARγ target genes, reduced adipogenicity *in vitro* and had an improved therapeutic index *in vivo*. Similar results were reported for the TZD PAT5a (**38**) [97,98]. Crystallographic evidence established that the SPPARγM BVT762 (**39**) binds in a distinct manner to the PPARγ LBD, forming a key hydrogen bond with Ser342; related analogs retain efficacy in *ob/ob* mice models [99,100]. The potent SPPARγM **40** [101], derived from a full agonist screening lead, evolved into benzoisoxazole **41** [102] which the authors established lacks the scissile *N*-acyl linkage of its progenitor. These SPPARγMs, while highly efficacious in the *db/db* models, lacked typical PPARγ adverse effects in rat toxicity studies (e.g. weight gain and cardiomegaly) and exhibited reduced adipogenesis (45% versus 183% for rosiglitazone) in this study. Interestingly, CLX-0921 (**42**), a modestly potent PPARγ agonist, exhibited significantly reduced co-activator recruitment and was weakly adipogenic both *in vitro* and *in vivo* (4-week studies) [103]. The biaryl pyrazole CRX000143 (**43**) also showed reduced weight gain (7.6%) versus rosiglitazone (17.8%) in SD rats (vehicle-treated animals, 10%) [104].

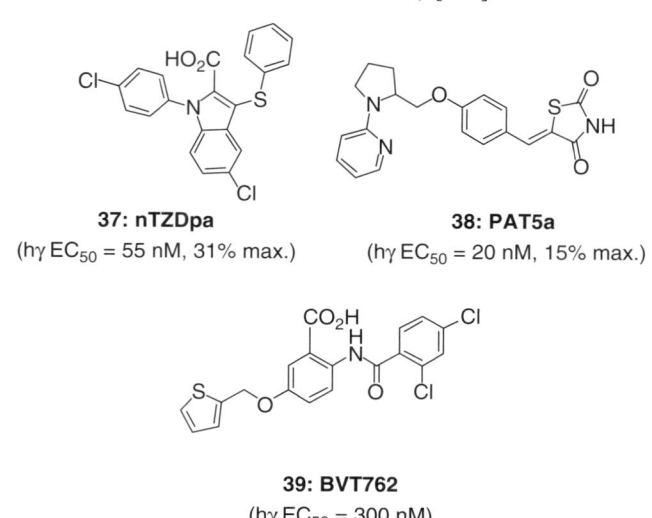

37: nTZDpa
(hγ EC$_{50}$ = 55 nM, 31% max.)

38: PAT5a
(hγ EC$_{50}$ = 20 nM, 15% max.)

39: BVT762
(hγ EC$_{50}$ = 300 nM)

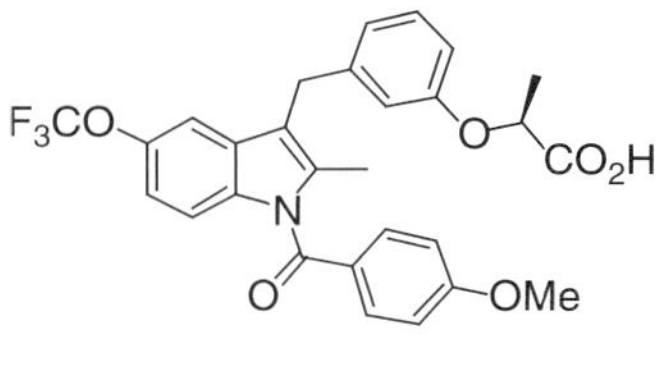

40
(hγ EC$_{50}$ = 2 nM, 21% max.)

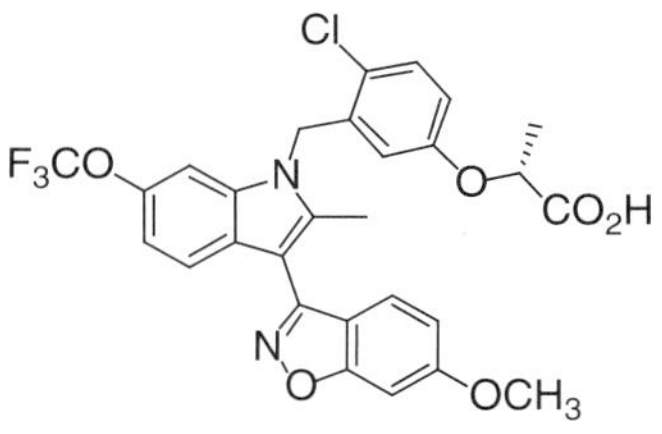

41
(hγ EC$_{50}$ = 3 nM, 24% max.)

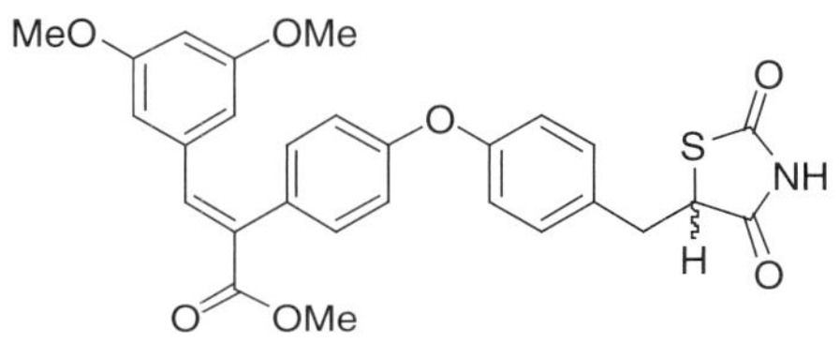

42: CLX-0921
(hγ EC$_{50}$ = 284 nM, 100% max.)

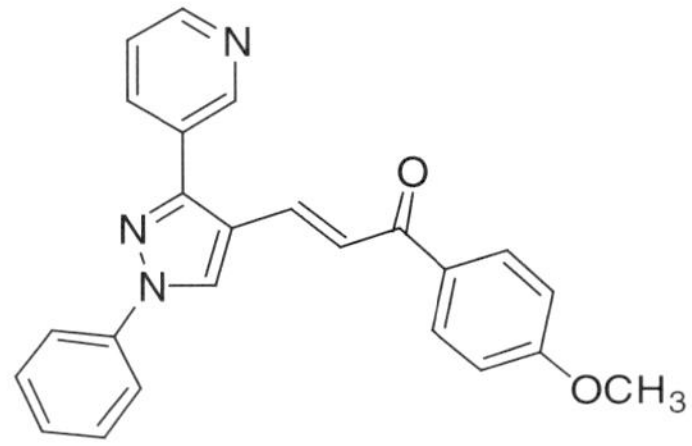

43: CRX000143
(hγ EC$_{50}$ = 141 nM)

3. RXR MODULATORS

RXR heterodimers with PPAR, LXR and FXR are known to activate gene transcription in the presence of an agonist for either of the partnered receptors and the diverse biological sequelae of RXR activation has been described in detail [105]. This summary will therefore focus upon the potential for antidiabetic effects of selective RXR agonists/modulators that have recently been reported. RXR agonists such as LG100268, based upon the rexinoid scaffold, have been shown to be insulin-sensitizing agents. Unfortunately, these compounds also cause undesirable side effects (dramatic increase in plasma TGs and suppression of the thyroid hormone axis) [106]. Recently reported were a series of new RXR modulators that selectively activate the RXR/PPAR heterodimer, but do not activate either RXR/RAR (retinoic acid receptor) or RXR/LXR [107]. Compound **44** afforded reductions in plasma glucose comparable to that of rosiglitazone in *db/db* mice but has distinct off-target activity relative to that of classic RXR agonists. Compound **44** suffers from poor PK, however, introducing a fluoro group onto the trienoic acid side chain (**45**) improved its PK profile substantially. Moreover, **45** retains desirable *in vitro* activity: potent binding to RXR (α/β/γ K_i = 7.8/11.5/13.2 nM) and high selectivity over RAR. In functional assays, **45** activates the RXR/PPAR heterodimer (EC$_{50}$ = 63 nM) and this activity is synergistic with rosiglitazone at 100 nM (EC$_{50}$ = 6.4 nM). These compounds possess little or no synergy with RAR agonists

(<2-fold). By replacing the trieneoic acid side chain with various benzofused heterocycles, agonists with varied constellations of activity were obtained. A closely related compound (**46**), in which a benzofuran was used to replace the trienoic acid, was found to be a full antagonist ($EC_{50} < 11$ nM, efficacy >90%) [108]. Alternatively, replacing the trienoic acid side chain with an indolyl group, (**47**), resulted in a compound with a potent EC_{50} (<11 nM) and very high efficacy (256%) in the RXR/PPAR heterodimer assay [109].

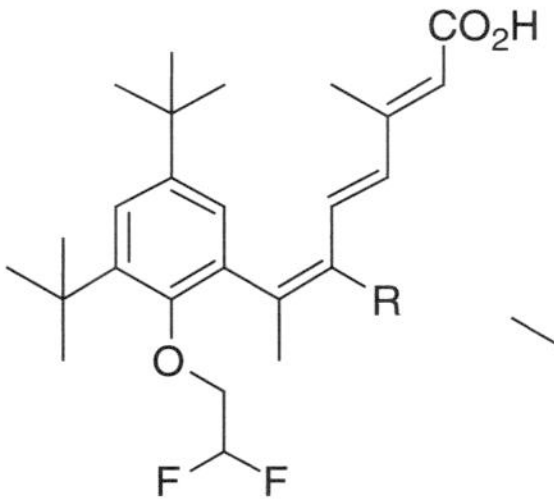

44: R = H
45: R = F

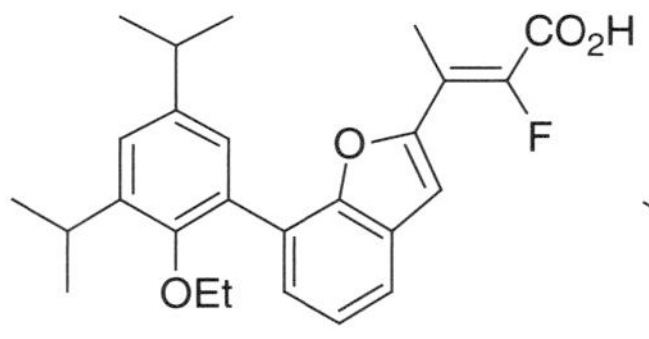

46

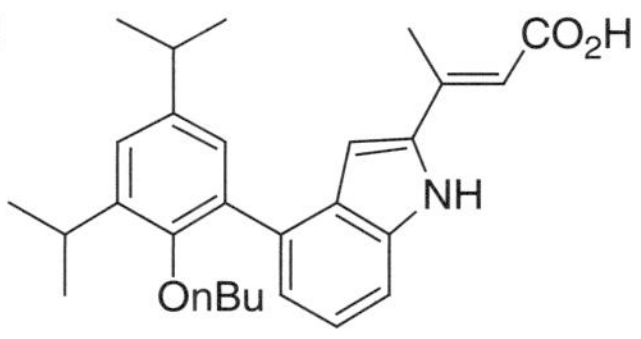

47

4. LXR

The NHR LXR is comprised of two subtypes, LXRα (NR1H3) and LXRβ (NR1H2) [110], which act as intracellular oxysterol sensors that regulate expression of genes involved in lipid metabolism [111]. These genes include many believed to be integral in the reverse cholesterol transport (RCT) pathway, specifically those regulating cellular cholesterol efflux, HDL metabolism and biliary cholesterol excretion. Of special interests, are the LXR-induced ATP-binding cassette (ABC) transporters ABCA1 and ABCG5/G8. ABCA1 encodes for a lipid pump which mediates cholesterol efflux from peripheral cells to the lipid-poor apolipoprotein acceptor (e.g. apoA-1) [112] as the first step in the RCT process, while ABCG5/G8 have been shown in mice to be required for the stimulation of biliary and fecal cholesterol excretion from the liver and small intestine [113,114]. Additionally, LXR regulates expression of apoE [115,116] and lipoprotein lipase (LPL). ApoE serves as a transport vehicle for cholesterol from peripheral cells to the liver, while liver LPL hydrolyzes TG (specifically in TG-rich lipoprotein chylomicrons and VLDL) leading to increased uptake of dissociated phospholipids and lipoproteins by HDL [117].

Agents acting on the LXR pathway have shown great promise in animal models of atherosclerotic cardiovascular disease, and this body of work is extensively summarized in recent reviews [118–120]. With a growing understanding of RCT and cholesterol efflux, it appears likely that their activation will provide improved CVD outcomes. Key proof of concept studies in rodent atherosclerosis models have demonstrated that LXR agonism can reduce progression of atherosclerotic lesion development [121,122].

Unfortunately, key genes involved in lipogenesis also are simultaneously activated with current LXR agonists, including those responsible for increased TG levels, such as (SREBP)1C (a master regulator of TG synthesis) and fatty acid synthetase (FAS) [123,124]. As an illustration of LXRα-mediated effects, chow-fed mice lacking LXRα, but not LXRβ, exhibit decreased expression of genes in the liver controlling TG and FA biosynthesis (i.e. (SREBP)1C, SCD-1, FAS and ACC) [125]. Consistent with these results, dosing of **48**, a synthetic LXR dual α/β agonist, in a KO mouse expressing only LXRβ leads to an increase in plasma HDL without the expected hypertriglyceridemia or resultant hepatic steatosis [126].

Naturally, the development of any successful LXR-based therapeutic must by necessity address agonist-driven lipogenesis and the accompanying steatosis. While it is clear that certain aspects of LXR-mediated gene regulation is species-specific, since (SREBP)1C and FAS are present in both rodents and humans the presumption is that known synthetic agents would produce unacceptable TG increases along with the desired lipid traffic effects. Promisingly, some synthetic derivatives of cholanoic acid when dosed in mice produced favorable lipid changes without increasing plasma TG or resulting in fatty liver [127], suggesting the potential to dissociate HDL increases from TG raising. As shown by these ligands, a strategy independent of receptor subtype selectivity would require agonists that activate LXR while simultaneously repressing SREBP processing downstream, akin to the endogenous oxysterols. [128].

More recently, the LXR pathway has been posited as a new avenue for treatment of conditions extending beyond atherosclerosis [129–132]. As LXR has broad brain expression [133] and LXRα/β agonists are capable of altering cholesterol levels in the CNS [134–136], potential applications include certain neurodegenerative diseases, including AD [137–139].

Since the development of the first-generation non-steroidal LXR agonists, T-1317 (**51**) and GW3965 (**52**), few second-generation tools have appeared in the literature [140]. New LXRα/β dual agonists with excellent selectivity versus other NHRs include a series of high-affinity maleimides [141,142] with the corresponding thiono-maleimides [143], which are typified by **53** (LXRα/β EC_{50} = 50/40 nM). While the *in vitro* genetic profiling for this family of compounds versus ABCA1 mRNA expression, cholesterol efflux, IL-6 secretion and TG biosynthesis was thoroughly discussed, no *in vivo* data were disclosed. A new series of potent heterocyclic LXR dual agonists was profiled for *in vitro* affinity and efficacy, PK characteristics and effect on HDL, cholesterol and TG levels *in vivo* [144]. Two of the most potent members, pyrazine **48** and oxyindole **49** (IC_{50} α/β 90/40 and 19/13 nM, respectively), also lack activity on other NHRs. Podocarpate dimers, such as **54** and **55**, were reported as high-affinity LXRα/β agonists (IC_{50} α/β 2/1 and 1/1 nM, respectively), though SAR exploration suggested limited modification of the subunits is tolerated [145]. In a subsequent report, replacement of one of the dimeric podocarpate subunits with simple alkyl or arene subunits, led to the representative and less potent ligand **56** (IC_{50} α/β 50/20 nM) [146].

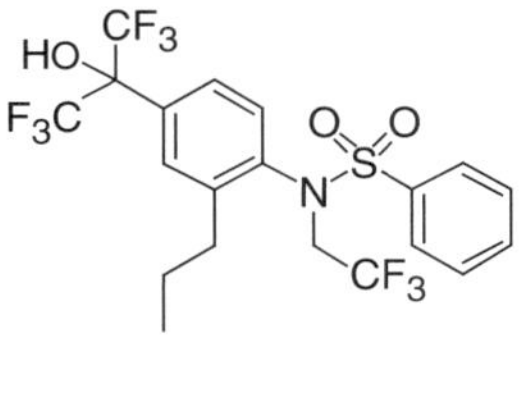

51: T-1317

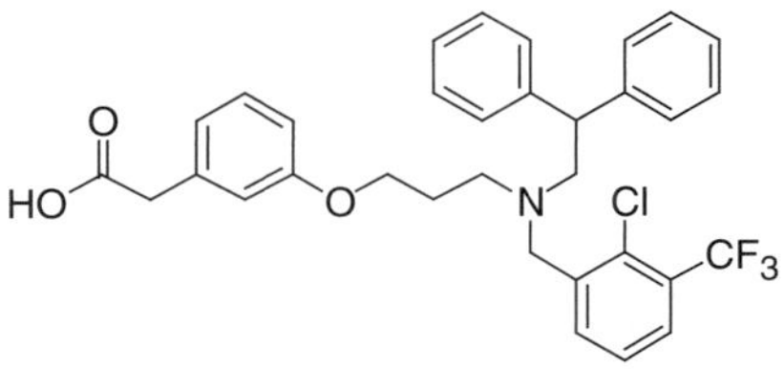

52: GW3965

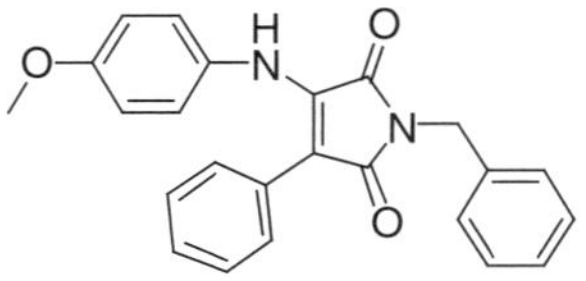

53

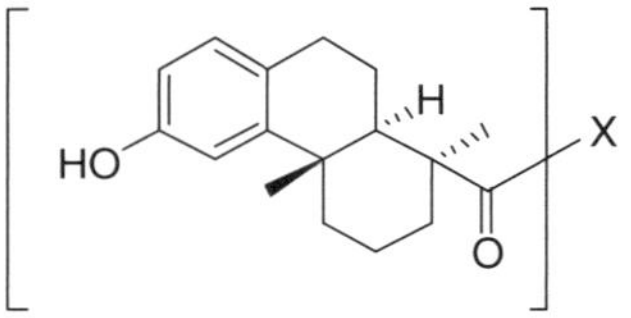

54: X = O
55: X = NH
56: X =

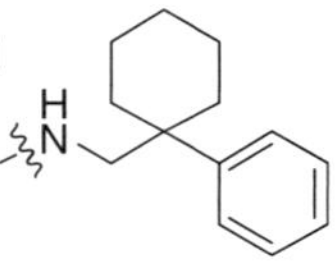

4.1. LXR subtype selective agonists

The apparent liabilities of LXRα/β dual agonists coupled with data garnered from LXR KO studies have led to a strong emphasis on the identification of LXR subtype selective agonists suitable for pharmacologic evaluation. Two natural products (**57, 58**) have been newly identified as LXRα-selective agonists with efficacy in cell-based assays (IC$_{50}$ α/β 0.12/>15 and 1.3/50 μM, respectively) [147]. Riccardin C (**59**) was identified as a LXRα agonist and a LXRβ antagonist [148]; unfortunately it exhibits weak affinity on LXRα (EC$_{50}$ = 30 μM). A similarly low-affinity LXRα selective agonist (**60**) was identified from the natural product *Gynostemma pentaphyllum* [149]. In a complementary approach, a recent report describes the selection of a series of DNA aptamers as LXRβ-selective agonists [150]. More recently, patent disclosures expanded on the SAR of benzisoxazole **48** by development of a potent class of urea agonists exemplified by **50** [151]. Urea **50** was reported to be a potent and highly selective LXRα ligand (IC$_{50}$ α/β 0.08/>50 μM) lacking activity on other NHRs [126]. Experiments in mice-deficient in LXRα demonstrated that **50** has no effect on plasma TG, liver cholesterol, liver TG or plasma HDL levels, whereas in WT mice increases in HDL and liver TG were noted. Interestingly, **61** and **62** have been reported as low-affinity LXRβ-selective partial agonists relative to T-1317 (no *in vivo* data disclosed) [152]. Also, a method was claimed for selecting LXR subtype specific peptides [153].

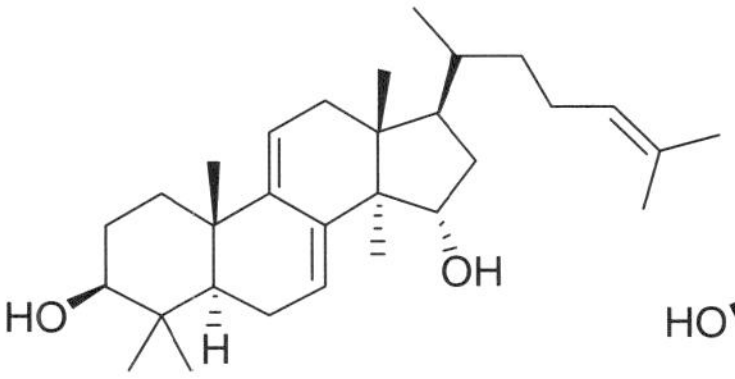

57 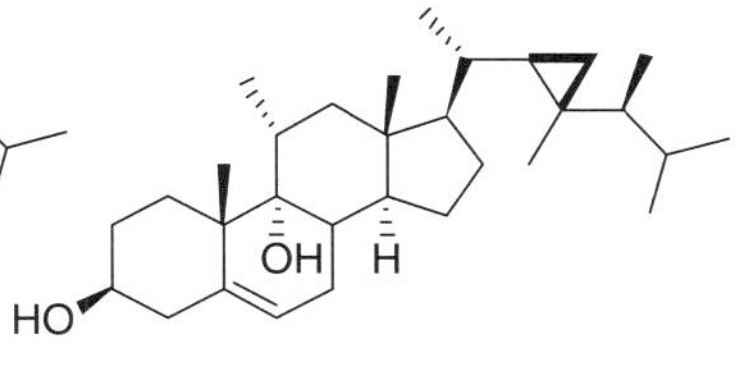 58 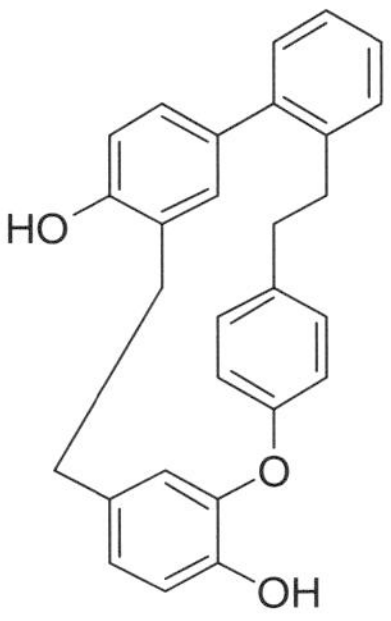 59

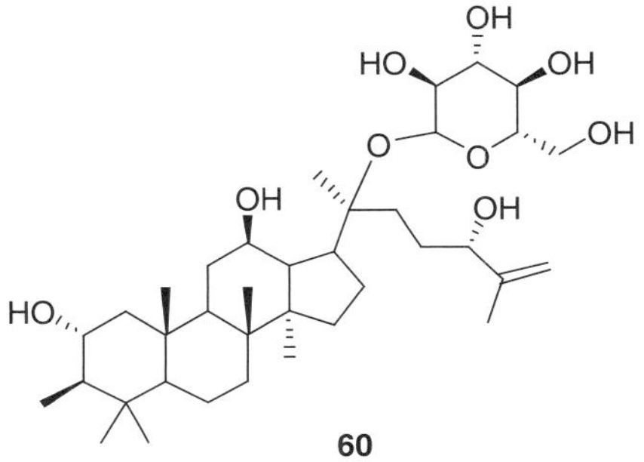

60 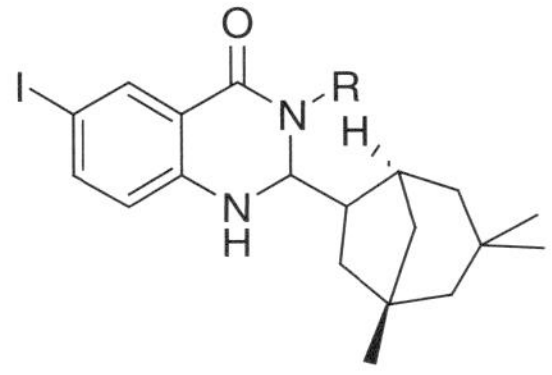 61: R = allyl 62: R = tBu

4.2. Current patent literature

Recent patent disclosures detail extensive efforts directed toward surrogates (**63–68**) for the phenylacetic acid moiety of GW3965, as exemplified by indole **63** [154], isoxazole **64** and pyrazole **65** [155], and encompasses quinolines or tetrahydroquinolines (i.e. **67**) [156] as the most preferred analogs. Interestingly, the carboxylic acid of GW3965 was also removed or converted into amines and amides. Significantly, the patent covering indoles such as **63** discloses new modifications for the amino terminus, arene substituent optimization, the replacement of the benzhydryl moiety with an α-phenethyl group, and exploration of specific methyl substitution (e.g. R_1, R_2 and R_3) throughout aliphatic linker region of **63** [154]. Interestingly, a recent patent application disclosed that GW3965 suppresses IL-13 overproduction and decreases IL-13 concentrations in bronchoalveolar lavage fluid, supporting a claim for the treatment of eosinophilia [157].

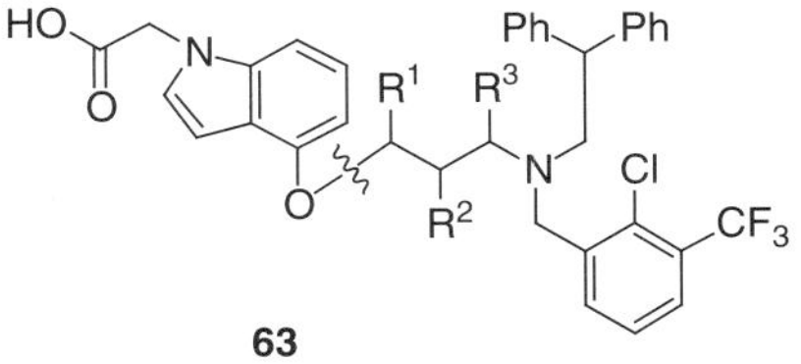

63 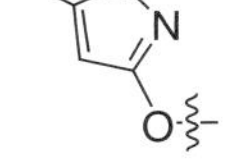 64 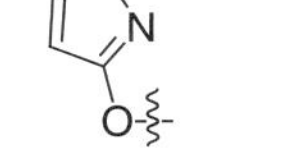 65

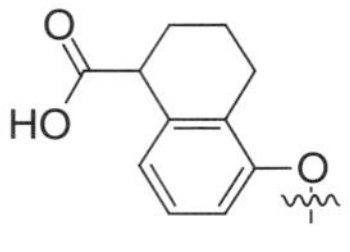

In extensions of the T-1317 aniline SAR, the sulfonamide was replaced with an aryl-substituted tertiary amine **69** [158]. A second patent further expands the scope where the most preferred analogs incorporate the aniline nitrogen into an indole ring (i.e. **70**) [159]. Separately, the nitrogen was also incorporated into an quinazolinone (**71**) [160]. In addition to these early structural templates, a host of new agonist classes with heteroaryl cores have been claimed (as exemplified by indazole **72** [161], quinoline **73** [162], pyrazole **74** [163] and thiadiazole **75** [164]. Substantial numbers of related analogs of **73** and **75** were synthesized, but the specific claims did not narrow the patent's scope. Amide agonists were reported with specific reference to their LXRβ activity as exemplified by **76** [165]. Separately, a set of adamantine amides were disclosed, as illustrated by **77**, with varied regioisomers of the furyl and pyridinyl rings claimed [166].

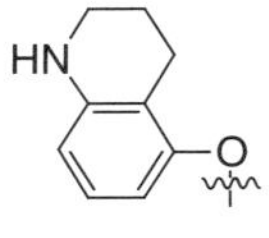

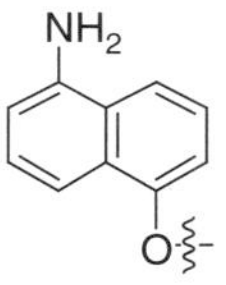

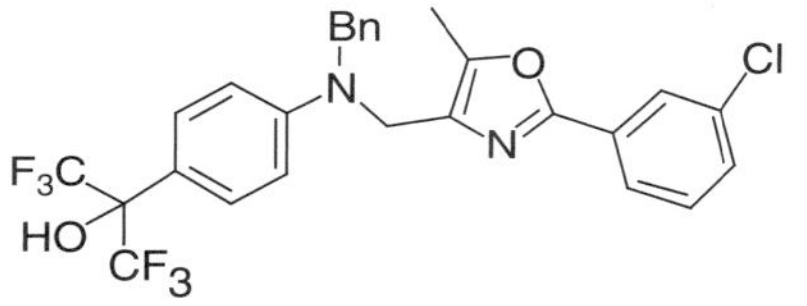

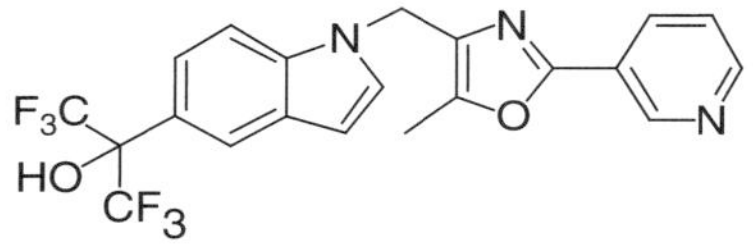

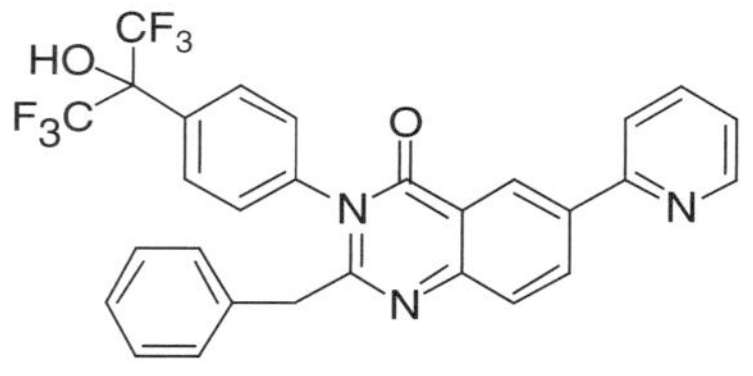

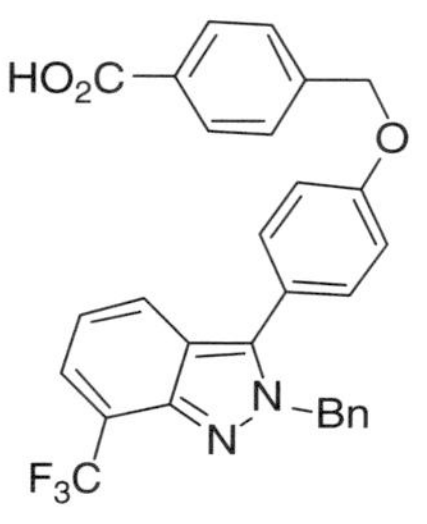

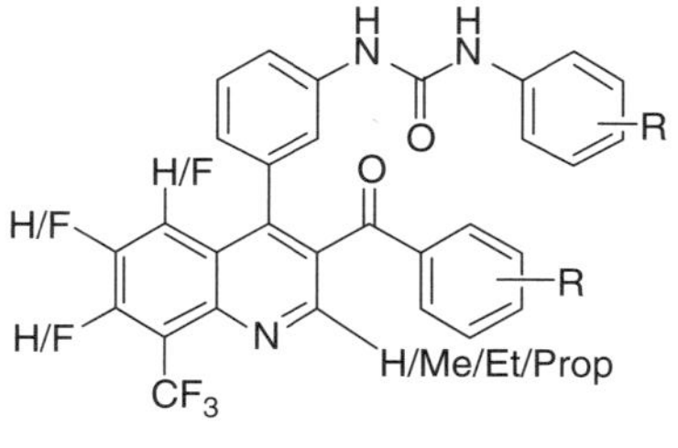

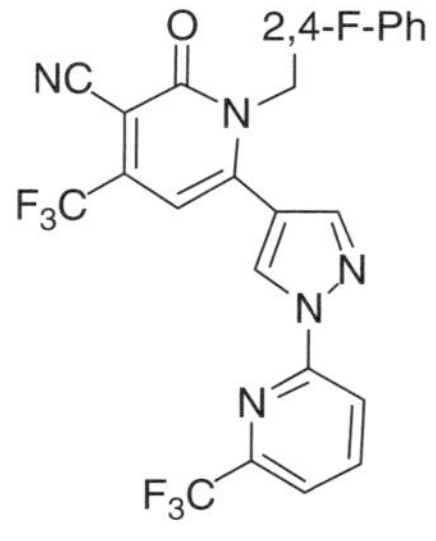

75

76

77

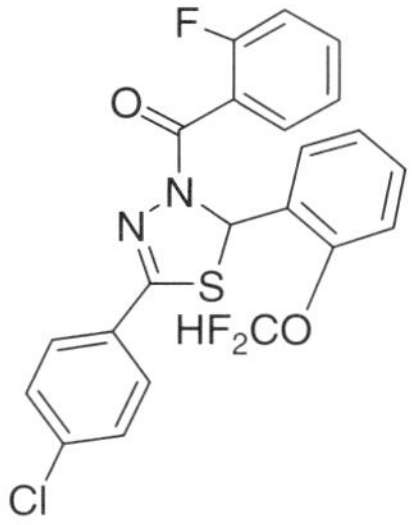

5. FXR

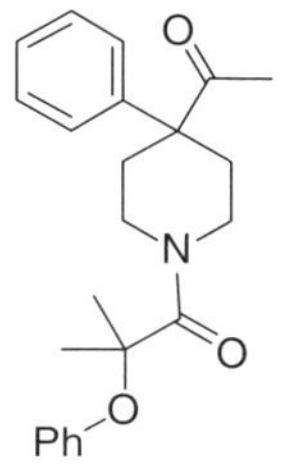

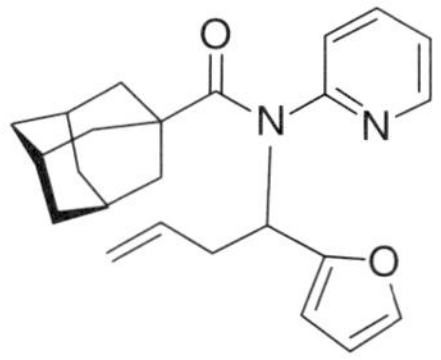

FXR is a NHR acting as a physiological sensor for bile acids by controlling genes regulating their metabolism and transport [167–169]. The repression of key genes, Cyp7A1 and Cyp8B1 by FXR hold much of its promise as a therapeutic target. These encode for cholesterol 7α-hydroxlyase and sterol 12α-hydrolyase, which are enzymes central to the synthesis of bile acids from cholesterol [170,171]. Mice deficient in FXR are characterized by elevated serum TG, cholesterol and bile acids, increased hepatic cholesterol and TG and a proatherogenic serum profile [172]. Early tools have been identified and utilized to probe the effect of FXR pharmacologic intervention: GW4064 (**78**) and guggulsterone (**79**). Additionally, potent new agonist templates continue to appear (**80**, $EC_{50} = 25$ nM) and the surrounding SAR is under active investigation [173,174].

78: GW4046

79: Guggulsterone

80

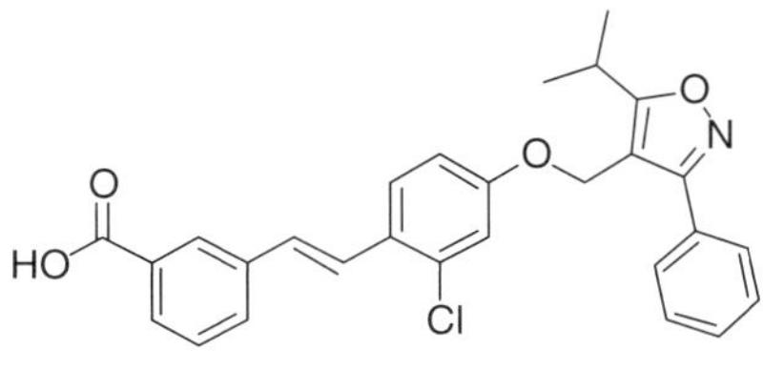

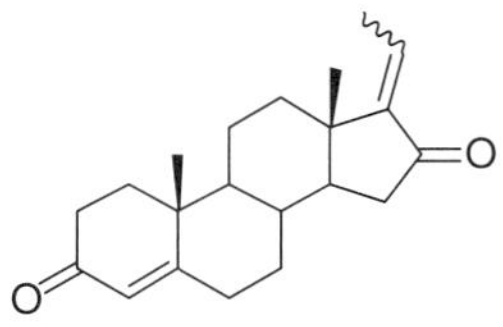

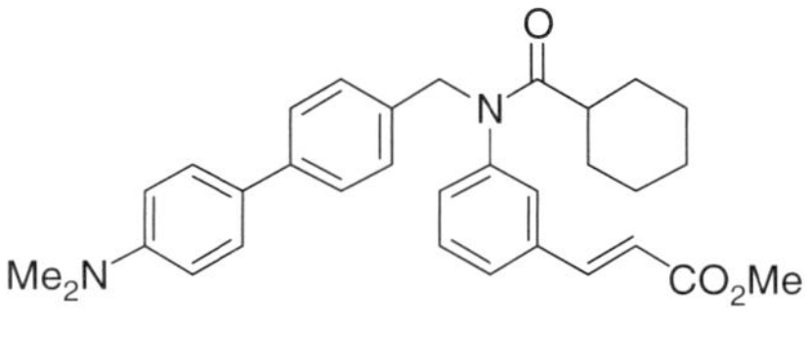

In mouse experiments, GW4064 has poor bioavailability (10%) and a moderate PK profile ($t_{1/2} = 3.5$ h). Dosing WT mice for 7 days with GW4064 (100 mg/kg, b.i.d.) produced a 45% decrease in plasma TG, with an $ED_{50} = 20$ mg/kg [175]. Additionally, dosing the antagonist guggulsterone (**79**) in WT mice, on a high cholesterol diet, lowered hepatic cholesterol levels. No effect was observed when guggulsterone was dosed in a FXR KO mouse [176].

It has been shown that guggulsterone regulates FXR target gene expression (e.g. Cyp7A1), thereby promoting cholesterol catabolism to bile acids and lowering cholesterol [177,178]. Unfortunately, even with a mechanistic basis for utility and animal efficacy as precedent, guggulsterone did not reduce cholesterol in randomized clinical trials [179]. Instead, it was observed that guggulsterone raised LDL cholesterol levels, highlighting the species differences inherent in drug discovery. As a consequence, further studies will be required to establish the relevance of FXR-based therapy for treatment of dyslipidemia in humans.

6. CONCLUSION

Clearly, the sedentary lifestyle and diet of current modern society has produced an unprecedented increase in the number of cases of diabetes and obesity requiring new therapies to preserve the state of human health. The prevalence of insulin resistance and dyslipidemia demands the development of new therapeutics acting via novel mechanisms to combat the increasing incidence of human disease. Modulators with a broad spectrum of activity on PPARα/γ/δ have the exciting potential to ameliorate certain aspects of the metabolic syndrome. Additionally, recent outcomes studies with PPAR ligands suggest that their utility may ultimately extend beyond frank control of glucose and lipid levels. In contrast to the extensive body of knowledge encompassing the PPARs, the RXR, LXR and FXR fields are comparatively nascent. In recent years, few reports describing novel, high-affinity RXR and FXR modulators have appeared, leaving their ultimate promise as therapeutic intervention points unclear. With respect to LXR, extensive research endeavors have generated potent agonists over a large variety of agonist chemotypes. These tools have greatly contributed to the rapid elucidation of the underlying associated biology, however to date, none of these agents are known to have advanced into clinical evaluation. Overall, the field of NHR modulators remains vibrant, with pharmaceutical discovery providing new tools to advance the understanding of the interplay between the pharmacology, physiology and functional genomics underpinning NHR modulation.

REFERENCES

[1] S. M. Rangwala and M. A. Lazar, *Trends Pharmacol. Sci.*, 2004, **25**, 331.
[2] J. Berger and D. E. Moller, *Ann. Rev. Med.*, 2002, **53**, 409.
[3] D. Sternbach, *Ann. Rep. Med. Chem.*, 2003, **38**, 71.

[4] T. M. Willson, P. J. Brown, D. D. Sternbach and B. R. Henke, *J. Med. Chem.*, 2000, **43**, 527.
[5] H. P. Guan, Y. Li, M. V. Jensen, C. B. Newgard, C. M. Steppan and M. A. Lazar, *Nat. Med.*, 2002, **8**, 1122.
[6] P. Tontonoz, E. Hu and B. M. Spiegelman, *Cell*, 1994, **79**, 1147.
[7] A. B. Goldfine, S. Crunkhorn, M. Costello, H. Gami, E. J. Landaker, M. Niinobe, K. Yoshikawa, D. Lo, A. Warren, J. Jimenez-Chillaron and M. E. Patti, *Diabetes*, 2006, **55**, 640.
[8] Y. Guan, C. Hao, D. R. Cha, R. Rao, W. Lu, D. E. Kohan, M. A. Magnuson, R. Redha, Y. Zhang and M. D. Breyer, *Nat. Med.*, 2005, **11**, 861.
[9] H. Zhang, A. Zhang, D. E. Kohan, R. D. Nelson, F. J. Gonzalez and T. Yang, *Proc. Natl. Acad. Sci. USA*, 2005, **102**, 9406.
[10] R. B. Goldberg, D. M. Kendall, M. A. Deeg, J. B. Buse, A. J. Zagar, J. A. Pinaire, M. H. Tan, M. A. Khan, A. T. Perez and S. J. Jacober, GLAI Study Investigators, *Diabetes Care*, 2005, **28**, 1547.
[11] P. D. Home, S. J. Pocock, H. Beck-Nielsen, R. Gomis, M. Hanefeld, H. Dargie, M. Komajda, J. Gubb, N. Biswas and N. P. Jones, *Diabetologia*, 2005, **8**, 1726.
[12] J. A. Dormandy, B. Charbonnel, D. J. Eckland, E. Erdmann, M. Massi-Benedetti, I. K. Moules, A. M. Skene, M. H. Tan, P. J. Lefebvre, G. D. Murray, E. Standl, R. G. Wilcox, L. Wilhelmsen, J. Betteridge, K. Birkeland, R. J. Golay, L. Heine, M. Koranyi, M. Laakso, A. Mokan, V. Norkus, T. Pirags, A. Podar, W. Scheen, G. Scherbaum, O. Schernthaner, J. Schmitz, U. Skrha and J. Smith, Taton and PROactive investigators, *Lancet*, 2005, **366**, 1279.
[13] A. H. Xiang, R. K. Peters, S. L. Kjos, A. Marroquin, J. Goico, C. Ochoa, M. Kawakubo and T. A. Buchanan, *Diabetes*, 2006, **55**, 517.
[14] F. J. Schopfer, Y. Lin, P. R. Baker, T. Cui, M. Garcia-Barrio, J. Zhang, K. Chen, Y. E. Chen and B. A. Freeman, *Proc. Natl. Acad. Sci. USA*, 2005, **102**, 2340.
[15] P. R. Baker, Y. Lin, F. J. Schopfer, S. R. Woodcock, A. L. Groeger, C. Batthyany, S. Sweeney, M. H. Long, K. E. Iles, L. M. Baker, B. P. Branchaud, Y. E. Chen and B. A. Freeman, *J. Biol. Chem.*, 2005, **280**, 42464.
[16] H. W. Lee, B. Y. Kim, J. B. Ahn, S. K. Kang, J. H. Lee, J. S. Shin, S. K. Ahn, S. J. Lee and S. S. Yoon, *Eur. J. Med. Chem.*, 2005, **40**, 862.
[17] P. J. Rybczynski, R. E. Zeck, J. Dudash, Jr., D. W. Combs, T. P. Burris, M. Yang, M. C Osborne, X. Chen and K. T Demarest, *J. Med. Chem.*, 2004, **47**, 196.
[18] P. J. Rybczynski, R. E. Zeck, D. W. Combs, I. Turchi, T. P. Burris, J. Z. Xu, M. Yang and K. T. Demarest, *Bioorg. Med. Chem. Lett.*, 2003, **13**, 2359.
[19] A. Sharon, R. Pratap, R. Vatsyayan, P. R. Maulik, U. Roy, A. Goel and V. J. Ram, *Bioorg. Med. Chem. Lett.*, 2005, **15**, 3356.
[20] P. Sauerberg, J. P. Mogensen, L. Jeppesen, L. A. Svensson, J. Fleckner, J. Nehlin, E. M. Wulff and I. Pettersson, *Bioorg. Med. Chem. Lett.*, 2005, **15**, 1497.
[21] T. Breidert, J. Callebert, M. T. Heneka, G. Landreth, J. M. Launay and E. C. Hirsch, *J. Neurochem.*, 2002, **82**, 615.
[22] L. Klotz, M. Schmidt, T. Giese, M. Sastre, P. Knolle, T. Klockgether and M. T. Heneka, *J. Immunol.*, 2005, **175**, 4948.
[23] G. S. Watson, B. A. Cholerton, M. A. Reger, L. D. Baker, S. R. Plymate, S. Asthana, M. A. Fishel, J. J. Kulstad, P. S. Green, D. G. Cook, S. E. Kahn, M. L. Keeling and S. Craft, *Am. J. Geriatr. Psychiatry*, 2005, **13**, 950.
[24] M. E. Risner, A. M. Saunders, J. F. Altman, G. C. Ormandy, S. Craft, I. M. Foley, M. E. Zvartau-Hind, D. A. Hosford and A. D. Roses, *Pharmacogenom. J.*, 2006, **6**, 246.
[25] G. S. Watson and S. Craft, *CNS Drugs*, 2003, **17**, 27.

[26] S. Kersten, J. Seydoux, J. M. Peters, F. J. Gonzalez, B. Desvergne and W. Wahli, *J. Clin. Invest.*, 1999, **103**, 1489.
[27] M. F. Linton and S. Fazio, *Curr. Atheroscler. Rep.*, 2000, **2**, 29.
[28] H. B. Rubins, S. J. Robins, D. Collins, C. L. Fye, J. W. Anderson, M. B. Elam, F. H. Faas, E. Linares, E. J. Schaefer, G. Schectman, T. J. Wilt and J. Wittes, *N. Engl. J. Med.*, 1999, **341**, 410.
[29] A. Keech, R. J. Simes, P. Barter, J. Best, R. Scott, M. R. Taskinen, P. Forder, A. Pillai, T. Davis, P. Glasziou, P. Drury, Y. A. Kesaniemi, D. Sullivan, D. Hunt, P. Colman, M. d'Emden, M. Whiting, C. Ehnholm and M. Laakso, FIELD Study Investigators, *Lancet*, 2005, **366**, 1849.
[30] A. Tenenbaum, M. Motro, E. Z. Fisman, D. Tanne, V. Boyko and S. Behar, *Arch. Intern. Med.*, 2005, **165**, 1154.
[31] Z. Li, C. Liao, B. C. Ko, S. Shan, E. H. Tong, Z. Yin, D. Pan, V. K. Wong, L. Shi, Z. Q. Ning, W. Hu, J. Zhou, S. S. Chung and X. P. Lu, *Bioorg. Med. Chem. Lett.*, 2004, **14**, 3507.
[32] K. Liu, L. Xu, J. P. Berger, K. L. Macnaul, G. Zhou, T. W. Doebber, M. J. Forrest, D. E. Moller and A. B. Jones, *J. Med. Chem.*, 2005, **48**, 2262.
[33] H. Koyama, J. K. Boueres, D. J. Miller, J. P. Berger, K. L. MacNaul, P. R. Wang, M. C. Ippolito, S. D. Wright, D. E. Moller and S. P. Sahoo, *Bioorg. Med. Chem. Lett.*, 2005, **15**, 3347.
[34] P. Sauerberg, P. S. Bury, J. P. Mogensen, H. J. Deussen, I. Pettersson, J. Fleckner, J. Nehlin, K. S. Frederiksen, T. Albrektsen, N. Din, L. A. Svensson, L. Ynddal, E. M. Wulff and L. Jeppesen, *J. Med. Chem.*, 2003, **46**, 4883.
[35] T. E. Johnson, X. Zhang, S. Shi and D. R. Umbenhauer, *Toxicol. Appl. Pharmacol.*, 2005, **208**, 210.
[36] R. Thomssen, S. Bonk, C. Propfe, K. H. Heermann, H. G. Kochel and A. Uy, *Med. Microbiol. Immunol. (Berl.)*, 1992, **181**, 293.
[37] S. Dharancy, M. Malapel, G. Perlemuter, T. Roskams, Y. Cheng, L. Dubuquoy, P. Podevin, F. Conti, V. Canva, D. Philippe, L. Gambiez, P. Mathurin, J. C. Paris, K. Schoonjans, Y. Calmus, S. Pol, J. Auwerx and P. Desreumaux, *Gastroenterology*, 2005, **128**, 334.
[38] N. Fujita, M. Kaito, H. Tanaka, S. Horiike and Y. Adachi, *Am. J. Gastroenterol.*, 2004, **99**, 2280.
[39] D. J. Kim, M. T. Bility, A. N. Billin, T. M. Willson, F. J. Gonzalez and J. M. Peters, *Cell Death Differ*, 2006, **13**, 53.
[40] R. L. Stephen, M. C. Gustafsson, M. Jarvis, R. Tatoud, B. R. Marshall, D. Knight, E. Ehrenborg, A. L. Harris, C. R. Wolf and C. N. Palmer, *Cancer Res.*, 2004, **64**, 3162.
[41] R. A. Gupta, D. Wang, S. Katkuri, H. Wang, S. K. Dey and R. N. DuBois, *Nat. Med.*, 2004, **10**, 245.
[42] Y. Yin, R. G. Russell, L. E. Dettin, R. Bai, Z. L. Wei, A. P. Kozikowski, L. Kopelovich and R. I. Glazer, *Cancer Res.*, 2005, **65**, 3950.
[43] Y. Barak, D. Liao, W. He, E. S. Ong, M. C. Nelson, J. M. Olefsky, R. Boland and R. M. Evans, *Proc. Natl. Acad. Sci. USA*, 2002, **99**, 303.
[44] T. Tanaka, J. Yamamoto, S. Iwasaki, H. Asaba, H. Hamura, Y. Ikeda, M. Watanabe, K. Magoori, R. X. Ioka, K. Tachibana, Y. Watanabe, Y. Uchiyama, K. Sumi, H. Iguchi, S. Ito, T. Doi, T. Hamakubo, M. Naito, J. Auwerx, M. Yanagisawa, T. Kodama and J. Sakai, *Proc. Natl. Acad. Sci. USA*, 2003, **100**, 15924.
[45] M. L. Sznaidman, C. D. Haffner, P. R. Maloney, A. Fivush, E. Chao, D. Goreham, M. L. Sierra, C. LeGrumelec, H. E. Xu, V. G. Montana, M. H. Lambert, T. M. Willson, W. R. Oliver Jr and D. D. Sternbach, *Bioorg. Med. Chem. Lett.*, 2003, **13**, 1517.

[46] J. M. Wallace, M. Schwarz, P. Coward, J. Houze, J. K. Sawyer, K. L. Kelley, A. Chai and L. L. Rudel, *J. Lipid Res.*, 2005, **46**, 1009.
[47] S. Weigand, H. Bischoff, E. Dittrich-Wengenroth, H. Heckroth, D. Lang, A. Vaupel and M. Woltering, *Bioorg. Med. Chem. Lett.*, 2005, **15**, 4619.
[48] T. L. Graham, C. Mookherjee, K. E. Suckling, C. N. Palmer and L. Patel, *Atherosclerosis*, 2005, **181**, 29.
[49] W. R. Jr. Oliver, J. L. Shenk, M. R. Snaith, C. S. Russell, K. D. Plunket, N. L. Bodkin, M. C. Lewis, D. A. Winegar, M. L. Sznaidman, M. H. Lambert, H. E. Xu, D. D. Sternbach, S. A. Kliewer, B. C. Hansen and T. M. Willson, *Proc. Natl. Acad. Sci. USA*, 2001, **98**, 5306.
[50] G. D. Barish, V. A. Narkar and R. M. Evans, *J. Clin. Invest.*, 2006, **116**, 590.
[51] A. C. Li, C. J. Binder, A. Gutierrez, K. K. Brown, C. R. Plotkin, J. W. Pattison, A. F. Valledor, R. A. Davis, T. M. Willson, J. L. Witztum, W. Palinski and C. K. Glass, *J. Clin. Invest.*, 2004, **114**, 1564.
[52] D. L. Sprecher, A. Johnson, G. Pearce, G. Watts and H. Barrett, *Circulation*, 2005, **112** (suppl., Abs), 1211.
[53] S. Seber, S. Ucak, O. Basat and Y. Altuntas, *Diabetes Res. Clin. Pract.*, 2006, **71**, 52.
[54] M. Guerre-Millo, P. Gervois, E. Raspe, L. Madsen, P. Poulain, B. Derudas, J. M. Herbert, D. A. Winegar, T. M. Willson, J. C. Fruchart, R. K. Berge and B. Staels, *J. Biol. Chem.*, 2000, **275**, 16638.
[55] J. M. Ye, P. J. Doyle, M. A. Iglesias, D. G. Watson, G. J. Cooney and E. W. Kraegen, *Diabetes*, 2001, **50**, 411.
[56] Q. Guo, S. P. Sahoo, P. R. Wang, D. P. Milot, M. C. Ippolito, M. S. Wu, J. Baffic, C. Biswas, M. Hernandez, M. Lam, N. Sharma, W. Han, L. J. Kelly, K. L. MacNaul, G. Zhou, R. Desai, J. V. Heck, T. W. Doebber, J. P. Berger, D. E. Moller, C. P. Sparrow, Y. S. Chao and S. D. Wright, *Endocrinology*, 2004, **145**, 1640.
[57] M. Nomura, S. Kinoshita, H. Satoh, T. Maeda, K. Murakami, M. Tsunoda, H. Miyachi and K. Awano, *Bioorg. Med. Chem. Lett.*, 1999, **9**, 533.
[58] T. W. Doebber, L. J. Kelly, G. Zhou, R. Meurer, C. Biswas, Y. Li, M. S. Wu, M. C. Ippolito, Y.-S. Chao, P.-R. Wang, S. D. Wright, D. E. Moller and J. P. Berger, *Biochem. Biophys. Res. Commun.*, 2004, **318**, 323.
[59] K. Decochez, R. K. Rippley, J. L. Miller, M. De Smet, K. X. Yan, K. Matthijs, K. A. Riffel, H. Song, H. Zhu, H. O. Maynor, W. Tanaka, A. O. Johnson-Levonas, M. J. Davies, K. M. Gottesdiener, B. Keymeulen and J. A. Wagner, *Drugs R. D.*, 2006, **7**, 99.
[60] R. Chakrabarti, R. K. Vikramadithyan, P. Misra, J. Hiriyan, S. Raichur, R. K. Damarla, C. Gershome, J. Suresh and R. Rajagopalan, *Br. J. Pharmacol.*, 2003, **140**, 527.
[61] L. C. Pickavance, C. L. Brand, K. Wassermann and J. P. H. Wilding, *Br. J. Pharmacol.*, 2005, **144**, 308.
[62] M. F. Saad, M. Greco, K. Osei, A. J. Lewin, C. Edwards, M. Nunez and R. R. Reinhardt, Ragaglitazar Dose-Ranging Study Group, *Diabetes Care*, 2004, **27**, 1324.
[63] P. V. Devasthale, S. Chen, Y. Jeon, F. Qu, C. Shao, W. Wang, H. Zhang, M. Cap, D. Farrelly, R. Golla, G. Grover, T. Harrity, Z. Ma, L. Moore, J. Ren, R. Seethala, L. Cheng, P. Sleph, W. Sun, A. Tieman, J. R. Wetterau, A. Doweyko, G. Chandrasena, S. Y. Chang, W. G. Humphreys, V. G. Sasseville, S. A. Biller, D. E. Ryono, F. Selan, N. Hariharan and P. T. Cheng, *J. Med. Chem.*, 2005, **48**, 2248.
[64] J. B. Buse, C. J. Rubin, R. Frederich, K. Viraswami-Appanna, K. C. Lin, R. Montoro, G. Shockey and J. A. Davidson, *Clin. Ther.*, 2005, **27**, 1181.
[65] J. A. Martin, D. A. Brooks, L. Prieto, R. Gonzalez, A. Torrado, I. Rojo, B. Lopez de Uralde, C. Lamas, R. Ferritto, M. Dolores Martin-Ortega, J. Agejas, F. Parra, J. R. Rizzo, G. A. Rhodes, R. L. Robey, C. A. Alt, S. R. Wendel, T. Y. Zhang, A.

Reifel-Miller, C. Montrose-Rafizadeh, J. T. Brozinick, E. Hawkins, E. A. Misener, D. A. Briere, R. Ardecky, J. D. Fraser and A. M. Warshawsky, *Bioorg. Med. Chem. Lett.*, 2005, **15**, 51.
[66] M. Prince, K. Spicer, J. Caro, E. Abu-Raddad, R. Konrad, M.-D. Wang, A. Negro-Vilar and J. Dananberg, 64th Scientific Sessions of the American Diabetes Association, Orlando, FL, June 4–8, 2004, Abstract 139 OR.
[67] P. Cronet, J. F. Petersen, R. Folmer, N. Blomberg, K. Sjoblom, U. Karlsson, E. L. Lindstedt and K. Bamberg, *Structure*, 2001, **9**, 699.
[68] B. Fagerberg, S. Edwards, T. Halmos, J. Lopatynski, H. Schuster, S. Stender, G. Stoa-Birketvedt, S. Tonstad, S. Halldorsdottir and I. Gause-Nilsson, *Diabetologia*, 2005, **48**, 1716.
[69] R. C. Desai, W. Han, E. J. Metzger, J. P. Bergman, D. F. Gratale, K. L. MacNaul, J. P. Berger, T. W. Doebber, K. Leung, D. E. Moller, J. V. Heck and S. P. Sahoo, *Bioorg. Med. Chem. Lett.*, 2003, **13**, 2795.
[70] R. C. Desai, D. F. Gratale, W. Han, H. Koyama, E. Metzger, V. K. Lombardo, K. L. MacNaul, T. W. Doebber, J. P. Berger, K. Leung, R. Franklin, D. E. Moller, J. V. Heck and S. P. Sahoo, *Bioorg. Med. Chem. Lett.*, 2003, **13**, 3541.
[71] F. Caijo, P. Mosset, R. Gree, V. Audinot-Bouchez, J. Bo3utin, P. Renard, D. H. Caignard and C. Dacquet, *Bioorg. Med. Chem. Lett.*, 2005, **15**, 4421.
[72] S. A. Kliewer, S. S. Sundseth, S. A. Jones, P. J. Brown, G. B. Wisely, C. S. Koble, P. Devchand, W. Wahli, T. M. Willson, J. M. Lenhard and J. M. Lehmann, *Proc. Natl. Acad. Sci. USA*, 1997, **94**, 4318.
[73] H. Koyama, D. J. Miller, J. K. Boueres, R. C. Desai, A. B. Jones, J. P. Berger, K. L. MacNaul, L. J. Kelly, T. W. Doebber, M. S. Wu, G. Zhou, P. R. Wang, M. C. Ippolito, Y. S. Chao, A. K. Agrawal, R. Franklin, J. V. Heck, S. D. Wright, D. E. Moller and S. P. Sahoo, *J. Med. Chem.*, 2004, **47**, 3255.
[74] Z. Cai, J. Feng, Y. Guo, P. Li, Z. Shen, F. Chu and Z. Guo, *Bioorg. Med. Chem.*, 2006, **14**, 866.
[75] D. B. Lowe, N. Bifulco, W. H. Bullock, T. Claus, P. Coish, M. Dai, F. E. Dela Cruz, D. Dickson, D. Fan, H. Hoover-Litty, T. Li, X. Ma, G. Mannelly, M. K. Monahan, I. Muegge, S. O'Connor, M. Rodriguez, T. Shelekhin, A. Stolle, L. Sweet, M. Wang, Y. Wang, C. Zhang, H. J. Zhang, M. Zhang, K. Zhao, Q. Zhao, J. Zhu, L. Zhu and M. Tsutsumi, *Bioorg. Med. Chem. Lett.*, 2006, **16**, 297.
[76] G. Q. Shi, J. D. Dropinski, B. M. McKeever, S. Xu, J. W. Becker, J. P. Berger, K. L. MacNaul, A. Elbrecht, G. Zhou, T. W. Doebber, P. Wang, Y.-S. Chao, M. Forrest, J. V. Heck, D. E. Moller and A. B Jones, *J. Med. Chem.*, 2005, **48**, 4457.
[77] A. D. Adams, Z. Hu, D. von Langen, A. Dadiz, A. Elbrecht, K. L. MacNaul, J. P. Berger, G. Zhou, T. W. Doebber, R. Meurer, M. J. Forrest, D. E. Moller and A. B. Jones, *Bioorg. Med. Chem. Lett.*, 2003, **13**, 3185.
[78] J. Kasuga, M. Makishima, Y. Hashimoto and H. Miyachi, *Bioorg. Med. Chem. Lett.*, 2006, **16**, 554.
[79] N. Mahindroo, C. C. Wang, C. C. Liao, C. F. Huang, I. L. Lu, T. W. Lien, Y. H. Peng, W. J. Huang, Y. T. Lin, M. C. Hsu, C. H. Lin, C. H. Tsai, J. T. Hsu, X. Chen, P. C. Lyu, Y. S. Chao, S. Y. Wu and H. P. Hsieh, *J. Med. Chem.*, 2006, **49**, 1212.
[80] N. Mahindroo, C. F. Huang, Y. H. Peng, C. C. Wang, C. C. Liao, T. W. Lien, S. K. Chittimalla, W. J. Huang, C. H. Chai, E. Prakash, C. P. Chen, T. A. Hsu, C. H. Peng, I. L. Lu, L. H. Lee, Y. W. Chang, W. C. Chen, Y. C. Chou, C. T. Chen, C. M. Goparaju, Y. S. Chen, S. J. Lan, M. C. Yu, X. Chen, Y. S. Chao, S. Y. Wu and H. P. Hsieh, *J. Med. Chem.*, 2005, **48**, 8194.

[81] J. P. Mogensen, L. Jeppesen, P. S. Bury, I. Pettersson, J. Fleckner, J. Nehlin, K. S. Frederiksen, T. Albrektsen, N. Din, S. B. Mortensen, L. A. Svensson, K. Wassermann, E. M. Wulff, L. Ynddal and P. Sauerberg, *Bioorg. Med. Chem. Lett.*, 2003, **13** (2), 257.
[82] K. G. Liu, J. S. Smith, A. H. Ayscue, B. R. Henke, M. H. Lambert, L. M. Leesnitzer, K. D. Plunket, T. M. Willson and D. D. Sternbach, *Bioorg. Med. Chem. Lett.*, 2001, **11**, 2385.
[83] Z. Zhao, X. Chen, J. Wang, H. Sun and J. S.-C. Liang, *WO Patent* 05080340 A1, 2005.
[84] J. Rosenstock, F. Flores-Losano, S. Schwartz, G. Gonzalez-Galvez and D. B. Karpf, *Diabetes*, 2005, **54**, A11, Abstract 44-OR.
[85] H. Minoura, S. Takeshita, M. Ita, J. Hirosumi, M. Mabuchi, I. Kawamura, S. Nakajima, O. Nakayama, H. Kayakiri, T. Oku, A. Ohkubo-Suzuki, M. Fukagawa, H. Kojo, K. Hanioka, N. Yamasaki, T. Imoto, Y. Kobayashi and S. Mutoh, *Eur. J. Pharmacol.*, 2004, **494**, 273.
[86] T. Fujimura, H. Sakuma, S. Konishi, T. Oe, N. Hosogai, C. Kimura, I. Aramori and S. Mutoh, *J. Pharmacol. Sci.*, 2005, **99**, 342.
[87] T. Fujimura, H. Sakuma, A. Ohkubo-Suzuki, I. Aramori and S. Mutoh, *Biol. Pharm. Bull.*, 2006, **29**, 423.
[88] L. R. McGee, J. B. Houze, S. M. Rubenstein, A. Hagiwara, N. Furukawa and H. Shinkai, *WO Patent* 0200633, 2002.
[89] K. Kersey, L. C. Floren, B. Pendleton, M. J. Stemien, J. Buchanon and F. Dunn, 64th ADA Scientific Sessions, Orlando, FL, 2004, Abstract 656-P.
[90] M. Schupp, M. Clemenz, R. Gineste, H. Witt, J. Janke, S. Helleboid, N. Hennuyer, P. Ruiz, T. Unger, B. Staels and U. Kintscher, *Diabetes*, 2005, **54**, 3442.
[91] S. C. Benson, H. A. Pershadsingh, C. I. Ho, A. Chittiboyina, P. Desai, M. Pravenec, N. Qi, J. Wang, M. A. Avery and T. W. Kurtz, *Hypertension*, 2004, **43**, 993.
[92] R. Clasen, M. Schupp, A. Foryst-Ludwig, C. Sprang, M. Clemenz, M. Krikov, C. Thone-Reineke, T. Unger and U. Kintscher, *Hypertension*, 2005, **46**, 137.
[93] H. A. Pershadsingh and T. W. Kurtz, *Diabetes Care*, 2004, **27**, 1015.
[94] Y. Miura, N. Yamamoto, S. Tsunekawa, S. Taguchi, Y. Eguchi, N. Ozaki and Y. Oiso, *Diabetes Care*, 2005, **28**, 757.
[95] J. P. Berger, A. E. Petro, K. L. Macnaul, L. J. Kelly, B. B. Zhang, K. Richards, A. Elbrecht, B. A. Johnson, G. Zhou, T. W. Doebber, C. Biswas, M. Parikh, N. Sharma, M. R. Tanen, G. M. Thompson, J. Ventre, A. D. Adams, R. Mosley, R. S. Surwit and D. E. Moller, *Mol. Endocrinol.*, 2003, **17**, 662.
[96] J. F. Dropinski, T. Akiyama, M. Einstein, B. Habulihaz, T. Doebber, J. P. Berger, P. T. Meinke and G. Q. Shi, *Bioorg. Med. Chem. Lett.*, 2005, **15**, 5035.
[97] P. Misra, R. Chakrabarti, R. K. Vikramadithyan, G. Bolusu, S. Juluri, J. Hiriyan, C. Gershome, A. Rajjak, P. Kashireddy, S. Yu, S. Surapureddi, C. Qi, Y. J. Zhu, M. J. K. Reddy and R. Ramanujam, *J. Pharmacol. Exp. Ther.*, 2003, **306**, 763.
[98] R. K. Vikramadithyan, R. Chakrabarti, P. Misra, M. Premkumar, S. K. Kumar, C. S. Rao, A. Ghosh, K. N. Reddy, C. Uma and R. Rajagopalan, *Metabolism*, 2000, **49**, 1417.
[99] M. Thor, K. Beierlein, G. Dykes, A. L. Gustavsson, J. Heidrich, L. Jendeberg, B. Lindqvist, C. Pegurier, P. Roussel, M. Slater, S. Svensson, M. Sydow-Backman, U. Thornstrom and J. Uppenberg, *Bioorg. Med. Chem. Lett.*, 2002, **12**, 3565.
[100] T. Ostberg, S. Svensson, G. Selen, J. Uppenberg, M. Thor, M. Sundbom, M. Sydow-Backman, A. Gustavsson L and L. Jendeberg, *J. Biol. Chem.*, 2004, **279**, 41124.
[101] J. J. Acton 3rd, R. M. Black, A. B. Jones, D. E. Moller, L. Colwell, T. W. Doebber, K. L. Macnaul, J. Berger and H. B. Wood, *Bioorg. Med. Chem. Lett.*, 2005, **15**, 357.

[102] K. Liu, R. M. Black, J. J. Acton 3rd, R. Mosley, S. Debenham, R. Abola, M. Yang, R. Tschirret-Guth, L. Colwell, C. Liu, M. Wu, C. F. Wang, K. L. MacNaul, M. E. McCann, D. E. Moller, J. P. Berger, P. T. Meinke, A. B. Jones and H. B. Wood, *Bioorg. Med. Chem. Lett.*, 2005, **15**, 2437.

[103] D. Dey, S. Medicherla, P. Neogi, M. Gowri, J. Cheng, C. Gross, S. D. Sharma, G. M. Reaven and B. Nag, *Metabolism*, 2003, **52**, 1012.

[104] J. Huck, R. Saladin and M. Sierra, *WO Patent* 043951 A1, 2004.

[105] I. G. Schulmann, *Handbook Exp. Pharmacol.*, 1999, **139**, 215.

[106] M. Stumvoll and H. Haring, *Horm. Res.*, 2001, **55**, 3.

[107] P. Y. Michellys, M. F. Boehm, J. H. Chen, T. A. Grese, D. S. Karanewsky, M. D. Leibowitz, S. Liu, D. A. Mais, C. M. Mapes, A. Reifel-Miller, K. M. Ogilvie, D. Rungta, A. W. Thompson, J. S. Tyhonas, N. Yumibe and R. J. Ardecky, *Bioorg. Med. Chem. Lett.*, 2003, **13**, 4071.

[108] P. Y. Michellys, J. D'Arrigo, T. A. Grese, D. S. Karanewsky, M. D. Leibowitz, D. A. Mais, C. M. Mapes, A. Reifel-Miller, D. Rungta and M. F. Boehm, *Bioorg. Med. Chem. Lett.*, 2004, **14**, 1593.

[109] D. L. Gernert, D. A. Neel, M. F. Boehm, M. D. Leibowitz, D. A. Mais, P. Y. Michellys, A. Rungta, A. Reifel-Miller and T. A. Grese, *Bioorg. Med. Chem. Lett.*, 2004, **14**, 2759.

[110] Nuclear Receptor Nomenclature Committee, *Cell*, 1999.

[111] P. Tontonoz and D. J. Mangelsdorf, *Mol. Endocrinol.*, 2003, **17**, 985.

[112] A. D. Attie, J. P. Kastelein and M. R. Hayden, *J. Lipid Res.*, 2001, **42**, 1717.

[113] J. A. Hubacek, K. E. Berge, J. C. Cohen and H. H. Hobbs, *Hum. Mutat.*, 2001, **18**, 359.

[114] L. Yu, J. York, K. von Bergmann, D. Lutjohann, J. C. Cohen and H. H. Hobbs, *J. Biol. Chem.*, 2003, **278**, 15565.

[115] P. A. Mak, B. A. Laffitte, C. Desrumaux, S. B. Joseph, L. K. Curtiss, D. J. Mangelsdorf, P. Tontonoz and P. A. Edwards, *J. Biol. Chem.*, 2002, **277**, 31900.

[116] B. A. Laffitte, J. J. Repa, S. B. Joseph, D. C. Wilpitz, H. R. Kast, D. J. Mangelsdorf and P. Tontonoz, *Proc. Natl. Acad. Sci. USA*, 2001, **98**, 507.

[117] I. J. Goldberg, *J. Lipid Res.*, 1996, **37**, 693.

[118] D. Bruemmer and R. E. Law, *Curr. Drug Targets Cardiovasc. Haematol. Disord.*, 2005, **5**, 533.

[119] M. Jaye, *Curr. Opin. Investig. Drugs*, 2003, **4**, 1053.

[120] E. G. Lund, J. G. Menke and C. P. Sparrow, *Arterioscler. Thromb. Vasc. Biol.*, 2003, **23**, 1169.

[121] G. U. Schuster, P. Parini, L. Wang, S. Alberti, K. R. Steffensen, G. K. Hansson, B. Angelin and J. A. Gustafsson, *Circulation*, 2002, **106**, 1147.

[122] S. B. Joseph, E. McKilligin, L. Pei, M. A. Watson, A. R. Collins, B. A Laffitte, M. Chen, G. Noh, J. Goodman, G. N. Hagger, J. Tran, T. K. Tippin, X. Wang, A. J. Lusis, W. A. Hsueh, R. E. Law, J. L. Collins, T. M. Willson and P. Tontonoz, *Proc. Natl. Acad. Sci. USA*, 2002, **99**, 7604.

[123] J. R. Schultz, H. Tu, A. Luk, J. J. Repa, J. C. Medina, L. Li, S. Schwendner, S. Wang, M. Thoolen, D. J. Mangelsdorf, K. D. Lustig and B. Shan, *Genes Dev.*, 2000, **14**, 2831.

[124] J. J. Repa, G. Liang, J. Ou, Y. Bashmakov, J. M. Lobaccaro, I. Shimomura, B. Shan, M. S. Brown, J. L. Goldstein and D. J. Mangelsdorf, *Genes Dev.*, 2000, **14**, 2819.

[125] D. J. Peet, S. D. Turley, W. Ma, B. A. Janowski, J. Lobaccaro, R. E. Hammer and D. J. Mangelsdorf, *Cell*, 1998, **93**, 693.

[126] E. G. Lund, L. B. Peterson, A. D. Adams, M. H. Lam, C. A. Burton, J. Chin, Q. Guo, S. Huang, M. Latham, J. C. Lopez, J. G. Menke, D. P. Milot, L. J. Mitnaul, S. E. Rex-Rabe, R. L. Rosa, J. Y. Tian, S. D. Wright and C. P. Sparrow, *Biochem. Pharmacol.*, 2006, **71**, 453.

[127] C. Song and S. Liao, *Steroids*, 2001, **66**, 673.
[128] J. D. Horton, J. L. Goldstein and M. S. Brown, *J. Clin. Invest.*, 2002, **109**, 1125.
[129] L. F. Michael, J. M. Schkeryantz and T. P. Burris, *Mini Rev. Med. Chem.*, 2005, **5**, 729.
[130] D. S. Lala, *Curr. Opin. Investig. Drugs*, 2005, **6**, 934.
[131] G. Cao, Y. Liang, X. C. Jiang and P. I. Eacho, *Drug News Perspect.*, 2004, **17**, 35.
[132] H. Gong and W. Xie, *Expert Opin. Ther. Targets*, 2004, **8**, 49.
[133] L. Wang, G. U. Schuster, K. Hultenby, Q. Zhang, S. Andersson and J. A. Gustafsson, *Proc. Natl. Acad. Sci. USA.*, 2002, **99**, 13878.
[134] K. D. Whitney, M. A. Watson, J. L. Collins, W. G. Benson, T. M. Stone, M. J. Numerick, T. K. Tippin, J. G. Wilson, D. A. Winegar and S. A. Kliewer, *Mol. Endocrinol.*, 2002, **16**, 1378.
[135] R. P. Koldamova, I. M. Lefterov, M. D. Ikonomovic, J. Skoko, P. I. Lefterov, B. A. Isanski, S. T. DeKosky and J. S. Lazo., *J. Biol. Chem.*, 2003, **278**, 13244.
[136] J. L. Collins, *Curr. Opin. Drug Discov. Devel.*, 2004, **7**, 692.
[137] R. D. Koldamova, *J. Biol. Chem.*, 2003, **278**, 13244.
[138] Y. Sun, J. Yao, T. W. Kim and A. R. Tall, *J. Biol. Chem.*, 2003, **278**, 27688.
[139] H. Fukumoto, A. Deng, M. C. Irizarry, M. L. Fitzgerald and G. W. Rebeck, *J. Biol. Chem.*, 2002, **277**, 48508.
[140] Recent LXR Medicinal Chemistry Reviews: (a). L. Michael, *Mini Rev. Med. Chem.* 2005, 5, 729. (b) J. Collins, *Curr. Opin. Drug Discov. Devel.* 2004, 7, 692.
[141] M. C. Jaye, J. A. Krawiec, N. Campobasso, A. Smallwood, C. Qiu, Q. Lu, J. J. Kerrigan, M. De Los Frailes Alvaro, B. Laffitte, W. S. Liu, J. P. Marino Jr, C. R. Meyer, J. A. Nichols, D. J. Parks, P. Perez, L. Sarov-Blat, S. D. Seepersaud, K. M. Steplewski, S. K. Thompson, P. Wang, M. A. Watson, C. L. Webb, D. Haigh, J. A. Caravella, C. H. Macphee, T. M. Willson and J. L. Collins, *J. Med. Chem.*, 2005, **48**, 5419.
[142] C. Boström, J. Brickmann, K. Holm, P. Sandberg, P. Swanson and M. Westerlund, *WO Patent* 005417 A1, 2005.
[143] P. Holm, *WO Patent* 005416 A1, 2005.
[144] J. W. Szewczyk, S. Huang, J. Chin, J. Tian, L. Mitnaul, R. L. Rosa, L. Peterson, C. P. Sparrow and A. D. Adams, *Bioorg. Med. Chem. Lett.*, 2006, **16**, 3055.
[145] S. B. Singh, J. G. Ondeyka, W. Liu, S. Chen, T. S. Chen, X. Li, A. Bouffard, J. Dropinski, A. B. Jones, S. McCormick, N. Hayes, J. Wang, N. Sharma, K. Macnaul, M. Hernandez, Y. S. Chao, J. Baffic, M. H. Lam, C. Burton, C. P. Sparrow and J. G. Menke, *Bioorg. Med. Chem. Lett.*, 2005, **15**, 2824.
[146] W. Liu, S. Chen, J. Dropinski, L. Colwell, M. Robins, M. Szymonifka, N. Hayes, N. Sharma, K. MacNaul, M. Hernandez, C. Burton, C. P. Sparrow, J. G. Menke and S. B. Singh, *Bioorg. Med. Chem. Lett.*, 2005, **15**, 4574.
[147] H. Jayasuriya, K. B. Herath, J. G. Ondeyka, Z. Guan, R. P. Borris, S. Tiwari, W. de Jong, F. Chavez, J. Moss D. W. Stevenson, H. T. Beck, M. Slattery, N. Zamora, M. Schulman, A. Ali, N. Sharma, K. MacNaul, N. Hayes, J. G. Menke and S. B. Singh, *J. Nat. Prod.*, 2005, **68**, 1247.
[148] N. Tamehiro, Y. Sato, T. Suzuki, T. Hashimoto, Y. Asakawa, S. Yokoyama, T. Kawanishi, Y. Ohno, K. Inoue, T. Nagao and T. Nishimaki-Mogami, *FEBS Lett.*, 2005, **579**, 5299.
[149] T. H. Huang, V. Razmovski-Naumovski, N. K. Salam, R. K. Duke, V. H. Tran, C. C. Duke and B. D. Roufogalis, *Biochem. Pharmacol.*, 2005, **70**, 1298.
[150] I. Surugiu-Warnmark, A. Warnmark, G. Toresson, J. A. Gustafsson and L. Bulow, *Biochem. Biophys. Res. Commun.*, 2005, **332**, 512.
[151] S. Huang and A. D. Adams, *US Patent* 0113419 A1, 2005.

[152] I. Deuschle, U. Kögl, M. Blume, B. Albers and M. Kober, *WO Patent* 04024161A1, 2004.
[153] S. Golz, U. Bruggemeier and A. Geerts, *WO Patent* 101011 A2, 2005.
[154] T. H. Hoang, S. K. Thompson and D. G. Washburn, *WO Patent* 05023196A2, 2005.
[155] T. H. Hoang, S. K. Thompson and D. G. Washburn, *WO Patent* 05023247A1, 2005.
[156] L. S. Kallander, J. P. Marino and S. K.. Thompson, *WO Patent* 05023188A2, 2005.
[157] H. Kikkawa, M. Kinoshita and O. Kurusu, *WO Patent* 05055998A1, 2005. (b) S. B. Joseph, A. Castrillo, B. A. Laffitte, D. J. Mangelsdorf and P. Tontonoz., *Nat. Med.*, 2003, **9**, 213.
[158] H. Dehmlow, R. Masciadri, N. Panday, H. Ratni and M. B. Wright, *US Patent* 0004068A1, 2006.
[159] H. Dehmlow, N. Panday, H. Ratni, T. Schulz-Gasch and M. B. Wright, *US Patent* 0245515, 2005.
[160] M. Arai, S. Kaneko, S. Shibuya, T. Watanabe and Y. Kumakura, *WO Patent* 023782 A1, 2005.
[161] R. J. Steffan,. E. M. Matelan, S. M. Bowen, J. W. Ullrich, J. E. Wrobel, E. Zamaratski, L. Kruger, A. L. Olsen, A. Cheng, T. Hansson, R. J. Uunwalla, C. P. Miller, P. P. Rhönnstad, *WO Patent* 017384A2, 2006.
[162] M. D. Collini, R. R.Singhaus, B. Hu, J. W. Jetter, R. L. Morris, D. H. Kaufman, C. P. Miller, J. W. Ullrich, R. J. Unwalla, J. E. Wrobel, E. Quinet, P. Nambi, R. C. Bernotas, M. Elloso, *WO Patent* 058834A2, 2005.
[163] C. D. Bayne, A. T. Johnson, S. Lu, R. Mohan, M. C. Nyman, E. J. Schweiger, W. C. Stevens, H. Wang and Y. Xie, *US Patent* 0080111 A1, 2005.
[164] V. Molteni, X. Li, J. Nanakka, D. A. Ellis, B. Anaclerio, E. Saez and J. Wityak, *WO Patent* 077124, 2005.
[165] P. Diaz, J. Bernardon and E. Thoreau, *WO Patent* 076418A1, 2004.
[166] L. Li, J. C. Medina and B. Shan, *US Patent* 6906069 B1, 2005.
[167] R. Pellicciari, A. Gioiello and G. Costantino, *Expert Opin. Ther. Patents*, 2006, **16**, 333.
[168] S. Westin, R. A. Heyman and R. Martin, *Mini Rev. Med. Chem.*, 2005, **5**, 719.
[169] R. Pellicciari, G. Costantino and S. Fiorucci, *J. Med. Chem.*, 2005, **48**, 5383.
[170] J. J. Repa, S. D. Turley, J. A. Lobaccaro, J. Medina, L. Li, K. Lustig, B. Shan, R. A. Heyman, J. M. Dietschy and D. J. Mangelsdorf, *Science*, 2000, **289**, 1524.
[171] Y. Yang, M. Zhang, G. Eggertsen and J. Y. Chiang, *Biochim. Biophys. Acta.*, 2002, **1583**, 63.
[172] C. J. Sinal, M. Tohkin, M. Miyata, J. M. Ward, G. Lambert and F. J. Gonzalez, *Cell*, 2000, **102**, 731.
[173] K. C. Nicolaou, R. M. Evans, A. J. Roecker, R. Hughes, M. Downes and J. A. Pfefferkorn, *Org. Biomol. Chem.*, 2003, **1**, 908.
[174] K. M. Honorio, R. C. Garratt and A. D. Andricopulo, *Bioorg. Med. Chem. Lett.*, 2005, **15**, 3119.
[175] P. R. Maloney, D. J. Parks, C. D. Haffner, A. M. Fivush, G. Chandra, K. D. Plunket, K. L. Creech, L. B. Moore, J. G. Wilson, M. C. Lewis, S. A. Jones and T. M Willson, *J. Med. Chem.*, 2000, **43**, 2971.
[176] N. L. Urizar, A. B. Liverman, D. T. Dodds, F. V. Silva, P. Ordentlich, Y. Yan, F. J. Gonzalez, R. A. Heyman, D. J. Mangelsdorf and D. D. Moore, *Science*, 2002, **296**, 1703.
[177] J. Cui, L. Huang, A. Zhao, J. L. Lew, J. Yu, S. Sahoo, P. T. Meinke, I. Royo, F. Pelaez and S. D. Wright, *J. Biol. Chem.*, 2003, **278**, 10214.
[178] J. Wu, C. Xia, J. Meier, S. Li, X. Hu and D. S. Lala, *Mol. Endocrinol.*, 2002, **16**, 1590.
[179] P. O. Szapary, M. L. Wolfe, L. T. Bloedon, A. J. Cucchiara, A. H. DerMarderosian, M. D. Cirigliano and D. J. Rader, *JAMA*, 2003, **290**, 765.

11β-Hydroxysteroid Dehydrogenase Type 1 Inhibitors

Craig D. Boyle, Timothy J. Kowalski and Lili Zhang

Schering-Plough Research Institute, 2015 Galloping Hill Road, Kenilworth, NJ 07033, USA

Contents

1. Introduction	127
2. 11β-HSD Isozymes	128
3. Potential therapeutic indications	130
4. 11β-HSD1 Inhibitors	131
4.1. Thiazole-based inhibitors	131
4.2. Triazole-based inhibitors	133
4.3. Amide-based inhibitors	133
4.4. Miscellaneous inhibitors	134
5. Conclusion	135
References	135

1. INTRODUCTION

The worldwide frequency of idiopathic obesity has increased significantly over the last decade [1], and the estimated annual direct and indirect cost in the US alone due to obesity is $117 billion [2]. Individuals classified as obese (body mass index (BMI) > 30) have a significantly higher risk of mortality and related comorbidities, such as insulin resistance and type 2 diabetes, dyslipidemia, and hypertension [1,3]. The clustering of obesity (abdominal obesity in particular) with insulin resistance, hypertension, and dyslipidemia has been termed the metabolic syndrome, and affects approximately 47 million Americans [4].

Glucocorticoids are important regulators of many physiological functions, particularly those associated with stress or inflammation and illness. The effects of glucocorticoids are mediated by interactions with intracellular receptors. Overexposure to glucocorticoids by excessive overproduction, as seen in Cushing's syndrome, or *via* exogenous administration (as seen in the treatment of inflammatory diseases) produces an increase in central (intra-abdominal) adiposity, insulin resistance, dyslipidemia, and hypertension [5,6]. When excess glucocorticoids are removed from Cushing's syndrome patients, many aspects of the phenotype are reversed.

Patients with metabolic syndrome are phenotypically similar to those with excess glucocorticoid (Cushing's syndrome), implicating a common underlying role for elevated glucocorticoid signaling in these diseases. Highlighting this are studies showing that removal of circulating glucocorticoids in rodent models of obesity by adrenalectomy attenuates the magnitude of obesity and insulin resistance [7].

Cortisol levels in obese individuals, however, are usually normal [8], although there is evidence of increased net cortisol production associated with a higher waist-to-hip ratio [9,10]. The discrepancy between high circulating cortisol in Cushing's syndrome and normal circulating cortisol levels with abdominal obesity and metabolic syndrome may be explained by enhanced tissue sensitivity to glucocorticoids. 11β-Hydroxysteroid dehydrogenase type 1 (11β-HSD1) catalyzes the conversion of the inactive 11-keto forms of glucocorticoids (cortisone, 11-dehydrocorticosterone) to their active forms (cortisol, corticosterone) in several tissues, thus amplifying local glucocorticoid signaling. Increased activity of this enzyme, and the resulting increase in tissue-specific glucocorticoid receptor activity, may be an underlying feature of metabolic syndrome. Indeed, studies have demonstrated elevated 11β-HSD1 mRNA and activity in adipose tissue of obese humans [10–14], supporting the involvement of glucocorticoid signaling in metabolic syndrome, and implicating 11β-HSD1 as a valid target for therapeutic intervention.

2. 11β-HSD ISOZYMES

Two 11β-HSD isozymes have been identified, 11β-HSD1 and 11β-HSD2 [15,16]. 11β-HSD2 is a nicotinamide adenine dinucleotide (NAD) dependent dehydrogenase that catalyzes the reverse reaction of 11β-HSD1 to inactivate cortisol to cortisone [17]. 11β-HSD2 is mainly localized in mineralocorticoid receptor (MR) target tissues such as kidney, colon, and salivary gland. The function of 11β-HSD2 is critical for protecting the MR from excessive exposure to cortisol and ensuring the normal activity of MR ligand aldosterone. 11β-HSD2 inhibition by genetic mutations or 18β-glycyrrhetinic acid, the active component of licorice, leads to sodium retention and hypertension [18–20]. Although 11β-HSD1 and 11β-HSD2 are isozymes in terms of their biological function, they share only 16% sequence homology [17]. The structural diversity between the two enzymes provides the potential to develop 11β-HSD1 selective inhibitors. Recent studies have identified several classes of potent, highly specific compounds targeting 11β-HSD1 [21–23], suggesting that it is possible to suppress 11β-HSD1 activity with minimal liability of 11β-HSD2 inhibition.

Both 11β-HSD1 and 11β-HSD2 belong to a large family of short-chain dehydrogenases and reductases (SDR) [24]. SDR enzymes share a conserved nucleotide binding motif and a consensus Ser-Tyr-Lys triad at the catalytic site that supports a general proton-transfer mechanism mediated by the Tyr residue. Crystal structures reveal a consensus α/β folding pattern but a diversified conformation at the substrate binding pocket that accommodates the large variety of substrates processed by SDRs. Several SDR family members are potential therapeutic targets for different disease indications [24]. The overall sequence homology shared among the SDRs is typically 15–30%, which again provides an advantage for developing target-specific inhibitors. Indeed, no significant off-target activity has been reported with the recently disclosed 11β-HSD1 inhibitors in subchronic and chronic studies [25,26].

Human 11β-HSD1 is a 292-amino-acid glycosylated protein with a single N-terminal transmembrane segment that anchors the catalytic domain in the lumen of the endoplasmic reticulum (ER) [27]. It is widely distributed with high expression in active glucocorticoid receptor (GR) tissues such as liver, adipose, and brain. In intact cells, 11β-HSD1 acts largely as a glucocorticoid reductase [28–30] driven by the redox potential maintained by the reduced nicotinamide adenine dinucleotide phosphate (NADPH) generating enzyme hexose-6-phosphate dehydrogenase (H6PDH) in the ER compartment [31]. *In vitro* biochemical characterization shows that 11β-HSD1 is an NADP(H)-dependent bidirectional enzyme that can function as both a reductase and a dehydrogenase, but that dehydrogenase activity predominates [32]. 11β-HSD1 has low affinity to its substrates, with a reported K_m in the submicromolar to low micromolar range. The wide affinity range may reflect an *in vitro* artifact associated with different isolation procedures. The local substrate concentration in the membrane can be much higher given the lipophilic nature of the substrate, which can facilitate the catalytic activity *in vivo*.

Rodent 11β-HSD1 utilizes 11-dehydrocorticosterone as a substrate, which is converted by the enzyme to the active form corticosterone. Recent studies also implicate the reductase activity of 11β-HSD1 in the conversion of 7-ketocholesterol to 7β-hydroxycholesterol, an important step in dietary oxysterol metabolism [33]. *In vitro*, the human enzyme can use both cortisone and dehydrocorticosterone as substrates with similar catalytic efficiency and the same is true for rodent enzymes [34]. The kinetic properties of 11β-HSD1 from species studied so far are comparable. However, significant species differences have been identified with several classes of 11β-HSD1 inhibitors [21,34]. The species selectivity presents a major challenge to evaluate 11β-HSD1 inhibitors in animal models.

The presence of a transmembrane domain renders 11β-HSD1 a difficult protein for purification and structural analysis. Only recently the crystal structures of 11β-HSD1 were reported using the N-terminus truncated soluble forms of the enzyme [35–37]. 11β-HSD1 structures from human, mouse, and guinea pig have the α/β backbone structure characteristic of SDR family members. The nucleotide binding site and catalytic site are also consistent with the common reaction mechanism shared by SDRs. 11β-HSD1 exists as a homodimer in crystals, which is likely the functional unit that has also been reported with SDRs, although tetramer formation has also been suggested for human 11β-HSD1 [35]. On the basis of the structural analysis, a model of enzyme orientation to the membrane is proposed where the substrate entry loop is in close proximity to the membrane bilayer. This model provides the lipophilic substrate easy access to the substrate binding and catalytic sites of the enzyme and it may explain the low *in vitro* substrate affinity of 11β-HSD1.

11β-HSD1 has been cloned from several species, and the sequence homology to human enzyme ranges between 75 and 95% [34,38]. Sequence alignment of 11β-HSD1 from different species identified several variable regions [38]. In a 3D structure of 11β-HSD1, these variable domains are all located around the active site and the substrate entry loop. This structural feature may explain the species selectivity of some 11β-HSD1 inhibitors. A recent study reported both competitive and mixed inhibition with arylsulfonamidothiazole inhibitors [38], however, much less is known about the mechanism of more recently discovered 11β-HSD1 inhibitors. Currently only the structures of enzyme-NADP(H) or enzyme-NADP(H)-substrate complexes have been solved for 11β-HSD1 [35–37]. Future studies of the 11β-HSD1 complex with different classes of inhibitors and the kinetic mechanism of these inhibitors will provide important insight for improving the profiles of current compounds to be eventually used as therapeutics in human.

3. POTENTIAL THERAPEUTIC INDICATIONS

Several lines of evidence implicate elevated 11β-HSD1 activity in the etiology and/or maintenance of type 2 diabetes and metabolic syndrome. There is a higher level of 11β-HSD1 mRNA and activity in adipose tissue of obese humans [11,12,14,39] and rodents [40]. Chronic high-fat feeding decreases 11β-HSD1 activity and mRNA in fat of C57Bl/6J mice, suggesting that this serves as an adaptive mechanism attempting to protect against the adverse metabolic consequence of high-fat feeding [41]. Interestingly, A/J mice chronically fed a high-fat diet become obese but are less hyperinsulinemic than C57Bl/6J mice, and this is associated with lower basal adipose tissue 11β-HSD1 activity and mRNA and a more pronounced decrease in activity with high-fat feeding [41]. This suggests that lower local synthesis of active glucocorticoid in A/J mice confers protection from dysregulated glucose homeostasis due to high-fat feeding. Transgenic overexpression of 11β-HSD1 in adipose tissue of mice produces intra-abdominal obesity, insulin resistance, dyslipidemia, and hypertension [42,43] in the absence of elevated corticosterone.

In contrast, mice with targeted disruption of the 11β-HSD1 gene are resistant to diet-induced obesity [44], have improved glucose tolerance [45] and an improved lipid profile [46]. Additionally, mice with adipose tissue-specific overexpression of 11β-HSD2, the enzyme that catalyzes the conversion of active glucocorticoid in to the inactive form, resist diet-induced obesity, have increased energy expenditure, improved glucose tolerance, and insulin sensitivity [47].

All these data suggest that pharmacological inhibition of 11β-HSD1 activity, particularly in adipose tissue, will reduce intracellular glucocorticoid signaling and improve aspects of metabolic syndrome. Indeed, experimental evidence exists that supports this hypothesis. In monogenic murine models of obesity and type 2 diabetes ($Lep^{ob/ob}$, KKA^y), oral administration (200 mg/kg bid for 4d) of a selective 11β-HSD1 inhibitor, BVT-2733 (*vide infra*), significantly reduced plasma glucose levels and improved glucose tolerance [25]. Treatment also produced a modest inhibition of food intake, and a decrease in hepatic glucose production with an accompanying decrease in the level of liver phosphoenolpyruvate carboxykinase mRNA [25]. Consistent with these findings in rodents, treatment of type 2 diabetic humans with the non-selective inhibitor carbenoxolone has been shown to improve hepatic insulin sensitivity [48,49].

A recent study reported that oral administration of a selective 11β-HSD1 inhibitor to DIO (diet-induced obese) mice (20 mg/kg bid for 11d) reduced food intake, body weight and adiposity, fasting glucose and insulin levels [23]. Administration of the same compound to high-fat-fed streptozotocin-treated mice (30 mg/kg bid for 9d) failed to affect feeding or body weight and adiposity, but reduced fasting and postprandial glucose and insulin and improved glucose tolerance. Lastly, treatment of apoE knockout mice with the selective 11β-HSD1 inhibitor (10 mg/kg/d for 8 weeks in diet) significantly attenuated the deposition of aortic plaque and cholesterol ester. This was accompanied by a decrease in circulating monocyte chemoattractant protein-1, suggesting that 11β-HSD1 inhibition has a direct anti-inflammatory effect at the blood-vessel wall.

Elevated glucocorticoids have been shown to have negative effects on hippocampal neurochemistry and function, and it has been proposed that exposure of the hippocampus to high levels of glucocorticoids contributes to aging-related cognitive impairment [50]. 11β-HSD1 is expressed throughout the brain, including the hippocampus [51], and studies with 11β-HSD1 knockout mice demonstrate that removal of glucocorticoid regeneration within the CNS has a protective effect against age-related cognitive impairment [51]. Indeed, a recent study in elderly men and type 2 diabetics reported that oral administration of carbenoxolone (100 mg tid for 4 weeks) improved verbal fluency in healthy men and verbal memory in type 2 diabetics [52]. These results suggest that 11β-HSD-1 inhibition may be effective in preventing or attenuating the decline of cognitive function with aging.

4. 11β-HSD1 INHIBITORS

Early inhibitors of 11β-HSD1 were natural product analogs that generally had low 11β-HSD1 potency and poor selectivity over 11β-HSD2. Compounds such as carbenoxolone demonstrated modest *in vivo* effects on markers of metabolic syndrome, but were generally poor pharmacological tools because of issues such as selectivity. These inhibitors, as well as early work on thiazole analogs, were reviewed in 2003 [53]. Subsequently, more *in vivo* results and several new chemical entities have been published in both the peer-reviewed and patent literature [54]. Thiazoles, triazoles, amides, and some miscellaneous derivatives have all demonstrated 11β-HSD1 potency as well as selectivity over 11β-HSD2. This review will focus on recent advances and compounds with published pharmacological data [55].

4.1. Thiazole-based inhibitors

One of the first classes of 11β-HSD1 inhibitors to demonstrate selectivity over 11β-HSD2 consisted of thiazole analogs. As depicted in structure **1**, the general structure of active thiazoles incorporates a sulfonamide moiety at the 2-position of the thiazole ring. Two preferred compounds that have been extensively studied are BVT-2733 (**2**) and BVT-14225 (**3**). Human isozyme *in vitro* studies showed that compounds **2** and **3** were reasonably potent 11β-HSD1 inhibitors (IC_{50} = 3.3 and 0.05 μM, respectively) and were selective vs. 11β-HSD2 (IC_{50} > 10 μM) [21]. Although **2** was less potent in the human assay, it was more potent than compound **3** against mouse 11β-HSD1 (IC_{50} = 96 and 284 nM, respectively) [21]. Both compounds were inactive at mouse 11β-HSD2 [21]. Compound **2** was tested in the hyperglycemic KKA^y mouse model of diabetes, and was shown to dose-dependently lower blood glucose levels, with a maximum glucose reduction of 53% (100 mg/kg po bid, 11 days) [21,56]. In addition to compounds **2** and **3**, several thiazole analogs have demonstrated *in vitro* 11β-HSD1 activity. Most of these derivatives of general structure **1** are represented in the patent literature, and the preferred compounds showed K_i values ranging from 10 to 500 nM [57–64].

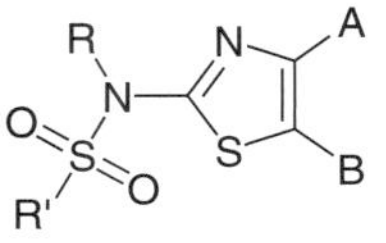

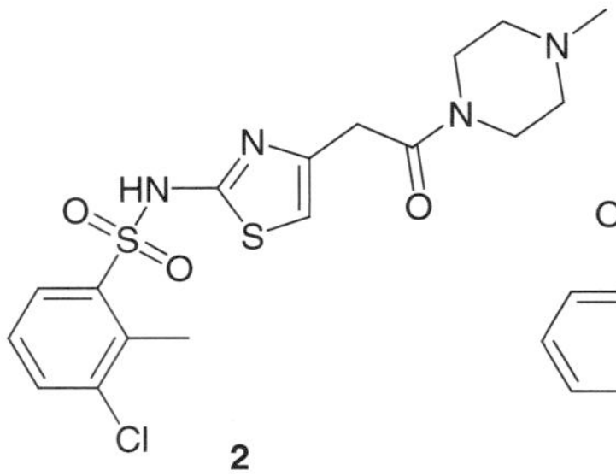

Minor modifications to the thiazole ring have also produced potent and selective 11β-HSD1 inhibitors. For example, the sulfonamidooxazole **4** had low micromolar 11β-HSD1 activity (IC$_{50}$ = 2.3 μM) with >100-fold selectivity vs. 11β-HSD2 [26]. Compound **4** had plasma exposure suitable for evaluation in *KKA^y* and *ob/ob* mice [26]. Related thiadiazoles have been reported as 11β-HSD1 inhibitors [65–68], and **5** had an 11β-HSD1 K_i of 219 nM [65,67]. Thienyl derivatives such as **6** have also been disclosed [69–71]. When dosed in *db/db* mice (50 mg/kg, po), compound **6** decreased plasma glucose by 42% and insulin by 61% over vehicle [69]. Thiazolones and oxazolones have also been disclosed as 11β-HSD1 inhibitors in patent applications [72,73], and compound **7** had a reported 11β-HSD1 K_i value of 107 nM [72].

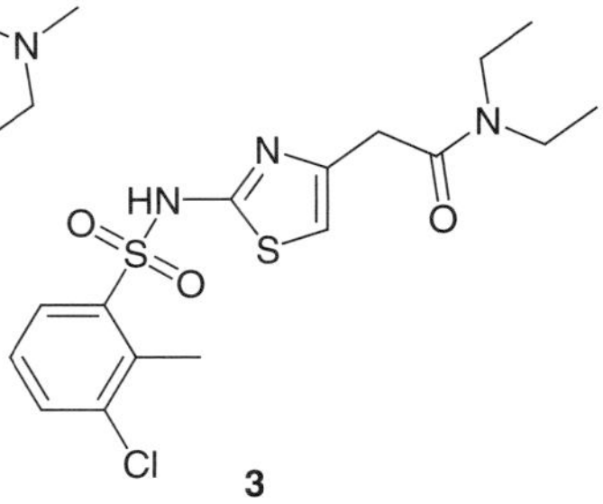

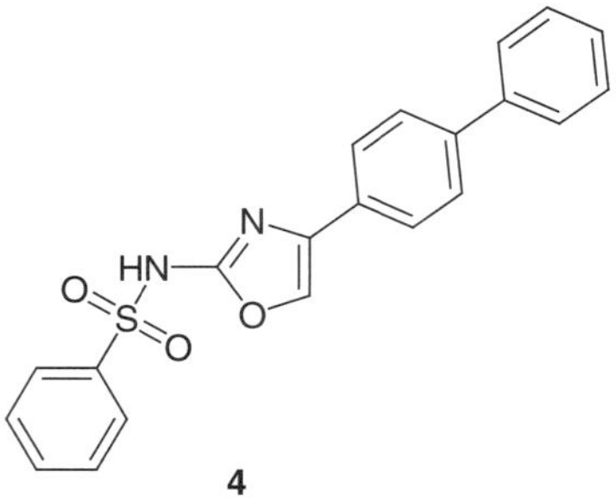

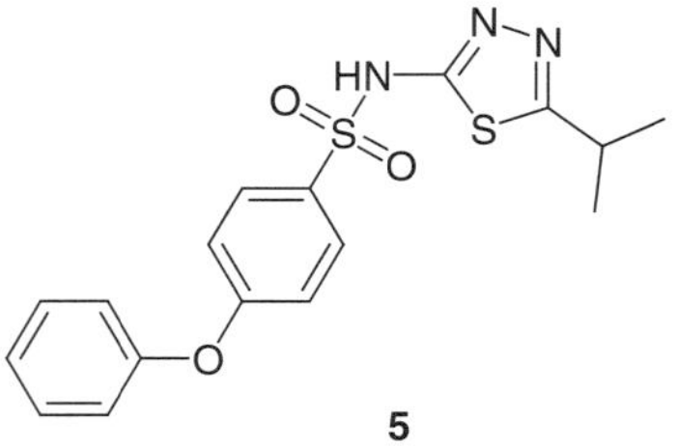

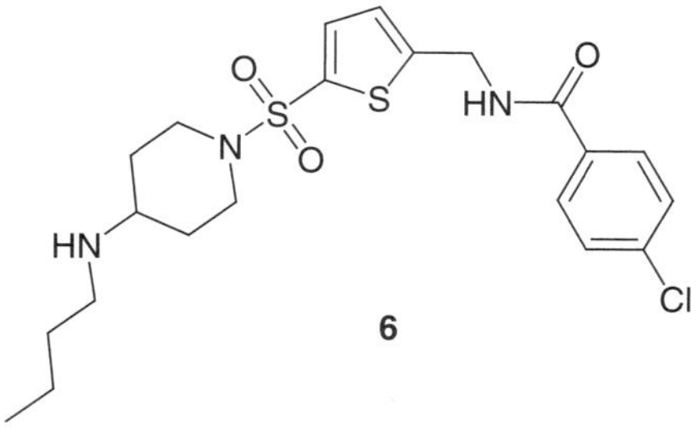

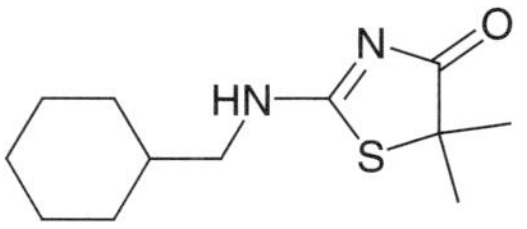

4.2. Triazole-based inhibitors

Triazoles are another highly studied class of 11β-HSD1 inhibitors. For example, adamantyl triazoles have been recently reported as potent and selective 11β-HSD1 inhibitors [74]. Compound **8** had a human 11β-HSD1 IC_{50} of 7.5 nM and a mouse IC_{50} value of 97 nM, and was inactive at both human and mouse 11β-HSD2 [23]. To study the effects of a triazole 11β-HSD1 inhibitor on metabolic syndrome, compound **8** was subjected to several *in vivo* models. In a pharmacodynamic mouse cortisone challenge assay, compound **8** inhibited cortisol formation by 60% at a 10 mg/kg oral dose after 1 h. Additionally, at 30 mg/kg, compound **8** showed efficacy up to 12 h [23]. Orally dosed DIO mice (20 mg/kg bid of **8**, 11d) demonstrated a lowering of body weight (7% vs. control), food intake (12% vs. control), and fasting serum glucose levels (15% vs. control) [23]. Compound **8** also lowered glucose levels and improved insulin resistance in the HF/STZ model of type 2 diabetes [23]. Serum lipid levels were also lowered in DIO and HF/STZ mice as well as in the atherosclerosis apoE KO mouse model [23]. Lastly, compound **8** slowed plaque formation in aortic lesions of apoE KO mice [23].

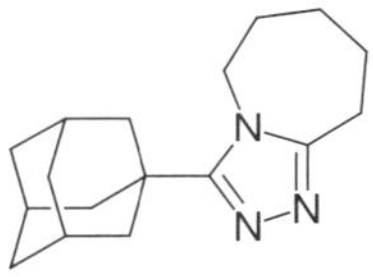

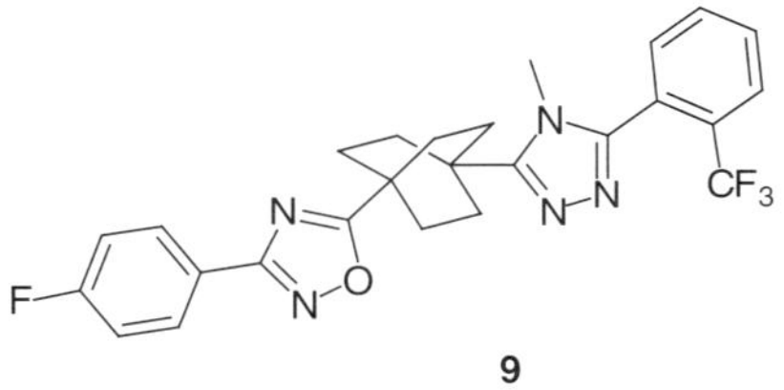

SAR development of the triazole series yielded several compounds with $IC_{50} < 10$ nM at the 11β-HSD1 isozyme. Oxadiazole **9** was equipotent in both human ($IC_{50} = 2.2$ nM) and mouse ($IC_{50} = 1.9$ nM) 11β-HSD1 [75]. Furthermore, compound **9** showed significant 11β-HSD1 inhibition at 10 mg/kg po in a pharmacodynamic cortisone challenge assay (86% at 4 h, 74% at 16 h) [75].

In addition to compound **8**, other triazole analogs have been covered in the patent literature as 11β-HSD1 inhibitors with reported IC_{50} values ranging from <10 to 200 nM [76–88]. In addition, pyrazole derivatives have demonstrated 11β-HSD1 potency. Compound **10** had a reported IC_{50} value of <10 nM [89].

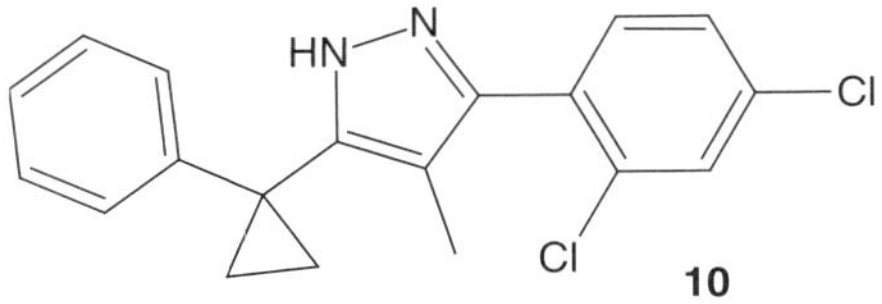

4.3. Amide-based inhibitors

Several different amide-based structures have been explored as 11β-HSD1 inhibitors. An extensive SAR study of benzamides yielded a number of potent and selective perhydroquinolylbenzamides [22]. Compounds **11**, **12**, and **13** exhibited submicromolar 11β-HSD1 activity (IC_{50} = 0.22, 0.56, and 0.10 μM, respectively) and good selectivity vs. 11β-HSD2 [22]. All three compounds also showed maximal *in vivo* activity in an ADX mouse model by a >70% lowering of liver corticosterone levels [22]. Other aryl amides have also been reported in the patent literature with IC_{50} values of preferred compounds ranging from 6 to 340 nM [90–94].

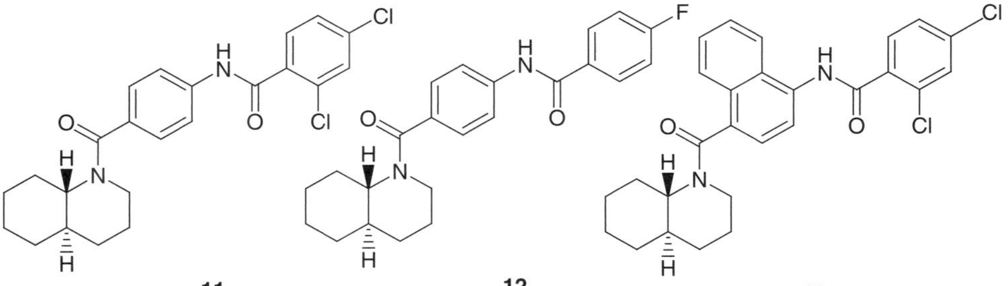

11   **12**   **13**

Similar to the triazole series, adamantyl amide derivatives exhibited 11β-HSD1 activity as well as selectivity over 11β-HSD2 [95–108]. Compound **14** had an 11β-HSD1 IC_{50} of 34 nM as well as >10 μM inhibition of 11β-HSD2 [101]. *In vivo*, compound **14** significantly lowered plasma corticosterone levels when dosed in a mouse cortisone challenge assay at 100 mg/kg [100]. Compound **14** also significantly lowered plasma glucose in an *ob/ob* mouse model of type 2 diabetes after 3 weeks of twice-daily dosing (30 and 100 mg/kg) [100]. Lactam analogs also demonstrated 11β-HSD1 potency [107–109]. The adamantyl lactam analog **15** had a reported 11β-HSD1 pIC_{50} > 6, and an 11β-HSD2 pIC_{50} < 5 [107]. The non-adamantyl analog **16** had the same pIC_{50} range as compound **15** for 11β-HSD1 and 2 [109]. Other amides have also been disclosed, but no pharmacological data was reported [110].

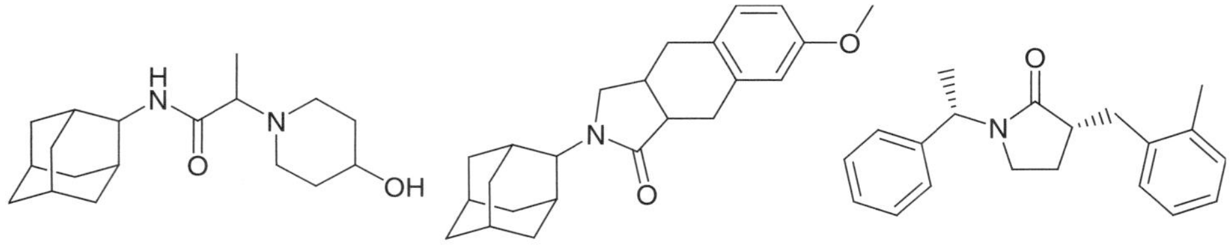

14
racemate

15
mixture of stereoisomers

16

4.4. Miscellaneous inhibitors

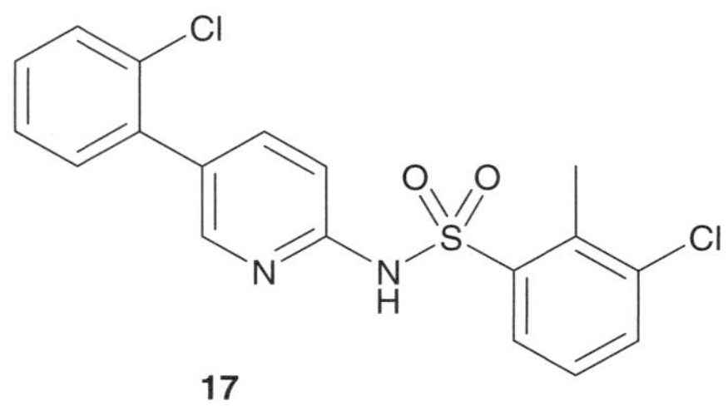

17

Various benzenesulfonamides have been disclosed in the patent literature [111–119]. For example, compound **17** had a reported *in vitro* 11β-HSD1 IC$_{50}$ of 3 nM [112]. Aryl and heterocyclic ketones have also demonstrated 11β-HSD1 activity [120–124]. Pyrrolidine **18** had an IC$_{50}$ of 70 nM [120], and the piperidine **19** had an IC$_{50}$ of 10 nM [121]. Finally, aryl sulfones have been disclosed as 11β-HSD1 inhibitors with IC$_{50}$ values ranging from subnanomolar to micromolar [125]. Interestingly, a byproduct of the synthesis of a derivative of oxazole **4** was the aryl sulfone **20**, which had an 11β-HSD1 IC$_{50}$ of 0.19 μM and selectivity over 11β-HSD2 [26].

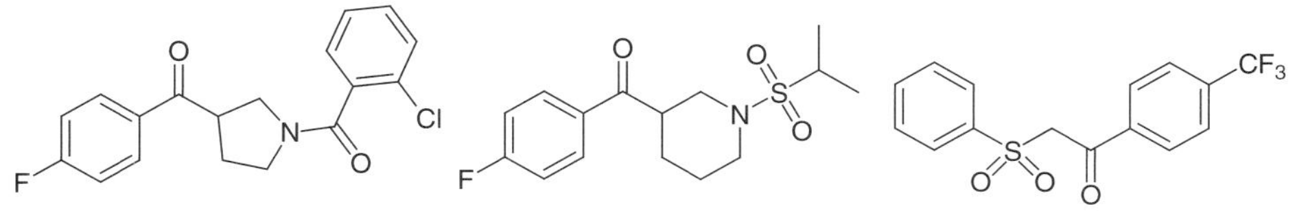

18
racemate

19
racemate

20

5. CONCLUSION

Several classes of compounds that are potent and selective 11β-HSD1 inhibitors have been identified. Representatives from the thiazole, triazole, and amide series have demonstrated activity in rodent models of diabetes and metabolic syndrome. However, no significant clinical data has been reported, and only two compounds (AMG-211 and INCB-13739, structures unknown [126]) are reported to be in clinical trials. Animal studies have suggested that 11β-HSD1 inhibition is a potential treatment for diabetes and metabolic syndrome, but the true benefit of 11β-HSD1 inhibitors will not be determined until clinical data are available.

REFERENCES

[1] A. H. Mokdad, E. S. Ford, B. A. Bowman, W. H. Dietz, F. Vinicor, V. S. Bales and J. S. Marks, *JAMA*, 2003, **289**, 76.
[2] Office of the Surgeon General, U.S. Department of Health and Human Services: Rockville, MD, 2001: www.surgeongeneral.gov/topics/obesity.
[3] A. Must, J. Spadano, E. H. Coakley, A. E. Field, G. Colditz and W. H. Dietz, *JAMA*, 1999, **282**, 1523.
[4] E. S. Ford, W. H. Giles and W. H. Dietz, *JAMA*, 2002, **287**, 356.
[5] W. Mayo-Smith, C. W. Hayes, B. M. Biller, A. Klibanski, H. Rosenthal and D. I. Rosenthal, *Radiology*, 1989, **170**, 515.
[6] M. Rebuffe-Scrive, M. Krotkiewski, J. Elfverson and P. Bjorntorp, *J. Clin. Endocrinol. Metab.*, 1988, **67**, 1122.
[7] Y. Shimomura, G. A. Bray and M. Lee, *Horm. Metab. Res.*, 1987, **19**, 295.
[8] G. W. Strain, B. Zumoff, J. J. Strain, J. Levin and D. K. Fukushima, *Metabolism*, 1980, **29**, 980.
[9] T. Ljung, B. Andersson, B. A. Bengtsson, P. Bjorntorp and P. Marin, *Obes. Res.*, 1996, **4**, 277.

[10] P. Marin, N. Darin, T. Amemiya, B. Andersson, S. Jern and P. Bjorntorp, *Metabolism*, 1992, **41**, 882.
[11] R. S. Lindsay, D. J. Wake, S. Nair, J. Bunt, D. E. Livingstone, P. A. Permana, P. A. Tataranni and B. R. Walker, *J. Clin. Endocrinol. Metab.*, 2003, **88**, 2738.
[12] O. Paulmyer-Lacroix, S. Boullu, C. Oliver, M. C. Alessi and M. Grino, *J. Clin. Endocrinol. Metab.*, 2002, **87**, 2701.
[13] E. Rask, B. R. Walker, S. Soderberg, D. E. Livingstone, M. Eliasson, O. Johnson, R. Andrew and T. Olsson, *J. Clin. Endocrinol. Metab.*, 2002, **87**, 3330.
[14] D. J. Wake, E. Rask, D. E. Livingstone, S. Soderberg, T. Olsson and B. R. Walker, *J. Clin. Endocrinol. Metab.*, 2003, **88**, 3983.
[15] J. R. Seckl, N. M. Morton, K. E. Chapman and B. R. Walker, *Recent Prog. Horm. Res.*, 2004, **59**, 359.
[16] Z. Krozowski, K. X. Li, K. Koyama, R. E. Smith, V. R. Obeyesekere, A. Stein-Oakley, H. Sasano, C. Coulter, T. Cole and K. E. Sheppard, *J. Steroid Biochem. Mol. Biol.*, 1999, **69**, 391.
[17] A. L. Albiston, V. R. Obeyesekere, R. E. Smith and Z. S. Krozowski, *Mol. Cell. Endocrinol.*, 1994, **105**, R11.
[18] T. Mune, F. M. Rogerson, H. Nikkila, A. K. Agarwal and P. C. White, *Nat. Genet.*, 1995, **10**, 394.
[19] P. M. Stewart, A. M. Wallace, R. Valentino, D. Burt, C. H. Shackleton and C. R. Edwards, *Lancet*, 1987, **2** (8563), 821.
[20] C. Monder, P. M. Stewart, V. Lakshmi, R. Valentino, D. Burt and C. R. Edwards, *Endocrinology*, 1989, **125**, 1046.
[21] T. Barf, J. Vallgårda, R. Emond, C. Häggström, G. Kurz, A. Nygren, V. Larwood, E. Mosialou, K. Axelsson, R. Olsson, L. Engblom, N. Edling, Y. Rönquist-Nii, B. Öhman, P. Alberts and L. Abrahmsén, *J. Med. Chem.*, 2002, **45**, 3813.
[22] G. M. Coppola, P. J. Kukkola, J. L. Stanton, A. D. Neubert, N. Marcopulos, N. A. Bilci, H. Wang, H. C. Tomaselli, J. Tan, T. D. Aicher, D. C. Knorr, A. Y. Jeng, B. Dardik and R. E. Chatelain, *J. Med. Chem.*, 2005, **48**, 6696.
[23] A. Hermanowski-Vosatka, J. M. Balkovec, K. Cheng, H. Y. Chen, M. Hernandez, G. C. Koo, C. B. LeGrand, Z. Li, J. M. Metzger, S. S. Mundt, H. Noonan, C. N. Nunes, S. H. Olson, B. Pikounis, N. Ren, N. Robertson, J. M. Schaeffer, K. Shah, M. S. Springer, A. M. Strack, M. Strowski, K. Wu, T. Wu, J. Xiao, B. B. Zhang, S. D. Wright and R. Thieringer, *J. Exp. Med.*, 2005, **202** (4), 517.
[24] U. Oppermann, C. Filling, M. Hult, N. Shafqat, X. Wu, M. Lindh, J. Shafqat, E. Nordling, Y. Kallberg, B. Persson and H. Jornvall, *Chem. Biol. Interact.*, 2003, **143–144**, 247.
[25] P. Alberts, C. Nilsson, G. Selen, L. O. Engblom, N. H. Edling, S. Norling, G. Klingstrom, C. Larsson, M. Forsgren, M. Ashkzari, C. E. Nilsson, M. Fiedler, E. Bergqvist, B. Ohman, E. Bjorkstrand and L. B. Abrahmsen, *Endocrinology*, 2003, **144**, 4755.
[26] J. Xiang, M. Ipek, V. Suri, W. Massefski, N. Pan, Y. Ge, M. Tam, Y. Xing, J. F. Tobin, X. Xu and S. Tam, *Bioorg. Med. Chem. Lett.*, 2005, **15**, 2865.
[27] A. Odermatt, P. Arnold, A. Stauffer, B. M. Frey and F. J. Frey, *J. Biol. Chem.*, 1999, **274**, 28762.
[28] P. M. Jamieson, K. E. Chapman, C. R. Edwards and J. R. Seckl, *Endocrinology*, 1995, **136**, 4754.
[29] S. C. Low, K. E. Chapman, C. R. Edwards and J. R. Seckl, *J. Mol. Endocrinol.*, 1994, **13**, 167.
[30] H. Duperrex, S. Kenouch, H. P. Gaeggeler, J. R. Seckl, C. R. Edwards, N. Farman and B. C. Rossier, *Endocrinology*, 1993, **132**, 612.

[31] A. G. Atanasov, L. G. Nashev, R. A. Schweizer, C. Frick and A. Odermatt, *FEBS Lett*, 2004, **571**, 129.
[32] U. C. Oppermann, B. Persson and H. Jornvall, *Eur. J. Biochem.*, 1997, **249**, 355.
[33] M. Hult, B. Elleby, N. Shafqat, S. Svensson, A. Rane, H. Jornvall, L. Abrahmsen and U. Oppermann, *Cell. Mol. Life Sci.*, 2004, **61**, 992.
[34] S. Arampatzis, B. Kadereit, D. Schuster, Z. Balazs, R. A. Schweizer, F. J. Frey, T. Langer and A. Odermatt, *J. Mol. Endocrinol.*, 2005, **35**, 89–101.
[35] D. J. Hosfield, Y. Wu, R. J. Skene, M. Hilgers, A. Jennings, G. P. Snell and K. Aertgeerts, *J. Biol. Chem.*, 2005, **280**, 4639.
[36] J. Zhang, T. D. Osslund, M. H. Plant, C. L. Clogston, R. E. Nybo, F. Xiong, J. M. Delaney and S. R. Jordan, *Biochemistry*, 2005, **44**, 6948.
[37] D. Ogg, B. Elleby, C. Norstrom, K. Stefansson, L. Abrahmsen, U. Oppermann and S. Svensson, *J. Biol. Chem.*, 2005, **280**, 3789.
[38] M. Hult, N. Shafqat, B. Elleby, D. Mitschke, S. Svensson, M. Forsgren, T. Barf, J. Vallgarda, L. Abrahmsen and U. Oppermann, *Mol. Cell. Endocrinol.*, 2006, **248**, 26.
[39] E. Rask, T. Olsson, S. Soderberg, R. Andrew, D. E. Livingstone, O. Johnson and B. R. Walker, *J. Clin. Endocrinol. Metab.*, 2001, **86**, 1418.
[40] D. E. Livingstone, G. C. Jones, K. Smith, P. M. Jamieson, R. Andrew, C. J. Kenyon and B. R. Walker, *Endocrinology*, 2000, **141**, 560.
[41] N. M. Morton, L. Ramage and J. R. Seckl, *Endocrinology*, 2004, **145**, 2707.
[42] H. Masuzaki, J. Paterson, H. Shinyama, N. M. Morton, J. J. Mullins, J. R. Seckl and J. S. Flier, *Science*, 2001, **294**, 2166.
[43] H. Masuzaki, H. Yamamoto, C. J. Kenyon, J. K. Elmquist, N. M. Morton, J. M. Paterson, H. Shinyama, M. G. Sharp, S. Fleming, J. J. Mullins, J. R. Seckl and J. S. Flier, *J. Clin. Invest.*, 2003, **112**, 83.
[44] N. M. Morton, J. M. Paterson, H. Masuzaki, M. C. Holmes, B. Staels, C. Fievet, B. R. Walker, J. S. Flier, J. J. Mullins and J. R. Seckl, *Diabetes*, 2004, **53**, 931.
[45] Y. Kotelevtsev, M. C. Holmes, A. Burchell, P. M. Houston, D. Schmoll, P. Jamieson, R. Best, R. Brown, C. R. Edwards, J. R. Seckl and J. J. Mullins, *Proc. Natl. Acad. Sci. USA*, 1997, **94**, 14924.
[46] N. M. Morton, M. C. Holmes, C. Fievet, B. Staels, A. Tailleux, J. J. Mullins and J. R. Seckl, *J. Biol. Chem.*, 2001, **276**, 41293.
[47] E. E. Kershaw, N. M. Morton, H. Dhillon, L. Ramage, J. R. Seckl and J. S. Flier, *Diabetes*, 2005, **54**, 1023.
[48] R. C. Andrews, O. Rooyackers and B. R. Walker, *J. Clin. Endocrinol. Metab.*, 2003, **88**, 285.
[49] B. R. Walker, A. A. Connacher, R. M. Lindsay, D. J. Webb and C. R. Edwards, *J. Clin. Endocrinol. Metab.*, 1995, **80**, 3155.
[50] S. J. Lupien, N. P. Nair, S. Briere, F. Maheu, M. T. Tu, M. Lemay, B. S. McEwen and M. J. Meaney, *Rev. Neurosci.*, 1999, **10**, 117.
[51] J. L. Yau, J. Noble, C. J. Kenyon, C. Hibberd, Y. Kotelevtsev, J. J. Mullins and J. R. Seckl, *Proc. Natl. Acad. Sci. USA*, 2001, **98**, 4716.
[52] T. C. Sandeep, J. L. W. Yau, A. M. J. MacLullich, J. Noble, I. J. Deary, B. R. Walker and J. R. Seckl, *Proc. Natl. Acad. Sci. USA*, 2004, **101**, 6734.
[53] B. R. Walker and J. R. Seckl, *Expert Opin. Ther. Targets*, 2003, **7** (6), 771.
[54] C. Fotsch, B. C. Askew and J. J. Chen, *Expert Opin. Ther. Pat.*, 2005, **15**, 289.
[55] Unless otherwise noted, the species for *in vitro* 11β-HSD data is human.
[56] P. Alberts, L. Engblom, N. Edling, M. Forsgren, G. Klingström, C. Larsson, Y. Rönquist-Nii, B. Öhman and L. Abrahmsén, *Diabetologia*, 2002, **45**, 1528.
[57] T. Barf, R. Emond, G. Kurz, M. Nilsson, J. Vallgårda and L. Zhang, *US Patent* 20050250776, 2005.

[58] H. Fukushima, M. Takahashi, T. Busujima and T. Kawaguchi, *WO Patent* 2005097764, 2005.
[59] T. Barf, R. Emond, G. Kurz, J. Vallgårda, M. Nilsson and L. Zhang, *WO Patent* 03043999, 2003.
[60] T. Barf, M. Nilsson and J. Vallgårda, *WO Patent* 0190094, 2001.
[61] M. Nilsson, *WO Patent* 0190093, 2001.
[62] G. Kurz and M. Nilsson, *WO Patent* 0190092, 2001.
[63] T. Barf, M. Nilsson, G. Kurz and J. Vallgårda, *WO Patent* 0190091, 2001.
[64] T. Barf, R. Emond, G. Kurz, J. Vallgårda and M. Nilsson, *WO Patent* 0190090, 2001.
[65] D. Pyring, M. Henriksson, J. Vagberg, M. Williams, C. Nilsson and C. Dreifeldt, *US Patent* 20050009821, 2005.
[66] C. Nilsson and C. Nilsson, *WO Patent* 2004113310, 2004.
[67] M. Williams, D. Pyring, J. Vågberg, M. Henriksson, C. Nilsson and C. Nilsson, *WO Patent* 2004103980, 2004.
[68] M. Williams, G. Kurz, M. Nilsson and J. Vallgårda, *WO Patent* 03044000, 2003.
[69] T. Rueckle, P. A. Vitte and J. P. Gotteland, *WO Patent* 2005025558, 2005.
[70] C. Nilsson and C. Dreifeldt, *WO Patent* 2004112779, 2004.
[71] M. Williams, G. Kurz, M. Nilsson and J. Vallgårda, *WO Patent* 03044009, 2003.
[72] M. Henriksson, E. Homan, L. Johansson, J. Vallgarda, M. Williams, E. Bercot, C. H. Fotsch, A. Li, G. Cai, R. W. Hungate, C. C. Yuan, C. Tegley, D. St. Jean, N. Han, Q. Huang, Q. Liu, M. D. Bartberger, G. A. Moniz and M. J. Frizzle, *WO Patent* 2005116002, 2005.
[73] L. Tedenborg, T. Barf, S. Nordin, J. Vallgårda, M. Williams and G. Kurz, *WO Patent* 2005075471, 2005.
[74] S. Olson, S. D. Aster, K. Brown, L. Carbin, D. W. Graham, A. Hermanowski-Vosatka, C. B. LeGrand, S. S. Mundt, M. A. Robbins, J. M. Schaeffer, L. H. Slossberg, M. J. Szymonifka, R. Thieringer, S. D. Wright and J. M. Balkovec, *Bioorg. Med. Chem. Lett.*, 2005, **15**, 4359.
[75] X. Gu, J. Dragovic, G. C. Koo, S. L. Koprak, C. LeGrand, S. S. Mundt, K. Shah, M. S. Springer, E. Y. Tan, R. Thieringer, A. Hermanowski-Vosatka, H. J. Zokian, J. M. Balkovec and S. T. Waddell, *Bioorg. Med. Chem. Lett.*, 2005, **15**, 5266.
[76] S. T. Waddell, G. M. Santorelli, M. M. Maletic, A. H. Leeman, X. Gu, D. W. Graham, J. M. Balkovec and S. D. Aster, *US Patent* 6 849 636, 2005.
[77] S. H. Olson, J. M. Balkovec and Y. Zhu, *US Patent* 6 730 690, 2004.
[78] S. D. Aster, J. M. Balkovec, D. W. Graham, X. Gu, N. J. Kevin, G. F. Patel and M. Ponpipom, *WO Patent* 2005097759, 2005.
[79] Y. Bereznitski, M. A. Huffman, J. E. Lynch and M. Zhao, *WO Patent* 2005073200, 2005.
[80] M. G. Cardozo, J. P. Powers, H. Goto, K. Harada, K. Imamura, M. Kakutani, I. Matsuda, Y. Ohe and S. Yata, *WO Patent* 2005044192, 2005.
[81] S. T. Waddell, G. M. Santorelli, M. M. Maletic, A. H. Leeman, X. Gu, D. W. Graham, J. M. Balkovec and S. D. Aster, *WO Patent* 2004106294, 2004.
[82] H. S. Andersen, G. C. T. Kampen, I. T. Christensen, J. P. Mogensen and A. R. Larsen, *WO Patent* 2004089380, 2004.
[83] H. S. Andersen, G. C. T. Kampen, I. T. Christensen, J. P. Mogensen and A. R. Larsen, *WO Patent* 2004089367, 2004.
[84] S. T. Waddell, G. M. Santorelli, M. M. Maletic, A. H. Leeman, X. Gu, D. W. Graham, J. M. Balkovec and S. D. Aster, *WO Patent* 2004058741, 2004.
[85] S. T. Waddell, G. M. Santorelli, M. M. Maletic, A. H. Leeman, X. Gu, D. W. Graham, J. M. Balkovec and S. D. Aster, *WO Patent* 2004058730, 2004.

[86] S. H. Olson, J. M. Balkovec and Y. Zhu, *WO Patent* 03104208, 2003.
[87] S. H. Olson, J. M. Balkovec and Y. Zhu, *WO Patent* 03104207, 2003.
[88] J. M. Balkovec, R. Thieringer, S. S. Mundt, A. Hermanowski-Vosatka, D. W. Graham, G. F. Patel, S. D. Aster, S. T. Waddell, S. H. Olson and M. Maletic, *WO Patent* 03065983, 2003.
[89] H. Goto, M. Kakutani, J. Nishiu, Y. Ohe, J. P. Powers and S. Yata, *WO Patent* 2005095350, 2005.
[90] Y. D. Gao, X. Gu, N. J. Kevin and S. T. Waddell, *WO Patent* 2005016877, 2005.
[91] A. Gundertofte, A. S. Jørgensen, G. C. T. Kampen, H. S. Andersen, I. T. Christensen and J. P. Kilburn, *WO Patent* 2004089896, 2004.
[92] H. S. Andersen, G. C. T. Kampen, I. T. Christensen, J. P. Mogensen and A. R. Larsen, *WO Patent* 2004089471, 2004.
[93] H. S. Andersen, G. C. T. Kampen, I. T. Christensen, J. P. Mogensen, A. R. Larsen and J. P. Kilburn, *WO Patent* 2004089470, 2004.
[94] G. M. Coppola, R. E. Damon, P. J. Kukkola and J. L. Stanton, *WO Patent* 2004065351, 2004.
[95] L. Jaroskova, J. T. M. Linders, L. J. E. Van Der Veken, G. H. M. Willemsens and F. P. Bischoff, *WO Patent* 2006024627, 2006.
[96] J. T. Link, Y. Chen, H.-S. Jae, J. R. Patel, M. A. Pliushchev, J. J. Rohde, Q. Shuai, B. K. Sorensen, M. Winn, D. Wodka and H. Yong, *US Patent* 20050277647, 2005.
[97] E. D. Hoff, J. T. Link, M. A. Pliushchev, J. J. Rohde and M. Winn, *US Patent* 20050261302, 2005.
[98] J. T. Link, M. A. Pliushchev, J. J. Rohde and D. Wodka, *US Patent* 20050245745, 2005.
[99] J. T. Link, M. A. Pliushchev, J. J. Rohde, D. Wodka, J. R. Patel and Q. Shuai, *US Patent* 20050245534, 2005.
[100] E. D. Hoff, J. T. Link, M. A. Pliushchev, J. J. Rohde and M. Winn, *US Patent* 20050245533, 2005.
[101] E. D. Hoff, J. T. Link, M. A. Pliushchev, J. J. Rohde and M. Winn, *US Patent* 20050245532, 2005.
[102] J. T. Link, M. A. Pliushchev, J. J. Rohde, D. Wodka, J. R. Patel and Q. Shuai, *WO Patent* 2005108368, 2005.
[103] H. Cheng, C. R. Smith, Y. Wang, T. J. Parrott, K. R. Dress, S. K. Nair, J. E. Hoffman, P. T. Q. Le, S. W. Kupchinsky, Y. Yang, S. J. Cripps and B. Huang, *WO Patent* 2005108359, 2005.
[104] J. T. M. Linders, G. H. M. Willemsens, R. A. H. J. Gilissen, C. F. R. N. Buyck, G. C. P. Vanhoof, L. J. E. Van Der Veken and L. Jaroskova, *WO Patent* 2004056745, 2004.
[105] J. T. M. Linders, G. H. M. Willemsens, R. A. H. J. Gilissen, C. F. R. N. Buyck, G. C. P. Vanhoof, L. J. E. Van Der Veken and L. Jaroskova, *WO Patent* 2004056744, 2004.
[106] S. T. Waddell, J. M. Balkovec, G. M. Santorelli, A. H. Leeman, M. Maletic and X. Gu, *WO Patent* 2006017542, 2006.
[107] L. Jaroskova, J. T. M. Linders, L. J. E. Van Der Veken and G. H. M. Willemsens, *WO Patent* 2006024628, 2006.
[108] L. Jaroskova, J. T. M. Linders, C. F. R. N. Buyck and L. J. E. Van Der Veken, *WO Patent* 2005108361, 2005.
[109] L. Jaroskova, J. T. M. Linders, C. F. R. N. Buyck, L. J. E. Van Der Veken, V. D. Dimitrov and T. T. Nikiforov, *WO Patent* 2005108360, 2005.
[110] W. Yao, C. Zhang, C. He and J. Zhuo, *US Patent* 20050288329, 2005.
[111] J. S. Xiang, J. C. McKew, S. Y. Tam, M. Ipek, V. Suri and T. S. Mansour, *US Patent* 20060025445, 2006.

[112] K. Amrein, D. Hunziker, B. Kuhn, A. Mayweg and W. Neidhart, *WO Patent* 2006010546, 2006.
[113] K. Amrein, D. Hunziker, B. Kuhn, A. Mayweg and W. Neidhart, *WO Patent* 2006000371, 2006.
[114] M. R. Degraffenreid, J. P. Powers, D. Sun and X. Yan, *WO Patent* 2005118538, 2005.
[115] N. Vicker, D. Ganeshapillai, A. Purohit, M. J. Reed and B. V. L. Potter, *WO Patent* 2005103023, 2005.
[116] M. R. Degraffenreid, X. He, J. P. Powers, D. Sun and X. Yan, *WO Patent* 2005063247, 2005.
[117] M. P. Edwards, T. O. Johnson, Jr., S. K. Nair, M. Siu, W. D. Taylor, S. J. Cripps, Y. Wang, H. Cheng and C. R. Smith, *WO Patent* 2005060963, 2005.
[118] N. Vicker, S. Xiangdong, D. Ganeshapillai, A. Purohit, M. J. Reed and B. V. L. Potter, *WO Patent* 2005042513, 2005.
[119] N. Vicker, X. Su, D. Ganeshapillai, A. Purohit, M. J. Reed and B. V. L. Potter, *WO Patent* 2004037251, 2004.
[120] P. J. Barton, R. J. Butlin and J. E. Pease, *WO Patent* 2005047250, 2005.
[121] P. J. Barton, R. J. Butlin and J. E. Pease, *WO Patent* 2005046685, 2005.
[122] P. J. Barton, D. S. Clarke, C. S. Donald and J. E. Pease, *WO Patent* 2004041264, 2004.
[123] P. J. Barton, P. J. Jewsbury and J. E. Pease, *WO Patent* 2004033427, 2004.
[124] P. J. Barton, D. S. Clarke, C. D. Davies, R. B. Hargreaves, J. E. Pease and M. T. Rankine, *WO Patent* 2004011410, 2004.
[125] P. Fan, H. Goto, X. He, M. Kakutani, M. Labelle, D. L. McMinn, J. P. Powers, Y. Rew, D. Sun and X. Yan, *WO Patent* 2005110980, 2005.
[126] Investigational Drugs Database. http://www.iddb3.com/ (accessed 7/24/06).

Glucokinase Activators for the Treatment of Type 2 Diabetes

Theodore O. Johnson and Paul S. Humphries

Pfizer Global Research and Development, San Diego, CA 92121, USA

Contents

1. Introduction	141
2. Biology of glucokinase	142
2.1. Features of glucokinase	142
2.2. Glucose-induced conformational change to explain GK cooperativity	142
2.3. Glucokinase in the liver	144
2.4. Glucokinase in the pancreatic β-cells	144
2.5. The mechanism of GK activation by allosteric-site binders	145
3. Small-molecule glucokinase activators	146
3.1. Carbon- and nitrogen-centered glucokinase activators	146
3.2. Aromatic ring-centered glucokinase activators	149
3.3. Miscellaneous	152
4. Conclusion and outlook	152
References	152

1. INTRODUCTION

Type 2 diabetes (T2D) affects an escalating percentage of populations in both the developed and developing parts of the world. According to the National Institute of Diabetes and Digestive and Kidney Diseases (NIDDK), 20.8 million Americans (~7% of the population) have diabetes, and nearly one-third of these cases are undiagnosed. Another 41 million have insulin resistance or pre-diabetes. Worldwide figures are especially astounding: in 2003 the World Health Organization (WHO) reported a global incidence of 194 million diabetes patients and this figure may possibly double by 2025. The underlying principle pathological origins of T2D include: (i) attenuated glucose-induced insulin release from pancreatic β-cells; (ii) insulin resistance in adipose tissue, skeletal muscle and liver; and (iii) disproportionate hepatic glucose production.

    Evidence gathered over the previous 40 years has ascertained an essential function for glucokinase (GK), also known as hexokinase IV or hexokinase D, in regulating glucose homeostasis. GK is expressed mainly in two tissues of high importance to glucose homeostasis and therapeutic strategies to treat T2D, pancreatic β-cells and hepatocytes. The mechanism of GK has been thoroughly studied and is well established [1–4]. In the liver, GK functions as a high-capacity enzyme that removes glucose from the blood and reacts it with adenosine triphosphate (ATP) to form glucose-6-phosphate (G-6-P), the first biosynthetic step in the conversion of glucose to its storage form, glycogen. In pancreatic β-cells, through its ability to sense glucose concentrations, GK operates as a glucose sensor to determine the threshold for glucose-stimulated insulin release (GSIR).

Because GK has a central role in glucose homeostasis, many researchers have attempted to develop drugs that augment GK activity to treat T2D. It is reasoned that an activator of GK would serve dual purposes of firstly decreasing blood glucose levels by increasing hepatic glucose uptake, and secondly, increasing insulin secretion by lowering the threshold of GSIR. In this review, the mechanism of enzymatic catalysis by GK, its function in the liver and pancreatic β-cells, key publications on small-molecule GK activators (GKAs) that report full characterization, and key compound disclosures from patents will be discussed.

2. BIOLOGY OF GLUCOKINASE

2.1. Features of glucokinase

The reaction catalyzed by GK is the Mg•ATP^{2-} mediated phosphorylation of glucose to produce G-6-P and adenosine diphosphate (ADP). GK is a 52-kDa protein, and is smaller compared with other hexokinase family members, which are typically around 100 kDa. GK is also unusual in that it is not inhibited by its product, G-6-P. This ensures that the activity of GK is regulated only by glucose concentration. Another key feature of GK is that it demonstrates slight positive cooperativity with respect to glucose. Its Hill slope coefficient is 1.5–1.7, and it has a slightly sigmoidal glucose saturation curve with a substrate concentration at half maximal velocity ($S_{0.5}$) of approximately 7.5 mM. The $S_{0.5}$ of GK is significantly higher than that of other hexokinases. The catalytic activity of GK is most sensitive to changes when the glucose concentration nears 4 mM, which represents the inflection point in the glucose saturation curve. This concentration is centered in the normal human physiological range of blood glucose concentrations and is optimally situated to produce maximum enzyme velocities when blood glucose concentrations rise above normal levels.

With respect to Mg•ATP^{2-}, GK kinetics are standard Michaelis–Menton type, with a K_m (Michaelis constant) of ATP around 0.4 mM, well below the intracellular physiologic concentration of ATP. This ensures that the enzyme rate is dictated by glucose concentration and not ATP concentration [1–4].

2.2. Glucose-induced conformational change to explain GK cooperativity

Two closely related models, the mnemonic model [5] and the slow transition model [6], were developed to explain the observed positive cooperativity of GK with respect to glucose. These models invoke two kinetically distinct conformational forms of the GK enzyme, a highly reactive form that is induced under high glucose conditions, and a less active form that is also a lower energy conformation, which forms under low glucose concentrations. The reactive form is associated with a fast catalytic cycle, and the less reactive form is associated with a slower catalytic cycle. Which form of the enzyme is operational at any particular time is entirely dependant on glucose concentration. Both models are consistent with the observed kinetic profile of GK.

Crystal structures of the apo form of GK and the liganded forms helped confirm the proposals regarding how cooperativity with respect to glucose is achieved [7]. The apo structure was found to exist in a low-energy "super-open" conformation, which was determined to be an inactive conformation because certain critical residues were absent from the active site. Upon binding glucose, a slow conformational change takes place to a higher-energy structure, the "open form". A subsequent fast conformational change to the "closed form" gives a complex that is able to bind the Mg•ATP^{2-} cofactor, leading to the occurrence of the catalytic reaction and generation and release of products, G-6-P and ADP. The catalytic cycle continues in the fast cycle as long as glucose concentrations remain high, bypassing the low-energy super-open conformation, and involves a reduced degree of protein dynamics relative to the slow cycle. When the glucose concentration drops, the conformation slowly relaxes back to the low-energy super-open form. If glucose concentration remains low, then the catalytic cycle operates in the slow cycle, where GK exhibits a larger, jaw-like range of motion with an angular movement of approximately 120° of the protein occurring after product dissociation (Fig. 1) [8].

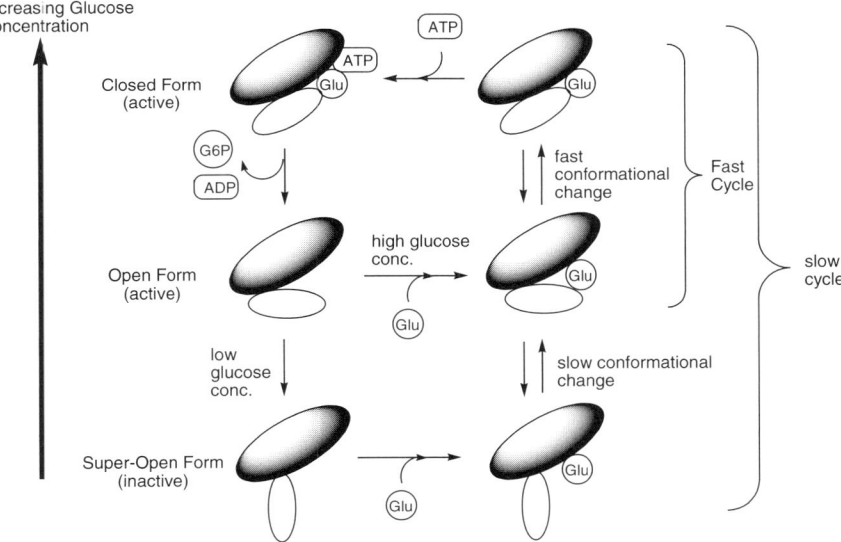

Fig. 1. Illustration of the kinetic mechanism model for GK demonstrating the closed and super-open conformations

2.3. Glucokinase in the liver

Glucose in the bloodstream is taken up mostly by muscle tissue and the liver. Hepatic GK is a high-capacity enzyme. It has been estimated that the human liver contains approximately 50–100 mg of GK enzyme, and this amount of enzyme has the capacity to remove and phosphorylate up to 1 g/min of glucose from the portal bloodstream and convert it into G-6-P, when the blood glucose concentration is 15 mM [3]. G-6-P can be converted into the final storage product, glycogen, by glycogen synthase, or to CO_2 by glycolysis. In addition, there are other minor pathways for glucose such as conversion into pentose units for nucleotide biosynthesis.

In hepatocytes, a 68-kDa GK regulatory protein (GKRP) is expressed, which forms a complex with GK in the presence of fructose-6-phosphate (F-6-P). When bound to GK, GKRP inhibits the kinase activity of GK, and also shields residues 300–310 of GK, a sequence known as the nuclear export sequence (NES). While the NES is shielded, GK remains localized in the nucleus. It is not clear whether GK performs other regulatory functions while sequestered in the nucleus, but GK's role in the metabolism of glucose is inoperational while in this configuration. When the concentration of fructose-1-phosphate (F-1-P) builds up as a result of fructokinase activity, which increases after meals, it displaces F-6-P from the GKRP-GK complex. Dissociation of GKRP from GK is then able to occur, exposing the GK NES. This allows GK to be exported from the nucleus to the cytoplasm, where it is localized while operational in its role in the metabolism of glucose. When the concentration of F-6-P again increases, GK is complexed with GKRP whereupon it is transported back to the nucleus. Since GK lacks a nuclear localization sequence, transport of GK into the nucleus is dependent on the GKRP nuclear localization sequence [2,9].

The homozygous knockout of GKRP shows 40% reduction in liver GK, suggesting that that complexation with GKRP may also protect the protein from proteolytic degradation. Mice deficient in GKRP have impaired glucose tolerance when given a bolus dose of glucose. It is thought that without GKRP, hepatocytes cannot store a reserve supply of GK in the hepatocyte nucleus, to be released under high glucose conditions [4].

2.4. Glucokinase in the pancreatic β-cells

In the pancreatic β-cells, GK establishes a threshold of GSIR by functioning as a glucose sensor to coordinate insulin secretion with blood glucose levels [10]. Glucose is transported into the β-cells by GLUT-2 transporters, where it is converted to G-6-P and ADP by GK. This reaction, along with glycolysis, establishes the critical ratio of ATP/ADP, which at a certain set point causes closure of ATP-sensitive potassium ion channels. Closure of this ion channel causes depolarization of the plasma membrane, and influx of calcium ions through voltage-sensitive calcium ion channels. This results in fusion of the insulin containing secretory granules with the plasma membranes, and release of insulin [4].

2.5. The mechanism of GK activation by allosteric-site binders

In a search for molecules that disrupt the GK-GKRP complex, compounds that directly activate GK by binding to an allosteric site of the enzyme were serendipitously discovered. The GKAs were shown to increase the affinity of GK for glucose (lower $S_{0.5}$), and to increase the V_{max}, although not all GKAs share the latter characteristic. The degree of positive cooperativity with respect to glucose decreased in the presence of a GKA, as reflected by a decrease in Hill slope from around 1.5–1.7 to approximately 1.2. The glucose saturation curve is also altered from being slightly sigmoidal to being very close to hyperbolic in the presence of an activator [1].

GKAs bind to the open form of GK in a small pocket remote from the active site. In the super-open form, this allosteric site does not exist. Upon binding of a GKA in the allosteric pocket, GK is prevented from relaxing to the lower energy super-open form, thereby locking the enzyme into the fast catalytic cycle.

In hepatocytes, where the inhibitory regulatory protein GKRP is involved in the regulation of GK activity, the binding of a GKA prevents relaxation to the super-open form to which GKRP binds (*vide supra*). Since the binding of GKRP is necessary for the sequestration of GK into the nucleus, GKAs have the effect of restricting the localization of GK to the cytoplasm [4].

To date, the X-ray co-crystal structures of GK with GKAs RO-27-5145 (**1**) [11], **2** [8] and with LY-2121260 (**3**) [12] have been reported. This work revealed a palm-shaped structure, consisting of a small and a large domain, separated by an interdomain cleft. Cleft residues Asn204 and Asp205, in combination with Glu256 and Glu290 of the large domain and small domain residues Thr168 and Lys169, are implicated in glucose binding. In addition, the allosteric site was identified at the interface connecting the two domains and is surrounded by linking region-I, the large domain (β1 strand and α5 helix), and the small domain (α13 helix) [8]. In each structure, the GKA binds at the allosteric site, which is 20 Å removed from the glucose binding site. In the case of **1**, the allosteric site is comprised of a ceiling created by residues 65–68, which are part of the first connection linking the two domains of GK, and a floor created by the hydrophobic residues Met235, Met210, Ile211, Val62, Val452, and Ile159. The crucial hydrogen-bond interactions of **1** with GK occur between the amide NH and the thiazole N of **1** to the Arg63 backbone carbonyl O and NH, respectively. Similarly **2** and **3** bind at the same location, and achieve analogous hydrogen-bonding interactions with Arg63. Moreover, **2** and **3** appear to make equivalent interactions with the other residues in the allosteric site.

Regulation of GK activity *via* the allosteric site facilitates the explanation of the activated kinetic properties of certain GK mutations that have been identified in patients with persistent hyperinsulinemic hypoglycemia of infancy (PHHI). For example, in patients with mutations in one GK allele in which methionine was substituted for valine at residue 455 (V455 M), tyrosine for cysteine at residue 214 (Y214C), or valine for alanine at residue 456 (A456 V), improved GK activity was exhibited [13]. These residues, which seem to be included in the allosteric site regulatory domain, are important for the activity of GK and also may aid the design of specific GK activators.

In vivo, GKAs increase glucose phosphorylation in the liver, thereby increasing glucose disposal from the blood, increasing hepatic glycogen production, and reducing hepatic gluconeogenesis. GKAs have been shown to be effective in reducing blood glucose levels in several animal models of T2D [7]. In pancreatic β-cells, both *in vitro* and in diabetic animal models, GKAs lowered the threshold of GSIR and increased the amount of secreted insulin. [1,7,14]. The fact that GKAs can favorably influence glucose homeostasis in the liver and pancreatic β-cells, two sites of action critical to diabetes, makes GK activation an attractive target for the potential treatment of T2D. Despite their potential, GKAs are still in the early stages of clinical development. This is probably because while historically the target appeared attractive for manipulation of liver glucose metabolism, the requirement of a kinase activator, and what was originally perceived as the necessity for the small molecule to interfere with a protein–protein interaction (GK–GKRP) seemed insurmountable.

3. SMALL-MOLECULE GLUCOKINASE ACTIVATORS

3.1. Carbon- and nitrogen-centered glucokinase activators

Investigation of small-molecule GKAs has rapidly turned into an extremely dynamic area of diabetes research and was recently reviewed [2]. Since the emergence of initial reports [7], numerous pharmaceutical companies have initiated drug discovery projects to identify small-molecule GKAs. To seek out small molecules that increase GK enzymatic activity, a library of 120,000 structurally diverse synthetic compounds was screened. A solitary hit was identified from this high-throughput screening (HTS) enzyme coupled assay, in the presence of GKRP. Ensuing kinetic analysis, in the absence of GKRP, confirmed the capacity of the hit molecule to directly bind and activate GK, rather than activating by disrupting the interaction of GKRP with GK. Extensive lead optimization efforts identified RO-28-0450 (**4**) as a preliminary lead. Ultimately, the corresponding R- and S-isomers, RO-28-1675 (**5**) and RO-28-1674 (**6**) respectively, were synthesized. Activation of GK was exquisitely sensitive to the chirality of the molecule: **5** was a potent activator ($EC_{50} = 0.75\,\mu M$), while **6** was inactive. Compound **5** amplified GK enzymatic activity in a dose-dependent fashion. At a concentration of $3\,\mu M$, **5** increased the V_{max} of GK 1.5-fold and increased the affinity for glucose, as signified by a decrease in the $S_{0.5}$ value from 8.6 to 2.0 mM. In addition to this, **5** reduced the threshold for GSIR and improved the degree of insulin release in isolated perfused rat pancreatic islets in a concentration-dependent manner [7].

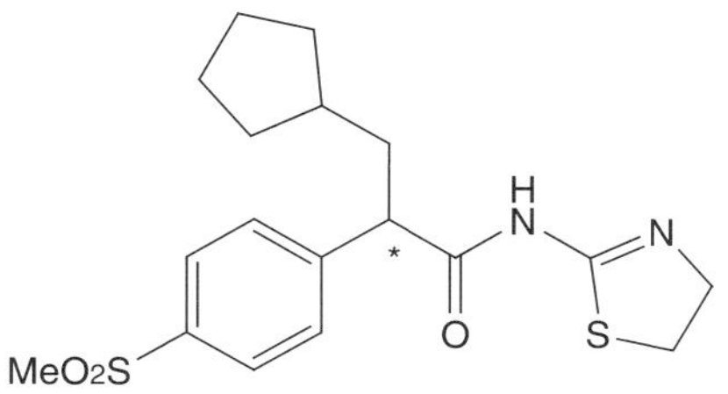

4 (racemic)
5 (R-isomer)
6 (S-isomer)

Compound **5** was assessed in a variety of rodent models of T2D and was effective in lowering basal blood glucose levels in addition to dampening glucose excursions during an oral glucose tolerance test (OGTT). Initial indications of participation of the liver became evident during OGTTs carried out in healthy and T2D mouse models. Compound **5** stimulated peak insulin levels at 45 min and an OGTT was carried out 120 min after oral administration of **5**. While **5** was equally efficacious regardless of when the OGTT was performed, sulfonylureas were generally more effective when an OGTT was carried out concurrently with the drug-induced peak insulin levels. These results implicated activation of hepatic GK by **5**, which was subsequently confirmed in a pancreatic clamp study in rats. These reports indicated improved glucose utilization in the drug-treated group. Therefore, GKAs were shown to increase the threshold of GSIR in the pancreas and also to improve glucose utilization in the liver. Furthermore, in a 40-week study, **5** prevented the development of hyperglycemia in mice fed on a high-fat diet, while having no effect on food intake or body weight relative to a vehicle-treated group [14]. Unfortunately, based on the cardiovascular risk (hERG $IC_{20} = 2.8\,\mu M$; Purkinje fiber $\Delta APD_{90} = 20\%$), further development of **5** was discontinued [15].

In a recent paper, the structure–activity relationship (SAR) and *in vivo* activity of a series of phenylurea derivatives was revealed [16]. In this series, the chiral carbon of **5** was substituted with an achiral nitrogen atom, resulting in a series of urea analogs, exemplified by **7**. Interestingly, **7** was ∼8- to 9-fold less potent than **5**, with GK EC_{50} values of 9.7 and 1.0 μM at 15 mM glucose for **7** and **5** respectively. Nevertheless, **7** had an effect equivalent to **5** on increasing V_{max} (221% control). The dramatic disparity in EC_{50}'s between **5** and **7** translated into *in vivo* activities. In overnight fasted C57BL/6J mice, **5** decreased basal blood glucose levels by 36% (compared to vehicle control) when administered orally at 50 mg/kg, while **7** demonstrated no blood glucose-lowering effects even at a higher dose of 100 mg/kg.

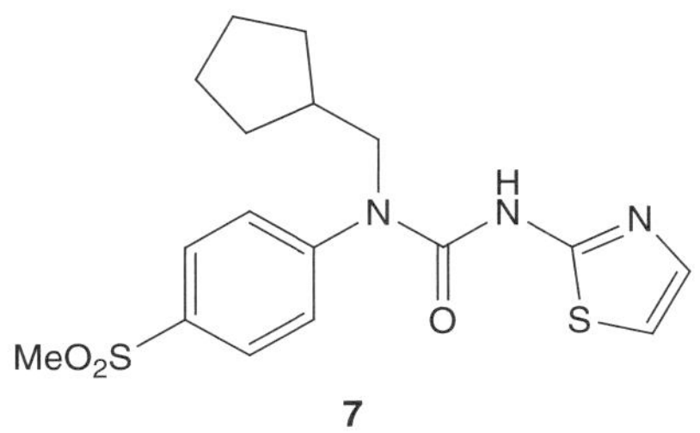

7 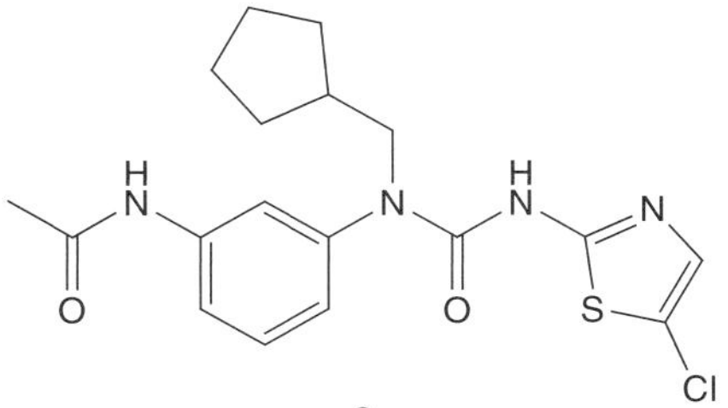   **8**

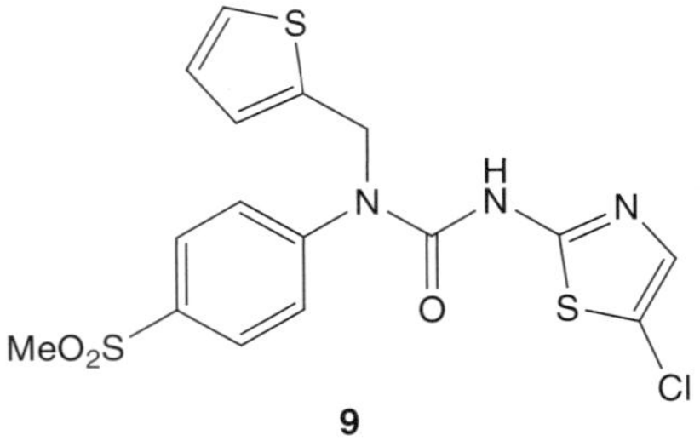

9

Additional optimization of the phenylurea derivatives led to **8** and **9**. Compound **9** reduced blood glucose levels by 34% (compared to vehicle control) when administered orally at 100 mg/kg to overnight fasted C57BL/6J mice.

Recently, the modulatory effects of a cyclopropane-based GKA, LY2121260 (**3**), in both pancreatic islet and hepatocyte assays were disclosed [12]. Compound **3**, like other GKAs, binds at the allosteric site of GK. The modulatory effects of **3** were also comparable to other GKAs in the pancreas and liver. At 10 µM, **3** increased V_{max} by 40%, and decreased the $S_{0.5}$ value from 6.8 to 0.4 mM. Compound **3** produced a 28% decrease in the area under the glucose curve following an OGTT in overnight-fasted Wistar rats, when administered orally at 50 mg/kg. In addition to this, incubation with **3** resulted in increased GK protein levels in pancreatic β-cells.

Numerous carbon- and nitrogen-centered structures have been published in the patent literature. Unfortunately, these disclose minimal biological data. Patent applications in this area invariably contain three different substitutions on a central carbon atom, with a hydrogen atom satisfying the fourth valence to create a chiral center (e.g. **10–12**) [17–21]. These patents explicitly state that the R-stereoisomers are preferred and at least one group has claimed sulfonamide GKAs (e.g. compound **11**) [20]. Carbon-centered compounds that contain an olefin (e.g. **13**) have also been claimed as GKAs [22–24]. Interestingly, additional patent applications also claim heterocyclic replacements for the earlier-claimed phenyl ring (e.g. compounds **14** and **15**) [25,26]. Spirocycles containing a quaternary carbon have also been claimed as GKAs by two companies (e.g. **16** and **17**) [27,28].

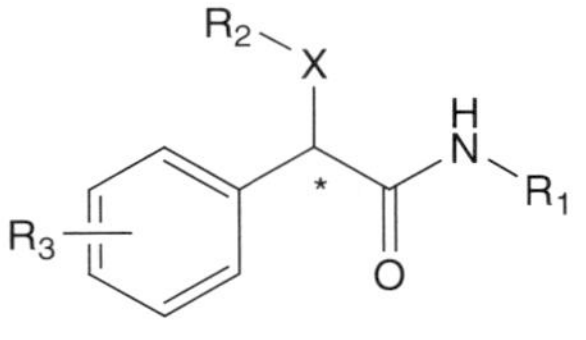

10 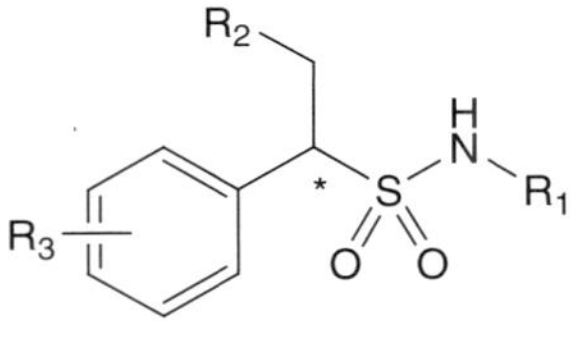   **11**

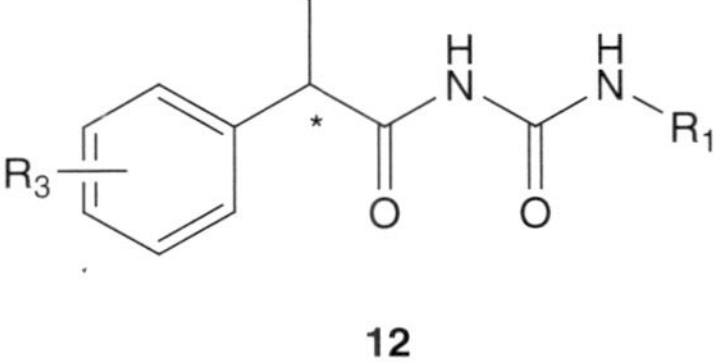

12 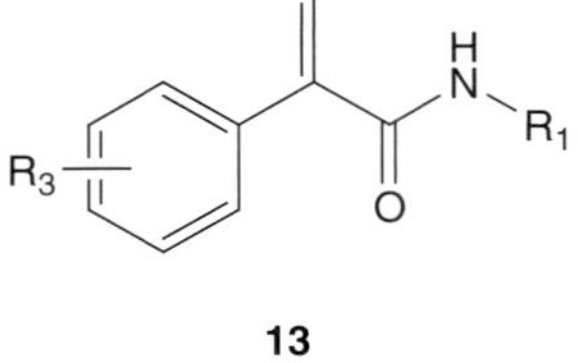   **13**

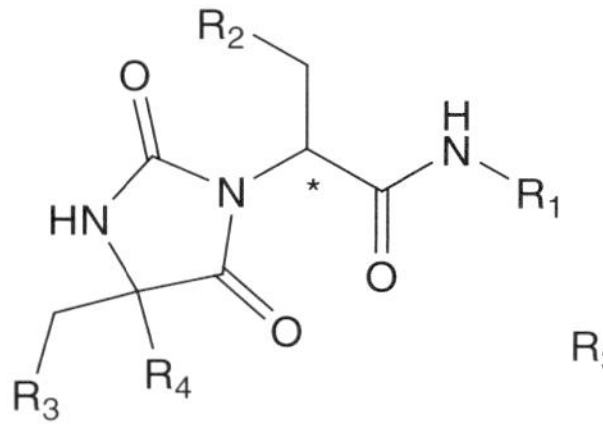

3.2. Aromatic ring-centered glucokinase activators

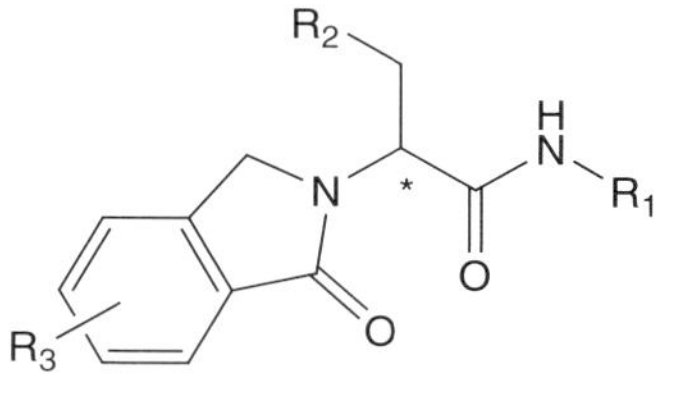

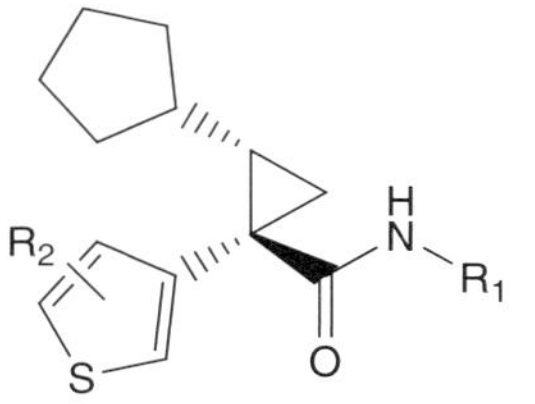

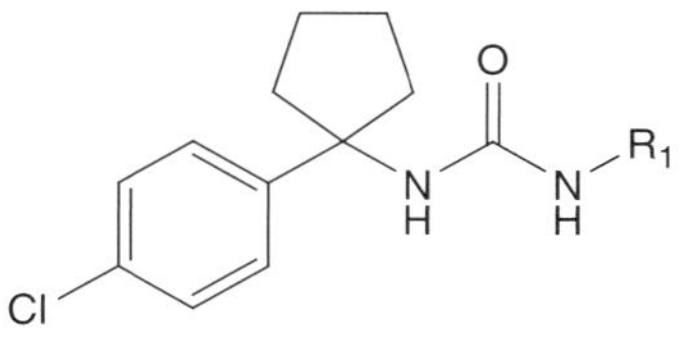

The discovery of a series of novel 1,3,5-trisubstituted benzenes as GKAs (e.g. **18** and **19**) has been reported. In an initial publication, the comprehensive mechanism of action of **18** and **19** was illustrated [29]. Compounds **18** and **19** improved the affinity of GK for glucose 2- and 4-fold, respectively, at 1 µM, and 4- and 11-fold, respectively, at 10 µM. Unlike **5**, these compounds exhibited an insignificant effect on V_{max}, an example of the fact that not all GKAs have the ability to increase the V_{max}. In hepatocytes, **18** and **19** increased free GK levels, glucose phosphorylation, glycolysis, and glycogen synthesis, with EC_{50} values of 2–3 and 1 µM, respectively. On the basis of these studies, the mechanism of action of **18** and **19** was hypothesized to be "glucose-like" in hepatocytes. Administration of a single acute oral dose of **18** to overnight fasted female mice produced a dose-dependent reduction in blood glucose levels [30].

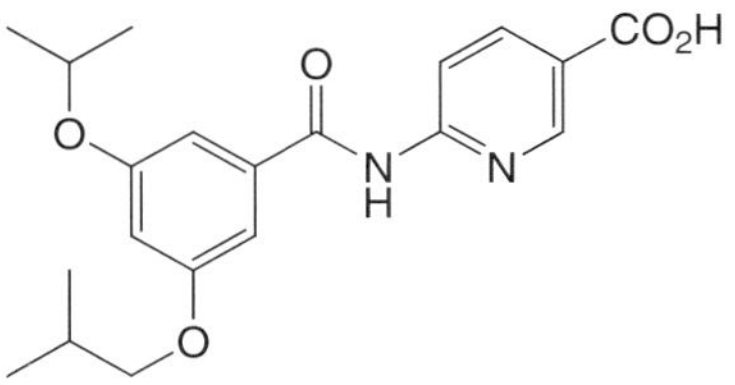

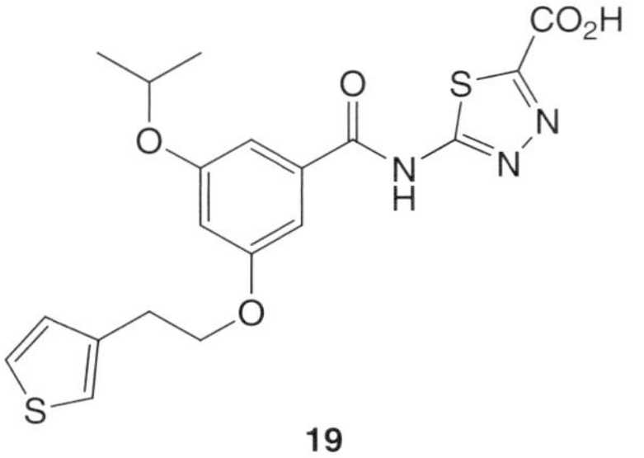

In a subsequent chemistry publication, the same group described the SAR of the trisubstituted benzene series, including **20** ($EC_{50} = 3.2$ µM), the original HTS hit [31]. Early optimization of **20** led to alteration of the olefin linkage to the nicotinic acid group and changing the substitution pattern from an unsymmetrical 1,2, 5-trisubstituted benzene to symmetrical 1,3,5-trisubstituted analogs. These endeavors afforded lead compound **21**, with an $EC_{50} = 0.09\,\mu M$. Additional optimization at positions 3 and 5 of the benzene ring led to the analogue **22** ($EC_{50} = 0.09\,\mu M$), deemed to have an appropriate balance of potency and physical properties (e.g. solubility and free drug level). Within this series there was a correlation between the potency and lipophilicity, as displayed in the reported cLogP values and high plasma protein binding. This relationship made the discovery of "drug-like" GKAs a challenge. Examination of the data also established a relationship between plasma free fraction of GKA and *in vivo* efficacy. This direct correlation led to a definition of a common free-drug exposure multiple of *in vitro* potency to observed *in vivo* efficacy [32]. In diet-induced obese Zucker rats, a 30 mg/kg oral dose of **22** reduced plasma glucose and produced a 14% decrease in the area under the glucose curve following an oral glucose challenge.

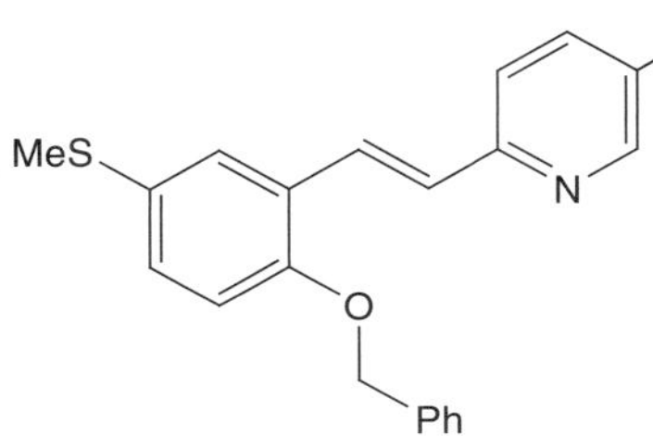

20

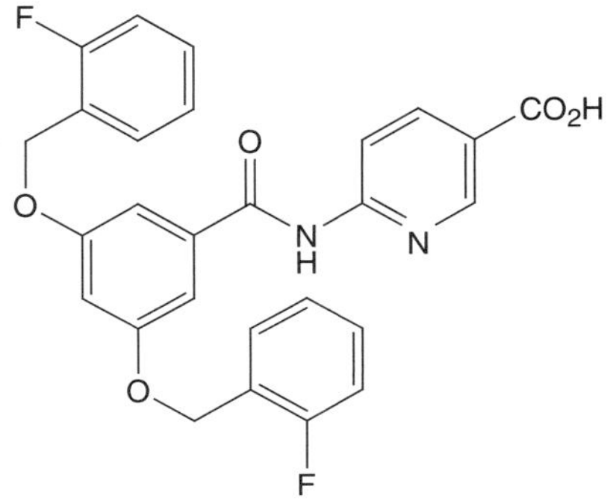

21

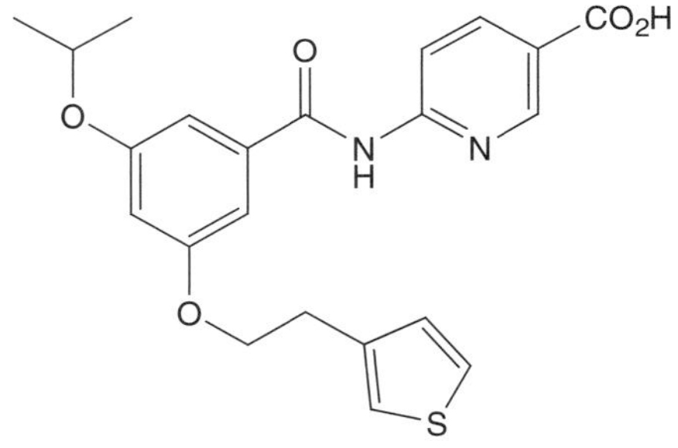

22

The first instance of an amino benzamide-containing GKA, **2**, and its effects on GK kinetics has been described [8]. At 30 μM, **2** reduced the $S_{0.5}$ value for glucose from 8 to 0.6 mM and increased V_{max} ~1.6-fold. Compound **2** lowered the Hill coefficient for glucose from 1.78 to 1.11, altering the shape of the glucose saturation curve from sigmoidal to close to a Michaelis–Menten hyperbolic curve.

Glucokinase Activators for the Treatment of Type 2 Diabetes

In this category, aromatic cores containing benzenes, pyridines, benzofurans, and indoles, have thus far been claimed. The examples cited in these applications comprise trisubstituted benzenes (e.g. **23**) [33], benzofurans (e.g. **24**) [34], and quinolines (e.g. **25** and **26**) [35]. Trisubstituted pyridines (e.g. **27**) [36] and aminobenzamides (e.g. **28**) [37], which demonstrated *in vivo* efficacy at 10 and 30 mg/kg doses, have also been claimed. However, the exact structures of *in vivo* active compounds are not unambiguously specified in these patent applications. Two groups have also claimed 1,3,6-trisubstituted indoles (e.g. **29**) as GKAs [38,39].

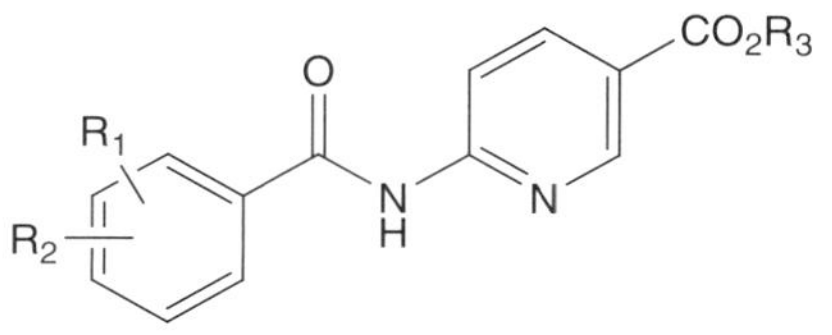

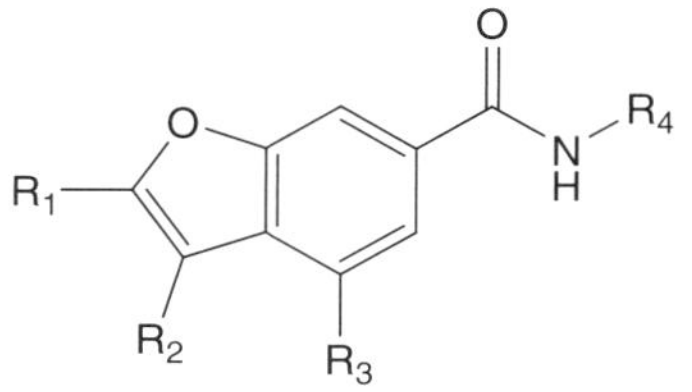

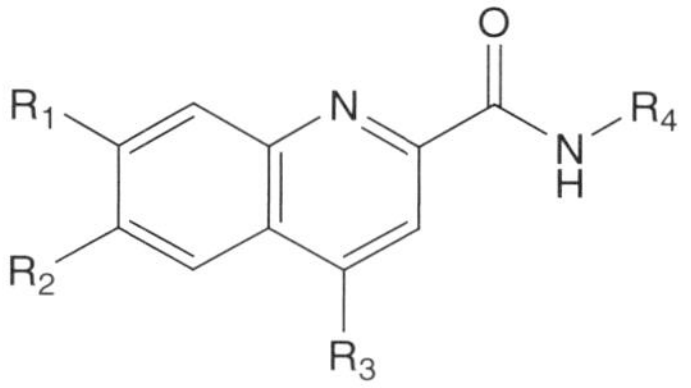

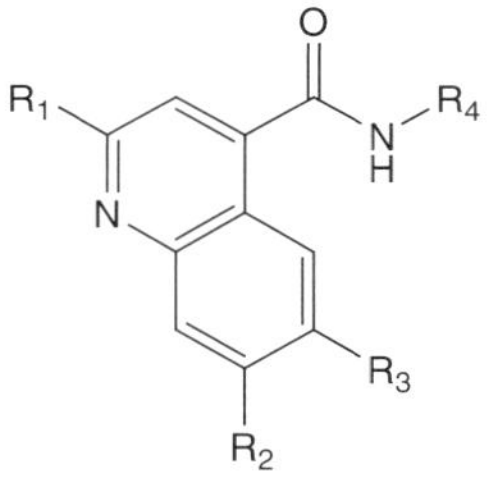

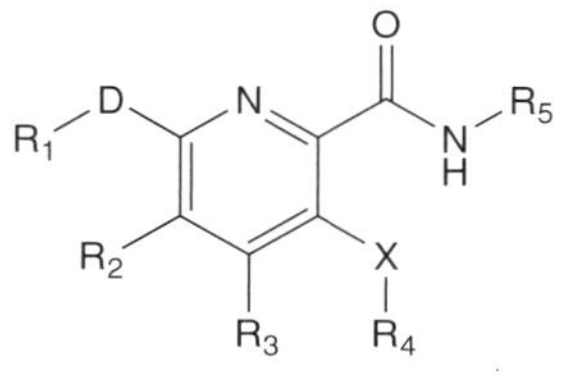

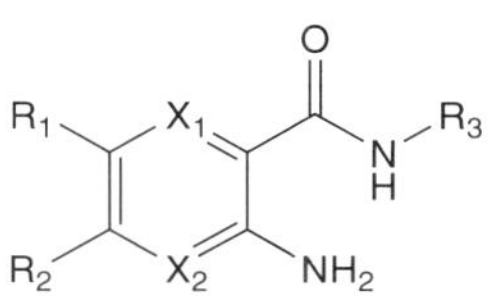

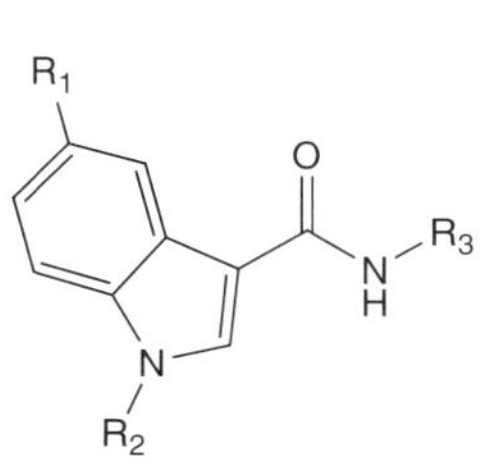

3.3. Miscellaneous

Recent patent applications have described novel chemical structures that do not fit into either of the above categories. Benzimidazoles (e.g. **30**) [40] and quinazolines (e.g. **31**) have been claimed as GKAs [41].

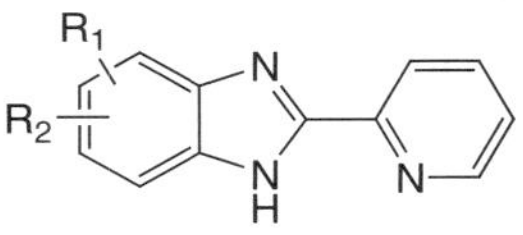

4. CONCLUSION AND OUTLOOK

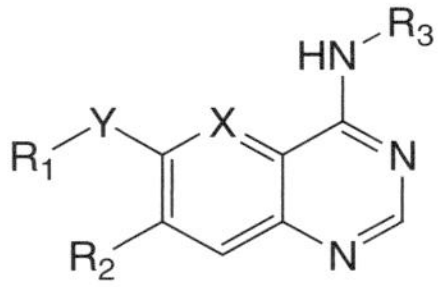

The vital role of GK in glucose homeostasis has been established for some time. Developments over the last decade, including the discovery of the regulatory function of GKRP, identification of maturity-onset diabetes of the young (MODY-2-), permanent neonatal diabetes (PNDM-) and (Persistent Hyperinsulinemic Hypoglycemia of Infancy) PHHI-related GK mutations, and the identification of novel GKAs and their co-crystal structures with GK, have radically improved our knowledge of GK structure and function. The structural categories of GKAs revealed thus far can be categorized as either carbon/nitrogen-centered, aromatic ring-centered activators, or a few special cases that do not fit into either category. These categories of GKAs seem to make comparable polar and non-polar interactions at the allosteric binding site, in spite of their considerable differences in structure. More significantly, in various animal models of T2D, GKAs have exhibited beneficial effects on glucose homeostasis, due to a dual effect of improving glucose utilization in the liver and glucose-dependent insulin secretion in pancreatic β-cells. Nevertheless, possible unwanted side effects, such as hypoglycemia, nausea, accumulation of fat in the liver, or even liver toxicity must be taken into consideration. The impact of GKA therapy on glycemic control in T2D patients and particularly their differentiation to the extensively prescribed sulfonylureas, is anxiously awaited.

REFERENCES

[1] F. M. Matschinsky, M. A. Magnuson, D. Zelent, T. L. Jetton, N. Doliba, Y. Han, R. Taub and J. Grimsby, *Diabetes*, 2006, **55**, 1.
[2] R. Sarabu and J. Grimsby, *Curr. Opin. Drug Disc. Dev.*, 2005, **8**, 631.
[3] F. M. Matschinsky, *Curr. Diabetes Rep.*, 2005, **5**, 171.
[4] T. Keitzmann and G. K. Ganjam, *Expert Opin. Ther. Patents*, 2005, **15**, 705.
[5] A. C. Storer and A. Cornish-Bowden, *Biochem. J.*, 1977, **165**, 61.
[6] K. E. Neet, R. P. Keenan and P. S. Tippett, *Biochem*, 1990, **29**, 770.
[7] J. Grimsby, R. Sarabu, W. L. Corbett, N. E. Haynes, F. T. Bizzarro, J. W. Coffey, K. R. Guertin, D. W. Hilliard, R. F. Kester, P. E. Mahaney, L. Marcus, L. Qi, C. L. Spence,

J. Tengi, M. A. Magnuson, C. A. Chu, M. T. Dvorozniak, F. M. Matschinsky and J. F. Grippo, *Science*, 2003, **301**, 370.
[8] K. Kamata, M. Mitsuya, T. Nishimura, J. Eiki and Y. Nagata, *Structure*, 2004, **12**, 429.
[9] E. Van Schaftingen, *Eur. J. Biochem.*, 1989, **179**, 179.
[10] M. D. Meglasson and F. M. Matschinsky, *Am. J. Physiol*, 1984, **246**, E1.
[11] P. Dunten, A. Swain, U. Kammlott, R. Crowther, C. M. Lukacs, W. Levin, L. Reik, J. Grimsby, W. L. Corbett, M. A. Magnuson, F. M. Matschinsky and J. F. Grippo, in *Glucokinase and Glycemic Disease – From Basics to Therapeutics* (eds F. M. Matschinsky and M. A. Magnuson), Karger AG, Basel, Germany, 2004, p. 145.
[12] A. M. Efanov, D. G. Barrett, M. B. Brenner, S. L. Briggs, A. Delaunois, J. D. Durbin, U. Giese, H. Guo, M. Radloff, G. S. Gil, S. Sewing, Y. Wang, A. Weichert, A. Zaliani and J. Gromada, *Endocrinology*, 2005, **146**, 3696.
[13] D. Zelent, H. Najafi, S. Odili, C. Buettger, H. Weik-Collins, C. Li, N. Doliba, J. Grimsby and F. M. Matschinsky, *Biochem. Soc. Trans.*, 2005, **33**, 306.
[14] J. Grimsby, F. M. Matschinsky and J. F. Grippo, in *Glucokinase and Glycemic Disease – From Basics to Therapeutics* (eds F. M. Matschinsky and M. A. Magnuson), Karger AG, Basel, Germany, 2004, p. 360.
[15] W. L. Corbett, 2nd Annual Mastering Medicinal Chemistry: Applying Organic Chemistry to Biological Problems, March, 2004, San Francisco, CA.
[16] A. L. Castelhano, H. Dong, M. C. T. Fyfe, L. S. Gardner, Y. Kamikozawa, S. Kurabayashi, M. Nawano, R. Ohashi, M. J. Procter, L. Qiu, C. M. Rasamison, K. L. Schofield, V. K. Shah, K. Ueta, G. M. Williams, D. Witter and K. Yasuda, *Bioorg. Med. Chem. Lett.*, 2005, **15**, 1501.
[17] F. T. Bizzaro, W. L. Corbett, J. F. Grippo, N. E. Haynes, G. W. Holland, R. F. Kester and R. Sarabu, *US Patent* 6 320 050 B1, 2001.
[18] F. T. Bizzaro, W. L. Corbett, J. F. Grippo, N. E. Haynes, G. W. Holland, R. F. Kester, P. E. Mahaney and R. Sarabu, *US Patent* 6 610 846 B1, 2003.
[19] G. R. Bebernitz, *WO Patent* 050645 A1, 2004.
[20] R. F. Kester and R. Sarabu, WO Patent, 08209 A1, 2002.
[21] F. T. Bizzaro, W. L. Corbett, A. Focella, J. F. Grippo, N. E. Haynes, G. W. Holland, R. F. Kester, P. E. Mahaney and R. Sarabu, *US Patent* 6 528 543 B1, 2003.
[22] W. L. Corbett, R. Sarabu and A. Sidduri, *US Patent* 6 353 111 B1, 2002.
[23] A. G. Weichert, D. G. Barrett, S. Heuser, R. Riedl, M. J. Tebbe and A. Zaliani, *WO Patent* 063194 A1, 2004.
[24] M. C. T. Fyfe, L. S. Gardner, M. Nawano, M. J. Procter, C. M. Rasamison, K. L. Schofield, V. K. Shah and K. Yasuda, *WO Patent* 072031 A2, 2004.
[25] R. A. Goodnow and L. E. Kang, *WO Patent* 83478 A2, 2001.
[26] K. R. Guertin, *WO Patent* 48106 A2, 2002.
[27] A. G. Weichert, D. G. Barrett, S. Heuser, R. Riedl, M. J. Tebbe and A. Zaliani, *WO Patent* 063179 A1, 2004.
[28] J. Lau, J. T. Kodra, M. Guzel, K. C. Santosh, A. M. Mjalli, R. C. Andrews and D. R. Polisetti, *WO Patent* 055482 A1, 2003.
[29] K. J. Brocklehurst, V. A. Payne, R. A. Davies, D. Carroll, H. L. Vertigan, H. J. Wightman, S. Aiston, I. D. Waddell, B. Leighton, M. P. Coghlan and L. Agius, *Diabetes*, 2004, **53**, 535.
[30] B. Leighton, A. Atkinson and M. P. Coghlan, *Biochem. Soc. Trans.*, 2005, **33**, 371.
[31] D. McKerrecher, J. V. Allen, S. S. Bowker, S. Boyd, P. W. R. Caulkett, G. S. Currie, C. D. Davies, M. L. Fenwick, H. Gaskin, E. Grange, R. B. Hargreaves, B. R. Hayter, R. James, K. M. Johnson, C. Johnstone, C. D. Jones, S. Lackie, J. W. Rayner and R. P. Walker, *Bioorg. Med. Chem. Lett.*, 2005, **15**, 2103.

[32] B. Leighton, A. M. Atkinson, D. Gill, G. J. Coope, H. Vertigan, P. Holme, C. Allott, C. Johnstone, D. McKerrecher and M. P. Coghlan, *Diabetes*, 2005, **54** (Suppl 1), Abs 2640-PO.
[33] B. R. Hayter, G. S. Currie, R. B. Hargreaves, R. James, C. D. Jones, D. McKerrecher, J. V. Allen, P. W. Caulkett, C. Johnstone and H. Gaskin, *WO Patent* 000267 A1, 2003.
[34] D. McKerrecher and J. W. Rayner, *WO Patent* 046139 A1, 2004.
[35] R. B. Hargreaves and C. D. Davies, *WO Patent* 045614 A1, 2004.
[36] M. Mitsuya, M. Bamba, F. Sakai, H. Watanabe, Y. Sasaki, T. Nishimura and J. Eiki, *WO Patent* 081001 A1, 2004.
[37] T. Nishimura, T. Iino, Y. Nagata and J. Eiki, *WO Patent* 080585 A1, 2003.
[38] W. L. Corbett, *US Patent* 0 067 939 A1, 2004.
[39] J. F. Lau, P. Vedso, J. T. Kodra, A. Murray and L. Jeppesen, *Eur. Patent EP* 1 532 980 A1, 2005.
[40] K. Nonoshita, Y. Ogino, M. Ishikawa, F. Sakai, H. Nakashima, Y. Nagae, D. Tsukahara, K. Arakawa, T. Nishimura and J. Eiki, *WO Patent* 063738 A1, 2005.
[41] M. Mitsuya, M. Bamba, Y. Sasaki, T. Nishimura, J. Eiki and K. Arakawa, *WO Patent* 090332 A1, 2005.

Renin Inhibitors

Colin M. Tice

Vitae Pharmaceuticals, 502 West Office Center Drive, Fort Washington, PA 19034, USA

Contents

1. Introduction	155
2. Biology	156
2.1. The renin–angiotensin system	156
2.2. Renin	156
2.3. Drugs targeting the RAS	157
2.4. Biomarkers and animal models	157
3. Classical renin inhibitors	158
4. Aliskiren	158
4.1. Aliskiren discovery	158
4.2. *In vivo* studies with aliskiren	160
5. Piperidines	161
6. Piperazines	162
7. Aminopyrimidines	163
8. Conclusion	164
References	164

1. INTRODUCTION

Hypertension is defined as systolic blood pressure of >140 mmHg or diastolic blood pressure of >90 mmHg. It affects about 25% of most populations and the prevalence is much higher in older people [1]. Hypertension occurs more frequently in patients suffering from insulin resistance, high LDL cholesterol, high triglycerides and obesity, and it is considered to be one facet of metabolic syndrome. Hypertension is also a risk factor for cardiovascular disease, including myocardial infarction, stroke and heart failure, and for renal disease [2–5]. Drugs available for the treatment of hypertension include diuretics, β-blockers, aldosterone receptor antagonists, angiotensin converting enzyme (ACE) inhibitors and angiotensin II receptor blockers (ARBs) alone and in combination. Nonetheless, hypertension is poorly controlled in many patients and the drugs prescribed may produce significant side effects. Renin has been recognized as a desirable target for antihypertensive drugs for almost four decades. Intensive efforts at many pharmaceutical companies in the 1980s, which led to the discovery of many potent inhibitors, have been reviewed [6–9]; however, no drug has reached the market. Over the past decade, new classes of renin inhibitors have been discovered and are progressing to market.

2. BIOLOGY

2.1. The renin–angiotensin system

In the classical rennin–angiotensin system (RAS), renin is synthesized in juxtoglomerular cells of the kidney in response to various signals and is secreted into the blood. Renin cleaves the protein angiotensinogen, produced primarily in the liver, liberating the physiologically inert decapeptide angiotensin I (Ang I). Ang I is further cleaved by ACE to the octapeptide angiotensin II (Ang II), which activates the Ang II type 1 receptor (AT_1). One of the downstream effects of AT_1 activation is a rise in blood pressure [10].

Further investigations of the RAS in recent years have uncovered greater complexity in the system. Ang II is subject to cleavage by a variety of peptidases to afford shorter peptides, which are also pharmacologically active. Furthermore, additional angiotensin receptors, AT_2, AT_3 and AT_4, which mediate different responses have been characterized. Ang II activates AT_2, and possibly AT_4, in addition to AT_1. Ang IV (the 3–8 peptide) activates AT_4. Evidence has also emerged for the presence of functioning local RASs in various tissues, including the heart [11–13]. Blocking these tissue RASs may be important in end-organ protection.

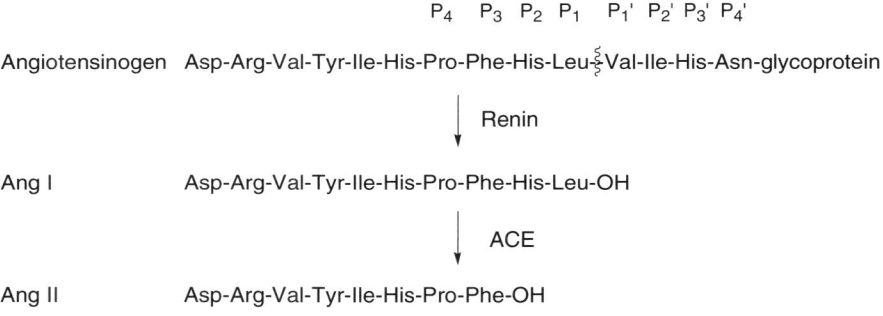

2.2. Renin

Renin [14] (EC 3.4.23.15) is a 44-kDa, 335-amino acid, glycosylated, monomeric aspartic protease. Renin is a member of the pepsin-like family [15]. Other members of this group include β-secretases 1 and 2, and cathepsins D and E. The renin gene is located on chromosome 1q32, spanning 12.5 kb of genomic DNA, and containing 10 exons and 9 introns. A number of SNPs have been identified but only one, in exon 9, yields an amino acid change (V351I). Two reports suggest an association between renin haplotype and essential hypertension [16,17]. The renin gene is translated into preprorenin, a protein with 401 amino acids. In the endoplasmic reticulum a 20-amino-acid signal peptide is cleaved to give prorenin, which is further processed by removal of an additional 46-amino-acid peptide in the Golgi apparatus to give renin itself [18].

Renin is a highly specific enzyme with no characterized activity other than cleavage of angiotensinogen to produce Ang I. Recently, one group has described a renin receptor but its physiological role is unknown [18,19]. Cathepsin D and tonin can also cleave angiotensinogen to Ang I; however, under normal conditions this activity is not pharmacologically relevant [20].

PDB contains two apo X-ray structures [21,22] of human renin (1bbs, 2ren) and four liganded structures [23–25] of human renin (1hrn, 1rne, 1bim, 1bil). Structure 1hrn has the best resolution at 1.8 Å. These structures reveal a binding site that interacts with residues P_4 to P_4' of angiotensinogen. Several of these structures show two conformations of the protein, termed open and closed, which differ mainly in the flap region (residues 72–81). Residues in the flap region form contacts with the P_3, P_2, P_1 and P_1' side chains of angiotensinogen [21]. Several additional X-ray structures of renin with bound inhibitors have also been described in the literature, but the coordinates are not publicly available [26–30].

2.3. Drugs targeting the RAS

Two prominent classes of antihypertensive drugs, the ACE inhibitors, e.g. benazepril, and the ARBs, e.g. valsartan, function by blocking the RAS. Despite the demonstrated efficacy of these drugs in controlling hypertension and reducing end-organ damage, both mechanisms of action have some disadvantages [10]. Complete inhibition of ACE does not prevent the conversion of Ang I to Ang II by other peptidases, including chymase. Indeed, in cardiac tissue, most Ang II is produced by enzymes other than ACE. In contrast, inhibition of bradykinin cleavage by ACE causes the side effects of coughing and angioedema in a substantial number of patients. Although the ARBs prevent binding of Ang II to AT_1, high levels of Ang II and its cleavage products remain in circulation and are available to activate AT_2, AT_3 and AT_4. Renin is anticipated to be a superior target for antihypertensives [31,32].

2.4. Biomarkers and animal models

In vivo studies of renin inhibitors have typically measured plasma renin activity and levels of Ang I and Ang II. Levels of all three components of the RAS should be reduced in a dose-dependent manner by an effective renin inhibitor.

Significant differences exist in the sequences and specificity of human and most non-primate renins and angiotensinogens. Human renin inhibitors are generally poor inhibitors of rat renin and this has precluded the use of traditional rat models of hypertension. Sodium-depleted primates, generally marmosets or cynomologous monkeys, have been the animal models of choice. The development of double transgenic rats (dTGR) [33] and mice [34], expressing both human renin and human angiotensinogen, has provided a valuable alternative. These animals are markedly hypertensive and suffer end-organ damage resulting in death in ~8 weeks. A range of tissue-specific mouse transgenes have been valuable in understanding the role of the RAS in end-organ damage. In addition to the RAS biomarkers mentioned above, studies in dTGR have also examined biomarkers for inflammation such as macrophage infiltration, which may be correlated with end-organ damage.

3. CLASSICAL RENIN INHIBITORS

Intensive programs during the 1980s led to the discovery of many potent peptidomimetic renin inhibitors [6–9,35] e.g. remikiren (RO 42-5892, **1**). During these efforts many of the now familiar peptide hydrolysis transition state isosteres, most of which position a hydroxyl group between the catalytic aspartates, were developed [36] and many potent inhibitors were discovered [6–9]. In **1**, the P_2 and P_3 side chains are identical to those in the natural substrate angiotensinogen, while the cyclohexylmethyl group replaces i-Bu at P_1. The molecule retains two peptidic secondary amides, has 4 stereocenters and its MW is 631. Remikiren has an IC_{50} of 0.7 nM against isolated renin and 0.8 nM in plasma; however, its oral bioavailability is $<1\%$. Stepwise progress was made in reducing the peptidic character of the inhibitors [37,38] but in most cases vestiges of the peptidic backbone and sidechains were readily apparent, molecular weights were >600 and oral bioavailability was inadequate. Compound **2** (BILA 2157 BS) is illustrative [39]: only a single secondary amide remains, the P_2 group has been modified, the P_3 group's attachment point and identity has been changed and the compound has 40% oral bioavailability in cynomologous monkeys; however, the molecular weight is 726 and the compound retains 4 stereocenters.

1

2

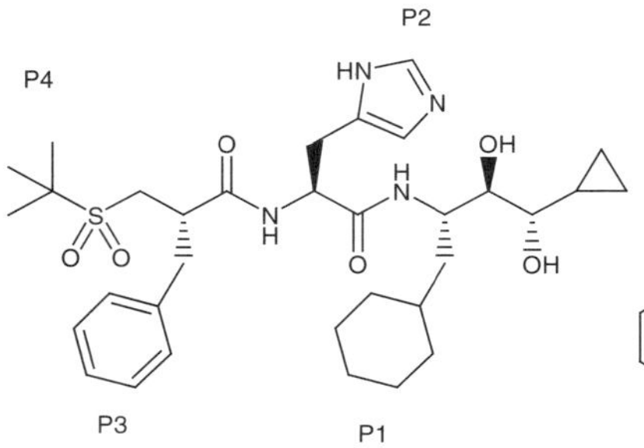

4. ALISKIREN

4.1. Aliskiren discovery

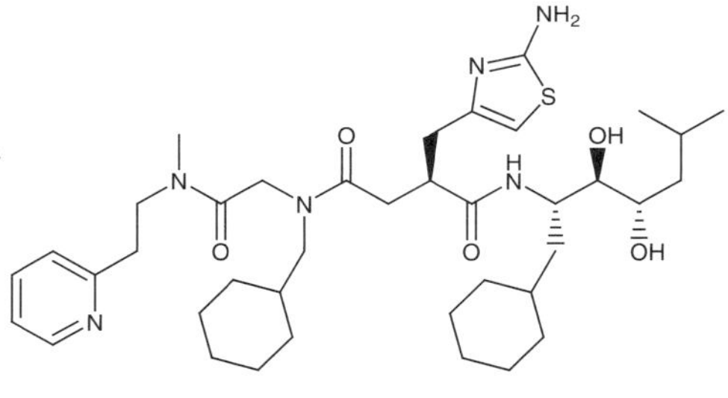

The publication of the first X-ray structures of renin revealed the presence of a subpocket adjacent to S_3, variously termed S_3^{aux} or S_3^{sp}, which is not filled by the natural substrate . SC-51106 (**3**) represents an early attempt to explore S_3^{sp} with a methyl group [26]. The binding mode was confirmed by X-ray crystallography. SC-51106 has an IC_{50} of 11 nM, compared to 70 nM for desmethyl analog SC-47921 (**4**).

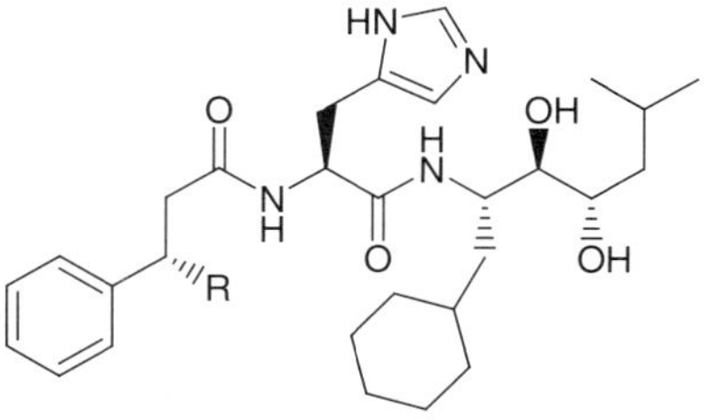

3 R = Me
4 R = H

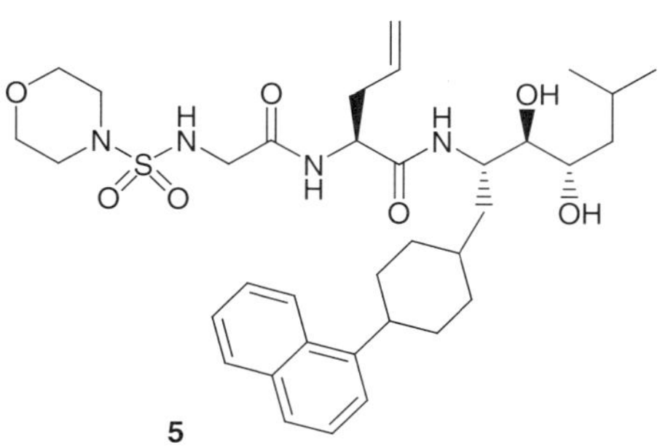

5

Examination of X-ray structures also made it clear that the S_1 and S_3 pockets of the enzyme are in fact contiguous and soon lead to attempts to fill S_3 by attaching a substituent to the P_1 group rather than from the peptide backbone [40–42]. Early compounds retained the peptide backbone with Gly replacing Phe. For example in **5**, which has an IC_{50} of 11 nM, S_1 and S_3 are filled by a (4-(1-naphthyl)cyclohexyl)methyl group [40,41].

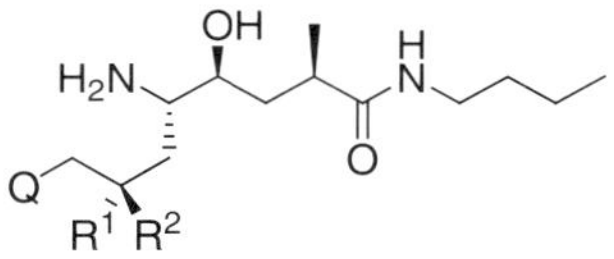

6a R^1 = H, R^2 = Me, Q = Ph
6b R^1 = Me, R^2 = Me, Q = 1-naphthylCH$_2$
6c R^1 = Me, R^2 = Me, Q = 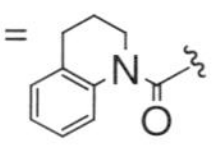
6d R^1 = Me, R^2 = Me, Q = 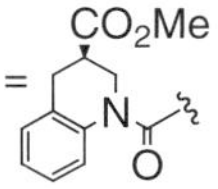
6e R^1 = H, R^2 = Me, Q =

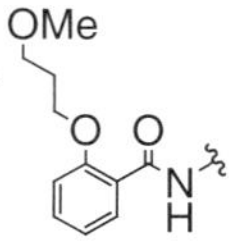

Subsequently, the peptide backbone was dispensed with entirely leading, initially, to weakly active compounds, e.g. **6a** $IC_{50} = 37 \mu M$ [43]. Optimization of this structural class, aided by X-ray crystallography, lead to several variations with improved potency: **6b** $IC_{50} = 700$ nM, **6c** $IC_{50} = 52$ nM, **6d** $IC_{50} = 8$ nM, **6e** $IC_{50} = 22$ nM [29,44,45]. Replacing the methyl group at R^2 in **6e** with an *i*-Pr group to give **7a** further increased potency to 0.9 nM. Application of this modification to **6a** and substitution of the phenyl ring with a 3-methoxypropoxy group at the 3-position and a methoxy group at the 4-position gave **7b** $IC_{50} = 0.4$ nM. Compounds **7a** and **7b** were less potent when assayed in the presence of plasma, with IC_{50}s of 4 and 3 nM, respectively. Replacement of the *n*-Bu group on the prime side with various polar groups was explored in an attempt to alleviate the loss of potency in the presence of plasma ultimately leading to the discovery of aliskiren (**8**, SPP100) which has $IC_{50}s = 0.6$ nM against both purified human renin and plasma renin [46].

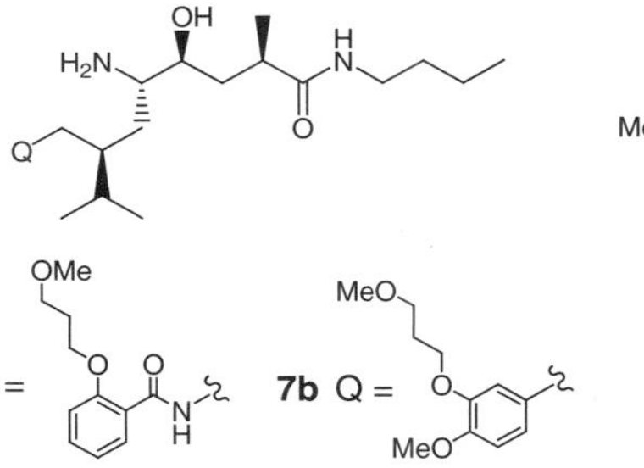

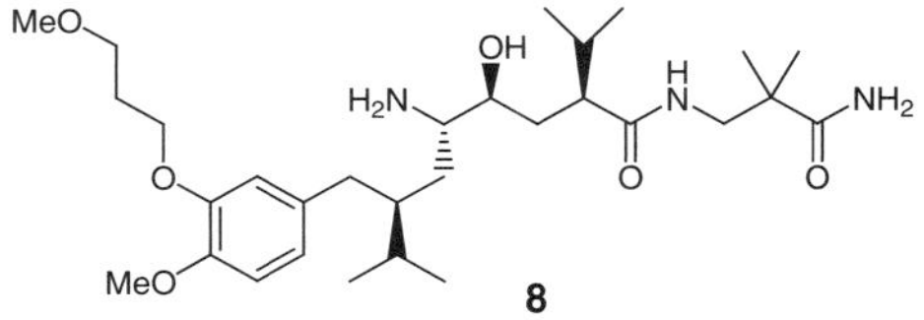

7a Q = [structure with OMe, benzamide linker]
7b Q = [structure with two MeO groups]

8 [aliskiren structure]

The MW of aliskiren is 552 and at pH 7.4 it has a log $P_{octanol/water}$ of 2.45 and water solubility of 350 mg/mL [46]. Its IC_{50} against porcine pepsin is >100 μM, a substantially greater margin of selectivity than was seen with classical peptidomimetic renin inhibitors. Aliskiren does not inhibit CYP isozymes at 20 μM but is a weak inhibitor of CYP3A4, CYP2C9 and CYP2D6 at 200 μM. The molecule has 4 acyclic stereocenters and is synthetically challenging. The original synthesis described in the patent literature is 17 steps [47]. A number of groups have published alternative routes [48–53].

More recently, prodrugs of aliskiren in which the alcohol is acylated have been described in the patent literature [54]. In addition, analogs based on a diaminopropanol core, e.g. **9a** $IC_{50} = 23$ nM and **9b** $IC_{50} = 71$ nM, have been reported by two groups [55,56].

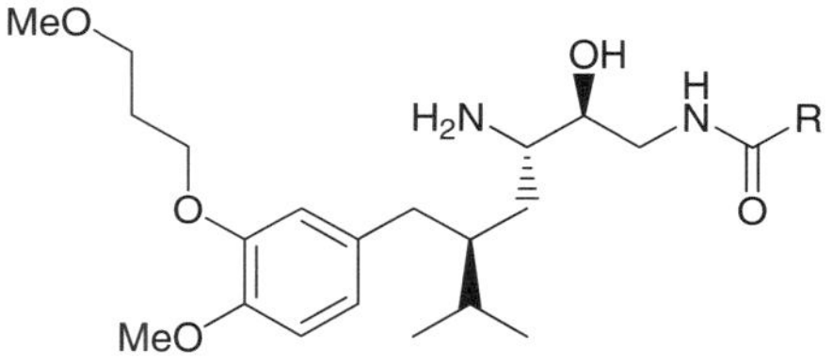

9a R = CH_2c-hex
9b R = CH_2Ph

4.2. *In vivo* studies with aliskiren

Aliskiren has been shown to lower blood pressure in sodium-depleted marmosets and in spontaneously hypertensive rats. In marmosets, a 10 mg/kg dose of aliskiren had an AUC of 49 mmol/l h and a mean $t_{1/2}$ of 2.3 h quantitated by biological activity [57]. Its calculated oral bioavailability was 16.3%. The compound's IC_{50} against marmoset renin is 2 nM and an oral dose of 3 mg/kg gave a 30-mmHg drop in blood pressure after 1 h and was more effective than the same dose of the older renin inhibitor **1**. An oral dose of 10 mg/kg of aliskiren was at least as effective as valsartan and benazepril. Combinations of aliskiren with valsartan or benazeprilat, designed to block the RAS at multiple sites, lowered blood pressure more effectively than the individual drugs but caused significant increases in heart rate. In spontaneously hypertensive rats, a 30 mg/kg dose of aliskiren had lower AUC, shorter $t_{1/2}$ and a calculated bioavailability of 2.4% [58]. Aliskiren is also effective in rats, however, higher doses were required, consistent with the aliskiren's reduced potency against rat renin ($IC_{50} = 80$ nM) and poorer pharmacokinetics. Subcutaneous administration of 0.3 mg/kg/day of aliskiren to dTGR lowered blood pressure by 63 mmHg and reduced albuminuria by 95% compared to untreated animals. All treated animals survived to 9 weeks and showed reductions in cardiac hypertrophy and other measures of cardiac damage and renal inflammation.

In healthy, normotensive volunteers on a constant sodium diet, aliskiren is well tolerated over 8 days of dosing and gives a dose-dependent reduction in plasma renin activity, Ang I and Ang II, while levels of renin in plasma increased 16- to 34-fold [59]. No effect on blood pressure was observed in these normotensive volunteers. In 226 patients with mild to moderate hypertension, alsikiren was again well tolerated and effected-dose dependent reductions in blood pressure and plasma renin activity [60]. A 150 mg/day dose of aliskiren afforded an 11-mm Hg mean reduction in daytime ambulatory systolic pressure and a 77% reduction in plasma renin activity. The effects on blood pressure were very similar to those seen with a 100-mg dose of the ARB losartan. In another study of patients with mild to moderate hypertension, 150 mg/day of alsikiren reduced mean sitting diastolic and systolic blood pressure by 9.3 and 11.4 mmHg, respectively [61]. The ARB irbesartan at 150 mg/day gave similar results. In studies of the effect of coadministration of aliskiren with warfarin [62], lovastatin, atenolol, celecoxib and cimetidine [63] no significant pharmacokinetic interactions were observed. The bioavailability of aliskiren has been reported to increase when cyclosporin D, an efflux protein inhibitor, is coadministered [64].

5. PIPERIDINES

High throughput screening uncovered trans-3,4-disubstituted piperidine **10a** as the first representative of a new structural class of inhibitors [27,28]. The (R,R) isomer of **10a** has an IC_{50} of 26 µM. A low resolution X-ray structure of the bromo analog **10b** bound to human renin suggested that the piperidine nitrogen was positioned near the two catalytic aspartates and that the 3-, 4- and 5-positions of the piperidine ring could be substituted. Analog synthesis led to the substantially more potent **10c**. The X-ray structure of **10c** bound to renin showed that the 2-naphthylmethyl group occupied the S_1 and S_3 pockets and the substituent at the 4-position of the piperidine ring occupied a new pocket formed by opening of the flap and substantial changes in the positions of Tyr75 and Trp39. Further modification of **10c** gave **11a** and **11b**, which has IC_{50}s of 87 pM and 12 nM against purified human renin and plasma renin, respectively. Introduction of a polar 3-methoxy-2-hydroxypropyl chain at the 5-position of the piperidine ring gave **11c** (RO0661132) with IC_{50}s of 67 pM and 8.9 nM against purified human renin and plasma renin, respectively. When dosed orally in sodium-depleted marmosets at 100 mg/kg, the compound lowered pressure about as effectively as cilazapril at 3 mg/kg. When dosed at 30 mg/kg/day in dTGRs, **11c** produced substantial blood pressure lowering and also reduced cardiac hypertrophy, albuminuria, Ang-II induced end-organ damage and pro-inflammatory responses.

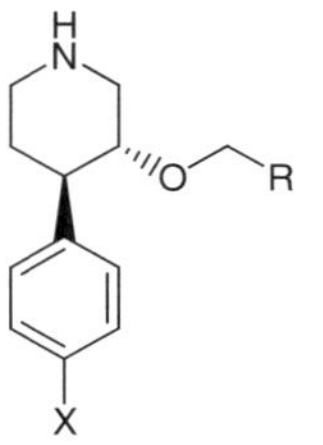

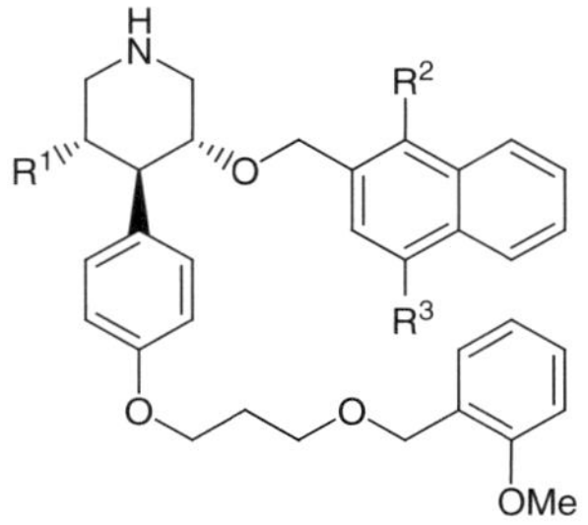

10a X = Cl, R = 4-MeO-Bn
10b X = Br, R = 4-MeO-Bn
10c X = 2-Cl-C_6H_4C(=O)CH_2, R = 2-naphthyl CH_2

11a R^1 = H, R^2 = H, R^3 = H
11b R^1 = H, R^2 = OMe, R^3 = OMe
11c R^1 = OCH_2CH(OH)CH_2OMe, R^2 = H, R^3 = OMe

Replacement of the naphthalene moiety of **11b** with a tetrahydroquinoline ring gave **12a** with IC_{50}s of 0.67 and 37 nM against purified human renin and plasma renin, respectively [65]. The nitrogen atom of the tetrahydroquinoline ring provided an attachment point for the 2-(acetylamino)ethyl chain of **12b** (RO0661168), which fills the S_3^{sp} pocket and improves potencies against both purified and plasma renin to 39 pM and 0.6 nM, respectively [66], albeit with pharmacokinetic properties inferior to **11c**. Both compounds were inhibitors of CYP3A4 [67]. Analogs of **11a** in which the substituted naphthylmethoxy substituent at the 3- position0 of the piperidine ring is replaced by a naphthylmethylamino [30] have been reported, e.g. **13** IC_{50} = 61 nM. Interestingly, in this series, the 3,4-*cis* stereochemistry is preferred. This series also inhibited CYP enzymes. Analogs in which the piperidine ring is bridged have been described in the patent literature: **14** is a preferred example [68].

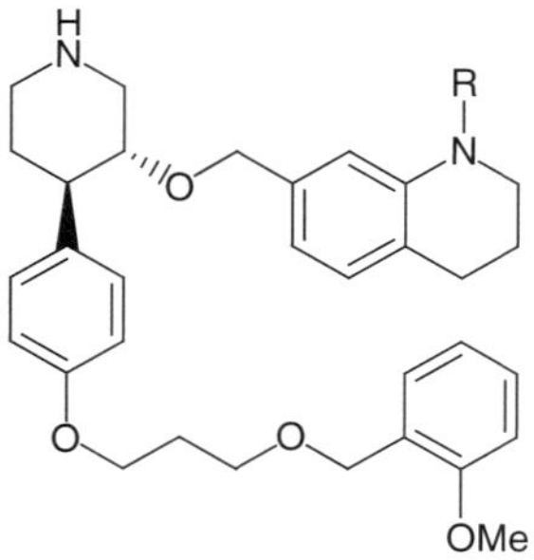

12a R = H
12b R = CH_2CH_2NHCOCH$_3$

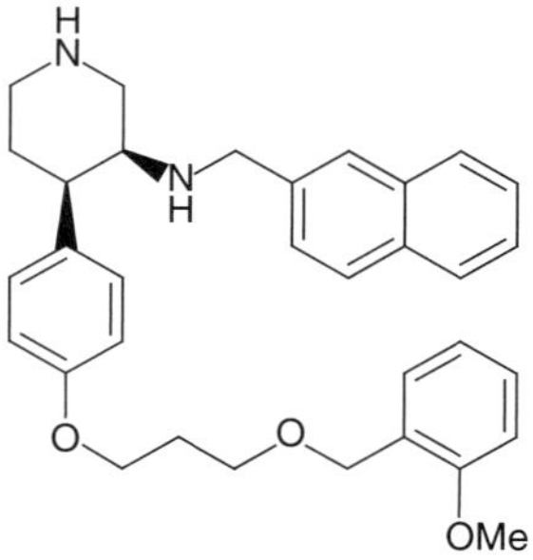

13

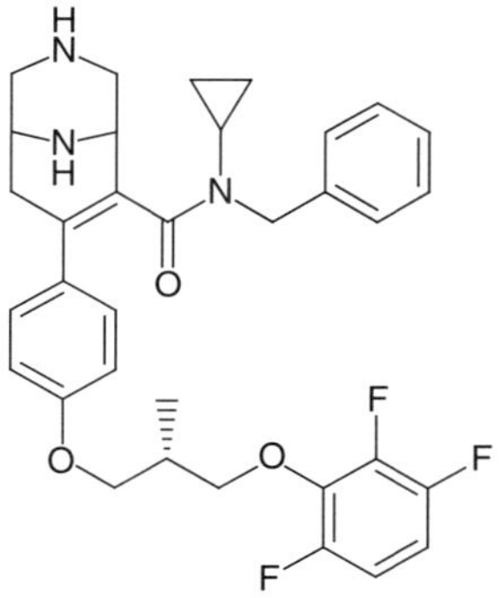

14

6. PIPERAZINES

Introduction of a second nitrogen into the piperidine ring of **11a** and moving the ether oxygen one position gave piperazine **15a** IC_{50} = 180 nM and ketopiperazine

15b $IC_{50} = 54$ nM [69,70]. Both these compounds were significantly less active than **11a** but provided a novel scaffold for optimization. Compound **15b** at 30 mg/kg was found to reduce blood pressure by up to 20 mmHg in double transgenic mice; however, blood pressure returned to baseline levels after 3 h. The most potent compounds in the series were **15c** and **15d** [71] with IC_{50}s of 0.30 and 0.18 nM respectively; however, at 3 μM, both **15c** and **15d** inhibit CYP3A4 >90%. In several cases, the R and S enantiomers in the ketopiperazine series were equipotent [72].

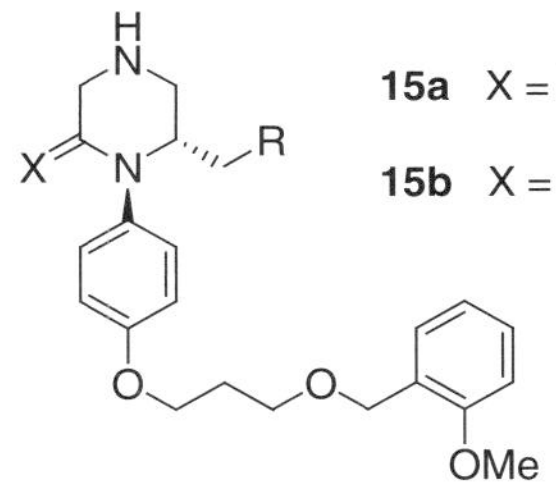

15a X = H_2, R = O-(2-naphthyl)
15b X = O, R = O-(2-naphthyl)
15c X = O, R = 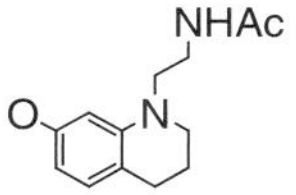
15d X = O, R =

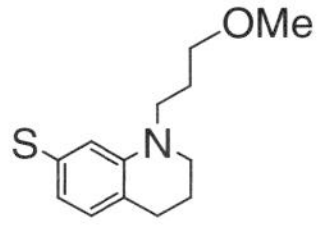

7. AMINOPYRIMIDINES

Optimization of a 2,4-diaminopyrimidine HTS hit using a combination of parallel synthesis and X-ray crystallography afforded lead compound **16** $IC_{50} = 655$ nM [73,74]. The 1-(3-methoxypropyl)-1,2,3,4-tetrahydroquinoline moiety of **16** was shown to occupy the S_3 and S_3^{sp} pockets of renin. Using NMR techniques, small molecules that could bind to the S_2 pocket were identified and linked to **16** to give **17** $IC_{50} = 1$ nM [75]. Further synthesis lead to **18** $IC_{50} = 1$ nM, $\%F = 8$), **19** ($IC_{50} = 48$ nM, $\%F = 74$) and **20** ($IC_{50} = 2$ nM, $\%F = 34$) [76,77].

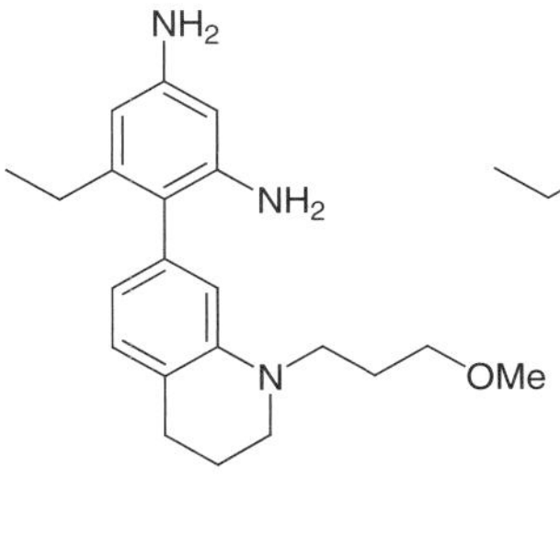

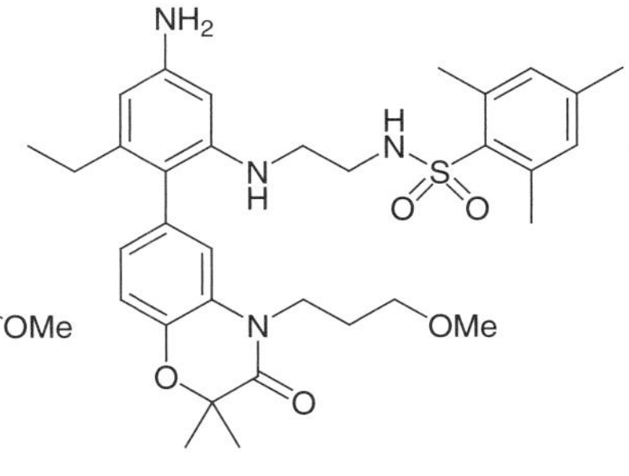

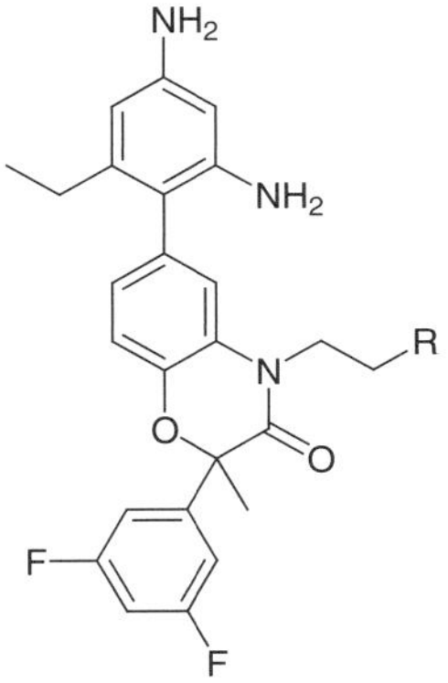

18 R = NHAc
19 R = CH_2CF_3, (S)isomer
20 R = $NHCO_2Me$

8. CONCLUSION

A decade after research into peptidomimetic renin inhibitors lost favor, several classes of non-peptidic inhibitors, including aliskiren, have been discovered. Potent examples of these classes contain basic amines that interact with the catalytic aspartates and structural elements that occupy S_3^{sp}. Structure-based drug design has played an influential role in the discovery of these compounds.

REFERENCES

[1] http://www.cdc.gov/nchs/data/hus/hus04trend.pdf#067.
[2] G. Cohuet and H. Struijker-Boudier, *Pharmacol. Ther.* 2006, in press.
[3] B. B. Scott, G. McGeehan and R. K. Harrison, *Curr. Protein Peptide Sci.*, 2006, in press.
[4] M. E. Cooper, *Am. J. Hypertens.*, 2004, **17**, 16S.
[5] K. Norris and C. Vaughn, *Expert Rev. Cardiovasc. Ther.*, 2003, **1**, 51.
[6] W. J. Greenlee, *Med. Res. Rev.*, 1990, **10**, 173.
[7] S. H. Rosenburg, *Prog. Med. Chem.*, 1995, **32**, 37.
[8] D. J. Hoover, B. A. Lefker, R. L. Rosati, R. T. Wester, E. F. Kleinman, J. S. Bindra, W. F. Holt, W. R. Murphy, M. L. Mangiapane and G. M. Hockel in *Aspartic Proteinases: Structure, Function, Biology and Biomedical Implications* (ed. K. Takahashi) Plenum Press, New York, 1995, pp 167–180.
[9] Rosenburg, S. H.; Boyd, S. A. in *Antihypertensive Drugs* (eds P. A. van Zweiten and W. J. Greenlee) 1997, pp 77–111.
[10] M. A. Zaman, S. Oparil and D. A. Calhoun, *Nat. Rev. Drug Disc.*, 2002, **1**, 621.
[11] R. M. Carey and H. M. Siragy, *Endoc. Rev.*, 2003, **24**, 261.
[12] K. Sakai and C. D. Sigmund, *Curr. Hypertens. Rep.*, 2005, **7**, 135.
[13] J. L. Lavoie and C. D. Sigmund, *Endocrinology*, 2003, **144**, 2179.
[14] H. L. Van Epps, *J. Exp. Med.*, 2005, **201**, 1351.
[15] B. M. Dunn, *Chem. Rev.*, 2002, **102**, 4431.
[16] B. Hasimu, T. Nakayama, Y. Mizutani, Y. Izumi, S. Asai, M. Soma, S. Kokubun and Y. Ozawa, *Hypertension*, 2003, **41**, 308.
[17] P. M. Frossard, A. M. Bokhari, S. H. Parvez and G. G. Lestringant, *Biogenic Amines*, 2005, **19**, 79.
[18] A. H. Jan Danser and J. Deinum, *Hypertension*, 2005, **46**, 1069.
[19] C. A. Burckle, A. H. Jan Danser, D. N. Muller, I. M. Carrelds, J. M. Jasc, E. Popova, R. Plehm, J. Peters, M. Bader and G. Nguyen, *Hypertension*, 2006, **47**, 552.
[20] R. H. Naseem, W. Hedegard, T. D. Henry, J. Lessard, K. Sutter and S. A. Katz, *Basic Res. Cardiol.*, 2005, **100**, 139.
[21] V. Dhanaraj, C. G. Dealwis, C. Frazao, M. Badasso, B. L. Sibanda, I. J. Tickle, J. B. Cooper, H. P. Driessen, M. Newman, C. Aguilar, S. P. Wood, T. L. Blundell, P. M. Hobart, K. F. Geoghegan, M. J. Ammirati, D. E. Danley, B. A. O'Connor and D. J. Hoover, *Nature*, 1992, **357**, 466.
[22] A. R. Sielecki, K. Hayakawa, M. Fujinaga, M. E. P. Murphy, M. Fraser, A. K. Muir, C. T. Carilli, J. A. Lewicki, J. D. Baxter and M. N. G. James, *Science*, 1989, **243**, 1346.
[23] L. Tong, S. Pav, D. Lamarre, L. Pilote, S. LaPlante, P. C. Anderson and G. Jung, *J. Mol. Biol.*, 1995, **250**, 211.
[24] J. Rahuel, J. P. Priestle and M. G. Grutter, *J. Struct. Biol.*, 1991, **107**, 227.

[25] L. Tong, S. Pav, D. Lamarre, B. Simoneau, P. Lavallee and G. Jung, *J. Biol. Chem.*, 1995, **270**, 29520.
[26] G. J. Hanson, M. Clare, N. L. Summers, L. W. Lim, D. J. Neidhart, H. S. Shieh and A. M. Stevens, *Bioorg. Med. Chem.*, 1994, **2**, 909.
[27] E. Viera, A. Binggeli, V. Breu, D. Bur, W. Fischli, R. Guller, G. Hirth, H. P. Marki, M. Muller and C. Oefner, *C. Bioorg. Med. Chem. Lett.*, 1999, **9**, 1397.
[28] C. Oefner, A. Binggeli, V. Breu, D. Bur, J. P. Clozel, A. D'Arcy, A. Dorn, W. Fischli, F. Gruninger, R. Guller, G. Hirth, H. P. Marki, S. Mathews, M. Muller, R. G. Ridley, H. Stadler, E. Vieira, M. Wilhelm, F. K. Winkler and W. Wostl, *Chem. Biol.*, 1999, **6**, 127.
[29] J. Rahuel, V. Rasetti, J. Maibaum, H. Rueger, R. Goschke, N. C. Cohen, S. Stutz, F. Cumin, W. Fuhrer, J. M. Wood and M. G. Grutter, *Chem. Biol.*, 2000, **7**, 493.
[30] W. L. Cody, D. D. Holsworth, N. A. Powell, M. Jalaie, E. Zhang, W. Wang, B. Samas, J. Bryant, R. Ostroski, M. J. Ryan and J. J. Edmunds, *Bioorg. Med. Chem.*, 2005, **13**, 59.
[31] A. Stanton, *J. Renin Angiotensin Aldosterone Syst.*, 2003, **4**, 6–10.
[32] N. D. L. Fisher and N. K. Hollenburg, *J. Am. Soc. Nephrol.*, 2005, **16**, 592.
[33] D. Ganten, J. Wagner, K. Zeb, M. Bader, J. B. Michel, M. Paul, F. Zimmermann, P. Ruf, U. Hilgenfeldt, U. Ganten, M. Kaling, S. Bachmann, A. Fukamizu, J. J. Mullins and K. Murakami, *Proc. Natl. Acad. Sci. USA*, 1992, **89**, 7806.
[34] A. Fukamizu, K. Sugimura, E. Takimoto, F. Sugiyama, M. S. Seo, S. Takahashi, T. Hatae, N. Kajiwara, K. Yagami and K. Murakami, *J. Biol. Chem.*, 1993, **268**, 11617.
[35] A. Himmelmann, A. Bergbrandt, A. Svensson, L. Hansson and M. Aurell, *Am. J. Hypertens.*, 1996, **9**, 517.
[36] M. G. Bursavich and D. H. Rich, *J. Med. Chem.*, 2002, **45**, 541.
[37] Y. Yamada, K. Ando, K. Komiyama, S. Shibata, I. Nakamura, Y. Hayashi, K. Ikegami and I. Uchida, *Bioorg. Med. Chem. Lett.*, 1997, **7**, 1863.
[38] G. L. Jung, P. C. Anderson, M. Bailey, M. Baillet, G. W. Bantle, S. Berthiaume, P. Lavallee, M. Llinas-Brunet, B. Thavonekham, D. Thibeault and B. Simoneau, *Bioorg. Med. Chem.*, 1998, **6**, 2317.
[39] B. Simoneau, P. Lavallee, P. C. Anderson, M. Bailey, G. Bantle, S. Berthiaume, C. Chabot, G. Fazal, T. Halmos, W. W. Ogilvie, M. A. Poupart, B. Thavonekham, Z. Xin, D. Thibeault, G. Bolger, M. Panzenbeck, R. Winquist and G. L. Jung, *Bioorg. Med. Chem.*, 1999, **7**, 489.
[40] M. Plummer, J. M. Hamby, G. Hingorani, B. L. Batley and S. T. Rapundalo, *Bioorg. Med. Chem. Lett.*, 1993, **1**, 2119.
[41] M. S. Plummer, A. Shahripour, J. S. Kaltenbronn, E. A. Lunney, B. A. Steinbaugh, J. M. Hamby, H. W. Hamilton, T. K. Sawyer, C. H. Humblet, A. M. Doherty, M. D. Taylor, G. Hingorani, B. L. Batley and S. T. Rapundalo, *J. Med. Chem.*, 1995, **38**, 2893.
[42] B. A. Lefker, W. A. Hada, A. S. Wright, W. H. Martin, I. A. Stock, G. K. Schulte, J. Pandit, D. E. Danley, M. J. Ammirati and S. F. Sneddon, *Bioorg. Med. Chem. Lett.*, 1995, **5**, 2623.
[43] S. Hanessian and S. Raghavan, *Bioorg. Med. Chem. Lett.*, 1994, **4**, 1697.
[44] V. Rasetti, N. C. Cohen, H. Rueger, R. Goschke, J. Maibaum, F. Cumin, W. Fuhrer and J. M. Wood, *Bioorg. Med. Chem. Lett.*, 1996, **6**, 1589.
[45] R. Goschke, N. C. Cohen and J. M. Wood, J Maibaum, *Bioorg. Med. Chem. Lett.*, 1997, **7**, 2735.
[46] J. M. Wood, J. Maibaum, J. Rahuel, M. G. Grutter, N. C. Cohen, V. Rasetti, H. Ruger, R. Goschke, S. Stutz, W. Fuhrer, W. Schilling, P. Rigollier, Y. Yamaguchi, F. Cumin, H. P. Baum, C. R. Schnell, P. Herold, R. Mah, C. Jensen, E. O'Brien, A. Stanton and M. P. Bedigian, *Biochem. Biophys. Res. Commun.*, 2003, **308**, 698.

[47] R. Goeschke, J. K. Maibaum, W. Schilling, S. Stutz, P. Rigollier, Y. Yamaguchi, N. C. Cohen and P. Herold, *Eur. Patent EP 678503-A1*, 1995.

[48] H. Rueger, S. Stutz, R. Goschke, F. Spindler and J. Maibaum, *Tetrahedron Lett.*, 2000, **41**, 10085.

[49] D. A. Sandham, R. J. Taylor, J. S. Carey and A. Fassler, *Tetrahedron Lett*, 2000, **41**, 10091.

[50] A. Dondoni, G. De Lathauwer and D. Perrone, *Tetrahedron Lett*, 2001, **42**, 4819.

[51] S. Hanessian, S. Claridge and S. Johnstone, *J. Org. Chem.*, 2002, **67**, 4261.

[52] R. Goschke, S. Stutz, W. Heinzelmann and J. Maibaum, *Helv. Chim. Acta*, 2003, **86**, 2848.

[53] H. Dong, Z. L. Zhang, J. H. Huang, R. Ma, S. H. Chen and G. Li, *Tetrahedron Lett*, 2005, **46**, 6337.

[54] H. Sellner, G. Gross, J. K. Maibaum and S. Cottens, *WO Patent 2005054177-A1*, 2005.

[55] Y. E. Bukhitiyarov, R. K. Harrison and B. B. Scott, PII 14, *4th General Meeting of the International Proteolysis Society,* Quebec, Canada, October 15–19, 2005.

[56] P. Herrold, S. Stutz, A. Stojanovic, V. Tschinke, C. Marti and M. Quirmbach, *WO Patent 2005070877-A1*, 2005.

[57] J. M. Wood, C. R. Schnell, F. Cumin, J. Menard and R. L. Webb, *J. Hypertension*, 2005, **23**, 417.

[58] B. Pilz, E. Shagdarsuren, M. Wellner, A. Fiebeler, R. Dechend, P. Gratze, S. Meiners, D. L. Feldman, R. L. Webb, I. M. Garrelds, A. H. J. Danser, F. C. Luft and D. N. Müller, *Hypertension*, 2005, **46**, 569.

[59] J. Nussberger, G. Wuerzner, C. Jensen and H. R. Brunner, *Hypertension*, 2002, **39**, e1.

[60] A. Stanton, C. Jensen, J. Nussberger and E. O'Brien, *Hypertension*, 2003, **42**, 1137.

[61] A. H. Gradman, R. E. Schmeider, R. L. Lins, J. Nussberger, Y. Chiang and M. P. Bedigian, *Circulation*, 2005, **111**, 1012.

[62] W. Dieterle, S. Corynen and J. Mann, *Brit. J. Clin. Pharm.*, 2004, **58**, 433–456.

[63] W. Dieterle, S. Corynen, S. Vaidyanathan and J. Mann, *Int. J. Clin. Pharm. Ther.*, 2005, **43**, 527.

[64] G. P. Camenisch, G. Gross, I. Ottinger and D. Wasmuth, *WO Patent 2006013094-A1*, 2006.

[65] R. Guller, A. Binggeli, V. Breu, D. Bur, W. Fischli, G. Hirth, C. Jenny, M. Kansy, F. Montavon, M. Muller, C. Oefner, H. Stadler, E. Vieira, M. Wilhelm, W. Wostl and H. P. Marki, *Bioorg. Med. Chem. Lett.*, 1999, **9**, 1403.

[66] H. P. Marki, A. Binggeli, B. Bittner, V. Bohner-Lang, V. Breu, D. Bur, P. Coassolo, J. P. Clozel, A. D'Arcy, H. Doebeli, W. Fischli, C. Funk, J. Foricher, T. Giller, F. Gruninger, A. Guenzi, R. Guller, T. Hartung, G. Hirth, C. Jenny, M. Kansy, U. Klinkhammer, T. Lave, B. Lohri, F. C. Luft, E. M. Mervaala, D. N. Muller, M. Muller, F. Montavon, C. Oefner, C. Qiu, A. Reichel, P. Sanwald-Ducray, M. Scalone, M. Schleimer, R. Schmid, H. Stadler, A. Treiber, O. Valdenaire, E. Vieira, P. Waldmeier, R. Wiegand-Chou, M. Wilhelm, W. Wostl, M. Zell and R. Zell, *Il Farmaco*, 2001, **56**, 21.

[67] H. P. Marki, MEDI-049, *ACS National Meeting*, New York, NY, September 7–11, 2003.

[68] O. Bezencon, D. Bur, W. Fischli, L. Remen, S. Richard-Bilstein, T. Weller and T. Sifferlen, *WO Patent 2004096116-A2*, 2004.

[69] D. D. Holsworth, N. A. Powell, D. M. Downing, C. Cai, W. L. Cody, J. M. Ryan, R. Ostroski, M. Jalaie, J. W. Bryant and J. J. Edmunds, *Bioorg. Med. Chem.*, 2005, **13**, 2657.

[70] N. A. Powell, E. H. Clay, D. D. Holsworth, J. W. Bryant, M. J. Ryan, M. Jalaie and J. J. Edmunds, *Bioorg. Med. Chem. Lett.*, 2005, **15**, 4713.

[71] D. D. Holsworth, C. Cai, X. M. Cheng, W. L. Cody, D. M. Downing, N. Erasga, C. Lee, N. A. Powell, J. J. Edmunds, M. Stier, M. Jalaie, E. Zhang, P. McConnell, M. J. Ryan, J. Bryant, T. Li, A. Kasani, E. Hall, R. Subedi, M. Rahim and S. Maiti, *Bioorg. Med. Chem. Lett.*, 2006, **16**, in press.
[72] N. A. Powell, E. H. Clay, D. D. Holsworth, J. W. Bryant, M. J. Ryan, M. Jalaie, E. Zhang and J. J. Edmunds, *Bioorg. Med. Chem. Lett.*, 2005, **15**, 2371.
[73] D. D. Holsworth, T. Belliotti, J. Bryant, C. Cai, W. L. Cody, D. M. Downing, M. Jalaie, A. Kasani, T. Li, S. Maiti, P. McConnell, N. A. Powell, M. Ryan, R. Subedi, E. Zhang and J. J. Edmunds, MEDI-087, ACS National Meeting, Atlanta, GA, March 26–30, 2006.
[74] J. J. Edmunds, T. Belliotti, J. Bryant, C. Cai, X. M. Cheng, F. L. Ciske, W. L. Cody, W. Collard, D. M. Downing, N. Erasga, S. Ferreira, E. Hall, D. D. Holsworth, M. Jalaie, A. Kasani, M. Kaufman, C. Lee, T. Li, S. Maiti, P. McConnell, K. Mennen, R. Ostroski, N. A. Powell, J. M. Rahim, M. Ryan, M. Stier, R. Subedi, C. Van Huis and E. Zhang, MEDI-195, ACS National Meeting, Atlanta, GA, March 26–30, 2006.
[75] D. D. Holsworth, M. Bury, W. L. Cody, D. Emerson, C. Van Huis, M. Jalaie, M. Kaufman, P. McConnell, N. A. Powell, P. Sahasrabudhe, R. W. Sarver, V. Thanabal and E. Zhang, MEDI-086, ACS National Meeting, Atlanta, GA, March 26–30, 2006.
[76] N. A. Powell, F. L. Ciske, C. Cai, W. L. Cody, D. M. Downing, D. D. Holsworth, K. Mennen, C. Van Huis, M. Jalaie, M. Ryan, J. Bryant, W. Collard, S. Ferreira, P. McConnell, E. Zhang and J. J. Edmunds, MEDI-089, ACS National Meeting, Atlanta, GA, March 26–30, 2006.
[77] N. A. Powell, F. L. Ciske, C. Cai, W. L. Cody, D. M. Downing, D. D. Holsworth, C. Van Huis, M. Jalaie, M. Ryan, J. Bryant, W. Collard, S. Ferreira and J. J. Edmunds, MEDI-090, ACS National Meeting, Atlanta, GA, March 26–30, 2006.

Antiarrhythmic Agents

Mark J. Suto and Douglas S. Krafte

Icagen Inc., 4222 Emperor Blvd, Suite 390, Durham, NC, USA

Contents

1. Introduction	169
2. Potassium channel modulators	169
2.1. I_{Kur} (Kv1.5) and atrial arrhythmias	170
2.2. I_{Ks} inhibitors	173
3. Other antiarrhythmic approaches	174
3.1. Sodium calcium exchanger	174
3.2. Adenosine A1 agonists and paroxysmal supraventricular tachycardia	175
4. Miscellaneous approaches	176
References	178

1. INTRODUCTION

The heart serves a relatively simple function in the body and acts as a mechanical pump to move blood throughout the vascular system. Underlying this simplicity, however, is a complex interplay of electrical activation, conduction and contraction that is necessary for normal cardiac function. Specialized cells in different regions of the heart have evolved to play specific roles ranging from pacemakers in the sinoatrial node to ventricular and atrial myocytes that regulate calcium entry and contraction. To achieve these specialized functions and maintain coordinated cardiac activity, cells express a range of different Na, Ca and K channels along with various ion exchangers and pumps [1]. While normal activity of these ion channels and pumps is essential for healthy functioning of the heart, aberrant activity can lead to a variety of arrhythmias, some of which can be life threatening [2]. A more complete understanding of the role these ion channels and pumps play in regulating normal and pathophysiological activity of the heart has led to new approaches in antiarrhythmic drug discovery that offer the possibility of better treatments in the coming years [3,4]. This review will focus on examples where new compounds have been identified and the approaches to arrhythmia management that are currently being investigated.

2. POTASSIUM CHANNEL MODULATORS

Potassium channel modulators have been investigated for years as a way to control both atrial and ventricular fibrillation. Many of the early clinical compounds were non-selective with respect to activity against I_{Kr}, I_{Ks} and I_{Kur} currents. Much of the recent work has focused on modulation of I_{Kur} through inhibition of Kv1.5. This approach lends itself to the observation that I_{Kur} is only found in atrial tissue and thus should produce a potential new and safer antiarrhythmic drug. However, there remains interest in continuing to explore more traditional Class III antiarrhythmic agents. A comprehensive review was published in 2003, which presents a historical perspective about cardiac voltage gated potassium channel modulators [5]. The paper highlights the structural differences in the various channels, the chemistry approaches pursued as well as the challenges these targets present. In addition, the emergence of Kv1.5 as the preferred target for the treatment of atrial fibrillation is discussed as well as the emerging structural studies that have been published and its relationship to the bacterial KcsA potassium channel. In addition, selectivity remains a challenge not only with the I_{Kr} and I_{Ks}, but also with other potassium channels.

2.1. I_{Kur} (Kv1.5) and atrial arrhythmias

Atrial fibrillation is the most common tachyarrhythmia in man and affects nearly 4% of the population over 60 years of age and 9% of the population over 80 years of age [4,6,7]. In data derived from a diverse population it was estimated that 1% of all adults are diagnosed with atrial fibrillation [5]. When atrial fibrillation is persistent, drug therapy is often required. However, most of the currently available drugs were not designed to treat atrial arrhythmias and can have effects on other regions of the heart [8]. Consequently, the search for newer and better molecular targets for drug discovery in this area continues.

An ideal molecular target for small molecule therapies to treat atrial fibrillation would be an ion channel important in regulating atrial electrical activity, but unlikely to play a similar role elsewhere in the heart. The ultrarapid delayed rectifier potassium current I_{Kur}, also known as Kv1.5, fits this description and is present in human atrial cells, but not in human ventricular myocytes [9]. This current activates rapidly with little inactivation and has been shown to play a role in regulating electrical activity of human atrial myocytes [10]. The Kv1.5 channel arises from the KCNA5 gene and expression studies confirm its selective expression in human atrial tissue [10,11]. Selective blockers of Kv1.5 would be expected to have little effect on ventricular electrical activity and offer the possibility for an improved therapeutic index compared to existing therapies.

The identification of Kv1.5 blockers for the treatment of atrial fibrillation has continued to dominate the scientific literature, particularly as the search for selective modulators progressed. Historically, it was found that a number of known drugs such as quinidine, clotrimazole, verapamil and nifedipine were Kv1.5 blockers. While all of these were non-selective, they did provide good chemical leads as evidenced by the fact that some of the first compounds developed were structurally related to these drugs.

A class of recent compounds that emerged from this early work was based upon a biphenyl core [12,13]. Several submicromolar compounds (**1–3**) were reported to be only threefold selective for Kv1.5 over Kv1.3, but had no significant activity against I_{Kr} currents (hERG). Compounds **2** and **3** were shown to prevent left atrial tachycardia (3 mg/kg, i.v.) without affecting QT-interval. An interesting structure–activity relationship observation was that the stereochemistry of the carbamate (**3**) had a dramatic effect on activity, whereby the S-enantiomer was 10-fold more potent than the R-enantiomer. No selectivity data of the enantiomers were given.

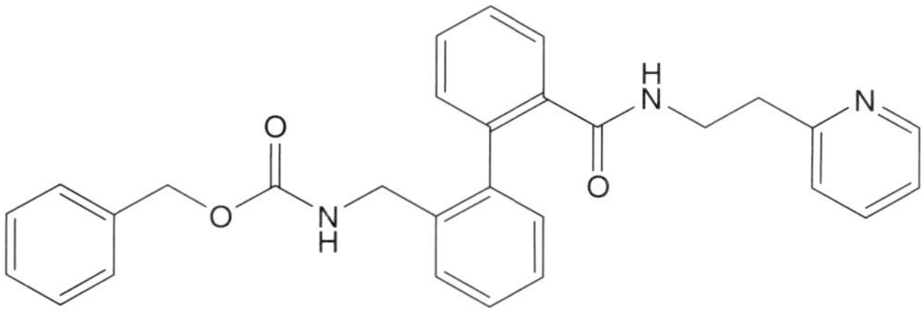

This work was extended to a series of anthranilic acid derivatives [14,15]. A pharmacophore model was developed, which was used to query an in-house compound data bank. The resulting compound **4** was less potent as a Kv1.5 blocker (IC$_{50}$ 5.5 µM), but had more favorable pharmacokinetic properties. A lead optimization program resulted in the 3-pyridinylmethyl analog **5**, which had submicromolar activity (IC$_{50}$ 0.7 µM) versus Kv1.5 and was inactive against hERG. In addition, **5** was 43% bioavailable in rats.

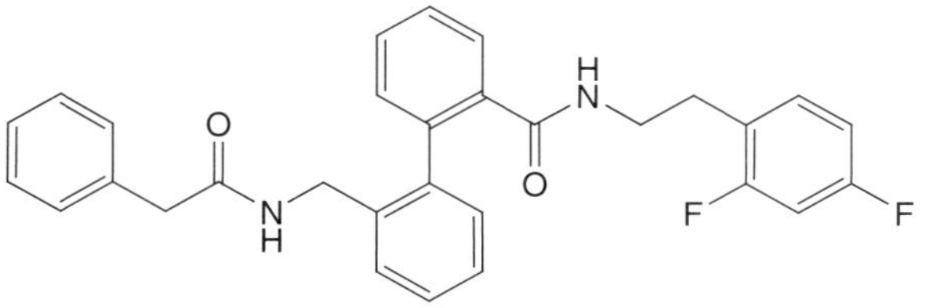

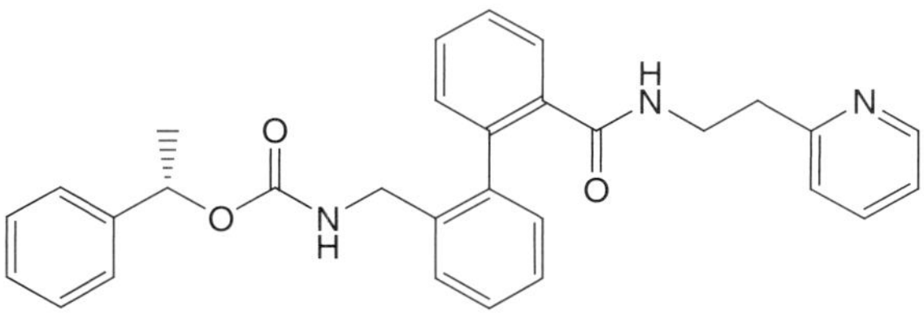

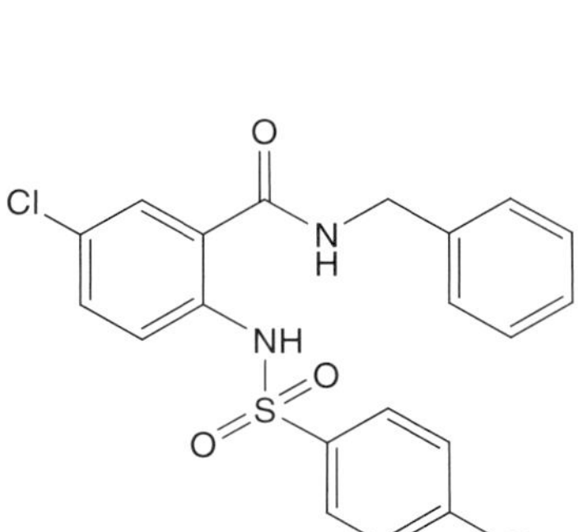

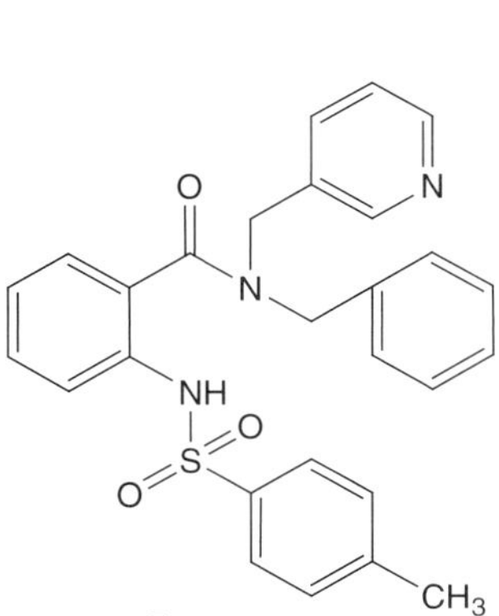

Computational models beyond the ligand and pharmacophore models were developed using the KcsA structure as a template [16]. The model was validated with historical data and then two putative binding sites were identified. The "internal site" was the focus based upon site-directed mutation studies. Virtual screening was then completed and 3102 hits were identified. The hits were then compared to a set of 29 known Kv1.5 blockers using topological descriptors. This resulted in identification of five new hits with significantly distinct molecular backbones, representing a successful scaffold hopping approach. The structures of these hits were not disclosed.

A series of substituted furanopyrimidine-based inhibitors of Kv1.5 were reported [17]. The most potent compounds (**6–9**) had submicromolar activity (>90% inhibition at 1 µM, in stably transfected CHO cells) and are structurally unique compared to previously reported Kv1.5 blockers.

Two related structural classes of indanes and tetrahydronaphthalenes (**10,11**) were reported as Kv1.5 inhibitors [18–20]. Multiple examples from each structural class were illustrated. All share a common motif and vary in the substituents found on the amide nitrogen. The activity for many of the compounds versus Kv1.5 was submicromolar with several analogs having IC_{50} values of <100 nM (>50% inhibition at 100 nM, CHO cells expressing human Kv1.5).

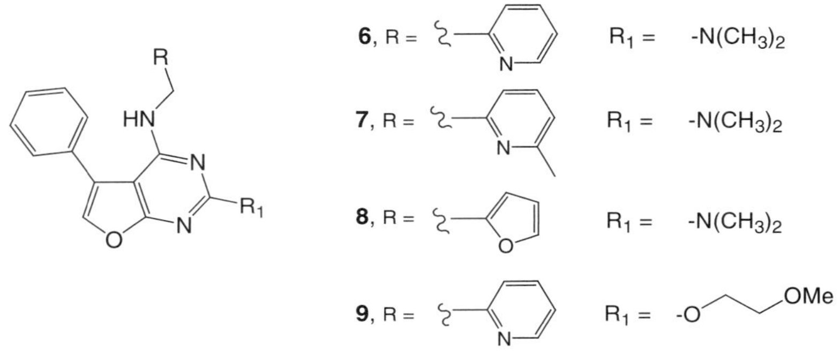

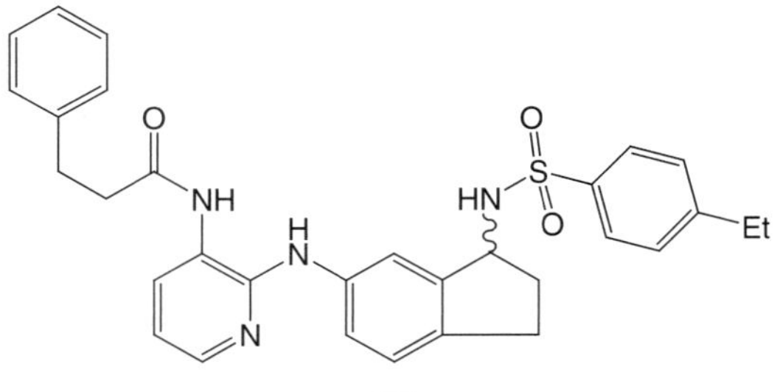

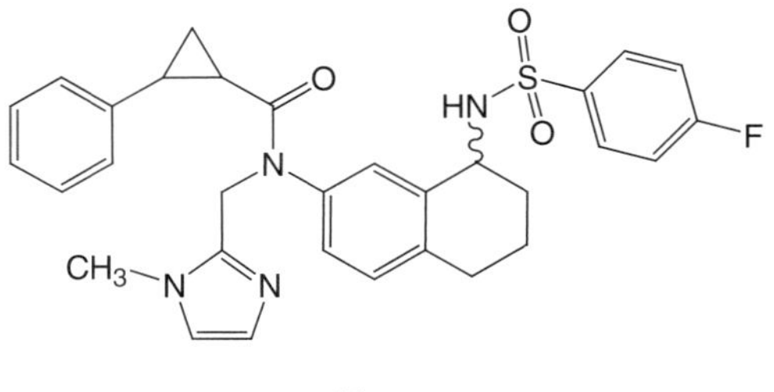

A series of compounds based upon chelidonine **12** was reported to inhibit Kv1.5 currents as well as I_{Kur} currents in human atrial myocytes [21]. The preferred compound, **13**, had an IC_{50} of 1.9 µM. The compounds were strongly voltage dependent between −30 and 0 mV, weakly voltage dependent between 0 and +60 mV and appeared to have stronger inhibitory activity versus the open state of the channel. In excised rabbit hearts, these compounds exhibited less antiarrhythmic activity at slower heart rates and therefore may provide antiarrhythmic efficacy with fewer side effects.

A limited series of isoquinolinones (**14,15**) were shown to block Kv1.5 as well as I_{Kur} currents [22]. The compounds all contained a 4-phenyl substituent and a 6-methoxy group, but contained a wide range of substituted amines and amides. The amide substituents ranged from simple dialkyl groups to complex alkyl and heteroalkyl amides. Of note is that a very simple N-methyl, N,N-dimethyl analog **14** was highlighted as most preferred. No data for the individual compounds were given, but several *in vitro* assays such as Kv1.5 expressed in CHO cells, human atrial myocytes, use-dependence and rate-dependent block were described.

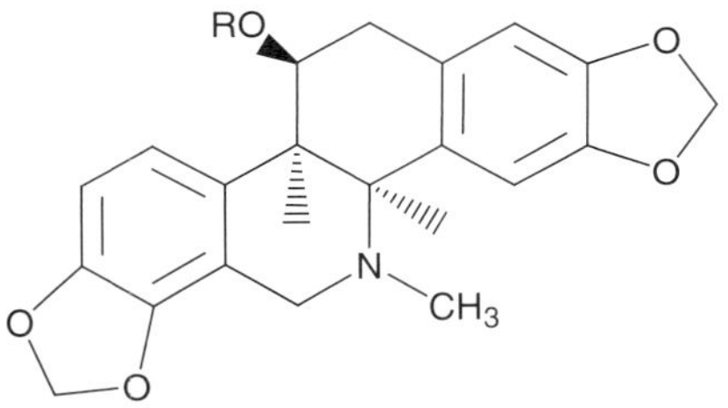

12 (Chelidonine), R=H
13, R = CONH$_2$

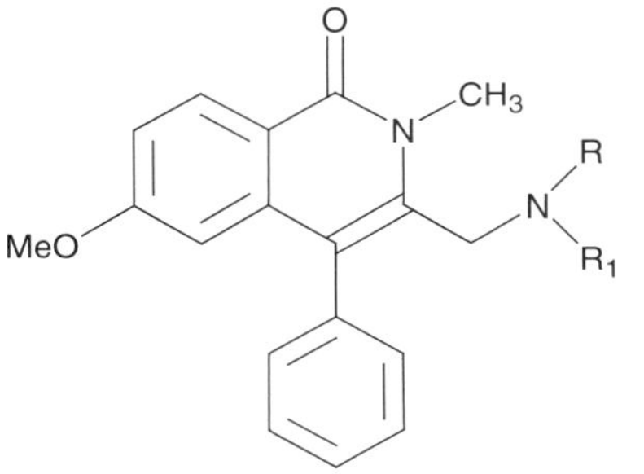

14, R$_1$ = R = CH$_3$
15, R$_1$ = H, R =

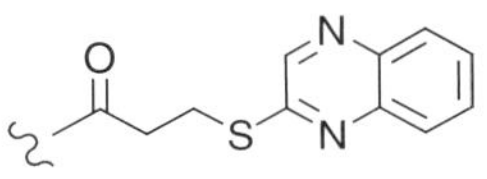

2.2. I_{Ks} inhibitors

Human genetics have helped to very clearly elucidate a number of key ion channels which regulate cardiac electrical activity, and mutations in several of the genes that encode for these ion channels form the basis of inherited arrhythmias [23]. Mutations in the KCNQ1 gene can lead to long QT syndrome (LQT1), as can mutations in the KCNH2 gene (LQT2). KCNQ1 and KCNH2 encode proteins responsible for I_{Ks} and I_{Kr} (hERG) currents, respectively, and these potassium channels have been studied in the cardiac electrophysiology arena for decades. Selective blockers of I_{Kr} have been shown to prolong both atrial and ventricular refractory periods, but are also pro-arrhythmogenic at low heart rates. I_{Kr} blockade is now routinely tested in all NCEs and is generally avoided. I_{Ks} blockers can also prolong cardiac action potential duration and QT intervals, and have been reported to show antiarrhythmic activity in canine models [24]. Despite the questions arising regarding the potential liabilities of any compound that prolongs QT interval as a function of a compound's mechanism of action, there continues to be additional investigation of combination blockers for I_{Kr} and I_{Ks} [25]. Such compounds may also have additional properties that could offset this liability, but it remains to be determined whether the benefits of this approach outweigh the risks.

A series of tetrahydronaphthalene derivatives were potent and selective blockers of I_{Ks} [26]. The series was derived from a screening hit **16**, which was further modified to provide **17**. This compound inhibits I_{Ks} with an IC_{50} of 0.037 μM and is 40-fold selective over I_{Kr}. Of note is that the *trans* derivative **17** is 10-fold more potent than the *cis* analog versus I_{Ks}. No activity versus I_{Kr} was reported for the *cis* isomer.

Another series of I_{Ks} blockers with excellent selectivity was reported [27]. The most potent compound, **18**, had an IC_{50} of 0.08 nM versus I_{Ks} and >100 nM versus I_{Kr}. This compound was also potent *in vivo* causing a significant increase in QT_c in anesthetized dogs at a dose of 0.001 mg/kg, i.v. Other parameters such as QRS interval, PR interval, conduction time and atrial excitation were unchanged. Some significant structure–activity relationships at the 5-position were observed, where the cyclopropyl analog was almost 100-fold more potent than the phenyl compound.

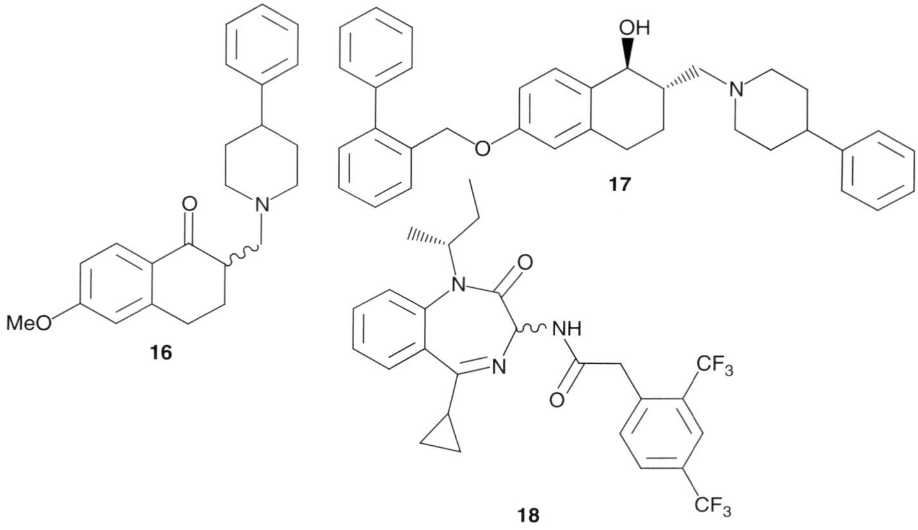

3. OTHER ANTIARRHYTHMIC APPROACHES

3.1. Sodium calcium exchanger

Heart failure results in decreased function of the heart as a pump. The prognosis is typically poor with approximately a 50% mortality rate over 5 years. In the more advanced stages of heart failure the prognosis is even worse [28]. Despite the progressive nature of the disease, half of all deaths occur suddenly presumably from the progression of ventricular tachycardia to ventricular fibrillation [29]. These arrhythmias may be triggered by delayed afterdepolarizations (DADs) or early afterdepolarizations (EADs) and both of these events are likely Ca-dependent.

Calcium handling is altered in myocytes from human heart failure patients [30] and one of the proteins that contribute to these alterations is the Na/Ca exchanger or NCX. The NCX families of proteins are the key exchangers in maintaining normal intracellular calcium levels. In the forward mode, NCX proteins transport calcium out of the cell and help maintain normal resting intracellular calcium levels. However, transport can be reversed under certain conditions and lead to enhanced calcium entry [31]. Complete block of NCX may be detrimental to the myocyte, but partial inhibition without calcium overload could have beneficial effects and this approach is being explored. NCX1 appears to be the dominant exchanger in the heart and, therefore, represents an interesting molecular target for the regulation of arrhythmias caused by aberrant calcium handling.

Compound **19** was shown to block the calcium exchanger, but suffered from activity against many of the critical cytochrome P450 enzymes [32]. Specific modifications to the pyridine moiety led to compounds with good inhibitory activity and reduced CYP liabilities. For example, **20** had an IC$_{50}$ value of 0.12 µM against NCX1.1 expressed in CCL39 cells and was orally active (ED$_{50}$ 2.9 mg/kg) in a rat model of ventricular tachycardia. The compound had a plasma half-life of 2.1 h and was 24% bioavailable. The addition of a 2-amino group to the pyridine improved the CYP inhibition profile considerably. Thus, compound **19** had IC$_{50}$ values against CYP 2C9, 2C19 and 2D6 of 0.1–0.6 µM. In contrast, the 2-amino compound **20**, had IC$_{50}$ values of > 50 µM for CYP 1A2 and 3A4, IC$_{50}$ values of 37 and 46 µM against 2C9 and 2C19 and an IC$_{50}$ value > 15 µM against 3A4, illustrating how small structural changes can lead to dramatic effects on CYP activity.

19, R = H
20, R = NH$_2$

3.2. Adenosine A1 agonists and paroxysmal supraventricular tachycardia

Paroxysmal Supraventricular Tachycardia (PSVT) has been reported to have a prevalence of over 500,000 patients in the general population [33]. Treatment is often achieved with AV-nodal blocking drugs to prevent spread to more life-threatening arrhythmias. Adenosine has been used clinically to terminate PSVT and also has the advantage of being very potent with a quick onset of action. However, since adenosine is an agonist for all adenosine receptor subtypes, A1, A2a, A2b and A3, its use can be associated with a variety of side effects. More recently, selective A1 agonists have been explored (e.g. [34,35]), which offer the potential for efficacy with reduction in side effects due to interactions with other adenosine receptor subtypes.

Adenosine, as well as selective A1 agonists, has a negative dromotropic effect on the AV node resulting in slowed conduction. This slowing conduction prevents the tachycardia from being transmitted from atrial to ventricular tissue. The ionic mechanism responsible for the action of adenosine and selective agonists has been reported to be G-protein coupled receptor-dependent stimulation of inward rectifier currents in AV-nodal cells [36]. Increases in inward rectifier currents would hyperpolarize the membrane potential and/or decrease the input conductance, both of which would dampen excitability in AV-nodal cells. The same investigators reported an indirect reduction in calcium currents, which could also contribute to the cellular effects of adenosine A1 agonists.

Adenosine receptor subtypes are widely distributed, and modulate many different processes [37]. In particular, there is a substantial body of pharmacological evidence to support the role of the A1 receptor in the regulation of a number of cardiovascular processes. For example, an A1 agonist can slow the electrical impulse propagation in the AV node resulting in a reduction of ventricular rate [38]. However, a full agonist may have associated liabilities such as high-grade AV block and therefore many newer compounds are partial A1 receptor agonists.

A series of 5'-carbamates and thiocarbamates of the full A1 agonist tecadenoson (**21**) such as compound **22** were prepared and shown to be partial A1 receptor agonists [38]. To improve metabolic stability and prevent *in vivo* generation of tecadenoson, a search for more stable derivatives resulted in compounds in which the 5'-hydroxy was substituted with a 3-fluorophenoxy substituent [39]. The resulting compound, **23**, had good affinity for A1 receptors and was a partial agonist in a guinea pig isolated heart model. It was shown that the type of substituent at the 5'-position had a dramatic effect on A1 affinity. *In vivo*, **23** was 81% bioavailable and had a plasma half-life of 1.6 h. However, upon further studies it was shown that a minor metabolite of **23** was indeed tecadenoson, however the consequences of the generation of this full agonist were not fully determined.

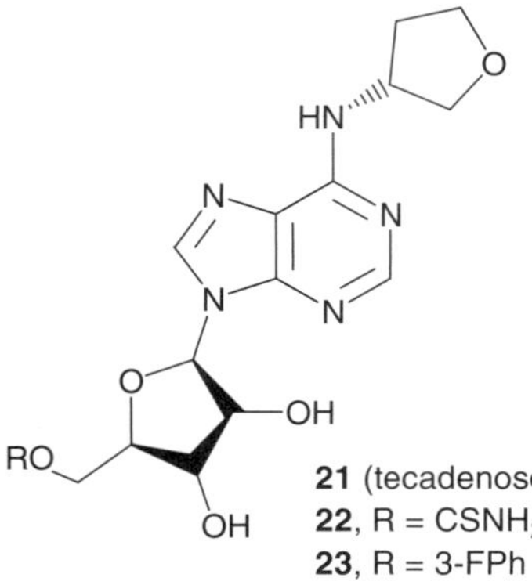

21 (tecadenoson), R = H
22, R = CSNH$_2$
23, R = 3-FPh

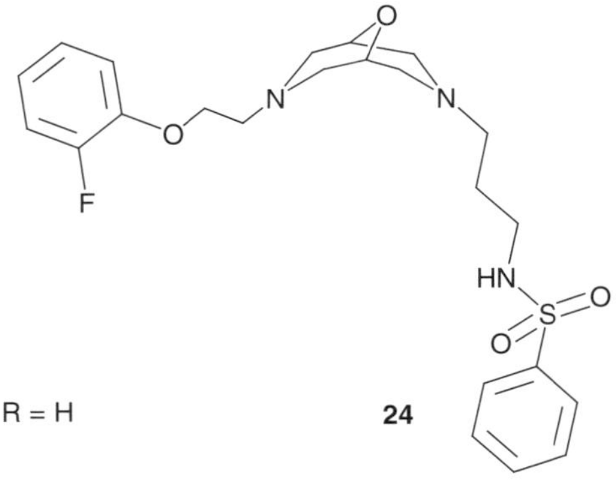

24

4. MISCELLANEOUS APPROACHES

Over the last year there have been several issued patents as well as applications that claim a variety of diverse structures having antiarrhythmic activity. Some fall into the Class III antiarrhythmic category and others have unknown or unreported

Antiarrhythmic Agents

mechanisms of action. However, from a medicinal chemistry perspective they provide some unique approaches for finding new drugs, several of which have been recently described in the literature [40,41].

The first examples are two patent applications describing a series of compounds based upon a 9-oxa-3,7-diazobicyclo[3.3.1]nonane backbone (oxabispidine) [42,43]. Interestingly, oxabispidine analogs were reported as antiarrhythmic agents several years ago, and have been used in metal coordination complexes. More recently reported compounds (i.e. **24**) are highly functionalized having a variety of substitutions on both nitrogen atoms. The antiarrhythmic data for these compounds are based upon *in vivo* studies looking at the prolongation of the monophasic action potential in anesthetized guinea pigs.

A group of benzopyran derivatives **25** related to cromakalim were disclosed, but the mechanism of their antiarrhythmic activity was not discussed [44]. However, the compounds were shown to prolong the refractory period for atrium muscle, but not ventricular muscle.

In several examples, simple chemical modifications to existing clinical compounds provided new antiarrhythmic agents. For example, evidence exists that tricyclic antidepressants such as imipramine and chlorpromazine induce self-defibrillation. However, these effects are seen at high doses resulting in therapeutic indices that are too low for use as antiarrhythmic agents. Modification of the side chain of imipramine to an amide (e.g. **26**) resulted in antiarrhythmic agents that were potent in a cat model of ventricular fibrillation (1–3 mg/kg i.v.) [45].

A recent US patent describes derivatives of amiodarone (**27**) for the treatment of ventricular fibrillation in patients with heart failure [46]. In the past, amiodarone has been reported to reduce arrhythmias in patients with congestive heart failure undergoing antihypertensive therapy. However, amiodarone has long-term toxicity related to its long half-life, drug–drug interactions and cardiac side effects. Therefore, analogs that are less lipophilic and contain a metabolically labile group should result in a Class III antiarrhythmic agent with significantly lower toxicity. Simple modification of the *n*-butyl group of amiodarone to incorporate an ester gave **28**, which was speculated to be quickly metabolized to a water-soluble compound resulting in a safer antiarrhythmic agent.

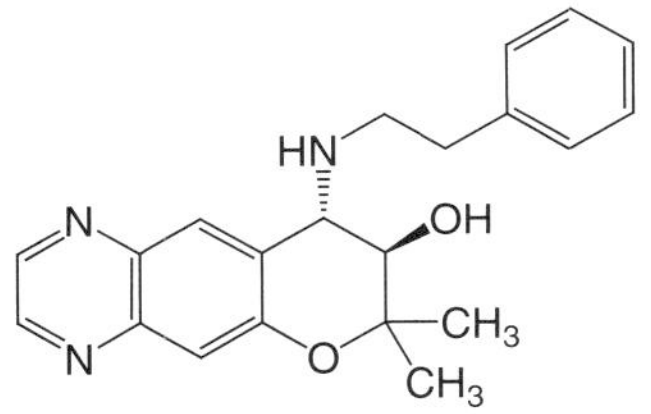

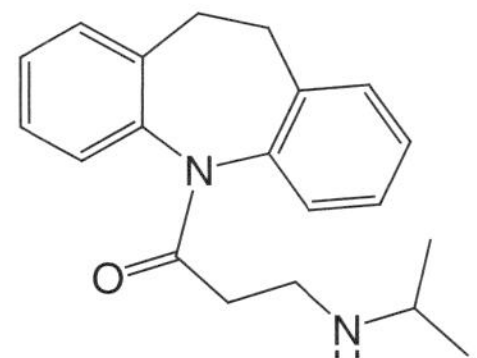

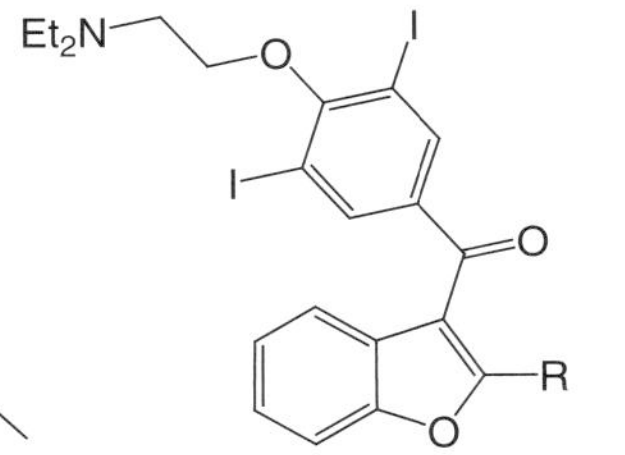

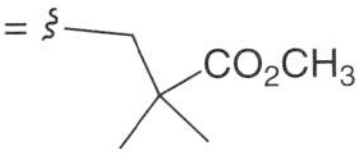

REFERENCES

[1] J. M. Nerbonne and R. S. Kass, *Physiol. Rev.*, 2005, **85**, 1205.
[2] M. Shah, F. G. Akar and G. F. Tomaselli, *Circulation*, 2005, **112**, 2517.
[3] A. Varro, P. Biliczki, N. Iost, L. Virag, O. Hala, P. Kovacs, P. Matyus and J. G. Papp, *Curr. Med. Chem.*, 2004, **11**, 1.
[4] R. L. Page and D. M. Roden, *Nat. Rev. Drug Discov.*, 2005, **4**, 899.
[5] J. Brendel and S. Peukert, *Curr. Med. Chem. Cardiovase. Hematol. Agents*, 2003, **1**, 273.
[6] A. S. Go, E. M. Hylek, K. A. Phillips, Y. Chang, L. E. Henault, J. V. Selby and D. E. Singer, *JAMA*, 2001, **285**, 2370.
[7] R. N. Goldstein and B. S. Stambler, *Prog. Cardiovasc. Dis.*, 2005, **48**, 193.
[8] V. Fuster, L. E. Ryden, R. W. Asinger, D. S. Cannom, H. J. Crijns, R. L. Frye, J. L. Halperin, G. N. Kay, W. W. Klein, S. Levy, R. L. McNamara, E. N. Prystowsky, L. S. Wann, D. G. Wyse, R. J. Gibbons, E. M. Antman, J. S. Alpert, D. P. Faxon, V. Fuster, G. Gregoratos, L. F. Hiratzka, A. K. Jacobs, R. O. Russell, S. C. Smith, Jr., W. W. Klein, A. Alonso-Garcia, C. Blomstrom-Lundqvist, G. de Backer, M. Flather, J. Hradec, A. Oto, A. Parkhomenko, S. Silber and A. Torbicki, *Circulation*, 2001, **104**, 2118.
[9] D. J. Snyders, M. M. Tamkun and P. B. Bennett, *J. Gen. Physiol.*, 1993, **101**, 513.
[10] M. M. Tamkun, K. M. Knoth, J. A. Walbridge, H. Kroemer, D. M. Roden and D. M. Glover, *FASEB J.*, 1991, **5**, 331.
[11] D. Fedida, B. Wible, Z. Wang, B. Fermini, F. Faust, S. Nattel and A. M. Brown, *Circ. Res.*, 1993, **73**, 210.
[12] S. Peukert, J. Brendel, B. Pirard, A. Bruggemann, P. Below, H. Kleemann, H. Hemmerle and W. Schmidt, *J. Med. Chem.*, 2003, **46**, 486.
[13] S. Peukert, J. Brendel, H. Hemmerle and H. Kleemann, *US Patent* 6,924,392, 2005.
[14] S. Peukert, J. Brendel, B. Pirard, C. Strubing, H. Kleemann, T. Bohme and H. Hemmerle, *Bioorg. Med. Chem. Lett.*, 2004, **14**, 2823.
[15] J. Brendel, T. Bohme, S. Peukert and H. Kleemann, *US Patent* 6,903,216, 2005.
[16] B. Pirard, J. Brendel and S. Peukert, *J. Chem. Info. Model*, 2005, **45**, 477.
[17] J. Ford, J. Palmer, F. Atherall and D. J. Madge, *US Patent* 0282829A1, 2005.
[18] S. Beaudoin, A. Reed and M. Gross, *US Patent* 6,849,634, 2005.
[19] M. Gross, A. Reed and S. Beaudoin, *US Patent* 6,858,610, 2005.
[20] A. Reed, M. Gross and S. Beaudoin, *US Patent* 6,858,611, 2005.
[21] J. Eun, Y. Kwak, D. Kim and S. Chae, *US Patent* 6,887,882, 2005.
[22] D. Claremon, C. McIntyre and N. Liverton, *US Patent* 6,870,055, 2005.
[23] D. M. Roden, *Trends Cardiovasc. Med.*, 2004, **14**, 112.
[24] U. Gerlach, *Curr. Med. Chem. Cardiovasc. Hematol. Agents*, 2003, **1**, 243.
[25] M. Carlson, *Expert Rev. Cardiovascu. Ther.*, 2005, **3**, 387.
[26] S. Ahmad, L. Doweyko, A. Ashfaq, F. Ferrara, S. Bisaha, J. Schmidt, J. DiMarco, M. Conder, T. Jenkins-West, D. Normandin, A. Russell, M. Smith, P. Levesque, N. Lodge, J. Lloyd, P. Stein and K. Atwal, *Bioorg. Med. Chem. Lett.*, 2004, **14**, 99.
[27] J. Butcher, N. Liverton, D. Claremon, R. freidinger, N. Jurkiewicz, J. Lynch, J. Salata, J. Wang, C. Dieckhaus, D. Slaughter and K. Vyas, *Bioorg. Med. Chem. Lett.*, 2003, **13**, 1165.
[28] D. W. Baker, D. Eisenstadter, C. C. Thomas and R. D. Cebul, *Am. Heart J.*, 2003, **146**, 258.
[29] S. M. Poqwizd and D. M. Bers, *Trends Cardiovasc. Med.*, 2004, **14**, 61.
[30] V. Piacentino, C. R. Weber, X. Chen, J. Weisser-Thomas, K. B. Margulies, D. M. Bers and S. R. Houser, *Circ. Res.*, 2003, **92**, 651.
[31] B. D. Quednau, D. A. Nicoll and K. D. Philipson, *Pflugers Arch*, 2004, **447**, 543.
[32] T. Kuramochi, A. Kakefuda, H. Yamada, I. Tsukamoto, T. Taguchi and S. Sakamoto, *Bioorg. Med. Chem.*, 2005, **13**, 4022.

[33] L. A. Orejarena, H. Vidaillet, F. DeStefano, D. L. Nordstrom, R. A. Vierkant, P. N. Smith and J. J. Hayes, *J. Am. Coll. Cardiol.*, 1998, **31**, 150.
[34] K. A. Ellenbogen, G. O'Neill, E. N. Prystowsky, J. A. Camm, L. Meng, H. D. Lieu, M. Jerling, R. Shreeniwas, L. Belardinelli and A. A. Wolff, *Circulation*, 2005, **111**, 3202.
[35] J. W. Cheung and B. B. Lerman, *Cardiovasc. Drug Rev.*, 2003, **21**, 277.
[36] A. C. Rankin, A. E. Martynyuk, A. J. Workman and K. A. Kane, *Can. J. Cardiol.*, 1997, **13**, 1183.
[37] S. Hutchinson and P. Scammells, *Curr. Pharm. Design*, 2004, **10**, 2021.
[38] V. Palle, V. Varkhedkar, P. Ibrahim, H. Ahmed, A. Li, Z. Gao, M. Ozeck, Y. Wu, D. Zeng, L. Wu, K. Leung, N. Chu and J. Zablocki, *Bioorg. Med. Chem. Lett.*, 2004, **14**, 535.
[39] C. Morrison, E. Elzein, B. Jiang, P. Ibrahim, T. Marquart, V. Palle, K. Shenk, V. Varkhedkar, T. Maa, L. Wu, Y. Wu, D. Xeng, I. Fong, D. Lustig, K. Leung and J. Zablocki, *Bioorg. Med. Chem. Lett.*, 2004, **14**, 3793.
[40] C. G. Wermuth, *J. Med. Chem.*, 2004, **47**, 1303.
[41] L. M. Lima and E. J. Barreiro, *Curr. Med. Chem.*, 2005, **12**, 23.
[42] A. Bjore, P. Bonn, U. Gran, J. Kajanus, C. Olsson and F. Ponten, *WO Patent* 05123748, 2005.
[43] A. Bjore, U. Gran and G. Stranlund, *WO Patent* 123747, 2005.
[44] K. Ohrai, Y. Shigeta, O. Uesugi, T. Okada and T. Matsuda, *WO Patent* 05090357, 2005.
[45] M. Erez, O. Levy and E. Keinan, *US Patent* 6 908 915, 2005.
[46] P. Druzgala, *US Patent* 6 869 972, 2005.

Section 3
Inflammatory, Pulmonary, and Gastrointestinal Diseases

Editor: Mark G. Bock
Novartis Institutes for BioMedical Research, Inc.
Cambridge
Massachusetts

Progress in Anti-SARS Coronavirus Chemistry, Biology and Chemotherapy

Arun K. Ghosh[1], Kai Xi[1], Michael E. Johnson[2], Susan C. Baker[3] and Andrew D. Mesecar[2]

[1]*Departments of Chemistry and Medicinal Chemistry, Purdue University, West Lafayette, IN, USA*
[2]*Center for Pharmaceutical Biotechnology and Department of Medicinal Chemistry and Pharmacognosy, University of Illinois at Chicago, IL, USA*
[3]*Department of Microbiology and Immunology, Loyola University Medical Center, IL, USA*

Contents

1. Introduction	183
2. SARS coronavirus proteases	184
3. SARS-CoV 3CLpro inhibitors	185
3.1. Covalent inhibitors of SARS-CoV 3CLpro	186
3.2. Non-covalent SARS-CoV 3CLpro inhibitors	187
3.3. SARS-CoV 3CLpro inhibitors from screening	190
4. PLpro inhibitors	191
5. Viral entry inhibitors	192
6. Miscellaneous inhibitors	193
7. Future outlook	194
Acknowledgments	194
References	194

1. INTRODUCTION

Severe acute respiratory syndrome (SARS) was first reported in Guangdong Province, China, in November 2002. Since then, it has spread to other Asian countries, North America and Europe. This outbreak reportedly affected more than 8000 individuals by July 2003, and resulted in 774 deaths. SARS is characterized by high fever, malaise, rigor, headache and non-productive cough or dyspnea and may progress to generalized interstitial infiltrates in the lung, requiring intubation and mechanical ventilation [1]. In an unprecedented response, the World Health Organization (WHO) called upon leading laboratories in the world and set up multi-center research to investigate the etiology of SARS and develop effective diagnostic tests. With the aid of modern information technologies, WHO set up a secured website to share emerging scientific information among several laboratories. The critical results including electron micrographs of viruses, sequences of genetic material, identification and characterization of the virus, virus isolates, samples from patients, and postmortem tissues were shared in real time. The goal was to pin down the causative agent for SARS and to develop diagnostic tools. Within months, a novel coronavirus was isolated from patients with SARS. Then within days, after comparison of the sequences of the coronavirus polymerase gene against all previously characterized strains, scientists concluded that this virus is distinct from all previously known human pathogens. Just over a month after the outbreak of the new illness, WHO announced that a new pathogen, a member of the coronavirus family never seen before in humans, is the cause of SARS [2–5]. In a remarkable scientific collaboration assembled by the WHO among teams of scientists from 13 laboratories in 10 countries, a novel coronavirus was identified as the etiological agent for SARS. Public health measures, including rapid identification of SARS cases and isolation of contacts, ultimately succeeded in controlling the 2002–2003 SARS epidemic. However, the identification of animal reservoirs for the virus and the possibility of re-emergence of epidemic or pandemic SARS provide strong motivation for the development of antiviral agents to treat this potentially fatal respiratory illness.

2. SARS CORONAVIRUS PROTEASES

Coronaviruses are a family of positive strand, enveloped RNA viruses that can cause respiratory, gastrointestinal and neurological diseases. Coronavirus virions are composed of a helical nucleocapsid surrounded by a lipid bilayer envelope studded with virus-specific glycoproteins [6]. The helical nucleocapsid structure contains the single-stranded, positive sense RNA genome surrounded by a nucleocapsid protein (N). The viral envelope contains the membrane glycoprotein (M), the envelope protein (E) and the spike glycoprotein (S).

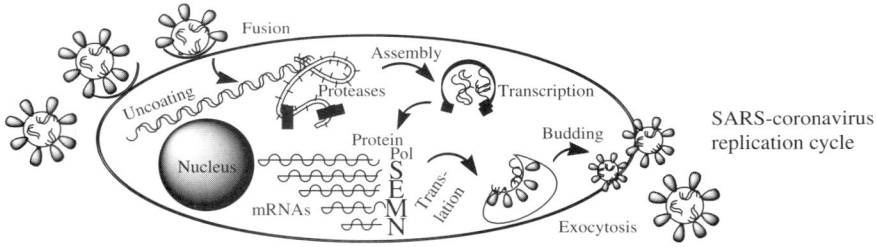

The SARS-CoV replicase is encoded in the 5′-most 21 kb of the ~29.7 kb viral genomic RNA. The genomic RNA is translated to produce two replicase polyproteins, termed pp1a and pp1ab [6,7]. The pp1a is a ~486 kilodalton (kDa) polyprotein that is predicted to contain a single papain-like protease (PLpro) analogous to the second murine coronavirus PLpro domain, a picornavirus 3C-like protease domain (3CLpro, also sometimes noted as Mpro or main protease), three putative membrane proteins, and several additional products of unknown function. The pp1ab (~790 kDa) is generated by ribosomal frameshifting and extends the pp1a product to include open reading frame 1b, which contains the core RNA polymerase and helical domains, and additional products of unknown function. The pp1a and pp1ab polyproteins are predicted to be processed to generate 16 protein products (termed non-structural proteins, nsp1–nsp16) [8], which assemble to form a membrane-associated viral replication complex.

3. SARS-COV 3CLPRO INHIBITORS

Proteolytic processing of the coronavirus replicase polyproteins is essential for on-going viral RNA synthesis. Therefore, the SARS-CoV proteases are attractive targets for the development of antiviral drugs to reduce viral replication and pathogenicity. The structure and activity of the coronavirus 3CLpro has already been elucidated and the design of inhibitors to 3CLpro as therapeutics has been proposed [9,10]. SARS-CoV 3CLpro has three domains: I (residues 8–101), II (residues 102–184), and III (residues 201–301). Domains I and II, which contain the active site region, are β-barrel domains and III is an α-helical domain. In the active site, a cysteine residue (Cys-145) acts as a nucleophile and a histidine residue (His-41) acts as the general acid base. The X-ray crystal structure of the related enzyme from porcine transmissible gastroenteritis coronavirus (TGEV 3CLpro) and a substrate-analogue hexapeptidyl chloromethyl ketone (CMK) inhibitor **1** (Cbz-Val-Asn-Ser-Thr-Leu-Gln-CMK) has been reported [8]. The sequence of this inhibitor was designed based upon P6 and P1 residues of the N-terminal autoprocessing site of TGEV 3CLpro. The corresponding sequences of SARS-CoV 3CLpro and HCoV-229E 3CLpro are Thr-Ser-Ala-Val-Leu and Tyr-Gly-Thr-Leu-Gln, respectively. The binding mode of this hexapeptidyl Gln inhibitor is similar to that which was observed for related human rhinovirus 3C protease (3Cpro) [9,11]. AG7088 (**2**), a prototype inhibitor of human rhinovirus 3Cpro [12] appears to bind to this enzyme in an orientation similar to the peptidyl CMK inhibitor in the binding site of TGEV 3CLpro [9,11]. Furthermore, substrate specificity of picornavirus 3Cpro for the P1-, P1'- and P4-sites is very similar to that of coronavirus 3CLpro. As a consequence, compounds **1** and **2** have become starting points for the design of SARS-CoV 3CLpro inhibitors.

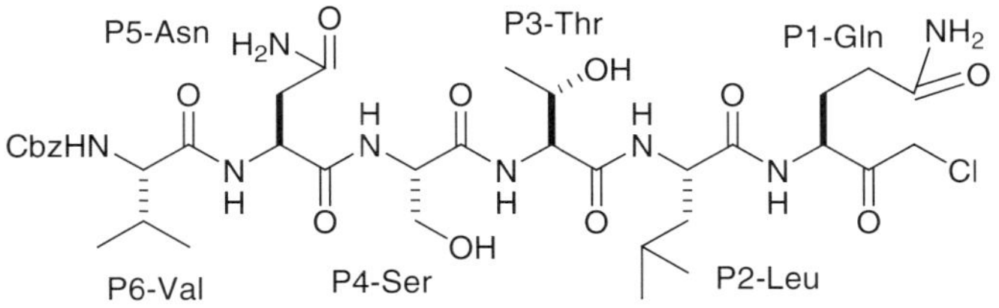

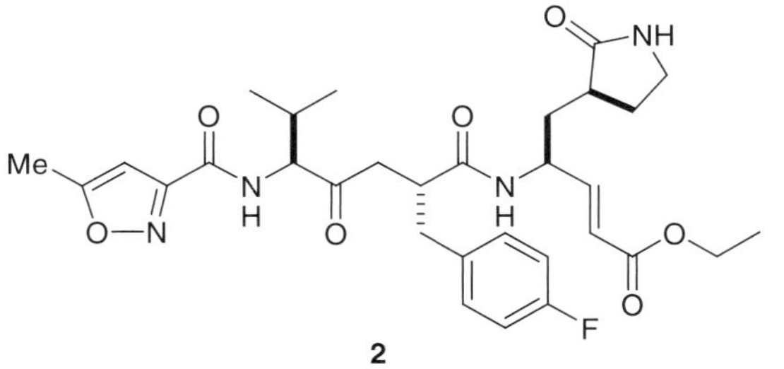

3.1. Covalent inhibitors of SARS-CoV 3CLpro

The design and synthesis of two analogues (**3,4**) of AG7088 (**2**) was recently reported [13]. Based upon the reported SARS-CoV 3CLpro structure [9,14], AG7088 was modified by changing the P2 side chain from a *p*-fluorobenzyl group to the smaller benzyl and prenyl groups. These inhibitors possess a P1/P1'- α,β-unsaturated ester functionality, which can covalently link to the Cys-145. Compound **2** is inactive against SARS-CoV in cell-culture assay. The antiviral activity for **2** was reported to be > 100 µg/ml [15]. The modified analogues are not only potent against SARS-CoV 3CLpro (k_{inact} values), but are effective in a SARS-CoV cell assay (IC_{50} values) as well. No toxicity was observed up to 100 µM. Moreover, an X-ray crystal structure of the SARS-CoV 3CLpro covalently linked with the synthetic small molecule inhibitor (**4**) was reported. In addition to the important covalent bond formed between **4** and the protease, the X-ray structure also showed crucial hydrogen bonding between **4** and His-164 and Glu-166. It revealed important insight into the molecular recognition of this type of inhibitor by SARS-CoV 3CLpro.

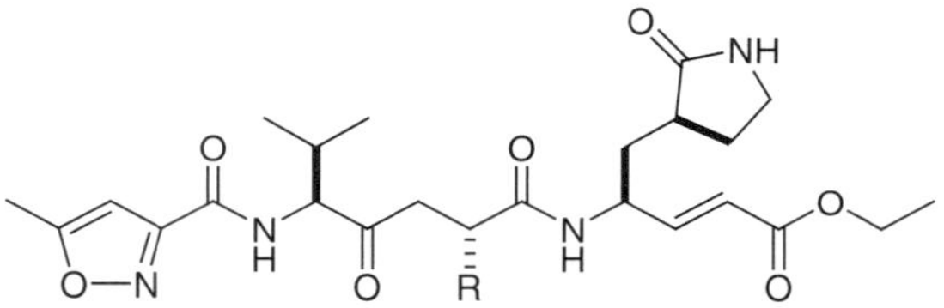

3 R = benzyl; K_i = 0.045 min^{-1}; IC_{50} = 45 µM
4 R = prenyl; K_i = 0.014 min^{-1}; IC_{50} = 70 µM

An X-ray crystal structure of a substrate-like aza-peptide epoxide (**5**) that inhibited the 3CLpro of SARS-CoV was published [16]. While these inhibitors are specific for clan CD cysteine peptidases [17], they are also lead candidates for the SARS-CoV 3CLpro, which has a Cys-145 and His-41 catalytic dyad in the active site. The best inhibition was obtained with the (*S,S*) diastereomer of compound **5** [k_{inact}/K_i = 1900($\pm$400) M/s]. The crystal structure revealed a covalent bond formed between the catalytic Cys-145 sulfur atom and the epoxide C3. Modeling studies of the four diastereomers binding to the SARS-CoV 3CLpro before nucleophilic attack by Cys-145 explained the necessity of the (*S,S*) configuration of the epoxide.

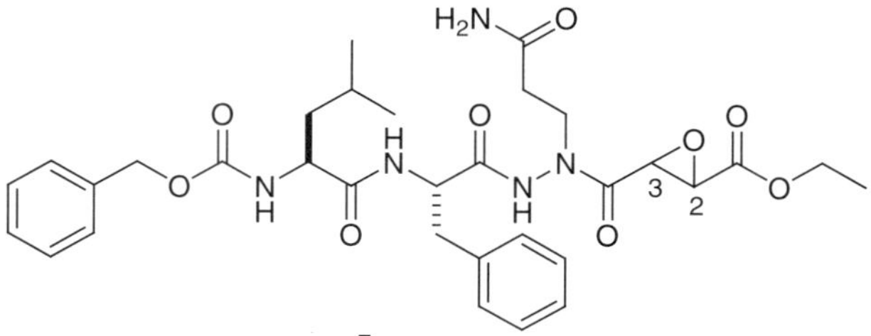

5

Another important series of covalent inhibitors was recently disclosed [18]. A series of tripeptide α,β-unsaturated esters and ketomethylene isosteres was assayed to target the SARS-CoV 3CLpro. The ketomethylene isosteres and tripeptide α,β-unsaturated esters containing both P1 and P2 phenylalanine residues show modest inhibitory activity (IC_{50} = 11–39 µM). The Phe–Phe dipeptide inhibitors were designed on the basis of computer modeling of the enzyme–inhibitor complex. The most potent inhibitor is compound **6** with an inhibition constant of 0.52 µM. The cell-based assays also indicate that this is a non-toxic anti-SARS agent with an EC_{50} value of 0.18 µM and an IC_{50} value of 1 µM. The computational study of structure–activity relationships shows that hydrogen bonding with the main chain Glu166 and the side chain Gln189 is crucial for inhibitory potency.

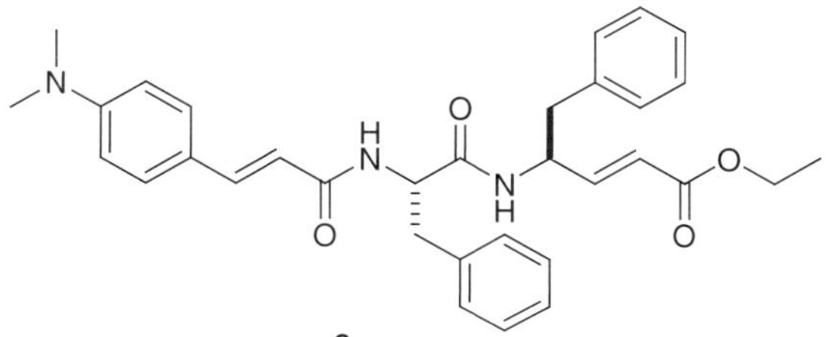

6

3.2. Non-covalent SARS-CoV 3CLpro inhibitors

Side effects and toxicity often arise with covalently bonded inhibitors which hinder or prevent their development as useful drug therapies [19,20]. To avoid such pitfalls, it is often desirable to design and develop non-covalent or reversible inhibitors as therapeutic agents.

A series of synthetic small molecule, non-covalent inhibitors of SARS-CoV 3CLpro was published in 2004 [21]. These investigators previously reported that keto-glutamine analogues with the phthalhydrazido group at α-position are reversible inhibitors of hepatitis A virus (HAV) 3C proteinase. The IC_{50} values of the inhibitors were in the low micromolar range [22,23]. They synthesized a series of keto-glutamine analogues with the phthalhydrazido group at the α-position and attachment of tripeptide (Ac-Val-Thr-Leu) as the inhibitors (**7–14**) for SARS-CoV 3CLpro. The combined effect of the β and β′ amino groups adjacent to the keto group and intramolecular hydrogen bonding to the carbonyl makes it more electrophilic. As a result, the carbonyl group can form a hemithioacetal with the sulfur of Cys-145. The K_i values of these reversible inhibitors remain to be determined.

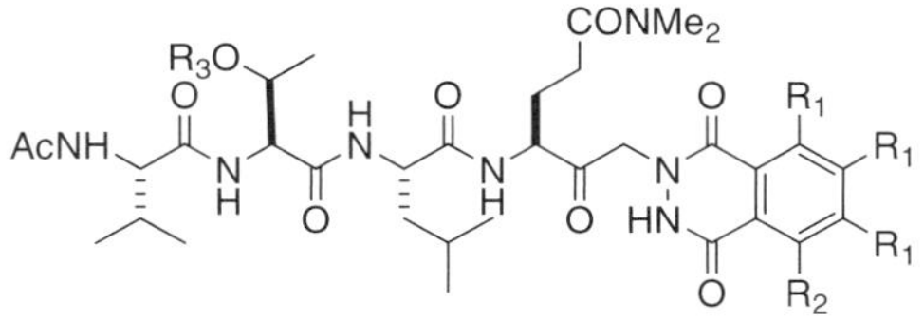

7 $R_1 = R_2 = H, R_3 = Bn$ $IC_{50} = 64$ µM
8 $R_1 = R_2 = R_3 = H$ $IC_{50} = 28$ µM
9 $R_1 = H, R_2 = NO_2, R_3 = Bn$ $IC_{50} = 70$ µM
10 $R_1 = R_3 = H, R_2 = NO_2$ $IC_{50} = 53$ µM

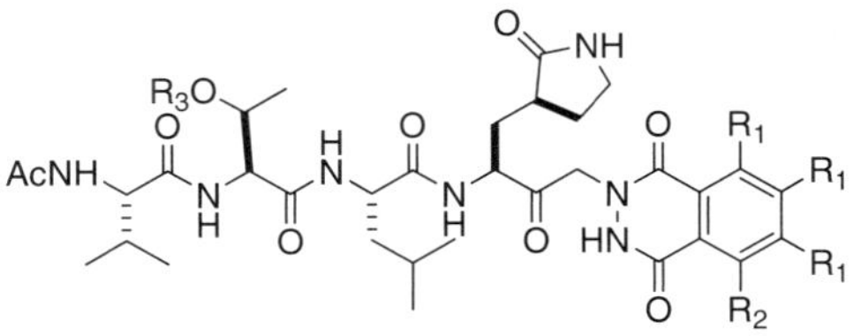

11 $R_1 = R_2 = H$, $R_3 = Bn$ $IC_{50} = 2.7$ μM
12 $R_1 = R_2 = R_3 = H$ $IC_{50} = 2.9$ μM
13 $R_1 = H$, $R_2 = NO_2$, $R_3 = Bn$ $IC_{50} = 0.60$ μM
14 $R_1 = R_3 = H$, $R_2 = NO_2$ $IC_{50} = 3.4$ μM

A diversified library of peptide anilides was prepared and their inhibitory activity was examined against the SARS-CoV 3CLpro by a fluorogenic tetradecapeptide substrate [24]. The most potent compound was **15** with a K_i value of 0.03 μM. Other analogues (**16–18**) were examined but showed substantially reduced inhibitory activity. The associated docking study showed that the dimethylamino, chloro and nitro groups that occupied the R_1, R_2 and R_3 positions, respectively, make important interactions with several residues of the SARS-CoV 3CLpro. These interactions are calculated to be responsible for a 9.1 kcal/mol energy difference between compound **15** and its analogues (**16–18**). Using the same assay methodology, the IC_{50} values of **15** against trypsin, chymotrypsin and papain were determined to be 110, 200 and 220 μM, respectively. Compound **15** is one of the most potent inhibitors of SARS-CoV 3CLpro reported to date and importantly, it is a competitive inhibitor which does not form covalent bonds with the protease.

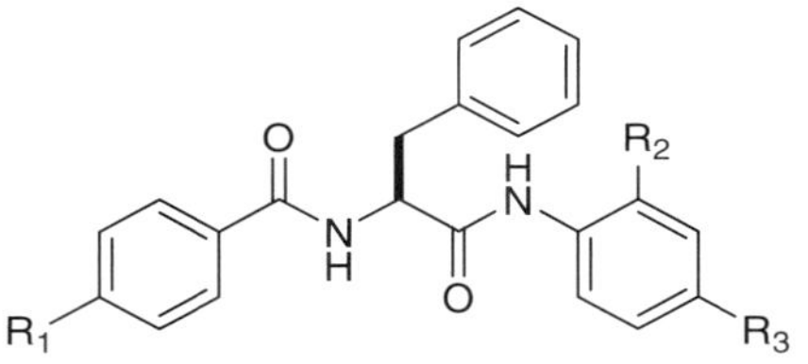

15 ~ 18

	R_1	R_2	R_3	IC_{50} (μM)
15	NMe₂	Cl	NO₂	0.06
16	H	Cl	NO₂	> 10
17	NMe₂	H	NO₂	> 10
18	NMe₂	Cl	H	> 10

A series of synthetic isatin derivatives were also reported as non-covalent SARS-CoV 3CLpro inhibitors [25]. It is known that certain isatin (2,3-dioxindole) compounds are potent inhibitors of rhinovirus 3Cpro [26]. Because SARS-CoV and rhinovirus have similar active sites and catalytic residues, isatin derivatives may be good candidates as SARS-CoV 3CLpro inhibitors. The SARS-CoV 3CLpro inhibition assays were conducted via fluorescence resonance energy transfer (FRET) according to the reported protocol [27]. These isatin derivatives inhibited SARS-CoV 3CLpro in the low micromolar range (0.95–17.5 μM). Among them, compounds **19** and **20** were the most potent inhibitors.

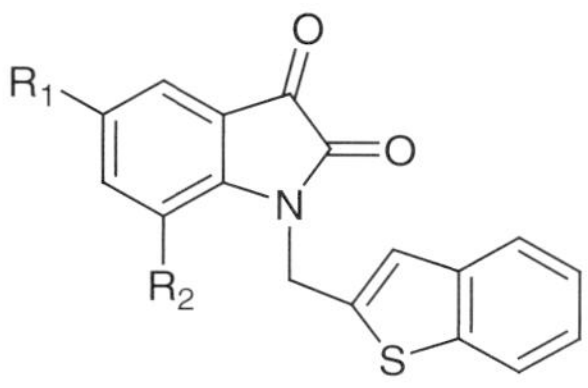

	R₁	R₂	IC$_{50}$ (µM)
19	H	Br	0.98
20	I	H	0.95

19, 20

Computer modeling showed that both compounds fit into the active pocket of SARS-CoV 3CLpro. The two carbonyl groups on isatin can form hydrogen bonds with the NH groups on Gly-143, Ser-144, Cys-145 and the His-41 side chain.

Analysis of the active site of SARS-CoV 3CLpro reveals the presence of a cluster of serine residues (Ser-139, Ser-144, Ser-147). A series of aryl boronic acid derivatives showed high binding affinities and have shown inhibition constants in low micromolar range [28]. Compound **21** with amide group linkages is the most potent.

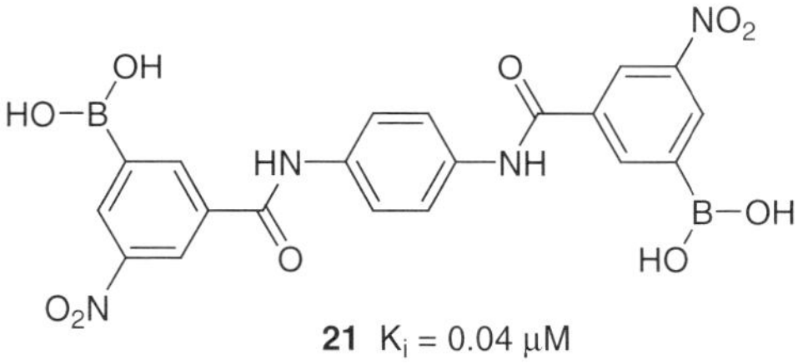

21 K$_i$ = 0.04 µM

Recently, a series of dipeptidyl fluoromethyl ketones were reported as SARS-CoV 3CLpro inhibitors [29]. The antiviral activity of these compounds was assessed by CPE inhibition in SARS-CoV infected Vero and CaCo-2 cultures. Compound **22** is the most potent inhibitor against SARS-CoV and showed low toxicity in cells. This compound exhibited a cellular-EC$_{50}$ value of 2.5 µM and a selectivity index of greater than 40.

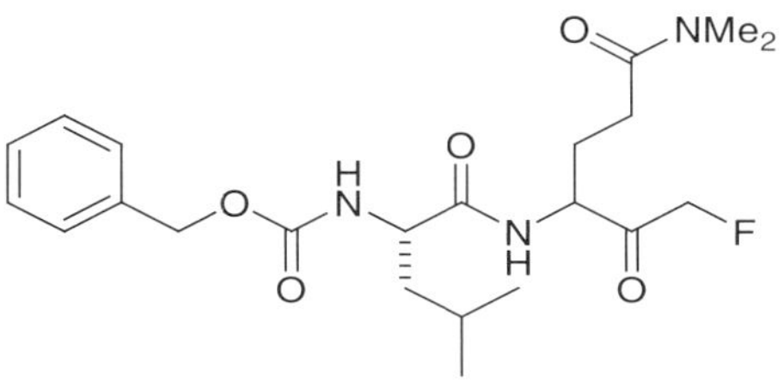

22

3.3. SARS-CoV 3CLpro inhibitors from screening

Extensive screening has been carried out in an effort to find structural leads against SARS-CoV 3CLpro from existing drugs. A major advantage is that approved drugs with minimal modifications may have the possibility of gaining accelerated approval by US Food and Drug Administration. It was reported that Kaletra, a mixture of protease inhibitors – Lopinavir and Ritonavir, approved for treating HIV in 2000, shows some effectiveness against the SARS virus [30]. Based on this observation, the binding affinities of six other drugs were investigated against SARS-CoV 3CLpro [31]. These include Lopinavir, Ritonavir, Niclosamide, Promazine and two other HIV inhibitors, PNU and UC2. The preliminary results indicated that these drugs could be useful as templates for designing SARS-protease inhibitors [32].

A collection of nearly 10,000 synthetic compounds and natural products was screened in an assay using SARS-CoV and Vero E6 cells [6,33]. For the SARS-CoV 3CLpro inhibiton assay, a C2-symmetric anti-HIV agent, compound **23** was found to inhibit SARS-CoV 3CLpro with a K_i value of 0.6 µM and showed a protective effect in the viral replication assay at a concentration of 10 µM [34]. The docking simulation of **23** showed that it is folded into a ring-like structure in the active site. Along similar lines, a compound library consisting of 960 commercially available drugs and biologically active substances was screened for inhibition of SARS-CoV 3CLpro [35]. Potent inhibition was achieved with the mercury-containing compounds thimerosal, phenylmercuric acetate and hexachlorophene in 1–10 µM range. Each compound inhibited viral replication in Vero E6 cell culture. Detailed mechanistic studies using a fluorescence-based protease assay demonstrated that the three compounds acted as competitive inhibitors (K_i = 0.7, 2.4 and 13.7 µM for phenylmercuric acetate, thimerosal and hexachlorophene, respectively). However, mercury-containing compounds pose toxicity problems. A panel of other metal ions including Zn^{2+} and its conjugates were also evaluated for their anti-SARS-CoV 3CLpro activities. Among these, 1-hydroxypyridine-2-thione zinc was shown to be the most potent competitive inhibitor (K_i = 0.17 µM). The addition of zinc-containing compounds such as (e.g. zinc acetate) as a supplement to the drug for Wilson's disease [36], suggests that zinc ion may be safe for human use.

23

Based on the concept of chemical genetics, 50,240 structurally diverse small molecules were screened yielding 104 compounds with anti-SARS-CoV activities [37]. Compound **24**, targeting SARS-CoV 3CLpro showed potent inhibitory activity with an IC_{50} = 2.5 µM and an EC_{50} of 7 µM in the Vero cell-based SARS-CoV plaque reduction assay. Another group of researchers, using a quenched FRET assay with a fully automated system screened 50,000 drug-like molecules, resulting in 572 hits [38]. After applying a series of virtual and experimental filters, five structurally novel molecules were identified which showed potent inhibitory activity (IC_{50} = 0.5–7 µM) against SARS-CoV 3CLpro. The inhibitory activities of the five compounds against four different proteases (HAV 3Cpro, NS3pro, chymotrypsin and papain) were also examined with the result that two compounds (**25** and **26**) showed apparent selectivity for SARS-CoV 3CLpro.

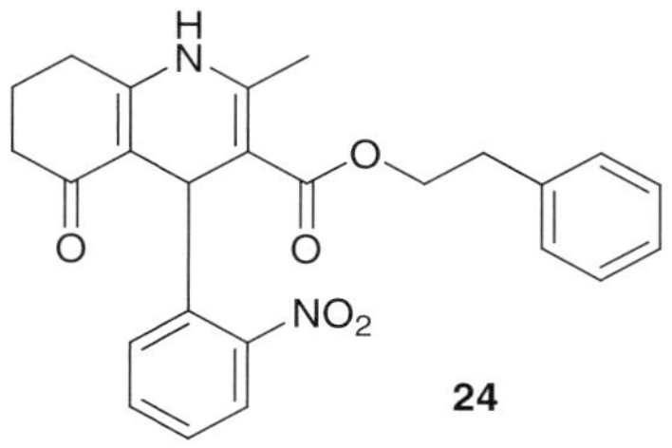

24

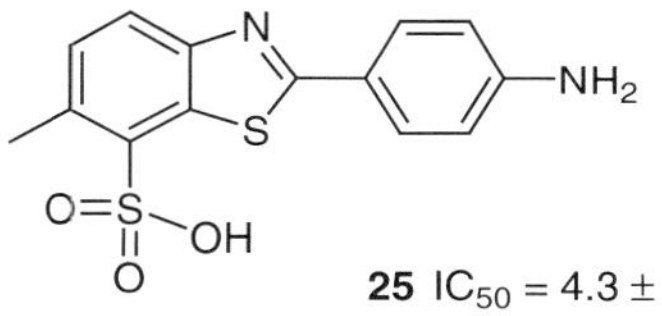

25 IC_{50} = 4.3 ± 0.5 µM

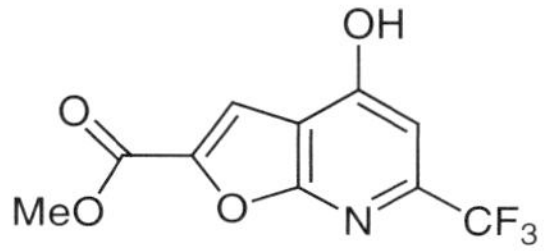

26 IC_{50} = 7 ± 2 µM

The screening of a database containing structural information for more than 8000 existing drugs identified the serotonin antagonist, cinanserin (SQ 10,643) [39,40]. Both a homology model and the crystallographic structure of the binding pocket of the SARS-CoV 3CLpro were utilized in these docking studies. Follow up experiments showed that both cinanserin and its hydrochloride bind to SARS-CoV 3CLpro (IC_{50} = 4.92 and 5.05 µM, respectively).

Forty compounds emerged from a virtual docking screen after postdock screening filters, including pharmacophore model, consensus scoring and "drug-like" filters were applied. Among the three compounds found to inhibit SARS-CoV 3CLpro was the calmodulin antagonist, calmidazolium (C3930) [41]. It showed a K_i value of 61 µM, moreover, calmidazolium is a non-covalent inhibitor.

Some ethacrynic acid derivatives were also tested as non-peptidic covalent inhibitors of the SARS-CoV 3CLpro [42]. An ethacrynic acid amide showed a K_i value of 35.3 µM in a fluorimetric assay using a novel FRET pair-labeled substrate.

Finally, three natural products contained in tea, tanic acid, 3-isotheaflavin-3-gallate (TF2B) and theaflavin-3,3'-digallate (TF3) also showed inhibitory properties against SARS-CoV 3CLpro with IC_{50} values <10 µM [43].

4. PLpro INHIBITORS

Numerous studies on the structural and mechanistic aspects of SARS-CoV 3CLpro have provided multiple avenues for structure-based design of antiviral compounds targeted against the 3CLpro active site [9,16,44,45]. On the other hand, structure-based design against the membrane-associated PLpro enzyme, either from SARS-CoV or related coronaviruses, has remained elusive due to lack of structural information. Unlike many coronaviruses that encode two PLpro paralogs (PLP1 or PLP2), SARS-CoV has a single copy of PLpro that cleaves pp1a at three sites at the N terminus to release nsp1, nsp2 and nsp3, respectively [10,46].

Interestingly, these cleavage sites bear strong resemblance to the C-terminal tail of ubiquitin (consensus sequence LXGG). As a result, it was hypothesized that SARS-CoV PLpro may have de-ubiquitinating activity [47]. Recently, the catalytic domain of PLpro was purified and it was shown that it efficiently disassembles di- and branched polyubiquitin chains, cleaves ubiquitin-7-amino-4-methylcoumarin substrates, and has de-ISGylating activity [48,49]. It has been reported that SARS-CoV PLpro can be inhibited by the specific de-ubiquitinating enzyme inhibitor, ubiquitin aldehyde with an inhibition constant of 0.21 µM [49]. However, the role of these de-ubiquitinating and de-ISGylating activities in the virus replication cycle remains unclear.

SARS-CoV PLpro is considered an equally viable target to 3CLpro for drug design because both are essential for viral replication. However, PLpro has likely not been pursued because of the paucity of structural information. Recently, the catalytic core of SARS-CoV PLpro was crystallized and its X-ray structure was determined to 1.9 Å [50]. The structure of SARS-CoV PLpro is the first to be elucidated for any coronavirus PLpro. This information should provide significant insight into its de-ubiquitinating function *in vitro* and expand the available structural templates of SARS-CoV enzymes that can be targeted for the discovery of novel therapeutic compounds that will halt the replication of SARS-CoV.

5. VIRAL ENTRY INHIBITORS

Since viral entry into a cell is the first step of viral infection, it is an attractive target for anti-SARS chemotherapy. Blocking the entry of a virus may effectively minimize the chance for the virus to evolve and acquire drug resistance [51,52]. The S1 domain of the SARS-CoV spike glycoprotein (S) can efficiently bind with a metallopeptidase, angiotensin-converting enzyme 2 (ACE2), at the virus entry step [53,54]. It was also proved *in vivo* that the binding of the S1 domain to ACE2 on host cell is responsible for SARS-CoV entry into the cells [55]. Anti-ACE2 antibody showed inhibitory ability toward viral replication on Vero E6 cells [56]. A human IgG1 form of 80R was found to bind the S1 domain of the SARS-CoV S protein ($K_d = 1.59$ nM) with a higher affinity comparable to that of ACE2 ($K_d = 1.70$ nM), which suggests that the 80R human monoclonal antibody is a useful viral entry inhibitor for SARS treatment [57].

27

Virtual screening aided in identifying structural leads for viral entry inhibitors of SARS-CoV. Compound **27**, which emerged from a 50,240 compound screen and inhibited pseudovirus entry and SARS-CoV plaque formation with EC_{50} values of 3 µM and 1.6 µM, respectively [37]. A two-step screening of Chinese herbal medicine-based, novel small molecules which bind avidly with the S protein was performed as well. Two virus entry inhibitors, tetra-*O*-galloyl-β- D-glucose (**28**) and luteolin (**29**) were identified and showed anti-SARS-CoV activities with EC_{50} values of 4.5 and 10.6 µM, respectively [58].

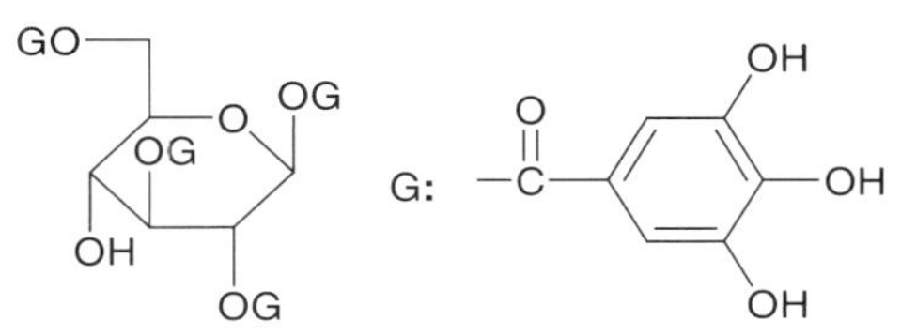

After binding with ACE2, SARS-CoV is taken up into a vesicle inside the cell. Special cellular enzymes (cathepsins) act in the acidic environment inside the vesicle, facilitating fusion of the viral membrane and the vesicle membrane, so that viral proteins and nucleic acids can enter the cell where viral replication occurs [59]. Thus, the cathepsin L inhibitor, MDL28170 (**30**) represents an attractive starting point for antiviral therapeutics targeting SARS-CoV entry.

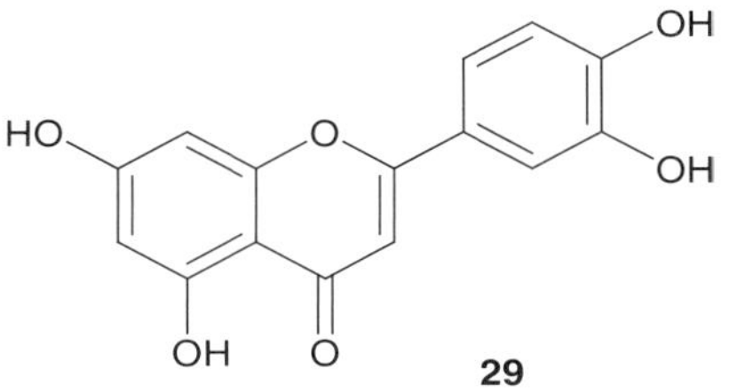

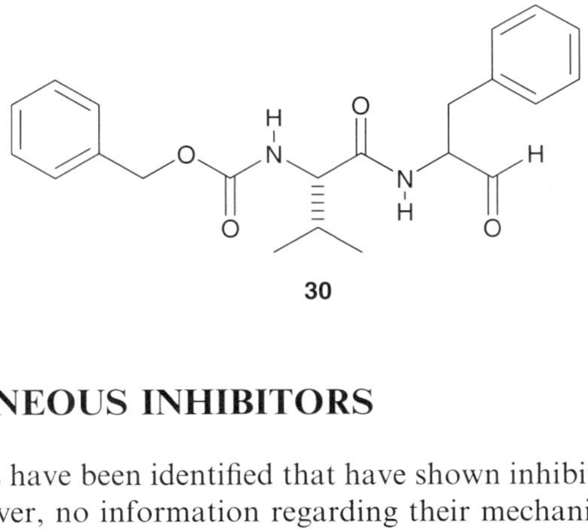

6. MISCELLANEOUS INHIBITORS

Several compounds have been identified that have shown inhibitory activity against SARS-CoV. However, no information regarding their mechanism of action or the corresponding target is known. **Glycyrrhizin** showed inhibitory activity for SARS-CoV replication with $EC_{50} = 300$ mg/l after virus absorption in Vero cells [60]. Some **glycyrrhizin** acid derivatives were found to inhibit SARS-CoV replication *in vitro* with EC_{50} values ranging from 5 to 50 µM. Unfortunately, these compounds show high cytotoxity [61]. The viral entry step was suspected to be inhibited by these derivatives. Nitric oxide (NO) has shown an inhibitory effect on some virus infections [62]. An organic NO donor, ***S*-nitroso-*N*-acetylpenicillamine** was shown to inhibit the replication cycle of SARS-CoV in a concentration-dependent manner, probably during the early steps of infection [63]. HIV protease inhibitor **nelfinavir** [64], antihelminthic drug **niclosamide** [65] and antimalarial agent **chloroquine** [66] all showed strong inhibitory activity ($EC_{50} = 0.048$ µM, $EC_{50} = 1–3$ µM, and $IC_{50} = 8.8 \pm 1.2$ µM, respectively) against SARS-CoV replication. However, no cytoprotective effect was found for **nelfinavir** in an independent study [67,68]. None of the foregoing compounds showed inhibitory activity against SARS-CoV 3CLpro or viral entry. **Ribavirin**, a broad-spectrum inhibitor of RNA and DNA viruses, was used for treatment of SARS patients [69], but it did not inhibit viral growth at concentrations attainable in human serum. In contrast, **interferon (IFN)-α** showed an *in vitro* inhibitory effect starting at concentrations of 1000 IU/ml [70]. Interestingly, the combination of **ribavirin** and **(IFN)-β** synergistically inhibited SARS-CoV replication [71].

7. FUTURE OUTLOOK

It is with unprecedented rapidity that a basic understanding of SARS-CoV life cycle has been achieved. Already, a number of targets including SARS-CoV 3CLpro and SARS-CoV PLpro appear very promising for anti-SARS-CoV chemotherapy. Since the global outbreak of SARS ended in 2003, only a small number of cases of SARS associated with laboratory exposures have been reported. However, with the identification of Chinese horseshoe bats as an animal reservoir for SARS-CoV, the potential danger of the transfer of this virus to humans still exists. To date, there is no effective therapy for the treatment of SARS in humans. While structure-based design and the screening of compounds have provided a number of promising structural leads for SARS-CoV 3CLpro, potent, low molecular weight inhibitors with less toxicity are needed for development. Interest in structure-based design and screening of SARS-CoV PLpro will increase since the X-ray structure of SARS-CoV PLpro was recently determined. Development of anti-SARS-CoV chemotherapy, based on the viral entry-step mechanism holds promise and requires further exploration. It will be of interest to determine if antivirals directed against SARS-CoV will be effective against the recently identified human coronaviruses NL-63 and HKU1, which cause respiratory infections and pneumonia in children and the elderly. Thus, it is likely that the quest for SARS-CoV chemotherapy will be of relevance to other coronavirus-related ailments.

ACKNOWLEDGMENTS

Our research work was supported by a grant from the National Institutes of Health (NIAID) P01 AI060915. We thank Dr. Geoff Bilcer for helpful comments.

REFERENCES

[1] N. Lee, *N. Engl. J. Med.*, 2003, **348**, 1986.
[2] T. G. Ksiazek and D. Erdman, *N. Engl. J. Med.*, 2003, **348**, 1953.

[3] C. Drosten and S. Gunther, *N. Engl. J. Med.*, 2003, **348**, 1967.
[4] T. Kuiken and R. A. Fouchier, *Lancet*, 2003, **362**, 263.
[5] J. S. Peiris and S. T. Lai, *Lancet*, 2003, **361**, 1319.
[6] M. A. Marra and S. J. Jones, *Science*, 2003, **300**, 1399.
[7] P. A. Rota and M. S. Oberste, *Science*, 2003, **300**, 1394.
[8] E. J. Snijder and P. J. Bredenbeek, *J. Mol. Biol.*, 2003, **331**, 991.
[9] K. Anand and J. Ziebuhr, *Science*, 2003, **300**, 1763.
[10] V. Thiel and K. A. Ivanov, *J. Gen. Virol.*, 2003, **84**, 2305.
[11] K. Anand and G. J. Palm, *EMBO J*, 2002, **21**, 3213.
[12] D. A. Matthews, *Proc. Natl. Acad. Sci. USA*, 1999, **96**, 11000.
[13] A. K. Ghosh and K. Xi, *J. Med. Chem.*, 2005, **48**, 6767.
[14] Z. H. Rao and H. T. Yang, *Proc. Natl. Acad. Sci. USA*, 2003, **100**, 13190.
[15] J. Seipelt and A. Guarne, *Virus Res.*, 1999, **62**, 159.
[16] M. N. G. James and T. W. Lee, *J. Mol. Biol.*, 2005, **353**, 1137.
[17] J. L. Asgian and K. E. James, *J. Med. Chem.*, 2002, **45**, 4958.
[18] J. M. Fang and P. H. Liang, *Bioorg. Med. Chem.*, 2005, **13**, 5240.
[19] M. J. McKeage, *Drug Saf.*, 1995, **13**, 228.
[20] K. M. Woessner, *Clin. Rev. Allergy Immunol.*, 2003, **24**, 149.
[21] J. C. Venderas and R. P. Jain, *J. Med. Chem.*, 2004, **47**, 6113.
[22] J. C. Venderas and R. P. Jain, *Bioorg. Med. Chem. Lett.*, 2004, **14**, 3655.
[23] Y. Ramtohul and J. C. Vederas, *J. Org. Chem.*, 2002, **67**, 3169.
[24] J. M. Fang and P. H. Liang, *J. Med. Chem.*, 2005, **48**, 4469.
[25] L. T. Liu, S. F. Chen and S. H. Juang, *Bioorg. Med. Chem. Lett.*, 2005, **15**, 3058.
[26] S. E. Webber and J. Tikhe, *J. Med. Chem.*, 1996, **39**, 5072.
[27] C.-J. Kuo and Y.-H. Chi, *Biochem. Biophys. Res. Commun.*, 2004, **318**, 862.
[28] E. Freire and U. Bacha, *Biochemistry*, 2004, **43**, 4906.
[29] S. X. Cai and H. Z. Zhang, *J. Med. Chem.*, 2006, **49**, 1198.
[30] B. Vastag, *JAMA*, 2003, **290**, 1695.
[31] X. W. Zhang and Y. L. Yap, *Bioorg. Med. Chem.*, 2004, **12**, 2517.
[32] N. Pattaribaman and R. V. Rajnayaranan, *Biochem. Biophys. Res. Commun.*, 2004, **321**, 370.
[33] C. Y. Wu and J. T. Jan, *Proc. Natl. Acad. Sci. USA*, 2004, **101**, 10012.
[34] A. Brik and Y. C. Lin, *Chem. Biol.*, 2002, **9**, 891.
[35] P. H. Liang and J. T. A. Hsu, *FEBS Lett.*, 2004, **574**, 116.
[36] G. J. Brewer and V. D. Johnson, *Hepatology*, 2000, **31**, 364.
[37] R. Y. Kao and W. H. W. Tsui, *Chem. Biol.*, 2004, **11**, 1293.
[38] E. D. Brown and L. D. Eltis, *Chem. Biol.*, 2004, **11**, 1445.
[39] S. Gunther and X. Shen, *J. Virol.*, 2005, **79**, 7095.
[40] B. Rubin and J. J. Piala, *Arch. Int. Pharmacodyn. Ther.*, 1964, **152**, 132.
[41] L. Lai and Y. Liu, *J. Chem. Inf. Model*, 2005, **45**, 10.
[42] T. Schirmeister and U. Kaeppler, *J. Med. Chem.*, 2005, **48**, 6832.
[43] C. N. Chen and K. P. C. Lin, *eCAM*, 2005, **2**, 209.
[44] L. D. Eltis and E. D. Brown, *Chem. Biol.*, 2004, **11**, 1445.
[45] H. Yang and Z. Rao, *Proc. Natl. Acad. Sci. USA*, 2003, **100**, 13190.
[46] B. H. Harcourt and S. C. Baker, *J. Virol.*, 2004, **78**, 13600.
[47] T. Sulea and R. Menard, *J. Virol.*, 2005, **79**, 4550.
[48] N. Barretto and S. C. Baker, *J. Virol.*, 2005, **79**, 15189.
[49] H. A. Lindner and R. Menard, *J. Virol.*, 2005, **79**, 15199.
[50] K. Ratia and A. D. Mesecar, *Proc. Natl. Acad. Sci. USA*, 2006, **103**, 5717.
[51] J. P. Moore and R. W. Doms, *Proc. Natl. Acad. Sci. USA*, 2003, **100**, 10598.

[52] J. A. Este, *Curr. Med. Chem.*, 2003, **10**, 1617.
[53] E. Jenwitheesuk and R. Samudrala, *BMC Struct. Biol.*, 2003, **3**, 2.
[54] S. R. Tipnis, *J. Biol. Chem.*, 2000, **275**, 33238.
[55] J. M. Penninger and K. Kuba, *Nat. Med.*, 2005, **11**, 875.
[56] W. Li and M. J. Moore, *Nature*, 2003, **426**, 450.
[57] W. A. Marasco and J. Sui, *Proc. Natl. Acad. Sci. USA*, 2004, **101**, 2536.
[58] H. K. Deng and X. J. Xu, *J. Virol.*, 2004, **78**, 11334.
[59] G. Simmons and P. Bates, *Proc. Natl. Acad. Sci. USA*, 2005, **102**, 11876.
[60] J. Cinati, Jr. and B. Morgenstern, *Lancet*, 2003, **361**, 2045.
[61] J. Cinati, Jr. and G. Hoever, *J. Med. Chem.*, 2005, **48**, 1256.
[62] T. E. Lane and A. D. Paoletti, *J. Virol.*, 1997, **71**, 2202.
[63] A. Mirazimi and M. Leijon, *J. Virol.*, 2005, **79**, 1966.
[64] N. Yamamoto and R. Yang, *Biochem. Biophys. Res. Commun.*, 2004, **318**, 719.
[65] J. T. A. Hsu and C. Wu, *Antimicrob. Agents Chemother.*, 2004, **48**, 2693.
[66] M. V. Ranst and E. Keyaerts, *Biochem. Biophys. Res. Commun.*, 2004, **323**, 264.
[67] E. L. Tan and L. W. Stanton, *Emerg. Infect. Dis.*, 2004, **10**, 581.
[68] Y. Liu, *Biochem Biophys Res Commun.*, 2005, **333**, 194.
[69] G. Koren and S. King, *CMAJ*, 2003, **168**, 1289.
[70] H. Feldmann and Y. Li, *J. Infect. Dis.*, 2004, **189**, 1164.
[71] J. Cinati, Jr. and B. Morgenstern, *Biochem. Biophys. Res. Commun.*, 2005, **326**, 905.

Inhibitors of the Expression of Vascular Cell Adhesion Molecule-1

Charles Q. Meng

AtheroGenics, Inc., 8995 Westside Parkway, Alpharetta, GA 30004, USA

Contents

1. Introduction	197
2. Therapeutic potential	198
3. Antioxidants	199
4. Cyclic depsipeptides	202
5. Chalcones	203
6. Other compounds	204
7. Conclusion	206
References	206

1. INTRODUCTION

In response to abnormal stimulation such as an injury or infection, endothelial cells become activated and express, among others, cell surface adhesion molecules, which, in turn, recruit leukocytes from the bloodstream into the affected blood vessels and adjacent tissues. Leukocyte recruitment is an essential physiological process for the removal of abnormal stimuli, tissue repair, and healing. If the stimuli linger for a prolonged time, massive amounts of leukocytes will be recruited to the affected area, which cause a dynamic complex of cytologic and chemical reactions and eventually lead to a chronic inflammatory state. In such cases, leukocyte recruitment turns from a beneficial process to a detrimental one. Blocking leukocyte recruitment into the endothelium is, therefore, a rational approach to drug discovery for chronic inflammatory diseases. Adhesion molecules expressed on endothelial cells as major players of the leukocyte recruitment process appear to be logical targets in this approach [1–3].

Vascular cell adhesion molecule-1 (VCAM-1), intercellular cell adhesion molecule-1 (ICAM-1), and selectins (e.g., E-selectin) represent the three major classes of cell surface adhesion molecules known to date. Selectins mediate the initial tethering of leukocytes to, and subsequent rolling along, the endothelium through binding to carbohydrate ligands on leukocytes. VCAM-1 and ICAM-1 are prominent members of the immunoglobin (Ig) superfamily and mediate, after initial tethering and rolling, firm attachment of leukocytes to the endothelium through binding to their counter receptors on leukocytes [2–5]. All three classes of cell adhesion molecules would seem to be reasonable targets for anti-inflammatory drug discovery. Although numerous selectin antagonists have been discovered and some have advanced to clinical stages [6], it is debatable whether intervention of the adhesion mediated by selectins is sufficient to be of therapeutic value since this process is reversible. The counter receptor of ICAM-1, lymphocyte function-associated antigen-1 (LFA-1), is expressed on neutrophils and chronic interference with the ICAM-1/LFA-1 interaction might lead to the impairment of these phagocytes and, thus, increased susceptibility to severe infection [7]. Inhibitors of ICAM-1 expression and LFA-1 antagonists have been reported [6,8,9]. The counter receptor of VCAM-1, very late antigen-4 (VLA-4; $\alpha 4\beta 1$ integrin), is not constitutively expressed on neutrophils and the leukocytes (e.g., T cells, monocytes, and eosinophils) that do express VLA-4 are key effector cells in various inflammatory disorders. Therefore, selective inhibitors of the expression of VCAM-1 appear to have the potential as therapeutic agents without interfering with the phagocytic property of neutrophils [7].

2. THERAPEUTIC POTENTIAL

First cloned in 1989 [10], VCAM-1 is an inducible cell surface protein of vascular endothelial cells. It is normally absent on resting cells, but is readily induced by a wide variety of inflammatory factors, such as interleukin-1 (IL-1), bacterial lipopolysaccharide (LPS), and tumor necrosis factor-α (TNF-α) [11,12]. At sites of inflammation, VCAM-1 can be cleaved from cell surface as soluble VCAM-1 (sVCAM-1) that can be detected in serum and other body fluids. VCAM-1 has been implicated in the pathology of numerous diseases since its discovery.

Atherosclerosis, perceived for over a century as a disorder solely caused by high levels of cholesterol in the circulation, is in fact an inflammatory disease [13,14]. Initial inflammatory stimulation on the endothelium induces VCAM-1 expression, which recruits monocytes from the circulation to adhere to the developing atherosclerotic lesion. VCAM-1 has been found to be specifically upregulated in arterial endothelial cells at lesion-prone areas in hypercholesterolemic animals [15–17]. Subsequent conversion of the recruited monocytes to foamy macrophages results in the synthesis of a wide variety of inflammatory cytokines, growth factors, and chemoattractants such as monocyte chemoattractant protein-1 (MCP-1) [18], all of which help to propagate the formation of mature atherosclerotic plaques from initial lesions and further induce the expression of inflammatory response genes, including VCAM-1. Studies using coronary arteries taken from explanted hearts concluded that the expression of cell adhesion molecules including VCAM-1 is an important element in the inflammatory component of human atherosclerosis [19]. It has also been shown that although both VCAM-1 and ICAM-1 are upregulated in atherosclerotic lesions, VCAM-1 but not ICAM-1 plays a dominant role in the initiation of atherosclerosis [20]. sVCAM-1 has been detected and proposed as an early indicator of human atherosclerosis [21].

Rheumatoid arthritis (RA) is a chronic inflammatory disease of the joints. Studies have shown that the highly specific VCAM-1/VLA-4 interaction could be the predominant adhesive interaction in the migration and adherence of leukocytes to inflamed synovium in RA, whereas the ICAM-1/LFA-1 pathway is less important [22–25]. Increased VCAM-1 expression has been found in RA synovial tissues compared to osteoarthritic and control tissues [4], and elevated serum levels of sVCAM-1 have been detected in RA patients [26]. Inflammatory cytokines such as TNF-α, IFN-γ, IL-1, and IL-4 upregulated VCAM-1 expression on synoviocytes *in vitro* [27]. VCAM-1 antibodies showed efficacy in rodent models of RA [28]. TNF-α blockers prevented TNF-α-induced upregulation of VCAM-1 on synovial fibroblasts *in vitro* [29,30] and reduced disease severity in RA patients [31,32].

Bronchial asthma is an inflammatory airway disease. Upon stimulation, adhesion molecules are expressed on the endothelial cells of the airways, which recruit leukocytes, primarily eosinophils, into the airways. Eosinophil accumulation eventually leads to airway edema, bronchoconstriction, airway wall remodeling and hyperresponsiveness. These events further induce the expression of inflammatory cytokines and adhesion molecules. Elevated expression of VCAM-1 has been detected in the lungs of asthmatic patients and anti-asthmatic treatment reduced the expression of VCAM-1 [33–35]. In animal models, VCAM-1 antibodies significantly protected against allergen-induced airway inflammation, inhibiting T-cell migration into the lungs and eosinophil accumulation in bronchoalveolar lavage fluid [36]. VCAM-1 is also implicated in other diseases such as multiple sclerosis [37–39], chronic transplant rejection [40–43], and diabetes [44,45].

This report briefly summarizes inhibitors of VCAM-1 expression with potential therapeutic value. A recent comprehensive review on this topic has appeared [46].

3. ANTIOXIDANTS

It was established in 1993 that activation of VCAM-1 expression on endothelial cells is partly regulated by a redox signal transduction pathway that is sensitive to inhibition by antioxidants. Pyrrolidine dithiocarbamate (PDTC), a known antioxidant, at 50 μM inhibited over 90% of IL-1β-induced VCAM-1 expression in cultured human umbilical vein endothelial cells (HUVEC) [47,48]. These findings led to the search for new compounds with antioxidant properties as inhibitors of VCAM-1 expression.

Probucol (**1**), a phenolic antioxidant, was once used worldwide as a lipid-lowering agent [49,50]. While probucol had moderate low-density lipoprotein (LDL)-lowering effects compared to statins, it also significantly lowered high-density lipoprotein (HDL) levels, and caused QTc prolongation. These adverse effects may explain the later withdrawal of probucol from most markets. However, the strong and unique antioxidant properties of probucol have been widely recognized in various settings and are believed to contribute to its anti-atherogenic effects [51–55]. Probucol inhibited neointima formation in animal models of artery balloon injury [56,57] and showed beneficial effects for the prevention of restenosis in humans after angioplasty [58–60]. In a clinical trial, probucol reduced carotid artery intima-media thickness in patients with hypercholesterolemia to the same extent as pravastatin and showed significantly lower incidence of cardiac events than the placebo group [61].

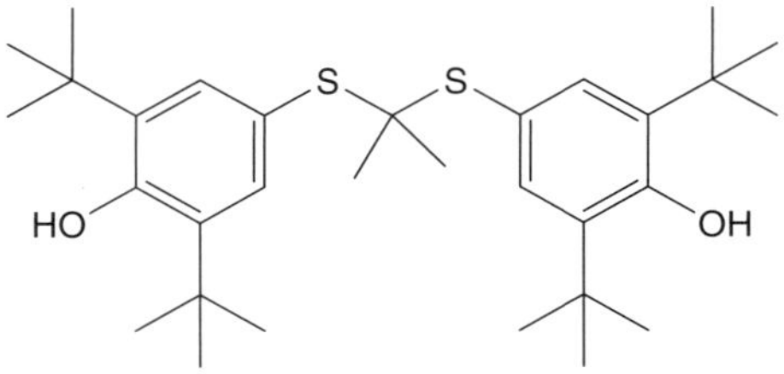

Probucol is metabolized to spiroquinone **2** [62]. The latter compound and/or its metabolites have been suspected to be responsible for causing QTc prolongation in some patients. Probucol derivatives have been designed with one of its phenol groups substituted to prevent the formation of the potentially harmful spiroquinone (**2**) and the other phenol group unsubstituted to retain the antioxidant property [63,64]. Some of these derivatives potently inhibited TNF-α-induced expression of VCAM-1 in cultured human endothelial cells, although probucol itself did not show any effect under the same condition. Hyperbolic correlations have been observed between inhibitory potency on VCAM-1 expression and lipophilicity of compounds among these probucol derivatives. Probucol, despite its strong antioxidant property, presumably does not have the right lipophilicity for potency while some of its derivatives do. Probucol was reported to inhibit VCAM-1 expression *in vivo* [65], but this effect could be indirect (as opposed to a direct effect in an *in vitro* cell-based assay) since probucol is known to be able to reduce the levels of oxidized LDL, which induces VCAM-1 expression [66].

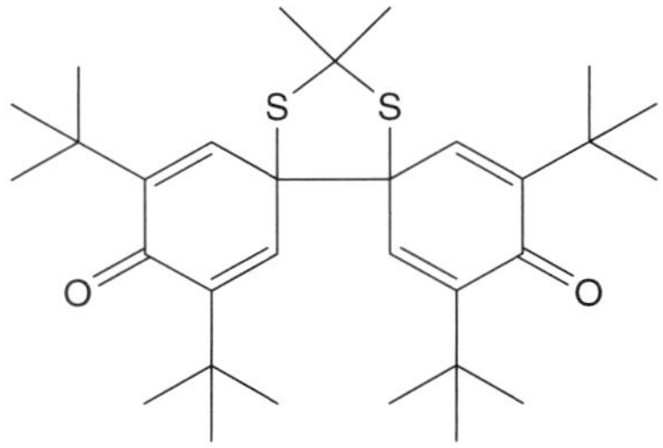

AGI-1067 (**3**), a clinical compound derived from this endeavor, exhibited many *in vitro* properties desirable in a molecule to treat atherosclerosis [67]. It showed similar antioxidant activity as probucol, potently and selectively inhibited VCAM-1 expression (IC50 = 6 µM) over ICAM-1, and also inhibited MCP-1 expression and human aortic smooth muscle cell proliferation. In animal models, AGI-1067 inhibited the progression of atherosclerosis and lowered LDL levels with neutral or elevating effects on HDL levels [68]. AGI-1067 was not metabolized to probucol in animals or humans, so it is not a prodrug of probucol. It did not exhibit the same pharmacological profile as probucol in animals or humans. AGI-1067 was well tolerated in a 1-year study in hypercholesterolemic monkeys. It lowered LDL levels by 90% and increased HDL levels by 107% at an oral dose of 150 mg/kg. Probucol only had modest LDL-lowering effect and decreased HDL levels in the same study.

Histopathological analysis of the aortas and coronary arteries revealed no atherosclerosis in the AGI-1067-treated group and minimal-to-moderate atherosclerosis in the vehicle and probucol groups [68]. In clinical trials, AGI-1067 improved lumen dimensions of reference segments of coronary arteries after angioplasty, suggesting a direct positive effect on atherosclerosis; furthermore, AGI-1067 did not cause QTc prolongation [69,70]. The current phase III studies with AGI-1067 will determine the merit of this novel, multifunctional drug in patients with coronary artery disease.

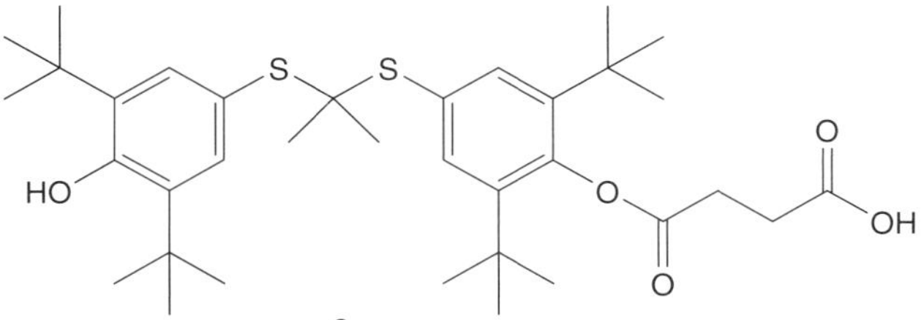

3

BO-653 (**4**), another phenolic antioxidant, is worth noting although there have been no reports on its inhibition of VCAM-1 expression. It was derived as a hybrid of probucol and vitamin E with the intention to retain their advantages and overcome their shortcomings as antioxidants [71–73]. Several key pharmacophores were incorporated in the design of BO-653: the phenol group present in both probucol and vitamin E as the antioxidant source; the di-*tert*-butyl groups from probucol to retain lipophilicity; and the 2,3-dihydrobenzofuran unit derived from a vitamin E analog, which had been known to have increased antioxidant activity compared to vitamin E and other 3,4-dihydro-2*H*-pyran analogs [74].

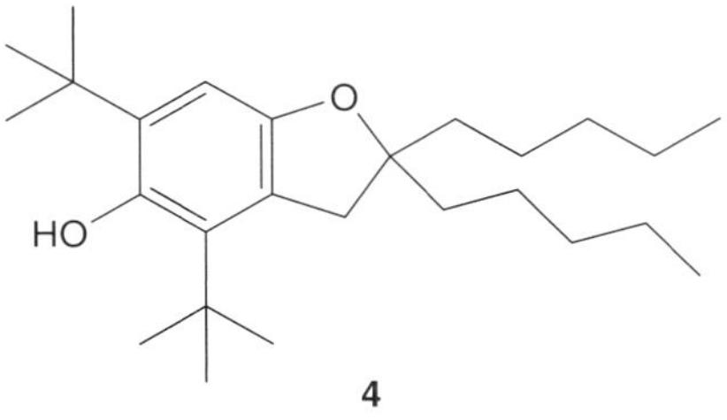

4

Although it showed weaker reactivity in quenching peroxyl radicals than vitamin E, BO-653 exhibited antioxidant property superior to that of vitamin E against lipid peroxidation [75–77]. In a study on the antioxidant activities of BO-653 against oxidative modification of human LDL particles, BO-653 was consumed faster than vitamin E and retarded consumption of vitamin E. Vitamin E was not consumed until most of the BO-653 was consumed. The formation of lipid hydroperoxides was effectively inhibited until almost all BO-653 was consumed. The superior antioxidant potency of BO-653 over vitamin E is likely due to the increased lipophilicity of the molecule and enhanced stability of its radical [77]. BO-653 has shown efficacy in animal models of atherosclerosis and restenosis; in some species the effects were superior to those of probucol. Also, it did not lower HDL levels in animals [78]. BO-653 was reported to be in clinical trials for the treatment of atherosclerosis. Data from these trials will help determine if its advantageous antioxidant property, lipophilicity, and anti-atherosclerotic effects provide superiority over probucol or vitamin E.

4. CYCLIC DEPSIPEPTIDES

HUN-7293 (**5**) is a cyclic depsipeptide originally isolated from a fungus [46,79]. It is the most potent inhibitor of VCAM-1 expression reported to date, with an IC_{50} value of 2 nM for inhibiting TNF-α-induced expression of VCAM-1 in human endothelial cells [80]. HUN-7293 showed efficacy in animal models of inflammation involving eosinophil infiltration and hypersensitivity reactions, indicating its potential for treating allergic asthma [81]. Compound **6**, also fungal-derived, showed similar potency as HUN-7293 to inhibit VCAM-1 expression, indicating that the cyano residue in HUN-7293 is not essential for inhibiting VCAM-1 expression [46].

Various derivatives and analogs of HUN-7293 have been synthesized to understand its structure–activity relationship (SAR) and to discover new, more potent inhibitors of VCAM-1 expression [46,82–85]. Several derivatives and analogs showed comparable potency to HUN-7293, but none have been identified which exhibited significantly better potency. For example, substitution of the ester function in the backbone of HUN-7293 with an amide bond yielded aza-HUN-7239, which had a 20-fold lower potency than HUN-7239 [85]. Compounds **7** and **8**, each replacing the cyano residue of HUN-7293 with a different group, showed similar potency as HUN-7293 for inhibition of VCAM-1 expression. Both compounds also inhibited the expression of vascular endothelial growth factor (VEGF) and significantly decreased retinal neovascularization and capillary non-perfusion in a murine model of ischemic retinopathy [46].

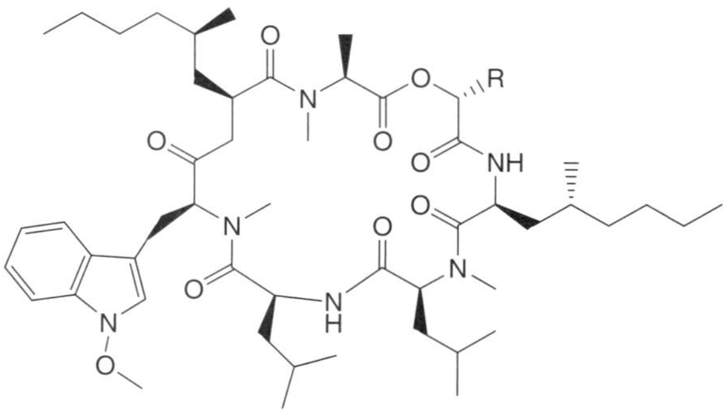

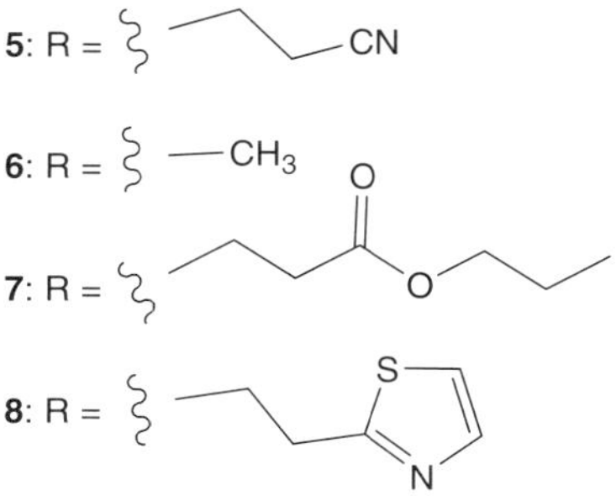

5: R = ⧝―⧸⧹―CN

6: R = ⧝―CH₃

7: R = ⧝―⧸⧹―C(=O)―O―⧸⧹

8: R = ⧝―⧸⧹―(thiazole)

5. CHALCONES

The chalcones, with a common 1,3-diphenyl-2-propen-1-one framework, have been known for over a century. Naturally occurring and synthetic chalcones have shown interesting biological activity as antioxidant, anti-inflammatory, anticancer, and anti-infective agents [86]. Based on the fact that some α,β-unsaturated carbonyl-containing natural products, such as halichlorine, are inhibitors of VCAM-1 expression, a screening of compounds with an α,β-unsaturated carbonyl led to the discovery of some chalcone compounds as potent inhibitors of VCAM-1 expression [87]. Hyperbolic lipophilicity-potency correlations were also observed with the chalcone compounds as inhibitors of VCAM-1 expression. Chalcone **9** showed an IC_{50} value of 1 μM in inhibiting VCAM-1 expression in cultured human endothelial cells, while its reduced analog, compound **10**, was inactive, indicating that the α,β-unsaturated carbonyl residue of the chalcone compounds is essential for inhibition of VCAM-1 expression [87] for electronic and/or steric reasons.

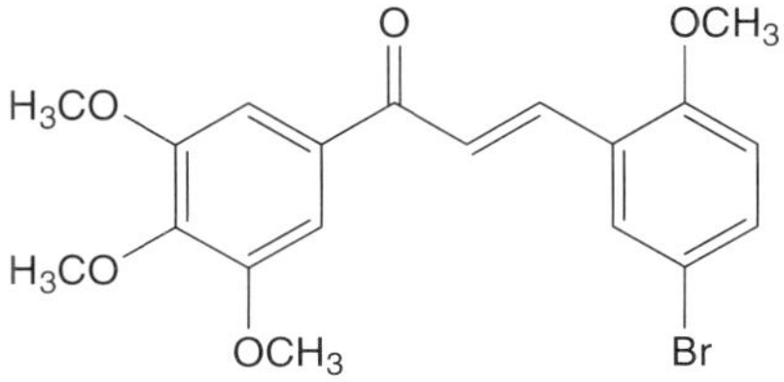

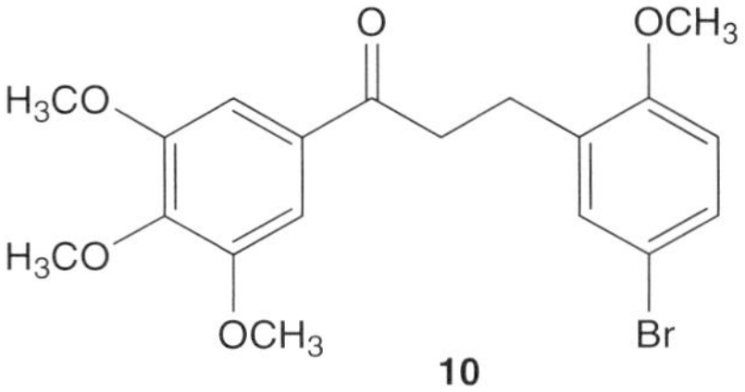

Chalcone derivative **11** inhibited VCAM-1 expression (IC50 = 0.6 μM) in cultured, TNF-α-induced human pulmonary arterial endothelial cells (HPAECs). It also inhibited E-selectin and MCP-1 expression under the same condition. In cultured, LPS-induced human peripheral blood mononuclear cells (HPBMCs), **11** completely inhibited IL-1β secretion at 5 μM. In a rat adjuvant arthritis model, prophylactic, subcutaneous dosing of **11** at 25 and 75 mg/kg, dose-dependently inhibited both paw swelling and splenomegaly. It also reached histological bone resorption and inflammation scores. These data indicate that compound **11** could be a potential disease-modifying anti-rheumatic drug (DMARD). Compound **11**

also inhibited paw swelling in a dose-dependent manner when dosed therapeutically. It also showed oral efficacy in the model [88].

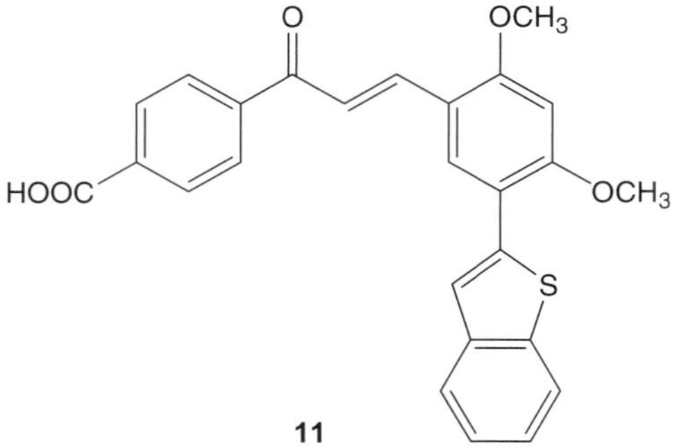

11

In an ovalbumin-sensitized/challenged mouse model of allergic asthma, compound **11** dose-dependently inhibited airway and tissue eosinophilia, reduced serum IgE levels, and reduced elevated lung mRNA levels of T helper 2 (Th2) cytokines, such as IL-4, IL-13, and IL-5 without affecting mRNA levels of Th1 cytokines such as IFN-γ and IL-2. Compound **11** also reduced airway hyper-responsiveness in the model [89].

6. OTHER COMPOUNDS

The cannabinoid receptor agonist, R-(+)-WIN-55,212-2 (**12**), has shown efficacy in controlling disease progression in animal models of multiple sclerosis (MS). This effect has been attributed to its ability to reduce migration of leukocytes into the central nervous system [90], in which adhesion molecules are believed to be involved. Compound **12**, but not its enantiomer, S-(-)-WIN-55,212-2, strongly inhibited IL-1-induced expression of VCAM-1 on astrocytoma and A-172 glioblastoma cells. Interestingly, S-(-)-WIN-55,212-2 showed no efficacy in models of MS. The inhibitory effect of **12** on VCAM-1 expression is not believed to be mediated via cannabinoid receptors, since both selective cannabinoid receptor antagonists and pertussis toxin failed to affect it. Experimental data suggest that **12** blocks IL-1 signaling by inhibiting the transactivation potential of nuclear factor-κB (NF-κB) [91].

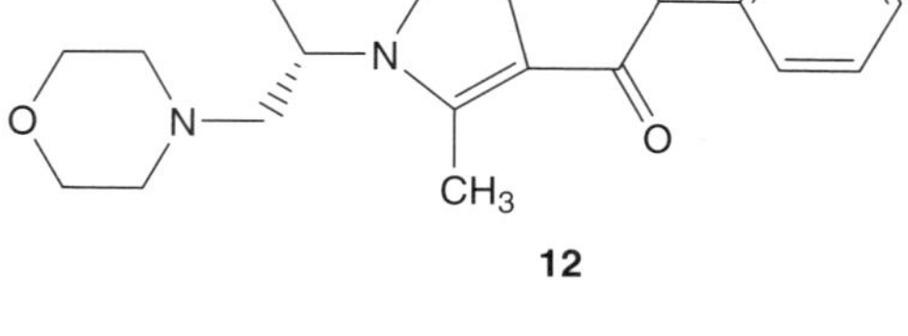

12

In recent years, dual 5-lipooxygenase (5-LOX) and cyclooxygenase (COX) inhibitors, interfering with both prostaglandin and leukotriene pathways in the arachidonic acid cascade, have emerged as possible alternatives to non-steroidal anti-inflammatory drugs (NSAIDs) which interact with the prostaglandin pathway only. The hope for dual 5-LOX/COX inhibition is to alleviate the side effects related to sole COX-inhibition, especially those exerted by selective COX-2 inhibitors. Licofelone (**13**), a dual 5-LOX/COX inhibitor, attenuated VCAM-1 expression in inflammatory endothelial cells *in vitro*. In a flow chamber assay, it dose-dependently decreased both the rolling and adhesion of leukocytes on endothelial cells. In contrast, no effects were found with the non-selective COX inhibitor indomethacin, the potent and selective 5-LOX inhibitor ZD-2138, or the selective COX-2 inhibitor celecoxib, or the combination of ZD-2138 with celecoxib. In a mouse peritonitis model, licofelone markedly reduced leukocyte accumulation [92].

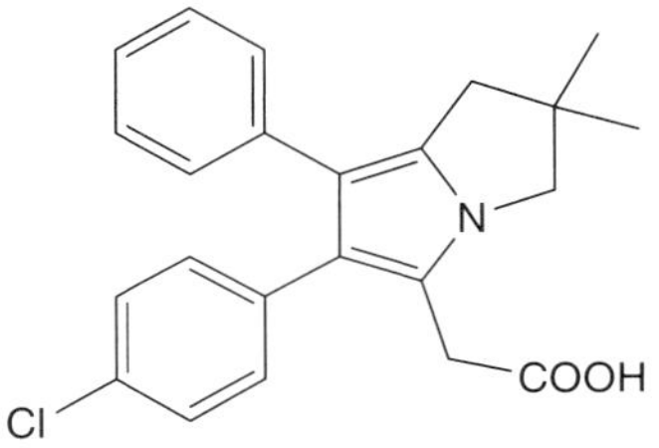

13

KR-31378 (**14**) [93] inhibited VCAM-1 expression and decreased adhesion and migration of monocytes in a dose-dependent manner. It has shown a neuroprotective effect for ischemia-reperfusion damage in rat brain. In LDL receptor-knockout mice which were fed a high-fat, high-cholesterol diet, treatment with **14** significantly inhibited fatty streak formation and macrophage accumulation on the artery wall, indicating its potential for treating atherosclerosis. Experimental data also showed that **14** might work by decreasing NF-κB activation [94].

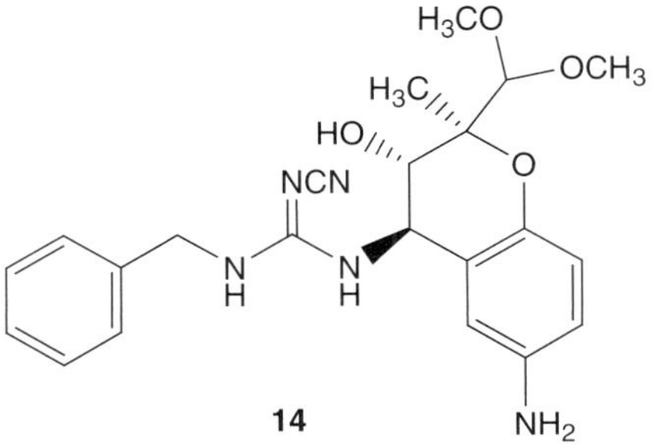

14

Methimazole (**15**) is a tautomeric cyclic thione used clinically for various autoimmune diseases. Several observations suggest that **15** could be an anti-inflammatory agent through inhibition of adhesion molecules [95]. For example, it reduced sVCAM-1 levels in the circulation of patients with Graves' disease [96]. A phenyl derivative (structure undisclosed) of methimazole [97] was far more effective in experimental models of lupus and diabetes than methimazole [98]. The phenyl analog inhibited TNF-α-induced VCAM-1 mRNA and protein expression in human airway epithelial cells (HAECs), reduced TNF-α-induced monocytic (U937) cell adhesion to HAECs under *in vitro* flow conditions, inhibited TNF-α-induced interferon regulatory factor-1 (IRF-1) binding to VCAM-1 promoter, and reduced TNF-α-induced IRF-1 expression in HAECs [95].

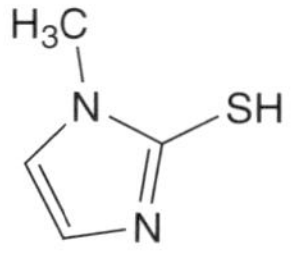

15

7. CONCLUSION

The diversity of chemical structures that have been reported [46] to be inhibitors of VCAM-1 expression suggests that there may be numerous upstream molecular targets regulating VCAM-1 expression with which different inhibitors interact. Antioxidants may interact with one target and chalcones with another. Even the antioxidants that have shown inhibition on VCAM-1 expression may interact with different targets of the VCAM-1 expression pathway, since PDTC, AGI-1067, and carvedilol [99] are different structurally. Therefore, VCAM-1 expression is not a single target in the traditional drug discovery sense. Rather, it is a collective marker of potentially many different upstream molecular targets.

VLA-4, the counter receptor of VCAM-1, is a single target, which has been pursued extensively for drug discovery [100,101]. An antibody of VLA-4, natalizumab, has shown beneficial effects in treating patients with MS and Crohn's disease [102,103]. The clinical trial with BIO 1211, a small molecule VLA-4 antagonist, was discontinued due to lack of efficacy for asthma [46]. This may mean that VLA-4 is not a viable target for asthma.

The most advanced inhibitor of VCAM-1 expression (AGI-1067) is currently in phase III clinical trials. Since there is no single, unique molecular target for VCAM-1 expression and inhibitors of VCAM-1 expression tend to be multifunctional, it will be challenging to unambiguously assess the outcomes of all inhibitors of VCAM-1 expression.

REFERENCES

[1] G. A. Koing, R. M. Schiffelers and G. Storm, *Endothelium*, 2002, **9**, 161.
[2] H. Yusuf-Makagiansar, M. E. Anderson, T. V. Yakovleva, J. S. Murray and T. J. Siahaan, *Med. Res. Rev.*, 2002, **22**, 146.
[3] M. P. Bevilacqua, R. M. Nelson, G. Mannori and O. Cecconi, *Annu. Rev. Med.*, 1994, **45**, 361.
[4] L. S. Wilkinson, J. C. Edwards, R. N. Poston and D. O. Haskard, *Lab. Invest.*, 1993, **68**, 82.

[5] T. M. Carlos and J. M. Harlan, *Blood*, 1994, **84**, 2068.
[6] H. Ulbrich, E. E. Eriksson and L. Lindbom, *Trends Pharmacol. Sci.*, 2003, **24**, 640.
[7] C. A. Foster, *J. Allergy Clin. Immunol.*, 1996, **98**, S270.
[8] Z. Pei, Z. Xin, G. Liu, Y. Li, E. B. Reilly, N. L. Lubbers, J. R. Huth, J. T. Link, T. W. von Geldern, B. F. Cox, S. Leitza, Y. Gao, K. C. Marsh, P. DeVries and G. F. Okasinski, *J. Med. Chem.*, 2001, **44**, 2913.
[9] G. D. Zhu, D. L. Arendsen, I. W. Gunawardana, S. A. Boyd, A. O. Stewart, D. G. Fry, B. L. Cool, L. Kifle, V. Schaefer, J. Meuth, K. C. Marsh, A. J. Kempf-Grote, P. Kilgannon, W. M. Gallatin and G. F. Okasinski, *J. Med. Chem.*, 2001, **44**, 3469.
[10] L. Osborn, C. Hession, R. Tizard, C. Vassallo, S. Luhowshkyj, G. Chi-Rosso and R. Lobb, *Cell*, 1989, **59**, 1203.
[11] M. Ahmad, N. Marui, R. W. Alexander and R. M. Medford, *J. Biol. Chem.*, 1995, **270**, 8976.
[12] G. E. Rice, J. M. Munro and M. P. Bevilacqua, *J. Exp. Med.*, 1990, **171**, 1369.
[13] R. Ross, *N. Engl. J. Med.*, 1999, **340**, 115.
[14] P. Libby, *Nature*, 2002, **420**, 868.
[15] H. Li, M. I. Cybulsky, M. A. Gimbrone, Jr. and P. Libby, *Am. J. Pathol.*, 1993, **143**, 1551.
[16] Y. Nakashima, E. W. Raines, A. S. Plump, J. L. Breslow and R. Ross, *Arterioscler. Thromb. Vasc. Biol.*, 1998, **18**, 842.
[17] K. Iiyama, L. Hajra, M. Iiyama, H. Li, M. DiChiara, B. D. Medoff and M. I. Cybulsky, *Circ. Res.*, 1999, **85**, 199.
[18] J. R. Harrington, *Stem Cells*, 2000, **18**, 65.
[19] M. J. Davies, J. L. Gordon, A. J. Gearing, R. Pigott, N. Woolf, D. Katz and A. Kyriakopoulos, *J. Pathol.*, 1993, **171**, 223.
[20] M. I. Cybulsky, K. Iiyama, H. Li, S. Zhu, M. Chen, M. Iiyama, V. Davis, J. C. Gutierrez-Ramos, P. W. Connelly and D. S. Milstone, *J. Clin. Invest.*, 2001, **107**, 1255.
[21] K. Peter, U. Weirich, T. K. Nordt, J. Ruef and C. Bode, *Thromb. Haemost.*, 1999, **82** (Suppl. 1), 38.
[22] A. C. van Dinther-Janssen, E. Horst, G. Koopman, W. Newmann, R. J. Scheper, C. J. Meijer and S. T. Pals, *J. Immunol.*, 1991, **147**, 4207.
[23] A. A. Postigo, R. Garcia-Vicuna, A. Laffon and F. Sanchez-Madrid, *Autoimmunity*, 1993, **16**, 69.
[24] C. Pitzalis, G. Kingsley and G. Panayi, *Ann. Rheum. Dis.*, 1994, **53**, 287.
[25] R. W. McMurray, *Semin. Arthritis Rheum.*, 1996, **25**, 215.
[26] M. N. Kolopp-Sarda, F. Guillemin, I. Chary-Valckenaere, M. C. Bene, J. Pourel and G. C. Faure, *Clin. Exp. Rheumatol.*, 2001, **19**, 165.
[27] J. Morales-Ducret, E. Wayner, M. J. Elices, J. M. Alvaro-Gracia, N. J. Zvaifler and G. S. Firestein, *J. Immunol.*, 1992, **149**, 1424.
[28] R. A. Carter, I. K. Campbell, K. L. O'Donnel and I. P. Wicks, *Clin. Exp. Immunol.*, 2002, **128**, 44.
[29] M. L. Blue, P. Conrad, D. L. Webb, T. Sarr and M. Macaro, *Lymphokine Cytokine Res*, 1993, **12**, 213.
[30] M. P. Bombara, D. L. Webb, P. Conrad, C. W. Marlor, T. Sarr, G. E. Ranges, T. M. Aune, J. M. Greve and M. L. Blue, *J. Leukoc. Biol.*, 1993, **54**, 399.
[31] J. D. Abbott and L. W. Moreland, *Expert Opin. Investig. Drugs*, 2004, **13**, 1007.
[32] M. C. Hochberg, M. G. Lebwohl, S. E. Plevy, K. F. Hobbs and D. E. Yocum, *Semin. Arthritis Rheum.*, 2005, **34**, 819.
[33] J. M. Pilewski and S. M. Albelda, *Am. J. Respir. Cell Mol. Biol.*, 1995, **12**, 1.

[34] S. Bazan-Socha, A. Bukiej, C. Marcinkiewicz and J. Musial, *Curr. Pharm. Des.*, 2005, **11**, 893.
[35] S. J. Wilson, A. Wallin, G. Della-Cioppa, T. Sandstrom and S. T. Holgate, *Am. J. Respir. Crit. Care Med.*, 2001, **164**, 1047.
[36] O. Kaminuma, H. Fujimura, K. Fushimi, A. Nakata, A. Sakai, S. Chishima, K. Ogawa, M. Kikuchi, H. Kikkawa, K. Akiyama and A. Mori, *Eur. J. Immunol.*, 2001, **31**, 2669.
[37] S. J. Lee and E. N. Benveniste, *J. Neuroimmunol.*, 1999, **98**, 77.
[38] A. Flugel, T. Berkowicz, T. Ritter, M. Labeur, D. E. Jenne, Z. Li, J. W. Ellwart, M. Willem, H. Lassmann and H. Wekerle, *Immunity*, 2001, **14**, 547.
[39] P. Rieckmann, N. Kruse, L. Nagelkerken, K. Beckmann, D. Miller, C. Polman, F. Dahlke, K. V. Toyka, H. P. Hartung and S. Sturzebecher, *J. Neurol.*, 2005, **252**, 526.
[40] G. M. Crews, L. Erickson, F. Pan, O. Fisniku, M. S. Jang, C. Wynn, H. Benediktsson, M. Kobayashi and H. Jiang, *Transplant. Proc.*, 2005, **37**, 1926.
[41] L. A. Robinson, L. Tu, D. A. Steeber, O. Preis, J. L. Platt and T. F. Tedder, *J. Immunol.*, 1998, **161**, 6931.
[42] M. D. Stegall, P. G. Dean, D. Ninova, A. J. Cohen, G. M. Shepard, C. Gup and R. G. Gill, *Transplantation*, 2001, **71**, 1549.
[43] C. L. Schlichting, W. D. Schareck and M. Weis, *J. Cardiovasc. Pharmacol.*, 2005, **46**, 250.
[44] B. Glowinska, M. Urban, J. Peczynska and B. Florys, *Metabolism*, 2005, **54**, 1020.
[45] F. Cipollone, F. Chiarelli, G. Davi, C. Ferri, G. Desideri, M. Fazia, A. Iezzi, F. Santilli, B. Pini, C. Cuccurullo, S. Tumini, A. Del Ponte, A. Santucci, F. Cuccurullo and A. Mezzetti, *Diabetologia*, 2005, **48**, 1216.
[46] E. P. Schreiner, B. Oberhauser and C. A. Foster, *Expert Opin. Ther. Patents*, 2003, **13**, 149.
[47] R. M. Medford, in *Cardiovascular Disease 2* (ed. L. L. Gallo), Plenum Press, New York, 1995, p. 121.
[48] N. Marui, M. K. Offermann, R. Swerlick, C. Kunsch, C. A. Rosen, M. Ahmad, R. W. Alexander and R. M. Medford, *J. Clin. Invest.*, 1993, **92**, 1866.
[49] J. W. Barnhart, D. J. Rytter and J. A. Molello, *Lipids*, 1977, **12**, 29.
[50] M. M. Buckley, K. L. Goa, A. H. Price and R. N. Brogden, *Drugs*, 1989, **37**, 761.
[51] D. Steinberg, S. Parthasarathy and T. E. Carew, *Am. J. Cardiol.*, 1988, **62**, 6B.
[52] P. D. Reaven, S. Parthasarathy, W. F. Beltz and J. L. Witztum, *Arterioscler. Thromb.*, 1992, **12**, 318.
[53] T. Kita, M. Yokode, K. Ishii, N. Kume, Y. Nagano, H. Arai, H. Otani, Y. Ueda and S. Hara, *Clin. Exp. Pharmacol. Physiol. Suppl.*, 1992, **20**, 37.
[54] D. Bonnefont-Rousselot, C. Segaud, D. Jore, J. Delattre and M. Gardes-Albert, *Radiat. Res.*, 1999, **151**, 343.
[55] S. Kondo, M. Shimizu, M. Urushihara, K. Tsuchiya, M. Yoshizumi, T. Tamaki, A. Nishiyama, H. Kawachi, F. Shimizu, M. T. Quinn, D. J. Lambeth and S. Kagami, *J. Am. Soc. Nephrol.*, 2006, **17**, 783.
[56] J. E. Schneider, B. C. Berk, M. B. Gravanis, E. C. Santoian, G. D. Cipolla, N. Tarazona, B. Lassegue and S. B. King, 3rd, *Circulation*, 1993, **88**, 628.
[57] C. L. Jackson and K. S. Pettersson, *Atherosclerosis*, 2001, **154**, 407.
[58] H. Yokoi, H. Daida, Y. Kuwabara, H. Nishikawa, F. Takatsu, H. Tomihara, Y. Nakata, Y. Kutsumi, S. Ohshima, S. Nishiyama, A. Seki, K. Kato, S. Nishimura, T. Kanoh and H. Yamaguchi, *J. Am. Coll. Cardiol.*, 1997, **30**, 855.
[59] J. C. Tardif, G. Cote, J. Lesperance, M. Bourassa, J. Lambert, S. Doucet, L. Bilodeau, S. Nattel and P. de Guise, *N. Engl. J. Med.*, 1997, **337**, 365.

[60] G. Cote, J. C. Tardif, J. Lesperance, J. Lambert, M. Bourassa, R. Bonan, G. Gosselin, M. Joyal, J. F. Tanguay, S. Nattel, R. Gallo and J. Crepeau, *Circulation*, 1999, **99**, 30.
[61] Y. Sawayama, C. Shimizu, N. Maeda, M. Tatsukawa, N. Kinukawa, S. Koyanagi, S. Kashiwagi and J. Hayashi, *J. Am. Coll. Cardiol.*, 2002, **39**, 610.
[62] J. W. Barnhart, E. R. Wagner and R. L. Jackson, in *Antilipidemic Drugs: Medicinal, Chemical and Biochemical Aspects* (eds D. T. Witiak, H. A. I. Newman and D. R. Feller), Elsevier, Amsterdam, 1991, p. 277.
[63] C. Q. Meng, P. K. Somers, C. L. Rachita, L. A. Holt, L. K. Hoong, X. S. Zheng, J. E. Simpson, R. R. Hill, L. K. Olliff, C. Kunsch, C. L. Sundell, S. Parthasarathy, U. Saxena, J. A. Sikorski and M. A. Wasserman, *Bioorg. Med. Chem. Lett.*, 2002, **12**, 2545.
[64] C. Q. Meng, P. K. Somers, L. K. Hoong, X. S. Zheng, Z. Ye, K. J. Worsencroft, J. E. Simpson, M. R. Hotema, M. D. Weingarten, M. L. MacDonald, R. R. Hill, E. M. Marino, K. L. Suen, J. Luchoomun, C. Kunsch, L. K. Landers, D. Stefanopoulos, R. B. Howard, C. L. Sundell, U. Saxena, M. A. Wasserman and J. A. Sikorski, *J. Med. Chem.*, 2004, **47**, 6420.
[65] J. Fruebis, V. Gonzalez, M. Silvestre and W. Palinski, *Arterioscler. Thromb. Vasc. Biol.*, 1997, **17**, 1289.
[66] L. Cominacini, U. Garbin, A. F. Pasini, A. Davoli, M. Campagnola, G. B. Contessi, A. M. Pastorino and V. Lo Cascio, *Free Radic. Biol. Med.*, 1997, **22**, 117.
[67] C. Kunsch, J. Luchoomun, J. Y. Grey, L. K. Olliff, L. B. Saint, R. F. Arrendale, M. A. Wasserman, U. Saxena and R. M. Medford, *J. Pharmacol. Exp. Ther.*, 2004, **308**, 820.
[68] C. L. Sundell, P. K. Somers, C. Q. Meng, L. K. Hoong, K. L. Suen, R. R. Hill, L. K. Landers, A. Chapman, D. Butteiger, M. Jones, D. Edwards, A. Daugherty, M. A. Wasserman, R. W. Alexander, R. M. Medford and U. Saxena, *J. Pharmacol. Exp. Ther.*, 2003, **305**, 1116.
[69] J. C. Tardif, J. Gregoire, L. Schwartz, L. Title, L. Laramee, F. Reeves, J. Lesperance, M. G. Bourassa, P. L. L'Allier, M. Glass, J. Lambert and M. C. Guertin, *Circulation*, 2003, **107**, 552.
[70] A. M. Franks and S. F. Gardner, *Ann. Pharmacother.*, 2006, **40**, 66.
[71] N. Noguchi and E. Niki, *Free Radic. Biol. Med.*, 2000, **28**, 1538.
[72] N. Noguchi, Y. Iwaki, M. Takahashi, E. Komuro, Y. Kato, K. Tamura, O. Cynshi, T. Kodama and E. Niki, *Arch. Biochem. Biophys.*, 1997, **342**, 236.
[73] C. Q. Meng, *Curr. Opin. Investig. Drugs*, 2003, **4**, 342.
[74] K. U. Ingold, G. W. Burton, D. O. Foster, M. Zuker, L. Hughes, S. Lacelle, E. Lusztyk and M. Slaby, *FEBS Lett*, 1986, **205**, 117.
[75] H. Kaise, M. Nakamura, Y. Takashima, O. Cynshi, N. Sakaguchi, M. Takeya, K. Takahashi, E. Niki and T. Kodama, *Atherosclerosis*, 1997, **134**, 203.
[76] A. Watanabe, N. Noguchi, A. Fujisawa, T. Kodama, K. Tamura, O. Cynshi and E. Niki, *J. Am. Chem. Soc.*, 2000, **122**, 5438.
[77] K. Muller, K. L. Carpenter, M. A. Freeman and M. J. Mitchinson, *Free Radic. Res.*, 1999, **30**, 59.
[78] O. Cynshi, Y. Kawabe, T. Suzuki, Y. Takashima, H. Kaise, M. Nakamura, Y. Ohba, Y. Kato. K. Tamura, A. Hayasaka, A. Higashida, H. Sakaguchi, M. Takeya, K. Takahashi, K. Inoue, N. Noguch, E. Niki and T. Kodama, *Proc. Natl. Acad. Sci. USA*, 1998, **95**, 10123.
[79] U. Hommel, H. P. Weber, L. Oberer, H. U. Naegeli, B. Oberhauser and C. A. Foster, *FEBS Lett*, 1996, **379**, 69.
[80] C. A. Foster, M. Dreyfuss, B. Mandak, J. G. Meingassner, H. U. Naegeli, A. Nussbaumer, L. Oberer, G. Scheel and E. M. Swoboda, *J. Dermatol.*, 1994, **21**, 847.

[81] C. A. Foster, J. Besemer, J. G. Meingassner, H. U. Naegeli, G. Schön, D. Bevec and T. Brend, *Skin Pharmacol*, 1996, **9**, 149.
[82] E. P. Schreiner, M. Kern and A. Steck, *J. Org. Chem.*, 2002, **67**, 8299.
[83] Y. Chen, M. Bilban, C. A. Foster and D. L. Boger, *J. Am. Chem. Soc.*, 2002, **124**, 5431.
[84] E. P. Schreiner, M. Kern, A. Steck and C. A. Foster, *Bioorg. Med. Chem. Lett.*, 2004, **14**, 5003.
[85] D. L. Boger, Y. Chen and C. A. Foster, *Bioorg. Med. Chem. Lett.*, 2000, **10**, 1741.
[86] L. Ni, C. Q. Meng and J. A. Sikorski, *Expert Opin. Ther. Patents*, 2004, **14**, 1669.
[87] C. Q. Meng, X. S. Zheng, L. Ni, Z. Ye, J. E. Simpson, K. J. Worsencroft, M. R. Hotema, M. D. Weingarten, J. W. Skudlarek, J. M. Gilmore, L. K. Hoong, R. R. Hill, E. M. Marino, K. L. Suen, C. Kunsch, M. A. Wasserman and J. A. Sikorski, *Bioorg. Med. Chem. Lett.*, 2004, **14**, 1513.
[88] R. B. Howard, A. M. McDonough, D. Stefanopoulos, J. Luchoomun, C. Q. Meng, J. A. Sikorski, M. A. Wasserman, C. Kunsch and C. L. Sundell, *FASEB J*, 2003, **17**, A666.
[89] C. L. Sundell, L. K. Landers, F. H. Qiu, L. Ni, A. Souder, K. Karu, M. McDonald, R. Howard, C. Q. Meng, J. Sikorski, C. Kunsch and M. A. Wasserman, *Am. J. Respir. Crit. Care Med.*, 2003, **167**, A357.
[90] T. Pfitzer, N. Niederhoffer and B. Szabo, *Br. J. Pharmacol.*, 2004, **142**, 943.
[91] N. M. Curran, B. D. Griffin, D. O'Toole, K. J. Brady, S. N. Fitzgerald and P. N. Moynagh, *J. Biol. Chem.*, 2005, **280**, 35797.
[92] H. Ulbrich, O. Soehnlein, X. Xie, E. E. Eriksson, L. Lindbom, W. Albrecht, S. Laufer and G. Dannhardt, *Biochem. Pharmacol.*, 2005, **70**, 30.
[93] S. E. Yoo, K. Y. Yi, S. Lee, J. Suh, N. Kim, B. H. Lee, H. W. Seo, S. O. Kim, D. H. Lee, H. Lim and H. S. Shin, *J. Med. Chem.*, 2001, **44**, 4207.
[94] J. Kim, K. H. Nam, S. O. Kim, J. H. Choi, H. C. Kim, S. D. Yang, J. H. Kang, Y. H. Ryu, G. T. Oh and S. E. Yoo, *FASEB J*, 2004, **18**, 714.
[95] N. M. Dagia, N. Harii, A. E. Meli, X. Sun, C. J. Lewis, L. D. Kohn and D. J. Goetz, *J. Immunol.*, 2004, **173**, 2041.
[96] C. Wenisch, D. Myskiw, A. Gessl and W. Graninger, *J. Clin. Endocrinol. Metab.*, 1995, **80**, 2122.
[97] L. D. Kohn, R. W. Curley and J. M. Rice, *US Patent* 6,365,616, 2002.
[98] D. S. Singer, L. Kohn, E. Mozes, M. Saji, J. Weissman, G. Napolitano and F. D. Ledley, *US Patent* 5,556,754, 1996.
[99] J. W. Chen, F. Y. Lin, Y. H. Chen, T. C. Wu, Y. L. Chen and S. J. Lin, *Arterioscler. Thromb. Vasc. Biol.*, 2004, **24**, 2075.
[100] D. M. Huryn, A. W. Konradi, S. Ashwell, S. B. Freedman, L. J. Lombardo, M. A. Pleiss, E. D. Thorsett, T. Yednock and J. D. Kennedy, *Curr. Top. Med. Chem.*, 2004, **4**, 1473.
[101] J. W. Tilley, L. Chen, A. Sidduri and N. Fotouhi, *Curr. Top. Med. Chem.*, 2004, **4**, 1509.
[102] L. Steinman, *Nat. Rev. Drug Discov.*, 2005, **4**, 510.
[103] K. A. Keeley, M. P. Rivey and D. R. Allington, *Ann. Pharmacother*, 2005, **39**, 1833, and references cited therein.

Recent Advances in Gastrointestinal Prokinetic Agents

David A. Sandham[1] and Hans-Jürgen Pfannkuche[2]

[1]Global Discovery Chemistry, Horsham Research Centre, Wimblehurst Road, Horsham RH12 5AB, UK
[2]Novartis Institutes of Biomedical Research, Gastrointestinal Diseases Area, Postfach, CH-4002 Basel, Switzerland

Contents

1. Introduction	211
2. Prokinetic agent target classes	212
2.1. Dopamine D2 receptor antagonists	212
2.2. Serotonin 5-HT$_4$ receptor agonists	213
2.3. Serotonin 5-HT$_3$ receptor agonists	214
2.4. Cholecystokinin CCK$_a$ receptor antagonists	215
2.5. Motilin receptor agonists	216
2.6. μ-opioid receptor antagonists	217
3. Conclusions	217
References	218

1. INTRODUCTION

Symptoms such as abdominal pain, nausea, vomiting, bloating and constipation impact on the quality of life of many patients suffering from functional gastrointestinal disorders (FGIDs). FGIDs present without overt biochemical or structural etiology and encompass conditions like functional dyspepsia (FD), symptomatic (non-erosive) gastro-esophagal reflux disease (sGERD), intestinal pseudo-obstruction, irritable bowel syndrome with constipation (IBS-C) and chronic idiopathic constipation. Various types of gastrointestinal (GI) dysmotility have been documented in patients with FGIDs, hence physicians have employed medications intended to modulate altered motility. Laxatives may be considered as progenitors of drugs that stimulate GI motility and are still in use for the treatment of patients with moderate to severe constipation. Laxatives comprise different classes of compound, and many of them increase intestinal electrolyte/water secretion, with stimulation of bowel motility being a secondary phenomenon. A compound that could be regarded as a first-generation prokinetic agent is metoclopramide (D2 anti-dopaminergic, 5-HT$_3$ antagonist, 5-HT$_4$ agonist). It stimulates gastric emptying through an improved antroduodenal coordination and can enhance propulsive activity in the upper GI tract. Metoclopramide is primarily used to treat symptoms of dyspepsia and nausea/vomiting of different origins. Cisapride (5-HT$_4$ agonist, 5-HT$_3$ antagonist) was developed as a second-generation agent that lacked overt CNS side effects and presented with greater prokinetic efficacy. It was used broadly, however, primarily applied to treat symptomatic GERD (e.g. nocturnal heartburn) and FD. Unfortunately, the use of this drug was associated with cases of cardiac arrhythmia and deaths due to QTc interval prolongation. Subsequently, cisapride was found to potently block a delayed rectifier potassium channel (I_{Kr}/hERG) in the heart and was withdrawn from most markets. Another compound that can act as a prokinetic agent is erythromycin, the macrolide antibiotic. Erythromycin has been introduced in the therapy of more severe dysmotility conditions affecting the upper GI tract. Acceleration of gastric emptying rate and induction of strong propagating contractions in the proximal small bowel are the most pronounced actions of erythromycin, which acts *via* stimulation of motilin receptors. A compound designed to act as a prokinetic agent throughout the GI tract is tegaserod (5-HT$_4$ agonist). This drug, which is now available in many countries, and several other compounds recently or currently in development as prokinetic agents are discussed below.

2. PROKINETIC AGENT TARGET CLASSES

2.1. Dopamine D2 receptor antagonists

Blockade of D2 receptors present on enteric neurons and/or those located at the chemoreceptor trigger zone may promote motility. In fact, D2 receptor antagonists such as metoclopramide and domperidone are in use for the treatment of dyspeptic symptoms, despite potential side effects such as hyperprolactinemia and extrapyramidal dystonic reactions [1].

Itopride **1**, which is available on the market in Japan, is a D2 receptor antagonist, which also exhibits acetylcholine esterase inhibition. The compound stimulated GI motility when dosed i.v. to conscious dogs and promoted colonic transit in rats and guinea pigs after oral administration [2]. After 8 weeks treatment in patients with FD, oral itopride t.i.d. was significantly more effective than placebo in reducing self-reported symptom scores including pain and fullness [3]. In a smaller trial in patients with GERD, itopride t.i.d. for 30 days significantly improved acid reflux and symptoms such as heartburn compared to pre-treatment [4]. In both studies, while there was some increase in prolactin levels, no significant adverse events occurred. However, most recent reports indicate that itopride did not meet expectations in clinical phase III studies in FD [5].

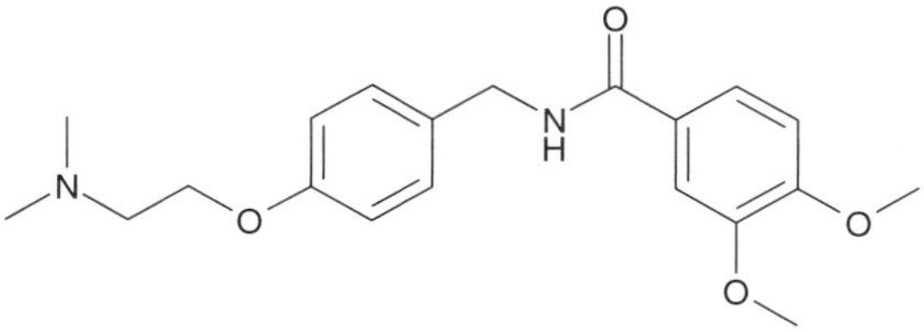

1

2.2. Serotonin 5-HT$_4$ receptor agonists

Serotonin (5-HT) is considered a key mediator/neurotransmitter in the GI tract. Large amounts of 5-HT are stored in enterochromaffin cells in the gut and released in a highly regulated fashion. Moreover, serotoninergic neurons are present in the enteric nervous system. Of all serotonin receptor subtypes found in the GI tract, both 5-HT$_3$ and 5-HT$_4$ receptors have been investigated most thoroughly. 5-HT$_4$ receptors located on nerve terminals of GI motor- and interneurons facilitate the release of neurotransmitters such as acetylcholine and, thereby, enhance motility. More recent data also suggest a role of 5-HT$_4$ receptors in the triggering of the peristaltic reflex [6]. Hence, agonism at the 5-HT$_4$ receptor is an established prokinetic mechanism [7].

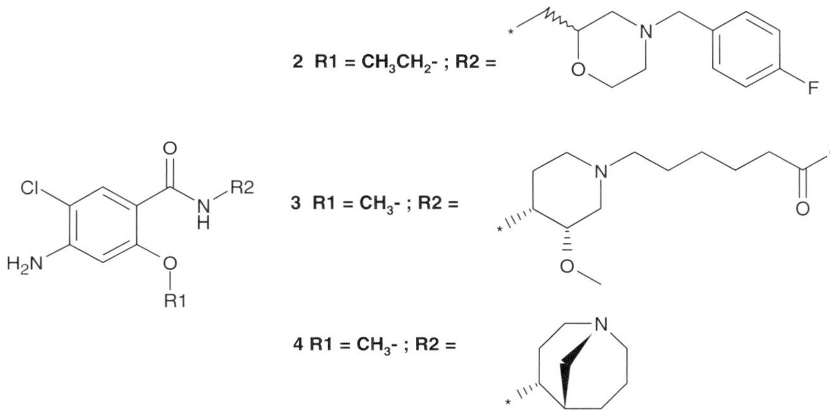

Mosapride **2** is a selective 5-HT$_4$ receptor agonist, which is marketed in Japan to treat gastric disturbances. While a close analogue of cisapride, mosapride shows no similar cardiac adverse effects. However, the overall pharmacodynamic profile of mosapride appears to be similar to that of cisapride, since the primary debenzylated metabolite of mosapride exhibits 5-HT$_3$ receptor antagonist activity. In GERD patients, mosapride t.i.d. for 7 days showed a significant but small effect on improvement in acid reflux variables and esophageal motor function compared to the pre-treatment period. These effects are comparable to those seen with cisapride [8]. In FD patients, mosapride t.i.d. for 4 weeks provided modest but significant improvements in self-reported symptom scores compared to the pre-treatment period [9].

ATI-7505 **3** is in development for the treatment of reflux disease. This compound, another 5-HT$_4$ receptor agonist, accelerated gastric emptying following dosing of 0.2 mg/kg i.v. to fed dogs. ATI-7505 exhibited prokinetic effects in the upper GI tract including small intestine rather than colonic motor activity [10].

Renzapride **4** is a dual 5-HT$_4$ agonist/5-HT$_3$ antagonist in development for the treatment of IBS-C. A recent mechanistic study demonstrated significantly improved colonic transit rates compared to placebo in patients with IBS-C after treatment for 11–14 days. Interestingly, the acceleration of colonic transit did not lead to any significant effects on bowel function (no improvement or satisfactory relief of symptom) [11].

Prucalopride **5** is a selective 5-HT$_4$ agonist which when dosed once daily for 4 weeks showed significant improvements compared to placebo in stool softening, decreased straining and time to first stool in patients with chronic constipation, who were refractory to laxatives. However, further development of this compound is on hold due to carcinogenicity issues [12].

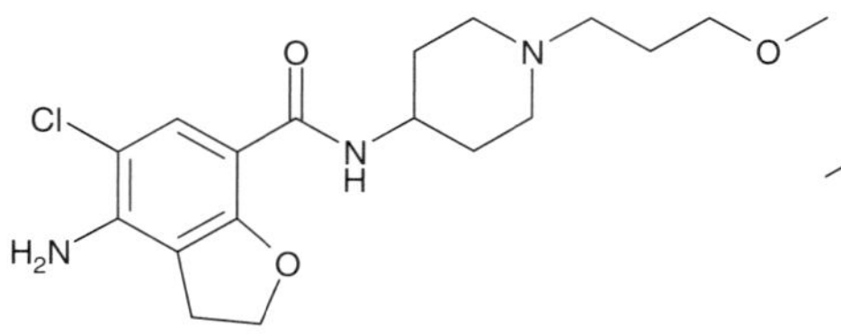

Tegaserod **6** is a 5-HT$_4$ receptor agonist which has been approved by the FDA for IBS-C and chronic idiopathic constipation and is marketed in the US and elsewhere. Tegaserod is rapidly absorbed in man and shows linear pharmacokinetics in a 2-12 mg oral dose range, with no significant differences found between healthy volunteers, young or elderly IBS patients [13]. In clinical phase III studies for IBS-C conducted in predominantly female patients, tegaserod b.i.d. at 6 mg for 12 weeks compared to placebo demonstrated significant improvements in self-reported symptom scores including abdominal pain, bloating and bowel function. Interestingly, stimulation of 5-HT$_4$ receptors with tegaserod has been shown to attenuate visceral sensitivity (visceral pain response) in both animals and humans. The drug is well tolerated, with transient diarrhea/soft stools being the most frequent adverse event found [14]. Phase III data in patients suffering from idiopathic chronic constipation indicated that tegaserod at 2 or 6 mg b.i.d. for 12 weeks significantly increased the number and quality of spontaneous bowel movements, compared to placebo [15]. A recent meta-analysis of all placebo-controlled trials of agents for chronic constipation ranked tegaserod with the osmotic laxative polyethylene glycol as the only two modalities for which there was a good level of evidence to recommend their use in this disorder [16].

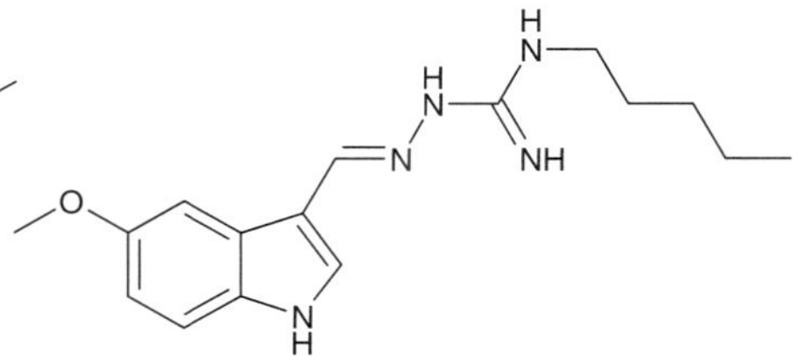

2.3. Serotonin 5-HT$_3$ receptor agonists

5-HT$_3$ receptors are located on vagal afferents and myenteric neurons. They are involved in both contractile and secretory responses and can trigger sensory signals through the vagus nerve. 5-HT$_3$ receptor antagonists are well known to attenuate chemotherapy-induced nausea and emesis through the inhibition of vagal signaling and to inhibit intestinal motility and secretion. In agreement with these activities, alosetron (5-HT$_3$ antagonist) has been developed for the treatment of IBS with diarrhea and conversely, a 5-HT$_3$ agonist might be expected to increase motility. In line with this hypothesis, a single 4 mg dose of the 5-HT$_3$ receptor agonist, pumosetrag **7**, significantly accelerated small intestinal transit compared to placebo in healthy volunteers. However, significant side effects of nausea, flushing and itching, all associated with systemic activation of 5-HT$_3$ receptors were also observed [17]. In a small exploratory study of patients with idiopathic constipation, pumosetrag given 0.5 mg once daily for 2 weeks showed significant improvement in bowel motility after a 1 week placebo run-in; the systemic side effects seen in the previous study were not noted at this lower dose [18].

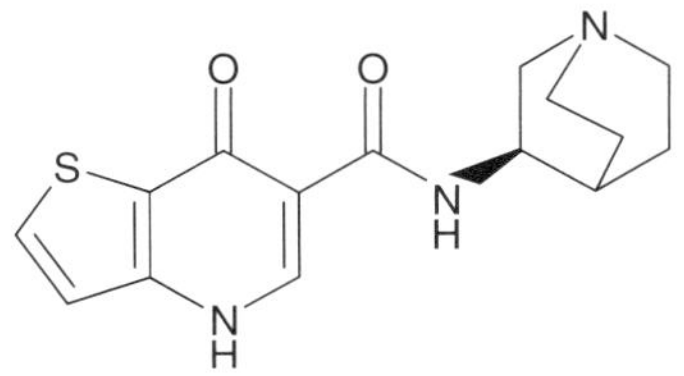

7

2.4. Cholecystokinin CCK$_a$ receptor antagonists

Chlolecystokinin (CCK) is released from endocrine cells in the duodenal and jejunal mucosa in response to certain food components and promotes inhibition of gastric emptying, among other effects. These effects of CCK are mediated through the CCK$_a$ (or CCK$_1$) receptor, blockade of which represents an approach to stimulate gut motility. In line with this, symptom improvements with the CCK$_a$ antagonist, loxiglumide, have been demonstrated in FD patients [7]. The active single enantiomer of loxiglumide, dexloxiglumide **8** t.i.d. for 7 days significantly accelerated gastric emptying in patients with IBS-C compared to placebo. However, overall colonic transit was unaffected and no significant relief of global IBS symptoms or overall bowel function were observed [19]. Recent information indicates that dexloxiglumide failed in clinical phase III studies in patients suffering from IBS-C [20].

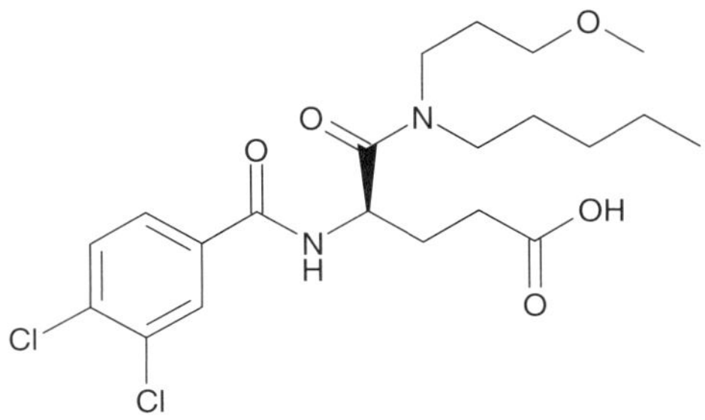

8

2.5. Motilin receptor agonists

Motilin is a 22 amino acid polypeptide, which primarily stimulates antral contractions and thus promotes gastric emptying. This effect is mediated through the motilin receptor, which is expressed throughout the gut. Many synthetic motilin agonists are derived from the macrolide antibiotic, erythromycin, which exhibits gastric motor stimulation effects due to motilin receptor activation, in addition to its antibiotic activity [21]. These compounds, such as alemcinal (ABT-229) **9**, exhibit improved potency at the motilin receptor and minimal antibiotic activity. However, **9** failed to show efficacy in relieving symptoms of GERD compared to placebo when dosed up to 10 mg t.i.d. for 8 weeks [22]. ABT-229 and another macrolide, KC11458, showed no improvement over placebo in affecting gastric emptying rates in patients with diabetic gastroparesis [23,24]. One explanation for the lack of efficacy with these compounds may be rapid induction of tachyphylaxis. Receptor trafficking studies using confocal microscopy have shown that the extent of motilin receptor downregulation varied from compound to compound [25,26], leaving open the possibility of selecting a clinically useful compound by early screening for tachylphylaxis effects. A recent patent highlights such an approach, identifying macrolides with reduced potential to induce tachyphylaxis compared with **9** or erythromycin, using an *in vitro* rabbit duodenal strip model [27].

In an alternative approach, a non-macrolide **10** optimized from a high-throughput screen for motilin agonism, was reported to show similar potency and reduced tachyphylactic potential compared to **9** [28].

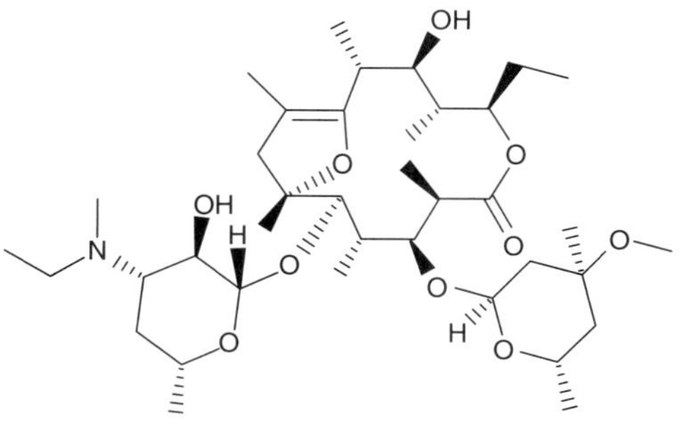

9

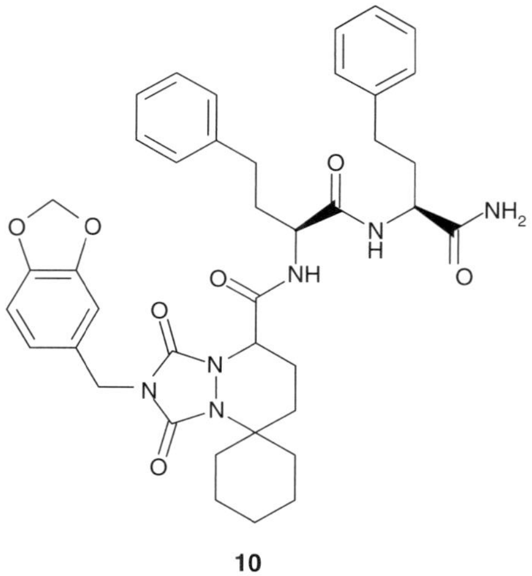

10

Very recently, atilmotin, a peptide analogue of the 1–14 amino acid fragment of motilin, has been reported, which remains a potent and selective motilin agonist. When dosed as a single i.v. bolus after three meals in healthy volunteers, atilmotin showed a relatively short-lived increase in gastric emptying rate compared to placebo, which may be related to the short half-life of the compound. However, the short half-life may also contribute to reduced tachyphylaxis effects of atilmotin [29].

2.6. μ-opioid receptor antagonists

Activation of peripheral μ-opioid receptors present in the enteric nervous system increases tone and phasic contractility leading to compromised motility throughout the GI tract. Opioid-induced bowel dysfunction is a condition for which no adequate treatment is currently available and, moreover, activation of opioid receptors might be a contributory factor to the development of post-operative ileus (POI). A peripheralized opioid receptor antagonist, which does not interfere with analgesia represents a possible treatment for these conditions. Methylnaltrexone **11** is a quaternary analogue of the μ-opioid antagonist naltrexone, which does not cross the blood–brain barrier and has limited oral absorption. After parenteral administration, methylnaltrexone reversed motility impairments caused by opioids in healthy volunteers and chronic methadone users [30]. The compound is presently in clinical development for various conditions associated with opioid-induced constipation.

Alvimopan **12** is a peripherally acting, poorly absorbed selective μ-opioid antagonist, which has received an FDA approvable letter for treatment of POI [31]. In a clinical phase III trial, alvimopan 6 mg and 12 mg, dosed orally prior to surgery and then b.i.d. for up to 7 days post-surgery in patients undergoing bowel resection or radical hysterectomy, significantly improved time to both recovery of GI motor function and hospital discharge compared with placebo [32]. In healthy volunteers, oral alvimopan reversed the inhibitory effects of codeine on small bowel and colonic transit of a ^{99}Tc labeled test meal, suggesting utility in opioid-induced bowel dysfunction [33]. In the same trial, alvimopan improved overall colonic transit compared with placebo alone; however, in a recent study in patients with chronic constipation, alvimopan failed to show clinical efficacy [34].

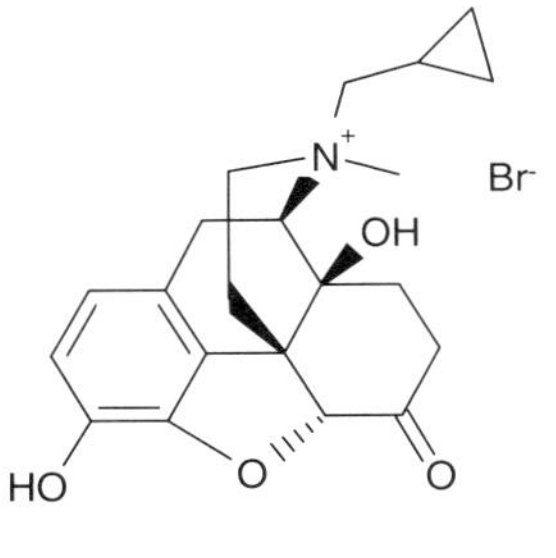

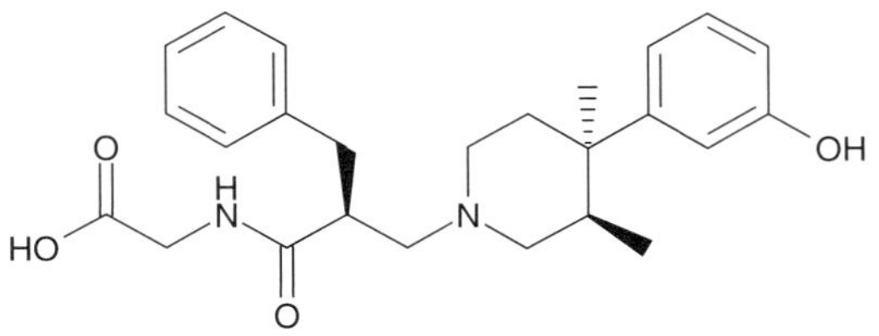

Alvimopan presents with an increased potency and duration of action compared with methylnaltrexone. The greater duration of action may be a consequence of a slower dissociation rate of alvimopan from the μ-opioid receptor [35].

3. CONCLUSIONS

GI disorders associated with impaired motility remain a therapeutic challenge. Although an increasing number of therapeutic options have become available during recent years, many patients are still in need of adequate treatment. Progress has been made with respect to prokinetic agents targeting the lower GI tract, whereas patients suffering from symptomatic GERD or FD still await drugs to effectively treat reflux and regurgitation, and to provide relief from symptoms due to gastric dysmotility.

In recent years, several prokinetic mechanisms have been identified which provided a strong scientific rationale for the treatment of conditions associated with GI motor dysfunction. Unfortunately, most of the mechanisms could not be validated since the corresponding compounds did not succeed in the clinical setting, despite demonstration of prokinetic activity in mechanistic studies. In fact, it is well recognized that vomiting, cramping, constipation, diarrhea and many other symptoms are due to disturbed GI motility. However, many altered GI motor activities are only weakly correlated with symptoms and are not sufficient to explain, for instance, chronic or recurrent abdominal pain/discomfort and bloating.

Research will continue to elucidate both the pathophysiology of FGIDs and mechanisms regulating GI motor function. Furthermore, a number of innovative agents are presently in development, some of which may eventually demonstrate both prokinetic activity and clinical benefit in patients.

REFERENCES

[1] M. Tonini, L. Cipollina, E. Poluzzi, F. Crema, G. R. Corazza and F. De Ponti, *Aliment. Pharmacol. Ther.*, 2004, **19**, 379.
[2] T. Tsubouchi, T. Saito, F. Mizutani, T. Yamauchi and Y. Iwanaga, *J. Pharmacol. Exp. Ther.*, 2003, **306**, 787.
[3] G. Holtmann, N. J. Talley, T. Liebregts, B. Adam and C. Parow, *N. Engl. J. Med.*, 2006, **354**, 832.
[4] Y. S. Kim, T. H. Kim, C. S. Choi, Y. W. Shon, S. K. Kim, S. Wook, S. G. Seo, Y. H. Nah, M. G. Choi and S. C. Choi, *World J. Gastroenterol.*, 2005, **11**, 4210.
[5] Anon, *Investigational Drugs Database: Drug Report for Itopride*, 2006.
[6] J. R. Grider, A. E. Foxx-Orenstein and J.-G. Jin, *Gastroenterology*, 1998, **115**, 370.
[7] J. J. Galligan and S. Vanner, *Neurogastroenterol. Motil.*, 2005, **17**, 643.
[8] M. Ruth, C. Finizia, L. Cange and L. Lundell, *Eur. J. Gastroenterol. Hepatol.*, 2003, **15**, 1115.
[9] M. Otaka, M. Jin, M. Odashima, T. Matsuhashi, I. Wada, Y. Horikawa, K. Komatsu, R. Ohba, J. Oyake, N. Hatakeyama and S. Watanabe, *Aliment. Pharmacol. Ther.*, 2005, **21**, 42.
[10] I. Irwin, M. Palme and C. Becker, *WO Patent* 068461, 2005.
[11] M. Camilleri, S. McKinzie, J. Fox, A. Foxx-Orenstein, D. Burton, G. Thomforde, K. Baxter and A. R. Zinsmeister, *Clin. Gastroenterol. Hepatol.*, 2004, **2**, 895.
[12] G. Coremans, R. Kerstens, M. De Pauw and M. Stevens, *Digestion*, 2003, **67**, 82.
[13] S. Appel-Dingemanse, *Clin. Pharmacokinet.*, 2002, **41**, 1021.
[14] M. Camilleri, *Aliment. Pharmacol. Ther.*, 2001, **15**, 277.
[15] J. F. Johanson, *Aliment. Pharmacol. Ther.*, 2004, **20**, 20.
[16] D. Ramkumar and S. S. C. Rao, *Am. J. Gastroenterol.*, 2005, **100**, 936.

[17] N. S. Coleman, L. Marciani, E. Blackshaw, J. Wright, M. Parker, T. Yano, S. Yamazaki, P. Q. Chan, K. Wilde, P. A. Gowland, A. C. Perkin and R. C. Spiller, *Aliment. Pharmacol. Ther.*, 2003, **18**, 1039.
[18] T. Fujita, S. Yokota, M. Sawada, M. Majima, Y. Ohtani and Y. Kumagai, *J. Clin. Pharm. Ther.*, 2005, **30**, 611.
[19] F. Cremonini, M. Camilleri, S. McKinzie, P. Carlson, C. E. Camilleri, D. Burton, G. Thomforde, R. Urrutia and A. R. Zinsmeister, *Am. J. Gastroenterol.*, 2005, **100**, 652.
[20] Anon, *Investigational Drugs Database: Drug Report for Dexloxiglumide*, 2006.
[21] T. L. Peters, *Neurogastroenterol. Motil.*, 2006, **18**, 1.
[22] C. L. Chen, W. C. Orr, M. H. Verlinden, A. Dettmer, H. Brinkhoff, D. Riff, S. Schwartz, R. Soloway, R. Krause, F. Lanza and R. J. Mack, *Aliment. Pharmacol. Ther.*, 2002, **16**, 749.
[23] N. J. Talley, M. Verlinden, D. J. Geenen, R. B. Hogan, D. Riff, R. W. McCallum and R. J. Mack, *Gut*, 2001, **49**, 395.
[24] A. Russo, J. E. Stevens, N. Giles, G. Krause, D. G. O'Donovan, M. Horowitz and K. L. Jones, *Aliment. Pharmacol. Ther.*, 2004, **20**, 333.
[25] V. Lamian, A. Rich, Z. Ma, J. Li, R. Seethala, D. Gordon and Y. Dubaquie, *Mol. Pharmacol.*, 2006, **69**, 109.
[26] L. Thielemans, I. Depoortere, J. Perret, P. Robberecht, Y. Liu, T. Thijs, C. Carreras, E. Burgeon and T. L. Peeters, *J. Pharmacol. Exp. Ther.*, 2005, **313**, 1397.
[27] Y. Liu, C. Carreras, D. Myles and Y. Chao, *WO Patent* 060693, 2005.
[28] J. J. Li, H.-G. Chao, H. Wang, J. A. Tino, R. M. Lawrence, W. R. Ewing, Z. Ma, M. Yan, D. Slusarchyk, R. Seethala, H. Sun, D. Li, N. T. Burford, R. H. Stoffel, M. E. Salyan, C. Y. Li, M. Witkus, N. Zhao, A. Rich and D. A. Gordon, *J. Med. Chem.*, 2004, **47**, 1704.
[29] M. I. Park, I. Ferber, M. Camilleri, K. Allenby, R. Trillo, D. Burton and A. R. Zinmeister, *Neurogastroenterol. Motil.*, 2006, **18**, 28.
[30] C.-S. Yuan, H. Doshan, M. R. Charney, M. O' Connor, T. Karrison, S. A. Maleckar, R. J. Israel and J. Moss, *J. Clin. Pharmacol.*, 2005, **45**, 538.
[31] I. A. Azodo and E. D. Ehrenpreis, *Curr. Opin. Investig. Drugs.*, 2002, **3**, 1496.
[32] C. P. Delaney, J. L. Weese, N. H. Hyman, J. Bauer, L. Techner, K. Gabriel, W. Du, W. K. Schmidt and B. A. Wallin, *Dis. Colon Rectum.*, 2005, **48**, 1114.
[33] J. Gonenne, M. Camilleri, I. Ferber, D. Burton, K. Baxter, K. Keyashian, J. Foss, B. Wallin, W. Du and A. R. Zinsmeister, *Clin. Gastroenterol. Hepatol.*, 2005, **3**, 784.
[34] Anon, *Investigational Drugs Database: Drug Report for Alvimopan*, 2006.
[35] J. A. Cassel, J. D. Daubert and R. N. Dehaven, *Eur. J. Pharmacol.*, 2005, **520**, 29.

PGD_2 Antagonists

Julio C. Medina and Jiwen Liu

Amgen Inc., South San Francisco, CA 94080, USA

Contents

1. Introduction	221
2. DP Antagonists	223
2.1. Early DP antagonists and their related analogs	223
2.2. 3-Indole acetic acid derivatives	223
2.3. 4-Indole acetic acid derivatives	224
2.4. Other indole derivatives	225
2.5. Benzimidazole derivatives	225
2.6. Phenyl acetic acid derivatives	225
3. CRTH2 Antagonists	226
3.1. 1-Indole acetic acid derivatives and related compounds	226
3.2. 3-Indole acetic acid derivatives and related compounds	228
3.3. Tetrahydroquinoline derivatives	229
3.4. Aryl acetic acid derivatives	230
3.5. Phenoxy acetic acid derivatives	230
3.6. Benzoic acid derivatives	231
4. Conclusion	231
References	232

1. INTRODUCTION

Prostaglandin D_2 (PGD_2, **1**) is believed to play a key role in mediating allergic reactions such as those seen in asthma, allergic rhinitis, atopic dermatitis and allergic conjunctivitis [1]. PGD_2 levels in bronchoalveolar lavage (BAL) fluid increase in response to antigen provocation. In fact, PGD_2 is the major cyclooxygenase product formed and secreted by activated mast cells [2–4] and, therefore, PGD_2 levels are used as a marker for activation of mast cells [5,6]. In animals, including humans, PGD_2 stimulates several responses observed in asthma, allergic rhinitis and other immune diseases (airway constriction, mucus secretion, increased microvascular permeability and recruitment of eosinophils) [7–12]. Mice that overexpress PGD_2 synthase, resulting in overproduction of PGD_2, experience increased levels of Th2 cytokines and chemokines accompanied by enhanced accumulation of eosinophils and lymphocytes in the lung following an allergic response to ovalbumin [11]. Thus, PGD_2 is thought to be involved in the acute and late phases of allergic reactions.

PGD_2 exerts its activity through two different G-protein-coupled receptors (GPCRs), termed DP (D prostanoid) and CRTH2 (chemoattractant receptor-homologous molecule expressed on Th2 cells).

DP, also know as DP_1, was the first PGD2 receptor discovered [13]. DP belongs to the prostanoid receptor family of GPCRs and is expressed on airway epithelium, smooth muscle and platelets. Upon stimulation, DP activates adenylate cyclase and increases the level of cAMP primarily *via* Gs-dependent pathways [14]. Genetic analysis of DP function using knock-out (KO) mice has shown that mice lacking DP do not develop asthmatic responses in an ovalbumin-induced asthma model [15]. Analysis of a selective DP antagonist in guinea-pig allergic rhinitis models demonstrated dramatic inhibition of early nasal responses, as assessed by sneezing, mucosal plasma exudation and nasal blockage, as well as late responses such as mucosal plasma exudation and eosinophil infiltration [16]. Moreover, DP antagonism alleviated allergen-induced plasma exudation in the conjunctiva in a guinea-pig allergic conjunctivitis model and antigen-induced eosinophil infiltration into the lung in a guinea-pig asthma model [16]. In addition, human genetic data suggest that DP may play a role in asthma [17]. These results demonstrate the importance of DP signaling for allergic responses.

The second PGD_2 receptor, CRTH2 (DP_2), is related to the N-formyl peptide receptor (FPR) subfamily of chemoattractant receptors [18]. CRTH2 is selectively expressed on Th2 cells, T cytotoxic type 2 (Tc2) cells, eosinophils and basophils [19–21]. CRTH2 activation induces an increase in intracellular Ca^{2+} mobilization *via* Gi–dependent pathways, allowing CRTH2 to transmit promigratory signals in response to PGD_2 [17,22,23]. In leukocytes, PGD_2 induces migration exclusively *via* CRTH2 [17,23–26]. The ability of PGD_2 to stimulate the migration of inflammatory cells has led to the hypothesis that CRTH2 may play a role in the recruitment of cellular components of the allergic response into diseased tissues. Studies in the rat of agonists and antagonists selective for CRTH2 over DP lend support to such a model. Selective activation of CRTH2 by intravenous injection of 13,14-dihydro-15-keto (DK)-PGD_2 into rats led to a dose- and time-dependent increase in the number of eosinophils in the peripheral blood. Pretreatment of the animals with Ramatroban (BAY u3405, 17), a CRTH2/thromboxane A2 receptor dual antagonist, completely abrogated DK-PGD_2-induced eosinophilia [12]. Furthermore, as with DP, human genetic data also suggest that CRTH2 may play a role in asthma [27].

These observations suggest that DP and CRTH2 play important and complementary roles in the physiological response of animals to PGD_2 and blockade of either one or both of these receptors may prove beneficial in alleviating allergic diseases triggered by PGD_2.

2. DP ANTAGONISTS

2.1. Early DP antagonists and their related analogs

2 BW A868C

3 S-5751

4

5

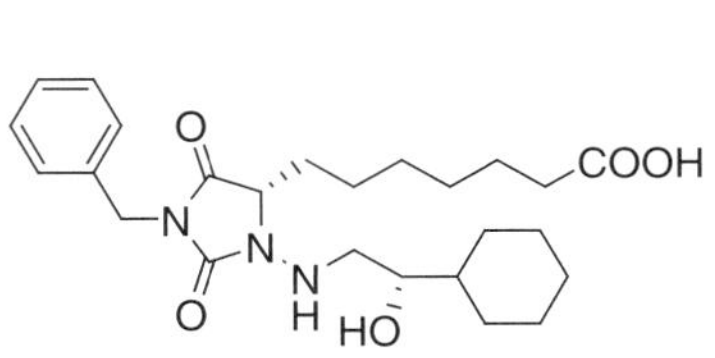

BW A868C (**2**) is one of the early DP antagonists reported in the literature. It binds the DP receptor with a K_d value of 1.45 nM (K_d was determined using human platelets) [28]. S-5751 (**3**) was the first DP antagonist reported to be under evaluation in clinical trials for rhinitis and asthma [29]. S-5751 demonstrated binding IC_{50} value of 1.9 nM in [^{3}H]-PGD_2 binding assays using human platelet membranes. It also inhibited PGD_2-induced cAMP formation in human platelets with an IC_{50} of 0.9 nM [30]. Several other DP antagonists, structurally similar to **3**, were described in the literature. Compound **4** had a binding IC_{50} value of 130 nM and an IC_{50} value of 70 nM in a functional assay [31].

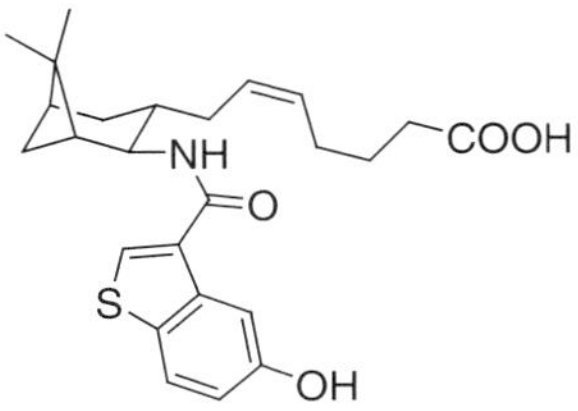

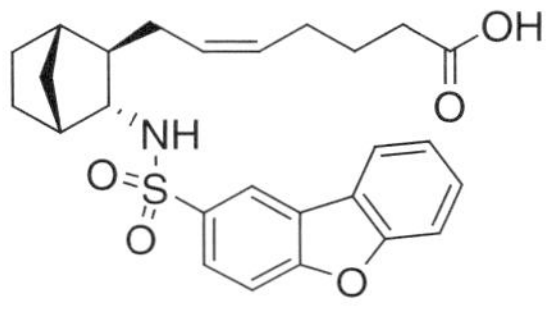

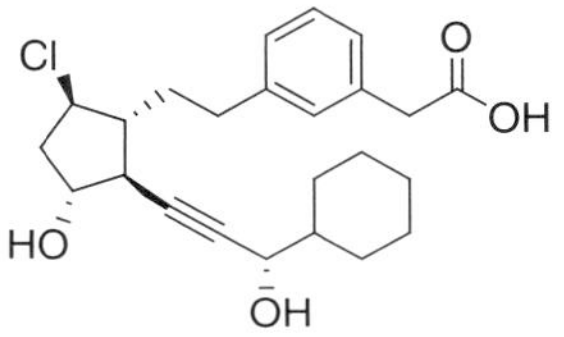

Another series of DP antagonists, exemplified by **5**, was reported. It displayed a [^{3}H]-PGD_2 binding K_i value of 0.8 nM at DP receptors in human platelet membranes and was shown to antagonize the PGD_2-induced increase in cAMP levels in EBTr cells (IC_{50} = 0.73 μM at 100 nM PGD_2; IC_{50} = 1.16 μM at 1 μM PGD_2) [32].

2.2. 3-Indole acetic acid derivatives

6

7a n = 1
7b n = 2
7c n = 3

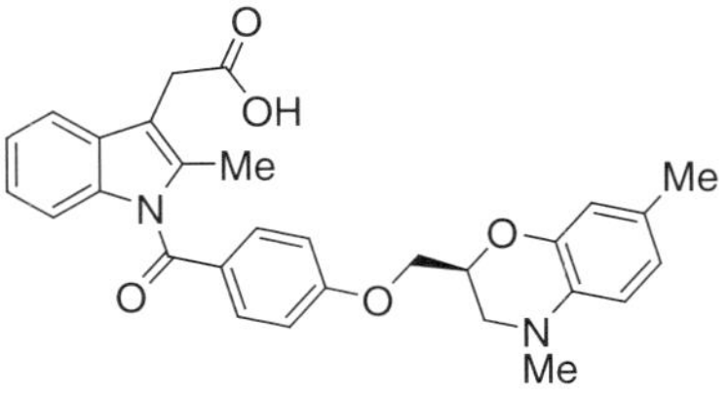

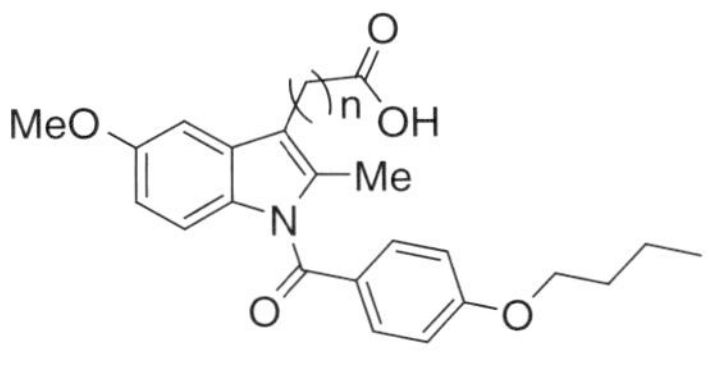

3-Indole acetic acid derivatives have been reported as both DP antagonists and CRTH2 antagonists (*vide infra*). While compounds **6** and **43** are structurally similar, compound **43** has been described as a CRTH2/DP dual antagonist, whereas compound **6** has been reported to be a DP antagonist [33]. It remains to be established if **6** has CRTH2 activity. The K_i for compound **6** in a [^{3}H]-PGD$_2$-binding assay was reported to be 7.4 nM. Compounds **7a**, **7b** and **7c** possess binding K_i values of 270, 280 and 190 nM, respectively, in a [^{3}H]-PGD$_2$ binding assays using mDP expressing CHO cell membranes. They also inhibited PGD$_2$-induced cAMP formation in mDP expressing CHO cells in the presence of 0.1% BSA at IC$_{50}$ values of 1.6, 1.5 and 4.5 µM, respectively [34,35]. Greater than 10-fold selectivity was achieved toward DP relative to mouse mEP1, mEP2, mEP3 and mEP4 receptors.

2.3. 4-Indole acetic acid derivatives

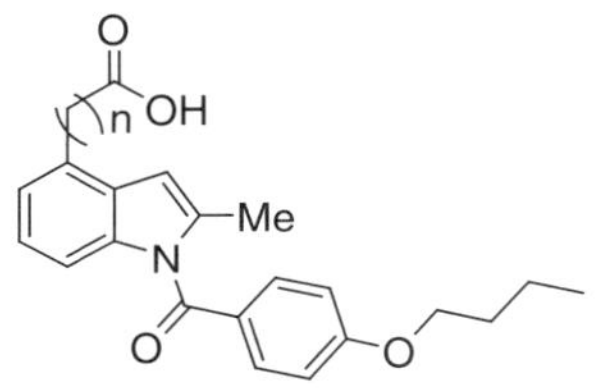

8a n = 1
8b n = 2
8c n = 3

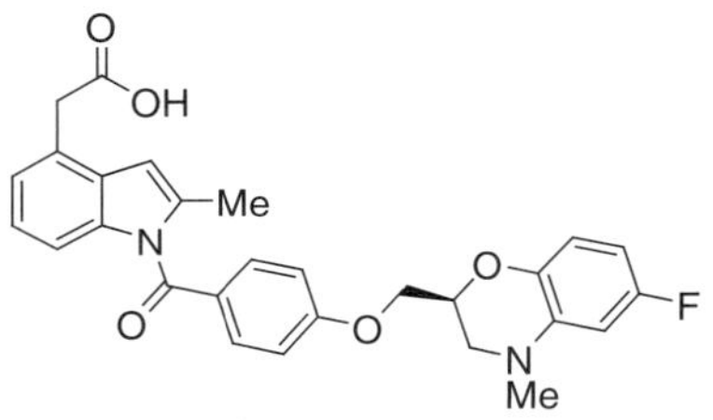

9

4-Indole acetic acid derivatives as DP antagonists, exemplified by **8a**, **8b** and **8c**, were derived from 3-indole acetic acids [35]. Compounds **8a**, **8b** and **8c** demonstrated binding K_i values of 10, 13 and 33 nM, respectively, in [^{3}H]-PGD$_2$ binding assays using mDP expressing CHO cell membranes. They also inhibited PGD$_2$-induced cAMP formation in mDP expressing CHO cells in the presence of 0.1% BSA at IC$_{50}$ values of 0.3, 0.43 and 0.57 µM, respectively [36]. Greater than 200-fold selectivity was achieved with **8a** and **8b** toward mDP relative to mEP1, mEP2, mEP3 and mEP4 receptors. Compound **8c** was reported to have some affinity for mEP2 (K_i = 46 nM) and >200-fold selectivity for mEP1, mEP3 and mEP4.

More potent DP antagonists were generated in this series, as exemplified by compound **9** [37,38]. The latter exhibited a binding K_i of 5.3 nM and an IC$_{50}$ of 0.8 nM in the same binding and functional assays. Evaluation *in vivo* showed a 60% inhibition of the increase in conjunctival vascular permeability induced by PGD$_2$ in guinea pigs (0.3 mg/kg, p.o.).

ONO-4127.Na is a DP antagonist in Phase I clinical trial for allergic rhinitis, but its structure has not been disclosed [39].

2.4. Other indole derivatives

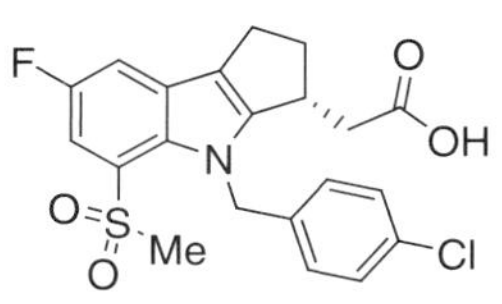

10 L-888839 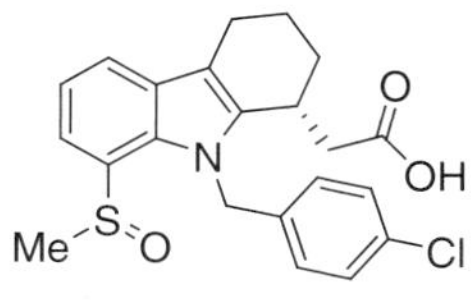   **11** 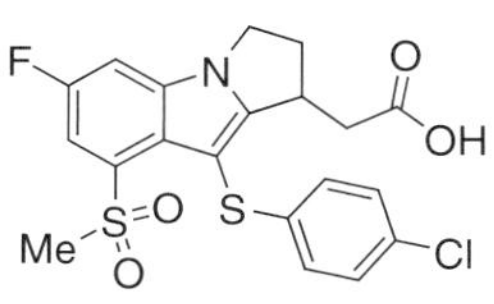   **12**

Compound **10** is a highly potent DP antagonist undergoing clinical evaluation for allergic rhinitis [40].

The related indole derivatives **11** and **12** were also disclosed as DP antagonists, but their DP activity was not reported [41–46]. Interestingly, very close structural analogs of **12** were also reported as CRTH2 agonists [47].

2.5. Benzimidazole derivatives

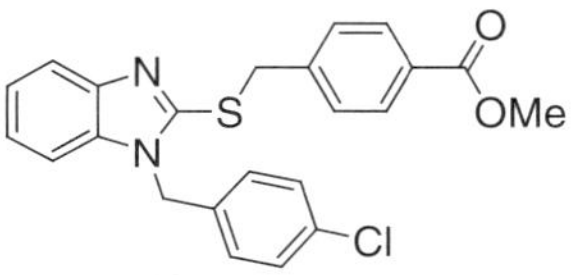

13 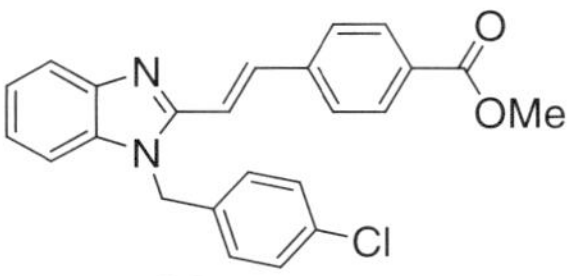   **14**

Benzimidazole derivatives have been evaluated as DP antagonists in [^{3}H]-PGD$_2$ binding assays using membranes from human DP (hDP)-expressing HEK293 (EBNA) cells. Compounds **13** and **14** are two of the most potent DP antagonists reported. Their DP binding K_i values were 73 and 38 nM, respectively. In the presence of 0.5% HSA, their affinities were decreased approximately by 2–4 fold. Compounds **13** and **14** have good selectivity. Their TP (thromboxane A2 receptor) binding K_i values were 9.1 and >75 μM, respectively. No functional data were reported [48].

2.6. Phenyl acetic acid derivatives

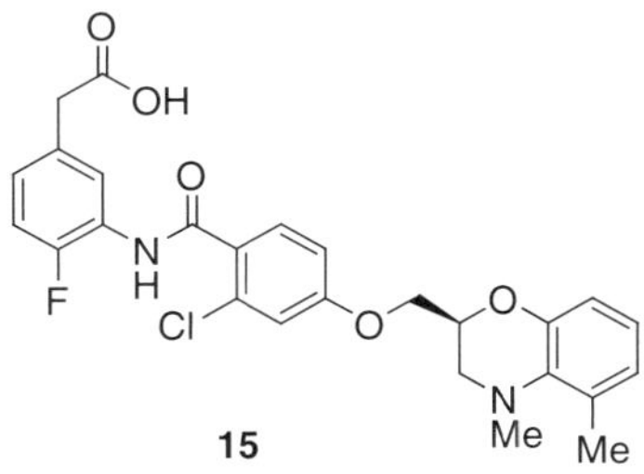

15 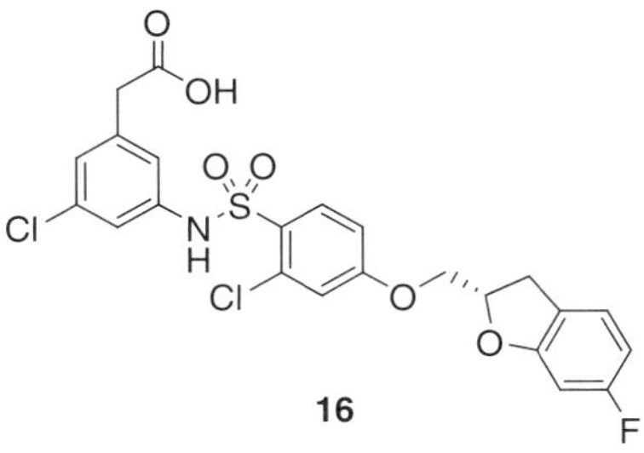   **16**

A series of phenylacetic acid derivatives, exemplified by **15** and **16**, were discovered as DP antagonists. The DP potencies were evaluated in [^{3}H]-PGD_2 binding and cAMP assays in hDP expressing CHO cells. The IC50 values were reported to be $<10\,\mu M$ [49,50].

3. CRTH2 ANTAGONISTS

3.1. 1-Indole acetic acid derivatives and related compounds

17 Ramatroban

18 R = H
19 R = Me

20

21

22

23

24

25

26

27

28

29

30

31

32

33

34

35

36

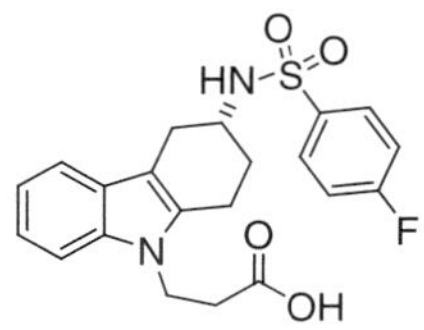

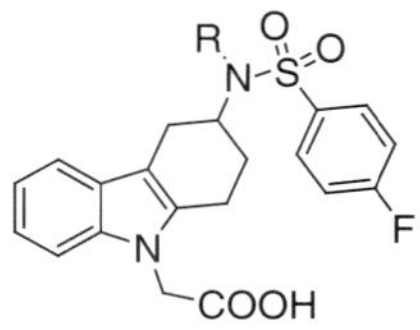

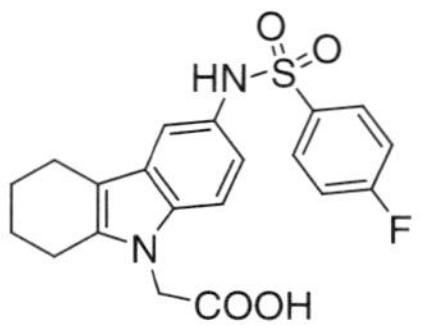

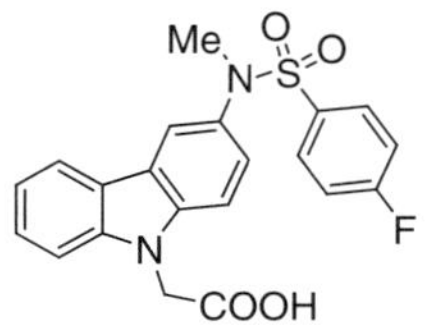

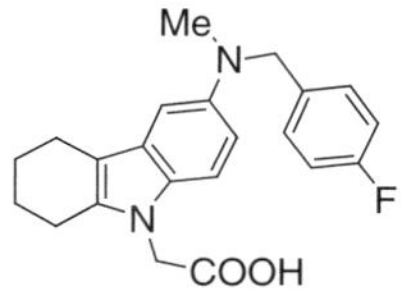

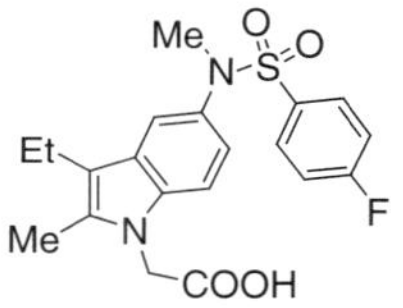

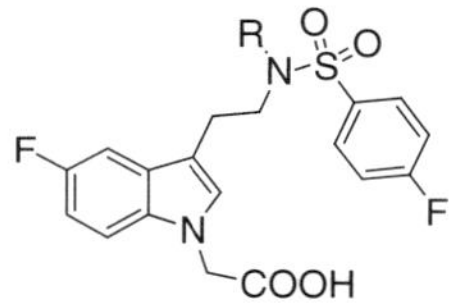

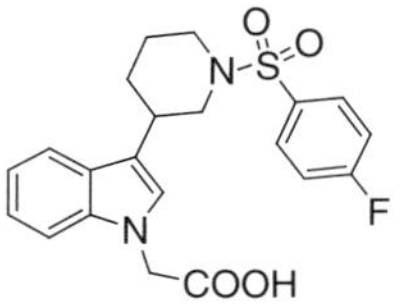

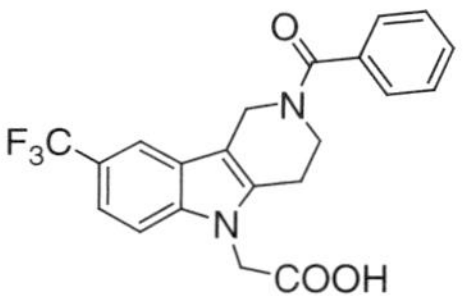

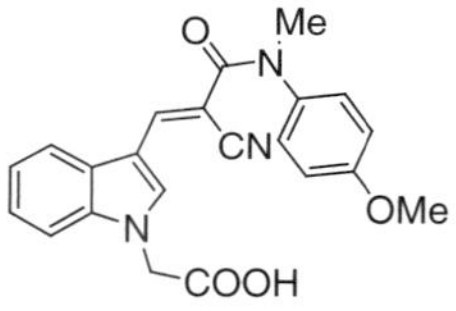

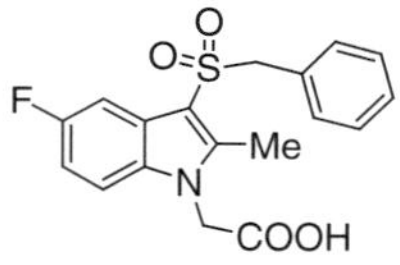

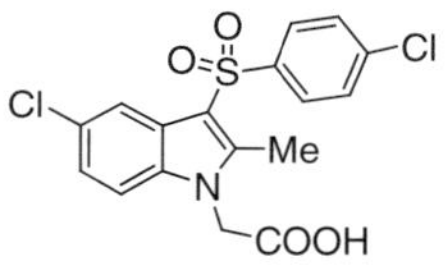

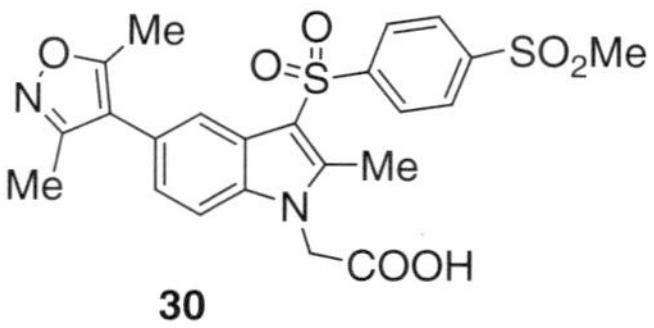

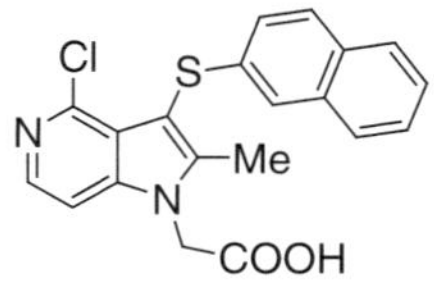

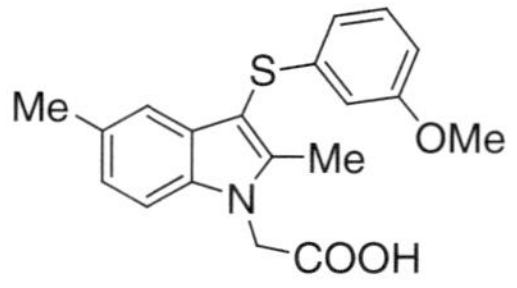

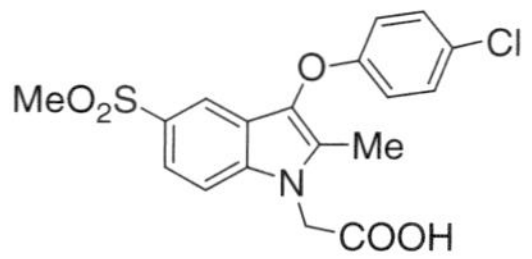

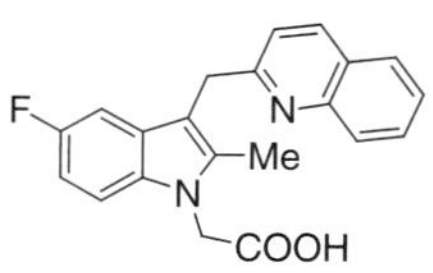

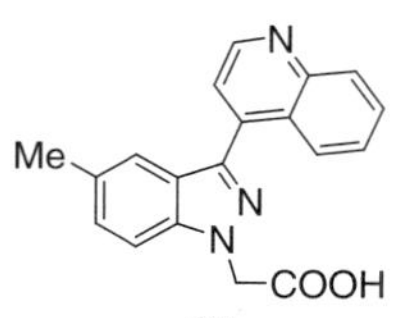

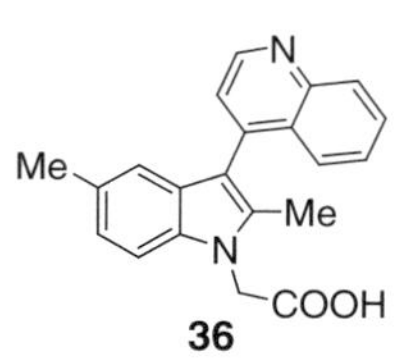

Numerous 1-indole acetic acid derivatives compounds have been reported as CRTH2 antagonists. These include the tetrahydrocarbazole derivative **17**, Ramatroban (BAY U3405), which is currently marketed in Japan for allergic rhinitis [51]. Ramatroban was developed as a thromboxane A2 receptor (TP) antagonist; later it was also shown to be a CRTH2 antagonist. Ramatroban has been reported to reduce antigen-induced early and late-phase allergic responses in mice, rats and guinea pigs. In humans, Ramatroban has been reported to attenuate PGD_2-induced bronchial hyper-responsiveness [52,53]. It has been postulated that the efficacy observed with Ramatroban in humans cannot be fully explained by its action on TP and that Ramatroban's efficacy in humans is due in part to its CRTH2 antagonist activity [23]. The affinities of Ramatroban and its close analogs, **18** and **19**, for the CRTH2 receptor were determined using a $[^3H]$-PGD_2 binding assay (K_i values for Ramatroban, **18** and **19** were 4.3, 0.5 and 0.6 nM, respectively) [54]. The antagonistic activity of these compounds was evaluated in a PGD_2-induced inositol phosphate production assay. Ramatroban and compounds **18** and **19** displayed IC_{50} values of 29, 3.8 and 1.2 nM, respectively. Thus, compounds **18** and **19**, analogs of Ramatroban featuring a shorter linker between the carboxylic acid and the indole, display significantly improved CRTH2 antagonistic activity. In addition, unlike Ramatroban, these compounds are reported to be selective for CRTH2 versus the TP receptor [54].

The CRTH2 binding affinity of compound **20** was evaluated together with Ramatroban in a $[^3H]$-PGD_2 binding assay using cell membranes from HEK293 cells expressing the CRTH2 receptor. While **20** was a more potent inhibitor than Ramatroban (K_i = 30 versus 290 nM) [55], unlike Ramatroban, **20** had no affinity for the TP receptor (K_i >20 µM) [55].

Compounds **19–25** are part of a large group of 1-indole acetic acid derivatives that were evaluated in a $[^3H]$-PGD_2 binding assay using CRTH2-expressing K562 cell membranes [56]. The IC_{50} values reported for compounds **19–25** are in the range of 3–20 nM. No functional data are available for these compounds.

Similarly, compounds **26** and **27** exemplify two indole series that are antagonists of the CRTH2 receptor. These analogs were evaluated in a $[^3H]$-PGD_2 binding assay using HEK293 cells expressing CRTH2 (IC_{50} = 1 and 7 nM, respectively) and in an intracellular calcium mobilization assay (FLIPR) [57,58]. No functional assay data were reported for compound **26**, while compound **27** had an IC_{50} value of 48 nM in the FLIPR assay.

Compounds **28–32** exemplify 3-sulfonyl and 3-sulfanyl 1-indole acetic acid derivatives antagonists of CRTH2. In addition, 3-aryloxy and 3-alkyl 1-indole acetic acid derivatives, exemplified by compounds **33** and **34**, have also been reported as antagonists of CRTH2 [59–65]. The most potent of these, analog **28**, had a K_i value of 6 nM in a $[^3H]$-PGD_2 binding assay using membranes from CRTH2-expressing CHO cells. Evaluation of these compounds in an intracellular calcium mobilization assay in CHO cells found **34** to be the most active (IC_{50} = 30 nM). In addition, compounds **31** and **34** were reported to have low affinity for the DP receptor [59–61]. Several compounds were also characterized in a $[^3H]$-PGD_2 binding assays using membranes from rhesus CRTH2 (rhCRTH2)-expressing HEK cells; compound **30** emerged as the most potent analog in this assay (pIC_{50} = 9.4) [62–65].

Finally, the 3-aryl-1-indazole acetic acid **35** and the 1-indole acetic acid **36** have been described as antagonists of CRTH2 (IC_{50} < 10 μM, rhCRTH2). A pIC_{50} value of 7.7 was reported for compound **36** [66,67].

3.2. 3-Indole acetic acid derivatives and related compounds

37

38

39

40

41

42

43

44

45

46

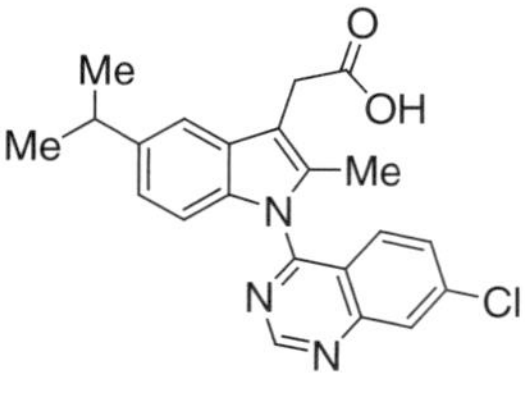

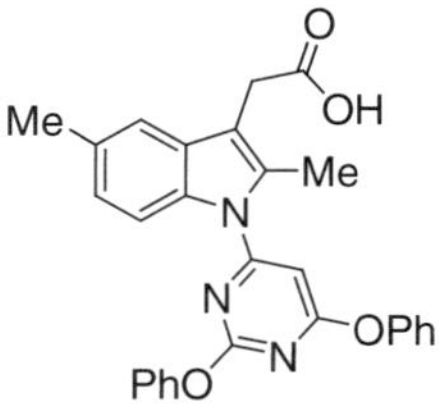

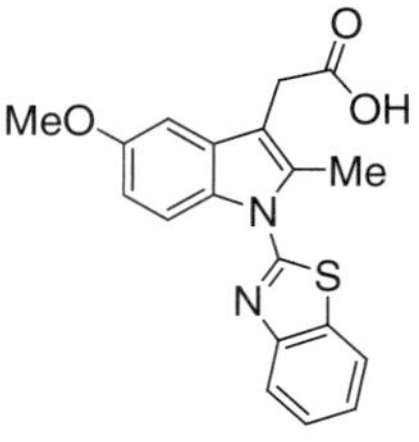

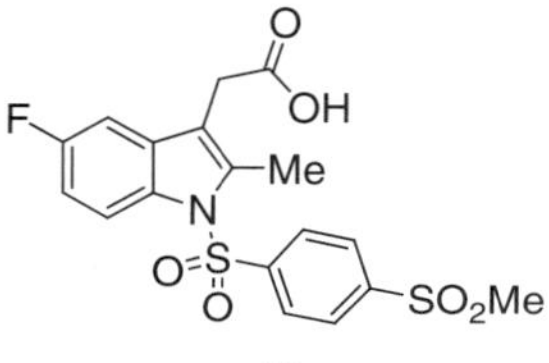

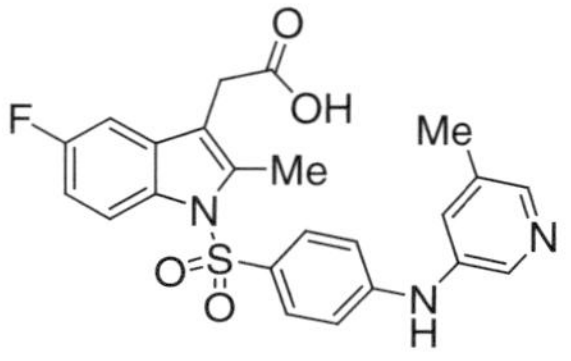

Multiple research groups have recently reported on 3-indole acetic acid derivatives as potent antagonists of the CRTH2 receptor. The antagonistic activity of the 1-aryl-3-indole acetic acid series, exemplified by **37–39**, was evaluated using an intracellular calcium mobilization assay in HEK cells transfected with the

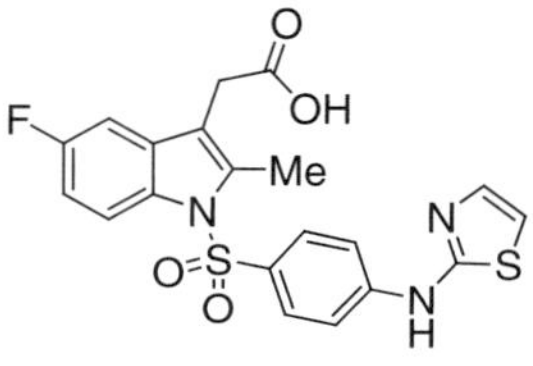

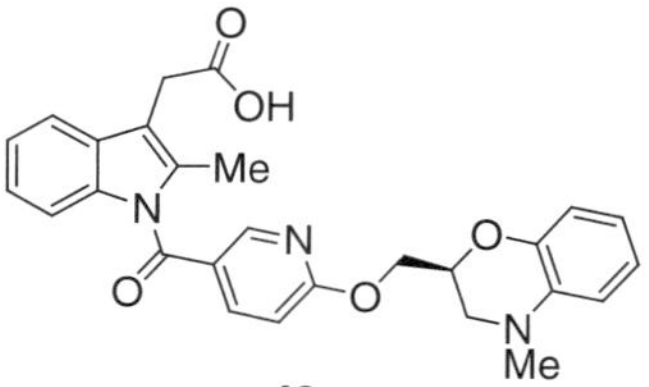

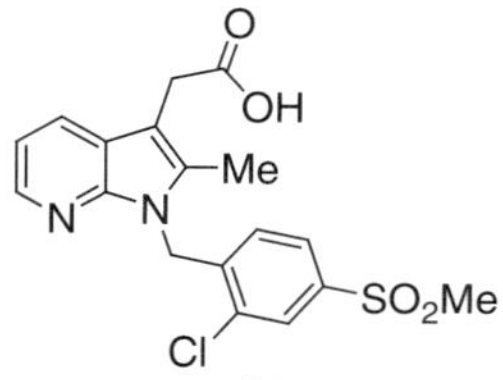

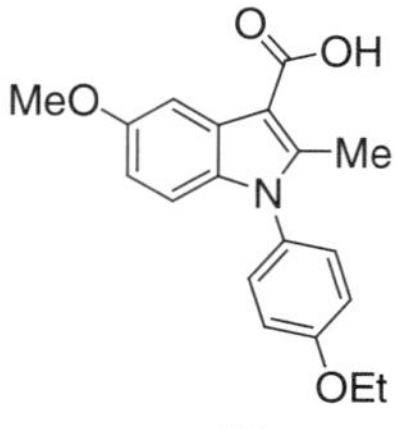

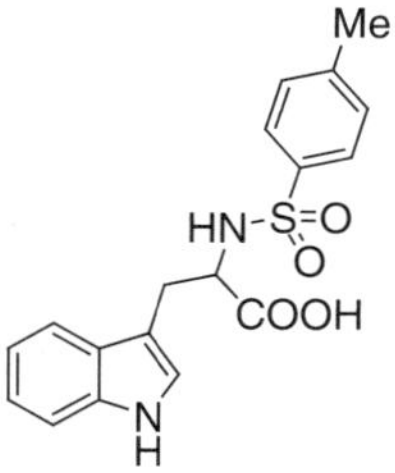

CRTH2 receptor [68,69]. The compounds were reported to possess IC_{50} values below 10 µM.

Likewise, 1-arylsulfonyl-3-indole acetic acids, exemplified by **40–42**, were disclosed as CRTH2 antagonists [70,71]. The compounds were evaluated in a [^{3}H]-PGD_2 binding assay where the most potent compound, **42**, displayed a K_i value of 15 nM. The CRTH2 antagonistic activity of these compounds was evaluated in a calcium mobilization assay yielding IC_{50} values between 19 and 200 nM. Interestingly, compound **41** was reported to also have good affinity for the DP receptor ($K_i = 150$ nM) in [^{3}H]-PGD_2 binding assays using membranes from CHO cells transfected with hDP. Both **40** and **42** were claimed to be selective for CRTH2 versus DP [71]. Compound **40** was reported to have a clearance in rat of 40 mL/min/kg after an i.v. dose of 1 mg/kg and a half life of 2.2 h [70]. The oral bioavailability of **40** was 56% after an oral gavage dose of 10 mg/kg [70].

3-Indole acetic acid derivatives, exemplified by **43**, have been reported to have affinity for both the CRTH2 and DP receptors [72]. Compound **43** was reported to possess IC_{50} values below 10 µM at both receptors.

3-Pyrrolopyridine acetic acid derivatives, exemplified by compound **44**, have also been reported as CRTH2 antagonists [73]. Most compounds were reported have K_i and IC_{50} values below 1 µM in binding and cAMP assays, respectively. Among the more potent members is compound **44** which had a K_i value of 3 nM in the binding assay and 14 nM in the cAMP assay.

Compounds **45** and **46** were reported as non-competitive CRTH2 antagonists [74]. They can bind to CRTH2 simultaneously with PGD_2. At the functional level, however, **45** and **46** do not interfere with PGD_2-mediated G-protein activation *via* CRTH2. Nonetheless, both compounds inhibit PGD_2-mediated arrestin translocation *via* a G-protein-independent mechanism.

3.3. Tetrahydroquinoline derivatives

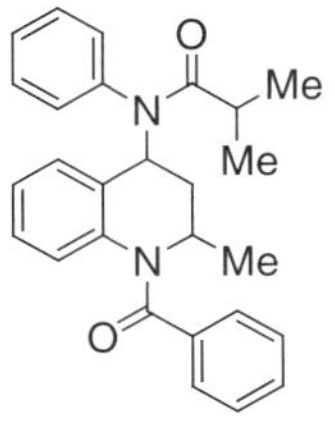

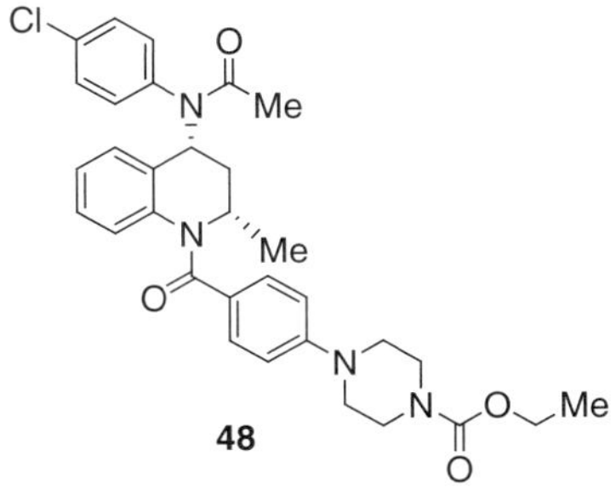

Several research groups have independently reported the discovery of tetrahydroquinoline derivatives, exemplified by compounds **47** and **48**, as CRTH2 antagonists [75–79]. This class of compounds is interesting because it contains a few CRTH2 antagonists that do not feature a carboxylic acid moiety. Compound **47**

3.4. Aryl acetic acid derivatives

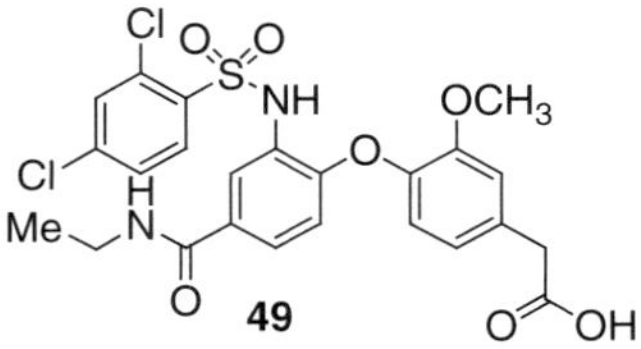

4-Aryloxy phenylacetic acid derivatives, such as compound **49**, have been reported as CRTH2 antagonists [81,82]. The activity of these compounds was evaluated in a [^{3}H]-PGD$_2$ binding assay in HEK293 cells transfected with CRTH2. Compound **49** and several related analogs were reported to have IC$_{50}$ values below 1 μM.

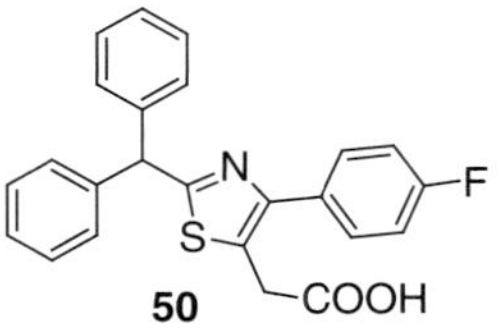

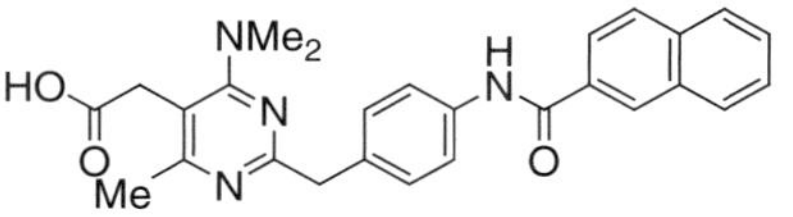

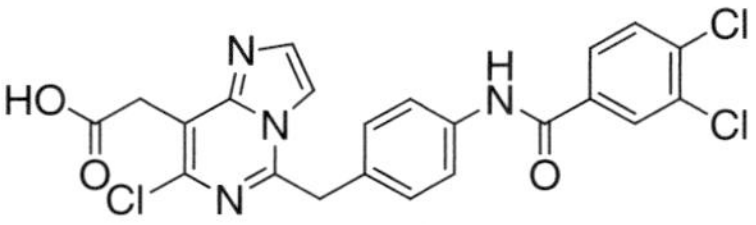

Similarly, substituted 5-thiazole acetic acid derivatives, exemplified by **50**, have been reported as CRTH2 antagonists [83]. This compound and its derivatives displayed IC$_{50}$ values below 0.5 μM in a [^{3}H]-PGD$_2$ binding assays using CRTH2-expressing COS-7 cells and in a functional bioluminescence resonance energy transfer (BRET) assay performed in HEK293 cells.

Finally, 5-pyrimidine acetic acid and 8-imidazopyrimidine acetic acid derivatives, exemplified by compounds **51** and **52**, were reported as CRTH2 antagonists [84,85]. Their CRTH2 activity was evaluated in a calcium mobilization assay using CRTH2-transfected L1.2 cells. Several compounds, including **51** and **52**, were reported to have IC$_{50}$ values below 10 nM.

3.5. Phenoxy acetic acid derivatives

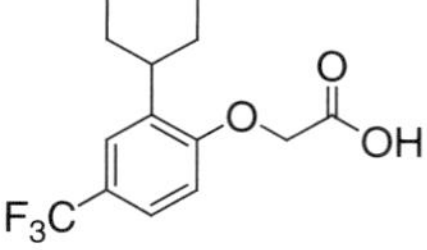

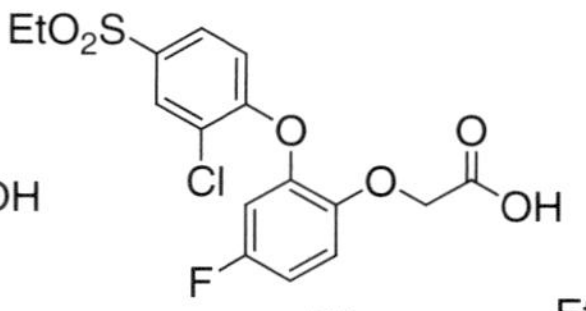

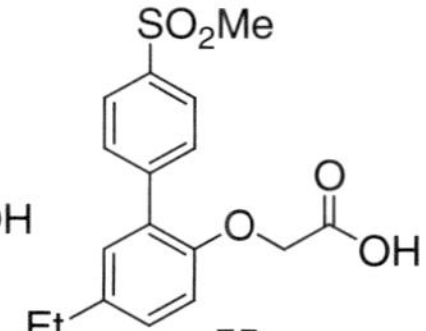

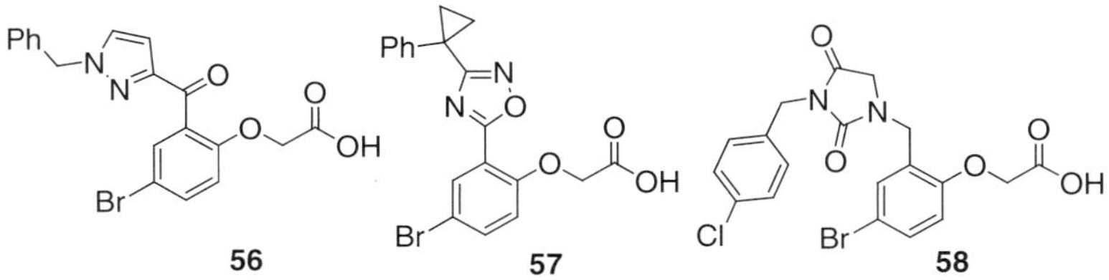

Several research groups have independently reported on phenoxyacetic acid derivatives as antagonists of the CRTH2 receptor. Compound **53** exemplifies a family of 2-cycloalkyl phenoxyacetic acid antagonists of CRTH2 [86]. This compound was reported to inhibit binding of [^{3}H]-PGD$_2$ to the CRTH2 receptor in membranes from K562 or CHO cells with a K_i of 21 nM. In addition, compound **53** was reported to have an IC$_{50}$ value of 139 nM in a cAMP assay using CHO cells expressing the CRTH2 receptor.

Researchers have also described 2-aryloxy and 2-aryl phenoxyacetic acid derivatives, exemplified by **54** and **55**, as CRTH2 antagonists [87–89]. The CRTH2 activity of these compounds was evaluated using a [^{3}H]-PGD$_2$ binding assay and pIC$_{50}$ values of 9 and 7.1 were reported for **54** and **55**, respectively. Compounds **56–58** exhibited IC$_{50}$ value below 0.5 μM in a [^{3}H]-PGD$_2$ binding assay using COS-7 cells expressing CRTH2 and in a functional BRET assay performed in HEK293 cells [90].

3.6. Benzoic acid derivatives

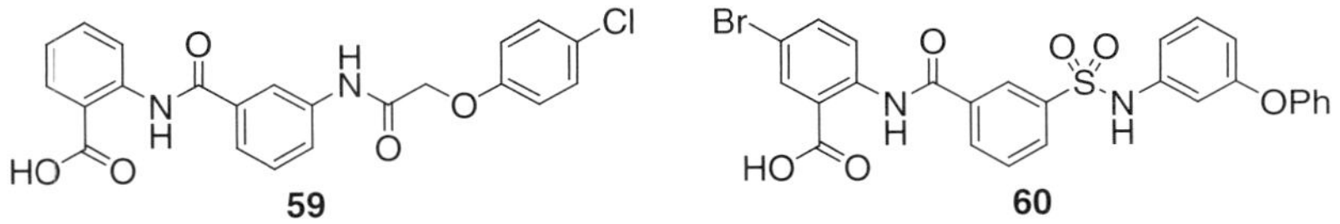

The benzoic acid derivatives have been reported as CRTH2 inhibitors (**59–60**) [91]. Both compounds were reported to inhibit [^{3}H]-PGD$_2$ binding to CRTH2 receptors expressed on COS-7 cells with IC$_{50}$ values below 0.5 μM. However, no functional activity was reported.

4. CONCLUSION

PGD$_2$ is the major cyclooxygenase product formed and secreted by activated mast cells during allergic reactions. However, the lack of suitable antagonists for its receptors, CRTH2 and DP, has limited progress toward elucidating the role that PGD$_2$ plays in allergic diseases. The recent discovery of potent and selective antagonists for CRTH2 and DP receptors, coupled with the genetic analysis of DP and CRTH2 function using KO mice, as well as the discovery of selective agonists for both PGD$_2$ receptors, are helping to clarify the crucial and complementary roles that these receptors play in mediating allergic reactions. Clinical data regarding the efficacy of PGD$_2$ antagonists are anticipated within the next few years.

REFERENCES

[1] S. Mitsumori, *Curr. Pharm. Des.*, 2004, **10**, 3533.
[2] R. A. Lewis, N. A. Soter, P. T. Diamond, K. F. Austen, J. A. Oates and L. J. Roberts, *J. Immunol.*, 1982, **129**, 1627.
[3] S. T. Holgate, G. B. Burns, C. Robinson and M. K. Church, *J. Immunol.*, 1984, **133**, 2138.
[4] R. H. Gundel, P. Kinkade, C. A. Torcellini, C. C. Clarke, J. Watrous, S. Desai, C. A. Homon, P. R. Farina and C. D. Wegner, *Am. Rev. Respir. Disord.*, 1991, **144**, 76.
[5] A. Miadonna, A. Tedeschi, C. Brasca, G. Folco, A. Sala and A. Murphy, *J. Allergy Clin. Immunol.*, 1990, **85**, 906.
[6] N. C. Turner, R. W. Fuller and D. M. Jackson, *J. Lipid. Mediat. Cell Signal*, 1995, **11**, 93.
[7] C. R. Beasley, C. Robinson, R. L. Featherstone, J. G. Varley, C. C. Hardy, M. K. Church and S. T. Holgate, *J. Clin. Invest.*, 1987, **79**, 978.
[8] D. L. Emery, T. D. Djokic, P. D. Graf and J. A. Nadel, *J. Appl. Physiol.*, 1989, **67**, 959.
[9] F. Pons, T. J. Williams, S. A. Kirk, F. McDonald and A. G. Rossi, *Eur. J. Pharmacol.*, 1994, **261**, 237.
[10] S. E. Sampson, A. P. Sampson and J. F. Costello, *Thorax*, 1997, **52**, 513.
[11] Y. Fujitani, Y. Kanaoka, K. Aritake, N. Uodome, K. Okazaki-Hatake and Y. Urade, *J. Immunol.*, 2002, **168**, 443.
[12] M. Shichijo, H. Sugimoto, K. Nagao, H. Inbe, J. A. Encinas, K. Takeshita, K. B. Bacon and F. Gantner, *J. Pharmacol. Exp. Ther.*, 2003, **307** (2), 518.
[13] B. Cooper and D. Ahern, *J. Clin. Invest.*, 1979, **64** (2), 586.
[14] K. Kabashima and S. Narumiya, *Prostaglandins Leukot. Essent. Fatty Acids*, 2003, **69**, 187.
[15] T. Matsuoka, M. Hirata, H. Tanaka, Y. Takahashi, T. Murata, K. Kabashima, Y. Sugimoto, T. Kobayashi, F. Ushikubi, Y. Aze, N. Eguchi, Y. Urade, N. Yoshida, K. Kimura, A. Mizoguchi, Y. Honda, H. Nagai and S. Narumiya, *Science*, 2000, **287**, 2013.
[16] A. Arimura, K. Yasui, J. Kishino, F. Asanuma, H. Hasegawa, S. Kakudo, M. Ohtani and H. Arita, *J. Pharmacol. Exp. Ther.*, 2001, **298**, 411.
[17] T. Oguma, L. J. Palmer, E. Birben, L. A. Sonna, K. Asano and C. M. Lilly, *New Eng. J. Med.*, 2004, **351** (17), 1752.
[18] H. Hirai, K. Tanaka, O. Yoshie, K. Ogawa, K. Kenmotsu, Y. Takamori, M. Ichimasa, K. Sugamura, M. Nakamura, S. Takano and K. Nagata, *J. Exp. Med.*, 2001, **193**, 255.
[19] K. Nagata, H. Hirai, K. Tanaka, K. Ogawa, T. Aso, K. Sugamura, M. Nakamura and S. Takano, *FEBS Lett.*, 1999, **459**, 195.
[20] K. Nagata, K. Tanaka, K. Ogawa, K. Kemmotsu, T. Imai, O. Yoshie, H. Abe, K. Tada, M. Nakamura, K. Sugamura and S. Takano, *J. Immunol.*, 1999, **162**, 1278.
[21] L. Cosmi, F. Annunziato, M. I. G. Galli, R. M. E. Maggi, K. Nagata and S. Romagnani., *Eur. J. Immunol.*, 2000, **30**, 2972.
[22] N. Sawyer, E. Cauchon, A. Chateauneuf, R. P. Cruz, D. W. Nicholson, K. M. Metters, G. P. O'Neill and F. G. Gervais, *Br. J. Pharmacol.*, 2002, **137**, 1163.
[23] H. Sugimoto, M. Shichijo, T. Iino, Y. Manabe, A. Watanabe, M. Shimazaki, F. Gantner and K. B. Bacon, *J. Pharmacol. Exp. Ther.*, 2003, **305**, 347.

[24] G. Monneret, S. Gravel, M. Diamond, J. Rokach and W. S. Powell, *Blood*, 2001, **98**, 1942.
[25] P. Gosset, F. Bureau, V. Angeli, M. Pichavant, C. Faveeuw, A. B. Tonnel and F. Trottein, *J. Immunol.*, 2003, **170**, 4943.
[26] A. N. Hata, R. Zent, M. D. Breyer and R. M. Breyer, *J. Pharmacol. Exp. Ther.*, 2003, **306**, 463.
[27] J.-L. Huang, P.-S. Gao, R. A. Mathias, T.-C. Yao, L.-C. Chen, M.-L. Kuo, S.-C. Hsu, B. Plunkett, A. Togias, K. C. Barnes, C. Stellato, T. H. Beaty and S.-K. Huang, *Hum. Mol. Genet.*, 2004, **13**, 2691.
[28] N. A. Sharif, G. W. Williams and T. L. Davis, *Br. J. Pharmacol.*, 2000, **131**, 1025.
[29] Development status summary for S-5751 from *Prous Science Integrity*, February, 2006.
[30] S. Mitsumori, T. Tsuri, T. Honma, Y. Hiramatsu, T. Okada, H. Hashizume, S. Kida, M. Inagaki, A. Arimura, K. Yasui, F. Asanuma, J. Kishino and M. Ohtani, *J. Med. Chem.*, 2003, **46**, 2446.
[31] S. Mitsumori, T. Tsuri, T. Honma, Y. Hiramatsu, T. Okada, H. Hashizume, M. Inagaki, A. Arimura, K. Yasui, F. Asanuma, J. Kishino and M. Ohtani, *J. Med. Chem.*, 2003, **46**, 2436.
[32] T. Tanami, N. Ono, M. Yagi, T. Seki and M. Sato, *WO Patent* 074240, 2004.
[33] K. Torisu, M. Iwahashi, K. Kobayashi and F. Nambu, *WO Patent* 022814, 2003.
[34] K. Torisu, K. Kobayashi, M. Iwahashi, H. Egashira, Y. Nakai, Y. Okada, F. Nanbu, S. Ohuchida, H. Nakai and M. Toda, *Bioorg. Med. Chem. Lett.*, 2004, **14**, 4557.
[35] K. Torisu, K. Kobayashi, M. Iwahashi, H. Egashira, Y. Nakai, Y. Okada, F. Nanbu, S. Ohuchida, H. Nakai and M. Toda, *Eur. J. Med. Chem.*, 2005, **40**, 505.
[36] K. Torisu, K. Kobayashi, M. Iwahashi, Y. Nakai, T. Onoda, T. Nagase, I. Sugimoto, Y. Okada, R. Matsumoto, F. Nanbu, S. Ohuchida, H. Nakai and M. Toda, *Bioorg. Med. Chem.*, 2004, **12**, 4685.
[37] K. Torisu, K. Kobayashi, M. Iwahashi, Y. Nakai, T. Onoda, T. Nagase, I. Sugimoto, Y. Okada, R. Matsumoto, F. Nanbu, S. Ohuchida, H. Nakai and M. Toda, *Bioorg. Med. Chem.*, 2004, **12**, 5361.
[38] K. Torisu, K. Kobayashi, M. Iwahashi, Y. Nakai, T. Onoda, T. Nagase, I. Sugimoto, Y. Okada, R. Matsumoto, F. Nanbu, S. Ohuchida, H. Nakai and M. Toda, *Bioorg. Med. Chem. Lett.*, 2004, **14**, 4891.
[39] Ono Pharmaceutical Press Release, September 11, 2005.
[40] Development status summary for L-888839 from *Prous Science Integrity*, February, 2006.
[41] L. Li, C. Beaulieu, D. Guay and C. Sturino, Z. Wang, *WO Patent* 103970, 2004.
[42] C. Beaulieu, D. Guay, C. Sturino, Z. Wang and R. Zamboni, *WO Patent* 056527, 2005.
[43] N. Lachance and C. Sturino, *WO Patent* 111047, 2004.
[44] Y. Leblanc, C. Dufresne and P. Roy, *WO Patent* 039807, 2004.
[45] C. Berthelette, N. Lachance, L. Li, C. Sturino and Z. Wang, *WO Patent* 062200, 2003.
[46] Z. Wang, C. Dufresne, D. Guay and Y. Leblanc, *WO Patent* 094830, 2002.
[47] F. G. Gervais, J. P. Morello, C. Beaulieu, N. Sawyer, D. Denis, G. Greig, D. Malebranche and G. P. O'Neill, *Mol. Pharmacol.*, 2005, **67**, 1834.
[48] C. Beaulieu, Z. Wang, D. Denis, G. Greig, S. Lamontagne, G. P. O'Neill, D. Slipetz and J. Wang, *Bioorg. Med. Chem. Lett.*, 2004, **14**, 3195.
[49] M. Iwahashi, K. Kobayashi and F. Nambu, *EP Patent* 1486491, 2004.
[50] A. Naganawa, M. Iwahashi, A. Kinoshita, A Shimabukuro, S. Ogawa, K. Yano, K. Kobayashi, Y. Okada, Y. Kishida, S. Kawauchi, K. Tsukamoto, Y. Matsunaga and F. Nambu, *WO Patent* 028455, 2005.

[51] Development status summary for Ramatroban from *Prous Science Integrity*, February, 2006.
[52] S. L. Johnson, P. G. Bardin, J. Harrison, W. Ritter, J. R. Joubert and S. T. Holgate, *Br. J. Clin. Pharmacol.*, 1992, **34**, 402.
[53] H. Magnussen, S. Boerger, K. Templin and A. R. Baunack, *J. Allergy Clin. Immunol.*, 1992, **89**, 1119.
[54] T. Ulven and E. Kostenis, *J. Med. Chem.*, 2005, **48**, 897.
[55] M. J. Robarge, D. C. Bom, L. N. Tumey, N. Varga, E. Gleason, D. Silver, J. Song, S. M. Murphy, G. Ekema, C. Doucette, D. Hanniford, M. Palmer, G. Pawlowski, J. Danzig, M. Loftus, K. Hunady, B. A. Sherf, R. W. Mays, A. Stricker-Krongrad, K. R. Brunden, J. J. Harrington and Y. L. Bennani, *Bioorg. Med. Chem. Lett.*, 2005, **15**, 1749.
[56] N. Tanimoto, Y. Hiramatsu, S. Mitsumori and M. Inagaki, *WO Patent* 097598, 2003.
[57] H. Fretz, A. Fecher, K. Hilpert and M. Riederer, *WO Patent* 095397, 2005.
[58] A. Fecher, H. Fretz, K. Hilpert and M. Riederer, *WO Patent* 094816, 2005.
[59] D. Middlemiss, M. R. Ashton, E. A. Boyd, F. A. Brookfield and R. E. Armer, *WO Patent* 040114, 2005.
[60] R. E. Armer, E. A. Boyd and F. A. Brookfield, *WO Patent* 121141, 2005.
[61] D. Middlemiss, M. R. Ashton, E. A. Boyd, F. A. Brookfield and R. E. Armer, *WO Patent* 044260, 2005.
[62] R. Bonnert, M. Dickinson, R. Rasul, H. Sanganee and S. Teague, *WO Patent* 007451, 2004.
[63] R. Bonnert and R. Rasul, *WO Patent* 106302, 2004.
[64] R. Bonnert, S. Brough, T. Cook, M. Dickinson, R. Rasul, H. Sanganee and S. Teague, *WO Patent* 101961, 2003.
[65] R. Bonnert, A. R. Cook, T. J. Luker, R. T. Mohammed and S. Thom, *WO Patent* 019171, 2005.
[66] R. Bonnert, R. T. Mohammed and S. Teague, *WO Patent* 054232, 2005.
[67] T. Birkinshaw, R. Bonnert, A. Cook, R. Rasul, H. Sanganee and S. Teague, *WO Patent* 101981, 2004.
[68] A. Baxter, J. Steele and S. Teague, *WO Patent* 066046, 2003.
[69] A. Baxter, J. Steele and S. Teague, *WO Patent* 066047, 2003.
[70] R. E. Armer, M. R. Ashton, E. A. Boyd, C. J. Brennan, F. A. Brookfield, L. Gazi, S. L. Gyles, P. A. Hay, M. G. Hunter, D. Middlemiss, M. Whittaker, L. Xue and R. Pettipher, *J. Med. Chem.*, 2005, **48**, 6174.
[71] D. Middlemiss, M. R. Ashton, E. A. Boyd, F. A. Brookfield and R. E. Armer, *WO Patent* 040112, 2005.
[72] M. Iwahashi, A. Naganawa, T. Nishiyama, T. Nagase, K. Kobayashi and F. Nambu, *WO Patent* 078719, 2004.
[73] K. Bala, C. Leblanc, D. A. Sandham, K. L. Turner, S. J. Watson, L. N. Brown and B. Cox, *WO Patent* 123731, 2005.
[74] J. M. Mathiesen, T. Ulven, L. Martini, L. O. Gerlach, A. Heinemann and E. Kostenis, *Mol. Pharmacol.*, 2005, **68**, 393.
[75] S. Ghosh, A. M. Elder, K. G. Carson, K. Sprott and S. Harrison, *WO Patent* 032848, 2004.
[76] S. Ghosh, A. M. Elder, K. G. Carson, K. Sprott, S. J. Harrison, F. A. Hicks, C. C. Renou and D. Reynolds, *WO Patent* 100321, 2005.
[77] C. Kuhn, F. Feru, M. Bazin, M. Awad and S. W. Goldstein, *EP Patent* 1413306, 2004.
[78] O. Kotera, E. Oshima, K. Ueno, T. Ikemura, H. Manabe, M. Sawada, H. Mimura, H. Miyaji and H. Nonaka, *WO Patent* 052863, 2004.
[79] W. Inman, J. Liu, J. C. Medina and S. Miao, *WO Patent* 007094, 2005.

[80] H. Mimura, T. Ikemura, O. Kotera, M. Sawada, S. Tashiro, E. Fuse, K. Ueno, H. Manabe, E. Oshima, A. Karasawa and H. Miyaji, *J. Pharmacol. Exp. Ther.*, 2005, **314**, 244.
[81] Expert Opin. Ther. Patents, 2005, **15**, 115.
[82] Z. Fu, X. A. Huang, J. Liu, J. C. Medina, M. J. Schmitt, L. H. Tang, Y. Wang and Q. Xu, *WO Patent* 058164, 2004.
[83] T. Ulven, T. Frimurer, O. Rist, E. Kostenis, T. Hogberg, J. M. Receveur and M. Grimstrup, *WO Patent* 116001, 2005.
[84] T. Ly, Y. Koriyama, T. Yoshino, H. Sato, K. Tanaka, H. Sugimoto, Y. Manabe, K. Bacon, K. Urbahns, M. Seki and T. Shintani, *WO Patent* 096777, 2004.
[85] T. Ly, T. Yoshino, Y. Takekawa, T. Shintani, H. Sugimoto, K. Bacon, K. Urbahns and M. Seki, *WO Patent* 073234, 2005.
[86] D. Sandham, K. L. Turner and C. Leblanc, *WO Patent* 105727, 2005.
[87] R. V. Bonnert, A. Patel and S. Thom, *WO Patent* 018529, 2005.
[88] G. Pairaudeau, R. Rasul and S. Thom, *WO Patent* 089884, 2004.
[89] R. Bonnert, S. Brough, A. Davies, T. Luker, T. McInally, I. Millichip, G. Pairaudeau, A. Patel, R. Rasul and S. Thom, *WO Patent* 089885, 2004.
[90] T. Ulven, T. Frimurer, O. Rist, E. Kostenis, T. Hogberg, J. M. Receveur and M. Grimstrup, *WO Patent* 115382, 2005.
[91] T. Ulven, T. Frimurer, O. Rist, E. Kostenis and T. Hogberg, *WO Patent* 115374, 2005.

Progress in the Development of Inhaled, Long-Acting β_2-Adrenoceptor Agonists

Paul A. Glossop and David A. Price

Pfizer Global Research and Development, Sandwich Laboratories, Ramsgate Road, Sandwich, Kent, CT13 9NJ, UK

Contents

1. Introduction	237
2. Quinolinones (8-hydroxycarbostyril)	238
3. Formamides	240
4. Saligenins	241
5. New head groups	244
6. Dual pharmacology	245
7. Conclusion	246
References	246

1. INTRODUCTION

Long acting β_2-adrenoceptor agonists are a highly precedented drug class used for the treatment of asthma and chronic obstructive pulmonary disease (COPD) [1,2]. There are currently two marketed long acting β_2-adrenoceptor agonists, salmeterol **1** [3], and formoterol **2** [4,5], neither of which provides a once daily dosing regimen. Therefore, an opportunity may exist for a once daily agent both as a mono-therapy and as part of combination products, particularly with steroidal or muscarinic ligands. This review is not intended to cover the current state of potential combination therapies; however, the final section of this review deals with an alternative paradigm of incorporating a second pharmacology within the same molecule as β_2-adrenoceptor agonism. This concept is referred to as dual pharmacology [6].

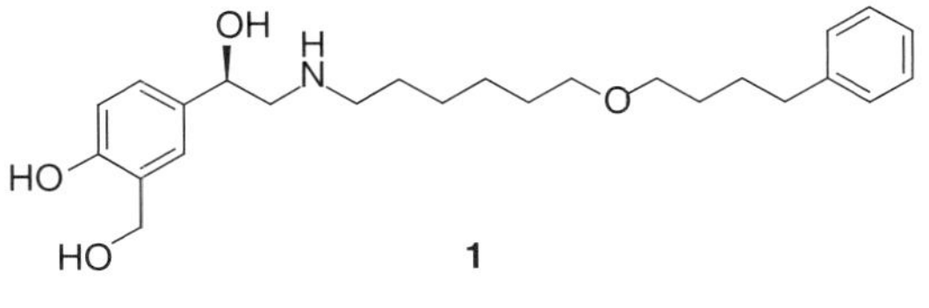

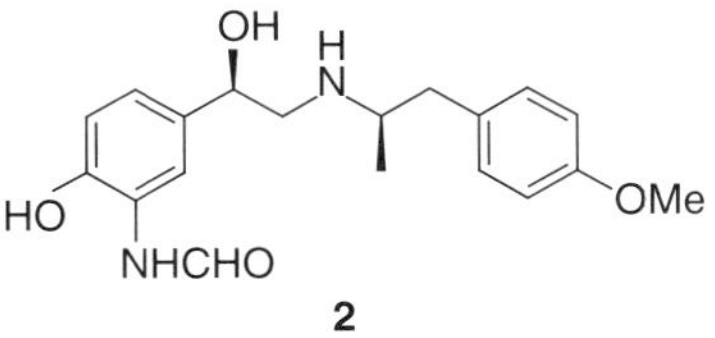

Asthma is a chronic inflammatory disorder of the airways causing recurrent episodes of wheezing, breathlessness, chest tightness, and coughing [7]. These symptoms often occur at night or in the early morning, impacting sleep patterns and so reducing overall quality of life. COPD is the fourth leading cause of death in the US and is characterized by airflow obstruction due to chronic bronchitis or emphysema and symptoms are typically breathing-related (e.g. chronic cough, exertional dyspnea, expectoration, and wheeze) [8]. It is believed that a once daily β_2-adrenoceptor agonist would be of tremendous value in the treatment of these conditions, with the added benefit of increased convenience for the patient population.

There has been considerable interest in the discovery of a once daily β_2-adrenoceptor agonist with a recent flurry of presentations, patent applications, and licensing agreements from a number of institutions. The aim of this review is to update the current state of the field, including clinical data for agents where available. The review is structured according to the various β_2-adrenoceptor agonist head groups employed by researchers and concludes with a discussion regarding the potential development of dual pharmacology agents.

2. QUINOLINONES (8-HYDROXYCARBOSTYRIL)

Indacaterol, 3 (QAB-149) is currently being developed as a once daily β_2-adrenoceptor agonist for the treatment of COPD and asthma [9]. Indacaterol contains the familiar quinolinone head group that appears in the marketed short-acting β_2-adrenoceptor agonist, procaterol, 4. This 8-hydroxycarbostyril moiety is known to have a high affinity for the β_2-adrenoceptor and has been postulated to give potential slow offset properties to ligands to drive duration of action [10]. A recently published patent suggests that the maleate salt of indacaterol is highly crystalline, stable, and may be the preferred form of this compound for development and clinical use [11].

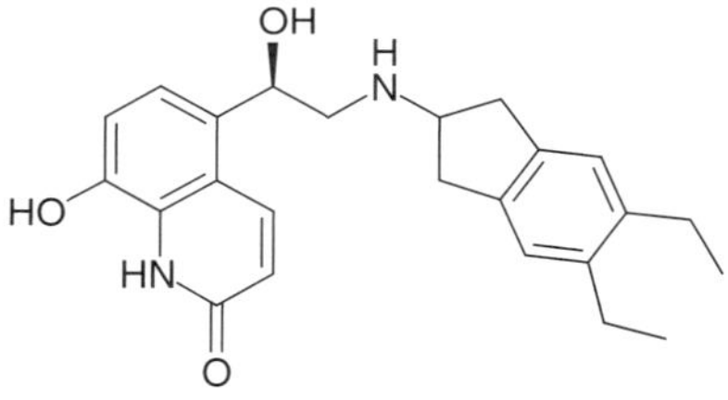

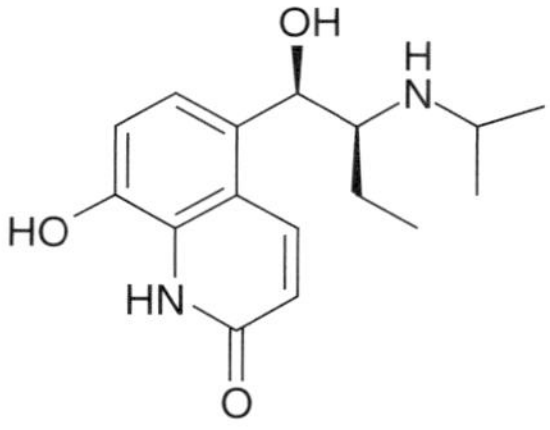

Data have recently been disclosed on the pharmacological profile of indacaterol. Using guinea-pig tissue preparations, potency data were generated in guinea-pig atria ($EC_{50} = 21.6\,\mu M$) vs. guinea-pig trachea ($EC_{50} = 45\,nM$), suggesting that the compound has selectivity over the β_1-adrenoceptor. Further studies using guinea-pig trachea tissue preparations with electrical field stimulation (EFS)-induced contractions suggest that indacaterol has lower potency than salmeterol ($EC_{50} = 45$ vs. $15\,nM$) but a more rapid onset of action. However, its duration of action was shorter than salmeterol (7 vs. $>12\,h$) when tested at $100\,nM$ concentrations [12]. Further profiling of indacaterol in isolated human bronchus using EFS-induced contractions again indicated that the compound was less potent than salmeterol ($EC_{50} = 131$ vs. $32\,nM$). In these experiments, indacaterol was highly efficacious as a bronchodilator and had similar duration of action relative to salmeterol [13].

Indacaterol was profiled using plethysmography in the conscious guinea pig with 5-hydroxytryptamine (5HT)-induced bronchoconstriction. When administered intra-tracheally as a dry powder, indacaterol was more potent than salmeterol ($ED_{50} = 0.22$ vs. $1.5\,\mu g/kg$) and also showed a superior duration of action at the ED_{80} dose.

Indacaterol was advanced to human trials using a hydrofluoroalkane-based multi-dose inhaler to deliver the drug in mild-to-moderate asthmatics. It showed long duration of action ($>24\,h$) and in some patients a rapid onset of bronchodilation at higher doses. The $400\,\mu g$ dose appeared to be statistically superior to placebo at all time points up to $26\,h$ post-dose without generating statistically significant safety concerns [14]. Further data have been released regarding multi-dose trials where an $800\,\mu g$ dose was tested against placebo in a 14-day trial in mild to moderate COPD patients. A $24\,h$ duration of action was achieved with no discontinuations due to adverse events. At the $800\,\mu g$ dose, the pharmacokinetic half-life was shown to be $48\,h$, with a twofold increase in systemic exposure relative to the $400\,\mu g$ dose [15]. Phase III studies are due to be completed in 2006 and regulatory submissions are planned for 2007 [16].

An alternative, once daily β_2-adrenoceptor agonist that has progressed to clinical trials is carmoterol, **5** (CHF-4226, TA-2005). Structurally, the compound resembles the marketed agent formoterol, while containing the 8-hydroxycarbostyril head group rather than the formamide. Extensive work to understand the preclinical pharmacology of carmoterol and studies using guinea-pig trachea preparations have shown the compound to possess a rapid onset of action and a long duration of action. It has been postulated that the long duration of action is linked to the tight binding of carmoterol to the β_2-adrenoceptor [17].

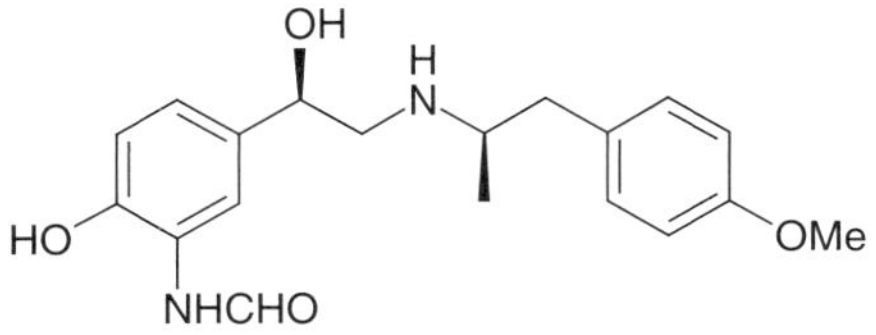

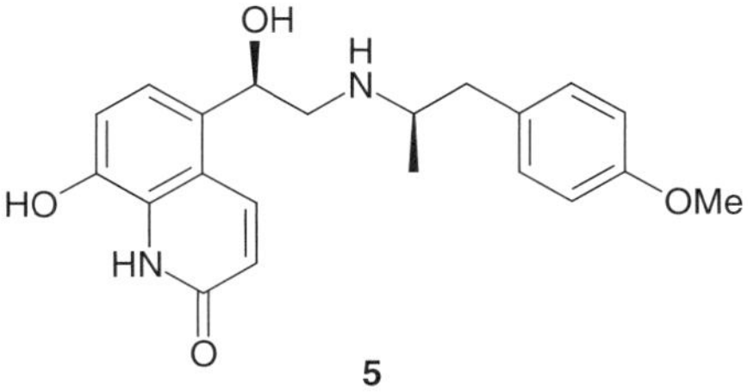

Carmoterol was previously in development and known to have entered Phase I trials; however, no development activities have been reported since 1994 [18]. The compound was then out-licensed for development and studies showed carmoterol to be an effective bronchodilator with a duration of action greater than $24\,h$ [19]. Interest in the compound was renewed and regulatory filings are now anticipated in 2006 [20].

The 8-hydroxycarbostyril ring system has been incorporated in structurally more complex molecules where increased lipophilicity may play a key role in providing duration of action. This is the postulated mechanism used to rationalize the duration of action of salmeterol where it is believed that better membrane partitioning underpins duration, in line with the diffusion micro kinetic theory [21].

Two patents claiming crystalline salt forms of two differing 8-hydroxycarbostyril-containing structures, **6** and **7**, have issued [22,23]. The filing of patents around specific salt forms of these compounds, including processes for their preparation and formulation suggests that these are significant compounds in the development pipeline as once daily β_2-adrenoceptor agonists.

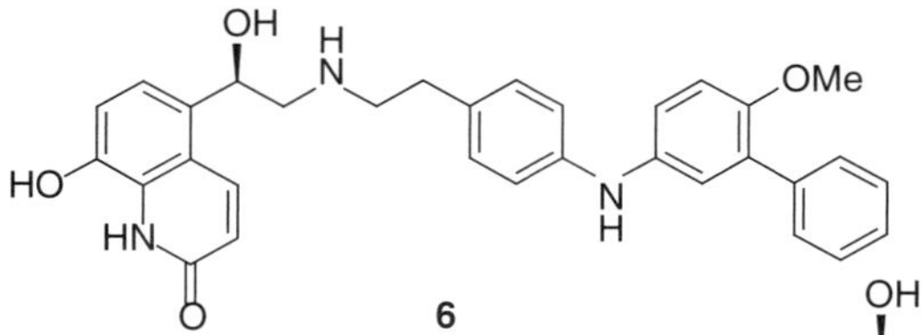

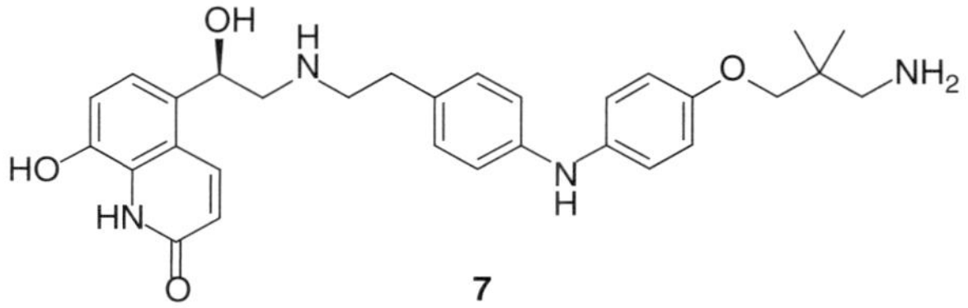

3. FORMAMIDES

A number of studies have been published wherein formoterol was used as a lead and the results suggest that the increase of lipophilicity of new analogues is related to the increase in their duration of action [24]. The authors suggest that the extended duration of action of compounds **8** and **9** is attributable to greater membrane partitioning of these analogues in line with the diffusion micro kinetic theory [21]. Both analogues **8** and **9** produced greater than twofold longer duration of action than formoterol in EFS-stimulated guinea-pig trachea preparations. Additionally, compound **8** also appeared to have a rapid onset of action. No further information regarding the development of either **8** or **9** has been reported.

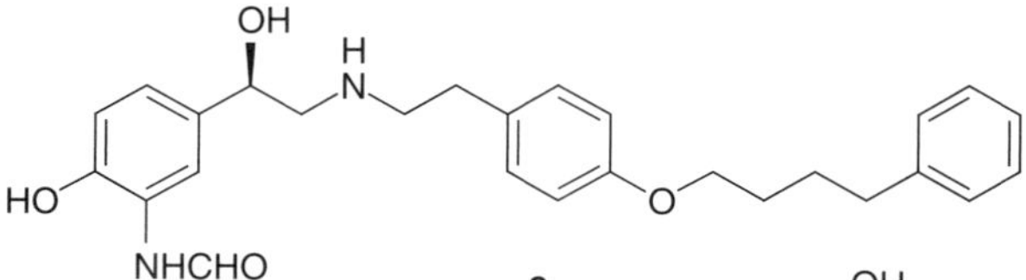

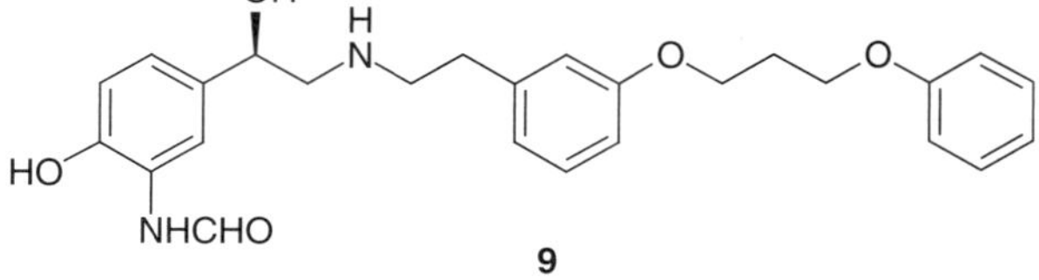

With the intense interest in developing once daily β_2-adrenoceptor agonists and with a heavy reliance on common structural leads, it is not surprising that more similarities than differences exist between final products from independent research efforts (cf. **8** and **10**).

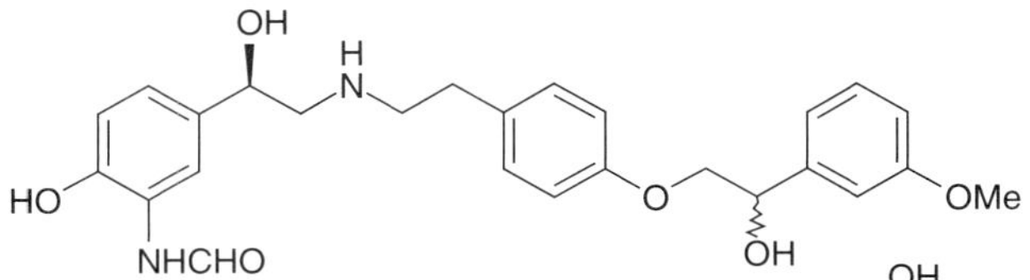

Extensive efforts to further develop the series exemplified by **10** [25] have delivered compounds **11** and **12**, which contain alternative heteroatom patterns [26,27]. For compound **12**, the preparation of a crystalline salt form was confirmed by powder X-ray diffraction studies.

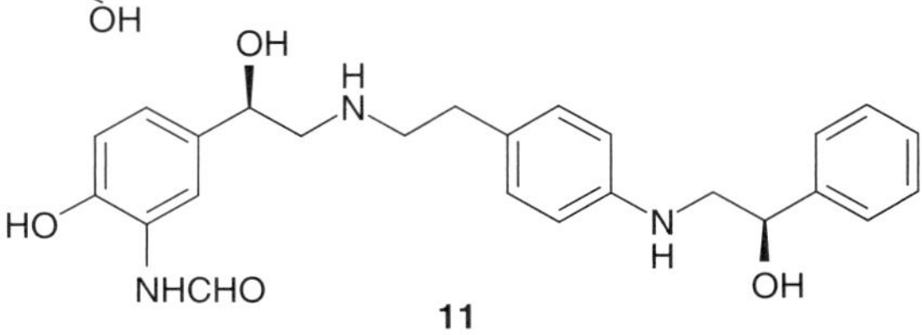

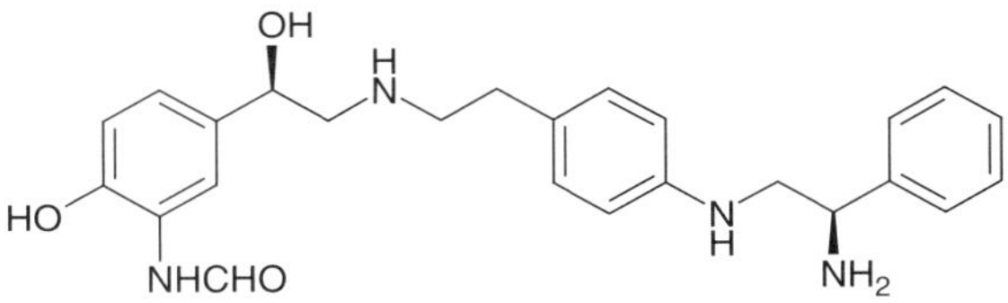

4. SALIGENINS

The saligenin head group has been used with success to prepare long acting β_2-adrenoceptor agonists with salmeterol playing a prominent role as a benchmark compound. In this connection, the combination agent Advair™ (salmeterol and fluticasone) is currently in third place globally in terms of sales, demonstrating the success of this approach.

A common tactic employed to initiate research in this area has been to emulate salmeterol and modify its scaffold to improve the duration of action of the corresponding derivatives. A number of patent applications have issued which claim compounds with enhanced duration of action. The design principles which were used to derive these new analogues build on the observations that increased lipophilicity leads to extended the duration of action. Since little supporting biological data are provided in these applications, it is difficult to assess any improvements relative to salmeterol or formoterol. The disclosure which describes compound **13** also claims specific salts with supporting physical data, suggesting a higher level of interest in this analogue [28]. Further evidence for interest in this dichloro expression is a second filing where alternative head groups to the saligenin are exemplified, in particular the formamide head group that appears in formoterol [29]. Another transformation that has been pursued to increase the lipophilicity of salmeterol is to add additional alkyl substituents to the ether linkage as exemplified in **14** [30].

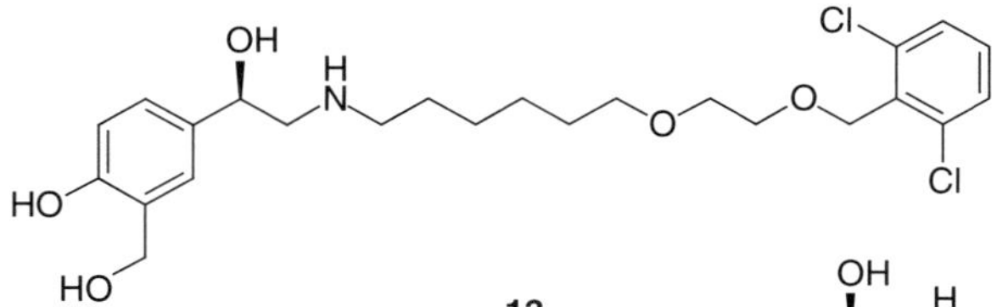

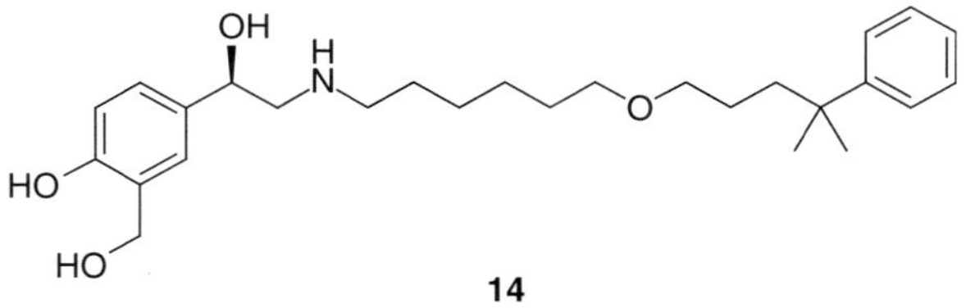

Researchers have been creatively replacing the alkyloxyalkyl linker of salmeterol with alternative expressions such as the phenethyl moiety in compound **15** and the conformationally constrained benzodioxane ring in analogue **16** [31,32].

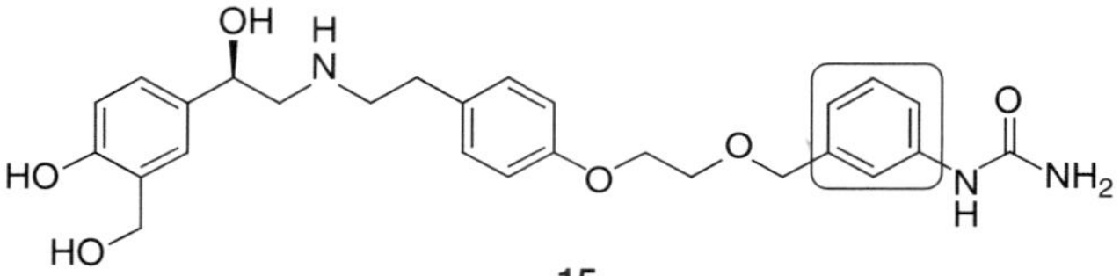

If compound **15** has a similar binding mode to that of **1** and **13**, it would suggest that the β_2-adrenoceptor is tolerant of different substituents of the boxed phenyl ring. There are a number of patent filings which claim a wide variety of heterocycles and polar substituents appended to this phenyl ring, such as compound **17** [33]. The application containing cyclopentylsulfone analogue **18** claims the preparation of seven different salts, suggesting a high level of interest in this particular compound [34].

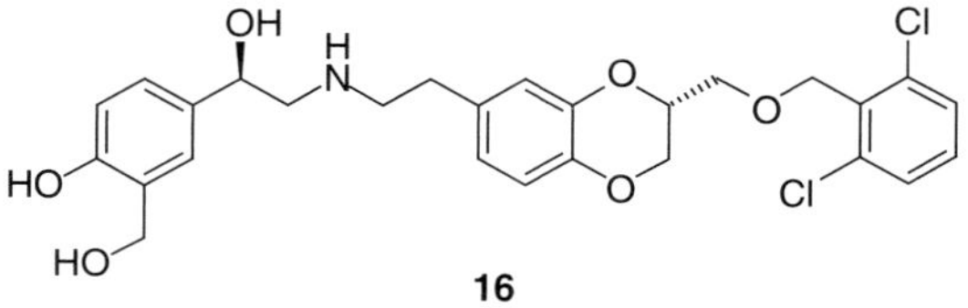

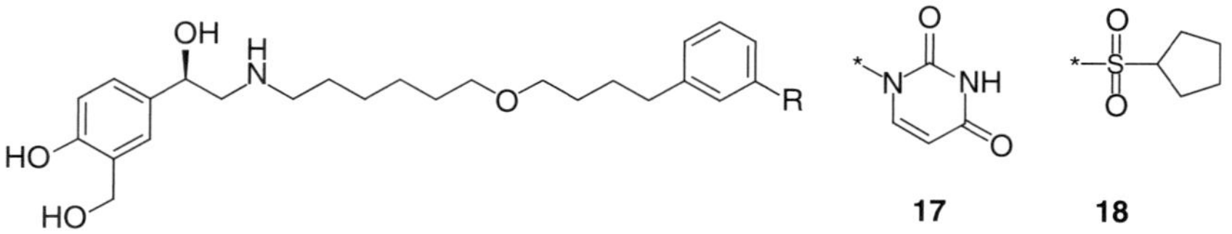

Diphenylanilines, exemplified by **7** have been the subject of follow-up studies. In particular, alternative substitution patterns around the diphenylaniline moiety have been discovered and these new findings have been extrapolated to other head groups. More complex substituents have also been appended, as in **19**; it also appears that these diphenylaniline expressions can be successfully combined with the saligenin head group [35].

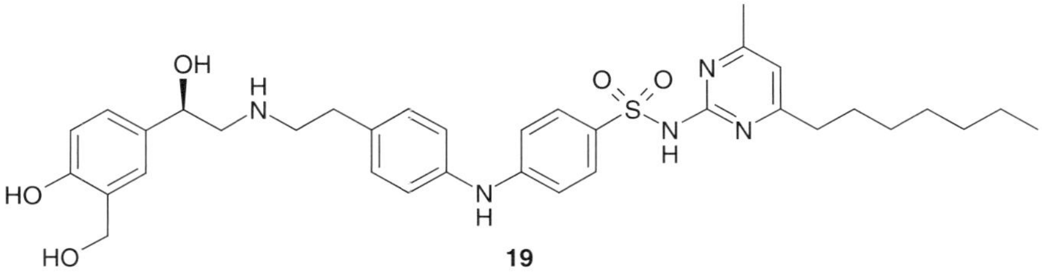

Recently published patent applications demonstrate the continuing interest in the saligenin expression, exemplified in **20**, a potent β_2-adrenoceptor agonist (EC$_{50}$ = 0.058 vs. 0.100 nM for **20** and **1**, respectively, in CHO cells recombinantly expressing the human β_2-adrenoceptor) that incorporates the saligenin head group and a novel indole linking group to a dimethoxy-substituted benzylamide [36,37]. A recent poster publication disclosed additional *in vitro* EFS guinea-pig trachea data for **20**, including potency relative to isoprenaline (RP = 0.021) and salmeterol-like duration of action [38]. The poster publication also speculated on the origins of the observed *in vitro* duration of action and concluded that exo-site binding and/or lipophilicity-driven tissue deposition may contribute toward the potential for long duration of action in humans [21].

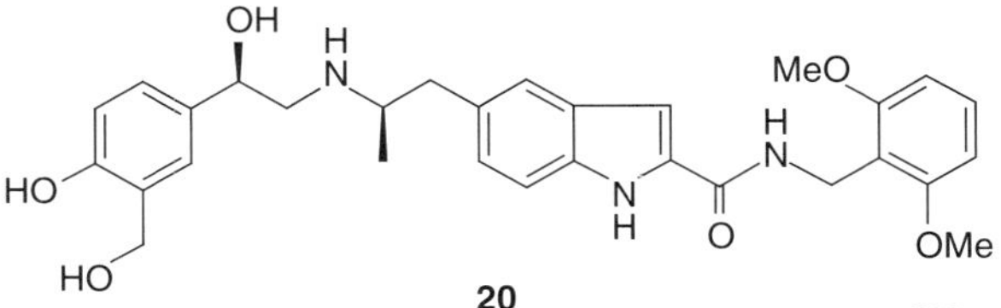

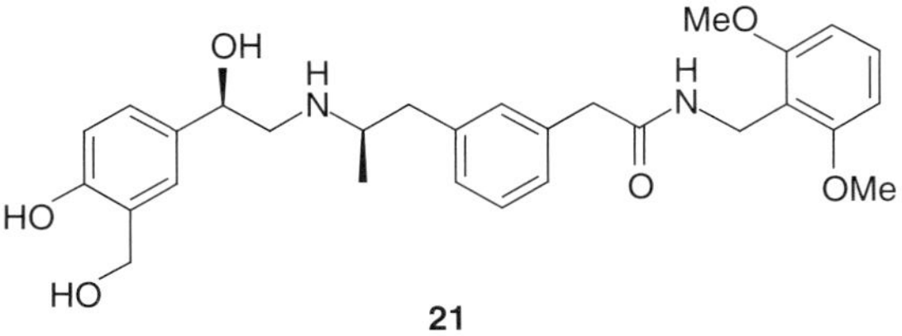

Compounds including **21** (EC$_{50}$ = 0.017 nM) have been identified that retain the saligenin head group, but replace the indole linker with a phenethylamine moiety that is substituted with a variety of aromatic and aliphatic amide tail groups [39–41].

5. NEW HEAD GROUPS

Instead of using head groups that have appeared in marketed inhaled β_2-adrenoceptor agonists, an alternative paradigm is to design bespoke head groups. One approach has been to utilize the 8-hydroxycarbostyril head group present in procaterol and introduce an oxygen into the heterocycle to generate compounds exemplified by **22** [42]. Within this filing, compound **23** is also exemplified where the heterocycle has been replaced by a benzylaniline derivative. In the latter example, the phenethyl expression present in **22** was also replaced with an *N*-linked benzimidazole and this theme is continued with triazole **24**. The head group is further developed in a separate filing in which the alternative *meta*-phenol regiochemistry is claimed as in **25** [43].

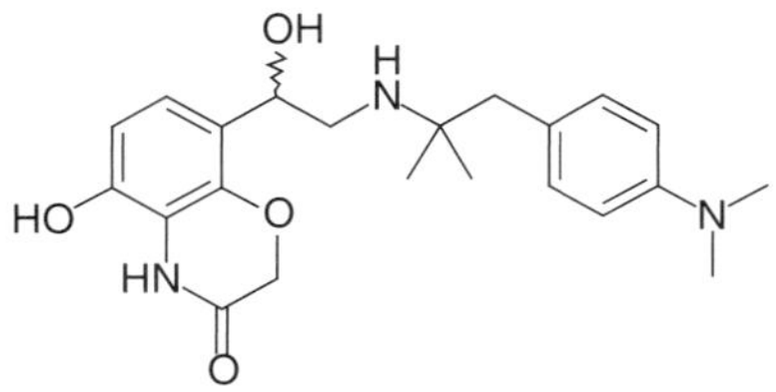

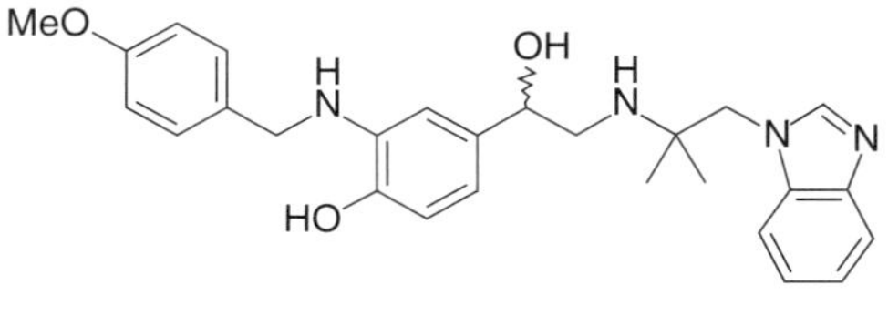

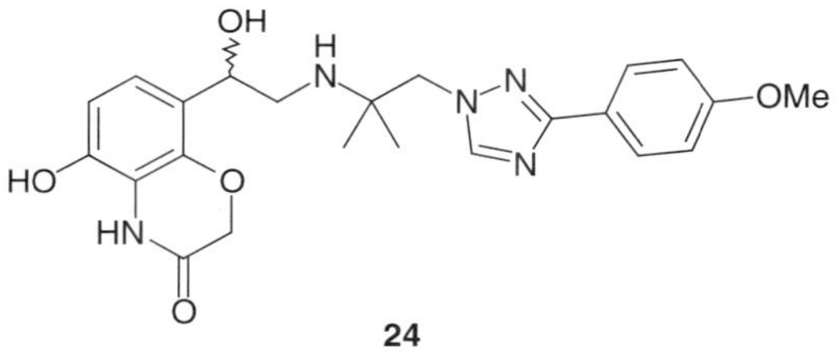

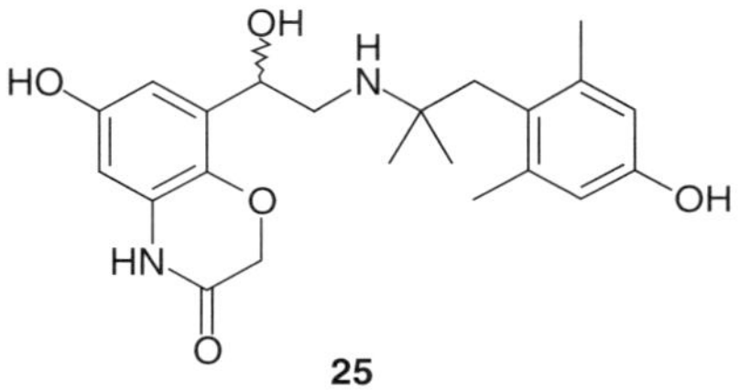

It has also been of interest to further develop the range of heterocyclic frameworks beyond triazoles and benzimidazoles and a recent patent describes further heterocycles such as **26** that are claimed to be long acting β_2-adrenoceptor agonists [44]. A recent disclosure also demonstrates a sulfonamide head group as exemplified in **27** that is claimed to have utility as a long acting β_2-adrenoceptor agonist [45].

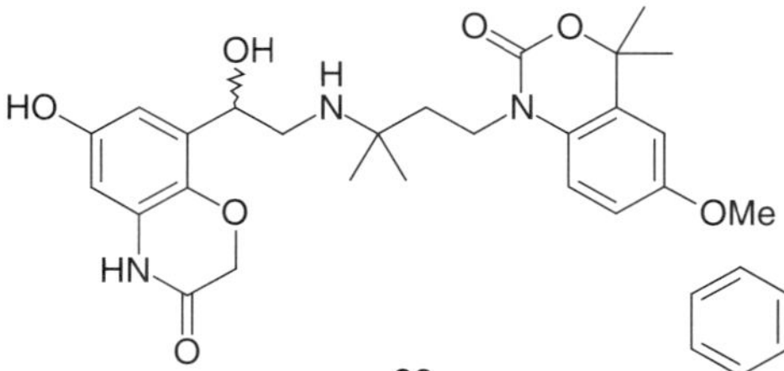

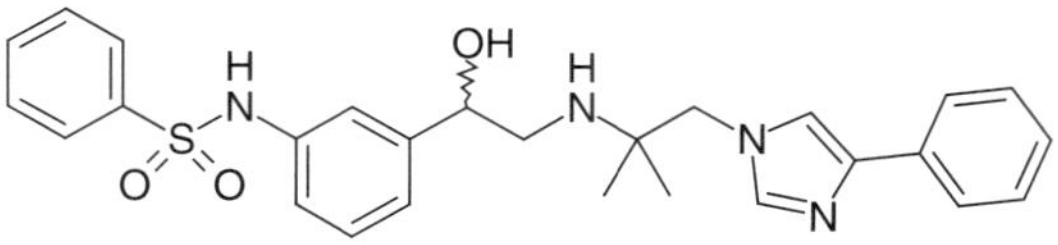

Researchers have also relied on results of early work in the β$_2$-adrenoceptor agonist arena by modifying the catechol group present in the endogenous ligand adrenaline [46]. This approach diverges from the more conservative methods which have relied on marketed head groups as starting points, and has led to the identification of compounds such as **28** and **29** [47,48].

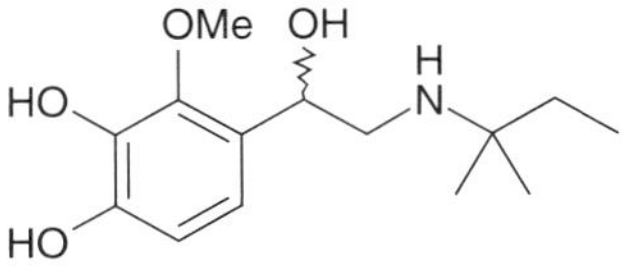

The structural features found in the saligenin **21** have been exploited by incorporating them into unprecedented β$_2$-agonist head groups such as the sulfonamide **30** or 2-aminopyridine **31** [49–52]. Literature precedent includes 2-aminopyridines as selective β$_3$-adrenoceptor agonists [53], whereas compound **31** is a potent and selective β$_2$-adrenoceptor agonist.

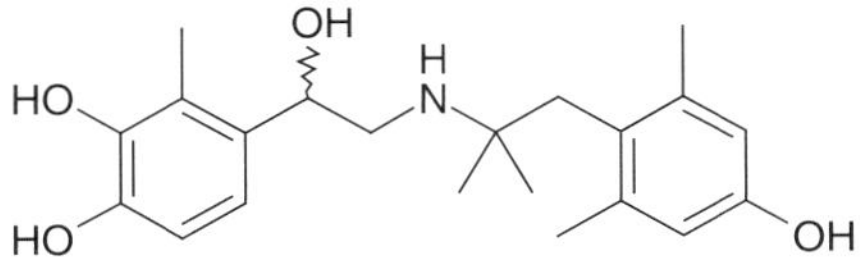

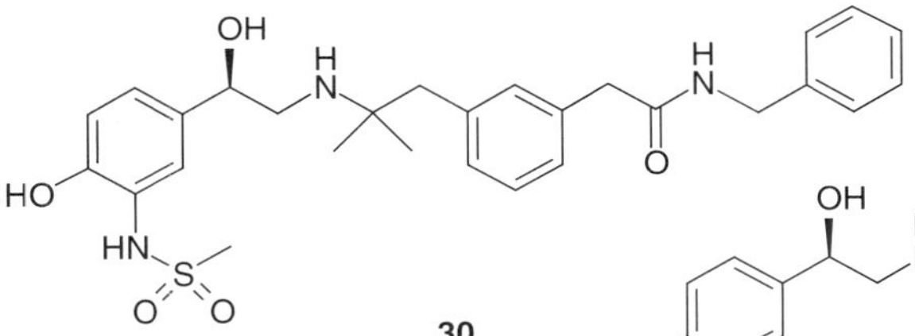

6. DUAL PHARMACOLOGY

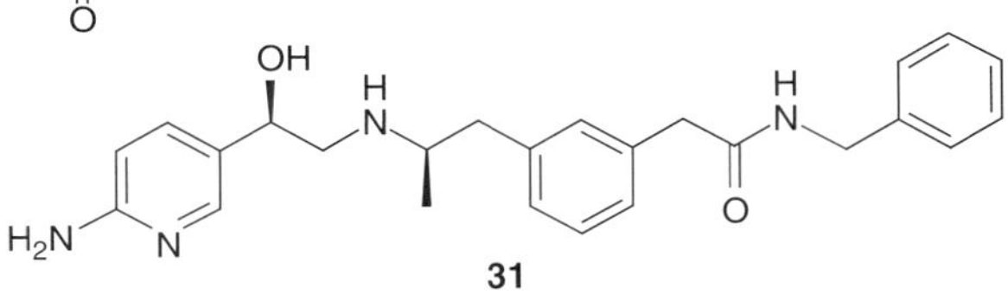

There is ample precedence in the long acting β$_2$-adrenoceptor agonist arena for combination therapy with inhaled corticosteroids (ICS). Notable examples include Advair^{TM}, which is the combination of salmeterol and fluticasone, and Symbicort^{TM}, which consists of formoterol and budesonide. There is also a proven market for combinations containing M$_3$ antagonists, such as Combivent^{TM}; the latter agent is a mixture of the short-acting β$_2$-adrenoceptor agonist, salbutamol, combined with ipratropium.

An alternative paradigm is to have the two pharmacologies incorporated within a single molecule. The dual pharmacology approach presents a number of potential advantages including ease of formulation, simpler pharmacokinetics, and the option of combining the dual agent with a third pharmacology in the inhaler.

The most notable effort to combine a second pharmacology with β$_2$-adrenoceptor agonism is the development of sibenadet (AR-C68397AA) **32**. Attempts were made to combine dopamine D$_2$ agonism with β$_2$-adrenoceptor agonism and this approach was highly encouraging with data being sufficiently attractive to submit sibenadet to

Phase III clinical trials [54,55]. Subsequently, the development of sibenadet was terminated due to the absence of sustained long-term benefits compared to other available treatments. This lack of efficacy was attributed to downregulation of the D_2 receptor over the course of the longer Phase III clinical trial [56].

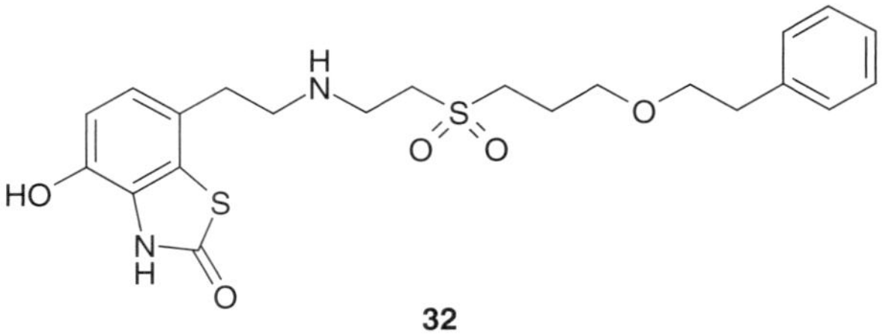

32

There has been a re-emerging interest of late in designing dual pharmacology agents and in particular, compounds that possess M_3 antagonism to complement long-acting β_2-adrenoceptor agonism. There have been a number of patent filings around this approach, including the recent disclosure that compound TD-5959 had progressed to preclinical trials [57]. The structure of this compound is currently unknown; however, a recent patent filing describes a crystalline salt form of **33** [58]. This application describes the preparation of the salt and claims that it can be used for the treatment of pulmonary conditions.

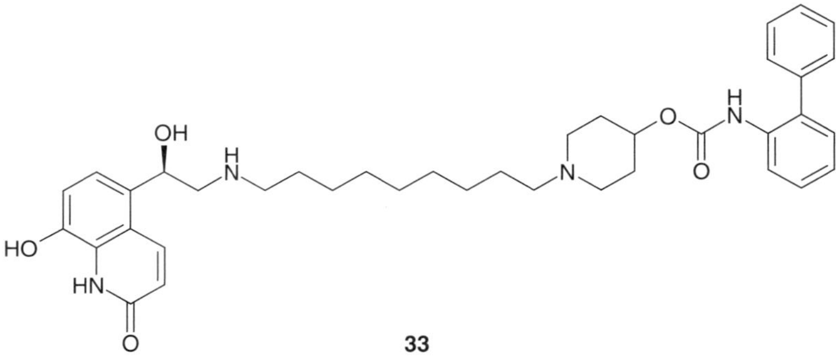

33

7. CONCLUSION

The interest in the pharmaceutical industry to develop long-acting β_2-adrenoceptor agonists with improved duration of action over salmeterol and formoterol is intense. The coming years will be pivotal as data for the most promising compounds will become available from clinical studies. There remains a continuing need for quality inhaled medicines for the treatment of asthma and COPD and it is a challenge to the industry to supply improved agents.

REFERENCES

[1] B. Waldeck, *Eur. J. Pharmacol*, 2002, **445**, 1.
[2] N. M. Siafakas, P. Vermeire and N. B. Price, *Eur. Respir. J.*, 1995, **8**, 1398.

[3] M. Johnson, *Med. Res. Rev.*, 1995, **15** (3), 225.
[4] G. P. Anderson, *Life Sci.*, 1993, **52**, 2145.
[5] M. Palmqvist, G. Persson, L. Lazer, J. Rosenborg, P. Larsson and J. Lotvall, *Eur. Respir. J.*, 1997, **10**, 2484.
[6] R. Morphy, C. Kay and Z. Rankovic, *Drug Discov. Today*, 2004, **9** (15), 641.
[7] S. T. Holgate, J. Holloway, S. Wilson, P. H. Howarth, H. M. Haitchi, S. Babu and D. E. Davies, *J. Allergy Clin. Immunol.*, 2006, **117** (3), 496.
[8] P. J. Barnes, *Pharm. Rev.*, 2004, **56** (4), 515.
[9] M. Cazzola, M. G. Matera and J. Lotvall, *Expert Opin. Investig. Drugs*, 2005, **14** (7), 775.
[10] H. Voss, D. Donnell and A. Bast, *Eur. J. Pharm.-Mol. Pharm. Sect.*, 1992, **227**, 403.
[11] B. Cuenoud, R. A. Fairhurst and N. Lowther, *WO Patent* 45703, 2002.
[12] C. A. Lewis, L. Jordan, D. Wyss, D. Bayley, J. Maas and C. Battram, *Proc. Am. Thorac. Soc.*, 2005, **2**, A355.
[13] E. Naline, M. Molimard, R. Fairhurst, A. Trifilieff and C. Advenier, *Proc. Am. Thorac. Soc.*, 2005, **2**, A356.
[14] Investigational Drugs Database, 603221.
[15] M. Aubier, X. Duval, H. Knight, S. Perry, J. Wood and L. Brookman, *Eur. Respir. J.*, 2005, **26** (49), A1920.
[16] Investigational Drugs Database, 624155.
[17] H. Voss, D. Donnell and A. Bast, *Eur. J. Pharm.-Mol. Pharm. Sect.*, 1992, **227**, 403.
[18] Investigational Drugs Database, 167306.
[19] Investigational Drugs Database, 257189.
[20] Investigational Drugs Database, 559622.
[21] G. P. Anderson, A. Linden and K. F. Rabe, *Eur. Respir J.*, 1994, **7**, 569.
[22] S. Axt and I. Stergiades, *US Patent* 0224982, 2004.
[23] R. M. McKinnell, J. R. Jacobsen, S. G. Trapp and D. R. Saito, *US Patent* 0159448, 2005.
[24] V. Alikhani, D. Beer, D. Bentley, I. Bruce, B. M. Cuenoud, R. A. Fairhurst, P. Gedeck, S. Haberthuer, C. Hayden, D. Janus, L. Jordan, C. Lewis and K. Smithies, *Bioorg. Med. Chem. Lett.*, 2004, **14**, 4705.
[25] E. J. Moran and E. Fournier, *WO Patent* 099764, 2003.
[26] M. S. Linsell, J. R. Jacobsen, D. Khossravi, M. Paborji and W. Zhang, *WO Patent* 011416, 2004.
[27] M. S. Linsell, J. R. Jacobsen and D. R. Saito, *WO Patent* 030678, 2005.
[28] P. C. Box, D. M. Coe, B. E. Looker and P. A. Procopiou, *WO Patent* 024439, 2003.
[29] K. Biggadike, P. C. Box, D. M. Coe, D. S. Holmes, B. E. Looker and P. A. Procopiou, *WO Patent* 071388, 2004.
[30] P. C. Box, *WO Patent* 037768, 2004.
[31] P. C. Box, D. M. Coe, B. E. Looker and P. A. Procopiou, *WO Patent* 091204, 2003.
[32] D. M. Coe and S. B. Guntrip, *WO Patent* 037807, 2004.
[33] B. S. Guntrip and P. A. Procopiou, *WO Patent* 022547, 2004.
[34] A. M. Chapman, S. B. Guntrip, B. E. Looker and P. A. Procopiou, *WO Patent* 037773, 2004.
[35] E. J. Moran, J. R. Jacobsen, M. R. Leadbetter, M. B. Modwell, S. G. Trapp, J. Aggen and T. J. Church, *US Patent* 0063755, 2004.
[36] A. D. Brown, J. S. Bryans, M. E. Bunnage, P. A. Glossop, C. A. L. Lane, R. A. Lewthwaite and S. J. Mantell, *WO Patent* 032921, 2004.
[37] A. D. Brown, J. S. Bryans, M. E. Bunnage, P. A. Glossop, C. A. L. Lane, R. A. Lewthwaite and S. J. Mantell, *WO Patent* 080964, 2004.

[38] A. D. Brown, J. S. Bryans, M. E. Bunnage, P. A. Glossop, C. A. L. Lane, R. A. Lewthwaite, S. J. Mantell and C. Perros-Huguet, Discovery of Novel Inhaled Long-Acting Beta-2 Agonists for Asthma & COPD, Poster P14, 13th RSC-SCI Medicinal Chemistry Symposium, September 4–7, 2005, Churchill College, Cambridge, UK.
[39] M. E. Bunnage, P. A. Glossop, C. A. L. Lane and R. A. Lewthwaite, *WO Patent* 100950, 2004.
[40] A. D. Brown, M. E. Bunnage, P. A. Glossop, K. James, C. A. L. Lane, R. A. Lewthwaite, G. Lunn and D. A. Price, *WO Patent* 090287, 2005.
[41] A. D. Brown, J. S. Bryans, P. A. Glossop, C. A. L. Lane and S. J. Mantell, *WO Patent* 090288, 2005.
[42] A. Walland, K. Schromm, K. H. Bozung and H. Schollenberger, *US Patent* 0137242, 2005.
[43] T. Bouyssou, F. Buettner, I. Konetzki, S. Pestel, A. Schnapp, H. Schollenberger, K. Schromm, C. Heine, K. Rudolf, P. Lustenberger and C. Hoenke, *WO Patent* 045618, 2004.
[44] T. Bouyssou, I. Konetzki, P. Lustenberger, J. Mack, A. Schnapp, D. Wiedenmayer, C. Hoenke and K. Rudolf, *US Patent* 0227975, 2005.
[45] T. Trieselmann and S. H. Bradford, *US Patent* 0245526, 2005.
[46] D. Jack, *Br. J. Clin. Pharmac.*, 1991, **31**, 501.
[47] T. Bouyssou, F. Buettner, I. Konetzki, S. Pestel, A. Schnapp, H. Schollenberger, K. Schromm and C. Heine, *WO Patent* 033412, 2004.
[48] T. Bouyssou, F. Buettner, I. Konetzki, S. Pestel, A. Schnapp, H. Schollenberger, K. Schromm and C. Heine, *WO Patent* 046083, 2004.
[49] M. H. Fisher, H. O. Ok and A. E. Weber, *US Patent* 5714506, 1998.
[50] A. D. Brown, C. A. L. Lane and R. A. Lewthwaite, *WO Patent* 108676, 2004.
[51] A. D. Brown, J. S. Bryans, C. A. L. Lane and S. J. Mantell, *WO Patent*108675, 2004.
[52] Investigational Drugs Database, 383469.
[53] A. Graul and J. Castaner, *Drugs Future*, 2000, **25** (2), 165.
[54] Investigational Drugs Database, 425928.
[55] Investigational Drugs Database, 590916.
[56] R. Chao, M. Rapta, P. J. Colson and J. Lee, *US Patent* 0209860, 2004.
[57] Investigational Drugs Database, 590916.
[58] R. Chao, M. Rapta, P. J. Colson and J. Lee, *US Patent* 0182092, 2005.

Section 4:
Cancer and Infectious Diseases

Editor: David Myles
Kosan Biosciences, Inc
Hayward
California

The Acyl Sulfonamide Antiproliferatives and Other Novel Antitumor Agents

Mary M. Mader

Lilly Research Laboratories, Eli Lilly and Company, Lilly Corporate Center, Indianapolis, IN 46285 USA

Contents

1. Introduction 251
2. The acyl sulfonamide antiproliferatives 252
3. SNS-595 255
4. Other novel antiproliferative agents 256
5. Summary and outlook 260
References 260

1. INTRODUCTION

In 2005, the FDA approved three new small molecules for cancer indications (Fig. 1): nelarabine **1** (Arranon®, GlaxoSmithKline; for acute lymphoblastic leukemia) [1], sorafenib **2** (Nexavar®, Bayer; for advanced renal cell carcinoma) [2], and lenalidomide **3** (Revlimid®, Celgene; for treatment of patients with transfusion-dependent anemia due to myelodysplastic syndrome) [3]. Each drug has a different mechanism of action (MOA). Nelarabine is a prodrug of the antimetabolite ara-G, sorafenib inhibits several kinases, and lenalidomide inhibits production of TNFalpha. All three drugs exhibit side effects to various degrees, ranging from nausea and alopecia to neutropenia [4,5], although the development of kinase inhibitors such as sorafenib, and earlier, erlotinib (Tarceva® for non-small cell lung cancer (NSCLC) and pancreatic cancer), imatinib (Gleevec® for gastrointestinal stromal tumors and chronic myeloid leukemia), and gefitinib (Iressa® for NSCLC) mark a strategy change aimed at reducing the cytotoxicity of cancer drugs. Kinase inhibitors are described as

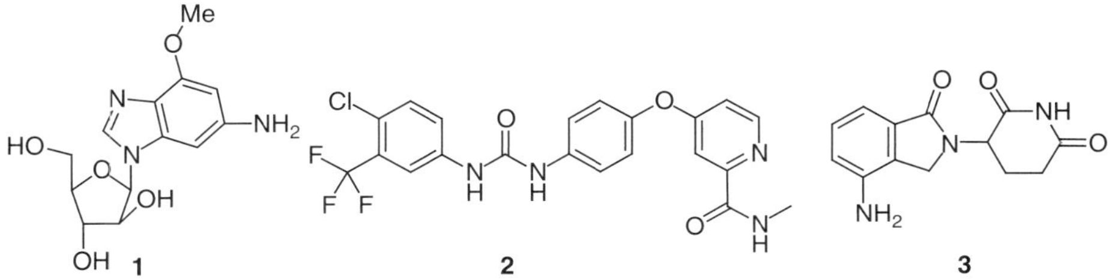

Fig. 1. New chemical entities approved by the US FDA for oncology indications in 2005.

"targeted therapies" because they are generally optimized for *in vitro* potency by assay in a specific enzyme which has been linked to a disease state. The targeted approach, in theory, should result in fewer of the side effects associated with the more traditional cytotoxic cancer therapies such as 5-fluorouracil, paclitaxel, and the platinum compounds [6]. Previously, oncology drugs were often found through *in vitro* screening for activity in cancer cell lines, and the specific molecular target of the drug was not necessarily known at the time of the drug's discovery, or even upon its entry into the clinic [7].

At present, a handful of new, non-natural product, small molecule drugs are identified through general cytotoxic screening methods and enter clinical trials. Although more than 180 industry-sponsored Phase I/II trials were underway in the US in the spring of 2006, investigations of new, small molecules as monotherapies represent less than one-third of the studies. The small molecules can be categorized further, as either enzyme-targeted therapies or natural products with anticancer activity. Only two new small molecules in Phase I or Phase II for oncology were identified through general cytotoxic screening methods: Lilly's acyl sulfonamide antiproliferative compound, LY573636, and Sunesis' SNS-595. Other compounds are described in the literature as having an "unknown" or "novel" MOA, and preclinical development is ongoing. The cell-based screening approach offers some advantages in early structure–activity relationships (SAR) optimization, despite the absence of a mechanistic understanding, thus we will examine the discovery and development of some of these compounds.

2. THE ACYL SULFONAMIDE ANTIPROLIFERATIVES

In 2004, Lilly reported [8] the discovery and SAR of acylsulfonamide antiproliferative (ASAP) compounds of general structure **4**. Subsequently, the activity of a second series **5** of thiophenyl acylsulfonamides was reported, and the compounds showed cellular activity comparable to the parent biaryl series (Fig. 2) [9]. The two initial leads **4a** and **4b**, one unsubstituted in the sulfonyl phenyl ring, and the other mono-substituted by chlorine in the *para*-position of the same ring, were identified in collaboration with researchers at Wayne State University through an *in vitro* screening effort that differentiated activity in solid tumor cell lines from normal fibroblasts and leukemia cell lines. Representative members of the two series were

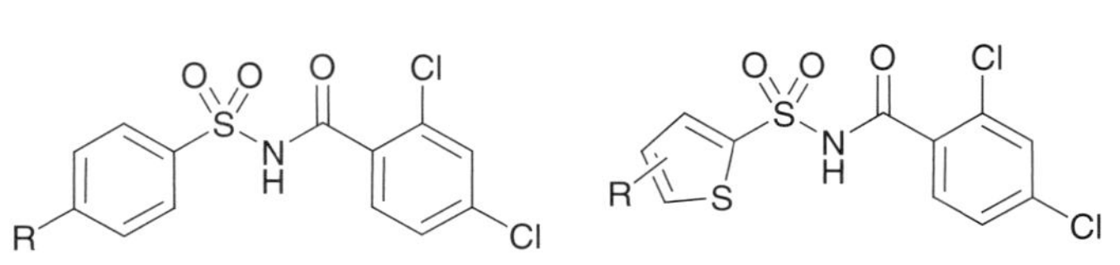

4a R = 4-H
4b R = 4-Cl

5

Fig 2. Lilly's antitumor antiproliferative compounds with unknown mechanism of action.

submitted to the 60-cell line COMPARE analysis of cell sensitivities of standard agents at the National Cancer Institute (NCI), and no correlations were found, suggesting a potentially unique MOA [10]. A Pearson correlation coefficient of >0.7 is considered significant for this type of analysis, but with one exception, chloroquinoxaline sulfonamide (PCC = 0.589), no compound in the database had a PCC>0.4. Although the ASAP compounds are observed to stimulate apoptosis, their MOA is not known.

Because the compounds showed cellular activity at the outset, rapid optimization of the ASAP compounds was possible. As a family, their "druggability" parameters met the criteria described by Lipinski's "rule of five" [11] with an average molecular weight of 360 and clogP of 4.0. In addition, the compounds were soluble in weakly basic bicarbonate buffer, due to the acidity of the amide proton ($pK_a = 2.8$), and were found to be highly stable to metabolism in liver microsomes. Thus, the medicinal chemistry efforts could be focused on optimizing potency and understanding the SAR of the molecules rather than addressing physicochemical properties, permeability, and metabolic instability.

The activity of the compounds was optimized in a vascular endothelial growth factor-stimulated human umbilical vein endothelial cell (VEGF-HUVEC) assay, and antiproliferative activity in this cell line was correlated to that in the colon cancer cell line, HCT116 [8]. In the bi-aryl series **4**, exploration of the substitution of the benzoyl ring revealed that the 2,4-disubstitution of the parent system was preferred (examples **4c–4h** in Table 1). The benzenesulfonamide accepted substitution at the *meta* position as well as *para*, and was slightly sensitive to electronic effects. The strongly electron donating *para*-methoxy- substituted compound **4l** was similar in activity to the weakly donating methyl analog **4k**, but introduction of a nitro (**4m**) or phenyl (**4n**) at the same position resulted in a fourfold decrease in activity. In comparison to the 35- to 100-fold losses observed by movement of substitution on the benzoyl side (**4b** vs. **4e–4g**), the benzenesulfonamide tolerated a wider range of substitution while retaining activity.

The observation that the benzenesulfonamide allowed a broader range of substitution led to the investigation of fused aromatic ring replacements as well as the thiophene series **5**. The thiophene analogs **5a–5e**, whether substituted at the 4- or 5-position of the thiophene, showed similar potency to the substituted phenyl analogs **4k–4m**. Small substituents such as halo or methyl were preferred, with IC_{50} values <1 μM. The 5/6 fused benzothiophene analog **5f**, attached at the C-2 position, was approximately threefold less active than the thiophenes, and is representative of the activity of other fused heterocycles that were prepared (Table 2).

In vivo testing was carried out in HCT116 tumor xenografts, and significant tumor growth delays were observed relative to both vehicle and CPT-11 for compounds **4a** and **4h**. Upon oral dosing, the average tumor growth delay was found to be 14 days for compound **4a** at 64 mg/kg, and 23 days for **4h** at the same dose. A separate study of the pharmacokinetic (PK) properties of **4a** and **4f** in mice revealed the compounds to be well absorbed with long half lives upon oral dosing (**4a**: AUC_{0-24} 11300 μg h/mL at 80 mg/kg; **4h**: AUC_{0-24} 10,700 μg h/mL at 80 mg/kg) with $t_{1/2}$ 27 and 132 hr for **4a** and **4h**, respectively. The length of the half-life may be associated with the protein-binding properties of the class.

Table 1. SAR of the biaryl acyl sulfonamide antiproliferatives [8]

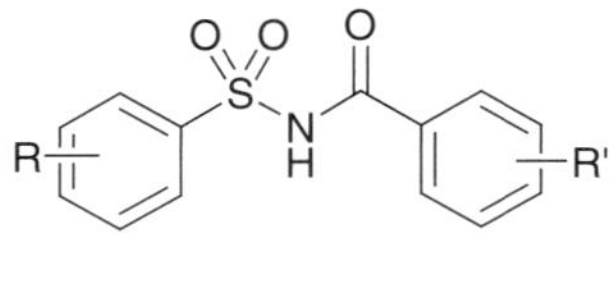

Cmpd no.	R	R'	VEGF-HUVEC IC$_{50}$ (µM)
4a	(H)	2,4-diCl	0.53
4b	4-Cl	2,4-diCl	0.17
4c	4-Cl	2-Cl	14
4d	4-Cl	4-Cl	6.3
4e	4-Cl	3,4-diCl	5.9
4f	4-Cl	2,5-diCl	13
4g	4-Cl	3,5-diCl	18
4h	4-Cl	2-Cl, 4-Br	0.20
4i	4-Cl	2-Me, 4-Cl	0.30
4j	3-Cl, 4-Me	2,4-diCl	0.36
4k	3-Cl	2,4-diCl	0.44
4l	4-Me	2,4-diCl	0.21
4m	4-OMe	2,4-diCl	0.35
4n	4-NO$_2$	2,4-diCl	1.3
4o	4-Ph	2,4-diCl	1.4

Table 2. SAR of the 2,4-dichlorobenzoyl thiophenyl sulfonamide antiproliferatives [9]

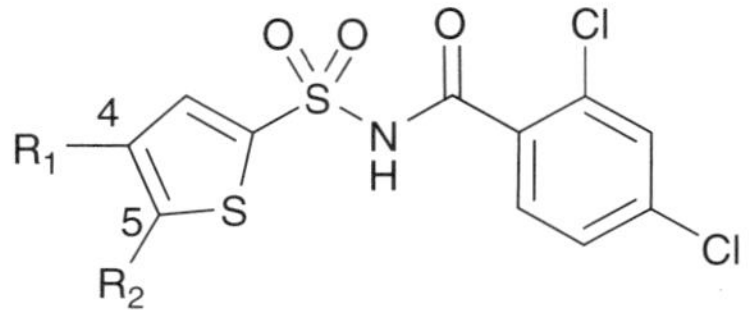

Cmpd no.	R1	R2	VEGF-HUVEC IC$_{50}$ (µM)
5a	H	Cl	0.36
5b	Cl	H	0.39
5c	H	Br	0.25
5d	H	Me	0.60
5e	H	OMe	0.68
5f	CH=CH–CH=CH		0.96

The sodium salt of LY573636 (LY573636.Na) entered clinical trials in 2003 in patients with solid tumors, and in 2005 preliminary results of the Phase I study were reported [12]. Patients were dosed intravenously for 2 h every 21 days, with doses ranging from 100 to 1400 mg. The tumor selective activity from the early screening studies in which fibroblasts were unaffected are borne out in the clinic, and toxicities have been minimal (Grade 1), with no hematological or gastrointestinal toxicity observed to date, nor have any dose-limiting toxicities been observed. The PK properties of the drug in humans were also reported in this Phase I study. The compound appears to be highly protein bound, as evidenced by a half-life of 15 days, which appears to correspond to the rate of albumin turnover. In addition, the total volume of distribution of LY573636.Na corresponds approximately to that of albumin (89 L).

3. SNS-595

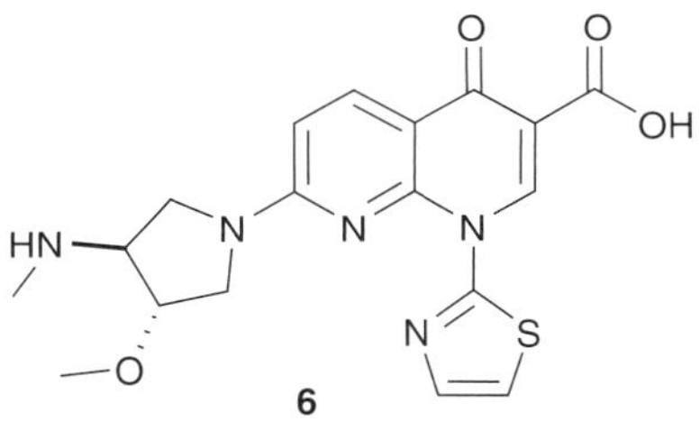

SNS-595 (**6**, SPC-595; formerly AG-7352) is a 1,8-disubstituted naphthyridine whose discovery and SAR were originally reported by Dainippon Pharmaceuticals [13,14]. The discovery came from screening antibacterial quinolines for antitumor activity, and finding that 1,8-naphthyridines possess cytotoxic activity against murine P388 leukemia. The rationale behind the screening for cytotoxic activity in tumors rather than bacteria was that the quinolone antibacterials are known to inhibit topoisomerase II enzymes, and that mammalian topoisomerase II is the target for DNA-active antitumor agents such as etoposide and doxorubicin. The hypothesis was supported by the antineoplastic activity of quinolines such as A-65282 [15], CP-115953 [16], and WIN 57294 [17].

As in the case of LY573636, the primary assay for the SAR studies was an *in vitro* cell assay, in this case P388 leukemia. In the SAR study of the series, several compounds inhibited the proliferation of murine P388 leukemia cells with $IC_{50} < 0.050\,\mu g/mL$ and were progressed to *in vivo* study in mice. SNS-595 demonstrated was dosed at three doses (3.13, 12.5, and 50 mg/kg) without toxicity at the top dose when administered intraperitoneally (ip) on days 1 and 5 after tumor implantation. In addition, **6** was found to have activity comparable or superior to etoposide in a panel of tumor cells *in vitro* (HL-60 leukemia, Hs746 T stomach, PANC-1 pancreas, SBC-3 lung, SK-OV-3 ovary, and SCaBER bladder) [14]. Earlier work on a closely related series also showed activity in solid tumors in addition to leukemia (AZ-512 stomach, HT-29 colon, A-427 lung, SK-OV-3 ovary, and SCaBER bladder), and the first Phase I studies with SNS-595 were carried out in patients with solid tumors [13].

Sunesis licensed the compound for clinical development in the US and has reported MOA studies. Although a specific protein target has not been identified, **6** acts in the S-phase of the cell cycle, and the data suggest that it causes selective double strand breaks during DNA synthesis. Many pathways exist to repair such damage, but **6** affects the DNA-damage sensing kinase (DNAPK) nonhomologous end-joining pathway. Even after removal of **6**, cells could not repair the strand breaks, resulting ultimately in apoptosis [18].

Phase I trials in patients with advanced malignancies progressed to Phase II trials in non-small cell lung cancer (NSCLC; in November 2005) as well as small cell lung cancer (SCLC; in March 2006). In the Phase I study, the dose-limiting toxicity was neutropenia, and the minimum toxic dose is between 12 and 15 mg/m^2 on a 28-day weekly schedule. A Phase I trial in acute myelogenous leukemia (AML) has been initiated as well [19].

4. OTHER NOVEL ANTIPROLIFERATIVE AGENTS

In the discovery phase, compounds continue to be identified and optimized by assay in a tumor cell line, although an MOA or biochemical target is not known. The compounds are identified from a variety of sources, including natural products, or from modification of structures that are known to be bioactive, or by *de novo* chemical synthesis. Authors often describe the compounds as "novel" because they have a previously unreported structural modification or they possess unique activity profiles in tumor cells. The data in the initial reports of these compounds are focused on SAR in an *in vitro* tumor cell assay, and may contain evidence of *in vivo* activity in a human tumor xenograft model. Information on ADME and PK properties is generally not available for the examples in Table 3.

An initial hypothesis that 2-phenylbenzothiazoles could mimic ATP and thus inhibit tyrosine kinases led to a series of compounds in which a 2-(dimethoxyphenyl)benzothiazole, GW610 (**7**), was found to be the most active member (Fig. 3) [20]. The initial *in vitro* assay was in four tumor cell lines (MCF-7, MDA 468, KM 12, and HCC2998), with follow up of the most active members in the NCI's full panel of 60 human tumor cell lines [21]. In the smaller set of tumor cells, **7** was most active in the breast cell line MDA 468. In the larger NCI panel, the compound exhibited its greatest activity in the colon and NSCLC subpanels. Interestingly, **7** appears to complement a more extensively studied 2-(aminophenyl)benzothiazole, in that breast tumor cell lines which acquire resistance to the (aminophenyl)benzothiazole retain sensitivity to the dimethoxyphenyl analog. A COMPARE analysis was not reported, a molecular target was not identified from this study, and *in vivo* studies were not undertaken, but the authors indicate such work will be reported at a later time.

A series of anthranilic acid analogs were prepared and assayed in the NCI's full panel of 60 tumor cell lines, and the most active compound was found to be entry **8**, a 2-trifluoromethylpyridyl derivative [22]. The compound was especially active in

Table 3. Novel antiproliferatives with new structures or biological characterization

Cmpd no.	Name	Source of structure	Assay for SAR	In vivo model	Mechanism
7	GW 610	Bioactive skeleton (benzothiazole)	MCF-7 (breast) MDA 468 (breast) KM 12 (colon) HCC 2998 (colon)	None reported	None reported
8	N-(2-(trifluoro-methyl)-pyridin-4-yl) anthranilic ester	Bioactive skeleton (anthranilic acid)	NCI's full panel of 60 human tumor cell lines	U251 (CNS)	COMPARE performed; possibly antimetabolite or topoisomerase II inhibitor
9	NSC-686288; TK-2339	Bioactive skeleton (flavone)	MCF7 (breast)	MCF7 (breast)	Induces DNA strand breaks
10	Cyclopenta-[e]azepine-4,10(1 H,5 H)-diones	Bioactive skeleton (benzazepine-diones)	In vitro cell line screen at NCI	None reported	COMPARE performed; possibly tyrosine kinases
11	Pyrrolo [3,2-f]quinolin-9-one	Bioactive skeleton (flavone)	Included: MCF7 (breast) H295R HT-29 (colon)	BNL 1ME A.7R.1 murine liver carcinoma	Antimitotic via tubulin depolymerization
12	EPC-2407; MX-116407	Synthetic library	T47D (breast), H1299 (NSCLC) DLD-1 (colon)	Calu-6 (lung)	Antimitotic via tubulin inhibition
13	Decursin	Natural product	DU145 (prostate) PC-3 (prostate) LNCaP (prostate)	Murine sarcoma 180	Cell cycle arrest, apoptosis via caspase-dependent and independent pathways

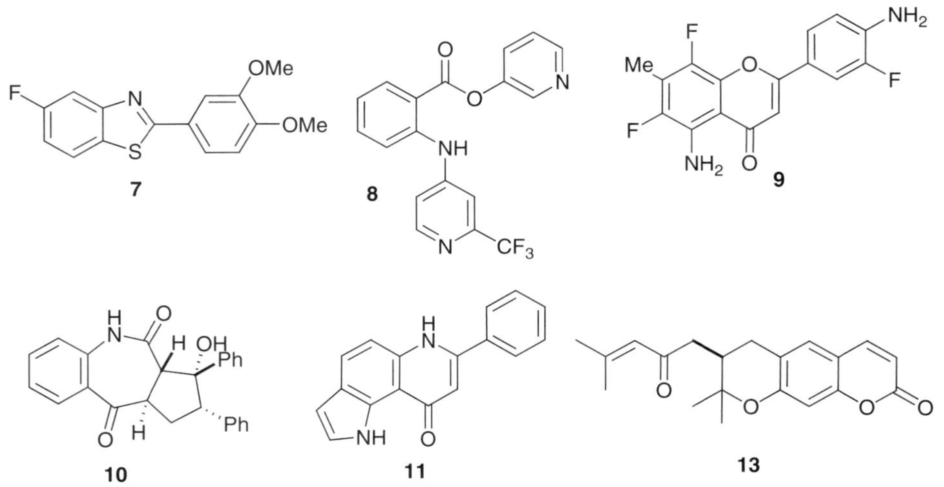

Fig. 3. Structures of novel antiproliferatives in discovery.

leukemia and colon tumor cell lines *in vitro*, but subsequent assay in hollow fibers and xenografts *in vivo* showed its greatest activity in a central nervous system tumor, U251, when dosed intraperitoneally (ip) at 100 mg/kg daily for 4 days. The growth inhibition was stated as "moderate" by the authors (67% of control under the assay conditions), but the PK properties of **8** were not described; potentially dosing on a different schedule or route could result in greater tumor growth inhibition. With respect to mechanism of action, the COMPARE PCC for **8** was greatest with the antimetabolite, morpholino-ADR, and the topoisomerase II inhibitor, vincristine. The conclusion drawn at present is that these anthranilates act by inhibiting cell cycle progression, but additional studies will be necessary to determine a specific target.

Natural and synthetic flavonoid derivatives exhibit a variety of biological activities, including antitumor activity, although the antiproliferative mechanisms are not fully understood. Researchers at Kyowa Hakko made two key observations: hydroxy-substituted flavonoids exhibit activity associated with breast cancer or estrogenic action, and that few analogs existed in which the hydroxyl group had been replaced by an amine. A series of modifications were analyzed, first confirming that an amino flavone inhibited *in vitro* antiproliferative activity in estradiol-stimulated MCF-7 breast cancer cells with an IC_{50} value of 7.2 nM [23]. Subsequently, fluorine was introduced to both the fused and pendant aromatic rings to enhance the metabolic stability of the scaffold, and this analog potently inhibited the growth of both estrogen receptor (ER) positive breast cancer cell lines and cells from ER-negative cancers of the breast, ovaries, endometrium, and liver [24]. The final iteration of the series explored the 7-position of the flavone ring, providing NSC-686288 (**9**), and it was found that methyl substitution was preferred at this position *in vitro* and *in vivo* in MCF7 cell and xenograft assays, respectively [25]. Mechanistic studies at the NCI have shown that **9** is metabolized by CYP1A1, and that it induces the expression of CYP1A1 in MCF-7 cells [26]. It additionally has been demonstrated that **9** involves the aryl hydrocarbon receptor (AhR) in the regulation of the CYP1A1 expression [27], and more recently **9** has been shown to induce DNA single strand breaks and DNA-protein cross-links [28].

The fused tricycle cyclopenta-[e]azepine-4,10(1 H,5 H)-dione **10** was discovered via modification of the synthesis which had provided cytotoxic antitumor benzazepines [29]. A tandem Michael–Aldol reaction in the presence of base led to a new compound class of 6/7/5 fused tricyclics as mixtures of diastereomers [30]. Separation and analysis of the diastereomers in the NCI's *in vitro* cell line screening project (IVCLSP) found that alteration of the substitution pattern on the saturated 5-membered ring (1,3-disubstitution vs. 2,3-disubstitution as shown in **10**) indicated distinctly different mechanisms of action. The highest correlations of **10** with known agents were with bleomycin and 4-nitro-estrone-4-methyl ester, which do not have a common molecular target. To further explore activity suggested by the NCI's database of molecular targets, the most active diastereomer of **10** was assayed at 10 µM concentration in 16 cancer-related kinases, but only moderate inhibitory activity was not found. Thus, it is proposed that **10** could interact with molecules of the signaling cascades in which kinases are involved, although it may not be a kinase inhibitor itself. Additional investigations will be necessary to establish the mechanism of action of this novel ring system.

The finding that some natural flavonoids have antimitotic cytotoxic activity by interfering with tubulin polymerization spurred the synthesis of another analog, 2-phenylpyrroloquinolin-4-one. Two structural variants have been reported, varying the fusion position of a pyrrole ring to the quinolone [31,32]. The SAR was developed in human and mouse tumor cell lines including those derived from liver, pancreas, colon, thyroid, ovary, and breast solid tumors. The daily ip injection of **11** (40 mg/kg) caused a significant (83%) inhibition of tumor growth relative to control groups in the hormone-sensitive syngeneic murine hepatocarcinoma BNL 1ME A.7R.1 [32]. Based on evidence of G2/M arrest in two cell lines by flow-activated cell sorting analysis and dose-dependent inhibition of tubulin polymerization similar to vincristin, the mechanism of cytotoxicity is hypothesized to be antimitotic via tubulin depolymerization, although a specific molecular target has not been identified.

A commercially obtained compound library yielded a novel series of 2-amino-4-(3-bromo-4,5-dimethoxy-phenyl)-3-cyano-4H chromenes as potent apoptosis inhibitors in high-throughput assay utilizing HL60 B-cell leukemia cancer cells [33]. Selected compounds were profiled further and found to have activity only toward proliferating cells and to interact with colchicines at the tubulin binding site, thereby inhibiting tubulin polymerization and leading to cell cycle arrest and apoptosis [34]. The most potent member of the family, MX-116407 (**12**), was found to be highly active in the human lung tumor xenograft, Calu-6, and is claimed as the lead candidate, although its structure has not been disclosed [35]. The SAR and structural variation within the family has been explicitly described in the literature, and the 7-dimethylamino analog, MX-58151, demonstrated potency similar to colchicines and vinblastine with an EC_{50} value of 19 nM in a caspase activation assay [33]. The compounds also are active in cells resistant to other antimitotic agents such as the taxanes and vinca alkaloids, and therefore might offer an advantage for the treatment of drug-resistant cancers.

Screening natural products for anticancer activity remains productive, and novel compounds continue to be identified in this manner. Decursin **13** is a coumarin compound that was isolated from Korean angelica gigas root and prepared via total synthesis in 2003 [36]. Like the flavonoids, it exhibits a variety of biological activities, including antibacterial, neuroprotective, and antioxidant. However, only recently has **13** been demonstrated to possess *in vivo* activity in murine sarcoma 180 [37] and to inhibit growth and induce cell death in human prostate cancer cell lines DU145, PC-3, and LNCaP [38]. Its inhibitory mechanism appears to be associated with G1 arrest in DU145 and LNCaP cells and its apoptotic effect is linked to both caspase-dependent and -independent pathways. Notably, it has no effect on normal PWR-1E prostate epithelial cells.

5. SUMMARY AND OUTLOOK

Antiproliferative agents with anticancer activity continue to be reported in the literature and to progress into clinical development. Active compounds identified in tumor cell-based assays tend to have good solubility and permeability characteristics, thus allowing rapid optimization of properties such as metabolic stability and absorption that are essential to successful drugs. The cell-based screening paradigm does not circumvent some key hurdles in the development of anticancer drugs, such as the poor correlation of *in vivo* tumor xenograft activity to clinical experience, and the off-target toxicities which are associated with agents that inhibit rapidly proliferating cells [39]. In the absence of a molecular target, the challenge of correlating of xenografts to clinical experience potentially could be overcome by monitoring not just for tumor growth inhibition, but also for biomarkers that predict the onset of apoptosis. In the pre-clinical setting, toxicities are currently predicted by monitoring for weight loss (or death) in treated animals, but the development of the Lilly ASAP series illustrates the use of counter-screening for selectivity vs. normal fibroblasts. In the final analysis, however, the need for effective anticancer drugs with improved side effect profiles remains, regardless of the initial method of identification of the drugs.

REFERENCES

[1] D. F. Kisor, *Ann. Pharmacother.*, 2005, **39**, 1056–1063.
[2] J. W. Clark, J. P. Eder, D. Ryan, C. Lathia and H.-J. Lenz, *Clin. Cancer Res.*, 2005, **11**, 5472–5480.
[3] A. List, S. Kurtin, D. J. Roe, A. Buresh, D. Mahadevan, D. Fuchs, L. Rimsza, R. Heaton, R. Knight and J. B. Zeldis, *New Eng. J. Med.*, 2005, **352**, 549–557.
[4] *Arranon Prescribing Information*. October 28, 2005 (cited April 12, 2005), available from: http://www.fda.gov/cder/foi/label/2005/021877lbl.pdf.
[5] *Revlimid Prescribing Information*. December 28, 2005 (cited April 12, 2006), available from: http://www.fda.gov/cder/foi/label/2005/021880lbl.pdf.

[6] A. D. Laird and J. M. Cherrington, *Expert Opin. Invest. Drugs*, 2003, **12**, 51–64.
[7] B. A. Chabner and T. G. J. Roberts, *Nat. Rev.: Cancer*, 2005, **5**, 65–72.
[8] K. L. Lobb, P. A. Hipskind, J. A. Aikins, E. Alvarez, Y.-Y. Cheung, E. L. Considine, A. De Dios, G. L. Durst, R. Ferritto, C. S. Grossman, D. D. Giera, B. A. Hollister, Z. Huang, P. W. Iversen, K. L. Law, T. Li, H.-S. Lin, B. Lopez, J. E. Lopez, L. M. Martin Cabrejas, D. J. McCann, V. Molero, J. E. Reilly, M. E. Richett, C. Shih, B. Teicher, J. H. Wikel, W. T. White and M. M. Mader, *J. Med. Chem.*, 2004, **47**, 5367–5380.
[9] M. M. Mader, C. Shih, E. L. Considine, A. De Dios, C. S. Grossman, P. A. Hipskind, H.-S. Lin, K. L. Lobb, B. Lopez, J. E. Lopez, L. M. Martin Cabrejas, M. E. Richett, W. T. White, Y.-Y. Cheung, Z. Huang, J. E. Reilly and S. R. Dinn, *Bioorg. Med. Chem. Lett.*, 2005, **15**, 617–620.
[10] J. N. Weinstein, T. G. Myers, P. M. O'Connor, S. H. Friend, A. J. Fornace, Jr., K. W. Kohn, T. Fojo, S. E. Bates, L. V. Rubinstein, N. L. Anderson, J. K. Buolamwini, W. W. Van Osdol, A. P. Monks, D. A. Scudiero, E. A. Sausville, D. W. Zaharevitz, B. Bunow, V. N. Viswanadhan, G. S. Johnson, R. E. Wittes and K. D. Paull, *Science*, 1997, **275**, 343–349.
[11] C. A. Lipinski, F. Lombardo, B. W. Dominy and P. J. Feeney, *Adv. Drug Delivery Rev.*, 1997, **23**, 3–25.
[12] G. Simon, M. Sovak, M. Wagner, E. Haura, S. Gerst, D. deAlwis, G. Bepler, D. Sullivan, A. Weitzman and D. Spriggs, *Eur. J. Cancer* (*Supplements*), 2004, **2**, 69, abs. 228.
[13] K. Tomita, Y. Tsuzuki, K.-i. Shibamori, M. Tashima, F. Kajikawa, Y. Sato, S. Kashimoto, K. Chiba and K. Hino, *J. Med. Chem.*, 2002, **45**, 5564–5575.
[14] Y. Tsuzuki, K. Tomita, K. Shibamori, Y. Sato, S. Kashimoto and K. Chiba, *J. Med. Chem.*, 2004, **47**, 2097–2109.
[15] W. E. Kohlbrenner, N. Wideburg, D. Weigl, A. Saldivar and D. T. Chu, *Antimicrob. Agents Chemotherap.*, 1992, **36**, 81–86.
[16] M. J. Robinson, B. A. Martin, T. D. Gootz, P. R. McGuirk and N. Osheroff, *Antimicrob. Agents Chemotherap.*, 1992, **36**, 751–756.
[17] M. P. Wentland, G. Y. Lesher, M. Reuman, M. D. Gruett, B. Singh, S. C. Aldous, P. H. Dorff, J. B. Rake and S. A. Coughlin, *J. Med. Chem.*, 1993, **36**, 2801–2809.
[18] J. Hyde, J. Wright, D. H. Walker, J. A. Silverman and M. R. Arkin, *Proc. Am. Assoc. Cancer Res.*, 2006, **47**, abs. 2074.
[19] S. Ebbinghaus, M. Gordon, R. Advani, H. Hurwitz, D. Mendelson, H. Wakelee, U. Hoch, J. A. Silverman, N. Havrilla and D. Adelman, *Proc. Am. Assoc. Cancer Res.*, 2006, **47**, abs. 2913.
[20] C. G. Mortimer, G. Wells, J.-P. Crochard, E. L. Stone, T. D. Bradshaw, M. F. G. Stevens and A. D. Westwell, *J. Med. Chem.*, 2006, **49**, 179–185.
[21] M. R. Boyd and K. D. Paull, *Drug Devel. Res.*, 1995, **34**, 91–109.
[22] C. Congiu, M. T. Cocco, V. Lilliu and V. Onnis, *J. Med. Chem.*, 2005, **48**, 8245–8252.
[23] T. Akama, Y. Shida, T. Sugaya, H. Ishida, K. Gomi and M. Kasai, *J. Med. Chem.*, 1996, **39**, 3461–3469.
[24] T. Akama, H. Ishida, Y. Shida, U. Kimura, K. Gomi, H. Saito, E. Fuse, S. Kobayashi, N. Yoda and M. Kasai, *J. Med. Chem.*, 1997, **40**, 1894–1900.
[25] T. Akama, H. Ishida, U. Kimura, K. Gomi and H. Saito, *J. Med. Chem.*, 1998, **41**, 2056–2067.
[26] M. J. Kuffel, J. C. Schroeder, L. J. Pobst, S. Naylor, J. M. Reid, S. H. Kaufmann and M. M. Ames, *Mol. Pharmacol.*, 2002, **62**, 143–153.

[27] A. I. Loaiza-Perez, S. Kenney, J. Boswell, M. Hollingshead, M. C. Alley, C. Hose, H. P. Ciolino, G. C. Yeh, J. B. Trepel, D. T. Vistica and E. A. Sausville, *Mol. Cancer Therap.*, 2004, **3**, 715–725.

[28] L.-h. Meng, G. Kohlhagen, Z.-y. Liao, S. Antony, E. Sausville and Y. Pommier, *Cancer Res.*, 2005, **65**, 5337–5343.

[29] A. Link and C. Kunick, *J. Med. Chem.*, 1998, **41**, 1299–1305.

[30] C. Kunick, C. Bleeker, C. Pruehs, F. Totzke, C. Schaechtele, M. H. G. Kubbutat and A. Link, *Bioorg. Med. Chem. Lett.*, 2006, **16**, 2148–2153.

[31] M. G. Ferlin, G. Chiarelotto, V. Gasparotto, L. Dalla Via, V. Pezzi, L. Barzon, G. Palu and I. Castagliuolo, *J. Med. Chem.*, 2005, **48**, 3417–3427.

[32] V. Gasparotto, I. Castagliuolo, G. Chiarelotto, V. Pezzi, D. Montanaro, P. Brun, G. Palu, G. Viola and M. G. Ferlin, *J. Med. Chem.*, 2006, **49**, 1910–1915.

[33] W. Kemnitzer, J. Drewe, S. Jiang, H. Zhang, Y. Wang, J. Zhao, S. Jia, J. Herich, D. Labreque, R. Storer, K. Meerovitch, D. Bouffard, R. Rej, R. Denis, C. Blais, S. Lamothe, G. Attardo, H. Gourdeau, B. Tseng, S. Kasibhatla and S. X. Cai, *J. Med. Chem.*, 2004, **47**, 6299–6310.

[34] S. Kasibhatla, H. Gourdeau, K. Meerovitch, J. Drewe, S. Reddy, L. Qiu, H. Zhang, F. Bergeron, D. Bouffard, Q. Yang, J. Herich, S. Lamothe, S. X. Cai and B. Tseng, *Mol. Cancer Therap.*, 2004, **3**, 1365–1374.

[35] H. Gourdeau, L. Leblond, B. Hamelin, C. Desputeau, K. Dong, I. Kianicka, D. Custeau, C. Boudreau, L. Geerts, S.-X. Cai, J. Drewe, D. Labrecque, S. Kasibhatla and B. Tseng, *Mol. Cancer Therap.*, 2004, **3**, 1375–1384.

[36] T. Nemoto, T. Ohshima and M. Shibasaki, *Tetrahedron*, 2003, **59**, 6889–6897.

[37] S. Lee, Y. S. Lee, S. H. Jung, K. H. Shin, B.-K. Kim and S. S. Kang, *Arch. Pharm. Res.*, 2003, **26**, 727–730.

[38] D. Yim, R. P. Singh, C. Agarwal, S. Lee, H. Chi and R. Agarwal, *Cancer Res.*, 2005, **65**, 1035–1044.

[39] A. Kamb, *Nat. Rev. Drug Discov.*, 2005, **4**, 161–164.

Progress on Mitotic Kinesin Inhibitors as Anti-cancer Therapeutics

Gustave Bergnes, Xiangping Qian and Andrew A. Wolff

Cytokinetics, Inc., 280 E Grand Ave, South San Francisco, CA 94080, USA

Contents

1. Introduction	263
2. Inhibitors of kinesin spindle protein	263
2.1. Ispinesib and related compounds	264
2.2. Monastrol analogs	267
2.3. Tetrahydro-β-carbolines	268
2.4. Dihydropyrroles and dihydropyrazoles	269
2.5. Tetrahydroisoquinolines	270
3. Conclusions	271
References	272

1. INTRODUCTION

Accurate completion of cell division requires the timely formation of a fully functional mitotic spindle. Success or failure in this endeavor is at least partly dependent on tubulin polymerization dynamics and the activities of multiple accessory proteins whose role is to associate with and manipulate the tubulin polymer [1,2]. Among these accessory proteins are kinesins [3], motor proteins that bind to microtubules and utilize the energy derived from ATP hydrolysis to drive many of the structural changes required for spindle formation and function. The therapeutic potential of mitosis-specific members of this protein family, known as mitotic kinesins [4], was first demonstrated in 1999 with the finding that inhibition of kinesin spindle protein (KSP, *Hs* Eg5) with a small molecule produces cell cycle arrest in prometaphase and eventual cell death [5]. Soon thereafter, CK0106023 (**1**) was reported to shrink tumors in a xenograft model via an anti-mitotic mechanism of action [6]. Since then, agents targeting KSP have entered clinical trials. The following pages describe some of the recent progress made in the discovery and development of mitotic kinesin-targeted therapeutics and ongoing studies that may define the future potential of this target class.

2. INHIBITORS OF KINESIN SPINDLE PROTEIN

KSP, also known as *Hs* Eg5, is a mitotic kinesin required in early mitosis for the formation of a bipolar mitotic spindle [7,8]. This protein exists as a homotetramer with an ability to bind to the oppositely polarized microtubules emanating from separate spindle poles. By coupling ATP hydrolysis with significant conformational mobility, KSP is able to move processively along microtubules via the coordinated action of paired motor domains arrayed at each end of the homotetramer. When bound to two microtubules, both ends of KSP attempt to move in the direction away from centrosomes, forcing spindle poles apart to create a bipolar mitotic spindle. Hence, inhibition of KSP causes mitotic arrest with the formation of monopolar mitotic spindles that result from failed centrosome separation. In addition, by acting on an accessory protein rather than on the microtubule cytoskeleton itself, agents targeting KSP would be expected to avoid the peripheral neuropathies associated with anti-microtubule agents. Mitotic arrest with small molecule inhibitors has been shown to result in cell death on a number of tumor cell lines both *in vitro* and *in vivo* that, along with an expectation of reduced neural toxicity, serves as the basis for several ongoing anti-mitotic drug discovery programs [9,10].

2.1. Ispinesib and related compounds

2.1.1. *Clinical progress on ispinesib and SB-743921*

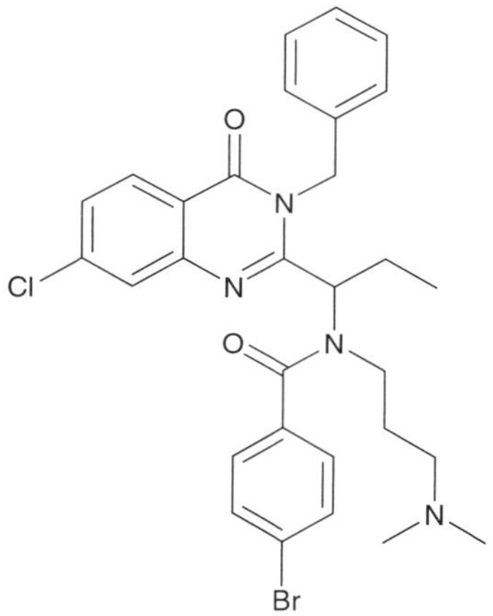

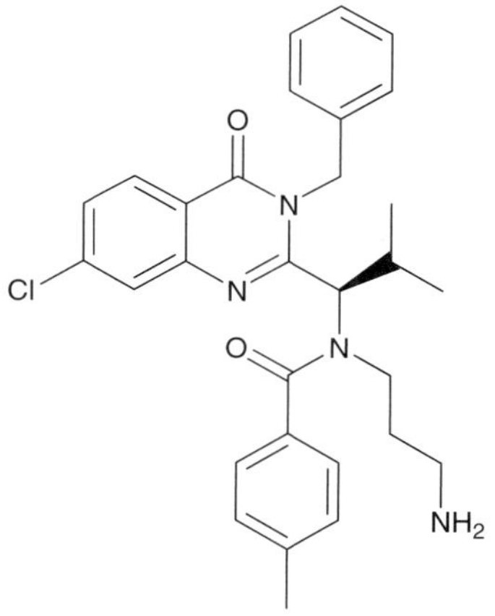

Ispinesib (SB-715992/CK0238273, **2**), one member of a KSP inhibitor series bearing a quinazolinone core, was the first mitotic kinesin inhibitor to enter clinical trials. Like its earlier reported analog **1**, ispinesib was reported to be an allosteric inhibitor of KSP that binds at the motor domain and inhibits its ATPase activity in an ATP uncompetitive manner. It was also shown to be >70,000-fold selective for KSP versus other members of the kinesin family. Results have been reported from two Phase I clinical trials with different dosing schedules. In trial KSP10001 [11], ispinesib was dosed i.v. once every 21 days (q21) in 45 patients. The drug was well tolerated without alopecia or prevalent neurotoxicity. At a dose of 21 mg/m^2, dose-limiting toxicities (DLT) were prolonged (≥ 5 days) neutropenia and febrile neutropenia. Prolonged disease stabilization, ranging from 7 to 43 weeks, was observed in 15 of 45 patients representing a variety of tumors. In addition, evidence of minimal response (<50% tumor mass decrease) was noted in three patients. Ispinesib had a mean (range) Cl of 98 (48–141) ml/min, Vss of 256 (162–514) l, and $t_{1/2}$ of 34 (17–56) h at a dose of 18 mg/m^2, which was the dose level selected for Phase II trials on the same q21-day schedule. In trial KSP 10002 [12], ispinesib was dosed i.v. on days 1, 8, and 15, one cycle every 28 days in 30 patients. The DLT observed at 8 mg/m^2 was grade 3 neutropenia preventing administration of the day 15 dose during cycle 1. The maximum tolerated dose (MTD) on this schedule was 7 mg/m^2. Prolonged disease stabilization ranging from 16 to over 32 weeks was evident in eight of 30 patients. At the MTD, median Cl was 94 ml/min and median $t_{1/2}$ was 33 h, very similar to results in the KSP10001 trial. The desired cellular phenotype, mitotic arrest with monopolar spindle formation, was observed in tumor biopsies in both Phase I trials.

Ispinesib has also been examined in Phase Ib clinical combination with docetaxel [13], capecitabine [14], and carboplatin [15]. The DLT for the combination of ispinesib and docetaxel was prolonged ($\geqslant$5 days) grade 4 neutropenia, and the optimally tolerated regimen (OTR) was defined as 10 mg/m^2 of ispinesib and 60 mg/m^2 of docetaxel, each administered once every 21 days. Ispinesib and docetaxel plasma concentrations were consistent with those previously reported when each drug was given as monotherapy, suggesting no pharmacokinetic interaction between them. A total of 13/24 patients (11 prostate, 1 renal, and 1 bladder) had a best response of stable disease (duration 2.25–7.5 months) [13].

For the combination of ispinesib capecitabine, the DLT and OTR had yet to be defined, although ispinesib had been administered at 18 mg/m^2 for every 21 days in the combination with capecitabine at 2000 mg/m^2 daily for 14 of 21 days. One DLT of prolonged ($\geqslant$5 days) grade 4 neutropenia had been observed at the time of the report. Ispinesib plasma concentrations did not appear to be affected by the presence of capecitabine. A total of eight patients (3 breast, and 1 each of tongue, colorectal cancer (CRC), bladder, thyroid, and salivary gland) had a best response of stable disease (duration 2–6.5 months) [14].

For the combination of ispinesib and carboplatin, DLTs included prolonged ($\geqslant$5 days) grade 4 neutropenia, grade 4 thrombocytopenia, and grade 3 febrile neutropenia. The OTR was ispinesib at 18 mg/m^2 (the Phase II dose) and a carboplatin target AUC of 6 (a commonly used monotherapy target exposure), both administered for q21 days. At the OTR, gastro intestinal (GI) toxicities were limited to grade 1/2 and minimal reports of grade 1 neuropathy were noted. The incidence of grade 3/4 thrombocytopenia and grade 3/4 neutropenia at the OTR were lower relative to full doses of carboplatin and ispinesib, respectively. At the OTR, ispinesib concentrations did not appear to be affected by carboplatin and systemic exposures of carboplatin were within 11% of predicted values, suggesting no interaction with ispinesib. One patient with breast cancer had the best response of partial response at cycle 2. A total of 13/28 (46%) patients had a best response of stable disease (duration 3–9 months) [15].

Ispinesib has been studied in eight Phase II clinical trials, including studies in patients with locally advanced or metastatic breast cancer, platinum-refractory and sensitive non-small cell lung cancer, ovarian cancer, hepatocellular cancer, colorectal cancer, head and neck cancer, hormone-refractory prostate cancer, and melanoma. In the ongoing breast cancer study, women with locally advanced or metastatic breast cancer, receive *ispinesib* as monotherapy at 18 mg/m^2 as a 1-h intravenous infusion for every 21 days. In an interim analysis of Stage 1 data from this two-stage trial, partial responses were reported in three of 33 evaluable patients. Maximum decreases in tumor size ranged from 46% to 68%, and the duration of response from 7.1 to 13.4 weeks. The overall response rate for all 33 evaluable patients was 9% with a median time to progression of 5.7 weeks. The adverse events were manageable, predictable, and consistent with the Phase I clinical trial experience with *ispinesib*, and *ispinesib* plasma concentrations were comparable to those observed in the Phase I clinical trial [11,16].

In mice with advanced MX-1 human breast carcinoma, ispinesib (30 mg/m^2) in combination with docetaxel (30 mg/m^2) or ispenisib (15 mg/m^2) with capecitabine (1500 mg/m^2) resulted in greater tumor growth delay than with either agent alone. Cisplatin was also shown to enhance the activity of ispinesib against murine P388 lymphocytic leukemia [17], with sub-MTD doses of both in combination being superior to either agent alone at their respective MTDs.

A second KSP inhibitor, SB-743921, entered clinical trials in 2004. Like ispinesib, it has been shown to be a very selective KSP inhibitor that is >40,000-fold more selective over other kinesins. Interim results were reported from a Phase I trial in 19 patients in which SB-743921 was administered intravenously q21 days [18]. DLT included prolonged grade 4 neutropenia, grade 3 febrile neutropenia, grade 3 elevated transaminases, grade 3 hyperbilirubinemia, and grade 3 hyponatremia. Also like ispinesib, SB-743921 did not cause neurotoxicity or alopecia. A dose of 4 mg/m^2 at q21 days was suggested for Phase II. At this dose median Cl was 21 ml/min, and $t_{1/2}$ was 28 h.

2.1.2. Quinazolinone core replacement

The potent pharmacophore represented by ispinesib has received a good deal of attention recently, particularly with respect to replacement of the quinazolinone core. Among the fused bicyclic examples to appear were **3–8** that encompass both 6,6 and 6,5 ring systems [19–25]. A potent version of **7**, known as BMS-601027, has been reported to have a KSP ATPase IC$_{50}$ of 86 nM and a cell IC$_{50}$ of 317 nM [22]. Several monocyclic analogs have also been reported including examples **9–14** that represent 6- and 5-membered replacements of the original pyrimidinone core [26–34].

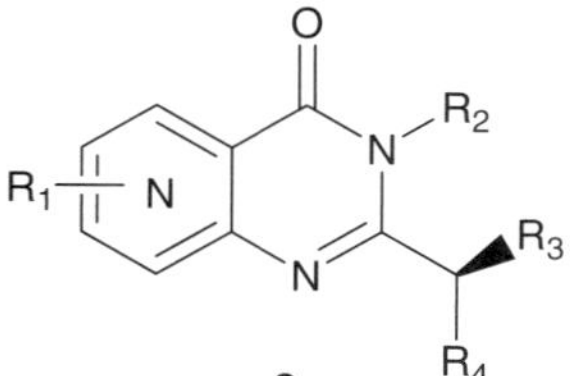

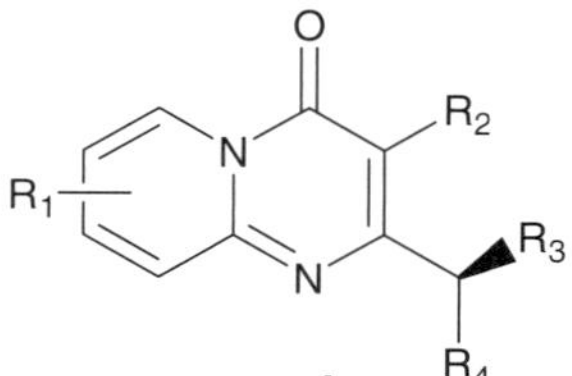

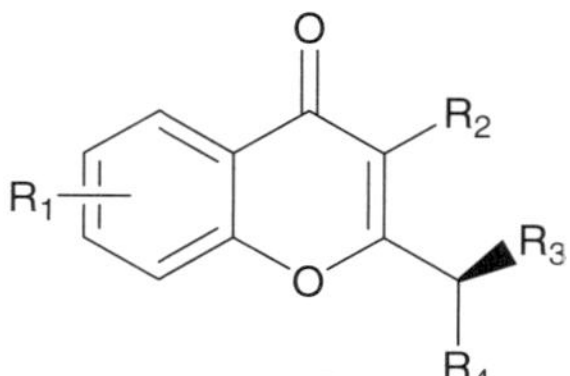

6

7

8

9

10

11

12

13

14

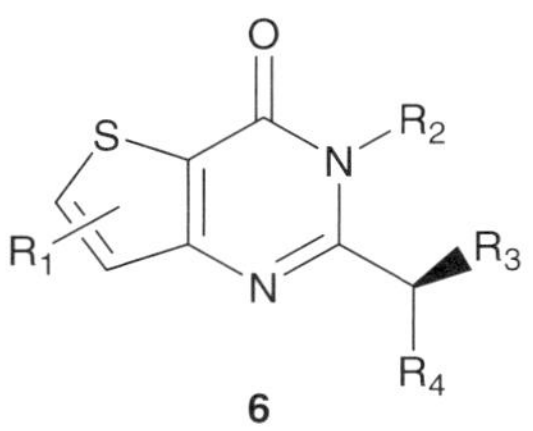

2.2. Monastrol analogs

15

16: R = H
17: R = Me

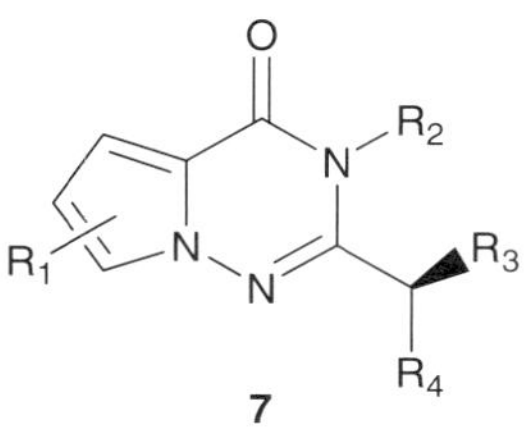

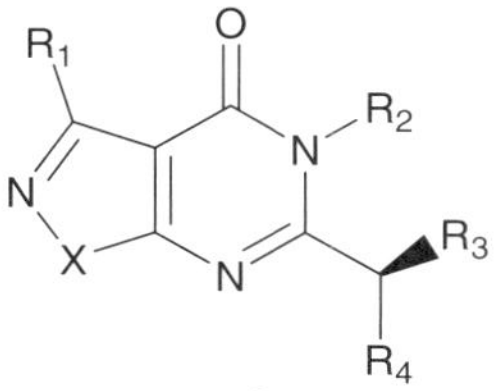

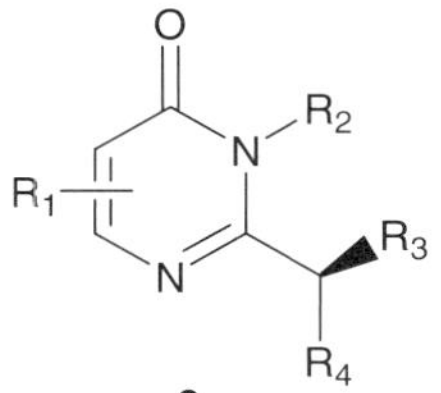

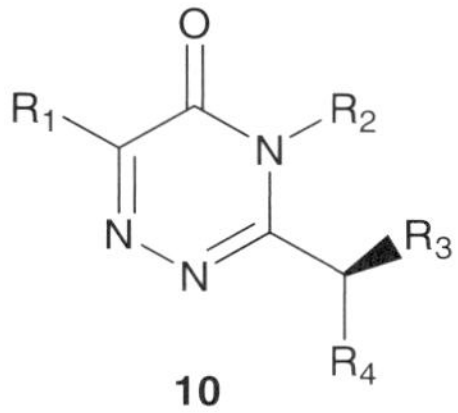

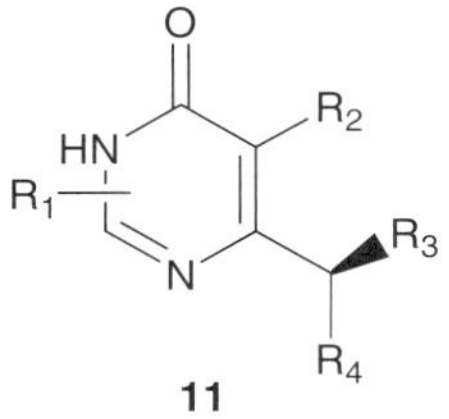

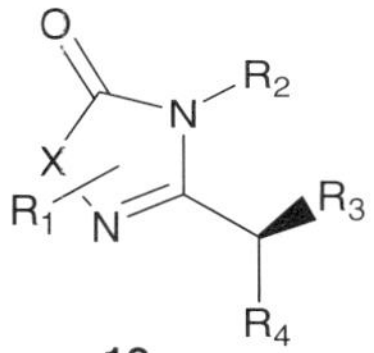

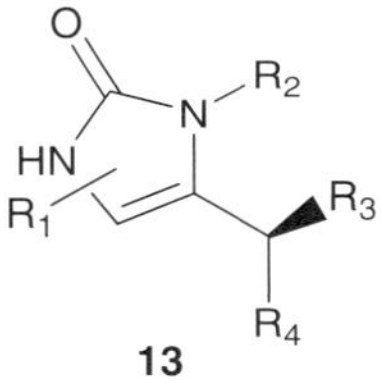

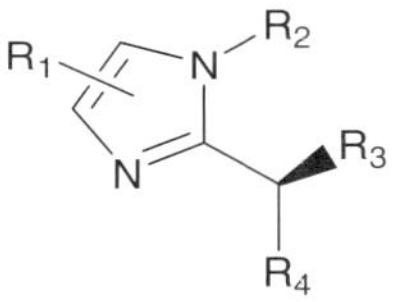

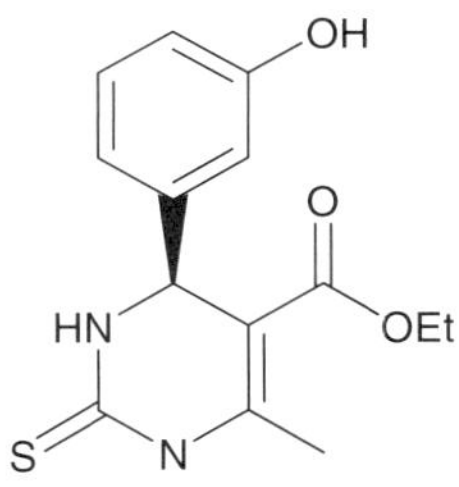

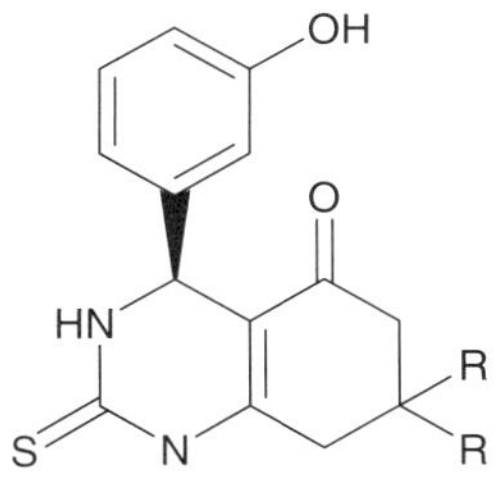

Monastrol (**15**) was the first reported specific inhibitor of KSP [5]. It was described as an allosteric inhibitor of motor function ($IC_{50} = 30\,\mu M$) that inhibits ADP release both in the presence and absence of microtubules. More recently, X-ray crystal structures of the KSP–ADP complex with monastrol provided insights into the structural basis for the inhibition by this agent [35,36]. Monastrol was found to induce dramatic conformational changes in helices α2 and α3 and the insertion loop (L5) that result in the formation of an induced-fit pocket not visible in the KSP–ADP structure [35]. Based on the subsequent kinetic and thermodynamic studies it was proposed that monastrol first binds weakly to the nucleotide-free state and/or the KSP–ATP collision complex which has an open conformation of loop L5 [37,38]. Upon tight ATP binding, a conformational change was suggested to occur that causes the inhibitor pocket to close, therefore increasing the binding affinity of monastrol to KSP. A series of spectroscopic probes were used to elucidate the pathway of structural changes in solution, and the results are consistent with the crystallographic model [39].

Analogs of monastrol have been pursued to improve biochemical and cellular potency for this scaffold [40]. Replacement of the phenyl ring or 3-OH group attenuated the activity as did replacement of the thiourea sulfur with oxygen. However, analogs with a fused bicyclic core showed improved potency. Compound **16** with an IC_{50} of 2 µM was 10 times more potent than monastrol. Introducing a gem-dimethyl group (compound **17**) further boosted the potency to 200 nM. Compounds **16** and **17** also had corresponding improvements in cellular potency with the desired cellular phenotype observed.

2.3. Tetrahydro-β-carbolines

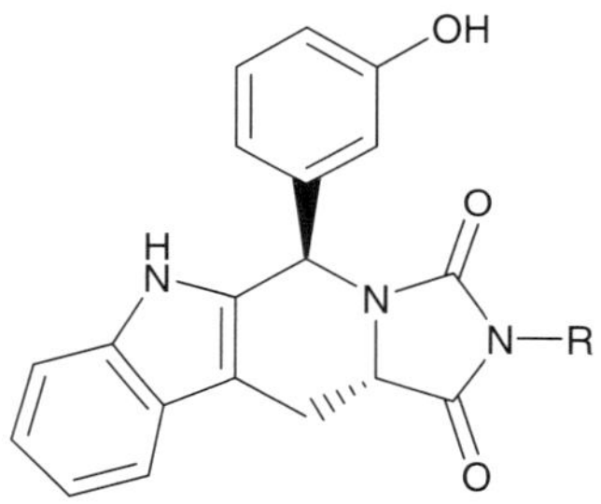

18: R = n-Bu
19: R = CH$_2$CH$_2$NH$_2$
20: R = Bn
21: R = n-Pr

Compound **18**, also known as HR22C16, was identified as a KSP inhibitor ($IC_{50} = 800$ nM) from a cell-based screen. Subsequent optimization yielded more potent compounds such as **19** ($IC_{50} = 90$ nM) [41]. A recent study was reported on the synthesis and biological evaluation of 60 analogs that included both *cis* and *trans*-isomers with different R groups [42]. *Trans*-isomers were found to be more active than *cis*-isomers in general, with the absolute stereochemistry shown being most prefered. Compound **20** was the most potent among the 60 analogs with an IC_{50} of 650 nM against KSP. HR22C16 and its analog **21** were shown to be active in both paclitaxel-sensitive (1A9) and paclitaxel-resistant (PTX10) ovarian cell lines [43]. Compound **21** had IC_{50} values of 0.8 and 2.3 µM in 1A9 and PTX10 cells, respectively. Compared to a 750-fold loss of activity for paclitaxel in PgP-overexpressing ovarian cell line A2780-D10, **21** had a 2.4-fold loss, suggesting that it is not a PgP substrate. The antiproliferative activity of **21** was attributed to mitotic arrest followed by cell death, which was mediated through an intrinsic apoptotic pathway. Interestingly, **21** was found to be antagonistic with paclitaxel treatment. Whether **21** (1 μM) was administrated concomitantly with paclitaxel (5 nM) or paclitaxel was dosed first (24 h), monopolar spindle phenotype was predominant. On the other hand, paclitaxel-type (i.e. multipolar) spindles were predominant when **20** was followed by paclitaxel [43].

2.4. Dihydropyrroles and dihydropyrazoles

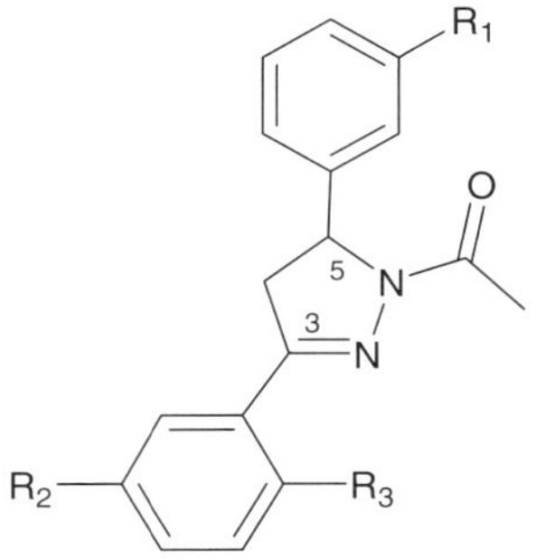

22: R_1 = H; R_2 = H; R_3 = Cl
23: R_1 = OH; R_2 = H; R_3 = H
24: R_1 = OH; R_2 = H; R_3 = Cl
25: R_1 = OH; R_2 = R_3 = F

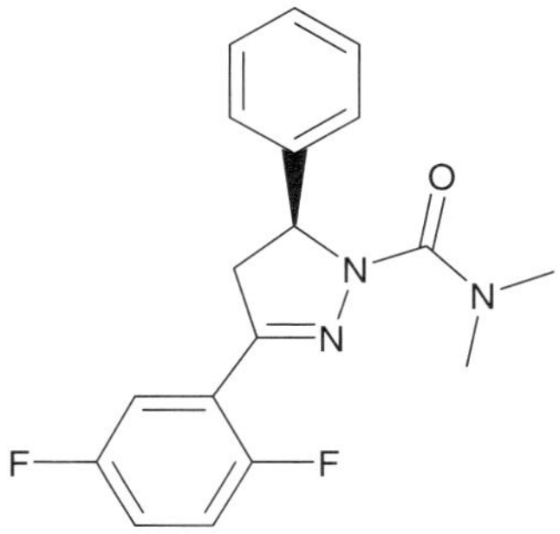

26

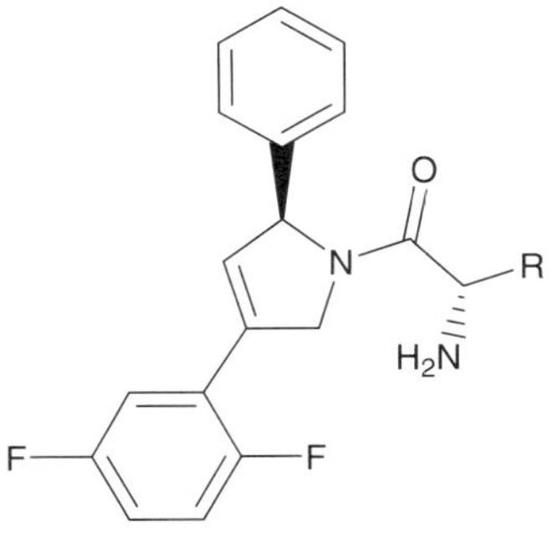

27: R = i-Pr
28: R = c-Pr

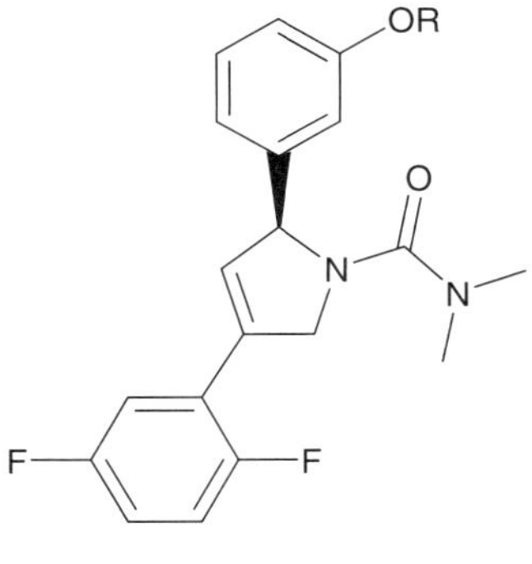

29: R = H
30: R = PO(OH)$_2$

3,5-Diaryl-4,5-dihydropyrazoles **22** and **23** were identified as screening hits with ATPase IC$_{50}$ values of 3.6 and 6.9 μM, respectively [44]. Combining chloro and hydroxy substituents in the same molecule yielded the more potent compound **24**, which had an IC$_{50}$ of 450 nM. Optimization work around the two phenyl groups demonstrated tolerance for 3-phenyl substitution leading to a more potent compound **25** (IC$_{50}$ = 51 nM), but almost no tolerance for variation on the 5-phenyl ring. Exploration of N1-substitution with larger acyl or alkyl groups generally resulted in analogs with lower potency, whereas compound **26** with a dimethyl urea had similar potency to compound **25**. The (S)-antipode of compound **25**, separated by chiral phase HPLC, was found to be the active stereoisomer with an ATPase IC$_{50}$

of 26 nM. It was also found to cause caspase-3 induction, a well-established marker of apoptosis, in A2780 human ovarian carcinoma cells with an IC_{50} of 15 nM [44].

Replacement of the dihydropyrazole core with a dihydropyrrole also yielded a viable scaffold in which the hybridization of the aryl-bearing carbon atoms was maintained. Introduction of basic amide and urea groups to the dihydropyrrole core led to enhanced potency. Compounds **27** and **28** had ATPase IC_{50} values of 3.6 and 2.0 nM, respectively. Both compounds caused mitotic arrest in A2780 cells with EC_{50}'s of 12 and 8.6 nM [43], respectively. X-ray crystal structures of KSP bound with inhibitors (S)-**25**, **26**, and **28** showed binding to the same allosteric pocket as monastrol [44,45].

Cellular response and the role of the spindle checkpoint in the induction of apoptosis by compound **27** were investigated in cancer cell lines. The results indicated that both activation of the spindle assembly checkpoint and mitotic slippage were required for the induction of apoptosis. This led to the suggestion that agents that promote mitotic slippage could act synergistically with KSP inhibitors in cancer cells with competent spindle checkpoints [46].

Compound **27** showed high aqueous solubility (>10 mg/ml) and moderate clearance (21–40 ml/min/kg) with $t_{1/2}$ ranging from 1–4 h in rat, dog, and monkey. Moderate human ether-à-go-go related gene (hERG) potassium channel binding was also observed for this dihydropyrrole series. Compounds **27** and **28** had hERG IC_{50} values of 2.4 and 3.5 µM, respectively [43], hence efforts to minimize hERG binding in this series were pursued [47]. Reintroduction of a 3-OH group to the northern phenyl ring in combination with neutral N1-acyl groups reduced hERG binding while maintaining potency. Compound **29** had an ATPase IC_{50} of 7 nM, cell mitotic EC_{50} of 22 nM, and hERG IC_{50} of 33 µM. With diminished solubility in the absence of the basic amino group, phosphate prodrugs of **29** were made to improve solubility. Prodrug **30** was shown to be inactive against KSP (IC_{50} > 1 µM), while its solubility was improved to >20 mg/ml at pH 7. It was cleaved rapidly in blood to produce parent **29**.

Some closely related KSP inhibitors have also been reported where the central dihydropyrrole and dihydropyrazole cores were replaced by tetrahydropyridine [48], dihydroisoxazole [49] and dihydrooxadiazole [50] groups.

2.5. Tetrahydroisoquinolines

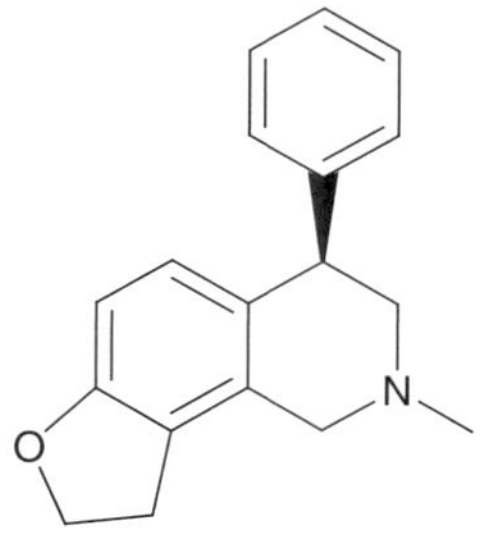

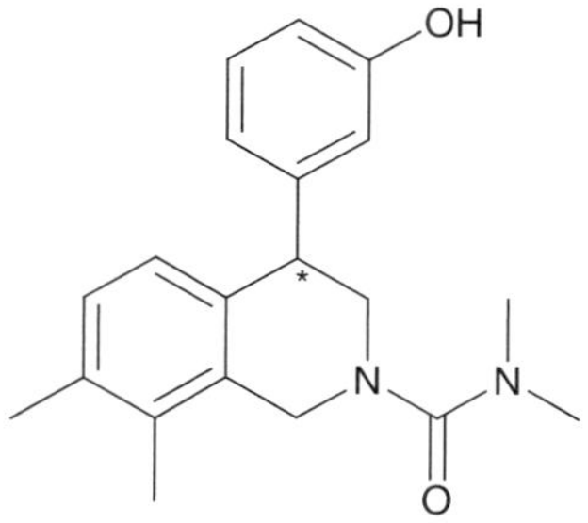

A new series of tetrahydroisoquinolines was identified from high-throughput screening. Representative compound **31** had an IC_{50} of 9.7 μM in an ATPase assay and an IC_{50} of 2.4 μM in a proliferation assay in A2780 human ovarian carcinoma cells [51]. Nuclear magnetic resonance (NMR) experiments showed compound **31** bound to the same allosteric site as monastrol. A binding model was constructed using NMR data and crystallographic data of co-crystal structures of KSP-monastrol and KSP-HR2216 [51]. The model was then used to help guide structure–activity relationship (SAR) development for the series.

In optimization studies, the fused dihydrofuran ring of **31** was found to be replaceable with 7,8-dimethyl groups without loss of potency. Substitution on N with alkyl groups larger than methyl was not well tolerated, but analogs with amino groups appended to 2-4 carbon atom chains were shown to have similar activity to **31**. This raised the possibility that the charged side chain may point toward solvent. In support of this, N-acylation products such as amides, sulfonamides, carbamates, and primary ureas were generally much less active. However, compound **32** (racemic) with a N,N-dimethyl urea, was slightly more potent than **31** (ATPase $IC_{50} = 2.75$ μM). In an attempt to recapitulate the H-bond between monastrol's phenol-OH group and the carbonyl of back-bone amide Glu 118 of KSP, a hydroxyl group was introduced to the 3-position of **32**. This was found to increase potency by about eight-fold in the ATPase assay ($IC_{50} = 306$ nM) and three-fold in the cellular assay ($IC_{50} = 376$ nM). For comparison, the 4-OH analog was completely inactive [51]. A co-crystal structure of KSP with the (R)-antipode of **32** was obtained, and as expected, the compound binds in the known allosteric site of KSP in the predicted orientation. An H-bond between the oxygen of the phenol and the carbonyl oxygen of Glu 118 and the side chain Arg 119 was evident. Van der Waals interactions between the tetrahydroisoquinoline, dimethyl urea, and phenyl ring with the protein were also observed.

3. CONCLUSIONS

The last several years have seen an increase in the number of targeted antimitotic drug discovery efforts as proteins critical for kinetocore, centrosome, and mitotic spindle function continue to be elucidated. The rapid progress made in progressing mitotic kinesin inhibitors into proof of concept clinical trials serves as a benchmark for the potential of this realm of anticancer drug discovery. The KSP inhibitor, ispinesib, has been studied in Phase II clinical trials against locally advanced or metastatic breast cancer (in which it has demonstrated anti-cancer activity), platinum-refractory and -sensitive non-small cell lung cancer, ovarian cancer, hepatocellular cancer, colorectal cancer, head and neck cancer, hormone-refractory prostate cancer, and melanoma. A second KSP inhibitor, SB-743921, has entered Phase I studies. In addition, a number of other potent and structurally diverse structures are in pre-clinical development, most if not all of which bind to the same induced-fit, allosteric pocket as monastrol. With another 13 proteins designated as mitotic kinesins, efforts against this target class are still at an early stage. Continued clinical development of KSP targeting agents should serve to bolster the interrogation of these unique proteins as anticancer targets.

REFERENCES

[1] K. W. Wood, W. D. Cornwell and J. R. Jackson, *Curr. Opin. Pharmacol.*, 2001, **1**, 370.
[2] K. W. Wood and G. Bergnes, *Ann. Rep. Med. Chem.*, 2004, **39**, 173.
[3] D. J. Sharp, G. C. Rogers and J. M. Scholey, *Nature*, 2000, **407**, 41.
[4] G. Bergnes, K. Brejc and L. Belmont, *Curr. Top. Med. Chem.*, 2005, **5**, 127.
[5] T. U. Mayer, T. M. Kapoor, S. J. Haggarty, R. W. King, S. L. Schreiber and T. J. Mitchison, *Science*, 1999, **286**, 971.
[6] R. Sakowicz, J. T. Finer, C. Beraud, A. Crompton, E. Lewis, A. Fritsch, Y. Lee, J. Mak, R. Moody, R. Turincio, J. C. Chabala, P. Gonzales, S. Roth, S. Weitman and K. W. Wood, *Cancer Res.*, 2004, **64**, 3276.
[7] A. Blangy, H. A. Lane, P. d'Herin, M. Harper, M. Kress and E. A. Nigg, *Cell*, 1995, **83**, 1159.
[8] V. Mountain, C. Simerly, L. Howard, A. Ando, G. Schatten and D. A. Compton, *J. Cell Biol.*, 1999, **147**, 351.
[9] P. J. Coleman and M. E. Fraley, *Expert Opin. Ther. Patents*, 2004, **14**, 1659.
[10] D. M. Duhl and P. A. Renhowe, *Curr. Opin. Drug Disc. Develop.*, 2005, **8**, 431.
[11] Q. Chu, K. D. Holen, E. K. Rowinsky, G. Wilding, J. L. Volkman, J. B. Orr, D. D. Williams, J. P. Hodge, C. A. Kerfoot and J. Sabry, *ASCO Annual Meeting*, New Orleans, LA, April 2004.
[12] H. Burris, III, P. LoRusso, S. Jones, T. Guthrie, J. B. Orr, D. D. Williams, J. P. Hodge, M. Bush and J. Sabry, *ASCO Annual Meeting*, New Orleans, LA, April 2004.
[13] S. Blagden, G. Seebaran, R. Molife, M. Payne, A. Reid, A. Protheroe, S. Kathman, D. Williams, C. Bowen, J. Hodge, M. Dar, J. de Bono and M. Middleton, *AACR-NCI-EORTC*, Philadelphia, PA, Nov. 2005.
[14] E. Calvo, Q. Chu, E. Till, E. Rowinsky, A. Patnaik, C. Takimoto, S. Kathman, D. Williams, C. Bowen, J. Hodge, M. Dar and A. Tolcher, *AACR-NCI-EORTC*, Philadelphia, PA, Nov. 2005.
[15] S. Jones, E. Plummer, H. Burris, A Razak, A. Meluch1, C. Bowen, D. Williams, L. Vasist, J. Hodge, M. Dar and A. Calvert, *ASCO Annual Meeting*, Atlanta, GA, June 2006.
[16] K. Miller, C. Ng, P. Ang, A. Brufsky, S. Lee, E. Dees, M. Piccart, M. Verrill, A. Wardley, J. Loftiss, J. Bal, S. Yeoh, J. Hodge, D. Williams, M. Dar, S. Kathman and P. Ho. *San Antonio Breast Cancer Symposium (SABCS)*, December 2005.
[17] D. Sutton, M. Diamond, J. Onori, S. Y. Zhang, M. Giardiniere, L. Faucette, L. Belmont, K. W. Wood, J. R. Jackson and P. Huang, *AACR-NCI-EORTC*, Philadelphia, PA, Nov. 2005.
[18] K. D. Holen, C. P. Belani, G. Wilding, S. Ramalingam, J. L. Volkman, R. K. Ramanathan, C. J. Bowen, D. D. Williams, J. P. Hodge, M. M. Dar and P. T. C. Ho, *ASCO Annual Meeting*, Orlando, FL, May 2005.
[19] P. J. Coleman, M. E. Fraley and W. F. Hoffman, *WO Patent 04039774-A2*, 2004.
[20] W. Wang, R. N. Constantine, L. M. Lagniton, S. Pecchi, M. T. Burger and M. C. Desai, *US Patent 0085490-A1*, 2005.
[21] A. I. McDonald, G. Bergnes, B. Feng, D. J. Morgans, Jr., S. D. Knight, K. A. Newlander, D. Dhanak and C. S. Brook, *WO Patent 03088903-A2*, 2003.

[22] W. Wang, L. M. Lagniton, R. N. Constantine and M. T. Burger, *WO Patent 04111058-A1*, 2004.
[23] L. J. Lombardo, R. S. Bhide, K. S. Kim and S. Lu, *WO Patent 03099286-A1*, 2003.
[24] D. L. Roussell, M. Ortega-Nanos, H. Shen, R. Batorsky, J. Ryder, R. Talbott, C. Raventos-Suarez, J. Naglich, C. Fairchild, R. Peterson, K. Johnston, A. Camuso, K. Menard, C. Flefleh, K. McGlinchey, D. Kan, S. Castaneda, I. Inigo, J. T. Hunt, R. Borzilleri, K. Kim, F. Lee, M. Gottardis, L. Lombardo and R. Wild, *AACR Annual Meeting*, Washington, DC, April 2006.
[25] B. Aquila, M. H. Block, A. Davies, J. Ezhuthachan, S. Filla, R. W. Luke, T. Pontz, M.-E. Theoclitou and X. Zheng, *WO Patent 04078758-A1*, 2004.
[26] P. J. Coleman, G. D. Hartman and L. A. Neilson, *WO Patent 03099211-A2*, 2003.
[27] X. Qian, G. Bergnes and D. J. Morgans, Jr., *WO Patent 04032879-A2*, 2004.
[28] X. Qian, G. Bergnes and D. J. Morgans, Jr., *WO Patent 04103282-A2*, 2004.
[29] H.-J. Zhou, A. I. Mcdonald, G. Bergnes, D. J. Morgans, Jr., J. C. Chabala, S. D. Knight and D. Dhanak, *WO Patent 04064741-A2*, 2004.
[30] W. Wang, R. Constantine and L. Lagniton, *WO Patent 05100357-A1*, 2005.
[31] X. Qian, G. Bergnes, D. J. Morgans, Jr., S. D. Knight and D. Dhanak, *WO Patent 04032840-A2*, 2004.
[32] X. Qian, H.-J. Zhou and G. Bergnes, *WO Patent 05013888-A2*, 2005.
[33] H.-J. Zhou and G. Bergnes, *WO Patent 04100873-A2*, 2004.
[34] W. Wang, P. A. Barsanti, Y. Xi, R. S. Boyce, S. Pecchi, N. Brammeier, M. Phillips, K. Mendenhall, K. Wayman, L. M. Lagniton, R. Constantine, H. Yang, E. Mieuli, S. Ramurthy, E. Jazan, A. Sharma, J. Rama, S. Sabramanian, P. Renhowe, K. W. Bair, D. Duhl, A. Walter, T. Abrams, K. Huh, E. Martin, M. Knapp and V. Le, Vincent. *WO Patent 06002236-A1*, 2006.
[35] J. Turner, R. Anderson, J. Guo, C. Beraud, R. Fletterick and R. Sakowicz, *J. Biol. Chem.*, 2001, **276**, 25496.
[36] Y. Yan, V. Sardana, B. Xu, C. Homnick, W. Halczenko, C. A. Buser, M. Schaber, G. D. Hartman, H. E. Huber and L. C. Kuo, *J. Mol. Biol.*, 2004, **335**, 547.
[37] J. C. Cochran, J. E. Gatial, III, T. M. Kapoor and S. P. Gilbert, *J. Biol .Chem.*, 2005, **280**, 12658.
[38] J. C. Cochran and S. P. Gilbert, *Biochemistry*, 2005, **44**, 16633.
[39] Z. Maliga, J. Xing, H. Cheung, L. J. Juszczak, J. M. Friedman and S. S. Rosenfeld, *J. Biol. Chem.*, 2006, **281**, 7977.
[40] M. Gartner, N. Sunder-Plassmann, J. Seiler, M. Utz, I. Vernos, T. Surrey and A. Giannis, *ChemBioChem*, 2005, **6**, 1.
[41] S. Hotha, J. C. Yarrow, J. G. Yang, S. Garrett, K. V. Renduchintala, T. U. Mayer and T. M. Kapoor, *Angew. Chem., Int. Ed. Engl.*, 2003, **42**, 2379.
[42] N. Sunder-Plassmann, V. Sarli, M. Gartner, M. Utz, J. Seiler, S. Huemmer, T. U. Mayer, T. Surrey and A. Giannis, *Bioorg. Med. Chem.*, 2005, **13**, 6094.
[43] A. I. Marcus, U. Peters, S. L. Thomas, S. Garrett, A. Zelank, T. M. Kapoor and P. Giannakakou, *J. Biol. Chem.*, 2005, **280**, 11569.
[44] C. D. Cox, M. J. Breslin, B. J. Mariano, P. J. Coleman, C. A. Buser, E. S. Walsh, K. Hamilton, H. E. Huber, N. E. Kohl, M. Torrent, Y. Yan, L. C. Kuo and G. D. Hartman, *Bioorg. Med. Chem. Lett.*, 2005, **15**, 2041.
[45] M. E. Fraley, R. M. Garbaccio, K. L. Arrington, W. F. Hoffman, E. S. Tasber, P. J. Coleman, C. A. Buser, E. S. Walsh, K. Hamilton, C. Fernandes, M. D. Schaber, R. B. Lobell, W. Tao, V. J. South, Y. Yan, L. C. Kuo, T. Prueksaritanont, C. Shu, M. Torrent, D. C. Heimbrook, N. E. Kohl, H. E. Huber and G. D. Hartman, *Bioorg. Med. Chem. Lett.*, 2006, **16**, 1775.

[46] W. Tao, V. J. South, Y. Zhang, J. P. Davide, L. Farrell, N. E. Kohl, L. Sepp-Lorenzino and R. B. Lobell, *Cancer Cell*, 2005, **8**, 49.
[47] R. M. Garbaccio, M. E. Fraley, E. S. Tasber, C. M. Olson, W. F. Hoffman, K. L. Arrington, M. Torent, C. A. Buser, E. S. Walsh, K. Hamilton, M. D. Schaber, C. Fernandes, R. B. Lobell, W. Tao, V. J. South, Y. Yan, L. C. Kuo, T. Prueksaritanont, D. E. Slaughter, C. Shu, D. C. Heimbrook, N. E. Kohl, H. E. Huber and G. D. Hartman, *Bioorg. Med. Chem. Lett.*, 2006, **16**, 1780.
[48] M. E. Fraley, R. M. Garbaccio, C. M. Olson and E. S. Tasber, *WO Patent 2004058700-A2*, 2004.
[49] M. E. Fraley, R. M. Garbaccio and G. D. Hartman, *WO Patent 06023440-A2*, 2006.
[50] C. D. Cox, M. E. Fraley and R. M. Garbaccio, *WO Patent 06031348-A2*, 2006.
[51] C. M. Tarby, R. F. Kaltenbach, III, T. Huynh, A. Pudzianowski, H. Shen, M. Ortega-Nanos, S. Sheriff, J. A. Newitt, P. A. McDonnell, N. Burford, C. R. Fairchild, W. Vaccaro, Z. Chen, R. M. Borzilleri, J. Naglich, L. J. Lombardo, M. Gottardis, G. L. Trainor and D. L. Roussell, *Bioorg. Med. Chem. Lett.*, 2006, **16**, 2095.

Developing Infectious Disease Strategies for the Developing World

Paul J. Lee and Leonard R. Krilov

Division of Pediatric Infectious Diseases, Winthrop University Hospital, Mineola, NY 11501, USA

Contents

1. Introduction	275
2. Avian influenza	276
2.1. Neuraminidase inhibitors	277
2.2. Avian influenza vaccine	278
3. MDR-TB	278
3.1. Quinolones	279
3.2. Nitroimidazoles	281
3.3. Pyrroles	282
3.4. Tuberculosis vaccine	283
4. Conclusion	284
References	284

1. INTRODUCTION

New infectious diseases continue to emerge and old ones have re-emerged in recent decades. Often these events begin and take root in developing countries. Infectious diseases of the developing world or emerging infections are defined by the Institute of Medicine as an infectious disease that has come to medical attention within the past two decades, or for which there is a potential that its prevalence will increase in the near future. Frequently, these diseases exist in nature as zoonoses and cross over to humans when people come into contact with formerly isolated animal population, such as monkeys in a rain forest that become less isolated with deforestation. Drug resistant organisms may also be considered as emerging infections since they result from human influence. Human immunodeficiency virus (HIV), Ebola virus, hantavirus pulmonary syndrome, monkey pox and multidrug-resistant *Mycobacterium tuberculosis* (MDR-TB), Severe Acute Respiratory Syndrome (SARS) associated coronavirus and avian influenza are examples of emerging infections. Many of these infections have arisen and become major health issues in the developing world, in part due to selection factors as noted above. Additionally, inadequate resource for public health surveillance and treatment of these illnesses plus poor sanitation and nutritional status may contribute to their flourishing. Once established, these illnesses then pose concerns for all mankind as we have truly become a global community. For these same reasons even common infections such as rotavirus gastrointestinal infection or respiratory syncytial virus (RSV) exert a greater toll in the developing world. Thus, even as molecular biology advances approaches to diagnoses, new challenges to prevention and treatment of infectious diseases continue to emerge. Through a discussion of avian influenza and MDR-TB, primarily, we will demonstrate the magnitude of the problems of infectious diseases in the developing world and discuss approaches that can be taken to in an attempt to monitor and contain these new threats as they emerge.

2. AVIAN INFLUENZA

Throughout history influenza has been a major infectious disease problem for man causing significant morbidity and mortality. Influenza viruses are members of the *Orthomyxoviridae* (myxo = mucous, Gr.) family and are classified into three types A, B, and C, based on the composition of their viral nuclear proteins. Influenza A viruses are further *categorized* into subtypes based on the antigenic structure of the two-surface membrane glycoproteins, hemagglutinin (H) and neuraminidase (N). Both of these proteins play critical roles in the viral life cycle. The hemagglutinin protein is involved in the process of viral attachment and entry to the host epithelial cell to initiate the process of viral infection. The neuraminidase is involved at the other end of the replication cycle as it functions to snip the cell membrane releasing newly made viral particles from the infected cell.

There are 15 distinct hemagglutinins (H1–H15) and nine neuraminidase subtypes (N1–N9) that have been described in nature. Until recently, human influenza viruses only contained H1, H2, H3 and N1 or N2 subtypes. Avian influenza viruses may have additional combinations of the 15 hemagglutinin and 9 neuraminidase proteins.

The influenza virus genome is composed of 8 segmented pieces of RNA with each coding for a separate viral protein. These viral proteins may change slowly (antigenic drift) through point mutations occurring during viral replication or suddenly (antigenic shift). The latter occurs only in influenza A strains and it results from a combination of the segmented genome described above and the ability of influenza A strains of human and avian origin to co-infect the same host (typically, pigs or swines are most receptive), and share their genetic information. During this process of reassortment a new strain acquiring one of the avian surface protein genes (H or N) may emerge. Such a novel strain then eludes pre-existing immunity in people previously infected with influenza and may spread rapidly among the entire population. This defines the potential for worldwide spread or an influenza pandemic. A number of influenza pandemics were documented over the 20th century with the most notable being the 1918–1919 Spanish or swine flu epidemic. During this pandemic, a new for the time H1NI strain arose leading to an estimated 20 million deaths worldwide. Scientists to this date, continue to study this strain attempting to discern virulence factors that contributed to the severity of illness caused by this strain. It has been suggested that the recently observed avian influenza H5N1 strain to be described below may share some of these same virulence properties which may contribute to the >50% mortality observed in the limited number of human cases reported to date [1]. Specifically, this virus predominately attaches to lower respiratory tract cells (type II pneumocytes, alveolar macrophages and nonciliated bronchiolar cells) leading to severe lower respiratory tract damage. However, it does not appear to attach well to upper respiratory tract cells, possibly explaining the low human-to-human risk from this strain [2].

The crowding of people, swine, and bird species in areas of the developing world such as Southeast Asia creates an environment that favors the origin of such strains. Public health vigilance has contributed to prompt recognition of new strains as they emerge and has contributed to control of their spread through destruction of bird and/or swine populations.

The H5N1 avian influenza that has attracted so much attention over the last several years does NOT presently meet the criteria of an antigenically shifted strain. It is presently an avian strain that has not undergone reassortment with a human strain and is not well adapted to humans. Only a small number of people ($n = 204$, with 113 deaths) have been confirmed as infected to date, and the vast majority of these individuals apparently acquired the virus through close contact with infected birds and/or consuming undercooked infected bird products [1,3]. Still the increasing geographic spread of infected birds beyond Asia to Africa and Europe, the transmission to felids (tigers and leopards) through consumption of raw infected chickens in Thailand, and the spread of reported human cases from Thailand and Viet Nam in 2004 to Cambodia, China and Indonesia in 2005 and Azerbaijan, Egypt, Iraq, and Turkey in 2006, raises concerns for a potential pandemic should further adaptation to humans occur.

2.1. Neuraminidase inhibitors

In the event of such a pandemic should arise, where do we stand in treating and preventing the spread of this virus? Since human influenza A (H5N1) isolates are resistant to the M2 inhibitors amantadine and rimantadine, these antivirals do not have a role for the treatment or prophylaxis against this strain. The neuraminidase inhibitors oseltamivir, **1**, and zanamivir, **2**, have *in vitro* activities against the human H5N1 isolates; however, recent data suggest that higher doses for longer periods (7–10 days vs. the prior recommendation of five days) may be required to be effective [4]. Oseltamivir is an oral agent approved for prophylaxis and treatment of influenza infections. Zanamivir is delivered topically to the respiratory tract with similar indications.

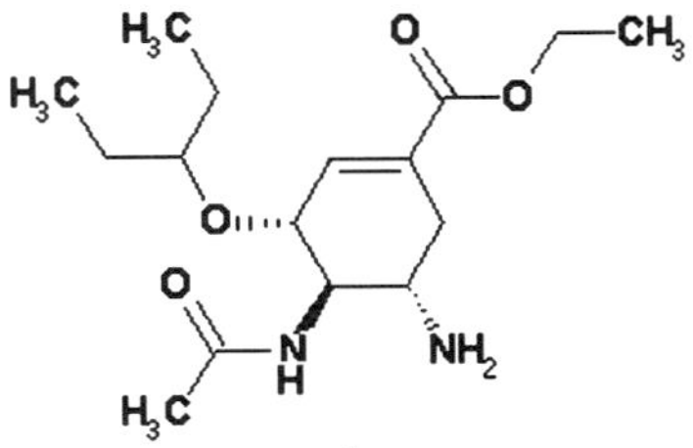

Oseltamivir resistance in human isolates from Viet Nam has been noted [5,6]. These strains may not be resistant to zanamivir [6]. Stockpiles of drug will be needed for prophylaxis and treatment if they are to be used in the event of a new pandemic.

A long-acting neuraminidase inhibitor, peramivir, **3**, has begun Phase I clinical trial to begin to determine whether it may be an intravenous option for this disease [7].

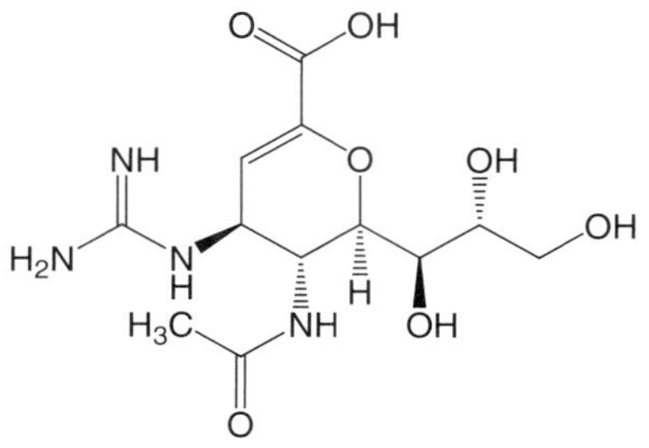

2 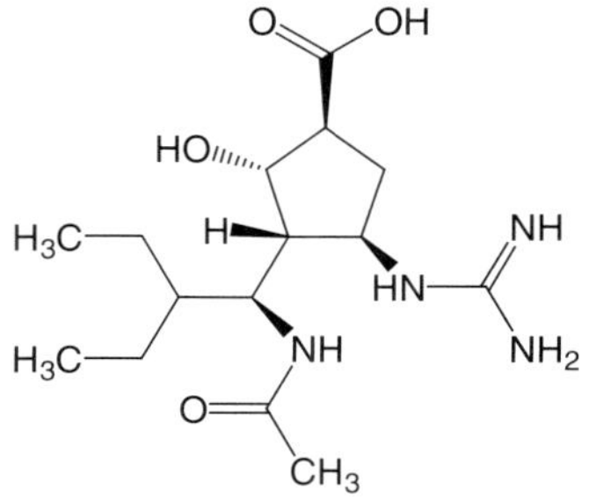   **3**

2.2. Avian influenza vaccine

Traditionally, vaccination has been the principal approach to protecting individuals against influenza. Currently, no influenza A H5N1 vaccine is available although several candidate vaccines are being developed. Preliminary data suggest that either higher concentrations of antigens than used in seasonal influenza vaccines and/or addition of adjuvants to these vaccines will be necessary to induce protective responses [8]. Gearing production, to rapidly make necessary quantities, of such a vaccine in the event of pandemic spread will be a great challenge to the vaccine industry.

Thus, although great progress has been made in understanding the biology of avian influenza H5N1 and although the virus is not presently capable of efficient human-to-human transmission there is reason for concern as noted above. Continued surveillance in animal and human populations and infection control measures to limit its spread are critical, while we learn to better combat this virus (or some other future strain) with pandemic potential.

3. MDR-TB

In 1952, when Selman Waksman received the Nobel Prize for his discovery of streptomycin, the first effective tuberculosis drug, he proclaimed that the conquest of the "Great White Plague", which was previously incurable and associated with up to 50% mortality, was in sight. Five decades later, tuberculosis remains a global epidemic and currently affects 2 billion people, or 1/3 of the world's population, killing 2 million of them annually. The World Health Organization (WHO) reports that new infections are increasing by 1% per year. By 2020, WHO predicts there will be 1 billion new infections, 200 million of who will develop active disease, and 35 million deaths caused by tuberculosis [9].

Tuberculosis is a particularly frustrating disease to cure because of a multitude of confounding factors. In fact, one of the greatest ironies is that the majority of antibiotics used to treat tuberculosis are still very effective. However, successful treatment requires a combination of multiple drugs for at least six months. The majority (95%) of tuberculosis occurs in the developing world, where the high relative cost, poor accessibility of medications, and weak health infrastructure causes poor compliance and subsequent selection of resistant strains of *M. tuberculosis* to the drugs it was exposed to. Two-thirds of people with TB do not have access to effective therapy. But even when therapy is available and properly utilized, it may be incompletely effective against these multi-drug resistant strains of tuberculosis. A Beijing/W genotype strain, which has developed resistance to both first and second line TB drugs, is of particular concern. A recent review combining >29,000 patients in 49 studies from 35 countries has found it to be epidemic in Cuba, the former Soviet Union, and South Africa, and endemic in East Asia [10]. And, estimates place the cost of treating MDR-TB at $250,000–$750,000 per patient, because of the extra isolation, monitoring, medical supervision, and medication costs required.

Hence, there has been renewed interest in developing new drugs for the treatment of tuberculosis. Current goals include compounds which can successfully treat active tuberculosis in a shorter period of time and/or with less frequent, intermittent dosing; compounds to deal with MDR-TB; and compounds to prevent tuberculosis in people at risk of MDR-TB. In fact, a recent Cochrane review covering appropriate studies to address the problem of treating latent tuberculosis infection in people exposed to MDR-TB found there were no randomized controlled trials in the database that have assessed the effectiveness of treatment [11].

There is also a need for relatively inexpensive drugs and medicines that can be given safely with antiretroviral (ARV) medications for HIV, because of the growing problem of tuberculosis and HIV co-infection. One-third of the 39 million people infected with HIV, are also infected with TB, mostly in sub-Saharan Africa. AIDS patients are particularly vulnerable to TB because of their weakened immune systems, which makes them more susceptible to contracting TB or allows previous infection with TB that has been suppressed and contained by the immune system to become active. TB is a leading cause of death for people living with HIV/AIDS, but treating both diseases simultaneously is difficult. Many of the first-line TB medications like rifampin cause adverse drug–drug reactions if given with ARV medications, because both utilize the same hepatic metabolic pathway.

Although more than 40 years has passed since the last novel TB drug, rifampin, was discovered, there are a large number of new drugs now in the pipeline. Many are still in early preclinical stages, but this discussion will focus on those already in clinical human trials and of particular relevance to the developing world.

3.1. Quinolones

3.1.1. Moxifloxacin

Moxifloxacin, **4**, is a methoxyfluoroquinolone which is already available and approved for the treatment of acute respiratory infections such as community-acquired pneumonia, intra-abdominal infections, acute sinusitis, and skin infections. It is an inhibitor of DNA gyrase, which is an enzyme important in bacterial growth and replication. It has an excellent safety profile and has already been used by 42 million patients worldwide. However, the drug has not been adequately evaluated in children and pregnant or lactating women.

In vitro and *in vivo* studies have demonstrated good activity against *M. tuberculosis* [12,13]. Moxifloxacin has a minimum inhibitory concentration (MIC) against *M. tuberculosis* fourfold lower than levofloxacin, a second-generation fluoroquinolone already in use for MDR-TB. (Older fluoroquinolones like levofloxacin and ofloxacin are only used for MDR-TB treatment because of initial studies showing lack of superiority to existing first-line drugs.) It also has a long half-life and high area under the curve concentration. A murine model study found that substituting moxifloxacin for isoniazid decreased the overall time necessary for TB eradication by two months. Moxifloxacin is currently in four different Phase II trials in eight countries – Brazil, Canada, South Africa, Spain, Tanzania, Uganda, the United States, and Zambia, with a goal of demonstrating efficacy as a first-line component of a shorter, standardized TB treatment regimen. The hypothesis is that moxifloxacin substitution for isoniazid or ethambutol in the usual four drug standard regimen (isoniazid, rifampin, ethambutol, and pyrazinamide) during the first two months of treatment will significantly increase the proportion of patients with culture-negative sputum at 8 weeks, compared with the standard regimen. Phase II trials, are expected to complete in 2007, with Phase III trials to follow. Moxifloxacin metabolism also does not utilize the hepatic cytochrome P-450 system, and does not lead to adverse reactions when taken alongside ARVs. However, one recent study in Italy found that 12/38 patients treated with long-term moxifloxacin for TB developed adverse reactions. None of the reactions was irreversible or fatal, and moxifloxacin was felt to be safe and result in treatment success in 31/38 patients. The success rate was only 5/14 in MDR-TB patients [14].

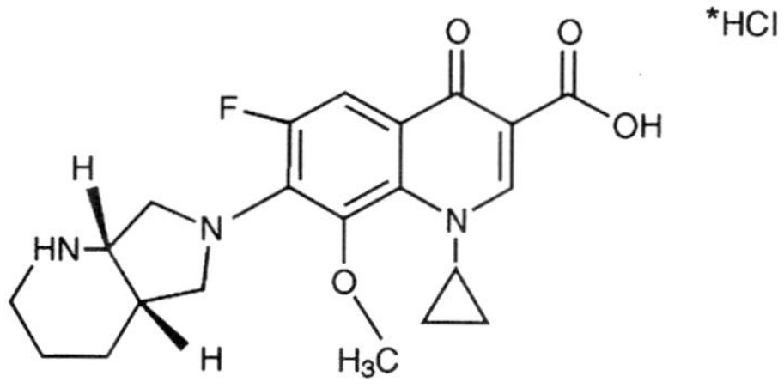

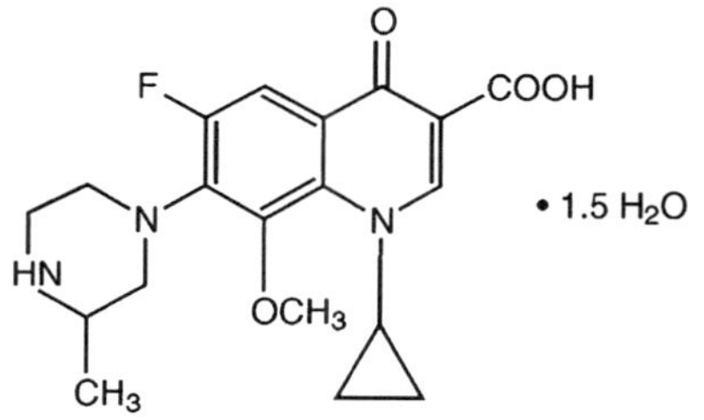

3.1.2. Gatifloxacin

Gatifloxacin, **5**, is another methoxyfluoroquinolone that is actually furthest along in clinical development for tuberculosis treatment. Like moxifloxacin, it is used to treat a variety of bacterial infections, has been used globally for many years, and has no interactions with ARV drugs. Gatifloxacin is a structurally similar to moxifloxacin and likely to have the same mechanism of action against bacterial DNA gyrase. This means cross-resistance is possible between these two drugs. Gatifloxacin's toxicity and drug–drug interaction profile is similar to moxifloxacin's.

Early clinical studies in Brazil have shown gatifloxacin to possess equivalent antituberculosis activity to moxifloxacin and similar, but slightly less potent, activity to isoniazid [15]. One murine study suggests that gatifloxacin in combination with ethionamide or pyrazinamide, could be an alternative regimen to cure active TB. This would be important in treating patients with drug-resistant tuberculosis where the standard isoniazid/rifampin-containing regimen would be ineffective [16].

A large multicenter phase III clinical trial is currently enrolling a planned 2000 patients in South Africa, Senegal, Kenya, Benin, and Guinea to evaluate whether using gatifloxacin instead of ethambutol can shorten the standard TB treatment in adults to four months. This study began in 2005, and is to be completed in 2009 [17].

3.1.3. TMC207

An additional quinolone derivative, TMC207, **6**, (formerly known as R207910) is currently being studied in Phase II trials in patients with tuberculosis. It is a diarylquinolone, with activity against both sensitive and MDR strains of TB. In a murine model it appeared equivalent to both isoniazid and rifampin at lower doses, and superior to the combination of isoniazid, rifampin, and pyrazinamide bactericidal activity by at least 1 log unit when combined at a higher dose with any two of the three. It even appeared to be superior to the combination as monotherapy. In combination with isoniazid or rifampin, TMC207 decrease TB sterilization time from four to less than two months. Its activity appears to be directed against mycobacterial ATP synthase, giving it a unique mechanism of action, compared with other tuberculosis drugs, including fluoroquinolones like moxifloxacin. Moreover, the half-life was greater than 24 hours, suggesting once weekly dosing is possible [18].

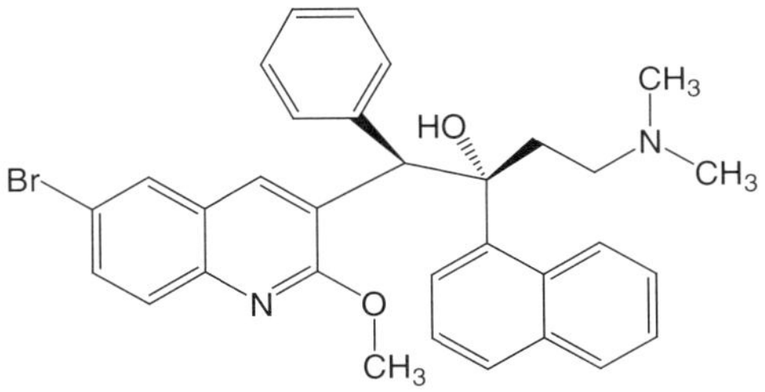

6

3.2. Nitroimidazoles

3.2.1. PA-824

PA-824, **7**, is a nitroimidazopyran discovered in 1995 as a potential cancer drug. In 2000, a study revealed its potency against *M. tuberculosis* [19]. It has a novel mechanism of action inhibiting protein and lipid synthesis, which is effective against both TB and tested clinical isolates of MDR-TB. It attacks the mycobacteria in both the initial, intensive phase of TB therapy, as well as the later, continuation phase. In the intensive phase, PA-824 has bactericidal activity comparable to isoniazid, and in the continuation phase, its bactericidal activity is close to the combination of isoniazid and rifampin. PA-824 also has activity against both actively replicating and the static, slow growing mycobacteria. This has the potential for shortening the course of therapy, since many current drugs are only active against actively replicating mycobacteria, and *M. tuberculosis* spends weeks to months in its static, slow-growing phase. In a murine model, PA-824 was more potent than rifampin when used in combination with moxifloxacin and pyrazinamide, and that when all four drugs were used together, the duration of treatment could be reduced to 3 months or less [20]. PA-824 successfully completed a phase I clinical trial in the U.S. in August 2005, and a phase II trial is planned for later in 2006, ending in 2008.

PA-824 has a bioavailability of about 40% and single oral dose rapidly travel to important target sites like the lung and spleen. PA-824 also does not inhibit the hepatic cytochrome P-450 isoenzyme system, suggesting it could be safely used with ARVs [21]. There is no evidence of genotoxicity, by such standard tests as Ames. The Global Alliance for TB Drug Development, a non-profit, public-private partnership has directed and funded PA-824's development and has obtained exclusive worldwide rights to PA-824 and its derivatives, so that the technology will be available royalty free in endemic countries.

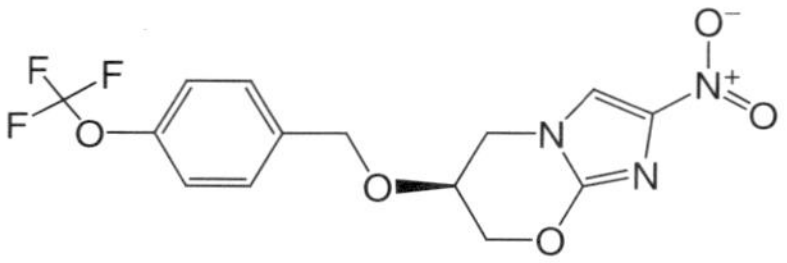

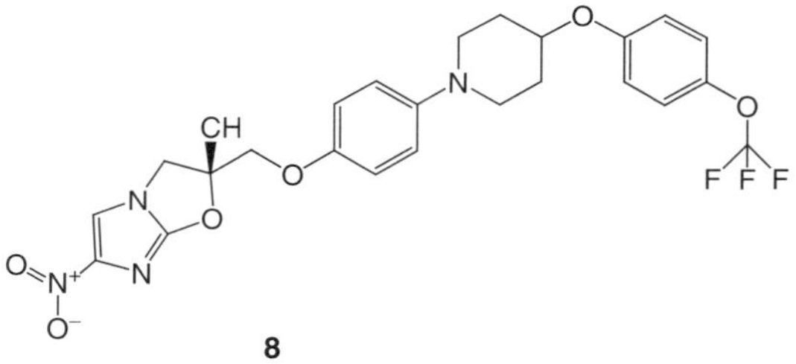

3.2.2. OPC-67683

Work has also begun on second-generation nitroimidazoles. In Japan, a nitroimidazo-oxazole OPC-67683, **8**, has entered phase I clinical trials and a backup nitroimidazo–oxazole is in early development. Preclinical data has been encouraging so far [22]. Second-generation nitroimidazopyran analogs with greater *in vivo* potency in animal models have also been identified.

3.3. Pyrroles

3.3.1. Sudoterb (LL3858, LL4858)

Sudoterb belongs to a class of compounds known as pyrroles, which are plant alkaloids. Sudoterb has been reported to have potent anti-TB activity *in vitro* and *in vivo. In vitro*, sudoterb has bactericidal activity similar to isoniazid and is synergistic with rifampin. The combination of sudoterb with isoniazid, rifampin, and pyrazinamide led to complete sterilization of both sensitive and MDR-TB strains in mice within two months, and in combination with rifampin and pyrazinamide, cured TB in all animals after three months. Sudoterb exhibits good oral bioavailability with once daily dosing. Phase I studies began in India in 2005 [23]. The structure of sudoterb has not been released.

3.4. Tuberculosis vaccine

Currently, BCG (Bacille Calmette-Guérin) is the only licensed tuberculosis vaccine. It has a number of shortcomings, among which is a relatively short period of protection, and poor antibody stimulation response beyond childhood. Since there are clearly many problems with treating TB at this point, a lot of effort and money has been focused on creation of a vaccine, as a means of preventing future infections in those at highest risk and putting a cap on the ever increasing TB epidemic.

Unfortunately, it appears the obstacles in the development of such a vaccine are no less challenging than those hindering drug development. It is well know that BCG varies in effectiveness among those who receive it. This has previously been attributed to different environmental factors or differences in the BCG strains used to create the vaccine. However, a recent gene analysis has found that *M. tuberculosis* is genetically distinct in different parts of the world and has evolved into 6 different lineages: East-African-Indian, East-Asian, Euro-American, Indo-Oceanic, and two West-African. These strains appear to have adapted over time to specific human populations, which suggest separate, wholly different vaccines might be required for each region [24].

Those who receive an effective vaccine may still not develop adequate protection. BCG is known to stimulate Th1 immune cells, which leads to the development of protective antibody and drives *M. tuberculosis* into latency in the body. In many areas of the developing world, parasitic helminthic infections are extremely common. Parasites stimulate a Th2 immune cell response, usually associated with an allergic response. This increase in background Th2 response is antagonistic to the Th1 cell response, and, in theory, will blunt or lead to an inadequate immune, as well as cause impaired bactericidal activity [25].

Nevertheless, work continues on a vaccine. A leading candidate, MVA85A, utilizes a recombinant vaccinia virus Ankara (MVA) expressing antigen 85A. MVA85A completed a Phase I clinical trial in the United Kingdom in 2004. When given with BCG, Th1 levels were upto 30-fold higher than groups given MVA85A or BCG alone [26]. Further clinical trials in the United Kingdom and Gambia are in progress. Another vaccine, Mtb72F/AS02A consisting of a recombinant fusion protein (Mtb72F) formulated in a proprietary adjuvant system (AS02A) has shown promising results in preclinical studies to induce strong, long lasting cellular and humoral immune responses. In addition, the vaccine has shown a good safety and immunogenicity profile in Phase I trials. Phase II trials are planned to being shortly [27]. Even with concerns with the potential efficacy of a vaccine, there may still be a role for its use. Combining the use of a vaccine with antituberculosis drugs seems to augment the bactericidal activity of the drugs, particularly with regard to the slow-growing mycobacteria [21].

We are still years away from actually seeing any of these developments in tuberculosis treatment actually enter clinical practice. But if their full potential is achieved, then it is likely that none of the current first-line drugs will be part of the new, optimized regimen, which should finally begin to decrease both the incidence of tuberculosis, and the development of MDR-TB.

4. CONCLUSION

It is clear that although drug and vaccine development for both avian influenza and MDR-TB is well underway, there is so much more that needs to be done. Currently, it is the developing world which is most affected by both of these infections; yet, the drugs and vaccines are not being optimized for these areas. Drug regimens need to be shorter and simpler, not only to make them affordable to the population at risk but also to prevent the development of drug resistance. Vaccines of more widespread efficacy are also needed to protect as well as prevent these infections from becoming global epidemics. Hopefully, this review has demonstrated the impact for all mankind of emerging infections in the developing world and will stimulate ongoing development of approaches aimed at addressing these issues for the people affected, and for the entire world.

REFERENCES

[1] http://www.who.int/csr/disease/avian_influenza, accessed on 4/21/2006.
[2] D. van Riel, V. J. Munster, E. de Wit, G. F. Rimmelzwaan, R. A. Fouchier, A. D. Osterhaus and T. Kuiken, *Science*, 2006, **312**, 399.
[3] J. H. Beigel, J. Farrar, A. M. Han, F. G. Hayden, R. Hyer, M. D. de Jong, S. Lochindarat, T. K. Nguyen, T. H. Nguyen, T. H. Tran, A. Nicoll, S. Touch and K. Y. Yuen, *N. Engl. J. Med.*, 2005, **353**, 1374.
[4] H. L. Yen, A. S. Monto, R. G. Webster and A. Govorkova, *J. Infect. Dis.*, 2005, **192**, 665.
[5] M. D. de Jong, T. T. Tran, H. K. Truong, M. H. Vo, G. J. Smith, V. C. Nguyen, V. C. Bach, T. Q. Phan, Q. H. Do, Y. Guan, J. S. Peiris, T. H. Tran and J. Farrar, *N. Engl. J. Med.*, 2005, **353**, 2667.
[6] R. K. Gupta and J. S. Nguyen-Van-Tam, *N. Engl. J. Med.*, 2006, **354**, 1423.
[7] http://www.clinialtrials.gov/show/NCT00297050, accesssed on 4/21/2006.
[8] J. J. Treanor, J. D. Campbell, K. M. Zangwill, T. Rowe and M. Wolff, *N. Engl. J. Med.*, 2006, **354**, 1343.
[9] C. Dye, C. J. Watt, D. M. Bleed, S. M. Hosseini and M. C. Raviglioni, *JAMA*, 2005, **293**, 2767.
[10] European Concerted Action on New Generation Genetic Markers and Techniques for the Epidemiology and Control of Tuberculosis, *Emerg. Infect. Dis.*, 2006, in press.
[11] A. Fraser, M. Paul, A. Attamna and L. Leibovici, *Cochrane Database Syst. Rev.*, 2006, CD005435.

[12] S. Sulochana, F. Rahman and C. N. Paramasivan, *J. Chemother.*, 2005, **17**, 169.
[13] U. G. Lalloo, R. Naidoo and A. Ambaram, *Curr. Opin. Pulm. Med.*, 2006, **12**, 179–185.
[14] L. R. Codecasa, G. Ferrara, M. Ferrarese, M. A. Morandi, V. Penati, C. Lacchini, P. Vaccarino and G. B. Migliori, *Respir. Med.,* 2006 (in press).
[15] http://www.cwru.edu/affil/tbru.trials.htm, accessed on 4/21/2006.
[16] M. H. Cynamon and M. Sklankey, *Antimicrob. Agents Chemother.*, 2003, **47**, 2442.
[17] http://www.clinicaltrials.gov/ct/show/NCT00216385, accessed on 4/21/2006.
[18] K. Andries, P. Verhasselt, J. Guillermont, H. W. Gohlmann, J. M. Neefs, H. Winkler, J. Van Gestel, P. Timmerman, M. Zhu, E. Lee, P. Williams, D. de Chaffoy, E. Huitric, S. Hoffner, E. Cambau, C. Truffot-Pernot, N. Lounis and V. Jarlier, *Science*, 2005, **307**, 223.
[19] C. K. Stover, P. Warrener, D. R. VanDevanter, D. R. Sherman, T. M. Arain, M. H. Langhorne, S. W. Anderson, J. A. Towell, Y. Yuan, D. N. Murray, B. N. Kreiswirth, C. E. Barry and W. R. Baker, *Nature*, 2000, **405**, 962.
[20] E. Nuermberger, S. Tyagi, K. N. Williams, I. Rosenthal, W. R. Bishai and J. H. Grosset, *Am. J .Respir. Crit. Care Med.*, 2005, **172**, 1452.
[21] A. J. Lenaerts, V. Gruppo, K. S. Marietta, C. M. Johnson, D. K. Driscoll, N. M. Tompkins, J. D. Rose, R. C. Reynolds and I. M. Orme, *Antimicrob. Agents Chemother.*, 2005, **49**, 2294–2301.
[22] M. Okakda, T. Tanka, K. Kita, N. Kanamaru, S. Hashimoto, H. Takai, Y. Fukunaga, Y. Sakaguchi, I. Furukawa, K. Yamada, Y. Muraki, S. Kuwayama, M. Izumiya, M. Matsumoto and M. Sakatani, *Abstract F-1467*, 45th ICAAC, Washington DC, 2005.
[23] R. K. Sinha, S. K. Arora, N. Sinha and V. M. Modak, *Abstract F-1116*, 44th ICAAC, Washington, DC, 2004.
[24] S. Gagneux, K. DeRiemer, T. Van, M. Kato-Maeda, B. C. de Jong, S. Narayanan, M. Nicol, S. Niemann, K. Kremer, M. C. Gutierrez, M. Hilty, P. C. Hopewell and P. M. Small, *Proc. Acad. Natl. Sci. U.S.A.*, **103**, 2869.
[25] G. A. Rook, K. Dheda and A. Zumla, *Vaccine*, 2005, **23**, 2115.
[26] H. McShane, A. A. Pathan, C. R. Sander, N. P. Goonetilleke, H. A. Fletcher and A. V. Hill, *Tuberculosis*, 2005, **85**, 47.
[27] S. Reed and Y. Lobet, *Microbes Infect*, 2005, **7**, 922.

Influenza Neuraminidase Inhibitors as Antiviral Agents

Y. Sudhakara Babu, Pooran Chand and Pravin L. Kotian

BioCryst Pharmaceuticals, Inc., 2190 Parkway Lake Dr., Birmingham, AL 35244, USA

Contents

1. Introduction	287
2. Inhibitors based upon six-membered heterocycles	288
2.1. Pyranose-based inhibitors	288
2.2. Cyclohexene-based inhibitors	290
2.3. Benzene-based inhibitors	290
3. Inhibitors based on five-membered ring structures	291
3.1. Furanose-based compounds	291
3.2. Cyclopentane-based inhibitors	292
3.3. Pyrrolidine-based inhibitors	293
4. Drugs approved or under clinical trial	293
5. Conclusions	294
References	294

1. INTRODUCTION

Influenza infections continue to be a significant worldwide health concern with yearly outbreaks affecting 10–20% of the world population. In the United States alone, the annual epidemic is associated with close to 200,000 hospitalizations and thousands of deaths [1]. Over the last few years, there has been considerable concern regarding the highly pathogenic H5N1 avian influenza virus, and its human pandemic potential. Outbreaks of avian H5N1 influenza among poultry have been reported in several parts of the world. As of March 13, 2006, there have been 177 human cases of avian influenza (H5N1) infections, 93 in Vietnam, 29 in Indonesia, 22 in Thailand, and some in Cambodia, China, Iraq, and Turkey, resulting in 98 deaths [2]. Existing options to control influenza have limitations. Immunization is effective only when administered annually in response to the seasonal antigenic variation of the virus. Vaccines that are specific for newly arising strains require several months of preparation. Although the development of a vaccine against H5N1 influenza is underway, none is presently available. Therefore, antiviral drugs are an important part of the strategy for treating/preventing an influenza epidemic or an influenza pandemic with a new influenza virus of any origin, including H5N1. There are presently two types of antiviral drugs to treat influenza:

- M2 ion channel blockers (amantadine and rimantadine) are effective only against influenza A and resistance strains to these two agents develops rapidly in the clinic [3,4].
- There has been a great deal of interest in targeting the neuraminidase (NA) enzyme on the surface of the influenza virus. This enzyme is required for the spread of the newly synthesized virus particles. The Food and Drug Administration (FDA) has approved two neuraminidase inhibitors, nebulized zanamivir (Relenza) and oral oseltamivir (Tamiflu), for treatment of influenza. Neuraminidase inhibitors have advantages over M2 ion channel blockers because they are effective against a wide range of influenza viruses unlike M2 ion channel blockers. Worldwide sales of Tamiflu were in excess of $1 billion in 2005, in large part due to the stock-piling of the drug by different governments in case of a pandemic avian influenza outbreak. Sales of Relenza are significantly lower due to the complexity of administering the drug.

2. INHIBITORS BASED UPON SIX-MEMBERED HETEROCYCLES

2.1. Pyranose-based inhibitors

2-Deoxy-2,3-didehydro-N-acetylneuraminic acid (DANA,**1**), a sialic acid analog, is a moderate inhibitor of NA *in vitro* ($IC_{50} = 10\,\mu M$). Efforts were made to modify the acetyl group in **1** to improve the potency. The trifluoroacetyl analog (**1a**) is a more potent inhibitor [$IC_{50} = 2\,\mu M$] than **1** [5,6]. The crystal structure of NA revealed a negatively charged pocket into which the C4-hydroxyl of DANA is positioned. Replacement of the C4-hydroxyl by basic groups resulted in significantly more potent compounds. The guanidino analog, zanamivir (**2**), is one of the most potent inhibitors of NA, and is almost 1000-fold more potent [$IC_{50} = 2$–$20\,nM$] than **1** [7–10]. Other modifications of hydroxyl, such as NHAc, N = C-CH$_3$, OCH$_2$CONH$_2$, OCH$_2$CN, OCH$_2$C(= NH)NH$_2$ at C-4 were not effective. The introduction of fluoro at C-2 and or C-3 position did not increase potency [11]. Other changes that were investigated included substituting the 9-position with azido, and the 5-position with COCF$_3$ [12,13]. From crystallographic studies, it appeared that the hydroxyl at C-7 of **2** does not interact with the protein. Hence, it was modified to F, H, O-alkyl, N$_3$, NH$_2$, NHC(O)-alkyl, OC(O)NH-alkyl, O(CO)N(alkyl-NH$_2$)$_2$ resulting in compounds of type **3** with activity similar to **2** [14–16]. The C-8 and C-9 hydroxyls were also removed to produce compounds with alkyl side chain. These compounds have only modest activity [17]. The replacement of the C-7 hydroxyl by an O-alkyl group with a polyglutamic acid backbone resulted in potent compounds [18]. A number of reports recently have appeared on di-, tri-, and tetra-meric compounds of **2** where the attachments are through the carbon atom at position 7 of zanamivir [19–21]. These novel structures are described in a recent review [22]. These compounds are about 100 times more potent than zanamivir. Some of the dimeric compounds **4a** and **4b** have shown long-lasting protective activities in mouse influenza infectivity experiments and very long residence times in the lungs of rats. The replacement of the entire glycerol side chain by hydrophobic groups (**5**) resulted in potent compounds against influenza A, but not against influenza B [23–27]. However, one compound, **6**, showed efficacy in mouse influenza models at 10 μM/kg [26]. Other heterocyclic cores have also been studied. For example, piperadine derivatives (**7a, 7b**) and 1,4,5,6-tetrahydropyridazine derivatives (**8a, 8b**) were prepared but had poor activity [28,29]. Sulfur analogs (**9a, 9b, 9c**) of DANA, 4-amino DANA and zanamivir were also reported [30,31].

1 R = CH_3
1a R = CF_3

2

3

3a R = F **3b** R = O-alkyl **3c** R = H,
3d R = N_3 **3e** R = NH_2
3f R = NHC(=O)alkyl

4a X = $C(=O)NH(CH_2)_{12}NHC(=O)$
4b X = $C(=O)NH(CH_2)_{14}NHC(=O)$

5

5a R = $OCH(alkyl)_2$ **5b** R = $C(=O)N(alkyl)_2$
5c R = $C(=O)CH(alkyl)_2$ **5d** R = $CH(OH)CH(alkyl)_2$

6

7a R = NH_2,
7b R = $NHCH(=NH)NH_2$

8a R = NH_2,
8b R = $NHCH(=NH)NH_2$

9a R = OH
9b R = NH_2,
9c R = $NHCH(=NH)NH_2$

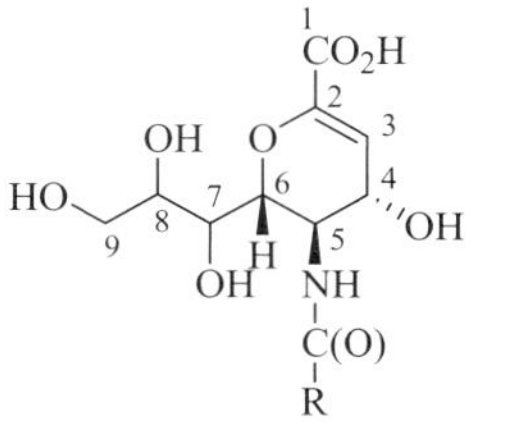

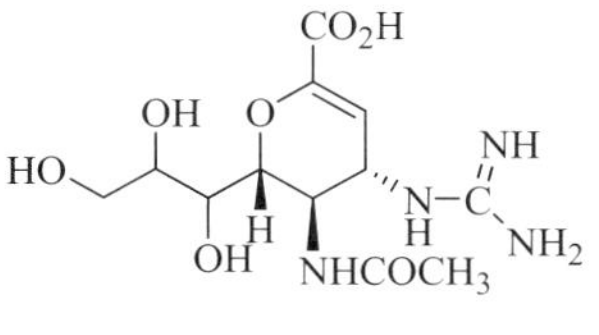

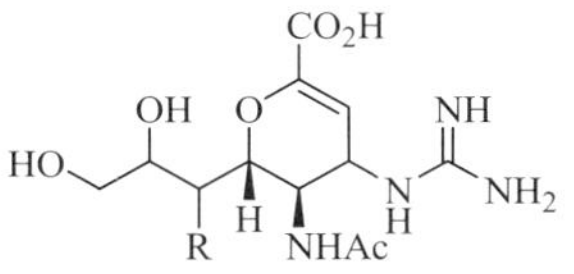

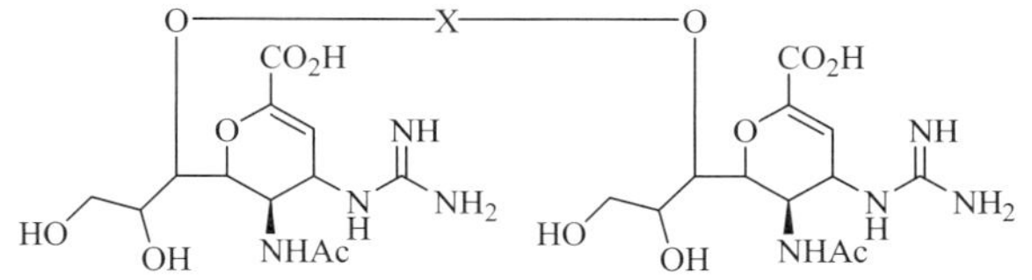

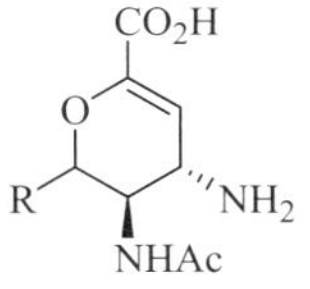

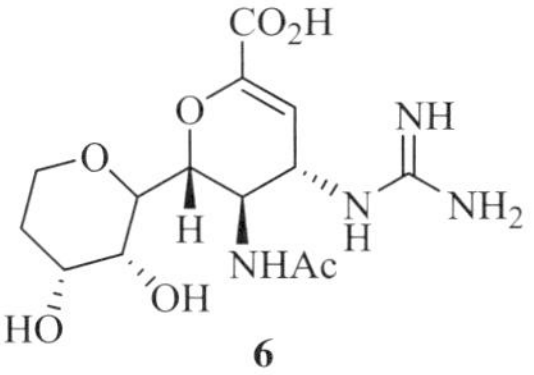

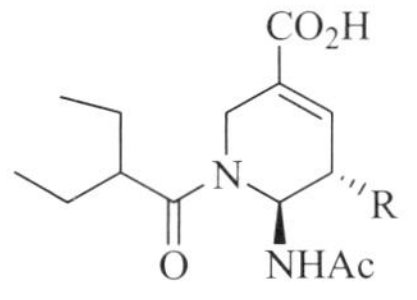

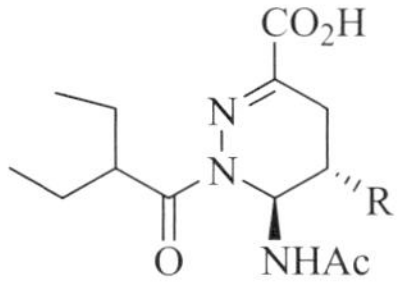

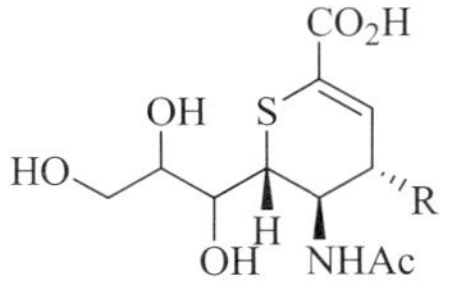

2.2. Cyclohexene-based inhibitors

Carbocyclic compounds based on the core structure of **10** have been investigated as potential neuraminidase inhibitors [32]. Sequential modification of the R group in **10** showed the *N*-propyl ether to be optimum. Further, SAR was conducted that led to **11**, the most potent compound in this series. This novel compound had an IC_{50} value of 1 nM against influenza A neuraminidase and 3 nM against influenza B neuraminidase [32,33]. When R = propoxy in **10**, the replacement of the oxygen of the propoxy group with CH_2, N or S resulted in compounds with poor activity [32–34]. Substituting $OCH(C_2H_5)_2$ with alkylamino, dialkylamino, acylamino, alkylacylamino, cyclicamides, etc. in **11** resulted in the loss of activity both against influenza A and influenza B [35–37]. However, 2-hydroxyethylazepine substitution (**12**) resulted in a compound with an IC_{50} of 8 and 14 nM against influenza A and B, respectively [36]. Replacement of NH_2 with OH in **11** resulted in decreased potency, while guanidino substitution increased the potency five times. Although it was not the most potent compound *in vitro*, compound **11** was selected for clinical development due to its superior bioavailability relative to the guanidine compound [34]. The replacement of the amino group with hydrophobic groups like vinyl, allyl, $CH_2CO_2CH_3$, CH_2CH_2OH produced compounds with poor activity [38]. Some isosteres of **11**, including compounds in which the double bond was either absent or isomerized, showed poor activity [34]. Some bicyclic structures, including **13** and **14**, were prepared but did not result in any compound having comparable activity to **11** [39,40]. Substitutions at the double bond carbon atom as shown in **15** were found to be very poor in activity [41]. The synthesis of **11** and its prodrug has been described in detail [42–44].

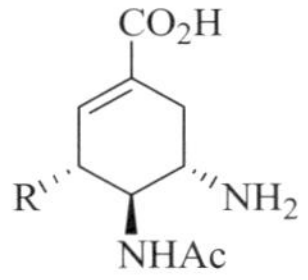

10
R = O-(alkyl)$_n$ (n = 1-9)

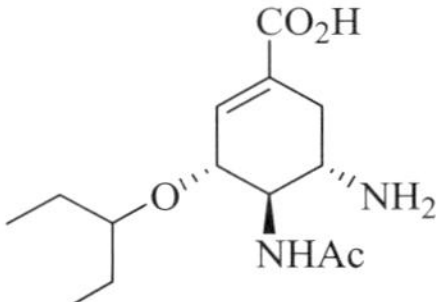

11

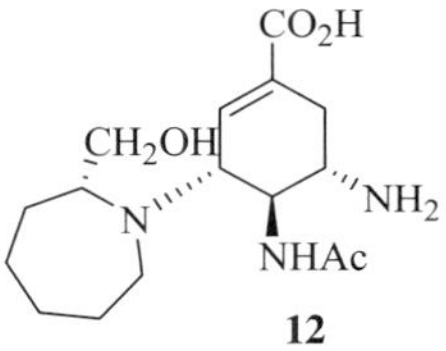

12

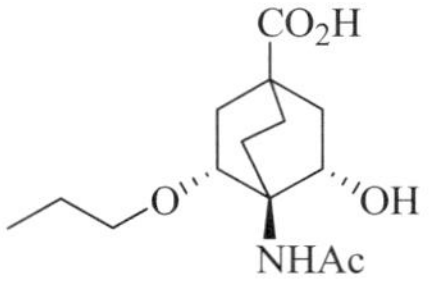

13

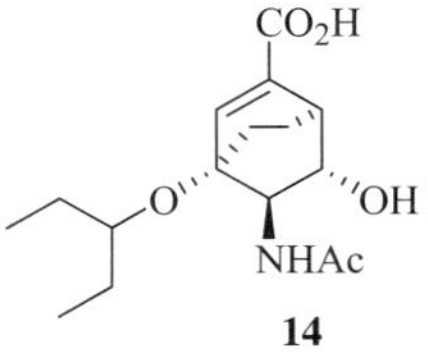

14

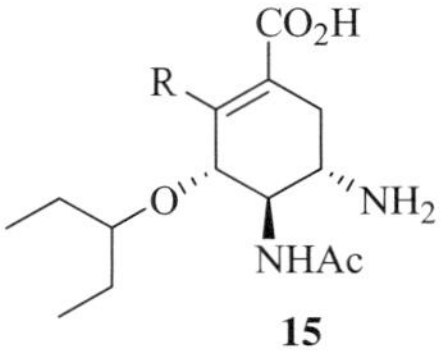

15
R = Cl, CH_3, SCH_3

2.3. Benzene-based inhibitors

The complex stereochemistry of the saturated carbocyclic NA inhibitors poses considerable synthetic challenge. Recent shortages of oseltamivir were due in part to the complex synthesis of this compound. Simplified analogs based on a benzene core have the potential to address this issue. Benzoic acid derivatives, **16**, **16a**, **16b** and **16c**, had promising activity against NA from both influenza A and B [45–47]. These findings encouraged a number of groups to investigate aromatic compounds. Ease of synthesis allowed the investigation of a number of substituents on 4-acetylaminobenzoic acid, including OH, NO_2, NH_2, guanidine, NHCN, $NHCONH_2$, CH_2NH_2, $CH_2CH_2NH_2$, CH_2-guanidine, etc. The best compound in this series, **16d**, showed an IC_{50} value of about 10 µM [45–48]. A crystal structure of **16d** with the NA enzyme revealed that the guanidine substituent occupies the same sub-site as the glycerol side chain of **2** in the active site [45]. The introduction of a fourth substituent on **16d**, such as another guanidine group, glycerol side chain, amino, CH_2OH, or CH_2CH_2OH resulted in less potent compounds [49–52]. Introduction of a hydrophobic group on the benzene ring resulted in compounds of type **17**. The lead compound in this series **17a** has an IC_{50} value of 15 µM [53,54]. Further optimization of this lead compound with R group being amino or guanidine improved the potency by about 10-fold [53,54]. Replacing the acetylamino group by a pyrrolidinone ring afforded compounds derived from the core **18** [55–58]. **18a** is the most potent compound in this series with an IC_{50} value of 0.8 µM against influenza A neuraminidase [57]. Screening of α and β amino acid libraries identified **19** with an IC_{50} value of 41 µM [59].

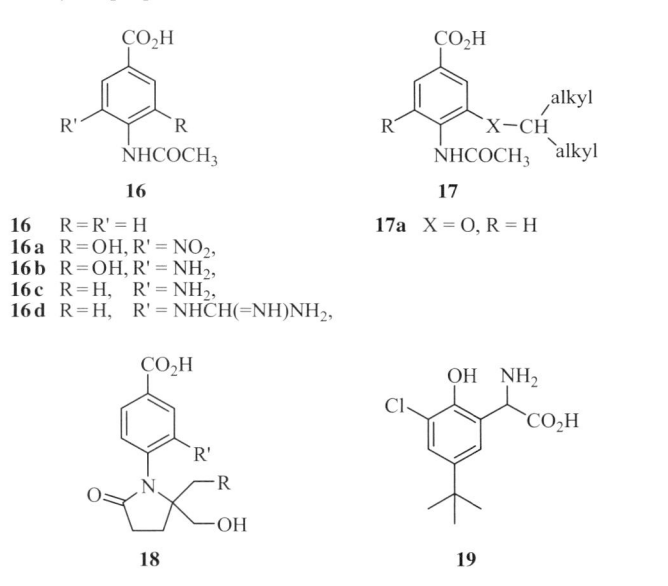

16 R = R' = H
16a R = OH, R' = NO_2,
16b R = OH, R' = NH_2,
16c R = H, R' = NH_2,
16d R = H, R' = NHCH(=NH)NH_2,

17a X = O, R = H

18a R = OH, R' = NHCH(=NH)NH_2

3. INHIBITORS BASED ON FIVE-MEMBERED RING STRUCTURES

3.1. Furanose-based compounds

A five-membered ring compound, furanose derivative **20**, having all the groups (COOH, OH, $NHCOCH_3$, and glycerol) necessary for interaction with the active site residues of NA, was observed to have activity similar to DANA [60]. No further modifications were reported on this ring structure, probably because of the synthetic difficulties. Crystal structure studies of **20** revealed that this compound binds in the active site in an orientation similar to that of DANA. A series of analogs in which the furanose ring system has been replaced by the more accessible cyclopentane ring has been reported [61]. Recently, two compounds of structure **21** in the furanose series were reported to have IC_{50} values in the range of 0.5–1.0 µM [62].

3.2. Cyclopentane-based inhibitors

The crystal structure of 20 and the activity of this compound against NA demonstrate that the relative position of the interacting groups is the most important factor for potent neuraminidase inhibition [61]. This observation suggested that a cyclopentane ring might be a suitable scaffold for a novel neuraminidase inhibitor. The first entry, **22**, in this class of compounds was initially synthesized as a racemate. Encouraged by the promising activity of this compound against the NA enzyme [61,63], compound **23** was synthesized to exploit the small hydrophobic surface in the active site of the NA enzyme. This compound was prepared initially as a mixture of eight isomers. The stereochemistry of the potent isomer from this mixture was identified from the crystal structure of **23** complexed with NA. The optimization of the hydrophobic side chains resulted in compounds **25a** and **25b**. Compounds **24a** and **24b**, obtained as intermediates during the synthesis of **25a** and **25b**, were found to possess activity similar to **25a** and **25b** [IC50 = 0.1–11 nM]. Among these compounds, **24a** was selected for clinical development because of ease of synthesis [61]. Attempts to replace the alkyl side chain with amide (**26**) yielded compounds with good activity against NA from influenza A but not against influenza B [64]. Some compounds of type **27** were also prepared, but did not show good activity against neuraminidase [65].

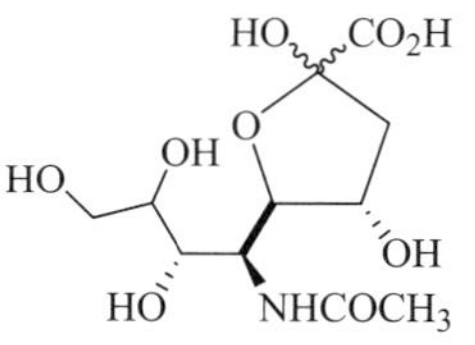

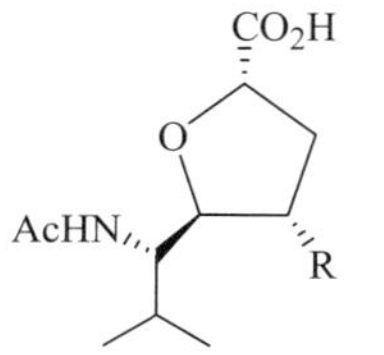

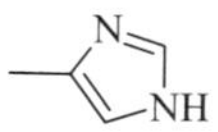

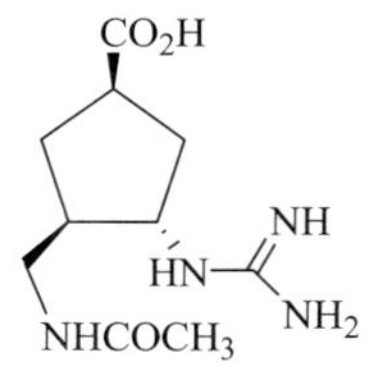

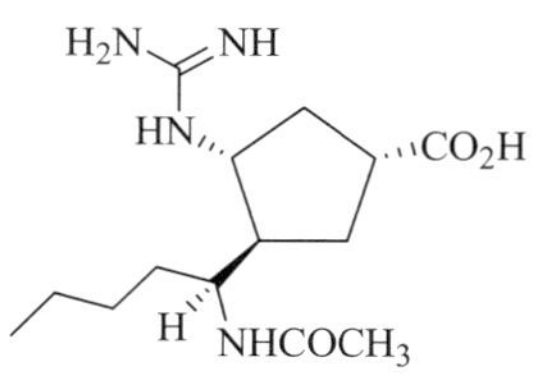

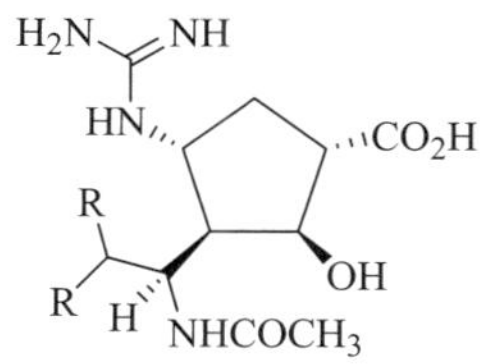

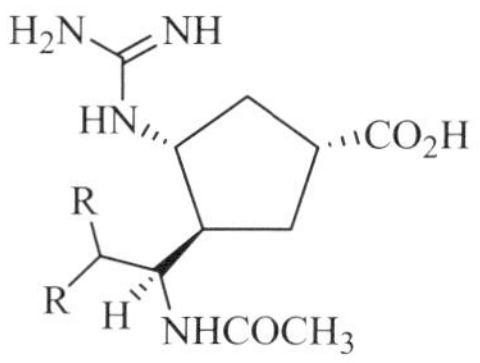

25a R = Et
25b R = n-Pr

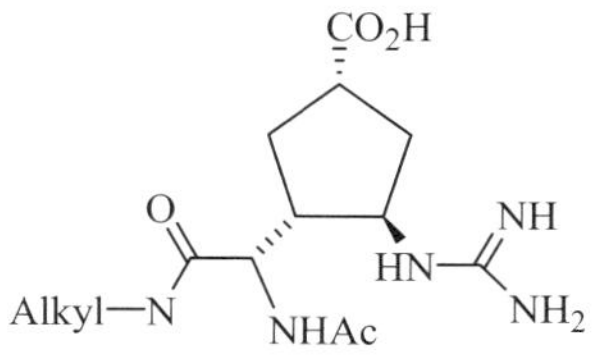

26

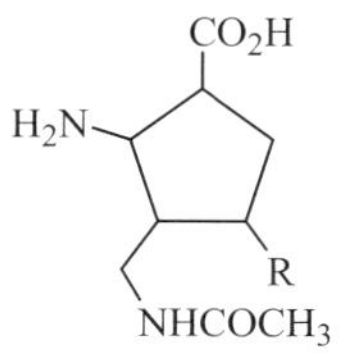

27 R = ester, amide

3.3. Pyrrolidine-based inhibitors

Screening a library of α and β amino acids resulted in the discovery of **28** as a competitive inhibitor with a K_i value of 58 µM [59]. Replacement of the *tert*-butoxycarbonyl group with various amides improved the potency significantly with the best compound in this series having a K_i value of 1 µM against the enzyme [66]. Further substitution at the 5 position with CH$_2$NHCOR (where R is Me, Et, CH=CH$_2$, i-Pr, CF$_3$) did not increase binding affinity [66]. Replacement of the amino group at the 4-position by alkylamino, amides, and guanidine also did not result in compounds with better inhibitory activity [66]. A different substitution on the pyrrolidine core resulted in **29** when R = ester and amine and R' = CH$_2$CH(CH$_3$ or C$_2$H$_5$)$_2$. These changes improved the potency in this series, with the best compound being R = CO$_2$CH$_3$ and R' = CH$_2$CH(C$_2$H$_5$)$_2$, and having a K_i value of 0.8 nM [65]. Additional modifications of the R group were investigated fixing R' as CH$_2$CH(CH$_3$)$_2$ as shown in **30** [67]. Compounds with various R groups including esters, ethers, ketones, amides, heterocyclics, and substituted vinyls were prepared. SAR identified *cis*-propene to be the optimum group. Compounds with this substituent showed K_i values of 0.02–0.8 µM [67]. Final optimization of the alkyl side chain resulted in the most active compound (**31**) in the series, with a K_i value of 0.2 nM. The detailed synthesis of **31** was reported [68–70]. Pyrrolidine compounds are well covered in a recent review article [71].

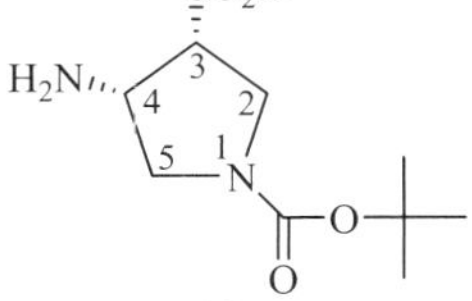

28

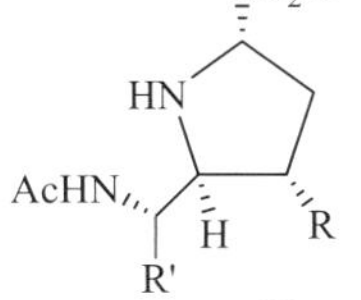

29

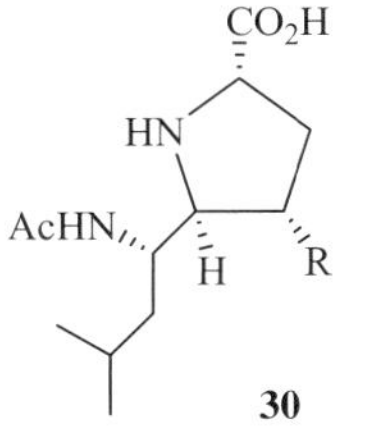

30

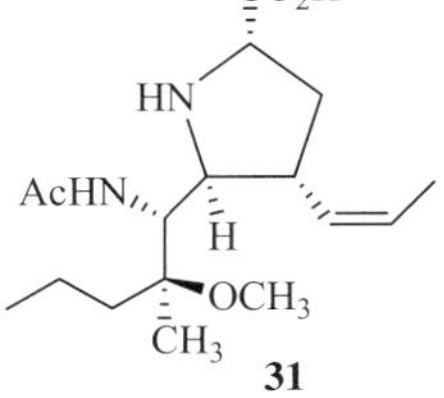

31

4. DRUGS APPROVED OR UNDER CLINICAL TRIAL

The FDA has approved two neuraminidase inhibitors, zanamivir (Relenza), and oseltamivir phosphate (Tamiflu) for the treatment of influenza. Both drugs are effective against a wide range of influenza viruses. Both compounds reduce the symptoms of the disease such as fever, headache, myalgia, and cough by 1.5–2 days if taken within 24–48 h of the onset of symptoms. The oral bioavailability of zanamivir is low. Hence, it is administered topically using a specifically designed inhaler at a dose of 10 mg twice daily for five days. Oseltamivir is given orally in 75 mg dose twice a day for five days. Zanamivir is approved only for treatment, while oseltamivir is approved both for treatment and prophylactic use.

Peramivir (BCX-1812, **24a**) is another potent neuraminidase inhibitor that has been shown to be very effective in mouse influenza models [72–77]. In phase III clinical trials, the compound did not meet its primary end point probably due to its limited oral bioavailability. Using different routes of administration, intravenous and intramuscular, Phase I clinical trials have been initiated recently [78].

5. CONCLUSIONS

Influenza virus infections and a potential influenza pandemic continue to be a major acute threat to the public health worldwide. The recent emergence of a highly pathogenic H5N1 avian influenza virus poses an increasing human pandemic threat. There is concern that it can mutate into a form that allows easy transmission between humans. Thus, there is an urgent need for more effective methods to control and treat influenza infections. Existing vaccine technologies require considerable lead-time in manufacturing a vaccine that is effective against a given strain of the virus. Neuraminidase inhibitors are attractive agents not only to treat illness caused by influenza infections but also to prevent infections. Tamiflu is widely prescribed to treat influenza infections, and is shown to be effective in reducing the time to alleviation of illness caused by influenza. However, its effectiveness in patients infected with H5N1 virus is limited, and virus strains resistant to Tamiflu are observed in the clinic [79]. Despite recent progress with the introduction of Relenza and Tamiflu, there is still urgent need for different classes of effective neuraminidase inhibitors to treat illness caused by influenza.

REFERENCES

[1] A. C. Schmidt, *Drugs*, 2004, **64**, 2031.
[2] World Health Organization Report March 13, 2006 on Cumulative Number of Confirmed Human Cases of Avian Influenza A/(H5N1) Reported to WHO.
[3] E. Arruda and F. G. Hayden, *Adv. Exp. Med. Biol.*, 1996, **394**, 175.
[4] S. Shigeta, *Nippon Rinsho*, 1997, **55**, 2758.
[5] P. Palese, J. L. Schulman, G. Bodo and P. Meindl, *Virology*, 1974, **59**, 490.
[6] J. L. Schulman and P. Palese, *Virology*, 1975, **63**, 98.
[7] P.-A. Driguez, B. Barrere, G. Quash and A. Doutheau, *Carbohydr. Res.*, 1994, **262**, 297.
[8] M. von Itzstein, W.-Y. Wu and B. Jin, *Carbohydr. Res.*, 1994, **259**, 301.
[9] M. Chandler, M. J. Bamford, R. Conroy, B. Lamont, B. Patel, V. K. Patel, I. P. Steeples, R. Storer, N. G. Weir, M. Wright and C. Williamson, *J. Chem. Soc. Perkin Trans.*, 1995, **1**, 1173.
[10] M. von Itzstein, J. C. Dyason, S. W. Oliver, H. F. White, W.-Y. Wu, G. B. Kok and M. S. Pegg, *J. Med. Chem.*, 1996, **39**, 388.

[11] T. Hagiwara, I. Kijima-Suda, T. Ido, H. Ohrui and K. Tomita, *Carbohydr. Res.*, 1994, **263**, 167.
[12] K. Ikeda, F. Kimura, K. Sano, Y. Suzuki and K. Achiwa, *Carbohydr. Res.*, 1998, **312**, 183.
[13] K. Ikeda, K. Sano, M. Ito, M. Saito, K. Hidari, T. Suzuki, Y. Suzuki and K. Tanaka, *Carbohydr. Res.*, 2001, **330**, 31.
[14] T. Honda, T. Masuda, S. Yoshida, M. Arai, Y. Kobayashi and M. Yamashita, *Bioorg. Med. Chem. Lett.*, 2002, **12**, 1921.
[15] T. Masuda, S. Yoshida, M. Arai, S. Kaneko, M. Yamashita and T. Honda, *Chem. Pharm. Bull.*, 2003, **51**, 1398.
[16] D. M. Andrews, P. C. Cherry, D. C. Humber, P. S. Jones, S. P. Keeling, P. F. Martin, C. D. Shaw and S. Swanson, *Eur. J. Med. Chem.*, 1999, **34**, 563.
[17] T. Honda, T. Masuda, S. Yoshida, M. Arai, S. Kaneko and M. Yamashita, *Bioorg. Med. Chem. Lett.*, 2002, **12**, 1925.
[18] T. Honda, S. Yoshida, M. Arai, T. Masuda and M. Yamashita, *Bioorg. Med. Chem. Lett.*, 2002, **12**, 1929.
[19] K. G. Watson, R. Cameron, R. J. Fenton, D. Gower, S. Hamilton, B. Jin, G. Y. Krippner, A. Luttick, D. McConnell, S. J. F. MacDonald, A. M. Mason, V. Nguyen, S. P. Tucker and W.-Y. Wu, *Bioorg. Med. Chem. Lett.*, 2004, **14**, 1589.
[20] S. J. F. Macdonald, R. Cameron, D. A. Demaine, R. J. Fenton, G. Foster, D. Gower, J. N. Hamblin, S. Hamilton, G. J. Hart, A. P. Hill, G. G. A. Inglis, B. Jin, H. T. Jones, D. B. McConnell, J. McKimm-Breschkin, G. Mills, V. Nguyen, I. J. Owens, N. Parry, S. E. Shanahan, D. Smith, K. G. Watson, W.-Y. Wu and S. P. Tucker, *J. Med. Chem.*, 2005, **48**, 2964.
[21] S. J. F. Macdonald, K. G. Watson, R. Cameron, D. K. Chalmers, D. A. Demaine, R. J. Fenton, D. Gower, J. N. Hamblin, S. Hamilton, G. J. Hart, G. G. A. Inglis, B. Jin, H. T. Jones, D. B. McConnell, A. M. Mason, V. Nguyen, I. J. Owens, N. Parry, P. A. Reece, S. E. Shanahan, D. Smith, W.-Y. Wu and S. P. Tucker, *Antimicrob. Agents Chemother.*, 2004, **48**, 4542.
[22] P. Chand, *Expert Opin. Ther. Patents*, 2005, **15**, 1009.
[23] P. W. Smith, J. E. Robinson, D. N. Evans, S. L. Sollis, P. D. Howes, N. Trivedi and R. C. Bethell, *Bioorg. Med. Chem. Lett.*, 1999, **9**, 601.
[24] N. R. Taylor, A. Clesaby, O. Singh, T. Skarzynski, A. J. Wonacott, P. W. Smith, S. L. Sollis, P. D. Howes, P. C. Cherry, R. Bethell, P. Colman and J. Varghese, *J. Med. Chem.*, 1998, **41**, 798.
[25] P. W. Smith, S. L. Sollis, P. D. Howes, P. C. Cherry, I. D. Starkey, K. N. Cobley, H. Weston, J. Scicinski, A. Merritt, A. Whittington, P. Wyatt, N. Taylor, D. Green, R. Bethell, S. Madar, R. J. Fenton, P. J. Morley, T. Pateman and A. Beresford, *J. Med. Chem.*, 1998, **41**, 787.
[26] T. Masuda, S. Shibuya, M. Arai, S. Yoshida, T. Tomozawa, A. Ohno, M. Yamashita and T. Honda, *Bioorg. Med. Chem. Lett.*, 2003, **13**, 669.
[27] P. G. Wyatt, B. A. Coomber, D. N. Evans, T. I. Jack, H. E. Fulton, A. J. Wonacott, P. Colman and J. Varghese, *Bioorg. Med. Chem. Lett.*, 2001, **11**, 669.
[28] E. Shitara, Y. Nishimura, K. Nerome, Y. Hiramoto and T. Takeuchi, *Org. Lett.*, 2000, **2**, 3837.
[29] L. Zhang, M. A. Williams, D. B. Mendel, P. A. Escarpe, X. Chen, K.-Y. Wang, B. J. Graves, G. Lawton and C. U. Kim, *Bioorg. Med. Chem. Lett.*, 1999, **9**, 1751.
[30] H. Mack and R. Brossmer, *Tetrahedron Lett.*, 1987, **28**, 191.
[31] G. B. Kok, M. Campbell, B. Mackey and M. von Itzstein, *J. Chem. Soc. Perkin Trans.*, 1996, **1**, 2811.

[32] C. U. Kim, W. Lew, M. A. Williams, H. Lin, L. Zhang, S. Swaminathan, N. Bischofberger, M. S. Chen, D. B. Mendel, C. Y. Tai, W. G. Laver and R. C. Stevens, *J. Am. Chem. Soc.*, 1997, **119**, 681.
[33] C. U. Kim, W. Lew, M. A. Williams, H. Wu, L. Zhang, X. Chen, P. A. Escarpe, D. B. Mendel, W. G. Laver and R. C. Stevens, *J. Med. Chem.*, 1998, **41**, 2451.
[34] W. Lew, X. Chen and C. U. Kim, *Curr. Med. Chem.*, 2000, **7**, 663.
[35] M. Hochgurtel, R. Biesinger, H. Kroth, D. Piecha, M. W. Hofmann, S. Krause, O. Schaaf, C. Nicolau and A. V. Eilseev, *J. Med. Chem.*, 2003, **46**, 356.
[36] W. Lew, H. Wu, X. Chen, B. J. Graves, P. A. Escarpe, H. L. MacArthur, D. B. Mendel and C. U. Kim, *Bioorg. Med. Chem. Lett.*, 2000, **10**, 1257.
[37] W. Lew, H. Wu, D. B. Mendel, P. A. Escarpe, X. Chen, W. G. Laver, B. J. Graves and C. U. Kim, *Bioorg. Med. Chem. Lett.*, 1998, **8**, 3321.
[38] S. Hanessian, J. Wang, D. Montgomery, V. Stoll, K. D. Stewart, W. Kati, C. Maring, D. Kempf, C. Hutchins and W. G. Laver, *Bioorg. Med. Chem. Lett.*, 2002, **12**, 3425.
[39] P. W. Smith, N. Trivedi, P. D. Howes, S. L. Sollis, G. Rahim, R. C. Bethell and S. Lynn, *Bioorg. Med. Chem. Lett.*, 1999, **9**, 611.
[40] P. S. Jones, P. W. Smith, G. W. Hardy, P. D. Howes, R. J. Upton and R. C. Bethell, *Bioorg. Med. Chem. Lett.*, 1999, **9**, 605.
[41] L. Zhang, M. A. Williams, D. B. Mendel, P. A. Escarpe and C. U. Kim, *Bioorg. Med. Chem. Lett.*, 1997, **7**, 1847.
[42] M. Federspiel, R. Fischer, M. Hennig, H.-J. Mair, T. Oberhauser, G. Rimmler, T. Albiez, J. Bruhin, H. Estermann, C. Gandert, V. Gockel, S. Gotzo, U. Hoffmann, G. Huber, G. Janatsch, S. Lauper, O. Rockel-Stabler, R. Trussardi and A. G. Zwahlen, *Org. Process Res. Dev.*, 1999, **3**, 266.
[43] J. C. Rohloff, K. M. Kent, M. J. Postich, M. W. Becker, H. H. Chapman, D. E. Kelly, W. Lew, M. S. Louie, L. R. McGee, E. J. Prisbe, L. M. Schultze, R. H. Yu and L. Zhang, *J. Org. Chem.*, 1998, **63**, 4545.
[44] M. Karpf and R. Trussardi, *J. Org. Chem.*, 2001, **66**, 2044.
[45] P. Chand, Y. S. Babu, S. Bantia, N. Chu, L. B. Cole, P. L. Kotian, W. G. Laver, J. A. Montgomery, V. P. Pathak, S. L. Petty, D. P. Shrout, D. A. Walsh and G. M. Walsh, *J. Med. Chem.*, 1997, **40**, 4030.
[46] M. J. Jedrzejas, S. Singh, W. J. Brouillette, W. G. Laver, G. M. Air and M. Luo, *Biochemistry*, 1995, **34**, 3144.
[47] M. Luo, M. J. Jedrzejas, S. Singh, C. L. White, W. J. Brouillette, G. M. Air and W. G. Laver, *Acta Cryst.*, 1995, **D51**, 504.
[48] E. A. Sudbeck, M. J. Jedrzejas, S. Singh, W. J. Brouillette, G. M. Air, W. G. Laver, Y. S. Babu, S. Bantia, P. Chand, N. Chu, J. A. Montgomery, D. A. Walsh and M. Luo, *J. Mol. Biol.*, 1997, **267**, 584.
[49] S. Singh, M. J. Jedrzejas, G. M. Air, M. Luo, W. G. Laver and W. J. Brouillette, *J. Med. Chem.*, 1995, **38**, 3217.
[50] P. Chand, P. L. Kotian, P. E. Morris, S. Bantia, D. A. Walsh and Y. S. Babu, *Bioorg. Med. Chem.*, 2005, **13**, 2665.
[51] A. Bianco, M. Brufani and C. Melchioni, *Il Farmaco*, 2001, **56**, 305.
[52] M. Williams, N. Bischofberger, S. Swaminathan and C. U. Kim, *Bioorg. Med. Chem. Lett.*, 1995, **5**, 2251.
[53] V. R. Atigadda, W. J. Brouillette, F. Duarte, Y. S. Babu, S. Bantia, P. Chand, N. Chu, J. A. Montgomery, D. A. Walsh, E. Sudbeck, J. Finley, G. M. Air, M. Luo and G. W. Laver, *Bioorg. Med. Chem.*, 1999, **7**, 2487.
[54] W. J. Brouillette, V. R. Atigadda, M. Luo, G. M. Air, Y. S. Babu and S. Bantia, *Bioorg. Med. Chem. Lett.*, 1999, **9**, 1901.

[55] V. R. Atigadda, W. J. Brouillette, F. Duarte, S. M. Ali, Y. S. Babu, S. Bantia, P. Chand, N. Chu, J. A. Montgomery, D. A. Walsh, E. A. Sudbeck, J. Finley, M. Luo, G. M. Air and G. W. Laver, *J. Med. Chem.*, 1999, **42**, 2332.

[56] J. B. Finley, V. R. Atigadda, F. Duarte, J. J. Zhao, W. J. Brouillette, G. M. Air and M. Luo, *J. Mol. Biol.*, 1999, **293**, 1107.

[57] W. J. Brouillette, S. N. Bajpai, S. M. Ali, S. E. Velu, V. R. Atigadda, B. S. Lommer, J. B. Finley, M. Luo and G. M. Air, *Bioorg. Med. Chem.*, 2003, **11**, 2739.

[58] B. S. Lommer, S. M. Ali, S. N. Bajpai, W. J. Brouillette, G. M. Air and M. Luo, *Acta Cryst.*, 2004, **D60**, 1017.

[59] W. M. Kati, D. Montgomery, C. Maring, V. S. Stoll, V. Giranda, X. Chen, W. G. Laver, W. Kohlbrenner and D. W. Norbeck, *Antimicrob. Agents Chemother.*, 2001, **45**, 2563.

[60] T. Yamamoto, H. Kumazawa, K. Inami, T. Teshima and T. Shiba, *Tetrahedron Lett.*, 1992, **33**, 5791.

[61] Y. S. Babu, P. Chand, S. Bantia, P. Kotian, Al Dehghani, Y. El-Kattan, T.-H. Lin, T. L. Hutchison, A. J. Elliott, C. D. Parker, S. L. Ananth, L. L. Horn, G. W. Laver and J. A. Montgomery, *J. Med. Chem.* **4**3 (2000) 3482.

[62] G. T. Wang, S. Wang, R. Gentles, T. Sowin, C. J. Maring, D. J. Kempf, W. M. Kati, V. Stoll, K. D. Stewart and G. Laver, *Bioorg. Med. Chem. Lett.*, 2005, **15**, 125.

[63] P. Chand, P. L. Kotian, A. Dehghani, Y. El-Kattan, T.-H. Lin, T. L. Hutchison, Y. S. Babu, S. Bantia, A. J. Elliott and J. A. Montgomery, *J. Med. Chem.*, 2001, **44**, 4379.

[64] P. Chand, Y. S. Babu, S. Bantia, S. Rowland, A. Dehghani, P. L. Kotian, T. L. Hutchison, S. Ali, W. Brouillette, Y. El-Kattan and T.-H. Lin, *J. Med. Chem.*, 2004, **47**, 1919.

[65] V. Stoll, K. D. Stewart, C. J. Maring, S. Muchmore, V. Giranda, Y.-g. Y. Gu, G. Wang, Y. Chen, M. Sun, C. Zhao, A. L. Kennedy, D. L. Madigan, Y. Xu, A. Saldivar, W. Kati, G. Lever, T. Sowin, H. L. Sham, J. Greer and D. Kempf, *Biochemistry*, 2003, **42**, 718.

[66] G. T. Wang, Y. Chen, S. Wang, R. Gentles, T. Sowin, W. Kati, S. Muchmore, V. Giranda, K. Stewart, H. Sham, D. Kempf and W. G. Laver, *J. Med. Chem.*, 2001, **44**, 1192.

[67] C. J. Maring, V. S. Stoll, C. Zhao, M. Sun, A. C. Krueger, K. D. Stewart, D. L. Madigan, W. M. Kati, Y. Xu, R. J. Carrick, D. A. Montgomery, A. Kempf-Grote, K. C. Marsh, A. Molla, K. R. Steffy, H. L. Sham, W. G. Laver, Y.-g. Gu, D. J. Kempf and W. E. Kohlbrenner, *J. Med. Chem.*, 2005, **48**, 3980.

[68] D. M. Barnes, M. A. McLaughlin, T. Oie, M. W. Rasmussen, K. D. Stewart and S. J. Wittenberger, *Org. Lett.*, 2002, **4**, 1427.

[69] S. Hanessian, M. Bayrakdarian and X. Luo, *J. Am. Chem. Soc.*, 2002, **124**, 4716.

[70] D. A. DeGoey, H.-J. Chen, W. J. Flosi, D. J. Grampovnik, C. M. Yeung, L. L. Klein and D. J. Kempf, *J. Org. Chem.*, 2002, **67**, 5445.

[71] G. T. Wang, *Expert Opin. Ther. Patents*, 2002, **12**, 845.

[72] E. A. Govorkova, I. A. Leneva, O. G. Goloubeva, K. Bush and R. G. Webster, *Antimicrob. Agents Chemother.*, 2001, **45**, 2723.

[73] D. Young, C. Fowler and K. Bush, *Phil. Trans. R. Soc. London. B*, 2001, **356**, 1905.

[74] D. F. Smee, J. H. Huffman, A. C. Morrison, D. L. Barnard and R. W. Sidwell, *Antimicrob. Agents Chemother.*, 2001, **45**, 743.

[75] R. W. Sidwell, D. F. Smee, J. H. Huffman, D. L. Barnard, K. W. Bailey, J. D. Morrey and Y. S. Babu, *Antimicrob. Agents Chemother.*, 2001, **45**, 749.

[76] G. R. Iyer, S. Liao and J. Massarella, *AAPS PharmSci 2002*, 2002, **4**, article 22.

[77] P. Chand, S. Bantia, P. L. Kotian, Y. El-Kattan, T.-H. Lin and Y. S. Babu, *Bioorg. Med. Chem.*, 2005, **13**, 4071.

[78] http://www.biocryst.com, Press release Nov. 28, 2005.

[79] M. D. de Jong, T. T. Thanh and T. H. Khanh, *N. Engl. J. Med.*, 2005, **353**, 2667.

Recent Developments in Antifungal Drug Discovery

Roberto Di Santo

Dipartimento di Studi Farmaceutici, Istituto Pasteur – Fondazione Cenci Bolognetti, Università di Roma "La Sapienza", P.le Aldo Moro 5, 00185 Rome, Italy

Contents

1. Introduction	299
2. Advances in antifungals with known mechanism of action	300
2.1. Azoles	300
2.2. Polyenes	305
2.3. Allylamines	306
2.4. Echinocandins	307
2.5. Pyrimidines	309
3. Emerging antigungal agents	310
3.1. Amino acids	310
3.2. Sordarins	310
3.3. Miscellaneous	311
4. Trends in antifungal drug discovery	312
4.1. Histone deacetylase inhibitors	312
4.2. Genome approach	312
4.3. Biofilm	312
References	313

1. INTRODUCTION

Fungi have emerged worldwide as an increasingly frequent cause of health-care associated infections [1,2]. Until the 1970s, fungal infections had generally been considered curable and thus the demand for new antifungal agents had been very low. To this day, superficial fungal infections of the skin and nails are common and for the most part treated successfully with existing antifungal agents. At the same time, serious invasive fungal infections are becoming a growing danger for human health [3,4]. A recent epidemiological survey in the US, reported that the incidence of fungal sepsis has increased threefold between 1979 and 2000 [5]. The most-common pathogenic agents that cause invasive fungal infections are *Candida* spp., *Aspergillus* (accounting for 70–90%, 10-20% of all invasive mycoses, respectively), *Criptococcus noeformans*, *Pneumocystis carinii* and *Histoplasma capsulatum* [3,4]. An alarming rise in life-threatening systemic fungal infections is reported in AIDS patients, who suffer from suppression of the immune system. Other conditions that have caused an increase in invasive fungal infections include the use of immunomodulatory agents for the prevention of rejection in bone marrow and organ transplants, the use of antineoplastic agents, the long-term use of corticoids and even the indiscriminate use of antibiotics [6–8]. Invasive medical procedures such as the use of vascular catheters, peritoneal dialysis, hemodialysis and parenteral nutrition have also contributed to the rise of fungal infections [9]. Further, there is an increasing emergence of fungal resistance to currently available antimycotic agents. Fungal resistance is one among many factors that justify the rapid increase of interest in research on new and more effective antifungal agents. As a consequence, the number of scientific publications reported in PubMed in the field of antimycotics in the last 10 years was double that of the previous decade (1986–1995).

2. ADVANCES IN ANTIFUNGALS WITH KNOWN MECHANISM OF ACTION

2.1. Azoles

Azole antifungal agents include imidazole and triazole derivatives that prevent the synthesis of ergosterol, a major component of fungal membranes, by inhibiting the cytochrome P-450-dependent enzyme 14α-lanosterol demethylase (CYP51) [10,11]. This enzyme contains an iron protoporphyrin unit located in its active site, which catalyzes the oxidative removal of the 14α-methyl group of lanosterol, by typical monooxygenase activity [12]. Azoles bind to the iron of the porphyrin [13] and cause the blockage of the fungal ergosterol biosynthesis pathway by blocking access of lanosterol to the active site of the enzyme. The depletion of ergosterol and the concomitant accumulation of 14α-methylated sterols alter the fluidity of the fungal membrane causing a reduction in the activity of membrane-associated enzymes and increased permeability. The net effect of these changes is to inhibit fungal growth and replication [14].

Imidazole derivatives, such as bifonazole, clotrimazole, econazole and miconazole were the first group of azole antifungal agents used in clinical practice in the 1970s. Recently, 1-[(aryl)(4-aryl-1*H*-pyrrol-3-yl)methyl]-1*H*-imidazole derivatives (such as compound **1**) related to bifonazole were reported, which showed potent antifungal activities both *in vitro* and *in vivo* experiments against *Candida albicans* and other *Candida* species. Interestingly, some derivatives were proven active *in vitro* against fluconazole resistant strains, with MIC_{50} ranging from 0.016 to 0.25 μg/ml [15–18]. Ketoconazole (**2**), a phenethylimidazole characterized by a dioxolane ring, was the first agent endowed with a wide spectrum of activity against both a variety of yeasts and dimorphic fungi. Further, **2** showed good bioavailability after oral administration and was used against serious invasive fungal infections. The clinical use of ketoconazole has been related to some adverse effects in healthy adults, especially local reactions, such as severe irritation, pruritus and stinging. Dispersions of ketoconazole in aqueous lipid nanoparticles were assessed as useful for targeting this drug into topical routes, minimizing the adverse side effects and providing a controlled release [19]. Ketoconazole has been given a teratogenic classification of C by the US Food and Drug Administration (FDA), thus leading research to be directed toward 1,2,4-triazole derivatives.

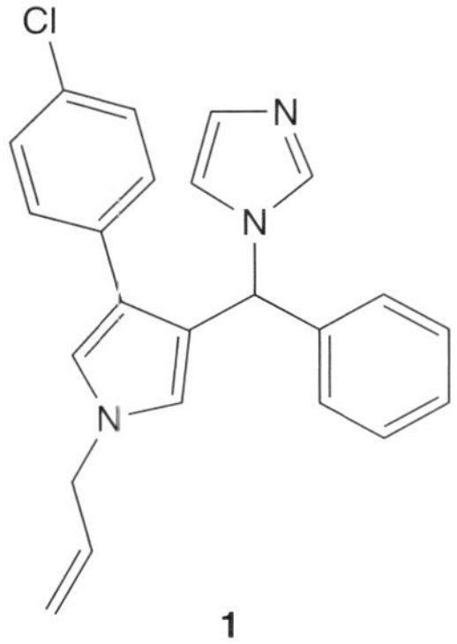

1

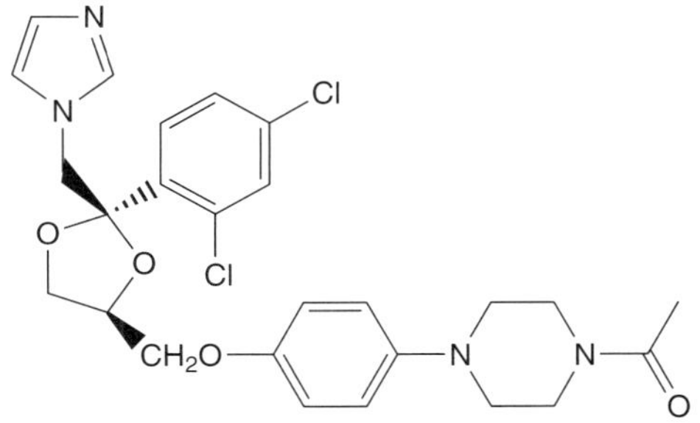

2

Itraconazole (**3**) is a triazole derivative related to **2**, which showed a broader spectrum of activity if compared to that of the parent compound. A recent study reported three itraconazole metabolites, hydroxy-itraconazole, keto-itraconazole and *N*-desalkyl-itraconazole that were competitive inhibitors of CYP3A4 [20]. A great enhancement of bioavailability of **3** after initial dosing, not influenced by fed/fasted state, was obtained with a self-emulsifying drug-delivery system composed of Transcutol®, Pluronic®, L64 and tocopherol acetate [21].

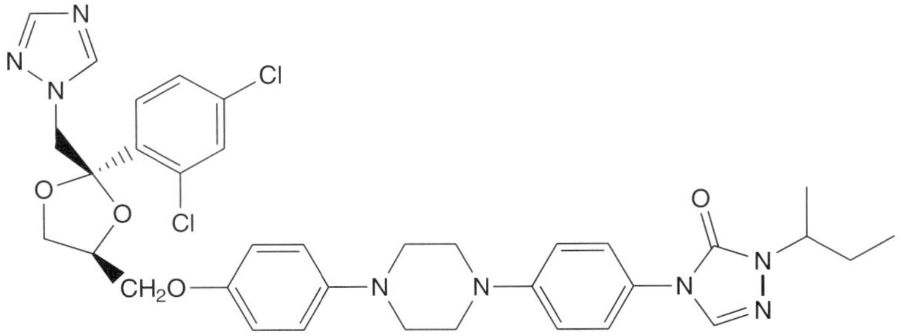

3

Posaconazole (**4**) is a second-generation triazole antifungal agent that is structurally related to itraconazole. The antifungal spectrum of posaconazole is broad and includes causative agents of invasive fungal infections such as *Candida* species, *Aspergillus* species, non-*Aspergillus* hyalohyphomycetes, phaeohyphomycetes, zygomycetes and endemic fungi. Posaconazole is presently involved in phase III trials. It is effective *in vitro* and *in vivo* against several fungi that are resistant to standard antifungals. It is the most active triazole against filamentous fungi, inhibiting 95% of isolates at concentrations of 1 µg/ml or below [22–25]. In a double-blinded, multicenter clinical trial for prophylaxis of invasive fungal infections, posaconazole was compared with fluconazole in 600 patients who had undergone hematopoietic stem cell transplant (HSCT) with graft-versus-host disease. Posaconazole (200 mg every 8 h) was significantly superior to fluconazole (400 mg once a day) in preventing aspergillosis [26]. Posaconazole demonstrated *in vitro* and *in vivo* activity against zygomycetes in two open-label, non-randomized, multicentered compassionate trials that evaluated oral posaconazole as salvage therapy for invasive fungal infections. It was given as an oral suspension of 200 mg four times a day or 400 mg twice a day. Overall, 19 of 24 subjects (79%) survived infection [27]. Posaconazole 5, 25 or 100 mg/kg as an oral suspension was effective in prolonging survival of mice in a murine model of central nervous system (CNS) aspergillosis [28]. A clinical trial was conducted to evaluate the safety and efficacy of posaconazole in subjects with invasive fungal infections who had refractory disease or who were intolerant of standard antifungal therapy. Subjects received posaconazole oral suspension 800 mg/day in divided doses for up to 1 year. Successful outcomes were observed in 48% subjects with cryptococcal meningitis, and 50% of subjects with CNS infections due to other fungal pathogens. Posaconazole was well tolerated, thus suggesting that as an oral medication it could provide a valuable alternative to parenteral therapy in patients failing existing antifungal agents [29].

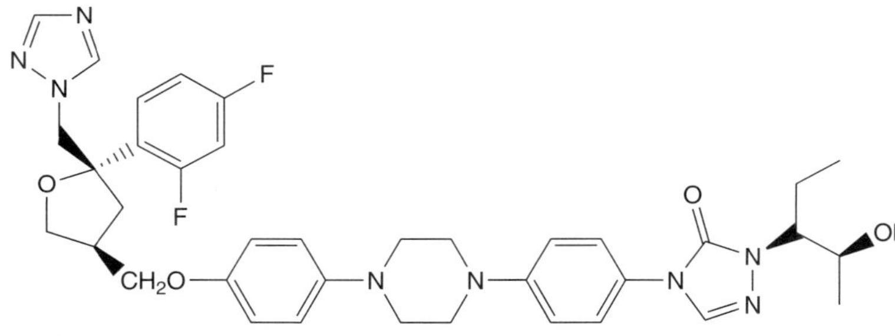

4

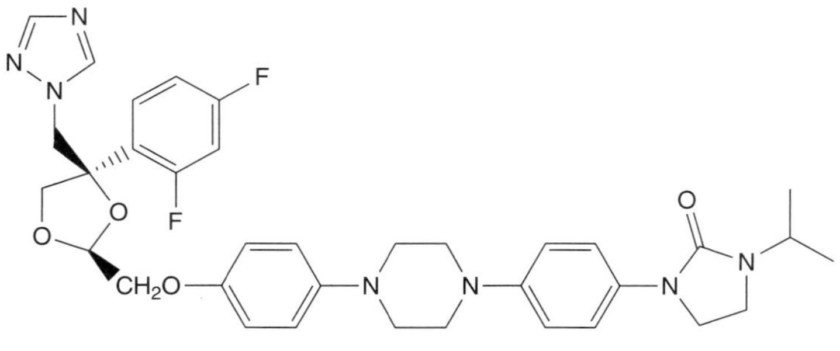

5

R126638 (**5**) was developed as a novel azole agent related to posaconazole, which showed very high levels of *in vitro* antifungal activity. Its potency was confirmed in *in vivo* models of cutaneous infections caused by *Microsporum canis*, and *Thricophyton mentagrophytes* in guinea pigs and mice. The ED_{50} of R126638 was calculated to be <0.63 mg/kg, superior to that of itraconazole used as a reference drug, which showed $ED_{50} = 3.02$ mg/kg [30,31].

Fluconazole (Diflucan®) (**6**) is a triazole derivative that shows both oral and parenteral fungistatic activity. Extensive clinical studies have demonstrated remarkable efficacy, favorable pharmacokinetics and a reassuring safety profile for **6**, all of which have contributed to its widespread use [32]. The efficacy, tolerability and safety of oral fluconazole given at 300 mg once weekly for 2 weeks was demonstrated for the treatment of *Tinea versicolor* [33]. Pulse fluconazole therapy, 300 mg once weekly, is effective against foot nail infections [34]. The upregulation of the ATP binding cassette (ABC) transporter-encoding gene AFR1 in *C. neoformans* was shown *in vivo* in a mouse model of systemic cryptococcosis treated with **6**. The reported findings indicate that the upregulation of the AFR1 gene is an important factor in determining the *in vivo* resistance to fluconazole [35]. Fosfluconazole, the phosphate prodrug of fluconazole, was developed as highly soluble form compared with the parent drug. It could be useful against fungal peritonitis in continuous ambulatory peritoneal dialysis patients given its high water solubility. After intravenous bolus injection of fosfluconazole 50–2000 mg, a rapid convertion to fluconazole was observed. It had a volume of distribution at the higher doses similar to the extracellular volume in man (0.2 l/kg) and was eliminated with a terminal half-life of 1.5–2.5 h [36,37].

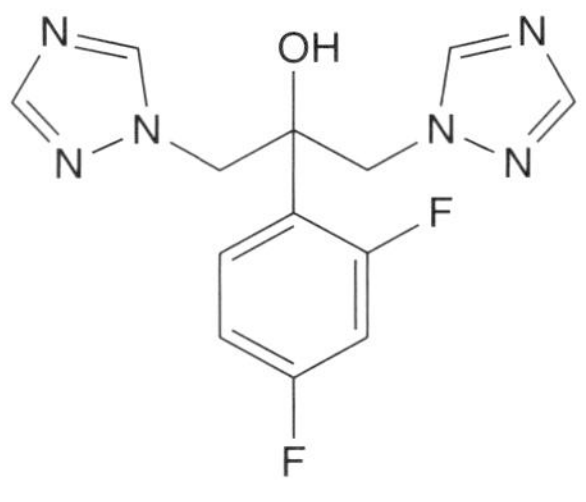

6

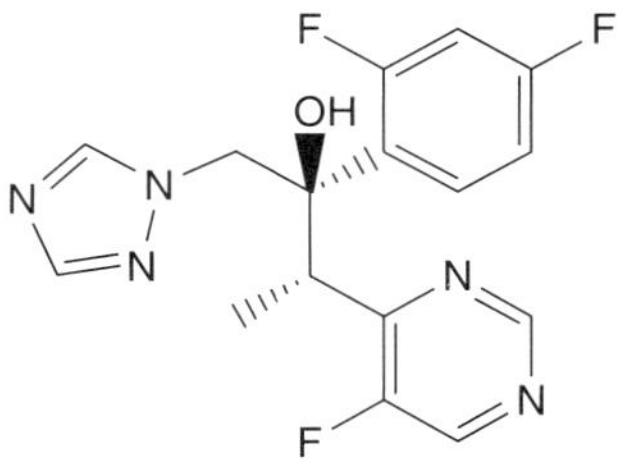

7

The major drawbacks of fluconazole were its poor bioavailability and the delayed development of an intravenous preparation. A second-generation of triazole antifungal agents was recently developed to address these limitations, including voriconazole (**7**), ravuconazole (**8**) and posaconazole (**4**). These new agents appeared to have expanded antifungal activity compared to prior azoles [38–40].

Voriconazole (**7**) is a synthetic derivative of fluconazole that was approved by FDA in 2002. Replacement of one of the triazole rings of **6** with a fluorinated pyrimidine and the addition of an α-methyl group resulted in expanded activity, compared with that of fluconazole. Voriconazole is broadly active against many species of *Aspergillus*, including *A. terreus*, and against all *Candida* species, including *C. krusei*, *C. glabrata* and strains of *C. albicans* resistant to fluconazole [41–43]. It is available in both intravenous (vials containing 20 mg), and oral formulation (50 and 200 mg film-coated tablets) [41]. Voriconazole at 0.125–0.5 mg/l, but not itraconazole and other azoles, inhibited conidiation in *A. fumigatus*, *A. flavus*, *A. niger* and *A. nidulans*. The authors suggest a mechanism unrelated to the inhibition of 14α-methyl lanosterol demethylase, perhaps directed against the formation of the conidia [44]. A comparison of the fungicidal activity against *A. fumigatus* hyphae, emphasized the higher potency of voriconazole versus amphotericin B. Approximately 99% killing of hyphae grown in peptone yeast extract glucose broth was obtained for voriconazole at 1 mg/l after 48 h of exposure, whereas amphotericin B at the same concentration yielded 82% killing after at the same time [45]. Topical application of voriconazole as therapy for *Paecilomyces lilacinus*, responsible for uncommon devasting fungal keratitis, was effective in infected eyes of rabbits. After 8 days of therapy hyphal masses were present in the control-infected eyes and absent in treated-infected eyes [46].

Ravuconazole (**8**) is a triazole antifungal related to voriconazole. Its bioavailability ranges from 48–74% in animals and it is highly protein bound. Metabolism is primarily hepatic, and there does not appear to be significant inhibition of CYP3A4 [38]. It is still under clinical development for human use. A phase I/II randomized, double-blind, placebo-controlled, dose-ranging study evaluated the efficacy, safety and pharmacokinetics of ravuconazole in the treatment of onychomycosis. A 200 mg/day for 12 weeks was the most effective of the regimens investigated [47]. The *in vitro* activity of ravuconazole was tested against a collection of 1796 clinical yeast isolates, including fluconazole-susceptible and -resistant strains. Ravuconazole exhibited potent activity with geometric mean MICs of 0.05 µg/ml. It was active against the majority of fluconazole-resistant isolates; but for 102 of 562 (18%) of the resistant isolates, mainly *C. tropicalis*, *C. glabrata*, and *Criptococcus neoformans*, the MICs were >1 µg/ml [48]. As ravuconazole demonstrates poor aqueous solubility, a water-soluble prodrug was developed for intravenous formulation. BMS-379224 is a phosphonooxymethyl ester derivative of ravuconazole that is rapidly converted into parent drug when infused in animals and healthy humans [49,50]. The pharmacokinetics of ravuconazole was investigated following administration of its produg BMS-379224. It fitted best to a three-compartment pharmacokinetic model. The compound revealed non-linear pharmacokinetics at higher dosages, indicating saturable clearance and/or protein binding. Ravuconazole displayed a long elimination half-life and achieved substantial plasma and tissue concentrations including in the brain.

Albaconazole (**9**) is a novel triazole derivative related to fluconazole. Its activity was determined *in vitro* against 12 isolates of *C. neoformans* and compared with that of fluconazole. Albaconazole was about 100-fold more potent than the reference drug, and showed MICs that ranged from ⩽0.0012–1.25 µg/ml, with MICs for most isolates being between 0.039 and 0.156 µg/ml. An *in vivo* test performed in infected rabbits showed that albaconazole at dosages ranging from 5 to 80 mg/kg of body weight a day was as effective as fluconazole [51].

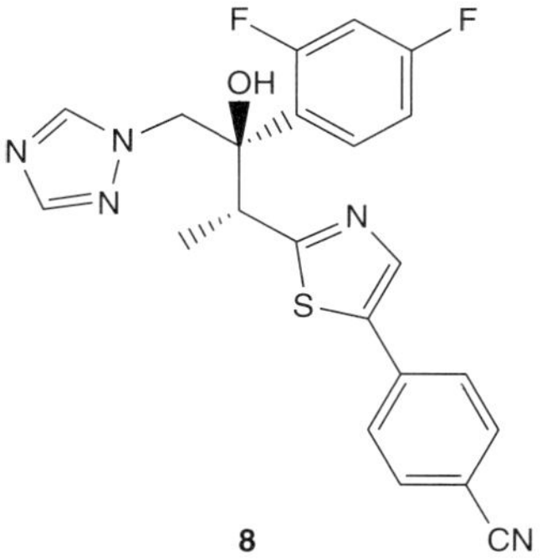

2.2. Polyenes

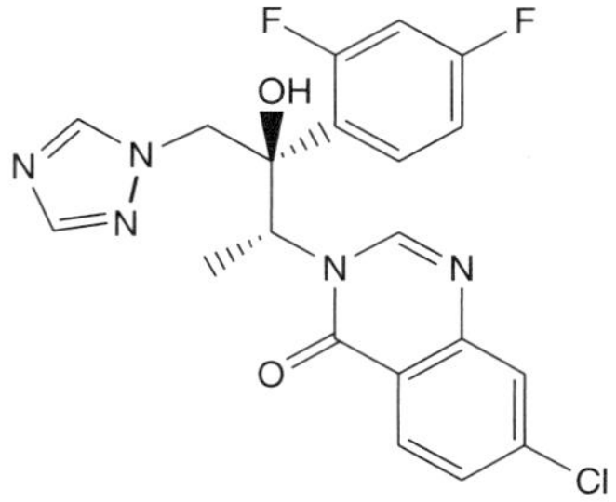

Polyene antibiotics and their semi-synthetic derivatives are macrocycle compounds characterized by the presence of a series of conjugated double bonds that are essential for antifungal activity. The chemical characteristics of these derivatives determine their mechanism of action. In fact, they bind primarily to ergosterol, resulting in alterations in membrane permeability, leakage of cellular components and cell death (fungicidal action). Amphotericin B (**10**) has been used since 1950, and is considered to be the gold standard of therapy for numerous serious, invasive fungal infections, due to its broad spectrum of activity, fungicidal properties and existing efficacy data [52]. Amphotericin B is used in clinical practice as deoxycholate (Fungizone®), lipid complex (Abelcet®), colloidal dispersion (Amphotec®) and (since 1997) as liposomal formulation [53]. Amphotericin B lipid complex was successful in the treatment of cryptococcosis. Its efficacy and renal safety were proven in 106 patients with cryptococcal infection (66% of clinical response) [54]. Amphotericin B lipid complex also gave positive response in 65% of 398 treated patients with invasive aspergillosis [55]. SPK-843 (**11**) is a water-soluble polyene that entered the phase II clinical trials for systemic mycosis in November 2004. It demonstrated activity against fungi in pre-clinical studies both *in vitro* and *in vivo*. **11** possesses broad-spectrum antifungal activity against yeast, dimorphic and filamentous fungi, and its eradicative efficacies in candidiasis were superior compared with amphotericin B [56]. SPK-843 was well tolerated in intravenous infusion in rabbits. The result of the test in this animal model indicate that doses up to 0.5 mg/kg can be administered for long periods of time [57].

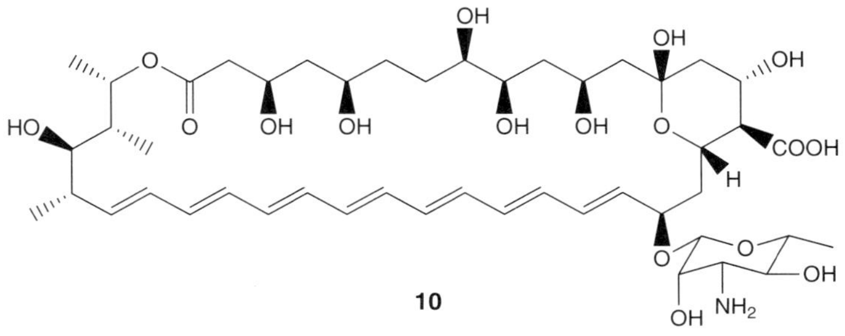

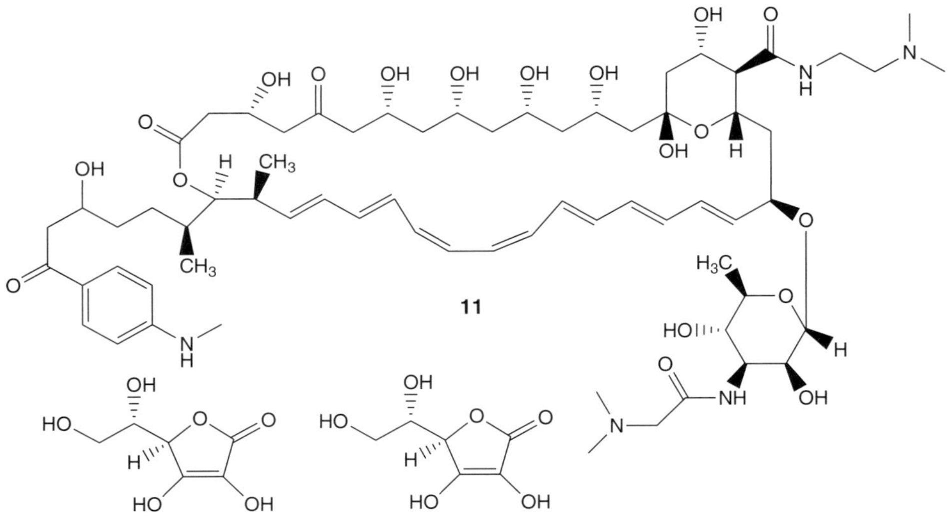

2.3. Allylamines

Allylamines are antifungal agents targeted to squalene epoxidase, an enzyme necessary for ergosterol biosynthesis. Naftifine (**12**) was the first allylamine agent introduced in therapy in the early 1980s as 1% cream or gel for topical use. It has fungicidal activity against dermatophytes and fungistatic activity against *Candida* species. Its sensitizing capacity seems to be greater than in the commonly used azoles [58]. Terbinafine (**13**) was approved in 1990s in the UK and USA for the treatment of onychomycosis. It is the most frequently prescribed oral antifungal agent in North America, for onychomycosis. Eighteen randomized controlled trials have shown terbinafine to be highly effective with mycological cure of 76%. **13** has an established safety profile and very low occurrence of drug interactions [59]. An improved antifungal composition for topical application to the skin and nails has been developed for allylamines (naftifine or terbinafine) [60]. A formulation to provide a product having a high therapeutic effect suitable for symptoms of weeping superficial mycosis was reported. It consists of an allylamine derivative (naftifine or terbinafine) blended with a fatty acid ester, a powdery component, an alcoholic solvent and an anti-itch component [61].

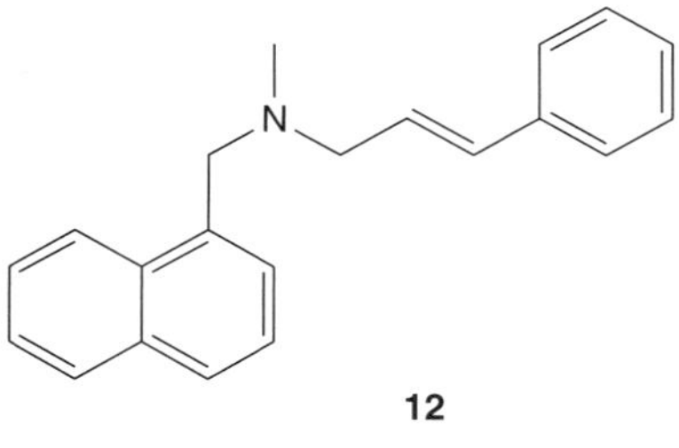

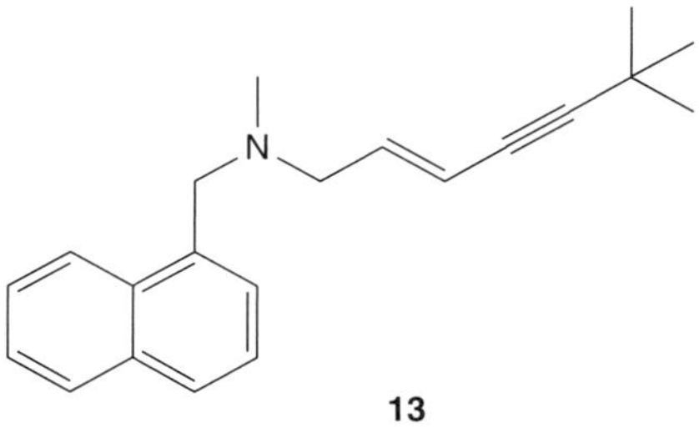

2.4. Echinocandins

Echinocandins are fungal secondary metabolites comprising a cyclic hexapeptide core with a lipid side chain responsible for antifungal activity against β-(1,3)-D-glucan synthesis. These compounds specifically act by inhibiting in a non-competitive manner the synthesis of the glucose omopolymer β-(1,3)-D-glucan. This inhibition leads to the depletion of cell wall glucan, osmotic instability and lysis of fungal cells. Echinocandins are fungicidal drugs that show broad spectrum of antifungal activity, and are an important addition to the antifungal armamentarium in the treatment of fungal infections in both immunocompromized patients and those with normal immunity [62]. Caspofungin (**14**) was the first-approved member of this class of agents with the proprietary name of Cancidas®. It is a semisynthetic derivative of pneumocandin B_0, isolated from the fungus *Glarea lozoyensis*. It is a well-tolerated agent that has proven highly effective in patients with *Candida* esophagitis as well as invasive aspergillosis [63,64]. Micafungin (**15**) is a new parenteral echinocandin that has proven efficacy in open non-comparative studies of esophageal candidiasis in HIV-infected patients and in comparative trials as antifungal prophylaxis in patients undergoing hematopoietic stem cell transplantation [65]. In HIV-positive patients, intravenous micafungin 50–150 mg/day dose dependently eradicated esophageal candidiasis [66]. Seventy patients with deep-seated mycosis were treated with micafungin doses of 12.5–150 mg/day intravenously for up to 56 days. The overall clinical response rates ranged from 55% to 100% depending on the pathogen species [67].

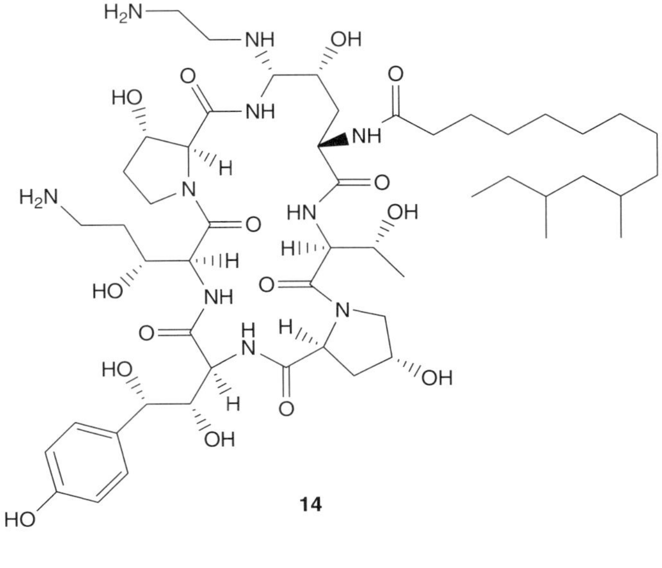

14

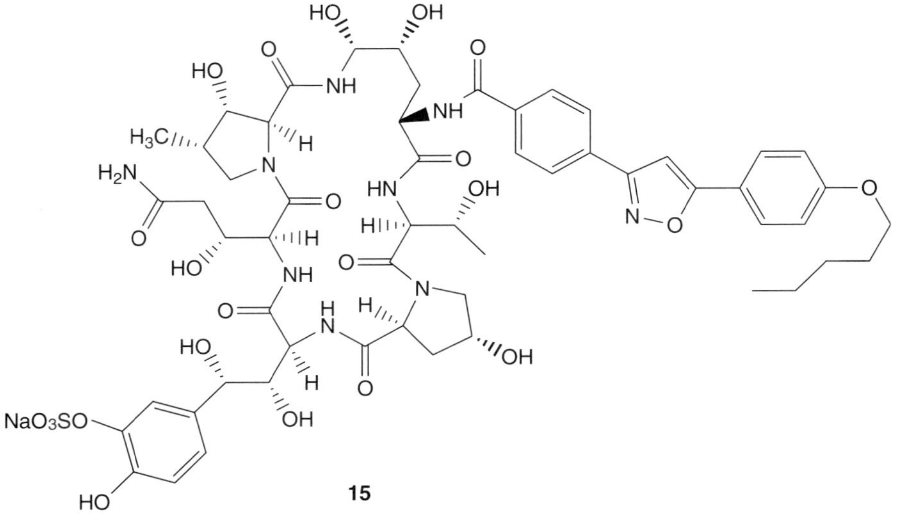

15

The third echinocandin introduced in the market was anidulafungin (**16**), with proved efficacy against *Candida* spp. and *Aspergillus* infections, and activity comparable to fluconazole [68]. A total of 123 elegible patients with invasive candidiasis were randomized to an intravenous regimen once daily. Success rates at the end of therapy were 89% at dose of 100 mg/day [69]. Aminocandin (IP960, **17**) is a new echinocandin with broad spectrum *in vitro* activity against *Aspergillus* and *Candida* spp. It was as effective as amphotericin B at doses $\geqslant 1.0$ mg/kg/day with >70% survival in an immunocompromized murine model of disseminated candidiasis [70].

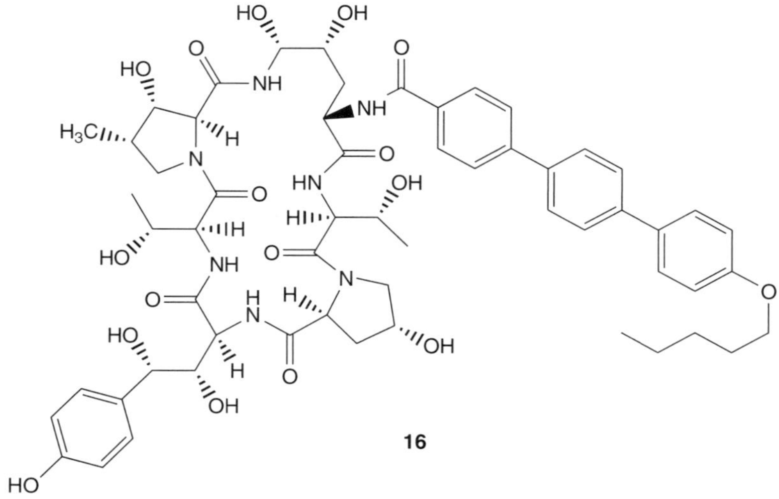

16

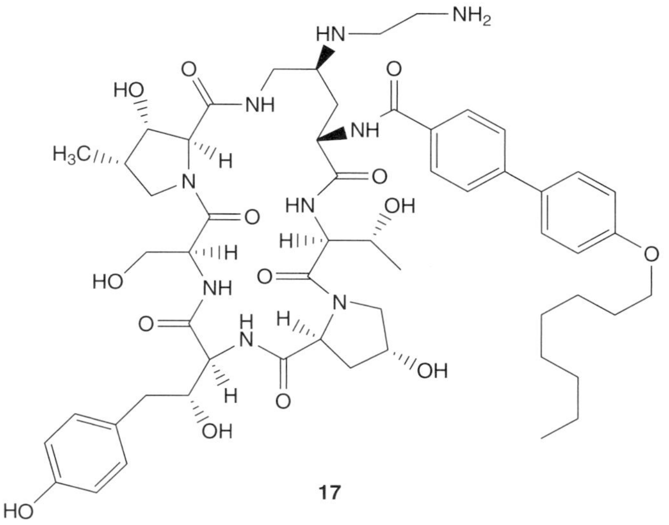

2.5. Pyrimidines

5-Fucitosine (5-FC, **18**) is a synthetic antifungal agent synthesized in 1957. *In vivo* it is converted into 5-fluorouracil (5-FU) within susceptible fungal cells, which is transformed into 5-fluorouridine triphosphate (FUTP). FUTP is incorporated into fungal RNA in place of uridylic acid. This alters the aminoacylation of tRNA, disturbs the amino acid pool and inhibits protein biosynthesis. A second mechanism of action is involved in the antifungal activity of 5-FC. 5-FU is metabolized into 5-fluorodeoxyuridine monophosphate, which is a potent inhibitor of thymidylate synthetase, a key enzyme in the biosynthesis of DNA. 5-FC monotherapy is used in cases of chromoblastomycosis and in uncomplicated lower tract candidosis and vaginal candidosis. In all other cases for treatment of systemic mycoses, 5-FC is usually used in combination therapy with amphotericin B. Efficacy of 5-FC monotherapy was recently reported in a neutropenic murine model of invasive aspergillosis. Survival rates of the treated mice with doses from 50 to 800 mg/kg of body weight/day administered at 6-, 12-, and 24-h intervals ranged from 40% to 90% [71].

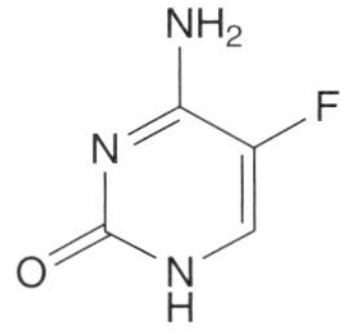

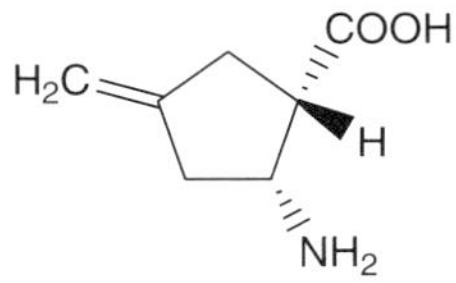

3. EMERGING ANTIGUNGAL AGENTS

3.1. Amino acids

β-Amino acids, 2-aminocyclohexenecarboxylic acid and cis-2-amino-cyclohexanecarboxylic acid (cispentacin) were recently identified as novel antifungal agents targeting protein biosynthesis through inhibition of isoleucyl tRNA synthetase. SAR studies of these lead compounds led to the discovery of (1S,2R)-2-amino-4-methylene-cyclopentanecarboxylic acid, (PLD-118, icofungipen, **19**). Icofungipen showed high antifungal activity against *C. albicans* in *in vitro* studies ($IC_{50} = 0.13\,\mu g/ml$). A successful outcome from a phase IIb study (started in November 2003) would indicate potential for salvage therapy in the oral treatment of azole-resistant oropharyngeal and esophageal candidiasis. Icofungipen was administered for 10 days starting 24 h after intravenous inoculation of *C. albicans* blastoconidia in rabbits. Treatment with icofungipen at 25 mg/kg/day in two divided dosages gave significant tissue clearance of *C. albicans* [72].

3.2. Sordarins

Sordarins are a novel class of antifungal agents that act by selectively inhibiting the protein synthesis elongation step of fungi. They interact with translation elongation factor 2 of fungi and large ribosomal subunit stalk rpP0. This multiple mechanism may explain the high degree of selectivity of this class of compounds between fungal and mammalian cells [73]. Sordarins exhibit potent antifungal activities with relatively broad-spectrum activities *in vitro* and some compounds exhibit good efficacy *in vivo*. Chemical modification of the natural sordarin zofimarin, which was isolated from the fungus *Zopfiella marina*, led to R-135853 (**20**). This compound exhibited potent activities against *C. albicans* including fluconazole-resistant strains, *C. glabrata*, *C. guilliermondii* and *C. neoformans*. It was highly absorbed by oral administration in mice and exhibited good efficacy in eradicating *C. albicans* from the esophagi of mice when it was administered at 50 mg/kg [74]. A novel sordarin derivative produced by *Morinia pestalozzioides* was discovered, which was named moriniafungin (**21**). This compound has a broad antifungal spectrum including *C. albicans*, *C. glabrata*, and *Saccharomyces cerevisiae*. The *in vivo* efficacy of moriniafungin was studied in a murine model of disseminated candidiasis with enhanced susceptibility to *C. albicans*. Moriniafungin did not produce any reduction in the number of fungal colonies in this model [75].

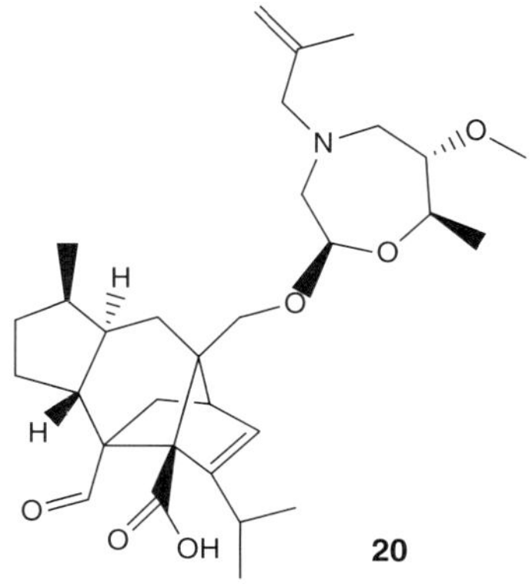

3.3. Miscellaneous

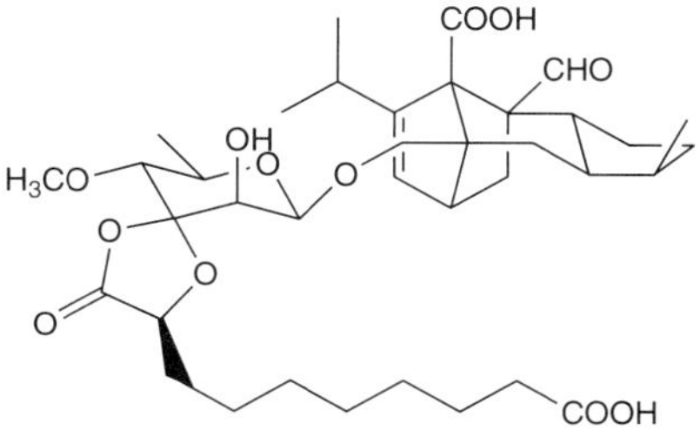

The novel benzothiazole derivative FTR1335 (**22**) was reported as potent antifungal agent against *C. albicans* isolates including fluconazole-resistant strains, and *C. tropicalis*. MICs were from 0.39 to 3.13 µM. Data reported in tests against *C. albicans* ATCC suggested that the antifungal activity of FTR1335 against this strain was fungicidal [76,77]. FTR1335 and related derivatives were proven to inhibit the *C. albicans* N-myristoyltransferase that catalyzes the transfer of the fatty acid myristate from myristoyl-CoA to the amino (*N*)-terminal glycine residue of a number of eukaryotic cellular proteins. Interesting antifungal activity was reported for some pyrrolo [1,2-*a*][1,4] benzodiazepine derivatives. Among them, compound **23** showed high antifungal activity *in vitro* against *A. fumigatus* and *C. parapsilosis*, but not against other *Candida* spp. After treatment with a 300 nM solution of **23** a strong accumulation of squalene was revealed, suggesting that squalene epoxidase is the biochemical target. Topical application of 1 g/day of a 2% **23** w/w carbowax cream on the skin of guinea pigs with cutaneous *M. canis* infection resulted in a cure after 2 weeks of treatment. No positive results were obtained with oral or intravenous administration [78].

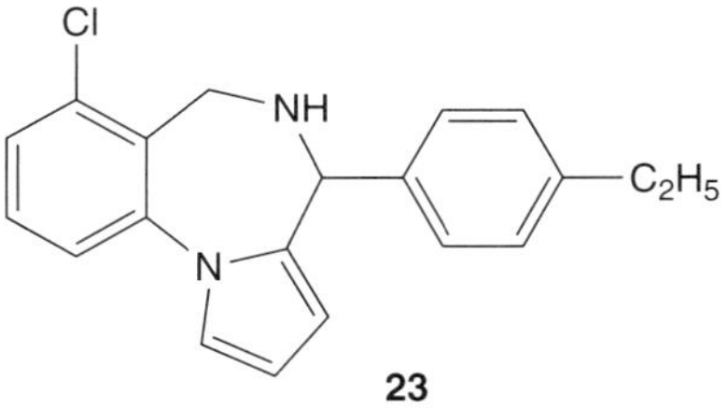

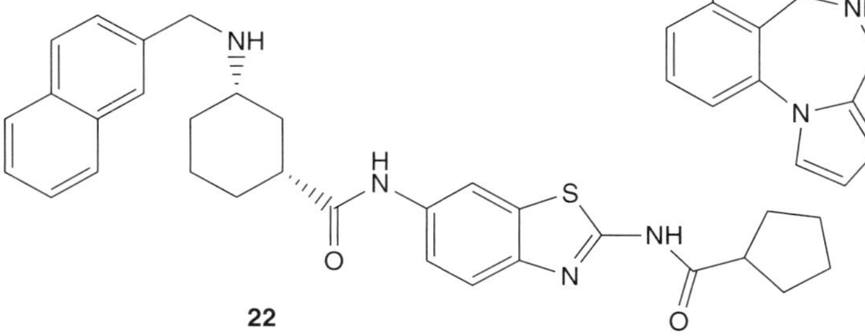

4. TRENDS IN ANTIFUNGAL DRUG DISCOVERY

4.1. Histone deacetylase inhibitors

Nuclear DNA is associated with proteins in complexes with several levels of higher-order structure. First-order packaging is at the level of nucleosome core particle, a structure that is composed of core histone proteins surrounded by DNA. In the light of the degree to which nuclear DNA is packaged, it is now appreciated that the structure of chromatin is necessarily highly dynamic, responding both to internal and external stimuli. Multiple post-translational modification of histones, including acetylation, phosphorylation, methylation ubiquitination, and ADP-ribosylation are now well documented. Of the post-translational modifications of histone proteins, acetylation has been the most extensively studied. Inhibitors of histone deacetylase (HDAC), the enzyme responsible for the reversible acetylation of histones, are proven antiproliferative agents. Recently, a role for HDAC inhibitors as antifungal agents was proposed, and directed or undirected effects of these agents on fungal growth were reported [79,80].

4.2. Genome approach

Genomic-based methodologies are increasingly used in drug development. The most-extensive applications have occurred in drug discovery due to advances in technologies that allow automated synthesis, analysis and for high-throughput screening against known targets. However, genomics can be used to identify previously uncharacterized pharmacologic actions. This tool provides the basis for the discovery and development of new classes of antifungal agents that should interfere with biological targets different to those described above. This could result in a wider chemotherapy agents armamentarium, which could be very useful to defeat fungal resistance [81–84].

4.3. Biofilm

The well-known high resistance of fungi species that frequently contaminate medical devices, led to the discovery of peculiar adherent microbial populations named biofilms. The pathogen produces a polymeric extracellular material, the synthesis of which markedly increases when developing biofilms are exposed to liquid flow. Recently it was demonstrated that in *C. albicans*, physical contact results in activation of the mitogen-activated protein kinase, Mkc1p. This enzyme is part of a signal transduction pathway known to be activated by cell wall stress. The role of Mkc1p for invasive hyphal growth and normal biofilm development was demonstrated [85]. The biofilm is scarcely penetrated by known antifungal agents. However, this poor antifungal penetration is not the major drug-resistance mechanism. Although the presence of matrix material decreases the diffusion of the antifungal agent, poor penetration does not account for the drug resistance of biofilm cells [86].

REFERENCES

[1] S. N. Banerjee, T. G. Emori, D. H. Culver, R. P. Gaynes, W. R. Jarvis, T. Horan, J. R. Edwards, J. Tolson, T. Henderson and W. J. Martone, *Am. J. Med.*, 1991, **91**, 86S.

[2] C. M. Beck-Sagué and W. R. Jarvis, *J. Infect. Dis.*, 1993, **167**, 1247.

[3] M. F. Vicente, A. Basilio, A. Cabello and F. Pelaez, *Clin. Microbiol. Infect.*, 2003, **9**, 15.

[4] A. Chakrabarti, *J. Postgrad. Med.*, 2005, **51**, 16–20.

[5] G. S. Martin, D. M. Mannino, S. Eaton and M. Moss, *N. Engl. J. Med.*, 2003, **348**, 1546.

[6] S. Zacchino, R. Yunes, V. Cechinel Filho, D. Enriz, V. Kouznetsov and C. Ribas, in *Plant-Derived Antimycotics Current Trends and Future Prospects* (eds M. Rai and D. Mares), Haworth Press, New York, NY, 2003, p. 1.

[7] S. Ablordeppey, P. Fan, J. Ablordeppey and L. Mardenborough, *Curr. Med. Chem.*, 1999, **6**, 1151.

[8] D. Li and R. Calderone, in *Pathogenic Fungi, Host Interaction and Emerging Strategies for Control* (eds G. San-Blas and R. Calderone), Caister Academic Press, Norfolk, 2004, p. 335.

[9] T. W. Boo, B. O'reilly, J. O'leary and B. Cryan, *Mycoses*, 2005, **48**, 251.

[10] M. A. Ghannoum, *Dermatol. Ther.*, 1990, **25**, 141–148.

[11] K. Asai, N. Tsuchimori, K. Okonogi, J. R. Perfect, O. Ogotoh and Y. Yoshida, *Antimicrob. Agents Chemother.*, 1999, **43**, 1163.

[12] J. H. Dawson and M. Sono, *Chem. Rev.*, 1987, **87**, 1255.

[13] F. Odds, A. J. P. Brown and N. A. R. Gow, *Trends Microbiol.*, 2003, **11**, 272.

[14] J. L. Adams and B. W. Metcalf, in *Comprehensive Medicinal Chemistry* (eds C. Hansch, P. G. Sammes and J. B. Taylor), Pergamon Press, Oxford, England, 1990, p. 333.

[15] M. Artico, R. Di Santo, R. Costi, S. Massa, A. Retico, M. Artico, G. Apuzzo, G. Simonetti and V. Strippoli, *J. Med. Chem.*, 1995, **38**, 4223.

[16] A. Tafi, J. Anastassopoulou, T. Theophanides, M. Botta, F. Corelli, S. Massa, M. Artico, R. Costi, R. Di Santo and R. Ragno, *J. Med. Chem.*, 1996, **39**, 1227.

[17] A. Tafi, R. Costi, M. Botta, R. Di Santo, F. Corelli, S. Massa, A. Ciacci, F. Manetti and M. Artico, *J. Med. Chem.*, 2002, **45**, 2720.

[18] R. Di Santo, A. Tafi, R. Costi, M. Botta, M. Artico, F. Corelli, M. Forte, F. Caporuscio, L. Angiolella and A. T. Palamara, *J. Med. Chem.*, 2005, **48**, 5140.

[19] E. B. Souto and R. H. Muller, *J. Microencapsul.*, 2005, **22**, 501.

[20] N. Isoherranen, K. L. Kunze, K. E. Allen, W. L. Nelson and K. E. Thummel, *Drug Metab. Dispos.*, 2004, **32**, 1121.

[21] J.-Y. Hong, J.-K. Kim, Y.-K. Song, J.-S. Park and C.-K. -Kim, *J. Control. Release*, 2006, **110**, 332.

[22] H. A. Torres, R. Y. Hachem, R. F. Chemaly, D. P. Kontoyiannis and I. I. Raad, *Lancet Infect. Dis.*, 2005, **5**, 775.

[23] R. Herbrecht, *Int. J. Clin. Pract.*, 2004, **58**, 612.

[24] E. S. Dodds Ashley and B. D. Alexander, *Drugs Today*, 2005, **41**, 393.

[25] G. M. Keating, *Drugs*, 2005, **65**, 1553.

[26] A. J. Ullmann, J. H. Lipton, D. H. Vesole, P. Chandrasekar, A. Langston, S. Taranolo, H. Greinix, C. Hardalo, H. Patino and S. Durrant, *Mycoses*, 2005, **48**, 26a.

[27] R. N. Greenberg, K. Mullane, J. A. van Burik, I. Raad, M. J. Abzung, G. Anstead, R. Herbrecht, A. Langston, K. A. Marr, G. Schiller, M. Schuster, J. R. Wingard, C. E. Gonzalez, S. G. Revankar, G. Concoran, R. J. Kryscio and R. Hare, *Antimicrob. Agents Chemother.*, 2006, **50**, 126.

[28] J. K. Imai, G. Singh, K. V. Clemons and D. A. Stevens, *Antimicrob. Agents Chemother.*, 2004, **48**, 4063.
[29] P. Pitisuttithum, R. Negroni, J. R. Graybill, B. Bustamante, P. Pappas, S. Chapman, R. S. Hare and C. J. Hardalo, *J. Antimicrob. Chemother.*, 2005, **56**, 745.
[30] F. Odds, J. Ausma, F. Van Gerven, F. Woestenborghs, L. Meerpoel, J. Heeres, H. Vanden Bossche and M. Borgers, *Antimicrob. Agents Chemother.*, 2004, **48**, 388.
[31] L. Meerpoel, L. J. J. Backx, L. J. E. Van der Veken, J. Heeres, D. Corens, A. De Groot, F. C. Odds, F. Van Gerven, F. A. A. Woestenborghs, A. Van Breda, M. Oris, P. van Dorsselaer, G. H. M. Willemsens, K. J. P. Vermuyten, P. J. M. G. Marichal, H. F. Vanden Bossche, J. Ausama and M. Borgers, *J. Med. Chem.*, 2005, **48**, 2184.
[32] R. Cha and J. D. Sobel, *Expert Rev. Anti-infect. Ther.*, 2004, **2**, 357.
[33] M. Karakas, M. Durdu and H. R. Memisoglu, *J. Dermatol.*, 2005, **32**, 19.
[34] L. G. Zisova, *Folia Med. (Plovdiv).*, 2004, **46**, 47.
[35] M. Sanguinetti, B. Posteraro, M. La Sorda, R. Torelli, B. Fiori, R. Santangelo, G. Delogu and G. Fadda, *Infect. Immun.*, 2006, **74**, 1352.
[36] T. Aoyama, K. Ogata, M. Shimizu, S. Hatta, K. Masuhara, Y. Shima, K. Rimura and Y. Matsumoto, *Drug Metab. Pharmacokinet.*, 2005, **20**, 485.
[37] S. Sobue, K. Sekiguchi, K. Shimatani and K. Tan, *J. Clin. Pharmacol.*, 2004, **44**, 284.
[38] P. Kale and L. B. Johnson, *Drugs Today*, 2005, **41**, 91.
[39] A. Chen and J. D. Sobel, *Expert Opin. Emerg. Drugs.*, 2005, **10**, 21.
[40] J. R. Wingard and H. Leather, *Biol. Blood Marrow Transplant.*, 2004, **10**, 73.
[41] R. Naithani and R. Kumar, *Indian Pediat*, 2005, **42**, 1207.
[42] J. R. Gaybill, *Lancet*, 2005, **366**, 1413.
[43] R. Herbrecht, *Expert Rev. Anti-infect. Ther.*, 2004, **2**, 485.
[44] N. L. Varanasi, I. Baskaran, G. J. Alangaden, P. H. Chandrasekar and E. K. Manavathu, *Int. J. Antimicrob. Agents*, 2004, **23**, 72.
[45] S. Krishnan, E. K. Manavathu and P. H. Chandrasekar, *J. Antimicrob. Chemother.*, 2005, **55**, 914.
[46] W. Sponsel, N. Chen, D. Dang, G. Paris, J. Graybill, L. K. Najvar, L. Zhou, K. W. Lam, R. Glickman and F. Scribbick, *Antimicrob. Agents Chemother.*, 2006, **50**, 262.
[47] A. K. Gupta, C. Leopardi, R. R. Stoltz, P. F. Pierce and B. Conetta, *J. Eur. Acad. Dermatol. Venereol.*, 2005, **19**, 437.
[48] M. Cuenca-Estrella, A. Gomez-Lopez, E. Mellado, G. Garcia-Effron and J. L. Rodriguez-Tuleda, *Antimicrob. Agents Chemother.*, 2004, **48**, 3107.
[49] Y. Ueda, N. Barbour and J. J. Bronson, 42nd Interscience Conferente on Antimicrobial Agents and Chemotherapy, San Diego September 27–30, 2002, Abstr. F-817.
[50] A. Bello, R. Ruso, D. Grasela and I. Salahudeen, 43rd Interscience Conferente on Antimicrobial Agents and Chemotherapy, Chicago September 14–17, 2003, Abstr. A-1567.
[51] J. L. Miller, W. A. Shell, E. A. Wills, D. L. Toffaletti, M. Boyce, D. K. Benjamin, Jr., J. Bartroli and J. R. Perfect, *Antimicrob. Agents Chemother.*, 2004, **48**, 384.
[52] L. Ostrosky-Zeichner, K. A. Marr, J. H. Rex and S. H. Cohen, *Clin. Infect. Dis.*, 2003, **37**, 415.
[53] W. J. Gibbs, R. H. Drew and J. R. Perfect, *Expert Rev. Anti-infect. Ther.*, 2005, **3**, 167.
[54] L. M. Baddour, J. R. Perfect and L. Ostrosky-Zeichner, *Clin. Infect. Dis.*, 2005, **40**, S409.
[55] P. H. Chandrasekar and J. I. Ito, *Clin. Infect. Dis.*, 2005, **40**, S392.
[56] N. Kasanah and M. T. Hamann, *Curr. Opin. Investig. Drugs*, 2005, **6**, 845.
[57] G. Mozzi, P. Benelli, T. Bruzzese, M. R. Galmozzi and A. Bonabello, *J. Antimicrob. Chemother.*, 2002, **49**, 321.
[58] M. Corazza, M. M. Lauriola and A. Virgili, *Contact Dermat.*, 2005, **53**, 302.

[59] A. K. Gupta, J. E. Ryder, L. E. Lynch and A. Tavakkol, *J. Drugs Dermatol.*, 2005, **4**, 302.
[60] M. E. Nimni, *US Patent 238672*, 2005.
[61] S. Sawamura, *WO Patent 032532 A1*, 2005.
[62] V. A. Morrison, *Expert Rev. Anti-infect. Ther.*, 2006, **4**, 325.
[63] G. Maschmeyer and A. Glasmacher, *Mycoses*, 2005, **48**, 227.
[64] R. Betts, A. Glasmacher, J. Maertens, G. Maschmeyer, J. A. Vazquez, H. Teppler, A. Taylor, R. Lupinacci, C. Sable and N. Kartsonis, *Cancer*, 2006, **106**, 466.
[65] P. L. Carver, *Ann. Pharmacother.*, 2004, **38**, 1707.
[66] B. Jarvis, D. P. Figgitt and L. J. Scott, *Drugs*, 2004, **64**, 969.
[67] S. Kohno, T. Masaoka, H. Yamaguchi, T. Mori, A. Urabe, A. Ito, Y. Niki and H. Ikemoto, *Scand. Infect. Dis.*, 2004, **36**, 372.
[68] R. H. Raasch, *Expert Rev. Anti-infect. Ther.*, 2004, **2**, 499.
[69] D. S. Krause, J. Reinhardt, J. A. Vazquez, A. Reboli, B. P. Goldstein, M. Wible and T. Henkel, *Antimicrob. Agents Chemother.*, 2004, **48**, 2021.
[70] P. A. Warn, A. Sharp, G. Morrissey and D. W. Denning, *J. Antimicrob. Chemother*, 2005, **56**, 590.
[71] D. T. te Dorsthorst, P. E. Verweij, J. F. Meis and J. W. Mouton, *Antimicrob. Agents Chemother.*, 2005, **49**, 4220.
[72] R. Petraitiene, V. Petraitis, A. M. Kelaher, A. A. Sarafandi, D. Mickiene, A. H. Groll, T. Sein, J. Bacher and T. J. Walsh, *Antimicrob. Agents Chemother.*, 2005, **49**, 2084.
[73] D. Gargallo-Viola, *Curr. Opin. Anti-Infect. Invest. Drugs*, 1999, **1**, 294.
[74] Y. Kamai, M. Kakuta, T. Shibayama, T. Fukuoka and S. Kuwahara, *Antimicrob. Agents Chemother.*, 2005, **49**, 52.
[75] A. Basilio, M. Justice, G. Harris, G. Bills, J. Collado, M. de la Cruz, M. T. Diez, P. Hernandez, P. Liberator, J. Nielsen kahn, F. Pelaez, G. Platas, D. Schmatz, M. Shastry, J. R. Tormo, G. R. Andersen and F. Vicente, *Bioorg. Med. Chem.*, 2006, **14**, 560.
[76] K. Yamazaki., Y. Kaneko, K. Suwa, S. Ebara, K. Nakazawa and K. Yasuno, *Bioorg. Med. Chem.*, 2005, **13**, 2509.
[77] S. Ebara, H. Naito, K. Nakazawa, F. Ishii and M. Nakamura, *Biol. Pharm. Bull.*, 2005, **28**, 591.
[78] L. Meerpoel, J. Van Gestel, F. Van Gerven, F. Woestenborghs, P. Marichal, V. Sipido, G. Terence, R. Nash, D. Corens and R. D. Richards, *Bioorg. Med. Chem. Lett.*, 2005, **15**, 3453.
[79] W. Lamar Smith and T. D. Edlind, *Antimicrob. Agents Chemother.*, 2002, **46**, 3532.
[80] J. Binoy, N. Norikazu and Y. Minoru, *Recent Res. Dev. Org. Bioorg. Chem.*, 2004, **6**, 11.
[81] R. E. Isaacson, *Current Pharm. Des.*, 2002, **8**, 1091.
[82] A. B. Parson, R. Geyer, T. R. Hughes and C. Boone, *Prog. Cell Cycle Res.*, 2003, **5**, 159.
[83] J. D. Cleary, L. A. Walker and R. L. Hawke, *Am. J. Pharmacogenom.*, 2005, **5**, 365.
[84] J. B. McAlpine, B. O. Bachmann, M. Piraee, S. Tremblay, A.-M. Alarco, E. Zazopoulos and C. M. Farnet, *J. Nat. Prod.*, 2005, **68**, 493.
[85] C. A. Kumamoto, *Proc. Natl. Acad. Sci. USA*, 2005, **102**, 5576.
[86] M. A. Al-Fattani and L. J. Douglas, *Antimicrob. Agents Chemother.*, 2004, **48**, 3291.

Section 5
Topics in Biology

Editor: Dalia Cohen
Rosetta Genomics Inc.
Technology Center of New Jersey
New Jersey

Genomic Data Mining and Its Impact on Drug Discovery

N.R. Nirmala

Genome and Proteome Sciences Department, Novartis Institutes for Biomedical Research, Cambridge, MA 02139, USA

Contents

1. Introduction	319
2. Gene expression profiling	319
3. Classification of human cancers using microRNA expression profiles	323
4. Pharmacogenetics and drug metabolism	325
5. Conclusion	328
Acknowledgements	329
References	329

1. INTRODUCTION

The sequence of the human genome and those of several other genomes have been available for a few years now. The availability of this information has held the pharmaceutical and scientific community in a state of suspended excitement about the potential that such information holds and the enormous impact that it would make, on our understanding of human biology, human disease and ultimately, drug therapy – from the discovery of new disease targets to the identification and functionalization of novel genes through the advances in genome-wide experimental and computational biology approaches [1]. There are a multitude of important applications of genomic data mining that have made an impact on our understanding of biology. We have chosen to highlight three specific examples here – two drawn from pre-clinical research areas and one example from studies most closely related to the treatment of human diseases. This article reviews one of the latest innovations in the analysis of gene expression profiling data, the use of computational biology and comparative genomics in the area of the discovery of microRNAs and in the prediction of their target genes and finally, applications in pharmacogenetics that have been brought about by our understanding of the human genome in the disease context.

2. GENE EXPRESSION PROFILING

Genom-wide RNA expression analysis has become a standard tool in the arsenal of the pharmaceutical researcher. Until recently, analysis of expression data consisted of the extraction of lists of genes that were statistically significantly up- or

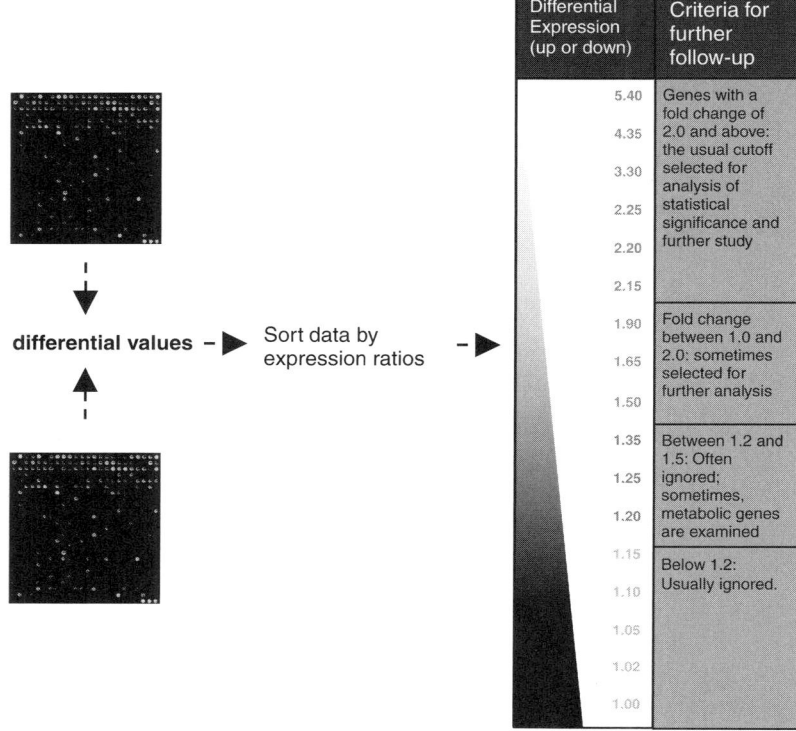

Fig. 1. A schematic of the initial workflow in the analysis of microarray data (For color version, see Color Plate Section).

downregulated in the experiment under study. Fig. 1 shows a typical workflow for the analysis of microarray data. Historically, the analysis used a gene-by-gene approach where expression levels across the two conditions (normal versus disease, time 1 versus time 2 and the like) were compared one gene at a time. Genes that are considered to be 'significantly regulated' are then further pursued experimentally. The criterion used for the selection of genes that are to be followed-up at the bench usually involve a fold change that is twofold or better in either direction (up or down) across the two states in the pairwise comparison. If the number of genes in this list is large, the strategy typically is to use an even higher cutoff, in order to reduce the number of genes to be followed-up to a tractable level. Consequently, the analysis of the same microarray data by different researchers may yield gene lists that show varying extent of overlap.

Subsequent to the generation of a statistically significant gene list [2,3] from the microarray data, the analysis focuses on attempting to place these genes in a biological context that is relevant to the set of experiments under study. Often, a number of the genes in the list may belong to a specific biological pathway and the up- or downregulation of these genes can be interpreted in this context. To the extent that such a pathway mapping is feasible, hypotheses are developed to explain and validate the analyses. It must also be pointed out that sometimes, it is simply not possible to project these genes on to any pathway to any significant extent. Such Approaches to interpret the gene expression profiling data are quite often inevitably influenced by the biological biases of the scientist analyzing the data.

A more agnostic approach toward such a pathway-based analysis was first described by Mootha *et al.* in a landmark paper in 2003 [4] followed by a more extensive implementation of the approach published in 2005 [5]. In this approach, which was dubbed as "gene set enrichment analysis" (GSEA), changes in expression levels for lists of genes in a number of known pathways were examined in the study to see which sets of genes changes coordinately in the same direction (up or down) in a statistically significant manner between muscle biopsies taken from normal and type II diabetes subjects. The general GSEA procedure as described in the paper is depicted schematically in Fig. 2.

Such a strategy enabled the detection of pathways in which the constituent genes changed by 20% or less between the normal and the disease (a set of genes comprising the oxidative phosphorylation pathway, in this case) conditions with statistical significance. Taken individually, a 20% change in expression levels of a gene in the oxidative phosphorylation pathway across conditions is usually ruled as "not interesting" and is often ignored (see Fig. 1). However, the biological significance of more than 90% of the genes in a particular gene set changing by even as little as 20% in a *coordinate* manner can hardly be ignored. Gene sets can be derived based on a number of factors – genes in a pathway, genes by chromosomal location, by the presence of common upstream *cis* motifs and also hand-curated sets of genes based on analyses of prior experiments. Several other reports have subsequently appeared in the literature [6,7] that use the same concept as the cornerstone of their approaches, albeit with different statistical tests to determine the significance of the influence of a gene set or pathway across a pair of conditions.

There are several immediately obvious advantages to the GSEA approach. GSEA considers all the genes in the experiment that are known to belong in a set, rather than a subset of the genes on the chip which are somewhat arbitrarily chosen based on a fold change or statistical significance cutoff. Also, since *all* gene sets are assessed for the level of their enrichment in the data, there is no *a priori* biological bias imposed on the analysis. In addition, GSEA is able to detect the significance of even those gene sets containing more subtle changes in expression levels of individual genes, which would otherwise be ignored in the more traditional gene-by-gene analysis. On the other hand, genes that are not known to be in a pathway will not be captured in such analyses and one has to use other methods such as *K*-means or hierarchical clustering and/or principal components analysis to categorize such genes. However, for genes that are not well characterized enough to be assigned to a known pathway, any algorithm can at best cluster them with other genes that have similar behavior. The cellular or biological context cannot be discerned for these, regardless of the analysis strategy used.

A powerful extension of this approach is to assemble a set of genes based upon the similarity of their expression profiles in one experiment and query the data of another profiling experiment to see if this set bears statistical significance [5] in the second experiment as well. Such an approach was attempted on three independent

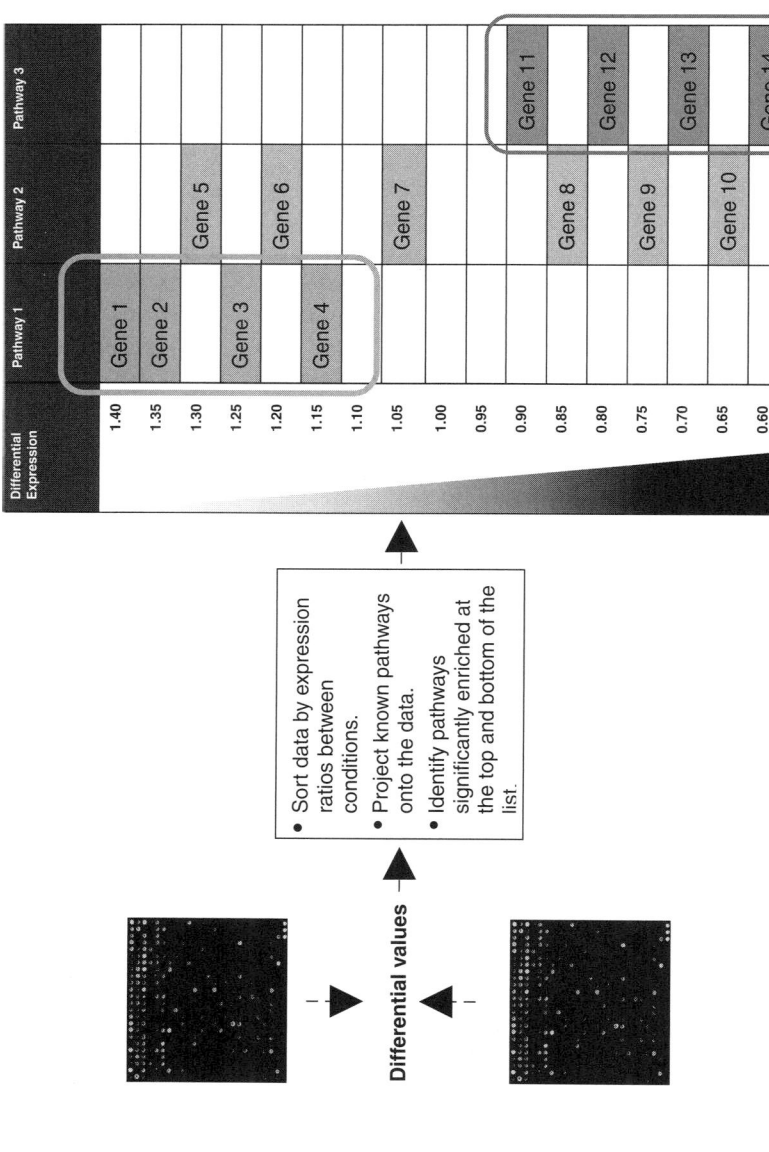

Fig. 2. A schematic overview of the Gene Set Enrichment Analysis approach (For color version, see Color Plate Section).

Color Plate Section

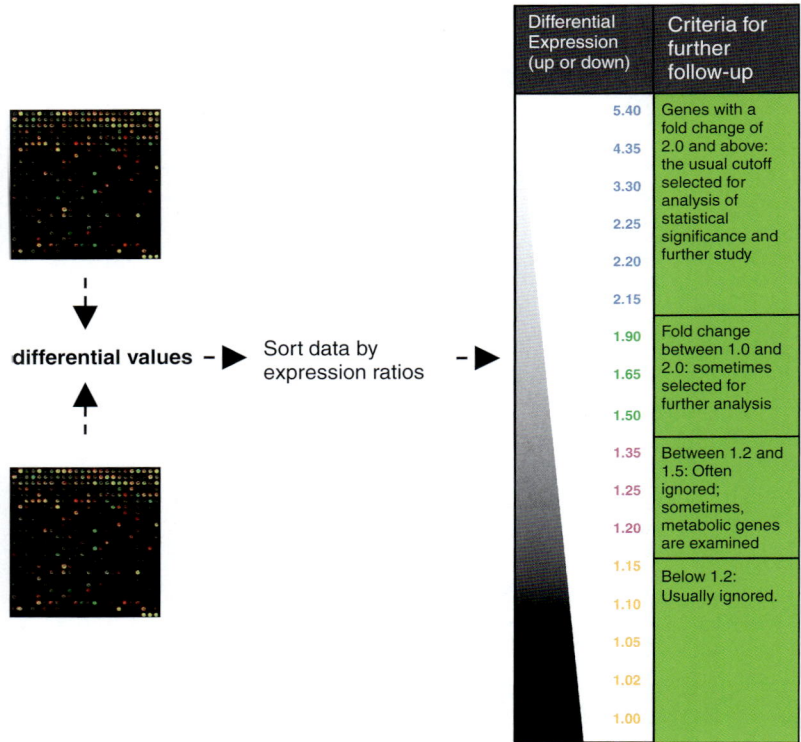

Plate 21.1. A schematic of the initial workflow in the analysis of microarray data.

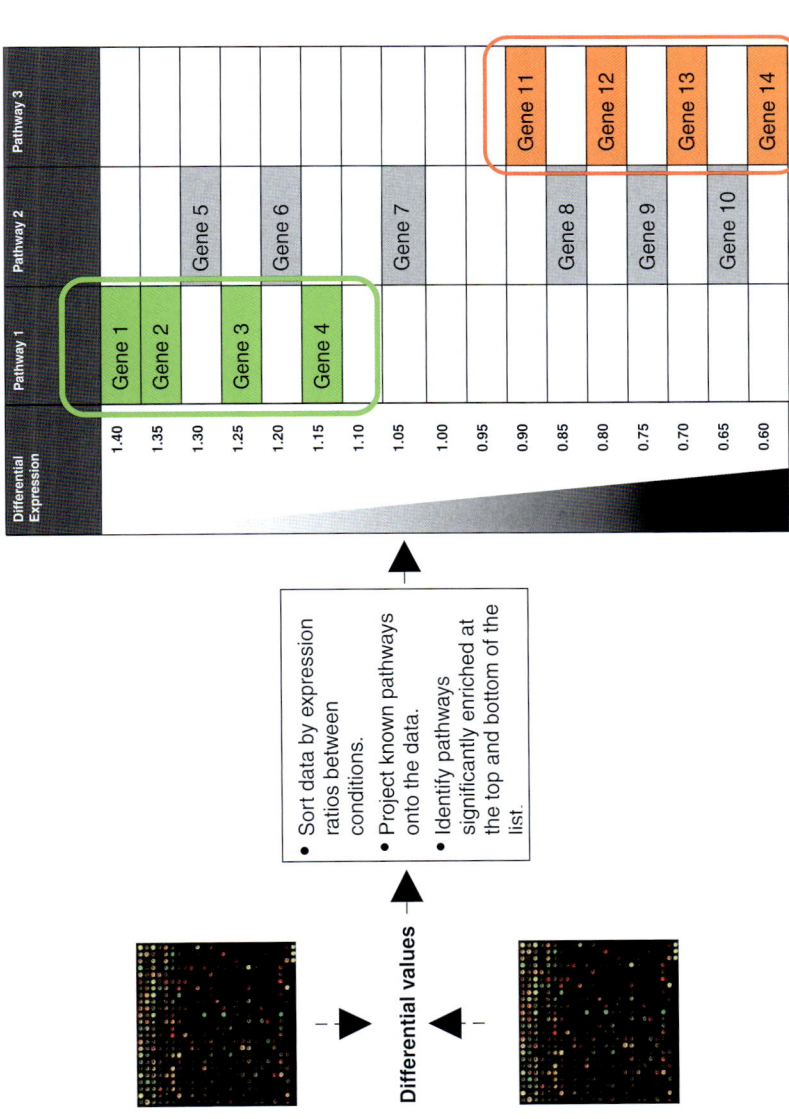

Plate 21.2. A schematic overview of the Gene Set Enrichment Analysis approach.

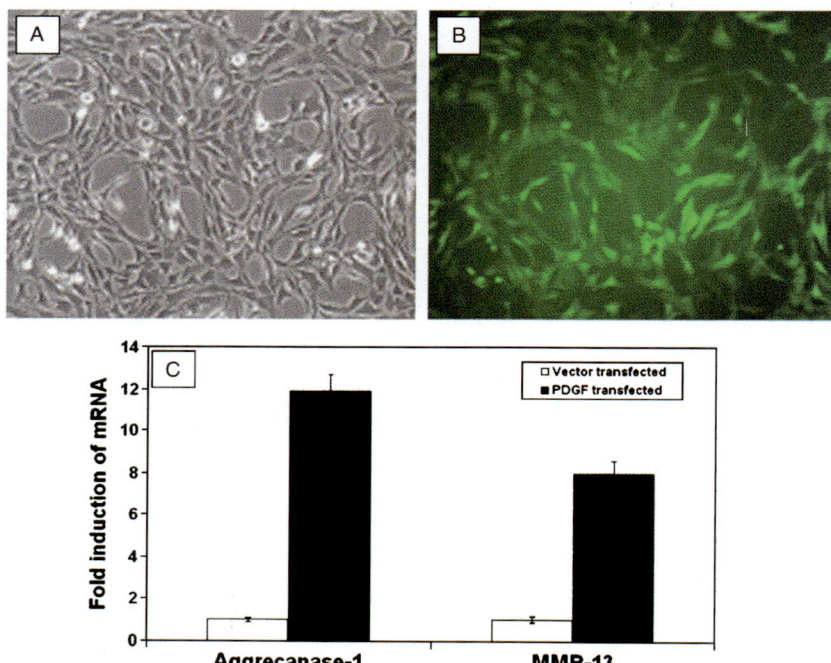

Plate 23.2. Quantitation of retroviral-mediated gene transfer efficiency using eGFP reporter plasmid and validation of the screen using PDGF as a test gene. Human articular chondrocytes were transfected with eGFP reporter plasmid by retroviral-mediated gene transfer. eGFP-expressing cells were visualized using an inverted fluorescent microscope (Olympus, $\times 1 \times 70$). (A) Represents bright light image and (B) represents fluorescent image. FACS analysis of transfected cells suggests $>70\%$ transduction efficiency. Using the same method, a PDGF cDNA clone was over-expressed in chondrocytes. (C) RNA was harvested 72 h post-transduction and changes in expression of MMP-13 and Agg-1 mRNA were detected by QPCR.

studies of lung adenocarcinomas [7] and it was found that a gene set containing the top genes in one study also tested to be statistically significant in the other two studies. This result has major implications – it is immediately evident that with this approach, it is possible to query across two different experiments with the same gene set to see if the gene set plays a similar role in different contexts. Previously underdetermined relationships between different experimental contexts could perhaps be discovered using such an approach.

While the traditional method of examining the expression profiles of one gene at a time is certainly useful, the GSEA approach has been convincingly shown to avoid the biases that the gene-by-gene approach inevitably contains. GSEA results are independent of the person analyzing the data and provides a number of pathway (or "gene set") contexts that serve as starting points for further analyses and follow-ups. GSEA should be a tool in the kit of all computational biologists who analyze gene expression profiling data. It is also quite likely that other high-throughput data types (proteomics data, for example) lend themselves to similar analyses.

3. CLASSIFICATION OF HUMAN CANCERS USING MICRORNA EXPRESSION PROFILES

MicroRNAs (miRNAs) are small, RNA molecules encoded in the genomes of plants and animals. These highly conserved, short (~22-mer) RNAs regulate the expression of genes by binding to the 3'-untranslated regions (3'-UTR) of their target mRNAs. They were first identified in *Caenorhabditis elegans* in 1993 [8] and it took several years before the first human miRNAs were discovered [8]. miRBase is a database that provides an integrated interface to miRNA sequence data, their annotation and predicted targets [9]. In the last 2 years, there has been a flurry of papers published describing miRNAs and their potential applications in drug discovery research. Several independent studies have shown that miRNAs could be key regulators in early development, cell proliferation and cell death [10]. Not surprisingly, there have been many studies of the expression and function of miRNAs in tumors [11].

There have been several lines of evidence suggesting that miRNAs function in a manner similar to that of siRNAs. In plants, miRNAs base pair with mRNAs with 100% complementarity and the target mRNA is directly cleaved and degraded using the RNA interference machinery in the cell [8]. miRNAs found in animals do not display the same level of complementarity to the mRNAs as that found in plants. As a result, suppression of the corresponding protein synthesis occurs via a mechanism that is not yet fully understood. Consequently, algorithms to identify the mRNAs that are the targets of miRNAs are being developed in many research labs. A month prior to the New York Academy of Sciences meeting in Feb 2005, six of the researchers who had developed algorithms to identify the targets of miRNAs were assigned "homework" where they were asked to predict the targets of two animal miRNAs [12]. In this blind test, it was found that there was virtually no overlap between the results of the four different algorithms that were reported. This startling outcome underlines the difficulty of the task of predicting *in silico* the targets of miRNAs in animals. Nevertheless, the task is one that is particularly well suited to bioinformatics since a major part of the algorithm involves matching miRNAs to the target mRNA at the primary sequence level. miRNAs usually do not bind to long complementary stretches in animals due to the lack of 100% complementarity with the target sequence. Consequently, loop-outs and non-Watson–Crick base pairing occur which is tackled differently at the computational level by different algorithmic approaches. Hence, the lack of overlap between the predictions made by the various algorithms.

The lack of 100% complementarity between a given miRNA sequence and the corresponding target mRNA may also indicate that the miRNA targets more than one mRNA, which potentially infers considerable regulatory functions to miRNAs. There is published evidence to suggest that on average, miRNAs have 100 target sites [13].

At the other end of the spectrum, several groups have been trying to identify new miRNA genes *in silico* [14]. In this analysis, a comparative genomics approach was taken. By looking at the 3'-UTRs across human, mouse, rat and dog genomes, Xie *et al.* discovered 106 conserved motifs. These motifs were reminiscent of many miRNA binding sites based on their length, distribution and the fact that they tended to end with the nucleotide "A", which is a characteristic of miRNA sequences. Further analysis of the 106 motifs by searching for complementary sequences in the miRNA database found 72 of the motifs to be complementary to known miRNAs. Additionally, 242 conserved and stable stem–loop sequences were found in the conserved sequences (across the four genomes) that were complementary to the 72 motifs mentioned above. These represented 113 sequences that encode already discovered miRNAs and 129 novel predicted miRNA genes. A representative set of 12 predicted new miRNA genes were tested in a set of pooled 10 adult human tissues (breast, prostate, pancreas, colon, stomach, uterus, lung, brain, liver and kidney). Although this set of 10 tissues is by no means complete, six of the 12 predicted miRNAs were found to be expressed in the pooled tissues. Considering that this analysis was limited to motifs conserved across four genomes and that the conserved motifs were stringently matched for complementarity to the known and predicted miRNAs, there exists the very real potential of discovering additional miRNAs either by relaxing the stringency of the sequence match or by limiting the analysis to only those genomes more closely related to humans (other primates, for example).

On the experimental side, an important publication in June 2005 [11] has described the characterization of miRNA expression profiles in human cancers across a diverse panel of human tissues (colon, kidney, lung, prostate, uterus and breast), both normal and cancerous. The analysis of the expression-profiling data suggests that the expression of the miRNAs distinguishes tumors of different origin, and that most of the tumors had lower levels of expression of the miRNA than the corresponding normal tissue. The data were so distinct that the classification of normal versus tumor could be done with 100% accuracy based on the differences in levels of expression between normal and tumor tissue. Compared to miRNA-expression levels, mRNA expression levels do not provide this level of specificity in distinguishing between normal and tumor tissue. This level of consistency in miRNA-expression level changes between normal and tumor tissue predicts that the monitoring of miRNA expression levels has diagnostic potential in the clinic.

Research in the field of miRNAs has not yet reached a steady state by any means. Several research labs are actively engaged in trying to discover additional miRNAs, in developing robust algorithms to predict the targets of miRNAs and in the understanding of their function and mechanism of action [10]. The multiple predictions of targets for miRNAs, the constant discovery of new miRNAs and the identification of their various functions suggest that we have much more to discover about the role of miRNAs in plant and animal biology.

4. PHARMACOGENETICS AND DRUG METABOLISM

Single-nucleotide polymorphisms (SNPs) form the basis of the inheritance of disease genes. SNPs occur in a genome when a single nucleotide (A, T, C or G) in the genome sequence is altered. For example a SNP might change the DNA sequence AATCCGAAT to AAACCGAAT. For a variation to be considered a SNP, it must occur in at least 1% of the population. SNPs, which make up about 90% of all human genetic variation, occur every 100-300 bases along the 3-billion base-long human genome. SNPs can occur in both coding (gene) and non-coding regions of the genome. If the SNP results in a change in the amino acid residue in the sequence, it is considered a non-synonymous SNP. A synonymous SNP occurs when the polymorphism does not give rise to a change in the amino acid residue of the protein sequence. SNPs are also evolutionarily stable – they do not change much through generations – thus making them easier to track in population studies. Ever since the sequencing of the human genome, there have been many studies that have attempted to identify disease-causing SNPs. The idea behind these studies is that if there is a link proven between the presence of a SNP and the occurrence of a disease, it might be possible to stratify the population according to the prevalence of this SNP and achieve greater therapeutic efficacy. While such an approach is indeed scientifically justifiable, it has met with some resistance in the scientific community due to the ethical ramifications of such studies[15].

Pharmacogenetics, on the other hand, deals with differing individual responses to drugs in the patient population and the ability to identify and forecast these responses based on genes responsible for these individual responses [16]. Figure 3 outlines the possible impact of the use of pharmacogenetics in the area of drug discovery and development.

As can be seen from the figure, mapping the polymorphisms in disease genes or genes implicated in pathways relevant to diseases may yield potential candidate therapeutic targets. These target genes then need to be further studied in order to establish a robust link between the genetic variant due to the polymorphism and any associated therapeutic implications. Such studies take a long time to do and in general are few and far between. The other prong of this approach involves studying polymorphisms in known drug-metabolizing enzymes (DMEs), which modulate drug responses in individuals [17]. This aspect of pharmacogenetics is much more tractable and will result in more tangible impact on drug development. In fact,

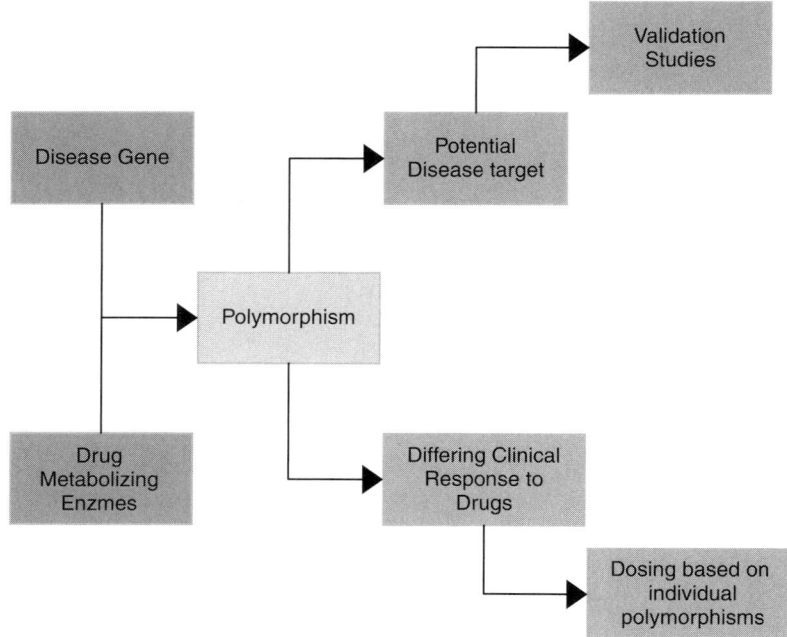

Fig. 3. Schematic representation showing the potential areas of impact using pharmacogenetics.

there have been many more successes in the efforts to elucidate the inherited basis of individual response to drugs compared to the discovery of drug targets using pharmacogenetics.

Potential consequences of polymorphic drug metabolism include: exacerbated therapeutic response, adverse reactions, toxicity and metabolism by alternative pathways [18]. Table 1 gives a brief overview of some of the DMEs that are known to have polymorphisms which affect drug response to several well-known therapeutic agents [19]. Several polymorphisms have been detected in DMEs, mainly in the cytochrome P450 (CYP450) family of heme-thiolate proteins, several of which are found to be expressed in high levels in the small intestine and liver. As a result, polymorphisms observed in these proteins can strongly influence the oral bioavailability and metabolic profile of drugs. The human CYP450 family consists of 57 members [20,21]. Sequence conservation among the P450's is low but they have a highly conserved three-dimensional structure with a conserved cysteine both at the sequence and structural level [22]. This residue forms the fifth ligand for the heme iron. P450's catalyze regiospecific and stereospecific oxidative attack of non-activated hydrocarbons at physiological temperatures. The active center for catalysis is the iron-protoporphyrin IX (heme) with the thiolate of the conserved cysteine residue as the fifth ligand [23].

Table 1. Common pharmacogenetic polymorphisms in human DMEs [19]

CYP450 polymorphism	Phenotype	Frequency in different ethnic groups	Known drug substrates	
			Total no. of drugs	Examples
CYP2D6	Poor metabolizer	White 6%, African-American 2%, and Oriental 1%	>100	Codeine, nortryptiline and dextromethorphan
	Ultra-rapid metaboliser	Ethiopian 20%, Spanish 7% and Scandinavian 1.5%		
CYP2C9	Reduced activity		>60	Tolbutamide, diazepam, ibuprofen and warfarin
CYP2C19	Poor metaboliser	Oriental 23% and White 4%	>50	Mephenytoin, omeprazole, proguanil and citalopram
N-Acetyl transferase (acetylation)	Poor metaboliser	White 60%, African-American 60%, Oriental 20% and Inuit 5%	>15	Isoniazid, procainamide, sulphonamides and hydralazines
Thiopurine methyltransferase	Poor metaboliser	Low in all populations	<10	6-Mercaptopurine, 6-thioguanine and azathioprine

The CYP3A subfamily of P450's has been implicated in the metabolism of structurally diverse compounds. CYP3A members are most abundantly expressed in the human intestine and liver tissues. Polymorphisms in the CYP 3A5 gene have recently been reported to invoke large inter-individual variations in the metabolism of tacrolimus, a calcineurin inhibitor used by renal, liver, lung and heart transplant recipients as a component of immunosuppressive therapy [24].

In this study, 40 men and 40 women renal transplant patients were recruited to investigate the correlation between polymorphisms in CYP3A5 and the dose response (concentration to dose ratio) to tacrolimus in these patients. The CYP3A5 gene was selected for the genotyping since tacrolimus is known to be a substrate of this gene. It was found that the dose range of tacrolimus given to these patients varied from 0.029 to 0.364 mg/kg/day, which represents a variation of 12-fold among these 80 patients. There was no correlation between the dose of the drug and the clinical outcome (delayed graft function or acute rejection episodes). Therefore, these variations in dose can be directly attributed to the oral bioavailability and/or metabolic clearance rates for tacrolimus in these patients. The patients were genotyped according to the CYP3A5 alleles that they carried, which revealed the presence of two alleles: the CYP3A5*1 allele (A at position 6986) and the CYP3A5*3 allele (G at position 6986). Subjects with the CYP3A5*1 allele were found to produce high levels of the protein. The CYP3A5*3 allele was found to display sequence variability in intron 3, which creates a cryptic splice site encoding for an aberrantly spliced mRNA with a premature stop codon.

The study found that the mean dose required to obtain the targeted concentration-to-dose ratio was far lower in the CYP3A5*3/*3 allele was 0.16 ± 0.07 mg/kg/day while the ratio for the CYP3A5*1/*1 genotype was 0.25 ± 0.06 mg/kg/day. This finding was explained by the fact that the latter allelic variant produced higher levels of the protein thereby reducing the bioavailability of tacrolimus in these subjects. Thus, based on the CYP3A5 genotype, one can reliably predict the amount of drug required. Such a direct application of pharmacogenetics in the clinical setting can help in the determination of appropriate dosing even prior to the transplantation surgery. A second independent study in pediatric heart transplant patients [25] corroborated the results of the renal transplant study, thus lending support to the strategy of genotyping patients for the CYP3A5 gene prior to the transplantation.

Several such studies are being conducted currently as it is becoming more and more widely recognized that pharmacogenetics can play an important role in the determination of individualized dosing provided that the DME for the drug under question is known, in addition to the response to the drug of the various genotypes. It is expected that this approach to patient stratification will become more and more widespread in the coming years.

5. CONCLUSION

As can be seen from the above examples, access to genomic sequences has far-reaching effects in our understanding of disease. The next few years will surely see the development of additional high-throughput technologies. To give a brief hint of what one might expect, technologies which measure levels of protein expression in tissues will certainly become more mature than they are today. While we have access to several techniques that measure mRNA expression levels in the cell, there is still not enough information when trying to understand the response of the cell to various contexts since there have been a number of studies that show that the levels of mRNA do not always correspond to protein levels. However, attempting to extend microarray technology to an analogous one of the protein domain is not so trivial. Some more time is needed for these technologies to evolve into robust methods for the analysis of proteomic data, especially in the clinical setting [26].

The advent of the genomic era has influenced the field of computational biology in several ways. First, the compilation of a catalog of the genes in the human genome itself involved the development of gene-prediction algorithms which have been primarily responsible for giving an estimate of the number of genes that are present in the human genome. The availability of the human genome has spawned several high-throughput technologies in order to get a comprehensive biological answer to one question or hypothesis. This has resulted not only in challenges for the associated analysis of high-throughput data, but it also has resulted in a view of biology that was simply not available prior to the human genome becoming publicly available. The sequencing of other genomes also has a high level of relevance in our understanding of the human genome. As can be seen by the example provided in the miRNA arena, the identification of regions of conserved sequence among various genomes using advanced computational biological techniques allows us to make many more reasonable inferences than would have been possible if access to those other genomes did not exist. Finally, there is still the challenge to meld diverse biological data together and get a unified view of the biological system under study. This represents a much larger computational problem which is being tackled at several levels in the field of systems biology. The integration of genome-wide biology, computational biology and data mining is sure to contribute to our understanding of biology, human disease and disease therapy in the years to come.

ACKNOWLEDGEMENTS

NRN gratefully acknowledges Charles Paulding and Steven Lewitzky for their input into this manuscript in the pharmacogenetics section. NRN also thanks Shaowen Wang, Delwood Richardson, Joseph Szustakowski and Som Wattanasin for a critical reading of the manuscript.

REFERENCES

[1] R. Kramer and D. Cohen, *Nat. Rev. Drug Disc.*, 2004, **3**, 965.
[2] B. Efron and R. Tibshirani, *Genet. Epidemiol.*, 2002, **23**, 70.
[3] M. B. Eisen, *Proc. Natl. Acad. Sci. USA*, 1998, **95**, 14863; S. Raychaudhuri, *Pac. Symp. Biocomput.*, 2000, **455**; P. Toronen, M. Kolehmainen, G. Wong and E. Castrén, *FEBS Lett*, 1999, **451**, 142.

[4] V. K. Mootha, C. M. Lindgren, K. F. Eriksson, A. Subramanian, S. Sihag, J. Lehar, P. Puigserver, E. Carlsson, M. Ridderstrale, E. Laurila, N. Houstis, M. J. Daly, N. Patterson, J. P. Mesirov, T. R. Golub, P. Tamayo, B. Spiegelman, E. S. Lander, J. N. Hirschhorn, D. Altshuler and L. C. Groop, *Nat. Genet.*, 2003, **34**, 267.
[5] A. Subramanian, P. Tamayo, V. K. Mootha, S. Mukherjee, B. L. Ebert, M. A. Gillette, A. Paulovich, S. L. Pomeroy, T. R. Golub, E. S. Lander and J. P. Mesirov, *Proc. Natl. Acad. Sci. USA*, 2005, **102**, 15545.
[6] S. Kim and D. Volsky, *BMC Bioinform*, 2005, **6**, 144.
[7] L. Tian, S. A. Greenberg, S. W. Kong, J. Altschuler, I. S. Kohane and P. J. Park, *Proc. Natl. Acad. Sci. USA*, 2005, **102**, 13544.
[8] R. C. Lee, R. L. Feinbaum and V. Ambros, *Cell*, 1993, **75**, 843; A. E. Pasquinelli, B. J. Reinhart, F. Slack, M. Q. Martindale, M. I. Kuroda, B. Maller, D. C. Hayward, E. E. Ball, B. Degnan, P. Muller, J. Spring, A. Srinivasan, M. Fishman, J. Finnerty, J. Corbo, M. Levine, P. Leahy, E. Davidson and G. Ruvkun, *Nature*, 2000, **408**, 86.
[9] S. Griffiths-Jones, R. J. Grocock, S. van Dongen, A. Bateman and A. J. Enright, *Nucl. Acids. Res.*, 2006, **34**, Database Issue, D140; S. Griffiths-Jones, *Nucl. Acids Res.*, 2004, **32**, Database Issue, D109. The microRNA database is accessible at http://microrna.sanger.ac.uk.
[10] J. Brennecke, D. R. Hipfner, A. Stark, R. B. Russell and S. M. Cohen, *Cell*, 2003, **113**, 25.
[11] J. Lu, G. Getz, E. A. Miska, E. varez-Saavedra, J. Lamb, D. Peck, A. Sweet-Cordero, B. L. Ebert, R. H. Mak, A. A. Ferrando, J. R. Downing, T. Jacks, H. R. Horvitz and T. R. Golub, *Nature*, 2005, **435**, 834.
[12] D. Steinberg, *The Scientist*, 2005, **19**, 14.
[13] V. Ambros, *Nature*, 2004, **431** 350; A. Stark, J. Brennecke, R. B. Russell and S. M. Cohen, *PLoS Biol*, 2003, **1**, 397.
[14] X. Xie, J. Lu, E. J. Kulbokas, T. R. Golub, V. Mootha, K. Lindblad-Toh, E. S. Lander and M. Kellis, *Nature*, 2005, **434**, 338.
[15] http://www.ornl.gov/sci/techresources/Human_Genome/faq/snps.shtml.
[16] D. Anglicheau, C. Legendre and E. Thervet, *Transplantation*, 2004, **78**, 311.
[17] F. P. Guengerich, in *Cytochrome P450: Structure, Mechanism, and Biochemistry* (ed. P. R. Ortiz de Montellano), 2nd edition, Plenum Press, New York, 1995, Chapter 14.
[18] C. R. Wolf, G. Smith and R. L. Smith, *BMJ*, 2000, **320**, 987–990.
[19] W. W. Weber, in *Pharmacogenetics*, Oxford University Press, 1997.
[20] H. Li and T. L. Poulos, *Cur. Top. Med. Chem.*, 2004, **4**, 1789.
[21] D. F. V. Lewis, *Pharmacogenomics*, 2004, **5**, 305.
[22] T. Omura, *Biochem. Biophys. Res. Commun.*, 1999, **266**, 690.
[23] D. Werck-Reichhart and R. Feyereisen, *Genome Biol.*, 2000, **1**, 3003.1.
[24] E. Thervet, D. Anglicheau, B. King, M. H. Schlageter, B. Cassinat, P. Beaune, C. Legendre and A. K. Daly, *Transplantation*, 2003, **76**, 1233.
[25] H. Zheng, S. Webber, A. Zeevi, E. Schuetz, J. Zhang, P. Bowman, G. Boyle, Y. Law, S. Miller, J. Lamba and G. J. Burckart, *Am. J. Transplant.*, 2003, **3**, 477.
[26] K. R. Coombes, J. S. Morris, J. Hu, S. R. Edmonson and K. A. Baggerly, *Nat. Biotechnol.*, 2005, **23**, 291.

HTS of cDNA and RNAi for Target Identification

Mark A. Labow

Genome and Proteome Sciences Department, Novartis Institutes for Biomedical Research Inc. 250 Massachusetts Avenue, Cambridge, MA 02139, USA

Contents

1. Introduction	331
2. High-throughput reverse genetics screens in cell-based systems	332
2.1. RNAi screens in drosophila cells	332
2.2. RNAi screens in mammalian cells	333
3. Gain of function screens using mammalian cDNAs	333
References	334

1. INTRODUCTION

The first tools of genomics, large scale gene sequencing and expression monitoring, provide extensive information on gene structure and quantitative measurement of gene activity. Complimenting this information has been the recent development of systematic, high-throughput genetic approaches that allow for the large-scale analysis of gene function in cultured cells. The core principles of this approach are (1) the analysis of genome-scale collection of biological reagents that can modulate the expression of virtually any gene, (2) quantitative measurement of the effect of each gene-specific reagent on a cellular process, and (3) the assembly of phenotypic data into searchable databases similar to that used for analysis of expression profiling. Genome-scale reagent collections are now available that allow for either systematic inhibition of gene expression using RNA interference (RNAi) technologies or overexpression of individual gene products using full length cDNAs, allowing both loss of function and gain of function genetic screens, respectively. The use of RNAi has been facilitated by several discoveries outlined in a number of reviews [1,2]. Briefly, RNAi in cell-based screens has been used extensively in drosophila and mammalian cell culture systems. In drosophila cell culture, RNAi is induced by double stranded RNAs (dsRNAs), produced by *in vitro* transcription that are added directly or through the use of simple chemical transfection techniques, to cultured fly cell lines (reviewed in [3]). In mammalian cells two techniques have been exploited. The first uses chemical transfection of small synthetic RNA duplexes of 21–29 bp in length (reviewed in [4]). These small interfering RNAs (siRNAs) can efficiently target for degradation endogenous messages carrying the siRNA sequence. In addition siRNAs can be produced intracellularly from an expression vector in mammalian cells and can be stably introduced into cells in a variety of ways, the most useful of which are retroviral or lentiviral vectors (reviewed in [5]). The use of cDNA overexpression in reverse genetic screens has been facilitated by the production of large collections of predicted or verified full length human and mouse cDNAs (as described in [6]). Genome-scale collections of these reagents have been extensively tested in a variety of phenotypic screens in cell culture, creating a large set of data on the role of known and previously uncharacterized genes in a number of biological settings as described below.

2. HIGH-THROUGHPUT REVERSE GENETICS SCREENS IN CELL-BASED SYSTEMS

2.1. RNAi screens in drosophila cells

In the past year a number of cell-based RNAi genetic screens in fly cells have been reported. While the first genome-scale RNAi screens in fly cells [7] focused only on cell growth, this technology has been applied to the study of a number of signal transduction pathways. Fly RNAi screens have been used to identify regulators of pathways induced by a variety of extracellular signaling proteins including wnt, hedgehog, and unpaired or Upd, a cytokine like molecule that activates the fly JAK/STAT pathway, [8–10]. Each of these screens was simple in nature; a reporter gene controlled by a promoter element inducible by one of each of the extracellular proteins was introduced into cultured fly cells. The transduced cells were combined with approximately 21,000 dsRNAi reagents representing most of the predicted drosophila genes, in individual wells of 384-well plates. Each fly cell-culture well was then tested for activation of the specific reporter genes. These studies suggested that large numbers of cellular genes can affect regulation of these pathways. For example, dasGupta *et al.*, reported that RNAi inhibition of 238 genes significantly affected fly wnt responses. Similarly another study identified 121 genes that appeared to significantly and specifically alter activation of the JAK/STAT signaling pathway in response to the Upd protein. It should also be noted that relatively few of the active RNAi reagents have been validated for their specificity and thus the accuracy of the screens are not yet known.

RNAi has been used in fly cells using number unique phenotypic assays in addition to those described above. For example, dsRNAi were used to identify genes that regulated uptake and replication of the intraceullar pathogens, *Mycobacterium fortuitum* and *Listerla monocytogenes* [11,12]. These experiments used as their phenotype the ability of drosophila S2 cells to support uptake and amplification of bacteria expressing a fluorescently tagged protein. These studies identified a number of potential proteins involved in replication of intracellular pathogens, the most interesting of which may be a CD36 family protein that is required in drosophila for bacterial uptake and is sufficient for uptake when expressed in cultured human cells. Finally another study utilized a thapsigargin-activated calcium entry assay in fly S2 cells to identify genes involved in capacitative calcium signaling [13]. After testing a collection of dsRNAi directed to a large set of candidate channel and membrane proteins one gene, STIM, was identified that was critical for calcium-release-activated Ca current (known as *Icrac*). Components of the *Icrac* channel have been sought for many years as this molecularly uncharacterized activity is central to a number of important processes including T-cell activation. The function of STIM appears to be conserved as RNAi depletion of human STIM in Jurkat T-cells was reported to also impair *Icrac*.

2.2. RNAi screens in mammalian cells

In accordance with the more difficult use of RNAi in mammalian cells, far fewer examples of large-scale genetic screens in mammalian cells have been reported. In the last two years large shRNA libraries have been produced and used in the study of regulation of protein degradation [14], the p53 pathway [15] as well as cell cycle control [16]. These studies have used sets of RNAi reagents smaller than that needed to screen the full genome, but they demonstrate the utility of systematic RNAi screening in mammalian systems. Widespread use of these reagents is likely as shRNA libraries re-renewable and publicly available (see for example http://www.openbiosystems.com and http://www.sigmaaldrich.com).

The use of synthetic siRNAs in the systematic evaluation of gene function is also becoming widespread. A large number of studies have been reported at scientific meetings and a number of papers have recently appeared. A recent study used a collection of siRNAs to 5000 human identified an important role for the PAK3 kinase in HIV replication *in vitro* [17]. Huesken et al. described the production and use of a siRNA collection to 23,000 predicted human genes [18]. This study was of note as it described the development of an *in silico* tool for prediction of highly effective siRNA sequences. In addition to using specific siRNA sequences, pools of siRNAs for large numbers of individual genes can be created by endonuclease digestion of dsRNA produced *in vitro* (known as esi pools). These pools were first used in screens for genes affecting cell division [19]. Interestingly an esi pool screen for calcium signaling regulators in human cells [20] identified the same protein, STIM, found in a drosophila cell culture screen described above. These observations suggest that RNAi screening can be highly reliable, identifying regulatory components long sought by other means.

While the promise of RNAi screening is great, it should be noted that a variety of studies suggest caution in interpreting the results of large scale screens. It is clear that the activity of siRNAs is not dependent on perfect nucleotide complementarity; thus a siRNA's specificity may be difficult to predict. Off-target effects of siRNAs (unintended inhibition of imperfectly matched mRNAs) have clearly been identified in functional screens. The most striking example of this is a recent report suggesting that active siRNAs from a small-scale hypoxia inducible gene expression screen all acted through small imperfect matches to the HIF1α mRNA, and not the genes containing exact 21 nucleotide siRNA sequence [21]. Thus interpretation of any screening result will require confirmation with multiple sequence-independent siRNAs directed against suspected target genes.

3. GAIN OF FUNCTION SCREENS USING MAMMALIAN CDNAS

Finally a number of reports have utilized the systematic overexpression of individual genes using arrayed, characterized full length cDNA libraries. These studies identified many novel regulators of a variety of cellular signaling pathways including that regulated by inflammatory cytokines, cAMP, wnt, and AP-1 proteins

[22–24]. While these initial reports used simple reporter gene assays to identify signaling proteins, a recent report [25] utilized a high-throughput microscopy assay to identify proteins capable of inducing the transport of [TORCs] (transducers of regulated CREB) from the cytoplasm to the nucleus. Interestingly, the TORC proteins were themselves previously identified in high-throughput screens using arrayed cDNA libraries. Another study utilized real-time microscopy to identify novel growth regulatory molecules [26]. While overexpression of genes through cDNAs is essentially how all expression cloning was done over the last 20 years, the systematic high-throughput use of this technique appears to be highly effective at identifying novel and specific regulatory proteins in a variety of mammalian systems.

REFERENCES

[1] J. Hall, *Nat. Rev. Genet.*, 2004, **5**, 552.
[2] R. K. Leung and P. A. Whittaker, *Pharmacol. Ther.*, 2005, **107**, 222.
[3] R. Dasgupta and N. Perrimon, *Oncogene*, 2004, **23**, 8359.
[4] P. Sandy, A. Ventura and T. Jacks, *Biotechniques*, 2005, **39**, 215.
[5] J. Silva, K. Chang, G. J. Hannon and F. V. Rivas, *Oncogene*, 2004, **23**, 8401.
[6] D. S. Gerhard, L. Wagner, E. A. Feingold, C. M. Shenmen, L. H. Grouse, G. Schuler, S. L. Klein, S. Old, R. Rasooly, P. Good, M. Guyer, A. M. Peck, J. G. Derge, D. Lipman and F. S. Collins, *Genome Res.*, 2004, **14**, 2121.
[7] M. Boutros, A. A. Kiger, S. Armknecht, K. Kerr, M. Hild, B. Koch, S. A. Haas, H. F. Consortium, R. Paro and N. Perrimon, *Science*, 2004, **303**, 832.
[8] R. Dasgupta, A. Kaykas, R. T. Moon and N. Perrimon, *Science*, 2005, **308**, 826.
[9] K. Nybakken, S. A. Vokes, T. Y. Lin, A. P. McMahon and N. Perrimon, *Nat. Genet.*, 2005, **37**, 1323.
[10] G. H. Baeg, R. Zhou and N. Perrimon, *Genes Dev.*, 2005, **19**, 1861.
[11] H. Agaisse, L. S. Burrack, J. A. Philips, E. J. Rubin, N. Perrimon and D. E. Higgins, *Science*, 2005, **309**, 1248.
[12] J. A. Philips, E. J. Rubin and N. Perrimon, *Science*, 2005, **309**, 1251.
[13] J. Roos, P. J. DiGregorio, A. V. Yeromin, K. Ohlsen, M. Lioudyno, S. Zhang, O. Safrina, J. A. Kozak, S. L. Wagner, M. D. Cahalan, G. Velicelebi and K. A. Stauderman, *J. Cell Biol.*, 2005, **169**, 435.
[14] P. J. Paddison, J. M. Silva, D. S. Conklin, M. Schlabach, M. Li, S. Aruleba, V. Balija, A. O'Shaughnessy, L. Gnoj, K. Scobie, K. Chang, T. Westbrook, M. Cleary, R. Sachidanandam, W. R. McCombie, S. J. Elledge and G. J. Hannon, *Nature*, 2004, **428**, 427.
[15] K. Berns, E. M. Hijmans, J. Mullenders, T. R. Brummelkamp, A. Velds, M. Heimerikx, R. M. Kerkhoven, M. Madiredjo, W. Nijkamp, B. Weigelt, R. Agami, W. Ge, G. Cavet, P. S. Linsley, R. L. Beijersbergen and R. Bernards, *Nature*, 2004, **428**, 431.
[16] J. Moffat, D. A. Grueneberg, X. Yang, S. Y. Kim, A. M. Kloepfer, G. Hinkle, B. Piqani, T. M. Eisenhaure, B. Luo, J. K. Grenier, A. E. Carpenter, S. Y. Foo, S. A. Stewart, B. R. Stockwell, N. Hacohen, W. C. Hahn, E. S. Lander, D. M. Sabatini and D. E. Root, *Cell*, 2006, **124**, 1283.
[17] D. G. Nguyen, K. C. Wolff, H. Yin, J. S. Caldwell and K. L. Kuhen, *J. Virol.*, 2006, **80**, 130.
[18] D. Huesken, J. Lange, C. Mickanin, J. Weiler, F. Asselbergs, J. Warner, B. Meloon, S. Engel, A. Rosenberg, D. Cohen, M. Labow, M. Reinhardt, F. Natt and J. Hall, *Nat. Biotechnol.*, 2005, **23**, 995.

[19] R. Kittler and F. Buchholz, *Cell Cycle*, 2005, **4**, 564.
[20] J. Liou, M. L. Kim, W. D. Heo, J. T. Jones, J. W. Myers and J. E. Ferrell, Jr. and T. Meyer,, *Curr.Biol.*, 2005, **15**, 1235.
[21] X. Lin, X. Ruan, M. G. Anderson, J. A. McDowell, P. E. Kroeger, S. W. Fesik and Y. Shen, *Nucleic Acids Res*, 2005, **33**, 4527.
[22] V. Iourgenko, W. Zhang, C. Mickanin, I. Daly, C. Jiang, J. M. Hexham, A. P. Orth, L. Miraglia, J. Meltzer, D. Garza, G. W. Chirn, E. McWhinnie, D. Cohen, J. Skelton, R. Terry, Y. Yu, D. Bodian, F. P. Buxton, J. Zhu, C. Song and M. A. Labow, *Proc. Natl. Acad. Sci. USA*, 2003, **100**, 12147.
[23] S. K. Chanda, S. White, A. P. Orth, R. Reisdorph, L. Miraglia, R. S. Thomas, P. DeJesus, D. E. Mason, Q. Huang, R. Vega, D. H. Yu, C. G. Nelson, B. M. Smith, R. Terry, A. S. Linford, Y. Yu, G. W. Chirn, C. Song, M. A. Labow, D. Cohen, F. J. King, E. C. Peters, P. G. Schultz, P. K. Vogt, J. B. Hogenesch and J. S. Caldwell, *Proc. Natl. Acad. Sci. U.S.A*, 2003, **100**, 12153.
[24] J. Liu, A. G. Bang, C. Kintner, A. P. Orth, S. K. Chanda, S. Ding and P. G. Schultz, *Proc. Natl. Acad. Sci. U.S.A*, 2005, **102**, 1927.
[25] M. A. Bittinger, E. McWhinnie, J. Meltzer, V. Iourgenko, B. Latario, X. Liu, C. H. Chen, C. Song, D. Garza and M. Labow, *Curr.Biol.*, 2004, **14**, 2156.
[26] J. N. Harada, K. E. Bower, A. P. Orth, S. Callaway, C. G. Nelson, C. Laris, J. B. Hogenesch, P. K. Vogt and S. K. Chanda, *Genome Res*, 2005, **15**, 1136.

Genomic Approaches to Identify Molecular Basis of Multi-Factorial Diseases

Chandrika Kumar

Novartis Institute for Biomedical Research, Cambridge, MA 02139, USA

Contents

1. Introduction	337
2. The complexities of the human genome	338
3. Current status of our understanding of human diseases at the molecular level	339
4. New methods to identify genes and pathways activated in complex multi-factorial diseases	340
5. Future directions	347
References	348

1. INTRODUCTION

Human disease is a malfunction of the body and is caused by one or more internal changes, usually in combination with external factors. Historically, systematic identification of causal mutations and aberrant signaling pathways associated with a disease took several years, even decades to unravel. In contrast, post genomic era provides tremendous opportunities for looking at disease in a more comprehensive manner. An explosion in the availability of new tools and technologies for genomic research has opened up important new avenues to understand the molecular basis of complex diseases. Some of these include: (1) availability of the complete sequence of human genome and that of other model organisms has enabled detailed characterization of the 22,000 predicted human proteins. Searching the genome methodically has enabled cataloging of all 'drugable' genes (kinases, enzymes, receptors, proteases, etc.) in the genome. Any chromosomal region of interest can be zoomed into and genes of interest can be picked out for analysis; (2) with every human gene available on an expression micro-array, the full extent of transcriptional changes that accompany several diseases can be investigated. For example, identification of unique transcriptional signature patterns characteristic of a tumor has enabled identification of precise signaling pathways activated in that specific tumor. This information provides a potential for choosing the most effective combination of therapies to kill the tumor; (3) comparative genome hybridizations can be used to search the entire genome of an individual for germ line variants, such as insertions, deletions, amplifications or translocations that might be associated with phenotypes such as mental disorders or congenital abnormalities; and (4) access to genome-wide complementary DNA (cDNA), RNAi/sh RNA reagents combined with high throughput screening capabilities using a multitude of functional assays have enabled scientists to identify the function of several novel genes of interest.

Thus, these developments have opened up a huge discovery space that can be exploited to get a comprehensive view of the signaling pathways activated in various diseases. This review describes some of the complexities of the human genome that is beginning to unravel in the post genomic era and the rapid strides that have been made in exploiting all the new information to get a better understanding of human disease at the molecular level.

2. THE COMPLEXITIES OF THE HUMAN GENOME

The foresight of sequencing a complete genome as opposed to only cataloging all the mRNAs (expressed part of the genome) came in recognizing that a genome is more than a bundle of genes and that the organization of genes in the context of surrounding information in the rest of the DNA is equally important [1]. An enormous amount of functionally important information has been found in the genome, in addition to the protein coding sequences. The human genome sequence contains non-coding RNA genes, regulatory sequences and structural motifs [2]. It also maintains short- and long -range spatial organization of sequences and it contains important evolutionary information. Only by going systematically along each chromosome from end to end could every piece of information be captured with certainty. This realization has initiated complete sequencing of genomes of several other organisms as well. The human genome contains over 10 million common polymorphic sites and an unlimited number of rarer variants. More than 7 million single-nucleotide polymorphisms (SNPs) have been mapped on the genome sequence and about 5 million have been validated [3–5]. This allows examination of each gene for variants that alter protein coding sequence or splice sites and test them for functional significance. One can also select polymorphisms for use as genetic markers, download the flanking sequence, determine the genotype of individual DNA samples and search for disease associations.

The finite number of protein coding genes hides a much greater diversity and extent of functional information in the human genome. For example, alternate splicing allows multiple functions encoded by the same gene to be selected in a cell specific manner [6,7]. Multiple promoters can confer diversity of inducible responses and substrate specificities on the same gene. The use of genomic information by each cell is governed by the interaction of multiple proteins with regulatory sequences that act as signal processors. As a result a response is initiated that takes into account all the information received from either inside or outside the cell. When analyzing the genome sequence, it is much harder to recognize regulatory sequences than protein coding sequences, because the rules are more complex and less obvious. Like protein coding regions, many regulatory sequences have been conserved during evolution, allowing one to use information from other organisms to try to find these functionally important elements of the human genome. Gene regulation is also governed by modifications to the DNA sequence (methylation) and to the proteins that bind to the DNA such as histones (acetylation) known as epigenetic changes [8]. Not much is known about the mechanisms and rules that govern this process or the replication of human chromosomes. Functional annotation of the protein coding region is relatively straightforward. Determining three dimensional structures, disrupting a gene sequence and correlating with the resulting phenotype, using model organisms for functional annotations are a few of the approaches that will enable one to harness the full potential of the human genome sequence.

3. CURRENT STATUS OF OUR UNDERSTANDING OF HUMAN DISEASES AT THE MOLECULAR LEVEL

More than 1400 human genes have been correlated with disease. In general these are single gene disorders which arise due to mutations that alter the function of a specific protein or lead to its complete absence in the disease phenotype. Such discoveries can lead to a precise test for monogenic disorders such as cystic fibrosis [9] or Huntington's disease [10,11] and also stimulate targeted research toward an effective cure by correction or replacement of the defective protein. The discovery that a chromosome translocation creates a new gene structure (the bcr-abl gene) in chronic myeloid leukemia (CML) led to the development of the drug imatinib (Gleevec, 12), which binds specifically to the bcr-abl protein and can alleviate the leukemia in patients for whom other treatments have failed. Effective cures are not necessarily guaranteed because the protein might be inaccessible or impossible to replace, the defect might be lethal too early in life, the translocated gene could undergo further mutations or there could be unexpected complication with therapy. Treatment of severe combined immunodeficiency by replacement of the defective adenosine deaminase gene [13,14] is proving to be a challenge to translate into clinical practice, illustrating that further research is needed to relate genomic data to patients.

What about more complex diseases such as diabetes, heart disease, arthritis, cancer or schizophrenia? Building up the etiology step by step by discovering genetic variants, and environmental factors is a monumental task. By identifying the key pathways nodes involved, one can pinpoint the most effective end points for intervention. New comprehensive, multi-parallel approaches are needed to identify such pathway nodes.

To find a pathway it is necessary to identify at least one of its components. For example, an enzyme in a metabolic pathway, a receptor or tranducer in a signaling pathway or a polymerase in a DNA repair pathway. Genetic approaches to this problem can benefit enormously from the human genome sequence. One can choose a gene, pick polymorphic markers from the sequence, test them for association with the disease and search the region for causative factors. If the disease has a familial mode of inheritance, it might be possible to use linkage analysis. The alternative is to use population -based association study and look for an imbalance in allele frequencies of a marker in a group of unrelated cases and compare with a matched control group. For example, an association study demonstrated the protective effect of a 32 base pair deletion in the cytokine receptor 5 (CCR5) gene against HIV-1 infection or AIDS progression [15] and two SNPs in the LTA-3 (dihydrolipoamide S-acetyltransferase) gene [16] showed a significant association with myocardial infarction. More detailed population studies will help to dissect out the genetic basis of complex traits including variable drug response.

One still lacks the understanding of how sets of proteins work together, the knowledge of the impact of the sequences outside of the coding regions as well as detailed understanding of the temporal expression of genes in response to various signals.

Acquiring this information will enable one to make much further use of the genome. Finding a genetic target would then allow quick investigation of a completely characterized biochemical pathway in order to understand the functional processes that are disrupted by a particular mutation, and to develop measures to discern the influence of non-genetic factors on these processes. For example, the gene for Huntington's disease is known and a specific alteration is sufficient to cause the disease. However the link with the pathology of the disease is still unknown. Point mutations in the peroxisome-proliferator-activated receptor gamma [17] have been associated with severe insulin resistance, but the mechanism by which this gene contributes to insulin sensitivity and glucose homeostasis is not understood. In the next section some of the novel strategies that were employed to delineate the molecular basis of human diseases will be reviewed.

4. NEW METHODS TO IDENTIFY GENES AND PATHWAYS ACTIVATED IN COMPLEX MULTI-FACTORIAL DISEASES

Most diseases result from an imbalance or perturbation in a signal transduction network. Signal transduction pathways are modular composites of functionally interdependent sets of proteins that act in a coordinated fashion to transform environmental information into a phenotypic response. Intracellular molecular signaling pathways are triggered by extracellular molecules that bind to receptors in the cell membrane, thereby switching on relay systems in the cell. The net result is gene activation or inactivation that affects a cell's behavior in a manner that it either undergoes cell division and growth or self- destruction (apoptosis). There are about 200 different signal transduction pathways known. To reliably discover medicines to correct diseases, it is important to acquire a comprehensive view of the complexities, interconnectivities and 'drugability' 'of major signaling pathways associated with disease. Cell-based pathway assays that are sensitized in a manner that model a disease can be used to elucidate the best therapeutic intervention points within these networks [18].

The availability of complete genome sequences from many organisms has yielded the ability to perform high-throughput, genome-wide screens of gene function. The major model systems used include yeast, flies, worms and mammals. There are many advantages to using these model organisms to identify signaling pathways activated or altered in disease and also for drug discovery. The genomes are less redundant and desired clone variants (mutants) can be generated and screened with both ease and speed. *Saccharomyces cerevisiae* is a model organism that has been used extensively to characterize cell cycle genes. Cell cycle studies performed in this organism have served as a guideline for understanding eukaryotic cell cycle progression. Classical screens by Pringle and Hartwell [19] identified temperature-sensitive mutants with specific arrest points throughout the cell division cycle (CDC) and have shed light on numerous aspects of the cell cycle. In an effort to complement this study and to identify novel CDC genes, Stevenson et al. [20] undertook a cDNA overexpression screen to identify genes that cause an alteration in cell cycle progression. A conditional over-expression approach with a *S. cerevisiae* cDNA library under the control of the GAL1 promoter was used because moderate over expression is more physiological than dramatic overproduction. In all, 113 genes including 19 hypothetical open reading frames (ORFs) that confer arrest or delay in specific compartments of the cell cycle were identified. Some of the genes identified in this approach overlap with those identified in loss of function screens, but a number of genes not previously implicated in cell cycle control were also identified. Through analysis of strains lacking these hypothetical ORFs, a variety of new CDC and checkpoint genes were identified [20]. This comprehensive list of cell cycle genes has given important clues to points in the cell cycle that is altered in diseases such as cancer even though mammalian cell cycle control is more intricate. Testing for synthetic lethal interactions, in which the combination of two mutations in the same cell causes death, is another approach that has been extensively used in yeast genetics to identify genes involved in the same biological process or pathway. Large-scale mapping of such interactions by synthetic genetic array (SGA) analysis have uncovered the key principles of genetic networks [21]. Synthetic lethal interactions occur frequently among different mutant alleles of two or more genes in out bred populations and may be the cause of common human diseases.

Even though classical mutagenesis approaches in the pre-genomic era provided definitive approaches to connect genes with molecular functions, it entailed working with one protein at a time. Fortunately RNA interference has come to the rescue, providing a simple method for depleting specific mRNAs and making it possible to assay for the effects of loss of function on a grand scale. Such studies in *C aenorhabditis elegans* has helped identification of genes involved in fat accumulation [22] transposon silencing [23] etc. The *Drosophila* community has followed suit and has used simple methods such as cell bathing, cell transfection or embryo injection to deliver siRNAs into *Drosophila* cells for systematic loss of function studies [24]. Such studies have helped identify genes involved in cell morphology [25], cytokinesis [26], hedgehog signaling [27] and cardiogenesis [28]. Thus genetic and biochemical analysis in model systems have successfully identified genes that play key regulatory roles in fundamental cellular and developmental processes. Understanding the normal function of these genes has provided significant insights into what goes awry in abnormal situations, such as cancer [29].

Use of genome -wide screens in complex mammalian cells is more challenging. Nevertheless, such studies are currently ongoing in different ways depending on the disease under investigation [30]. A wide variety of genomic approaches have been used to get insights into the molecular basis of tumorogenesis and predict clinical outcomes. Oncogenesis occurs through the acquisition and selection of multiple somatic mutations, each contributing to the growth, survival and spread of the cancer. Key attributes of the malignant phenotype, such as unchecked proliferation and cell survival, can often be "reversed" by the selective diminution of dominant oncogenes by chemical or genetic means (e.g. β-catenin in colorectal carcinomas; bcr-abl in CML). These observations suggest that the products of oncogenes, or of secondary genes that mediate and maintain tumor phenotypes, might be revealed through the systematic disruption of each and every gene in tumor-derived cells. Some of these genes may encode proteins amenable to therapeutic intervention, thus fueling the cancer drug discovery process. However, a functional assessment of each known or predicted gene in mammalian cells is a daunting task and represents the rate-limiting step in drug target identification and validation. For example, mutations in the p53 tumor suppressor protein are associated with a number of cancers. In order to get a comprehensive list of modifiers of the p53 activity, a genome-scale gain-of-function cellular screen was carried out in HCT116 tumor cells [31]. Seven new proteins, that enhance p53 activity and two proteins that inhibit the tumor suppressor were identified. Among the activators, five act at the posttranslational level, and two function at the transcriptional level. A number of the activators belong to the basic helix–loop–helix (bHLH) superfamily of transcription factors, three of which (HEY1, HES1 and TFAP4) are shown to activate p53 through transcriptional modulation of HDM2 expression. Furthermore, HEY1 and HES1 consistently recapitulate p53-induced phenotypes when ectopically expressed in zebrafish and chick embryos and can function as p53-dependent suppressors of transformation in mouse embryo fibroblasts. Thus these new approaches have enabled the identification of novel players that play critical roles in specific cell survival pathways.

Another phenomenon associated with cancer is called oncogene 'addiction' [32]. This is the genetic reprogramming of cells that occurs in the presence of a causal oncogenic mutation. These oncogene-driven cells become survival dependent ('addicted') to the presence of oncoprotein. Clinically exploited examples include the inhibition of c-abl kinase activity in BCR-abl transformed 32D tumor cells which lead to loss of viability with no effect on parental cells. Systematic gene silencing in cancer cells with large number of siRNAs has revealed unforeseen cellular dependencies that may be pharmaceutically tractable.

CML is one of the rare tumor types in which precise molecular abnormality which leads to transformation is known [12]. In most tumors, multiple abnormalities are responsible for their malignant behaviors. Current approaches to the design of drugs against cancer assume that almost all tumors escape normal growth regulation by usurping a few of the dozen or so key cell-signaling pathways. However, pathways can be activated at different points, so it is not always easy to tell which signaling mechanism has been activated by looking for mutations in known cancer-associated genes (oncogenes or tumor-suppressor genes). If the gene at the top of a signaling cascade is unaffected, for instance, one cannot assume that the pathway is not involved, as a factor further downstream might have been activated. It can thus be hard to predict the best treatment for a particular tumor. This problem has been addressed in two ways.

Tumor cells prone to 'oncogenic addiction' rely heavily on the activation of only one or two signaling pathways. The RAS pathway, for example can be activated either by mutation of RAS oncogene itself (seen in 40% of lung tumors) or by mutation of the BRAF oncogene, the next factor in the pathway (mutated in 60% of melanomas). One step of the RAS pathway that is being targeted by candidate drugs is MEK, an enzyme that is directly activated by BRAF, and is responsible for much of the down stream signaling from RAS. To see whether MEK inhibitors could be useful for treating all tumors with aberrant RAS signaling, Solit et al. [33] tested human tumor cell lines carrying mutations in BRAF or RAS for sensitivity to these drugs. Cells bearing an activating BRAF mutation were extremely sensitive to MEK inhibitors, both *in vitro* and when transplanted into immunodeficient mice. By contrast, cells with an activating RAS mutation showed much lower and more variable sensitivity to these inhibitors. The fact that the mutational status of BRAF predicts sensitivity to inhibition of MEK suggests that all oncogenic signaling from BRAF is mediated through the activity of its direct target MEK. However, signaling from RAS bifurcates to several downstream targets in addition to BRAF and MEK, including key pathways such as that involving phosphatidylinositol 3-kinase. Thus, inhibition of MEK may not be sufficient to inhibit cell proliferation triggered by these pathways, at least in some situations. The clear implication is that the MEK inhibitors being developed as potential drugs should be tested on tumors bearing activating BRAF mutations, such as melanomas, and not on tumors with activating RAS mutations, such as pancreatic and lung carcinomas. Indeed, the mutational status of BRAF should be used to stratify patients in any such trials.

In a situation where multiple oncogenic mutations lead to transformation and metastasis and different approach is necessary. Bild et al. [34] used microarrays to analyse the gene expression profiles of human mammary epithelial cells in which five key oncogenic pathways had been activated, by mutational activation of the MYC, RAS, SRC or beta-catenin proteins, or by loss of the Rb tumor-suppressor gene. In each case, the authors defined a signature of a hundred or so genes whose expression correlated with activation of the specific pathway. Similar individual signatures have been characterized before [35], but Bild et al. used them simultaneously to analyse the activation state of each of the pathways in a range of human and mouse tumors. The signatures successfully predicted the activating mutation in several mouse models of cancer and in human lung cancers bearing RAS mutations. In addition, predictions of the degree of deregulation of each pathway could be used as a basis for categorizing tumors into clusters that showed marked correlations with clinical outcome. For example, in lung cancer, deregulation of MYC, RAS, SRC and β-catenin together correlated with particularly poor patient survival. This information can then be used to rationalize the combination of drugs that are necessary to treat a particular type of cancer.

Another example is the use of genomic approaches to understand a complex multi-factorial disease such as osteoarthritis (OA). OA is a very prevalent joint disease and most individuals over age 60 have radiographic or histologic signs of OA [36,37]. Current therapeutic approaches are directed toward symptomatic relief because pharmacologic interventions that prevent disease progression are not available. Many patients thus progress to advanced disease where total joint replacement surgery is indicated. The major pathological change and source of subjective symptoms and joint dysfunction is the progressive loss of articular cartilage.

Other joint structures are also involved in the OA process. OA is associated with synovial inflammation, osteophyte formation and remodeling of the subchondral bone. Although considered a degenerative disorder, the OA process is driven by activation of chondrocytes. This cell activation is manifested by increased cell proliferation or cluster formation and cell death. Chondrocytes produce a broad spectrum of extracellular matrix-degrading enzymes and inflammatory mediators including cytokines, prostaglandins, nitric oxide and oxygen radicals. The formation of new tissue that is not hyaline articular cartilage but fibro cartilage also occurs. The presence of fibro cartilage and the expression of type X and IIA collagen are related to aberrant chondrocyte differentiation or de-differentiation. These observations suggest that OA pathogenesis involves a complex series of gene expression changes leading to chondrocyte activation and release of inflammatory mediators. Previous approaches to develop OA therapies were primarily focused on extracellular matrix degradation. Identification of new drug targets has recently been pursued by the analysis of gene expression patterns that are characteristic of OA. Several gene profiling studies demonstrate that a large number of genes are differentially expressed in OA vs normal cartilage and support the notion that OA is associated with chondrocyte activation and de-differentiation. The full potential of gene-profiling studies can only be realized if efficient and biologically meaningful assays for the analysis of function of differentially expressed genes are available.

Daouti et al. [37] have described the development of two novel gain of function screens in primary chondrocytes with read outs representing the major changes that are characteristic of OA pathogenesis.

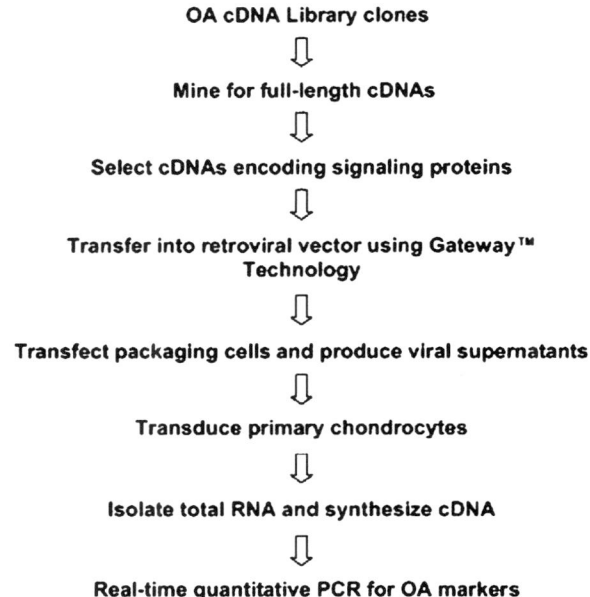

Fig. 1. Flow chart of high-throughput real-time PCR screen. The screen was run using 1200 cDNAs expressed in OA cartilage.

In the first approach (Figs. 1 and 2), cDNA libraries were constructed from OA cartilage RNA, sequenced and full-length clones primarily expressed in cartilage were selected (about 1200). These cDNAs were transferred into a retroviral vector and overexpressed in human articular chondrocytes (HAC) by retroviral-mediated gene transfer. The induction of OA-associated markers, including aggrecanase-1 (Agg-1), matrix metalloproteinase-13 (MMP-13), inducible nitric oxide synthase (iNOS), cyclooxygenase-2 (COX-2), collagen IIA and collagen X was measured by quantitative real-time polymerase chain reaction (QPCR). In the second approach (Figs. 3 and 4), whole cDNA libraries were transduced into chondrocytes and screened for chondrocyte cluster formation, another feature characteristic of OA, in three-dimensional agarose cultures. Using green fluorescent protein (eGFP) as a marker gene, it was shown that the retroviral method has a transduction efficiency

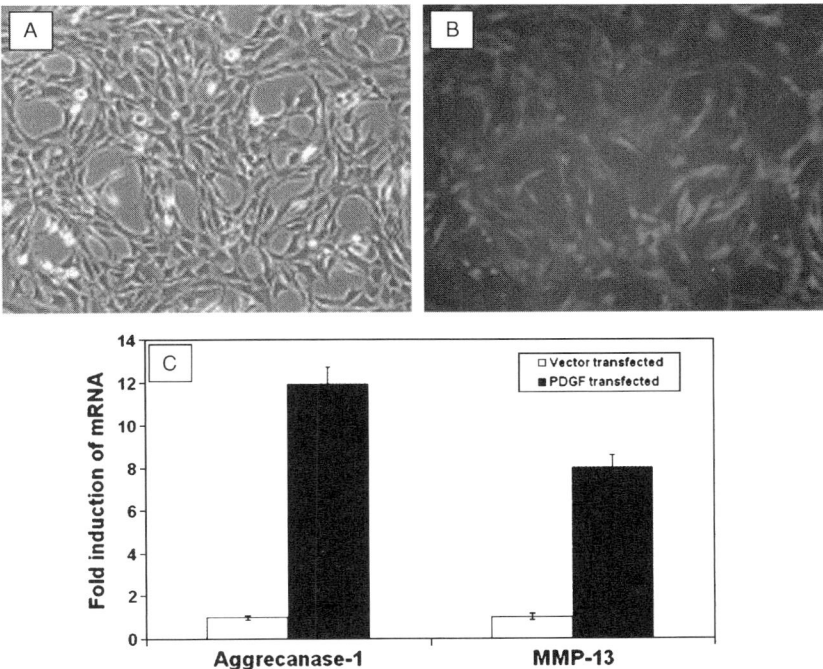

Fig. 2. Quantitation of retroviral-mediated gene transfer efficiency using eGFP reporter plasmid and validation of the screen using PDGF as a test gene. Human articular chondrocytes were transfected with eGFP reporter plasmid by retroviral-mediated gene transfer. eGFP-expressing cells were visualized using an inverted fluorescent microscope (Olympus, $\times 1 \times 70$). (A) Represents bright light image and (B) represents fluorescent image. FACS analysis of transfected cells suggests $>70\%$ transduction efficiency. Using the same method, a PDGF cDNA clone was overexpressed in chondrocytes. (C) RNA was harvested 72 h post-transduction and changes in expression of MMP-13 and Agg-1 mRNA were detected by QPCR (For color version, see Color Plate Section).

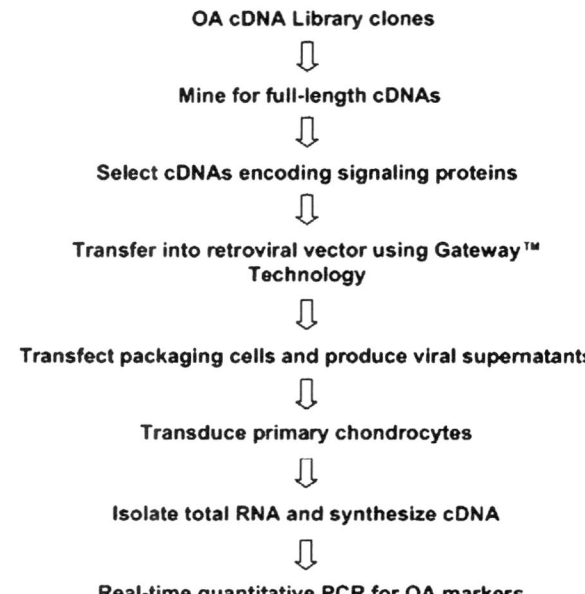

Fig. 3. Flow chart of high-throughput screen for chondrocyte cloning. An OA cDNA library was transduced into human articular chondrocytes and the tranduced chondrocytes were assayed in agarose cultures for 3-4 weeks for cluster formation.

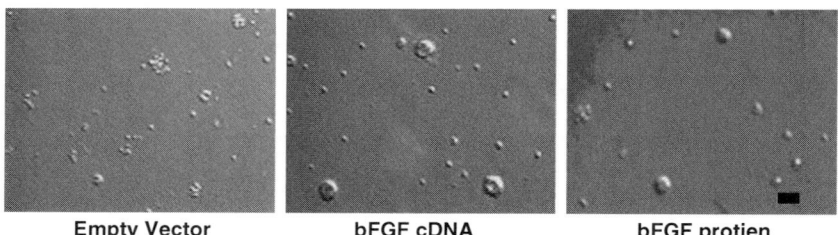

Fig. 4. Validation of the screen using basic FGF cDNA and protein. Chondrocytes were retrovirally transduced with bFGF or with empty vector and maintained in agarose cultures for four weeks. Non-transduced cells were stimulated with bFGF (100 ng/ml). The figure shows colonies > 50 µm after four weeks in culture. First image represents chondrocytes transduced with empty vector; second image represents chondrocytes transduced with bFGF cDNA clone; third image represents normal chondrocytes that were cultured in the presence of bFGF protein (100 ng/ml) added exogenously.

of > 70%. The most potent hits in both screens were the tyrosine kinases, Axl and Tyro-3 as well as members of the PI3 kinase pathway. Thus the key gene and pathway (PI3 K) involved in chondrocyte activation and de-differentiation was identified. This study demonstrates that systematic coordinated functional genomic approaches can be used to identify genes and pathways activated in complex human diseases.

5. FUTURE DIRECTIONS

The completion of whole-genome sequencing of various model organisms and the recent explosion of new technologies in the field of Functional Genomics and Proteomics is poised to revolutionize the way scientists identify and characterize gene function.

As shown in Fig. 5, there is great promise in integrating proteomics with high content imaging and small molecule chemical genetic screens to improve the understanding and treatment of cancer. Combining genetic linkage studies with high-throughput genotyping and transgenics is beginning to unravel the molecular basis of hypertension, a complex polygenetic disease with a major impact on health worldwide [38]. Recently by combining expression profiling with pathway analysis using a novel gene clustering strategy [39], pathways involved in myogenic differentiation and insulin sensitivity were identified. The free availability of human sequence information together with information on the novel approaches that are being used to understand the biochemical pathways that are fundamental to cellular homeostasis and what perturbations occur in disease will allow for cross fertilization of ideas, insights and methods that will lead to a more integrated view of human disease. This will also provide many opportunities for all scientists to continue to contribute their vision, expertise and innovation in the future.

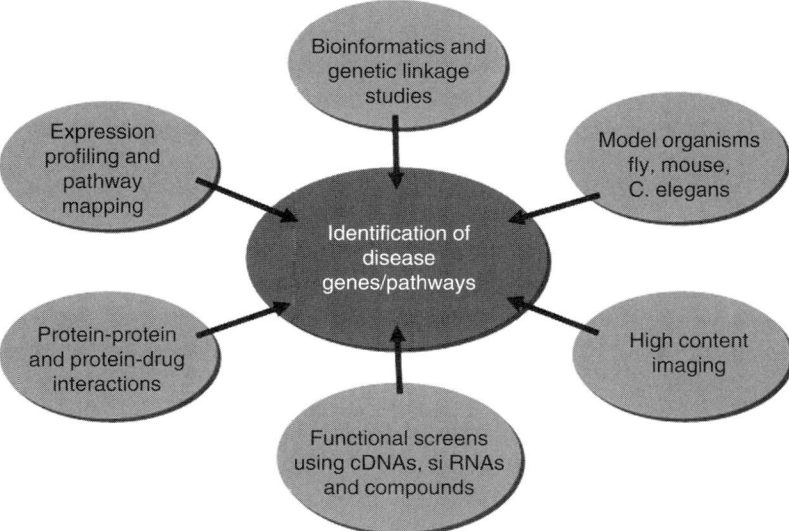

Fig. 5. Paradigm for identification of pathways activated in human disease: a multi-disciplinary approach.

REFERENCES

[1] D. R. Bentley, *Nature*, 2004, **429**, 440.
[2] D. Kampa, J. Cheng, P. Kapranov, M. Yamanaka, S. Brubaker, S. Cawley, J. Drenkow, A. Piccolboni, S. Bekiranov, G. Helt, H. Tammana and T. R. Gingeras, *Genome Res.*, 2003, **14**, 331.
[3] L. Kruglyak and D. A. Nickerson, *Nat. Genet.*, 2001, **27**, 234.
[4] The International SNP Map working group., *Nature*, 2001, **409**, 928.
[5] The International Hap Map Consortium., *Nature*, 2003, **426**, 789.
[6] A. Orr-Urtreger, M. T. Bedford, T. Burakova, E. Arman, Y. Zimmer, A. Yayon, D. Givol and P. Lonai, *Dev. Biol.*, 1993, **158**, 475.
[7] Q. H. Gong, *Pharmacogenetics*, 2001, **11**, 357.
[8] D. Rodenhiser and M. Mann, *CMAJ*, 2006, **174**, 341.
[9] F. Becg, *Curr. Pharm Des.*, 2006, **12**, 471.
[10] A. Van Dellen and A. J. Hannan, *Neurogenetics*, 2004, **5**, 9.
[11] N. Georgiou-Karistianis, *Brain Res. Bull.*, 2003, **59**, 331.
[12] B. J. Druker, *Semin. Hematol*, 2003, **40**, 50.
[13] A. Aiuti, F. Ficara, F. Cattaneo, C. Bordignon and M. G. Roncarolo, *Curr. Opin. Allergy Clin. Immunol.*, 2003, **3**, 461.
[14] S. Hacein-Bey-Abina, *N. Engl. J. Med.*, 2003, **348**, 255.
[15] M. Dean, M. Carrington, C. Winkler, G. A. Huttley, M. W. Smith, R. Allikmets, J. J. Goedert, S. P. Buchbinder, E. Vittinghoff, E. Gomperts, S. Donfield, D. Vlahov, R. Kaslow, A. Saah, C. Rinaldo, R. Detels and S. J. O'Brien, *Science*, 1996, **273**, 1856.
[16] K. Ozaki, Y. Ohnishi, A. Iida, A. Sekine, R. Yamada, H. Tsunoda, H. Sato, H. Sato, M. Hori, Y. Nakamura and T. Tanaka, *Nat. Genet.*, 2002, **32**, 650.
[17] M. Barroso, V. E. Gurnell, F. Crowley, M. Agostini, J. W. Schwabe, M. A. Soos, G. LI Maslen, T. D. M. Williams, H. Lewis, A. J. Schafer, V. K. K. Chatterjee and S. O'Rahilly, *Nature*, 1999, **402**, 880.
[18] M. C. Fishman and J. A. Porter, *Nature*, 2005, **437**, 2005.
[19] J. R. Pringle and L. H. Hartwell, in *Molecular Biology of the yeast Saccharomyces* (eds J. N. Southern, E. W. Jones and J. R. Broach), Cold Spring Harbor Laboratory Press, Plainview, NY, 1981, Vol. 1, p. 97.
[20] L. F. Stevenson, B. K. Kennedy and E. Harlow, *Proc. Natl. Acad. Sc. USA*, 2001, **98**, 3946.
[21] H. Friesen, J. S. Millman and B. J. Andrews, *Genome Biol.*, 2003, **4**, 352.
[22] K. Ashrafi, F. Y. Chang, J. L. Watts, A. G. Fraser, R. S. Kamath, J. Ahringer and G. Ruvkun, *Nature*, 2003, **421**, 268.
[23] N. L. Vastenhouw, S. E. Fischer, V. J. Robert, K. L. Thijssen, A. G. Fraser, R. S. Kamath, J. Ahringer and R. H. Plasterk, *Curr Biol.*, 2003, **13**, 1311.
[24] C. Echeverri and N. Perrimon, *Nat. Rev. Genet.*, 2006, **7**, 373.
[25] A. Kiger, B. Baum, S. Jones, M. Jones, A. Coulson, C. Echeverri and N. Perrimon, *J. Biol.*, 2003, **2**, 27.
[26] U. S. Eggert, A. A. Kiger, C. Richter, Z. E. Perlman, N. Perrimon, T. J. Mitchison and C. M. Field, *PLoS Biol.*, 2004, **2**, 379.
[27] R. Dasgupta, A. Kaykas, R. T. Monn and N. Perrimon, *Science*, 2005, **308**, 826.
[28] Y. O. Kim, S. J. Park, R. S. Balaban, M. Nirenberg and Y. Kim, *Proc. Natl. Acad. Sc. USA*, 2004, **101**, 159.
[29] R. Dasgupta and N. Perrimon, *Oncogene*, 2004, **23**, 8359.
[30] R. Kramer and D. Cohen, *Nat. Rev.*, 2004, **3**, 965.

[31] Q. Huang, A. Raya, P. DeJesus, S. Chao, K. C. Quon, J. S. Caldwell, S. K. Chanda, J. C. Izpisua-Belmonte and P. G. Schultz, *Proc. Natl. Acad. Sc., USA*, 2004, **101**, 3456.
[32] J. Downward, *Nature*, 2006, **439**, 274.
[33] D. B. Solit, L. A. Garraway, C. A. Pratilas, A. Sawai, G. Getz, A. Basso, Q. Ye, J. M. Lobo, Y. She, I. Osman, T. R. Golub, J. Sebolt-Leopold, W. R. Sellers and N. Rosen, *Nature*, 2006, **439**, 358.
[34] A. H. Bild, G. Yao, J. T. Chang, Q. Wang, A. Potti, D. Chasse, M. Joshi, D. Harpole, J. M. Lancaster, A. Berchuck, J. A. Olson, Jr., J. A. Marks, H. K. Dressman, M. West and J. R. Nevins, *Nature*, 2006, **439**, 353.
[35] D. R. Rhodes, *Proc. Natl. Acad. Sci. USA*, 2004, **101**, 9309.
[36] J. Y. Reginster, *Rheumatology (Oxford)*, 2002, **41** (Suppl 1), 3.
[37] S. Daouti, B. Latario, S. Nagulapalli, F. Buxton, S. Uziel-Fusi, G. W. Chirn, D. Bodian, C. Song, M. Labow, M. Lotz, J. Quitavalla and C. Kumar, *Osteoarthritis Cartilage*, 2005, **13**, 508.
[38] M. W. McBride, D. Graham, D. C. Delles and A. F. Dominiczak, *Curr. Opin. Nephrol. Hypertens.*, 2006, **15**, 145.
[39] J. D. Szustakowski, J. Lee, C. A. Marrese, P. A. Kosinski, N. R. Nirmala and D. M. Kemp, *Genomics*, 2006, **87**, 129.

Section 6
Topics in Drug Design and Discovery

Editor: Manoj C. Desai
Gilead Sciences Inc.
Foster City
California

Structure–Activity Relationships for *In vitro* and *In vivo* Toxicity

Julian Blagg

Pfizer Global Research and Development, Sandwich, Kent CT13 9NJ, UK

Contents

1. Introduction	353
2. Structure–toxicity relationships where outcomes may be associated with specific structural fragments	354
2.1. General themes	354
2.2. Toxicophores which do not require metabolism to manifest their adverse events or toxicity	355
2.3. Toxicophores which require metabolism to manifest their adverse events or toxicity	356
2.4. Screening methods to derisk compounds for toxicity related to specific structural fragments	360
3. Toxicities linked to descriptors for the overall properties of the molecule	362
4. Predicting structure–toxicity relationships	363
5. Conclusion	365
Acknowledgment	365
References	365

1. INTRODUCTION

Adverse safety or toxicity profiles for compounds of therapeutic interest can be indicated by a range of *in silico* methods, *in vitro* screens, *in vivo* animal studies or through subsequent human exposure [1]. Many toxicological or safety outcomes are manifest as a result of the coincidence of a number of factors; for example, the observed QT prolongation effects of terfenadine are the result of a combination of high affinity for the HERG channel coupled with high exposure in the presence of a CYP3A4 inhibitor that prevents metabolism to the principle active metabolite of lower HERG affinity [2]. Thus, individual compounds may have the same originating factors for a potential toxicology or safety outcome but a difference in dose size, route of administration, pharmacokinetic profile, genetic risk factors, tissue distribution or metabolic pathway may result in different observed outcomes. Thus, the genesis of any toxicological or safety outcome is multifactorial and complex. For this reason it can be difficult for medicinal chemists to factor the avoidance of toxicity outcomes into their target design; a focus on readily measurable parameters (e.g. primary potency) is easier to handle. However, the fundamental origins of adverse safety or toxicity findings can be considered as deriving from four parameters, all of which are within the control of the medicinal chemist.

- Primary mechanism of action of the therapeutic agent.
- Secondary pharmacology of the therapeutic agent (off target pharmacology including interaction with DNA).
- The presence of a well-defined structural fragment in the molecule with evidence that associates the fragment with adverse outcomes (a structural alert) henceforth referred to as a toxicophore (see Section 2).
- The overall physicochemical properties of the molecule, which may predispose it to adverse outcomes (see Section 3).

Links between primary mechanism of action and adverse outcome; for example the well-known link between PDE-4 inhibition and emesis in animals [3] and human [4], is a concern of every drug discovery team and tactics to understand and mitigate these risks are pivotal to any drug discovery program. Understanding the links between secondary pharmacology of potential therapeutic agents and adverse outcomes is also a field of huge scope and importance. A primary pharmacology for one therapeutic indication can be an unwanted side effect profile in another therapy area. There is growing understanding of the link between *in vitro* pharmacology and adverse outcome. Companies such as CEREP are at the forefront of building an understanding of the links between pharmacology and clinical adverse outcomes through provision of broad *in vitro* screening on marketed drugs, failed candidates and novel compounds; the BioPrint^{TM} approach [5]. The optimal use of *in vitro* safety pharmacology profiling to establish the secondary pharmacology credentials of potential therapeutic agents [6] and the use of predictive *in silico*, *in vitro* and *in vivo* systems to better characterize risk profiles [1,7] have both been the subject of useful recent reviews.

This review will focus on literature that highlights the links between adverse safety or toxicity findings, which can be directly correlated to the structure or physicochemical properties of a small molecule. The links between adverse outcomes and structures of protein, antibody or antisense therapeutic agents are out of scope of this review as is the considerable amount of work published on structural links to environmental toxicity (e.g. aquatic toxicity). This review will also cover recent advances in methods for the prediction of toxicity from chemical structure. Understanding these aspects of structure–toxicity is an important weapon in the medicinal chemists armory. Once a molecule is synthesized, all the properties of that molecule are fixed apart from those that can be changed by salt and physical form variation. Thus, awareness of structural fragments and physicochemical properties that have a higher risk of attrition through safety or toxicity findings is key to optimal target design and reducing attrition in the pharmaceutical industry.

2. STRUCTURE–TOXICITY RELATIONSHIPS WHERE OUTCOMES MAY BE ASSOCIATED WITH SPECIFIC STRUCTURAL FRAGMENTS

2.1. General themes

The generic pathways by which structural fragments may link to adverse outcomes are summarized in Fig. 1. Of particular note are idiosyncratic drug responses which

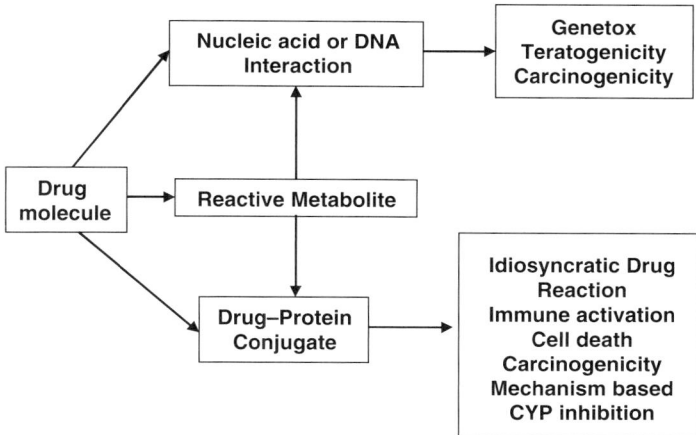

Fig. 1. Pathways linking toxicophores to adverse outcome.

often remain undetected until late stage clinical trials or post marketing when a large enough patient population is exposed to the drug to detect such a response. Such reactions are, by their very nature, difficult to screen for and adverse reactions of this type have lead to the costly withdrawal of many marketed agents [8]. Several hypotheses have been proposed to explain the cause of idiosyncratic drug reactions; the most relevant to the medicinal chemist being the formation of reactive species and subsequent protein adducts which trigger immune or hypersensitivity reactions (see Section 2.3). A number of excellent reviews over the last few years cover advances in the understanding of idiosyncratic drug reactions [9,10].

Toxicophores which have been associated with adverse outcomes can be subdivided into two classes; those fragments which do not require metabolic transformation to manifest an adverse event or toxicity and those where the evidence points to a requirement for metabolic or photochemical activation (see below). A number of excellent and comprehensive reviews have been published which catalogue the literature evidence behind fragments of both classes. Much of the evidence relates observed outcomes to specific structural fragments through biotransformation of that fragment to a reactive metabolite [11–13]. A recent paper uses an extensive Ames dataset of 4337 compounds and demonstrates that most mutagens can be associated with 29 simple toxicophore fragments with a classification error of 18% [14]. Although the toxicophores highlighted in this paper signpost an increased potential for mutagenicity, many of these toxicophores have also been associated with adverse clinical outcomes *via* metabolic activation and are discussed in the sections below. Irrespective of the method of correlation (e.g. to clinical outcome or to mutagenicity outcome), the resultant lists of toxicophores are highly consistent.

2.2. Toxicophores which do not require metabolism to manifest their adverse events or toxicity

Toxicophores which do not require metabolic activation encompass those associated with direct effects on DNA, frank electrophiles and groups associated with cytochrome P450 (CYP) inhibition through metal chelation. Table 1 highlights some important fragments where there is significant literature weight of evidence for their association with such outcomes.

For compounds whose mechanism of toxicity involves the formation of covalent bonds, chemical reactivity is obviously a key parameter. Knowledge of organic chemistry principles can educate design teams to the risks involved in incorporating such functionality into target molecules, alternatively quantum mechanical calculations have been used in this setting to predict chemical reactivity [23]. In cases where chemical reactivity results in the irreversible modulation of the protein of therapeutic interest then exquisite *in vivo* selectivity for the desired protein is paramount to the avoidance of toxicological sequelae.

Compounds which interact with CYP enzymes may be reversible through interaction with the heme as discussed above or irreversible through covalent modification of the enzyme either directly, or *via in situ* reactive metabolite formation as discussed in Section 2.3. A thorough recent review covers the clinical manifestations of both these classes of interaction by psychotropic drugs as well as information on the central regulation of CYP expression by these compounds [24].

2.3. Toxicophores which require metabolism to manifest their adverse events or toxicity

Toxicophores which require metabolism have been extensively covered in two recent reviews. These reviews also contain a comprehensive listing of drug molecules susceptible to metabolic activation that are associated with an adverse outcome [25,26]. They lay out the evidence that associates certain toxicophores with reactive metabolites and downstream clinical outcomes. Table 2 summarizes the structural fragments where the evidence for downstream adverse safety or toxicity outcomes is particularly strong. Many of these fragments are also associated with higher incidence of mechanism-based inhibition of CYP enzymes. A CYP enzyme often catalyses metabolite formation, thus the probability of covalent modification of CYP enzymes is high, particularly by short-lived reactive metabolites which do not escape the CYP active site. This topic has been recently reviewed in a useful paper that contains a comprehensive listing of 128 mechanism-based human CYP inactivators and 106 non-human CYP isoform inactivators [27]. A recent review specific to mechanism-based inactivators of CYP3A4 is also available [28].

The fragments in Table 2 serve as a signpost to medicinal chemists to be aware of the potential for metabolism to reactive species; however, each fragment has its own structure activity relationship associated with activation, a so-called structure–toxicity relationship or STR. The literature evidence for some common STRs (aromatic amines, nitro- and nitroso-aromatics) has recently been reviewed in a study of toxicophores linked to mutagenicity, hepatotoxicity and cardiotoxicity [29]. In a more specific example, 3-methylindole is susceptible to metabolic activation to a 3-methylene-indolenine reactive metabolite as outlined in Table 2. However, this metabolic activation is less well documented for indoles bearing substitution on the 3-methyl group. The recent observation of metabolic activation of the 3-benzylindole

Table 1. Toxicophores which do not require metabolic activation

Class	Fragment name	Fragment	References
Frank electrophile	Alkyl halide	R—X R = sp³ carbon X = Cl, Br, I	[15]
Frank electrophile	Michael acceptors	 CH=CH–EWG EWG = Electron withdrawing group	[16]
Frank electrophile	Reactive heterocycles	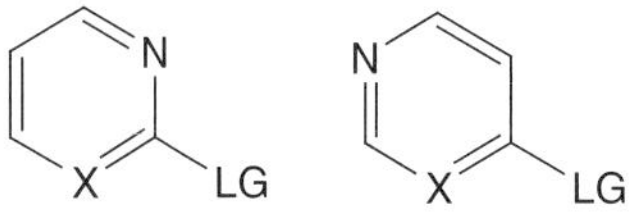 pyridine-X-LG ; pyridine-LG X = N or CH LG = Leaving group	[17]
Frank electrophile	Epoxides	 epoxide	[18]
Cross linking agent	Thiols and disulfides	 R–S–S–R ; R–S–H R = Any carbon	[19]
Potential metal chelators and/or CYP inhibitors	Imidazoles, triazoles and pyridines	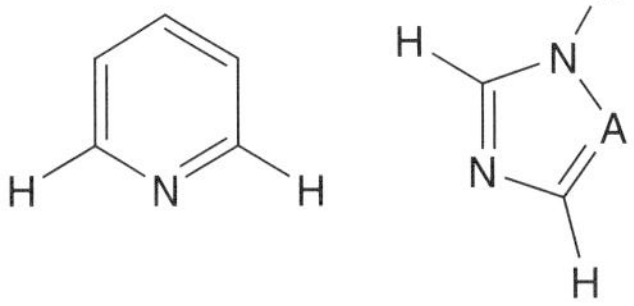 imidazole/triazole ; pyridine	[20,21]
Potential DNA-damaging agents	Furocoumarins	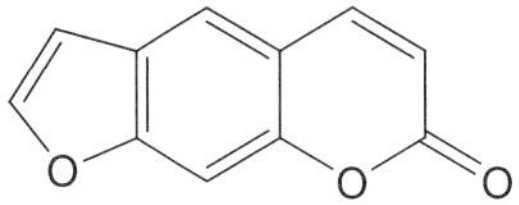 furocoumarin A = CH or N R = Any carbon	[22]

Table 2. Toxicophores requiring metabolic activation

Fragment name	Fragment structure	Examples of principle reactive intermediates	Key references
Alkynes	 $R_1-\equiv-R_2$ R_1 = Any carbon R_2 = H or any carbon	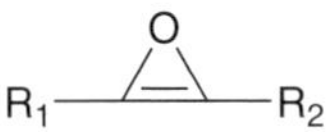 epoxide with R_1, R_2	[35]
Aminothiazoles	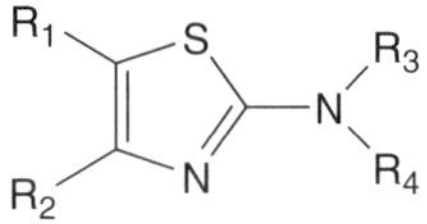 2-aminothiazole with R_1, R_2, NR_3R_4 R_1 to R_4 = H or Alkyl	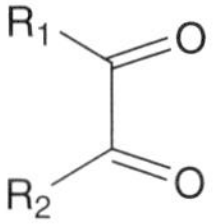 1,2-dicarbonyl (R_1CO-COR_2) and thiourea ($H_2N-C(S)-NR_3R_4$) 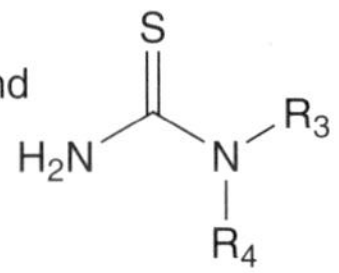	[36]
Anilines	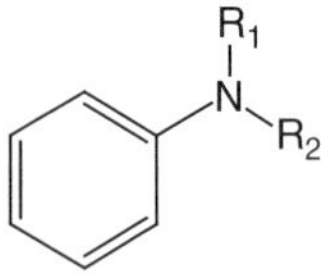 R_1R_2N–phenyl R_1, R_2 = H, alkyl, acyl, acyloxy, phenyl	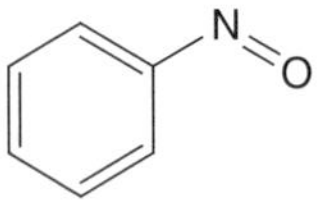 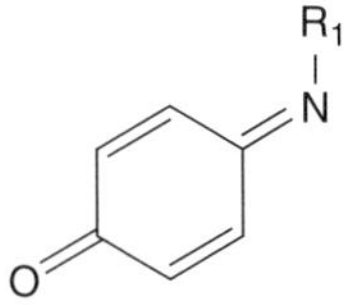 quinone imine ($R_1-N=$cyclohexadienone) and nitrosobenzene (Ph–N=O)	[37]
Benzodioxolanes	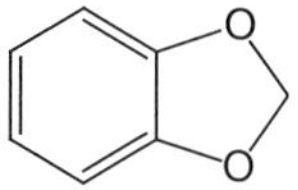 benzo[1,3]dioxole	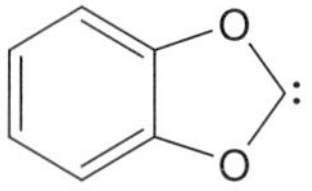 carbene (benzodioxole carbene) Carbene	[38]
Dibenzazepines	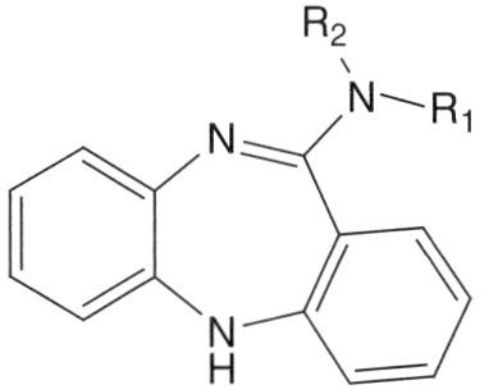 dibenzazepine with R_2R_1N- R_1, R_2 = H, cycloalkyl	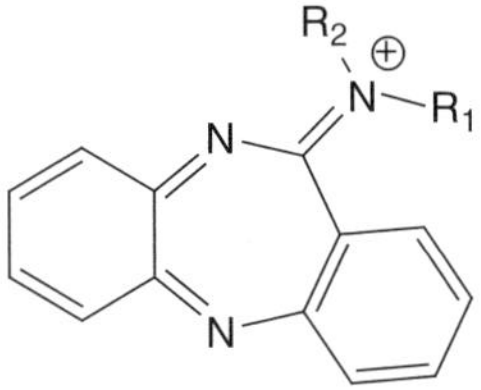 iminium ($R_2\overset{\oplus}{N}-R_1$) dibenzazepine	[39]

358   J. Blagg

Structure–Activity Relationships for *In vitro* and *In vivo* Toxicity

Class	Structure	Notes	Ref
Hydrazines	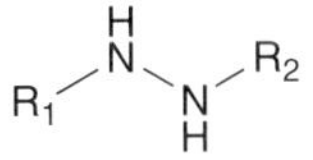 $R_1-\overset{H}{N}-\overset{H}{N}-R_2$; $R_1-N=N-R_2$	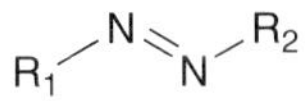 R_1, R_2 = Any carbon. Subsequent radical formation at R_1 and/or R_2	[37]
Furans and thiophenes	(furan/thiophene ring with X); (epoxide)	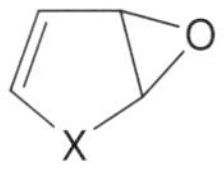 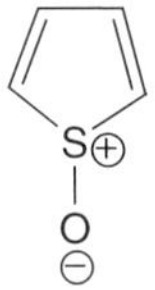 $X = S$ or O	[36]
3-Methylindoles	3-methylindole (Me at 3, R on N); 3-methyleneindolenium cation		[40]
Nitro-aromatics	$R-NO_2$; $R-N=O$	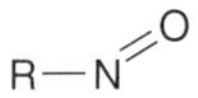 $R = H$, Any carbon	[41]
Ortho or para-hydroquinones	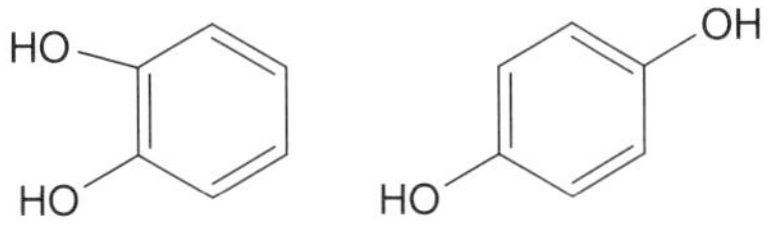 catechol (o-dihydroxybenzene); hydroquinone (p-dihydroxybenzene); o-benzoquinone; p-benzoquinone	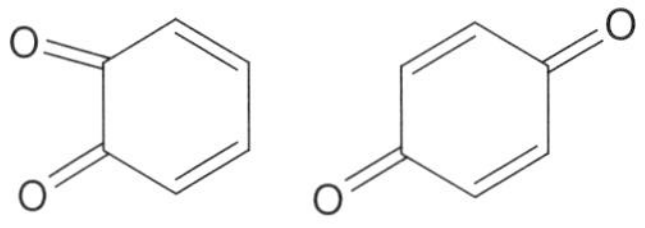 R = Aryl or heteroaryl	[42]
Thioamides and Thioureas	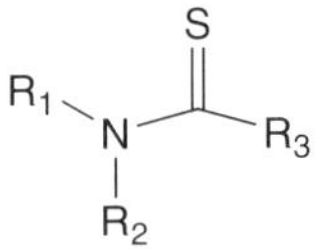 $R_1\underset{R_2}{\overset{S}{-N}}-R_3$; $R_1\underset{R_2}{\overset{SO_xH}{\overset{\oplus}{-N}}}-R_3$	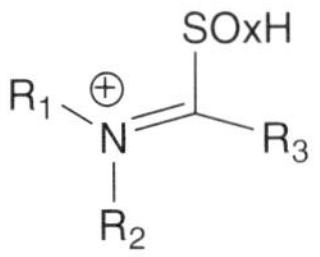 $R_1, R_2 = H$, Any carbon; R_3 = Any nitrogen or any carbon; $x = 1$ to 3	[43]

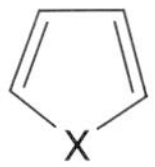

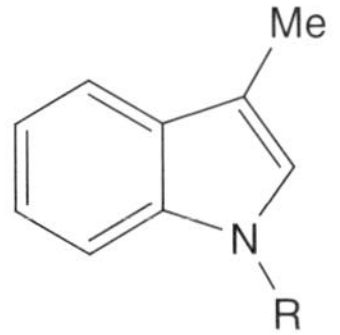

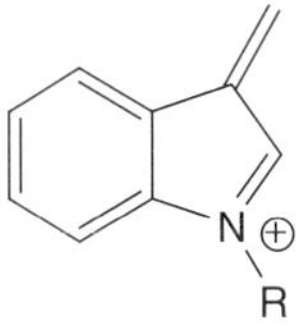

moiety in zafirlukast (1) to give the glutathione adduct (2) is a noteworthy extension to the SAR of the 3-methylindole activation pathway [30].

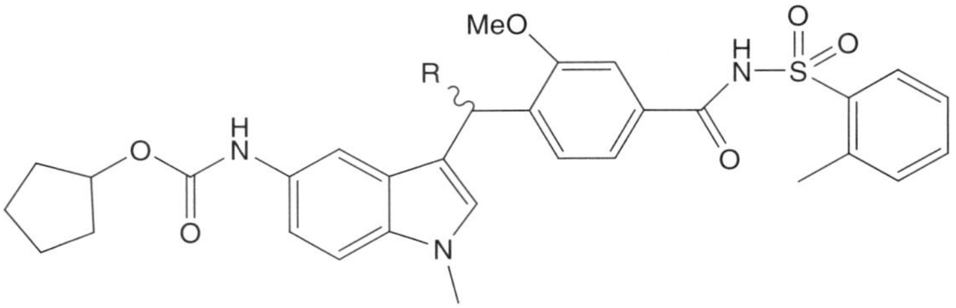

(1) R = H: Zafirlukast
(2) R = Glutathione adduct

The thiophene fragment has similarly been associated with metabolic activation in a number of publications (Table 2); however, there has been less evidence for activation of benzothiophene. A recent paper highlights metabolic activation of 2-substituted benzothiophenes [31]. Metabolism of the benzothiophene moiety in the 5-lipoxygenase inhibitor zileuton (3) has also recently been reported [32]. Formation of the reactive intermediate (4) is followed by trapping *in vitro* and *in vivo* to yield the adduct (5)

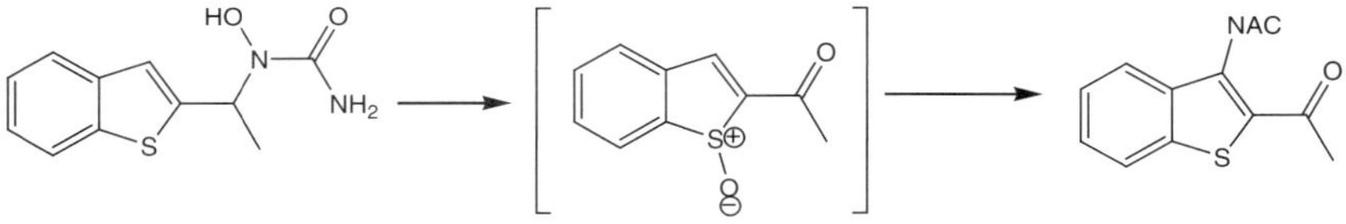

(3)   (4)   (5) NAC= N-Acetylcysteineadduct

Many of the fragments included in Table 2 are subject to activation by CYP enzymes; however the role of peroxidases in the metabolism and side effects of drugs has recently been reviewed [33]; for example, myeloperoxidase is unique in catalyzing chloride oxidation to produce hypochlorite and hypochlorous acid in activated neutrophils that has, in turn, been implicated in the oxidation of ticlodipine to its reactive metabolite [34]. Even if parent compounds are not peroxidase substrates, peroxidases may be implicated in the metabolic processing of metabolites; for example phenoxy radical formation from phenolic metabolites [33].

2.4. Screening methods to derisk compounds for toxicity related to specific structural fragments

Some molecules incorporating a toxicophore may never generate an adverse toxicological or CYP inhibition outcome. Many compounds generate an adverse outcome without a recognized toxicophore in the molecule through, for example, primary or secondary pharmacology. Some safe marketed drugs contain a toxicophore. Thus a toxicophore is not, on its own, predictive of an adverse outcome.

Many other factors such as the clinical dose and exposure, route of drug clearance or the presence of enzyme polymorphisms or metabolic routes that divert metabolism away from the toxicophore need to be considered. Preclinical screening of compounds to quantify the risk associated with the presence of postulated toxicophores in the context of the factors outlined above is therefore desirable. This subject has been recently reviewed [44–47] and is also the subject of a comprehensive review in this volume. This section will focus on a brief summary of recent advances in the principle assays; *in silico* predictive methods are covered in Section 4; Chapter 25.

A widely used qualitative indicator of the presence or absence of reactive metabolites is the *in vitro* reactive metabolite screen in which compounds are incubated with the endogenous trapping agent glutathione in the presence a metabolic activation medium [48,49]. This screen has several drawbacks; for example, the enzymes responsible for activation *in vivo* may not be present in the *in vitro* activation system, or the reactive metabolite may abstract a hydrogen atom faster than the rate of trapping by glutathione. In addition, the detection of glutathione adducts depends upon their stability and the power of the detection system, thus quantitation of the amount of adduct formation and the overall relevance to the *in vivo* setting is difficult. A recent paper describes the use of dansylglutathione as a trapping agent which facilitates use of the reactive metabolite screen as a quantitative estimation of reactive metabolite formation [50]. However, this methodology requires HPLC mediated separation of the fluorescent conjugate from unreacted starting material for quantitative purposes. The use of quaternary ammonium glutathione analogues to quantify reactive metabolite levels has recently been published [51]. This method has the advantage of direct and high throughput analysis.

The use of gene expression profiles as markers of downstream events due to reactive metabolite formation is a developing field [52]. Recently a 69 gene set has been proposed to maximally separate compounds which are macrophage activators, peroxisome proliferators and oxidative stressors from control compounds [53]. The gene expression profiles produced by the human carcinogens 4-aminobiphenyl and benzo[a]pyrene have recently been published; gene expression consistant with generalized stress response, antioxidant, glutathione transferase and DNA repair pathways were found; however, a statistically significant dose response was not uncovered [54]. The use of gene expression profiles from cultured human hepatocytes to successfully differentiate quinolone antibiotics has also recently been published [55].

The use of a radiolabeled compound to detect covalent binding to endogenous proteins is an alternative method for detection of reactive metabolite formation; however, the investment required to prepare radiolabeled material prevents use of this method as an early screening tool [56]. Importantly, advanced mass spectral techniques have greatly enhanced the ability to detect the adducted proteins; for example, in the case of detection of ^{14}C-napathalene adducts in the lung [57]. The use of animal toxicology studies to highlight toxicity or safety issues is a regulatory requirement. Until recently, these studies have not shown the potential to predict idiosyncratic drug reactions apart from rare cases where animal models of hypersensitivity reactions specific to a particular chemical class have been developed [58].

However, recent papers highlight the use of liver gene expression analysis in rats cotreated with the test compound and an inflammatory stimulus (lipopolysaccharide) [59,60]. This *in vivo* method differentiates the hepatic outcomes resulting from trovafloxacin (associated with human idiosyncratic drug reactions in liver) and levofloxacin (a fluoroquinolone antibiotic without idiosyncratic associations). Accelerated cytotoxic mechanism screening (ACMS) is another general *in vitro* technique that uses isolated rat hepatocytes primed with non cytotoxic concentrations of peroxidase and H_2O_2 to flag compounds at risk of idiosyncratic hepatotoxicity associated with sporadic inflammatory episodes [61]. Many compounds which have been withdrawn from the market for safety reasons or have hepatotoxic or GI safety issues (e.g. troglitazone, tolcapone and diclofenac) were identified in this system.

3. TOXICITIES LINKED TO DESCRIPTORS FOR THE OVERALL PROPERTIES OF THE MOLECULE

Sections 2.1–2.3 summarize the weight of evidence that links specific structural fragments (toxicophores) to adverse outcomes. Literature evidence linking overall molecular properties and adverse outcomes is less prevalent. One approach is to examine the link between overall physicochemical parameters and toxic endpoints and has been the subject of an excellent recent review which discusses the physicochemical parameters commonly associated with toxicity endpoints [62]. Lipophilicity, usually described by the octanol–water partition coefficient, (log P) or the calculated value, (clog P) appears to be the most general requirement in quantitative structure–toxicity relationships (QSTRs) reflective of the link between lipophilicity and access to tissue and cell compartments through membrane permeability and lipophilicity driven non-specific binding to hydrophobic protein sites or membranes. This paper also highlights the literature evidence linking lipophilicity to metabolism, including binding to and inhibition of CYP450 enzymes. Other physicochemical parameters linked to toxicity outcomes include bond dissociation energies (e.g. the homolytic cleavage of a phenolic O–H bond) to produce free radicals [62,63], HOMO–LUMO energy gap and phototoxicity [62,64], the link between oxidation potential and aromatic amine carcinogenicity [62,65] and the association of lipophilic basic compounds with phospholipidosis [66]. Structure–toxicity relationships associated with mitochondrial toxins have been recently reviewed. Mitochondrial uncoupling mechanisms are common toxic pathways: for example, a correlation of mitochondrial toxicity with pKa and clog P for phenolic compounds is observed [67]. A recent paper discusses QSTRs of a series of non-steroidal anti-inflammatory agents (NSAIDs); a clear correlation between lipophilicity and cytotoxicity in rat hepatocytes was established. In addition, for the benzoic acid class of NSAID, a correlation of HOMO–LUMO gap with cytotoxicity was also noted, probably reflective of the ease of oxidation of the diphenylamine template of many compounds in this class [68]. This paper highlights the use of QSTR analysis on subsets of compounds within a class to establish parameters correlating with cytotoxic outcomes and subsequent use of this information to link to possible chemical mechanisms of toxicity.

4. PREDICTING STRUCTURE–TOXICITY RELATIONSHIPS

In silico predictions of QSTRs have been reviewed [69]. The methods used fall into two broad categories; those which are knowledge-based expert systems built upon literature data of structural fragments linked to adverse outcomes and those which are probabilistic methods based upon statistical analysis of relationships between structure and/or property descriptors and toxicity endpoints. Many of the current methods are built upon carcinogenicity or mutagenicity datasets and there is a paucity of methods linking to other toxicity endpoints or correlating with an overall risk of adverse outcomes. DEREK (deductive estimation of risk from existing knowledge) [70] and TOPKAT (Toxicity prediction by Komputer-assisted technology) [71] are representative of the two approaches. The former is a rule-based system based upon a compilation of structural fragments that have been associated with toxic endpoints; the latter computes a probability of adverse outcomes based upon attributes of the chemical structure. Comparison of the performance of the two methods in predicting bacterial mutagenicity highlights an overall 65% and 73% concordance for DEREK and TOPKAT, respectively [72]. A recent paper highlights the potential advantage of a multiple computer automated structure evaluation (MCASE) methodology [73] in the prediction of mutagenic potential [74]. This analysis was based upon a larger dataset (2513 compounds drawn from multiple *Salmonella* test strains) and resulted in a better prediction of active mutagens. However, it should be recognized that the concordance of *Salmonella* mutagenicity with the regulatory endpoint of rodent carcinogenicity outcomes has been found to be as low as 60% [75] reflecting the important point that the overall relevance of any predictive model is only as good as the quality and relevance of the training set. MCASE is a statistical correlation system which evaluates a chemically diverse learning set to correlate structural features with adverse outomes; it subsequently applies modulators for the observed activity which may be physicochemical (e.g. clog P) or structure related. A recent relevant review comparing the sensitivity, specificity and concordance of the DEREK, TOPKAT and MCASE approaches to genotoxicity prediction has been published [76]. It is clear that the overall specificity of prediction of positive Ames outcomes by all three methods is poor (43–52% of positives flagged) a reflection of the fact that structures at risk of genotoxic outcomes that are outside the well-known training set of toxicophores (see Sections 2.1–2.3) are difficult to predict. The MCASE methodology has also been applied to a number of other toxicology endpoints [77]. The FDA have recently reported a database of adverse events from marketed drugs in a format suitable for QSTR modeling by abstracting relevant information from the FDA/CDER spontaneous reporting system (SRS) which contains 1.5 million adverse drug reaction reports [78]. The group then attempted to link adverse events to molecular structure using MCASE/MC4PC QSAR software; unfortunately, they did not report the alerted structures which resulted from this analysis. An alternative approach to mutagenicity prediction is common reactivity patterns analysis (COREPA), which identifies the parameters that best discriminate groups of chemicals with the same biological activity and uses a Bayesian decision tree to classify new structures. One advantage of this method is that it considers all conformers of a molecule and recognizes that different activities may be associated with different conformers. Application of this approach to a mutagenicity dataset correctly predicted 82% of the primary mutagens that do not rely upon metabolic activation for activity [79].

A recent interesting paper describes a powerful toxicity alerting system for compounds based upon computational analysis of toxicity outcomes from the registry of toxic effects of chemical substances (RTECS) database and the world drug index (WDI). The authors have built a database of structural fragments "at risk" of mutagenic, tumorigenic, irritant or reproductive adverse outcomes based upon their relative incidence in the RTECS toxicity database versus the corresponding incidence in the WDI [80]. A free web-based Java applet permits toxicity-risk assessment for any structure [81]. Of particular note in this work is the visualization of potentially toxic areas of chemical space through construction of a "self-organizing map" where clusters of compounds at risk in diverse chemical space can be highlighted. Libraries of novel compounds superimposed onto the map highlight areas of structure–toxicity risk in the library design [80].

Although many *in silico* approaches have focused on mutagenicity and carcinogenicity datasets, other toxicity endpoints have been the subject of study. *In silico* methods have been developed to predict HERG activity [82] and a variety of other ADMET outcomes [83]. QSAR relationships for skin sensitization have been recently reviewed with the conclusion that skin sensitizers are best classified according to their mechanism of activation and that, within each class, outcomes are best related to electrophilic chemical reactivity [84]. One interesting recent report covers a computational prediction of overall toxicity profile (predicted LD_{50}) by correlating chemical similarity (daylight fingerprints) with rat oral LD_{50} data from 13,645 compounds [85]. This method relies upon the assumption that similar compounds will have a similar toxicological profile and that there will be significant near neighbors to facilitate a robust prediction. However, there are clear exceptions to these assumptions, some of which are quoted in this paper.

In addition to *in silico* prediction of structural links to toxic endpoints based upon databases of specific toxic outcomes, there are efforts to establish databases of genome, tissue, organ and cellular pathway responses to external stimuli, for example the study of expressed proteins and peptides in response to chemical intervention (proteomics). To date, there is little literature precedent for predictive association of structural types to particular profiles or from characteristic profiles to *in vivo* outcomes in these systems [86]. One example of an initiative to tackle this challenge is the chemical effects in biological systems knowledge base (CEBS) run by the National Center for Toxicogenomics (NCT). One aim of this challenging initiative is to provide scholarship on structure–toxicity relationships by building an integrated dataset of gene, protein and metabolite changes collected in context of exposure, time, target organ and severity of outcome [87]. An overarching issue in building predictive structure–toxicity relationships is the need for large consistent datasets of safety or toxicity outcomes which are amenable to statistically meaningful structural investigation. One example of an initiative to develop consistent information, accessible to all, is the DSSTox database which is designed to facilitate data integration and interrogation across biology and chemistry disciplines and to support the discovery of new structure–toxicity relationships [88].

5. CONCLUSION

It is important for the medicinal chemist to be aware of structural features that have a higher than normal propensity for adverse outcomes. For certain structural fragments outlined in Section 2, there is a gathering weight of evidence that their incorporation into target molecules increases the risk of attrition. In addition, each of these fragments has its own structure–toxicity relationship which helps define the at risk structural features. One strategy is to avoid these well defined structural fragments which together represent only a small fraction of the huge diversity of chemical space. An alternative is to incorporate them and rely on *in vitro* and *in vivo* screens to build confidence through the development process; however, as discussed in Section 3.4, the ability of these screens to educate clear and quick decisions is poor and the risk may not be fully resolved until completion of expensive Phase III clinical trials. Less scholarship is available linking overall structural properties to outcomes; although relationships have been demonstrated for some physicochemical parameters (in particular clog *P*) versus some specific outcomes. Placing all of these structural insights into the context of dosing protocol, predicted efficacious exposure and routes of metabolism is important [89]. *In silico* prediction is becoming more robust for prediction of Ames mutagenicity and CYP-inhibition outcomes; however, *in silico* prediction of other adverse outcomes requires significant development. Consistent datasets large enough to support statistically meaningful analyses are required and deconvolution of the possible causes of adverse events is necessary to highlight those specifically related to chemical structure as opposed to those originating from primary or secondary pharmacology. This review highlights the ever increasing opportunity for drug-design teams to educate themselves on certain well-researched toxicophores and areas of physicochemical property space associated with adverse outcomes and thereby facilitate the delivery of safer therapeutic agents with reduced likelihood of attrition in drug development.

ACKNOWLEDGMENT

With thanks to Deepak Dalvie, Nigel Greene, David Hepworth, Amit Kalgutkar and David J. Walsh for their assistance in the preparation of this manuscript.

REFERENCES

[1] N. Bhogal, C. Grindon, R. Combes and M. Balls, *Trends Biotechnol*, 2005, **23** (6), 299.
[2] P. K. Honig, D. C. Wortham, K. Zamani, D. P. Conner, J. C. Mullin and L. R. Cantilena, *J.A.M.A.*, 1993, **269**, 1513.
[3] A. Robichaud, C. Savoie, P. B. Stamatiou, F. D. Tattersall and C. C. Chan, *Neuropharmacology*, 2001, **40**, 262.
[4] V. Boswell-Smith, D Spina, *Curr. Opin. Inv. Drugs*, 2005, **6**, 1136.
[5] C. M. Krejsa, D. Horvath, S. L. Rogalski, J. E. Penzotti, B. Mao, F. Barbosa and J. C. Migeon, *Curr. Opin. Drug Discov. Dev.*, 2003, **6**, 470.

[6] S. Whitebread, J. Hamon, D. Bojanic and L. Urban, *Drug Discov. Today*, 2005, **10**, 1421.
[7] D. E. Johnson and G. H. I. Wolfgang, *Drug Discov. Today*, 2000, **5**, 445.
[8] J. F. Waring and M. G. Anderson, *Curr. Opin. Drug Discov. Dev.*, 2005, **8**, 59.
[9] C. Ju and J. P. Utrecht, *Curr Drug Metab.*, 2002, **3**, 367.
[10] N. Kaplowicz, *Nat. Rev. Drug Discov.*, 2005, **4**, 489.
[11] S. D. Nelson, *J. Med. Chem.*, 1982, **25** (7), 753.
[12] S. D. Nelson, *Adv Exp. Med. Biol.*, 2001, **500**, 33.
[13] D. P. Williams and D. J. Naisbitt, *Curr. Opin. Drug Discov. Dev.*, 2002, **5**, 104.
[14] J. Kazius, R. McGuire and R. Bursi, *J. Med. Chem.*, 2005, **48**, 312.
[15] M. W. Anders, *Adv. Exp. Med. Biol.*, 2001, **500**, 113.
[16] T. W. Schultz and J. W. Yarbrough, *SAR QSAR Environ. Res.*, 2004, **15** (2), 139.
[17] Z. Zhao, K. A. Koeplinger, T. Peterson, R. A. Conradi, P. S. Burton, A. Suarato, R. L. Heinrikson and A. G. Tomasselli, *Drug Metab. Disp.*, 1999, **27**, 992.
[18] R. L. Melnick, *Ann NY. Acad. Sci.*, 2002, **982**, 177.
[19] J. H. Yeung, A. M. Breckenridge and B. K. Park, *Biochem. Pharmacol.*, 1983, **32** (23), 3619.
[20] M. R. Franklin, *Curr. Drug Metab.*, 2002, **3**, 599–607.
[21] W. Zhang, Y. Ramamoorthy, T. Kilicarslan, H. Nolte, R. F. Tyndale and E. M. Sellers, *Drug Metabol. Dispos.*, 2002, **30**, 314.
[22] L. Dalla Via, O. Gia, S. M. Magno, L. Santana, M. Teijeira and E. Uriarte, *J. Med. Chem*, 1999, **42**, 4405, and references therein.
[23] W. White, in *Alternative Toxicological Methods*, (eds H. Salem and S. A. Katz), CRC Press, Boca Raton, Fla., 2003, Chapter 41, p. 533.
[24] A. D. Wladyslawa, *Expert. Opin. Drug. Metab. Toxicol.*, 2005, **1**, 203.
[25] A. S. Kalgutkar and J. R. Soglia, *Expert. Opin. Drug. Metab. Toxicol.*, 2005, **1**, 91.
[26] A. S. Kalgutkar, I. Gardner, R. S. Obach, C. L. Shaffer, E. Callegari, K. Henne, A. E. Mutlib, D. K. Dalvie, J. S. Lee, Y. Nakai, J. P. O'Donnell, J. Boer and S. P. Harriman, *Curr Drug Metab.*, 2005, **6**, 161.
[27] E. Fontana, P. M. Dansette and S. M. Poli, *Curr. Drug Metab.*, 2005, **6**, 413.
[28] S. Zhou, S. Y. Chan, B. C. Goh, W. Duan, M. Huang and H. L. McLeod, *Clin. Pharmacokinet.*, 2005, **44**, 279.
[29] G. H. Hakimelahi and G. A. Khodarahmi, *J. Iranian Chem. Soc.*, 2005, **2** (4), 244.
[30] K. Kassahun, K. Skordos, I. McIntosh, D. Slaughter, G. A. Doss, T. A. Baillie and G. S. Yost, *Chem. Res. Toxicol.*, 2005, **18**, 1427.
[31] C. A. Evans, H. E. Fries and K. W. Ward, *Chem.–Biol. Interact.*, 2005, **152**, 25.
[32] E. M. Joshi, M. D. Chordia, B. H. Heasley and T. L. MacDonald, *Chem. Res. Toxicol.*, 2004, **17**, 137.
[33] S. Tafazoli and P. J. O'Brien, *Drug Discov. Today*, 2005, **10** (9), 617.
[34] Z. C. Liu and J. P. Utrecht, *Drug Metab. Dispos.*, 2000, **28**, 726.
[35] K. He, T. F. Woolf and P. F. Hollenberg, *J. Pharmacol. Exp. Ther.*, 1999, **288** (2), 791.
[36] D. K. Dalvie, A. S. Kalgutkar, S. C. Khojasteh-Bakht, R. S. Obach and J. P. O'Donnell, *Chem. Res. Toxicol.*, 2002, **15** (3), 269.
[37] A. S. Kalgutkar, D. K. Dalvie, J. P. O'Donnell, T. J. Taylor and D. C. Sahakian, *Curr. Drug. Metab.*, 2002, **3** (4), 379.
[38] M. Murray, *Curr. Drug Metab.*, 2000, **1** (1), 67.
[39] Z. C. Liu and J. P. Utrecht, *J. Pharmacol. Exp. Ther.*, 1995, **275** (3), 1476.
[40] K. W. Skordos, J. D. Laycock and G. S. Yost, *Chem. Res. Toxicol.*, 1998, **11** (11), 1326.
[41] G. L. Kedderis and G. T. Miwa, *Drug. Met. Rev.*, 1988, **19** (1), 33.
[42] J. L. Bolton, *Toxicology*, 2002, **177** (1), 55.

[43] G. J. Stevens, K. Hitchcock, Y. K. Wang, G. M. Coppola, R. W. Versace, J. A. Chin, M. Shapiro, S. Suwanrumpha and B. L. Mangold, *Chem. Res. Toxicol.*, 1997, **10** (7), 733–741.
[44] A.-E. F. Nassar and A. Lopez-Anaya, *Curr. Opin., Drug Discov. Dev.*, 2004, **7** (1), 126.
[45] A.-E. F. Nassar, A. M. Kamel and C. Clairmont, *Drug Discov. Today*, 2004, **9** (24), 1055.
[46] J. Utrecht, *Drug Discov. Today*, 2003, **8** (18), 832.
[47] K. Park, D. P. Williams, D. J. Naisbitt, N. R. Kitteringham and M. Pirmohamed, *Toxicol. App. Pharmacol.,*, 2005, **207**, S425.
[48] W. G. Chen, C. Zhang, M. J. Avery and H. G. Fouda, *Adv. Exp. Med. Biol.*, 2001, **500**, 521.
[49] J. Castro-Perez, R. Plumb, L. Liang and E. Yang, *Rapid Commun. Mass Spectrom.*, 2005, **19**, 798.
[50] J. Gan, T. W. Harper, M.-M. Hseuh, Q. Qu and W. G. Humphreys, *Chem. Res. Toxicol.*, 2005, **18**, 896.
[51] J. R. Soglia, L. G. Contillo, A. S. Kalgutkar, S. Zhao, C. E. C. A. Hop, J. G. Boyd and M. J. Cole, *Chem. Res. Toxicol.*, 2006, **19**, 480.
[52] D. P. Williams and C. O'Donnell, *Comp. Clin. Path.*, 2003, **12**, 1.
[53] M. McMillian, A. Y. Nie, J. B. Parker, A. Leone, S. Bryant, M. Kemmerer, J. Herlich, Y. Liu, L. Yieh, A. Bittner, X. Liu, J. Wan and M. D. Johnson, *Biochem. Pharmacol.*, 2004, **68**, 2249.
[54] W. Luo, W. Fan, H. Xie, L. Jing, E. Ricicki, P. Vouros, L. P. Zhao and H. Zarbl, *Chem. Res. Toxicol.*, 2005, **18**, 619.
[55] M. J. Liguori, M. G. Anderson, S. Bukofzer, J. McKim, J. F. Pregenzer, J. Retief, B. B. Spear and J. F. Waring, *Hepatology*, 2005, **1** (41), 177.
[56] D. C. Evans, A. P. Watt, D. A. Nicoll-Griffith and T. A. Baillie, *Chem. Res. Toxicol.*, 2004, **17**, 3.
[57] M. A. Isbell, D. Morin, B. Boland, A. Buckpitt, M. Salemi and J. Presley, *Proteomics*, 2005, **5**, 4197.
[58] J. Utrecht, *Toxicology*, 2005, **209**, 113.
[59] J. F. Waring, M. J. Liguori, J. P. Luyendyk, J. F. Maddox, P. E. Ganey, R. F. Stachlewitz, C. North, E. A. G. Blomme and R. A. Roth, *J. Pharmacol. Exp. Ther.*, 2006, **316** (3), 1080.
[60] P. E. Ganey, J. P. Luyendyk, J. F. Maddox and R. A. Roth, *Chem-Biol. Interact*, 2004, **150**, 35.
[61] S. Tafazoli, D. D. Spehar and P. J. O'Brien, *Drug Met. Rev.*, 2005, **37**, 311.
[62] A. G. Siraki, T. Chevaldina, M. Y. Moridani and P. J. O'Brien, *Curr. Opin. Drug Discov. Dev.*, 2004, **7** (1), 118.
[63] R. Garg, A. Kurup and C. Hansch, *Crit. Rev. Toxicol.*, 2001, **31** (2), 223.
[64] S. Dong, P. P. Fu, R. N. Shirsat, H. M. Hwang, J. Leszczynski and H. Yu, *Chem. Res. Toxicol.*, 2002, **15** (3), 400.
[65] R. Benigni and L. Passerini, *Mutat., Res.*, 2002, **511** (3), 191.
[66] J.-P. H. Ploemen, J. Kelder, T. Hafmans, H. van de Sandt, J. A. van Burgsteden, P. Salemink and E. van Esch, *Exp. Toxicol. Pathol.*, 2004, **55** (5), 347.
[67] P. Kovacic, R. S. Pozos, R. Somanathan, N. Shangari and P. J. O'Brien, *Curr. Med. Chem.*, 2005, **12**, 2601.
[68] A. G. Siraki, T. Chevaldina and P. J. O'Brien, *Chem.– Biol. Interact.*, 2005, **151**, 177.
[69] M. T. D. Cronin and D. J. Livingstone, *Predicting Chemical Toxicity and Fate*, CRC Press, Boca Raton, Fla, 2004.
[70] http://www.chem.leeds.ac.uk/luk/derek/index.html.

[71] http://www.accelrys.com/products/topkat/index.html.
[72] N. F. Cariello, J. D. Wilson, B. H. Britt, D. J. Wedd, B. Burlinson and V. Gombar, *Mutagenesis*, 2002, **17** (4), 321.
[73] http://www.multicase.com/.
[74] G. Klopman, H. Zhu, M. A. Fuller and R. D. Saiakhov, *SAR QSAR Environ. Res.*, 2004, **15** (4), 251.
[75] Y. Lee, B. G. Buchanan, G. Klopman, M. Dimayuga and H. S. Rosenkranz, *Mutat. Res.*, 1996, **358**, 37.
[76] R. D. Snyder and M. D. Smith, *Drug Discov. Today*, 2005, **10** (16), 1119.
[77] G. Klopman, S. Chakravarti, J. Ivanov and R. Saiakhov, in *Predictive Toxicology* (ed. C. Helma), Taylor & Francis, Boca Raton, Fla, 2005, p. 423.
[78] E. J. Matthews, N. L. Kruhlak, J. L. Weaver, R. D. Benz and J. F. Contrera, *Curr. Drug. Disc. Technol.*, 2004, **1**, 243.
[79] O. Mekenyan, S. Dimitrov, P. Schmeider and G. Veith, *SAR QSAR Environ. Res.*, 2003, **14**, 361.
[80] M. von Korff and T. Sander, *J. Chem. Inf. Model*, 2006, **46** (2), 536.
[81] http://www.actelion.com/uninet/www/www_main_p.nsf/Content/Technologies+Property+Explorer.
[82] R. Pearlstein, R. Vaz and D. Rampe, *J. Med. Chem.*, 2003, **46** (11), 2017.
[83] H. van der Waterbeemd and E. Gifford, *Nat. Rev. Drug Discov.*, 2003, **2**, 192.
[84] A. O. Aptula, G. Patlewicz and D. W. Roberts, *Chem. Res. Toxicol.*, 2005, **18**, 1420.
[85] S. M. Muskal, S. K. Jha, M. P. Kishore and P. Tyagi, *J. Chem. Inf. Compu. Sci.*, 2003, **43**, 1673.
[86] A. Bugrim, T. Nikolskaya and Y. Nikolsky, *Drug Discov. Today*, 2004, **9** (3), 127.
[87] M. Waters, G. Boorman, P. Bushel, M. Cunningham, R. Irwin, A. Merrick, K. Olden, R. Paules, J. Selkirk, S. Stasiewicz, B. Weis, B. Van Houten, N. Walker and R. Tennant, *Environ. Health Persp.*, 2003, **111** (6), 811.
[88] A. M. Richard, *Preclinica*, 2004, **2** (2), 103.
[89] A. P. Li, *Drug Discov. Today*, 2004, **9** (16), 687.

Importance of Early Assessment of Bioactivation in Drug Discovery

Cornelis E.C.A. Hop, Amit S. Kalgutkar and John R. Soglia

Pharmacokinetics, Dynamics and Metabolism, Pfizer Global Research and Development, Eastern Point Road, MS8118D-2026, Groton, CT, USA

Contents

1. Introduction	369
2. Assays to monitor formation of reactive metabolites	370
2.1. Covalent binding	370
2.2. Detection of glutathione adducts	373
2.3. Mechanism-based inactivation of cytochrome P450 enzymes	378
3. Strategic directions	379
References	379

1. INTRODUCTION

Safety-related issues continue to be a significant contributor to overall attrition statistics in the pharmaceutical industry. Recently, it was shown that toxicity is responsible for at least 20% of the attrition in drug development [1]. Moreover, attrition due to toxicity has increased since 1991. Therefore, increased emphasis has been placed on the identification of indicators of adverse drug reactions in animals and humans as early as possible in the overall discovery/development process. Cumulative evidence gathered from over 50 years of research has implicated chemically reactive, electrophilic metabolites as mediators of toxicity associated with several drugs. Frequently, these metabolites are formed by the cytochrome P450 (CYP) family of enzymes. Cytochrome P450 enzymes are involved in the oxidation of drugs, other xenobiotics and endogenous chemicals. The major function of these enzymes is to facilitate the clearance of drugs by generating metabolites, which are more polar and, therefore, can be readily excreted by way of the bile as well as the urine. However, in some cases the enzymatic reactions can give rise to metabolites, which are reactive toward cellular and/or circulating macromolecules. Idiosyncratic adverse drug reactions (e.g., rashes) and more distinctly dose-related toxicity, such as hepatobiliary dysfunction and cholestatis, have been reported for several drugs; it has been suggested that reactive metabolites play a causative role [2–4]. Reactive metabolites can give rise to covalent modification of macromolecules, such as DNA or proteins. Modification of DNA can have genotoxic and carcinogenic consequences, whereas modification of proteins can interfere with normal physiology and can give rise to (idiosyncratic) an immune response. Evidence has been presented that the response may be more pronounced if there is an underlying inflammatory condition – a danger signal [5]. Although there is no formal proof of a *causal* relationship between reactive metabolite formation and (idiosyncratic) adverse drug reactions, circumstantial evidence exists for myriad examples of drugs displaying toxicity that are also bioactivated to reactive metabolites and react with endogenous nucleophiles, including proteins. Bromobenzene and acetaminophen are probably the best document cases. Brodie *et al.* [6,7] showed that metabolites of these hepatotoxins covalently modify hepatic proteins and that binding is highest in the necrotic areas. Formation of the reactive quinone-imine, NAPQI, is proposed to mediate the hepatotoxicity of acetaminophen by covalent modification of critical cellular proteins. Acetaminophen is not considered to cause idiosyncratic hepatotoxicity, because the response is dose-dependent and largely predictable. Several excellent reviews have appeared describing the bioactivation of drugs and the tentative link with (idiosyncratic) toxicity [2–5].

Formation of reactive metabolites has received considerable attention in drug discovery and development, and *in vitro* assays have been established to monitor formation of reactive metabolites. Although each of these assays has relevance, none of the available assays provides an unambiguous link to a direct *in vivo* toxicological observation. A relatively simple screen to assess reactive metabolite formation is covalent binding to macromolecules using radiolabeled compounds. Alternatively, reactive, electrophilic metabolites can be trapped by the addition of exogenous nucleophiles, such as glutathione and cysteine, and the stable conjugates can be characterized by liquid chromatography–tandem mass spectrometry (LC–MS/MS) and/or nuclear magnetic resonance (NMR) spectroscopy. Advantages of the latter methodology are that no radiolabeled material is required and that it is amenable to higher throughput screening. Furthermore, the elucidation of structure of these conjugates can provide indirect information about the structure of the electrophilic species, thereby providing insight into the bioactivation mechanism and hence a rationale on which to base subsequent chemical intervention strategies. The bioactivation of compounds by CYP enzymes is often accompanied by mechanism-based inactivation of these enzymes. The mechanism is distinct from that of competitive drug–drug interactions in that it results in prolonged reduction of the metabolic capacity. In this article, the assays and their advantages, disadvantages and limitations will be described in detail. Finally, context will be provided to the general phenomenon of bioactivation and how to position its relevance in drug discovery and development.

2. ASSAYS TO MONITOR FORMATION OF REACTIVE METABOLITES

Formation of reactive metabolites has received considerable attention and several *in vitro* assays have been established to monitor and address this phenomenon.

2.1. Covalent binding

Availability of a radiolabeled compound allows a quantitative assessment of the amount of covalent binding either *in vitro* or in tissue or blood/plasma obtained from preclinical *in vivo* studies. To assess the impact of metabolic bioactivation, the compound is usually incubated with human liver microsomes. Following incubation, the solution is centrifuged and the supernatant is removed. The macromolecular pellet (in particular proteins) has to be washed extensively (i.e., multiple dissolution-centrifugation-supernatant removal cycles) to ensure that non-covalently bound radiolabeled material (starting material and metabolites) is removed. Finally, the non-extractable amount of radioactivity can be counted. Important experimental parameters are the *in vitro* compound concentration and the duration of the incubation; both parameters should be based on anticipated clinical relevance. Some companies have adopted a limit of 50 pM/mg microsomal protein as the cutoff for further development [8,9]. This limit is based on the commonly observed amount of covalent binding detected in the livers of animals receiving a prototypic hepatotoxin (e.g., acetaminophen, bromobenzene, furosemide or 4-ipomeanol), which is associated with overt hepatotoxicity (about 1000 pM/mg microsomal protein) and a 20-fold safety margin. However, it should be noted that the 50 pM/mg microsomal protein is not a hard cutoff. The rigor with which teams adhere to this limit depends on multiple factors such as the therapeutic area, the duration of therapy (acute versus chronic), the target population, first in class or best in class and, of course, the anticipated human pharmacokinetics and dose.

Covalent-binding studies can be performed *in vivo* as well. Either tissue or blood/plasma can be examined for the degree of covalent binding. However, covalent binding may require multiple dosing to establish the true impact of the compound. Reactive metabolites formed after the first dose may be efficiently trapped by nucleophiles such as glutathione and eliminated from the body (e.g., via the bile). Once glutathione is depleted, the extent of covalent binding with cellular or circulating proteins may increase rapidly, which could result in an adverse drug reaction.

The advantage of covalent-binding studies is that they directly measure covalent binding of reactive metabolites to macromolecules, which could cause an adverse immunological response or direct organ toxicity. Nevertheless, no information is available about the nature of the covalently modified proteins. Many drugs display a degree of covalent modification of proteins, but only a fraction thereof cause toxicity (the critical target hypothesis). Thus, a direct link with a toxicological endpoint is not guaranteed. Indeed, the mechanism of action of some drugs (e.g., aspirin, finasteride, clopidogrel and omeprazole) involves covalent binding to the target. For finasteride, the binding *specifically* involves 5α-reductase [10] and therefore no toxicity due to off-target covalent binding is observed. Acetaminophen is one of the few drugs for which detailed information is available about the types of proteins covalently modified by reactive metabolites [4]. Extensive covalent modification of microsomal, cytosolic, mitochondrial and cell membrane proteins was observed *in vitro* and *in vivo* at high acetaminophen concentrations. These findings are remarkable considering the over-the-counter status of acetaminophen.

An added disadvantage of this approach is that covalent-binding experiments are laborious. Recently, a semiautomated method has been described in which a cell harvester is used to capture and wash macromolecules [11]. The results obtained with this method agreed with those obtained with the traditional dissolution-centrifugation-supernatant removal method. Finally, radiolabeled material is not routinely available in early drug discovery at most pharmaceutical companies.

A classical example to illustrate the use of covalent-binding data is the proposed link between reactive metabolite formation and the agranulocytosis observed for clozapine [12]. The incidence of agranulocytosis in humans was about 1% after 1

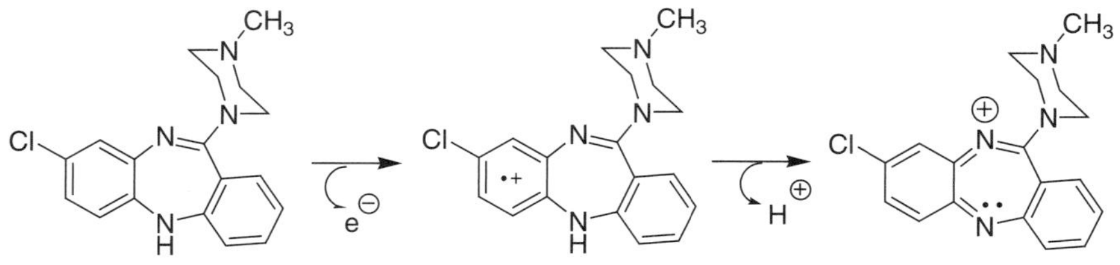

Fig. 1. Bioactivation of clozapine.

year of clozapine administration. Idiosyncratic hepatoxicity was observed for clozapine as well. Studies with anti-clozapine antiserum detected covalent binding of clozapine to multiple human neutrophil proteins *in vitro*. The molecular weights of the modified proteins were determined, but their identification is still outstanding. Proteins covalently modified with clozapine were also observed in neutrophils of patients being treated with clozapine, which reaffirms the relevance of the *in vitro* studies. The bioactivation pathway described in Fig. 1 was observed in neutrophils and has been associated with agranulocytosis although the link is still speculative and is largely derived from structure–toxicity relationship analysis with clozapine analogs. Thus, replacement of the nitrogen that connects the two aryl rings on clozapine with oxygen or sulfur results in analogs incapable of undergoing bioactivation to the reactive iminium species, and the resultant drugs are devoid of the toxicological consequences associated with clozapine. For example, quetiapine is a clozapine analog and marketed drug that contains sulfur instead of the bridging nitrogen. Despite administration at doses comparable to clozapine, cases of agranulocytosis with quetiapine are extremely rare. Finally, it is interesting to note that no agranulocytosis has been reported with the close-in analog, olanzapine. Covalent binding was observed for olanzapine *in vitro* in neutrophils, but not *in vivo*. The critical difference between clozapine and olanzapine is the dose: 300–450 mg/day for clozapine and ⩽20 mg/day for olanzapine.

A few examples have appeared in the literature illustrating the use of covalent binding for SAR purposes. For example, Samuel *et al.* [13] combined covalent-binding studies with mechanistic metabolism studies to identify the molecular liability giving rise to formation of reactive metabolites. Table 1 shows the data obtained for five analogs. The first compound, **1**, displayed extensive covalent binding (3870 pM/mg microsomal protein) and the result was rationalized by 3,4-epoxide formation; the resulting quinones were implied to cause covalent binding. To support the latter argument, addition of glutathione to the incubation reduced the amount of covalent binding to 647 pM/mg microsomal protein. The glutathione adducts were identified by LC–MS and the structures of the glutathione adducts were indicative of quinone reactive intermediates. Replacement of the phenyl ring with a pyridine ring and blocking the 4-position reduced the amount of covalent binding by more than 10-fold. However, this iterative process required synthesis of several radiolabeled compounds, which was time-consuming, and it is generally not compatible with the speed of drug discovery. Overall, reduction of covalent binding is viewed as desirable, but it has to be assessed in light of other preclinical toxicological observations and the overall risk-benefit of the drug.

Table 1. Covalent binding of compounds 1–5 in human liver microsomes [13]

Structure	Covalent binding to human liver microsomes (pmol equiv. mg^{-1} protein/1h incubation)
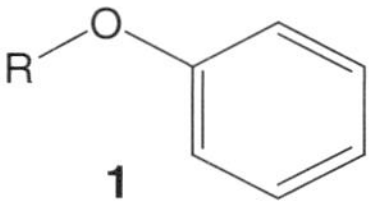 1 (R–O–phenyl)	3870 ± 303
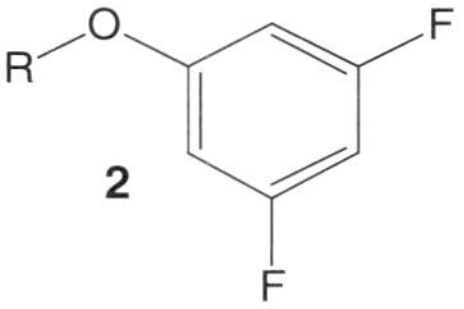 2 (R–O–3,5-difluorophenyl)	1690 ± 315
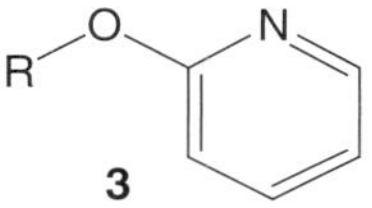 3 (R–O–pyridyl)	911 ± 109
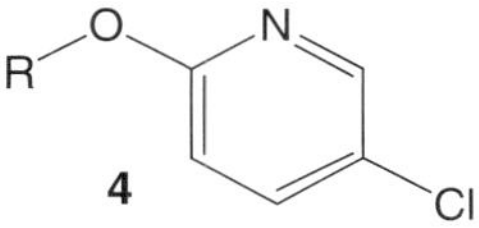 4 (R–O–pyridyl–Cl)	303 ± 81
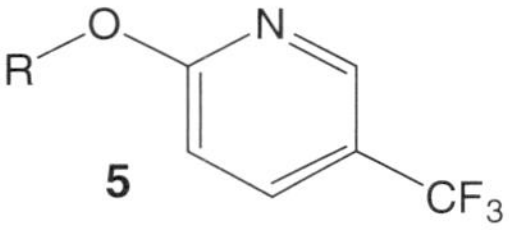 5 (R–O–pyridyl–CF$_3$)	88 ± 4

2.2. Detection of glutathione adducts

Reactive metabolites are frequently electrophilic in nature and can be trapped both *in vitro* and *in vivo* with effective nucleophiles such as glutathione and *N*-acetylcysteine. *In vivo*, the liver is endowed with about 5 mM of glutathione to prevent adverse effects caused by drugs or food components. Gluthathione adducts are relatively easy to detect by LC–MS/MS. Baillie *et al.* [14] showed that glutathione adducts display a characteristic loss of 129 Da upon collisional activation in MS/MS experiments of the [M+H]$^+$ ions. Since 129 Da corresponds to the loss of the glutamate moiety of glutathione, monitoring the loss of 129 Da in a constant neutral loss experiment on a triple quadrupole mass spectrometer provides a generic and selective means to identify glutathione adducts. The constant neutral loss scan will provide the molecular weight of the glutathione adduct and subsequent product-ion scans can provide structural information about the nature of the adduct. Recently, promising data for the detection of glutathione adducts using LC–MS/MS in the negative-ion mode were presented [15]. The adducts were detected *via* a specific precursor ion scan of *m/z* 272 (the deprotonated γ-glutamyl-dehydroalanyl-glycine anion). Although the data obtained by LC–MS/MS are valuable, NMR is frequently required to obtain detailed structural information (*vide infra*). This information will allow chemists to synthesize compounds without the offending moiety or with the site of reactivity blocked by an appropriate substituent.

To address formation of reactive intermediates early in the drug discovery process Soglia *et al.* [16] devised a higher throughput assay for detection of glutathione adducts. New chemical entities were incubated in human liver microsomes fortified with glutathione-ethyl ester; a microbore liquid chromatography–microelectrospray ionization-triple quadrupole mass spectrometer with multiple reaction monitoring (MRM) was used for detection of the adducts. A standard battery of MRM transitions was used, which was based on (1) the most common types of glutathione addition and (2) the ubiquitous loss of 129 Da for glutathione adducts. For example, the MRM battery included transitions such as $(M+323) \rightarrow (M+323-129)$ for quinones and $(M+306) \rightarrow (M+306-129)$ for thiophenes. (M stands for the molecular weight of the parent compound.) The advantages of this method include increased sensitivity, speed, and no prior knowledge of the type of glutathione adduct. Although detection of glutathione adducts by these MS/MS methods provides useful information, the experiments are not quantitative because of the variable ionization efficiency of compounds. This makes SAR studies aimed at circumventing the issue harder. Recently, different trapping reagents have been proposed to facilitate quantitation. Gan *et al.* [17] used glutathione with a fluorescent dansyl tag. Analysis involved liquid chromatography with quantitative detection via fluorescence and mass spectrometry for determination of the molecular weight of the adduct. Soglia *et al.* [18] proposed the use of glutathione derivatized with a quaternary ammonium group to reduce differences in ionization efficiencies of the glutathione adducts. Occasionally, endogenous compounds may display collision-induced loss of 129 Da as well, which would interfere with the detection of glutathione adducts. Castro-Perez *et al.* [19] developed a similar assay on a quadrupole time-of-flight mass spectrometer, but the accurate mass capability increased the selectivity of the assay and significantly reduced interference from endogenous material.

As described above, detection of glutathione adducts can be converted into an assay that can be incorporated as a routine ADME screen in the drug discovery process. The detection of glutathione adducts is indicative of the formation of reactive metabolites. However, the absence or presence of glutathione adducts has to be interpreted with caution.

1 In its current format, the glutathione assay is predominantly non-quantitative. However, it is likely that the degree of bioactivation is an important factor *in vivo*.
2 Not all reactive metabolites can be trapped with glutathione. Other trapping agents, such as cyanide and methoxylamine, may be required to get a more complete picture about metabolic bioactivation.
3 Glutathione adducts observed with human liver microsomes may not be observed with human hepatocytes or *in vivo* where other clearance pathways may be operational. For example, glucuronidation in the gut and liver [20] constitutes the principal clearance pathways of raloxifene in humans (as opposed to bioactivation [21,22]).

4 Some marketed drugs are bioactivated and form glutathione adducts, but have no associated clinical adverse events or have manageable adverse events (e.g., raloxifene [21,22] and pioglitazone [23]). The *in vitro* glutathione adduct findings have to be viewed in light of the prescribed or predicted human dose and pharmacokinetics. (The doses of raloxifene and pioglitazone are low: 60 and ⩽45 mg, respectively.)
5 Finally, the evidence linking reactive metabolites with direct organ toxicity and/or idiosyncratic toxicity is not conclusive.

Diclofenac has been studied extensively. Mild, reversible elevation of liver enzymes occurs in about 15% of patients taking diclofenac and significant hepatotoxicity has been observed for only 1–5 per 100,000 patients [3]. Two bioactivation pathways have been proposed for diclofenac. First, the carboxylic acid group of diclofenac is conjugated to an acyl glucuronide and data have been presented showing the ability of the acyl glucuronide to covalently modify proteins [24], in particular dipetidyl peptidase IV in bile canaliculi [25]. Second, several bioactivation pathways have been proposed based on phase I metabolic reactions (Fig. 2). Tang *et al.* [26] demonstrated *in vitro* that diclofenac is hydroxylated at the 5- and 4'-positions, which leads to quinone-imines. The quinone-imines are reactive and can either be trapped by glutathione or can react with macromolecules. The same glutathione adducts were detected in urine samples from humans [27], which confirms the relevance of the *in vitro* findings. Recently, a novel glutathione adduct, 4'-hydroxy-2'-glutathione-deschlorodiclofenac (4'-OH-2'-GS-DDF), was observed, which suggests that bioactivation accompanied by chlorine displacement occurs [28]. The structure of the metabolite was confirmed by ^{1}H-NMR. The relative contributions of these reactive metabolites to covalent modification of proteins and ultimately toxicity remain to be determined. Thus, sometimes multiple bioactivation pathways may occur, and it may not be possible to establish a strong link between a particular mechanism and the observed toxicity. Consequently, no clear guidance can be provided to the synthetic chemists (aside from avoiding multiple functional moieties).

A second example is the bioactivation and formation of glutathione adducts for the non-tricyclic antidepressant nefazodone. Severe hepatotoxicity resulting in death or the need for a liver transplant has been reported for nefazodone [29] and the incidence (29 cases of hepatic injury per 100,000 patient years) far exceeds that associated with other antidepressants on the market (generally less than four cases of hepatic injury per 100,000 patient years). Recently, nefazodone was voluntarily withdrawn from the market. Following incubation of nefazodone with human liver microsomes fortified with glutathione, a glutathione adduct associated with the 3-chlorophenyl piperazine moiety (Fig. 3) was detected [30]. The authors suggested that formation of this glutathione adduct was mediated via para-hydroxylation of the chlorophenyl ring followed by two-electron oxidation to the reactive quinone-imine (Fig. 3, pathway A). A novel N-dearylated metabolite was detected as well and it was speculated that its formation should be accompanied by release of 2-chloro-1,4-benzoquinone. Indirect evidence for release of 2-chloro-1,4-benzoquinone was obtained via trapping of this intermediate with glutathione (Fig. 3, pathways B and C) and detection of the

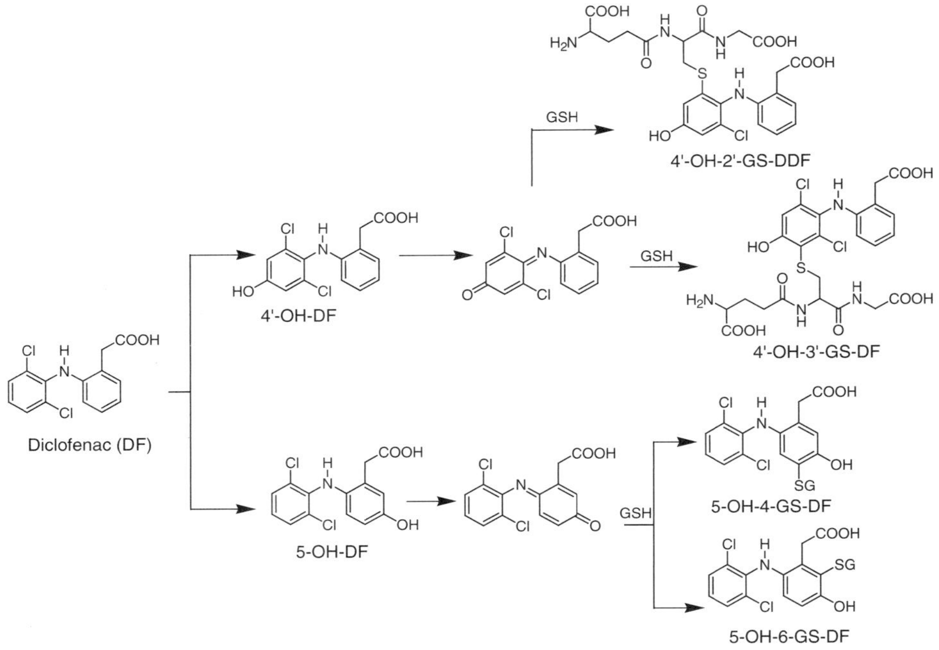

Fig. 2. Bioactivation of diclofenac initiated by para-hydroxylation.

adduct by LC–MS/MS. Although these data point to a potential liability for nefazodone, it must be emphasized that an unambiguous link between the formation of reactive metabolites and the observed hepatotoxicity has not been established. Moreover, recent evidence suggests that the parent compound itself is responsible for the clinical hepatotoxicity [31]. *In vitro* and *in vivo* studies were performed, which indicate that nefazodone inhibits the human bile salt export pump (BSEP) in the liver. Interestingly, two related compounds, buspirone and trazodone, which are not associated with overt hepatotoxicity, do not inhibit BSEP. *In vivo* rats studies showed an increase in serum bile acids following administration of nefazodone, but not for buspirone. *In vitro* hepatocyte studies in the presence of 1-aminobenzotriazole (ABT), a non-specific mechanism-based inhibitor of CYPs, increased the cytotoxicity. ABT inhibited the hydroxylation of nefazodone, which is the first step in the formation of glutathione adducts (Fig. 3). These two observations combined suggest that the metabolic bioactivation is not the mechanism responsible for nefazodone toxicity. This example highlights the difficulty associated with finding the root mechanistic cause for toxicity.

A third example, presented in Fig. 4, is also instructive. Following incubation of a potent compound targeting obesity, compound **A** (for proprietary reasons only a partial structure can be revealed), a large amount of a glutathione adduct was observed by LC–MS/MS. The proposed mechanism involves para-hydroxylation of the indole ring and subsequent formation of a quinone-imine, which was trapped effectively with glutathione (Fig. 4). Next, compounds were synthesized with

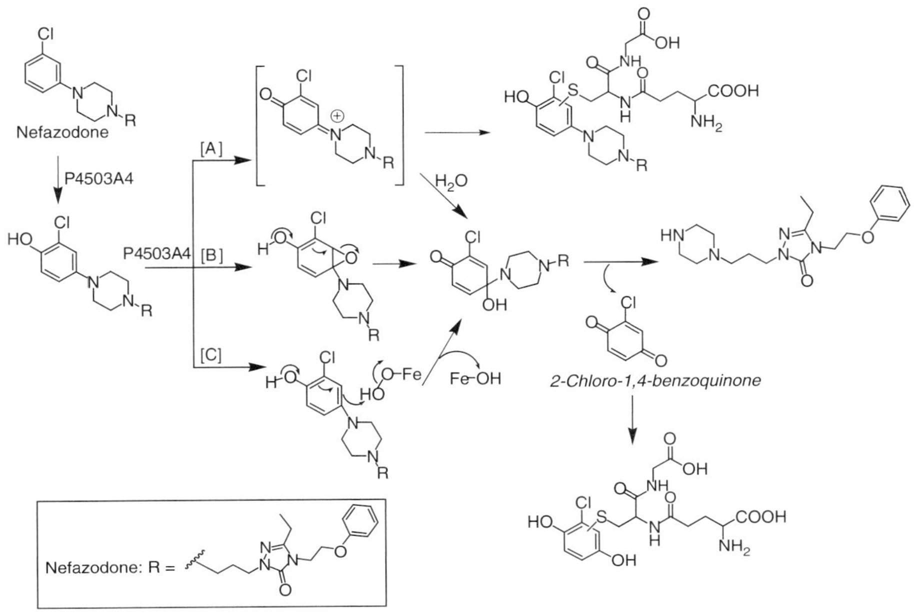

Fig. 3. Bioactivation of nefazodone.

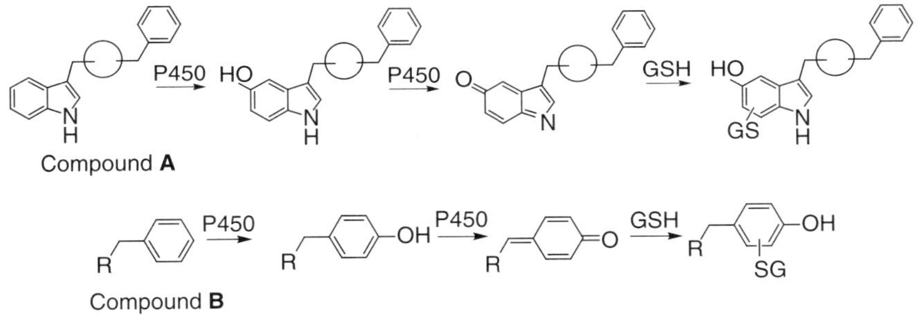

Fig. 4. Bioactivation of compounds **A** and **B**.

suitable substituents on the indole ring (e.g., a halogen or alkyl substituent in the para-position). However, glutathione adducts were still detected albeit in a lower quantity. Furthermore, compounds with the indole group removed altogether still displayed formation of glutathione adducts. More detailed LC–MS/MS studies indicated that removal of the indole group resulted in compounds (e.g., compound **B**), which were bioactivated at the terminal benzyl group (Fig. 4). Thus, metabolic switching occurred, and this complicated the efforts to circumvent bioactivation. In the end, a compound was nominated for development that formed glutathione adducts, but the mitigating circumstances were the anticipation of a low dose and very low (free) systemic exposure (free $C_{max} \leqslant 0.1$ nM).

At this stage, formation of reactive metabolites is seen as a liability, and if structural modifications can be made without sacrificing potency, pharmacokinetics, etc., this should be considered. A common approach for prevention of bioactivation is avoiding incorporation of certain structural moieties, e.g., 3-methylindoles, furans, quinones, and anilines. Kalgutkar *et al.* [32] have presented a comprehensive listing of bioactivation pathways associated with specific functional groups. If formation of reactive metabolites cannot be avoided, one should proceed with caution, especially if it is a first in class drug and/or targets a life-threatening disease.

2.3. Mechanism-based inactivation of cytochrome P450 enzymes

Another assay that points toward bioactivation is mechanism-based inhibition of cytochrome P450 enzymes. Competitive inhibition has been studied in great detail and has been recognized as an important parameter to study and optimize in drug discovery [33]. A large number of human CYP enzymes have been identified with the CYPs of greatest metabolic significance being 3A4, 2D6, 2C8, 2C9, 2C19, 1A1/2, 2A6, 2B6, and 2E1. If a drug is an inhibitor of one particular isozyme, it could affect (i.e., increase) the level of a co-administered drug, if the latter drug is metabolized by this particular isozyme. This drug–drug interaction could be particularly serious if the affected drug has a narrow therapeutic index. The effects of mechanism-based inactivation can be more severe, because the reactive metabolites bind covalently to the apoprotein or the heme group, which permanently inactivates the enzyme. The only way to restore the metabolic capacity is to resynthesize the enzyme. The effect may be more pronounced *in vivo* if the dose and exposure are high and if the half-life is long. One way to assess mechanism-based inactivation is to pre-incubate liver microsomes with the compound under investigation prior to addition of the probe substrate (which is known to be metabolized by a specific CYP enzyme). A decrease in the turnover of the probe substrate with increasing pre-incubation time is indicative of mechanism-based inaction of CYP enzymes. Formation of reactive metabolites can manifest itself in mechanism-based inactivation of CYPs if the reactive metabolites do not leave the active site of the CYP enzyme and covalently modify the apoprotein or the heme group.

A significant fraction of the compounds with the most severe drug–drug interactions are mechanism-based inhibitors of CYPs, and it can severely restrict their use. Ritonavir, nelfinavir, saquinavir, mibefradil, nefazodone, troleandomycin, erythromycin, and clarithromycin are all potent mechanism-based inhibitors, and they increase the pharmacokinetic area-under-the-curve of midazolam (the prototypical clinical CYP3A4 probe substrate) by fourfold or more. Some of these drugs have been associated with severe adverse reactions, including hepatotoxicity, but their presence on the market may be justified by their life-saving potential (e.g., ritonavir) or the temporary nature of the treatment (e.g., clarithromycin). Occasionally, a mechanism-based inhibitor may counteract its inhibitory effects by also acting as an inducer of CYPs (e.g., ritonavir). Not surprisingly, many compounds that are bioactivated and form glutathione adducts are also mechanism-based inactivators of a particular CYP (e.g., nefazodone (*vide supra*), raloxifene [21] and verapamil [34]). For example, Shen *et al.* [35] showed that the male rat-specific cytochrome P4502C11 metabolizes diclofenac into a highly reactive metabolite that covalently binds to the enzyme before it can diffuse out of the reactive site, and the adduct formation resulted in a 72% drop in the activity of CYP2C11. A detailed review of mechanism-based inhibition of CYPs and its link with common substructures and their reactivity has been presented by Fontana *et al.* [36].

3. STRATEGIC DIRECTIONS

Overall, formation of reactive metabolites and subsequent covalent modification of macromolecules is a potential liability in drug discovery and a potential source of attrition. However, it must be emphasized that in most cases there is no conclusive evidence linking formation of reactive metabolites and (idiosyncratic) toxicity. Indeed, for some of the examples presented above alternative mechanisms have been proposed to explain the toxicity, which did not involve bioactivation. Nevertheless, there is evidence to suggest that bioactivation increases the risk of attrition. What to do when bioactivation (as defined by covalent binding, formation of glutathione adducts or mechanism-based inactivation of CYPs) is observed continues to be debated. First of all, not all *in vitro* observations translate into *in vivo* liabilities. A low dose and low systemic exposure are mitigating circumstances; in a recent paper Smith and Schmid present a compelling case that aiming for a low dose is a good strategy to avoid toxicity [37]. In addition, the availability of alternative clearance pathways *in vivo* can prevent bioactivation by shunting the metabolism to innocuous pathways. Other factors to consider are the availability of other lead series, the therapeutic area (drugs targeting life-threatening diseases vs lifestyle drugs), the target population, the duration of therapy, and whether it is a prototype for a new target or a follow-up. When nominating a drug with this liability, one should keep in mind that bioactivation can result in unpredictable and unmanageable toxicity and the liability may not show up until late in development or when the drug is on the market. Thus, some measures to prevent bioactivation are appropriate. The most pragmatic approach is the avoidance of certain structural moieties that have been associated with bioactivation. Chapter 24 in this volume by J. Blagg provides more details about this approach.

REFERENCES

[1] I. Kola and J. Landis, *Nature Rev. Drug Disc.*, 2004, **3**, 711–715.
[2] D. P. Williams and B. K. Park, *Drug Disc. Today*, 2003, **8**, 1044–1050.
[3] J. L. Walgren, M. D. Mitchell and D. C. Thompson, *Critical Rev. Toxicol.*, 2005, **35**, 325–361.
[4] B. K. Park, N. R. Kitteringham, J. L. Maggs, M. Pirmohamed and D. P. Williams, *Annu. Rev. Pharmacol. Toxicol.*, 2005, **45**, 177–202.
[5] R. A. Roth, J. P. Luyendyk, J. F. Maddox and P. E. Ganey, *J. Pharmacol. Expt. Therap.*, 2003, **307**, 1–8.
[6] B. B. Brodie, W. D. Reid, A. K. Cho, G. Sipes, G. Krishna and J. R. Gillette, *Proc. Natl. Acad. Sci. USA*, 1971, **68**, 160–164.
[7] D. J. Jollow, J. R. Mitchell, W. Z. Potter, D. C. Davis, J. R. Gillette and B. B. Brodie, *J. Pharmacol. Exp. Ther.*, 1973, **187**, 195–202.

[8] D. C. Evans, A. P. Watt, D. A. Nicoll-Griffith and T. A. Baillie, *Chem. Res. Toxicol.*, 2004, **17**, 3–16.
[9] D. C. Evans and T. A. Baillie, *Curr. Opin. Drug Disc. Dev.*, 2005, **8**, 44–50.
[10] G. Tian, S.-Y. Chen, K. L. Facchine and S. R. Prakash, *J. Am. Chem. Soc.*, 1995, **117**, 2369–2370.
[11] S. H. Day, A. Mao, R. White, T. Schulz-Utermoehl, R. Miller and M. G. Beconi, *J. Pharmacol. Toxicol. Meth.*, 2005, **52**, 278–285.
[12] I. Gardner, J. S. Leeder, T. Chin, N. Zahid and J. Uetrecht, *Mol. Pharmacol.*, 1998, **53**, 999–1008.
[13] K. Samuel, W. Yin, R. A. Stearns, Y. S. Tang, A. K. Chaudhary, J. P. Jewell, T. Lanza, Jr., L. S. Lin, W. K. Hagmann, D. C. Evans and S. Kumar, *J. Mass Spectrom.*, 2003, **38**, 211–221.
[14] T. A. Baillie and M. R. Davis, *Biol. Mass Spectrom.*, 1993, **22**, 319–325.
[15] C. M. Dieckhaus, C. L. Fernandez-Metzler, R. King, P. H. Krolikowski and T. A. Baillie, *Chem. Res. Toxicol.*, 2005, **18**, 630–638.
[16] J. R. Soglia, S. P. Harriman, S. Zhao, J. Barberia, M. J. Cole, J. G. Boyd and L. G. Contillo, *J. Pharm. Biomed. Anal.*, 2004, **36**, 105–116.
[17] J. Gan, T. W. Harper, M.-M. Hsueh, Q. Qu and W. G. Humphreys, *Chem. Res. Toxicol.*, 2005, **18**, 896–903.
[18] J. R. Soglia, L. G. Contillo, A. S. Kalgutkar, S. Zhao, C. E. C. A. Hop, J. G. Boyd and M. J. Cole, *Chem. Res. Toxicol.*, 2006, **19**, 480–490.
[19] J. Castro-Perez, R. Plumb, L. Liang and E. Yang, *Rapid Commun. Mass Spectrom.*, 2005, **19**, 798–804.
[20] D. C. Kemp, P. W. Fan and J. C. Stevens, *Drug Metab. Dispos.*, 2002, **30**, 694–700.
[21] Q. Chen, J. S. Ngui, G. A. Doss, R. W. Wang, X. Cai, F. P. DiNinno, T. A. Blizzard, M. L. Hammond, R. A. Stearns, D. C. Evans, T. A. Baillie and W. Tang, *Chem. Res. Toxicol.*, 2002, **15**, 907–914.
[22] L. Yu, H. Liu, W. Li, F. Zhang, C. Luckie, R. B. van Breemen, G. R. J. Thatcher and J. L. Bolton, *Chem. Res. Toxicol.*, 2004, **17**, 879–888.
[23] T. M. Baughman, R. A. Graham, K. Wells-Knecht, I. S. Silver, L. O. Meyer, M. Wells-Knecht and Z. Zhao, *Drug Metab. Dispos.*, 2005, **33**, 733–738.
[24] A. Kretz-Rommel and U. A. Boelsterli, *Toxicol. Appl. Pharmacol.*, 1993, **120**, 155–161.
[25] S. J. Hargus, B. M. Martin, J. W. George and L. R. Pohl, *Chem. Res. Toxicol.*, 1995, **8**, 993–996.
[26] W. Tang, R. A. Stearns, S. M. Bandiera, Y. Zhang, C. Raab, M. P. Braun, D. C. Dean, J. Pang, K. H. Leung, G. A. Doss, J. R. Strauss, G. Y. Kwei, T. H. Rushmore, S.-H. L. Chiu and T. A. Baillie, *Drug Metab. Dispos.*, 1999, **27**, 365–372.
[27] G. K. Poon, Q. Chen, Y. Teffera, J. S. Ngui, P. R. Griffin, M. P. Braun, G. A. Doss, C. Freeden, R. A. Stearns, D. C. Evans, T. A. Baillie and W. Tang, *Drug Metab. Dispos.*, 2001, **29**, 1608–1613.
[28] L. J. Yu, Y. Chen, M. P. DeNinno, T. N. O'Connell and C. E. C. A. Hop, *Drug Metab. Dispos.*, 2005, **33**, 484–488.
[29] A. Carvajal Garcia-Pando, J. Garcia del Pozo, A. Sanchez Sanchez, A. Velaso Martin, A. M. Rueda de Castro and M. I. Lucena, *J. Clin. Psychiatry*, 2002, **63**, 135–137.
[30] A. S. Kalgutkar, A. D. N. Vaz, M. E. Lame, K. R. Henne, J. Soglia, S. X. Zhao, Y. A. Abramov, F. Lombardo, C. Collin and Z. S. Hendsch, and C. E. C. A. Hop,, *Drug Metab. Dispos.*, 2005, **33**, 243–253.
[31] S. E. Kostrubsky, S. C. Strom, A. S. Kalgutkar, S. Kulkarni, J. Atherton, R. Mireles, B. Feng, R. Kubik, E. Urda and A. E. Mutlib, *Tox. Sci.*, 2006, **90**, 451–459.

[32] A. S. Kalgutkar, I. Gardner, R. S. Obach, C. L. Shaffer, E. Callegari, K. R. Henne, A. E. Mutlib, D. K. Dalvie, J. S. Lee, Y. Nakai, J. P. O'Donnell, J. Boer and S. P. Harriman, *Curr. Drug Metab.*, 2005, **6**, 161–225.
[33] G. Zlokarnik, P. D. J. Grootenhuis and J. B. Watson, *Drug Disc. Today*, 2005, **10**, 1443–1450.
[34] M. Walles, P. Grude, P. E. Morin and J. Ducharme, Proceedings of the 53rd ASMS Conference on Mass Spectrometry and Allied Topics, San Antonio, Texas, June 5–9, 2005.
[35] S. Shen, S. J. Hargus, B. M. Martin and L. R. Pohl, *Chem. Res. Toxicol.*, 1997, **10**, 420–423.
[36] E. Fontana, P. M. Dansette and S. M. Poli, *Curr. Drug Metab.*, 2005, **6**, 413–454.
[37] D. A. Smith and E. F. Schmid, *Curr. Opin. Drug Disc. Dev.*, 2006, **9**, 38–46.

The Application of Pulmonary Inhalation Technology to Drug Discovery

A.R. Clark, R.K. Wolff, M.A. Eldon and S.K. Dwivedi

Nektar Therapeutics, San Carlos, CA, USA

Contents

1. Introduction	383
2. The lungs	384
3. Absorption and bioavailability	385
4. Pharmacokinetics	386
5. Safety	386
5.1. Toxicology studies	387
5.2. Clinical studies	387
6. Inhalation dosage form and delivery system considerations	390
7. Inhalation product development	391
8. Inhalation products	391
9. Conclusions	391
References	392

1. INTRODUCTION

"Pharmaceutical inhalers" have been used to treat respiratory diseases for centuries. However, "real" pulmonary delivery technology is an invention of the 20th century. At the turn of the 19th century, liquid nebulizers were invented. These devices use either mechanical energy or compressed gas to atomize an aqueous solution or suspension. In 1956, the chlorofluorocarbon (CFC)-propelled pressured metered dose inhaler (pMDI) was introduced, and in the late 1950s the dry powder inhaler (DPI) was added to the arsenal of aerosol delivery techniques. Over the last two decades, CFCs have been replaced with more environmentally friendly hydroflouralkane propellants and a large number of dry powder inhalers have been developed [1]. However, most of these technologies have been used to deliver drugs to the lung to treat local lung diseases such as asthma, chronic obstructive pulmonary disease (COPD) and cystic fibrosis (CF). Recently, much progress has been made in understanding the lung as an organ and its associated absorption and clearance mechanisms allowing the development of more efficient and reproducible pulmonary delivery systems. It is now possible to contemplate using the lungs as a portal of entry into the systemic circulation for proteins and peptides [2] and small molecules [3]. In the case of small molecules, pulmonary delivery can be advantageous where solubility, oral absorption and first-pass metabolism may limit drug development, or where rapid onset pharmacokinetics are required. Indeed, inhaled human insulin (Exubera® – insulin human [rDNA origin] inhalation powder; Pfizer

Inc., New York, NY; Nektar Therapeutics, San Carlos, CA), a 6-KD peptide, for the treatment of hyperglycemia in adults with type 1 or 2 diabetes mellitus has recently received approval from both the Food and Drug Administration and the European Agency for the Evaluation of Medicinal Products. A number of systemically targeted small molecules such as morphine, fentanyl, di-hydroergotamine and tetrahydrocannabinoid are also being developed as inhaled products. The inhaled route is becoming a practical route of delivery for both local and systemic therapies across a range of molecules with diverse physicochemical characteristics.

Inhalation drug delivery may offer a new avenue to scientists for understanding and enabling compounds hitherto difficult to deliver via the traditional routes. Molecules, with limited oral bioavailability, may be delivered systemically through the lungs. The unique attributes of drug's absorption, distribution, metabolism and excretion (ADME) through the pulmonary route may also offer a means of attaining rapid pharmacokinetic profiles and novel pharmacokinetic/pharmacodynamic relationships.

2. THE LUNGS

Although the lungs are an internal organs, they represent the largest surface (*ca.* 80 m^2) in the body that is in direct contact with the external environment (Fig. 1). The pulmonary epithelium, where gas exchange takes place, is thin (200–300 nm [4]) and permeable enough to allow absorption of proteins, peptides and small molecules. The lungs are also enzymatically less harsh than the gut. Although lung fluid and tissues do contain peptidases, proteases and cytochrome P450 enzymes, they are at far lower levels than in the gut [5] and 5–20 times lower than in the liver [6]. As a result, the lack

Epithelium cells	Clearance		Deposition	Anatomy
		Upper Airways		
Ciliated columnar Mucus (goblet) Serous Clara (non-ciliated)	Ingestion		> 10 μm	Mouth, pharynx, larynx
Brush Basal Intermediate	Mucociliary clearance and ingestion	Conducting Airways	3 – 10 μm	1-16 generations bifurcating structure Surface area ~ 1 m² Epithelial thickness 100μm to 10μm
Neuroendocrine Alveolar type I Alveolar type II Alveolar Macrophages Mast	Phagocytosis	Peripheral Airways	1 – 3 μm	16-23 generation of alveolated airways Surface area ~ 80 m² Epithelial thickness 100nm to 200nm

* Clearance: Mechanism for insoluble particles / Soluble material will also undergo some absorption in all airway compartments

Fig. 1. Airway structure and biology.

of substantial first-pass metabolism means that metabolic profiles can be strikingly different for this route of delivery. For example, steroids used for respiratory therapy such as budesonide and fluticasone have lung bioavailability close to unity but have very poor oral bioavailability due to poor absorption and first-pass effects [7,8].

However, to complement these low-level biological defense mechanisms, the lungs have an anatomical defense built into their structure (Fig. 1) [9]. The head is the first barrier to an inhaled aerosol that is trying to reach the absorptive surface of the airways and lungs. The oral cavity and the pharynx together present a tortuous path to an inhaled aerosol and large aerosol particles greater than 5–7 μm are mainly captured in the head and swallowed. The conducting airways that lead from the pharynx to the lung periphery consist of bifurcating tubes of ever-decreasing diameter and length. This labyrinth of tubes results in inhaled particles greater than 3–5 μm in aerodynamic diameter being deposited in these upper airways [10]. Particles less than 0.5–1 μm have a tendency to remain suspended in the inhaled air and are exhaled during expiration. The general requirement for efficient delivery to the lung and subsequent absorption is, therefore, to generate an aerosol with the majority of particles in the 1–5 μm range. The requirement for proteins and peptides is a little more stringent, 1–3 μm. The evidence suggests that small molecules can be absorbed throughout the airways [7] while the common belief is that proteins and peptides are mainly absorbed from the peripheral lung [2].

3. ABSORPTION AND BIOAVAILABILITY

The ability of modern aerosol technologies to generate fine aerosols, in addition to the physicochemical properties of lung fluid (approximately 10–25 mL of fluid containing lipids and proteins [11,12] with low enzyme content) and the large surface area of the lungs, point to a novel portal of entry for poorly soluble small molecules as well as biological molecules.

However, both the rate and extent of absorption show a dependency on the physical chemistry of the molecule being delivered. For proteins and peptides, molecular weight (MW) appears to be an overriding factor with absorption half-lives ($t_{1/2}$) being >24 h for larger biological molecules (MW > 10^5 kD) and a few minutes for small peptides [2–3]. For small molecules, lipid and aqueous solubility are important. Lipid-insoluble materials show a marked dependence of absorption kinetics on molecular weight and lipid-soluble molecules show virtually no dependence [13]. In the molecular weight range 50–1000 kD, small molecules show a $t_{1/2}$ ranging from 1 to 100 min. Prolonged absorption over several hours is seen in only two cases: highly insoluble molecules, such as fluticasone, amphotercin B and all-trans retinoic acid [3], or highly positively charged molecules, such as pentamidine and tobramycin [14]. In the latter case, retention may be due to a charge interaction between the molecule and the negative charges on the surface of the cell membrane.

As stated above, even though lung fluid and tissue possess substantially less enzymatic activity than the gut or liver, enzymes *are* present, and as a result the bioavailability of biologics is usually considerably <1. Insulin, for example, appears to have an intrinsic bioavailability (fraction absorbed from the fraction of compound deposited in the lung) of 0.3 [15]. The relationship between bioavailability and molecular weight for a number of macromolecules as determined by rat intratracheal instillation suggests that a "sweet spot" can be seen for proteins in the molecular weight range 10–40 kD [2]. However, most small molecules, with the exception of molecules such as ciclesonide [16], which is designed to undergo lung metabolism, are not degraded in the lung to any large extent and their intrinsic bioavailabilities appear to be close to 1.

4. PHARMACOKINETICS

The amount of drug reaching and depositing in the lungs and the amount filtered by the mouth is a function of the aerodynamic particle size characteristics of the inhaled aerosol and the patients' inhalation pattern. The amount of drug reaching the systemic circulation (central compartment) is a function of the extent of first-pass elimination that occurs in the lungs and gastrointestinal tract (GI). (First pass in the lungs is relatively low for most small molecules and GI absorption from an inhaled product can only be the result of oral deposition and subsequent swallowing.) After reaching the systemic circulation, the drug is distributed and is eliminated in the same manner as it would if it had been administered by any other route. Excessive oropharyngeal deposition can result in a large fraction of the drug being swallowed and not reaching the lungs until it has been absorbed and redistributed, typically at much lower concentrations than after local delivery. Peptide and protein drugs, or drugs susceptible to oral first-pass elimination or solubility-limited absorption, will only reach the systemic circulation provided that adequate deposition in the lungs occurs, i.e. oropharyngeal losses are minimized. Examples of the impact of inhaler and formulation design on oropharyngeal losses and lung deposition can be found elsewhere [17–19], however, lung deposition can range from 5% to 60% of the nominal dose depending upon inhaler and formulation design.

A typical example of the rapid absorption kinetics attainable *via* the pulmonary route with some small molecules is illustrated in the data presented [20] for inhaled morphine where the kinetics are essentially the same as intravenous (IV) delivery. For neutral small molecules, this type of profile is typical. Initial absorption is rapid and complete, and the elimination half-life is governed by systemic clearance rather than rate-limiting absorption. However, for some small molecules exhibiting prolonged absorption, flip–flop kinetics can be seen and the terminal phase can be governed by absorption [21].

5. SAFETY

Safety of inhalation delivery has been investigated for both small and large molecules. Beta-adrenergic agonists, corticosteroids, and anticholinergics have been inhaled for decades and reviews related to patient management do not reveal any meaningful adverse effects on the lung. Adverse events are associated most predominantly with systemic effects resulting from absorption of the drug into the general circulation and action at sites distant from the lung [22].

There have been fewer studies on the safety of inhaled biologics, since this is still a relatively new area [23].

5.1. Toxicology studies

Table 1 presents a short summary of some of the animal toxicology studies of inhaled pharmaceuticals. These studies provide data on higher dose exposures than are given clinically. With respect to insulin, it is encouraging to note that 6-month studies in both rats and monkeys have shown no adverse effects on lung histopathology [30]. The most pronounced findings are immunologically mediated effects because animals have been dosed with human proteins. Significant immunologically mediated reactions appear to be a low probability for most native therapeutic proteins that might be inhaled by humans. The limited data available suggest that if subcutaneous delivery has been well tolerated immunologically, there is a reasonable expectation that inhalation delivery will also be well tolerated.

Recently, studies of inhaled cytotoxic agents have been described [32]. The fact that lung toxicity was relatively modest even with these highly active molecules suggests that lung epithelium is relatively resilient and as a corollary, dosing with a wide range of agents can be contemplated. An aspect that works in favor of intrinsic safety of compounds delivered to the lung is that they need to be more pharmacologically potent than an orally administered compound because less dose can be delivered to the lung from a practical perspective (doses ranging from micrograms to tens of milligrams).

5.2. Clinical studies

Table 2 lists a number of macromolecules (proteins, peptides) and other high-molecular-weight molecules that are currently in use or are being investigated for clinical use *via* lung delivery. These trials were designed to investigate both efficacy and safety. Many of these inhaled proteins have shown efficacy as noted in other reviews [38]; only the safety aspects are emphasized here. There have been virtually no reports to date of adverse lung reactions, anaphylaxis, or immune reactions. There are several years of clinical data for DNase, tobramycin and inhaled insulin. In addition, a number of trials have been carried out for other agents (e.g., up to 6 months for leuprolide acetate [27], 1.3 years for heparin [40] and up to 1 year for interferon-alpha [35]). The lack of adverse respiratory effects in these trials is encouraging for other biologics that may be delivered *via* inhalation.

The experience with DNase and tobramycin is noteworthy because they are marketed products and hence have the most extensive databases. DNase has been on the market for over 9 years, and thousands of patients have used it for this length of time with a low frequency of reports of adverse respiratory effects

Table 1. Summary of toxicology studies of inhaled pharmaceutical candidates

Compound	Aerosol concentration (mg/m³)	Duration	Results	Reference
Human growth hormone (rats)	0.27	0.3 h/day 7 and 28 days	No airway or alveolar inflammation, mild perivascular cuffing	[24]
Bovine growth hormone (rats)	0.27	0.3 h/day 7 d	No apparent lung reaction	[24]
Human growth hormone (monkeys)	50, 150, 500	0.25 h/day 28 days	No apparent lung reaction	[24]
Alpha-1-antitrypsin (rats)	31, 162, 655	6 h/day 28 days	Concentration-dependent increase in round cells; lung weight increase on high dose only	[25]
Albumin (rats)	719	6 h/day 28 days	Lung weight increase	[26]
Leuprolide acetate (dogs)	100	0.1, 0.2, 0.5 h/day 6 m	No apparent lung reaction	[27]
DNase (rats and monkeys)	70	0.1, 0.2, 0.5 h/day 28 d	Dose-related alvelitis and bronchiolitis	[28]
Tobramycin (rats)	1000	0.3 h/day 1 h/day 3 h/day 6 months	Dose-related increase in lung weights; dose-related hypreplasia of bronchial/bronchiolar epithelium, alveolar macrophage influx at all doses	[29]
Insulin (rats and monkeys)	Not quoted	6 months	No adverse lung effects	[30]
IgE (rats and monkeys)	116 µg/day (rats) 67 µg/day	7 days	No adverse lung effects reported	[31]
Chemotherapeutic agents (dogs)	1–480	Dose escalation protocol	Mixed results No lung toxicity • carboplatin (up to 45 mg/m³) Dose-dependent toxicity • cisplatin (up to 140 mg/m³) • doxorubicin (up to 120 mg/m³)	[32]

Note: Ig, immunoglobulin.

attributable to the inhaled protein. The only adverse effects of the formulated drug on the respiratory system noted in the summary basis of approval [28] were voice alteration, laryngitis and pharyngitis. Considering that DNase is an active enzyme, such effects are not entirely unexpected. The experience with tobramycin has been similar. Treatment of CF patients with tobramycin in clinical trials showed an improvement in pulmonary function and also a decrease in the incidence of several adverse experiences. No safety problems were noted in the summary basis of approval [29] with the exception of some tinnitus and voice alteration.

Table 2. Clinical use of inhaled biomolecules

Biomolecule	No. of patients	Length of treatment	Reference
DNase	Thousands	Up to 9 years	[28]
Tobramycin	Thousands	Up to 7 years	[29]
Insulin	Thousands	Up to 4 years	[33,34,35,36]
Leuprolide acetate	Hundreds	6 months	[27]
Antibiotics	Hundreds	Up to 2 years	[37]
Interferon-alpha	16		[38]
Interferon-gamma	5		[39]
Alpha-1-antitrypsin	12		[40]
Calcitonin	Not stated	1 day	[41]
Human growth hormone	12		[24]
Anti-IgE	15	Up to 42 days	[42]

Note: Ig, immunoglobulin.

There have also been extensive studies of antibiotics other than tobramycin. Although not proteins, these high-molecular-weight molecules have solubility and absorption properties within the range of therapeutic proteins. Hall [34] has reviewed the use of inhaled antibiotics, noting that there have been several studies conducted with sizable numbers of patients (50–100) for periods extending up to 2 years. The aerosolized antibiotic gentamicin [44] and polypeptides such as polymixin [42] achieved the highest exposure among the inhaled compounds studied. Typically, this is on the order of several tens to hundreds of milligrams as compared to the 2.5-mg dose for DNase and other proteins or 1–100 µg for most local respiratory compounds. This is another example of human inhalation exposure to large therapeutic molecules in which virtually no adverse reactions have been reported, except minor irritation in some cases.

Recent experience with insulin has been of considerable interest as the first approved inhaled protein intended for systemic delivery. The most recent data from 24 months of use in patients treated either with inhaled human insulin (Exubera®), or oral agents, provide the longest evaluation of pulmonary safety. Both groups experienced declines in pulmonary function from baseline measured as FEV_1 (forced expired volume in 1 s) and DL_{CO} (carbon monoxide diffusing capacity) [33]. It was found that mean treatment group differences were small, occurred early after treatment initiation, had no clinical relevance and did not progress with up to 2 years of continued treatment. It was observed that serum antibody levels were higher in patients receiving inhaled human insulin compared with those taking subcutaneous injections. Studies have shown that these antibodies are not neutralizing and appear to have no impact on blood glucose control and therefore are of no real clinical significance [34,35]. Studies also have shown higher incidence of cough in subjects treated with inhaled insulin, which is predominantly mild and rarely productive, occurring soon after dose administration [36].

A number of trials have been carried out for significant periods of time for other agents (e.g., up to 6 months for leuprolide acetate [27], 1.3 years for heparin [43] and up to 1 year for interferon-alpha [38]). The lack of adverse respiratory effects in a significant number of subjects is encouraging for other macromolecules that may be delivered *via* inhalation.

The above discussion suggests that safety of inhaled molecules will continue to have to be investigated on a case-by-case basis. However, the general conclusion is that for a wide range of inhaled small and large molecules of pharmaceutical interest, the lung appears to be relatively tolerant and should be investigated as viable route of delivery when warranted.

6. INHALATION DOSAGE FORM AND DELIVERY SYSTEM CONSIDERATIONS

The science and technology of drug delivery through the lungs has advanced rapidly in recent decades. Inhalation product development has become a significant new discipline [45,46], with expanding types of products addressing various disease targets. As described above, most of these pulmonary products have been focused on lung or lung-related diseases, but newer products, especially those intended for the systemic circulation, are proliferating.

Inhalation products, whether intended for intra-pulmonary or systemic delivery, have one fundamental attribute at the basis of dosage form design: the aerodynamic particle size distribution. Drugs can be converted to pulmonary formulations by simple size reduction processes such as jet-milling, often followed by admixture with bulking agents such as lactose. More sophisticated approaches involve spray-drying of drug solution or suspension in compositions comprising one or more excipients, such as simple sugars, amino acids, phospholipids, buffering agents and surfactants. While delivering excipients to the lung can require extensive toxicology assessment, there are now a wide range of excipients used in approved inhalation products. The inhalation product manufacturing systems employ technologies and unit operations customized to handling these fine particles, and the pharmaceutical industry is now familiar with a broad range of inhalation product platforms for pulmonary drug delivery.

The delivery of pulmonary formulations requires the development of appropriate devices or delivery systems. Nebulizers for liquid or suspension formulations, single- or multiple-dose inhalers for dry powder formulations, metered dose inhalers for solution or suspension formulations and many ancillary devices for delivery assurance and compliance enhancement are available on the market [47,48].

Inhalation products deliver to the lungs anywhere from 5% to 60% of the nominal dose contained in primary packages. For high potency compounds, this could mean delivery of just a few micrograms to the lungs. For other compounds requiring large doses, lung doses as high as 50–100 mg may be considered. The adjustment of the product profile is achieved by an integrated design effort on formulation, packaging and delivery system.

7. INHALATION PRODUCT DEVELOPMENT

While the *in vitro* science of inhalation drug delivery is well established and diverse means exist for *in vivo* study of inhalation exposure, the integrated development experience of inhalation products is limited to just over the past two decades. This limited experience has required the regulatory authorities to adopt a conservative approach toward establishing and finalizing guidance for development of inhalation products [49].

8. INHALATION PRODUCTS

The advancements in design and control of high-performance inhalation products are complemented by the advancements in the tools necessary for understanding drugs for pulmonary drug delivery. At the early stages of development, the pulmonary ADME of compounds can be explored rapidly and inexpensively by the use of custom delivery systems for small and large animal species. Precise animal experiments are often able to accurately predict development paths. However, the efforts in inhalation product design in response to chemical properties and potencies are delivered to the inhalation product development scientists by the discovery chemist. Very few drug design efforts have focused on pulmonary drug delivery [15,50,51].

As scientists begin to appreciate the potential of the pulmonary route, it is expected that compounds specifically designed to leverage the biopharmaceutics of this route will be advanced into the clinic for existing and new targets. It is expected that the compounds previously abandoned due to delivery-related problems might be revived. Besides leveraging the technical potential of this route, the discovery chemists may find themselves in uncharted intellectual property territory expanding the business potential of both inhalation products and pharmacological targets. The future growth of pulmonary drug delivery is as much in the hands of discovery experts, by considering it early in their drug design efforts for all targets, as it is in the hands of development experts by effectively converting the discovered candidates into commercial products.

9. CONCLUSIONS

Over the past few decades, much progress has been made in understanding lung biology and the physical and biological barriers associated with pulmonary delivery. Technology has been developed to efficiently and reliably deliver pharmaceuticals to the lung in doses ranging from micrograms to several tens of milligrams. The lungs exhibit novel pharmacokinetics with substantially reduced first-pass metabolism and the potential for rapid absorption. Delivery to the lungs appears to be generally safe. The combination of these features makes the lung an obvious and valid route of delivery for macro- and small molecules alike and establishes pulmonary delivery as an additional option to treat human diseases.

REFERENCES

[1] A. R. Clark, in *Inhalation Aerosols: Physical and Biological Basis for Therapy* (ed. A. J. Hickey), 2nd edition. Taylor & Francis Inc., New York, 2004.
[2] J. S. Patton, *Adv. Drug Deliver. Rev.*, 1996, **19**, 3–36.
[3] J. S. Patton, S. C. Fishburn and J. G. Weers, *Proc. Am. Thorax Soc.*, 2004, **1**, 338–344.
[4] E. R. Weibel, *Morphometry of the Human Lung*, Springer-Verlag, Berlin, 1963.
[5] J. R. Bend, G. E. R. Hook and T. E. Gram, *Drug Metab. Dispos.*, 1972, **1**, 358–367.
[6] T. Matsusbara, R. A. Prough, M. D. Burke and R. W. Estabrook, *Cancer Res.*, 1974, **34**, 2196–2203.
[7] L. Borgstrom and M. Nilsson, *Pharm. Res.*, 1990, **7**, 1068.
[8] Consensus statement from a workshop of the British Association for Lung Research, *Respir. Med.*, 1999, **93**, 123–133.
[9] E. R. Weibel, *Respiratory Physiology: An Analytical Approach*, Marcel Dekker, New York, 1989, 1–56.
[10] G. Rudolf, R. Kobirch and W. Stahlhofen, *J. Aerosol. Sci.*, 1990, **21**, S403–S406.
[11] G. E. Hatch, in: *Treatise on Pulmonary Toxicology: Comparative Biology of the Normal Lung* (ed. R. A. Parent), CRC Press, Boca Raton, FL, 1992, 617–632.
[12] R. H. Stephens, A. R. Benjamin and D. V. Walters, *J. Appl. Physiol.*, 1996, **80**, 1911–1920.
[13] L. S. Shanker, *Biochem. Pharmacol.*, 1978, **27**, 381–385.
[14] J. P. Monk and P. Benfield, *Drugs*, 1990, **39**, 741–756.
[15] J. S. Patton, J. Bukar and S. Nagarajan, *Adv. Drug Deliver. Rev.*, 1999, **35**, 235–247.
[16] M. Stoeck, R. Riedel, G. Hocchaus, D. Häfner, J. M. Masso, B. Schmidt, D. Hatzelmann and D. S. Bundschuh, *J. Pharmacol. Exp. Ther.*, 2004, **309**, 249–258.
[17] L. Borgstrom, H. Bisgaard, C. O'Callaghan and S. Pedersen, in *Drug Delivery to the Lungs* (eds. H. Bissgard, C. O'Callaghan and G. C. Samldone), Marcel Dekker, New York, 2002, 421–448.
[18] C. O'Callaghan and P. Wright, in *Drug Delivery to the Lungs* (eds. H. Bissgard, C. O'Callaghan and G. C. Samldone), Marcel Dekker, New York, 2002, 337–370.
[19] A. L. Adjei, Y. Qiu and P. Gupta, in *Inhalation Aerosols* (ed. A. J. Hickey), Marcel Dekker, New York, 1996, 197–228.
[20] S. J. Farr, J. A. Schuster, P. Lloyd, L. J. Lloyd, J. K. Okikawa and R. M. Rubsamen, in *Respiratory Drug Delivery VIII* (eds. R. N. Dalby, P. R. Byron and S. J. Farr), Interpharm Press, Tucson, AZ, 1996, 175–185
[21] A. M. Taburet and B. Schmit, *Clin. Pharmacokinet.*, 1994, **26** (5), 396–418.
[22] M. R. Sears and J. Lovall, *Respir. Med.*, 2005, **99** (2), 152–170.
[23] R. K. Wolff, *J. Aerosol Med.*, 1998, **11**, 197–219.
[24] R. K. Wolff, D. Clarke, M. Carfagna, M. Stoiff, H. Smith, M. Shaw, L. Johnston, B. Jackson and D. Edwards, *Congress of International Society of Aerosol in Medicine*, Baltimore, MD, June 14–18, 2003.
[25] J. Pauluhn and T. Martins, Proceedings of the Fourth International Aerosol Conference, Los Angeles, CA, 1994, p. 683 (abstract).
[26] J. D. Green, *Hum. Exp. Toxicol.*, 1994, **13** (suppl), Sl–S42.
[27] A. L. Adjei, *J. Aerosol Med.*, 1995, **8**, 131.
[28] United States Food and Drug Administration, Summary Basis of Approval for Dornase Alfa, recombinant Pulmozyme Inhalation Solution, 1993, FDA, Washington, DC.
[29] United States Food and Drug Administration, Summary Basis of Approval for Tobramycin, 1997, FDA Washington, DC.

[30] J. S. Patton and R. M. Platz, in *Respiratory Drug Delivery IV* (eds. P. R. Byron, R. N. Dalby and S. J. Farr), Interpharm Press, Buffalo Grove, IL, 1994, 65–74.
[31] T. D. Sweeney, M. Marian, K. Achilles, J. Bussiere, J. Ruppel, M. Shoenhoff and R. J. Mrsny, in *Respiratory Drug Delivery VII.* (eds. R. N. Dalby, P. R. Byron, S. J. Farr and J. Peart), Serentec Press, Tarpon Springs, FL, 2000, 59–66.
[32] M. E. Placke, W. C. Zimlich, J. Y. Ding, D. J. Westaway and A. R. Imodi, in *Respiratory Drug Delivery VIII* (eds. R. N. Dalby, P. R. Byron, J. Peart and S. J. Farr), Davis Horword International Publishing, Tucson, AZ, 2002, 15–23.
[33] M. Dryer, *Diabetologia*, 2004, **47** (suppl 1), A44.
[34] T. Heise, S. Bott, C. Tusek, J.-A. Stephan, T. Kawabata, D. Finco-Kent, C. Liu and A. Krasner, *Diabetes Care*, 2005, **28**, 2161–2169.
[35] S. E. Fineberg, T. Kawabata, D. Finco-Kent, C. Liu and A. Krasner, *J. Clin. Edocrinol. Metab.*, 2005, **90**, 3280–3294.
[36] Exubera® [Package insert], 2006, Pfizer Inc., New York, NY.
[37] C. B. Hall, *J. Aerosol Med*, 1989, **2**, 221–231.
[38] V. Kinnula, K. Cantell and K. Mattson, *Eur. J. Cancer*, 1990, **26**, 740–741.
[39] R. J. Martin, M. Boguniewicz, J. E. Henson, A. C. Celniker, M. Williams, R. C. Giorno and D. Y. Leung., *Am. Rev. Respir. Dis.*, 1993, **148**, 677–1682.
[40] N. G. McElvaney, R. C. Hubbard, P. Birrer, M. S. Chernick, D. B. Caplan, M. M. Frank and R. G. Crystal, *Lancet*, 1991, **337**, 392–394.
[41] J. S. Patton and R. M. Platz, *Adv. Drug Deliv. Rev.*, 1992, **8**, 179–196.
[42] L. P. Boulet, K. R. Chapman, J. Cote, S. Kalra, R. Bhagat, V. A. Swystun, M. Laviolette, L. D. Cleland, F. Deschesnes, J. Q. Su, A. DeVault, R. B. Vick and D. W. Cockcroft, *Am J. Respir. Crit. Care Med.*, 1997, **155**, 1835–1840.
[43] D. Kohler, *J. Aerosol Med.*, 1994, **7**, 307–314.
[44] J. S. Ilowite, J. D. Gorvoy and G. C. Smaldone, *Am. Rev. Respir. Dis.*, 1987, **136**, 1445–1449.
[45] H. M. Courrier, N. Butz and T. F. Vandamme, *Crit. Rev. Ther.*, 2002, **19**, 425–498.
[46] D. A. Groneberg, C. Witt, U. Wagner, K. F. Chung and A. Fischer, *Resp. Med.*, 2003, **97**, 382–387.
[47] N. R. Labiris and M. B. Dolovich, *Brit. J. Clin. Pharmacol.*, 2003, **56**, 600–612.
[48] D. W. Combs, M. S. Rampulla, R. K. Russel, R. A. Rampulla, D. H. Klaubert, D. Ritchie, A. S. Meeks and T. Kirchner, *Drug Des. Deliv.*, 1990, **6**, 241–254.
[49] FDA draft guidance for nasal spray and inhalation solutions, suspensions and spray drug products, May 1999.
[50] D. Jack, *Drug Safety*, 1990, **5** (suppl. 1), 4–23.
[51] I. Szelenyi, G. Hocchaus, S. Heer, S. Küsters, D. Marx, H. Poppe and J. Engel, *Drugs Today*, 2000, **36**, 313–320.

Recent Advances in Oral Prodrug Discovery

Aesop Cho

Gilead Sciences, Inc. 333 Lakeside Drive, Foster City, CA, USA

Contents

1. Introduction	395
2. Prodrugs of poorly soluble compounds	395
3. Prodrugs of polar compounds	398
4. Prodrugs targeting active transporters	400
5. Prodrugs for targeted tissue delivery	401
6. Prodrugs with other benefits	403
7. Conclusions	405
References	405

1. INTRODUCTION

In general, prodrug research begins with careful selection of pro-moieties to be conjugated to the parent molecule. A number of structural classes of pro-moieties are available for prodrug derivatization with information on enzymatic or non-enzymatic activation of prodrugs to the parent drugs [1]. Once a prodrug derivative that affords a promise in correcting the pharmacokinetic limitation of the parent drug is found, further optimization efforts may follow for most favorable physico-chemical and pharmacokinetic properties.

During the past few years, significant progress has been achieved in oral prodrug discovery research. In particular, rational prodrug design has been more frequently practiced because of increased understanding of the complex functions of the intestinal epithelium in the drug absorption process. Furthermore, selective delivery of active parents to target sites has been pursued through the exploitation of tissue-specific transporters and metabolic enzymes for prodrugs. Other innovative applications of prodrugs, such as prolongation of plasma half-life, have also been realized. Structurally, a number of new useful pro-moieties have been discovered from these recent efforts, broadening the scope of prodrug derivatizations. This chapter will provide an update on the progress of oral prodrug research, with illustration of recent examples of prodrug derivatives. Several previous reviews concerning oral prodrug approaches can also be consulted for further information [2–7].

2. PRODRUGS OF POORLY SOLUBLE COMPOUNDS

Compounds are considered to be poorly soluble when their oral bioavailability is limited by the rate and extent of dissolution in the gastrointestinal (GI) tract.

A typical prodrug strategy in such cases focuses on chemical derivatization to improve water solubility. Highly polar, ionizable pro-moieties are often required to obtain an adequate level of solubility. Paradoxically, however, this tactic compromises intestinal permeability. One way to get around this issue is to design solubilizing pro-moieties so as to utilize hydrolytic enzymes bound to the intestinal mucosal membrane. Activation of the prodrug in such proximity to the membrane results in absorption of the parent compound [8]. Less polar pro-moieties are also useful for compounds whose solubility is limited by intermolecular interactions, affording an appropriate balance of lipophilicity and solubility for intestinal absorption. A list of pro-moieties commonly used for poorly soluble compounds is available in a previous review on this topic [5]. More recent examples of new prodrugs and pro-moieties are included here.

Amprenavir (**1**), a clinically used HIV protease inhibitor, has low-aqueous solubility (~0.04 mg/ml) that limits oral bioavailability. Accordingly, the clinical dosage form is a soft-gel capsule where a high ratio of excipients to drug (~8:1) is necessary to aid solubilization in the GI tract and thus absorption. As a more compact dosage form was desirable, a water-soluble prodrug was sought. Out of 60 prodrug forms that were evaluated *in vitro* and *in vivo* screens, the phosphate prodrug **2** (fosamprenavir) was chosen for development. Compound **2** is highly soluble ($\geqslant$3 mg/ml) and affords, upon oral administration to dogs as hard-gelatin capsules, 64% of the systemic drug blood levels attained with the clinical formulation of amprenavir [9]. The data suggested that the prodrug **2** itself does not cross the intestinal epithelium, but that the parent **1** is generated upon hydrolysis by alkaline phosphatase on the apical membrane and then crosses the epithelium, thereby reaching the systemic circulation. A crystalline calcium salt of **2** entered the clinic in 2003.

1 R = H
2 R = -P(O)(OH)$_2$

3 R = H
4 R = -CH$_2$OP(O)(OH)$_2$

Another example of a phosphate-based prodrug class is that of BMS-488043 (**3**), a HIV-1 attachment inhibitor. The prodrug **4** is a structural variant that utilizes a spacer group (–CH$_2$O–) between the parent molecule and the phosphate pro-moiety. Owing to the poor solubility of **3**, oral absorption becomes saturated at high doses. For this reason, doses of 800–1800 mg of **3** b.i.d. must be taken in conjunction with high-fat meals to achieve sufficient plasma exposure for significant antiviral activity in humans. These constraints limit the clinical utility of the drug. Compound **4** is highly soluble (>18 mg/ml at pH 6.5) and is converted to the parent rapidly and completely by alkaline phosphatase both *in vitro* and *in vivo*. The absolute oral bioavailability of **3** from **4** in animals was 62–94%, which is comparable to that of **3** itself when formulated as a solution in PEG (60–90%). Comparative oral pharmacokinetic studies in monkeys found that use of **4** gave shorter T_{max} (0.83 h vs. 2.7 h) and higher C_{max} (16 µM vs. 5.8 µM) of **3** than those resulting from dosage of the parent itself, indicating more rapid and efficient absorption. The C_{24h} (0.067 µM vs. 0.074 µM) and terminal $t_{1/2}$ (4.8 h vs. 4.2 h) were about the same. In addition, dose-linearity of the plasma exposure was observed even at high doses [10].

An antifungal triazole BAL8514 (**5**) was derivatized to the prodrug BAL8557 (**6**) initially for parenteral formulation. The pro-moiety including the spacer group was developed by rational iterative optimization [11]. The incorporation of a quaternary triazolium salt, along with sarcosine and pyridine groups, gave a huge increase in solubility (>100 mg/ml vs. <0.0001 mg/ml). The pro-moiety was designed in such a way that it is rapidly degraded to release the parent **5** immediately upon enzymatic hydrolysis of the sarcosine ester group followed by chemical breakdown. The oral bioavailability of **5** from **6** in monkeys was good (87%) [11], and after oral or intravenous administration of **6** the total plasma AUC of **5** in humans was comparable [12]. Another prodrug form, **7**, was also orally bioavailable [13]. A similar approach involving quaternarization of a tertiary amine was applied to the poorly soluble compound siramesine (Lu28-179, **8**). Prodrugs such as **9** and **10** showed ~10^6-fold increase in solubility as well as efficient enzyme-mediated activation to the parent. However, the oral bioavailability of **8** was not improved through use of these prodrugs. This is most likely due to precipitation of the parent, resulting from over-rapid hydrolysis of the prodrugs [14].

A new structural class of prodrugs was devised for the arylsulfonamide GW678248 (**11**), a HIV-1 non-nucleoside reverse transcriptase inhibitor. A series of *N*-acylated analogs were evaluated, and GW695634 (**12**) was identified among them as optimal, exhibiting improved aqueous solubility (92 mg/ml as the sodium salt vs. 0.00018 mg/ml for **11**) and oral bioavailability in rats and dogs (>40%). The oral bioavailability in monkeys was low (10–20%) due to less-efficient activation to the parent [15,16]. An *N,N*-dimethylglycine ester of the lipophilic alcohol CEP-5214 (**13** to CEP-7055, **14**), a Mannich base–type prodrug of the indoline-2-one semaxanib (**15,16**) and an acyloxymethylcarbamate derivative of an oxazolidinone (**17,18**) are other examples of recent prodrugs displaying improved oral bioavailability [17–19]. Notably, compound **18** has a neutral, non-ionizable pro-moiety that still affords three to four fold increase in oral bioavailability (75% vs. 21%) in dogs.

8

9 R =
10 R =

11 R = H
12 R =

13 R = H
14 R =

15 R = H
16 R =

17 R = H
18 R =

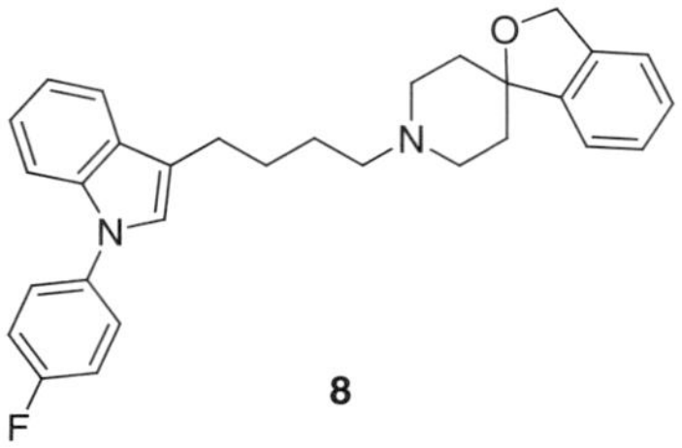

3. PRODRUGS OF POLAR COMPOUNDS

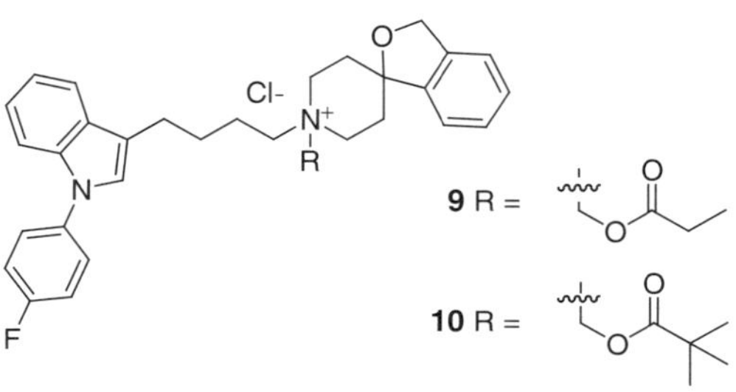

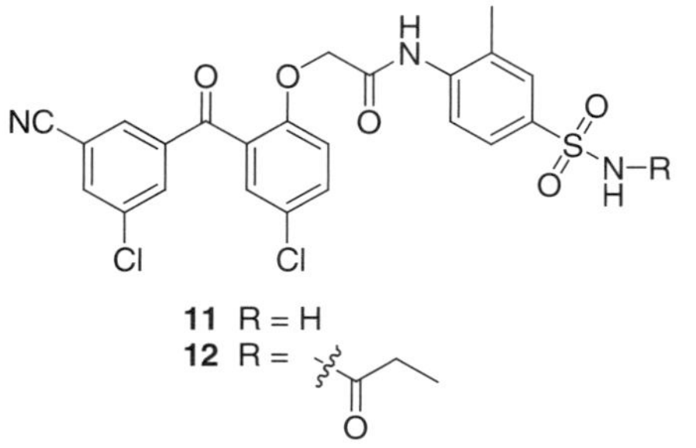

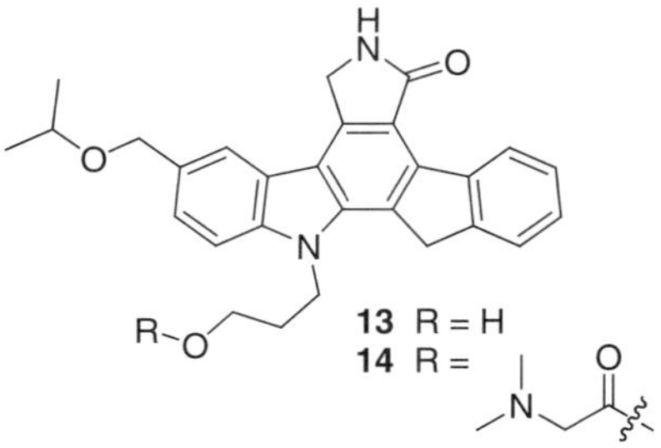

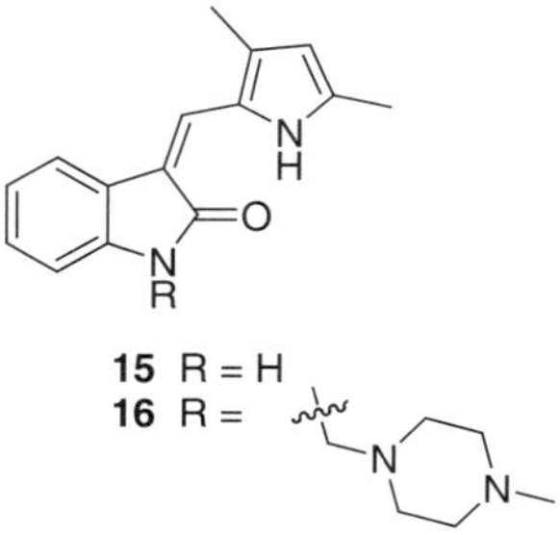

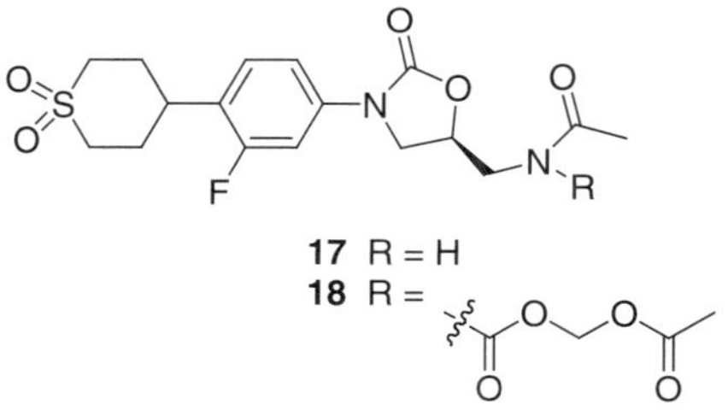

Prodrug derivatization of polar compounds is primarily directed toward increasing lipophilicity to enhance passive transcellular absorption, while retaining an acceptable level of solubility. Accordingly, the design involves masking ionizable or hydrogen-bonding groups with appropriate pro-moieties. This strategy has been generally successful and a number of prodrugs in this category are being used clinically [4]. However, it has become apparent that an increase in lipophilicity alone does not necessarily lead to improvement in oral bioavailability. This is due to the diverse function of intestinal mucosal cells (enterocytes) as the first-line of barrier to absorption. Enterocytes can metabolize prodrugs, and actively control the trafficking of prodrugs and their parents (generated inside cells) between the intestinal lumen and the plasma by influx and efflux transporters embedded on both the apical and basolateral membranes. Therefore, the influence of all these factors upon oral bioavailability has been an important issue in prodrug design and optimization. Prodrugs of polar compounds and analysis of the mechanisms of their intestinal absorption were previously reviewed [5]. More recently published work is covered here.

The polar carboxylic acid **19**, a beta-lactamase inhibitor, is very poorly absorbed in rodents when administered orally. Several ester-type prodrugs with known pro-moieties including **20** and **21** were evaluated initially, but their oral bioavailability was not satisfactory. Further exploration of new pro-moieties led to the discovery of highly orally bioavailable (>93%) prodrugs such as **22** and **23**. Compound **22** was crystalline, with a melting point (136°C) suitable for tableting, milling, and purification [20].

The diacid **24** is poorly orally bioavailable and was derivatized to the corresponding diethyl ester **25**, which showed high oral bioavailability (50%) in rats [21]. In contrast, the diester prodrug ME3229 (**27**) afforded low oral bioavailability (10%) of the parent diacid ME3277 (**26**) in rats due to an efflux transporter expressed on the apical membrane of enterocytes. Compound **27** crossed the apical membrane efficiently, but was metabolized to **26**, **28**, and **29** inside cells. These metabolites were then expelled back into the intestinal lumen by breast cancer resistant protein (BCRP) efflux transporters [22]. Various active transporters influencing pharmacokinetics of drugs were recently reviewed [23]. An opposite case was reported where an efflux transporter helps improve oral absorption. Experiments with Caco-2 cells indicated that pivampicillin (**31**) loaded into the apical chamber crosses the apical membrane and is converted to ampicillin (**30**), which is actively transported out to the basolateral medium [24].

Similar to the discovery of capacitabine (see Section 5 for details), a nucleoside antimetabolite **32** was converted to a series of lipophilic carbamate-type prodrug derivatives. Compound **33**, with a log *D* of 2.0, afforded the most improved oral bioavailability (43% vs. 11% for the parent control) in rats [25].

19 R = H

20 R =

21 R =

22 R =

23 R =

24 R = H
25 R = Et

26 R_1, R_2 = H
27 R_1, R_2 = nBu
28 R_1 = H, R_2 = nBu
29 R_1 = nBu, R_2 = H

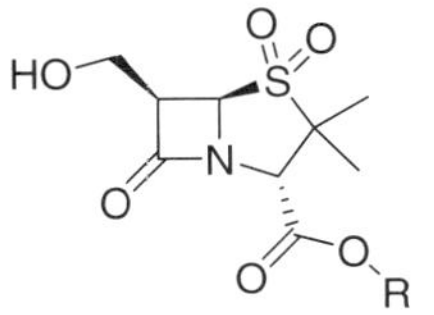

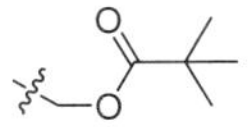

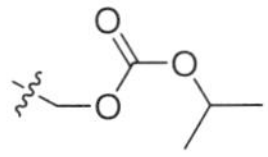

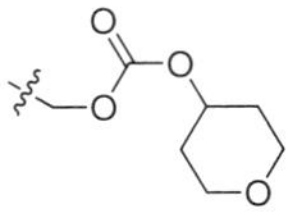

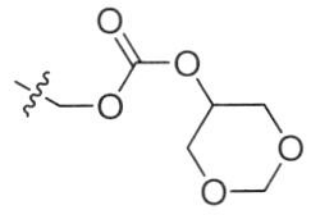

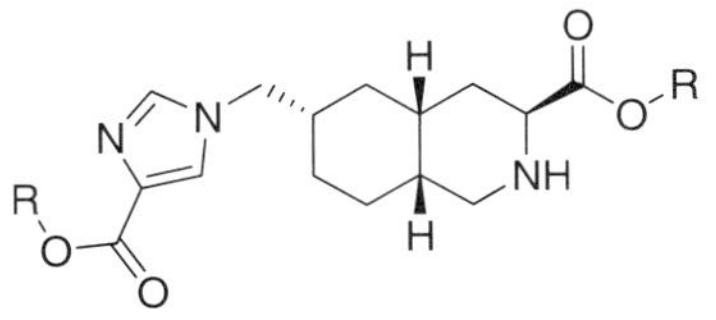

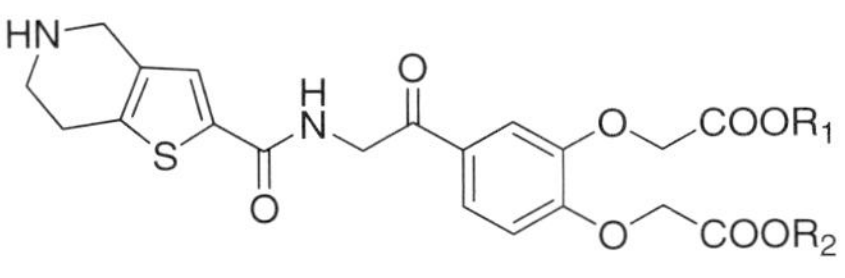

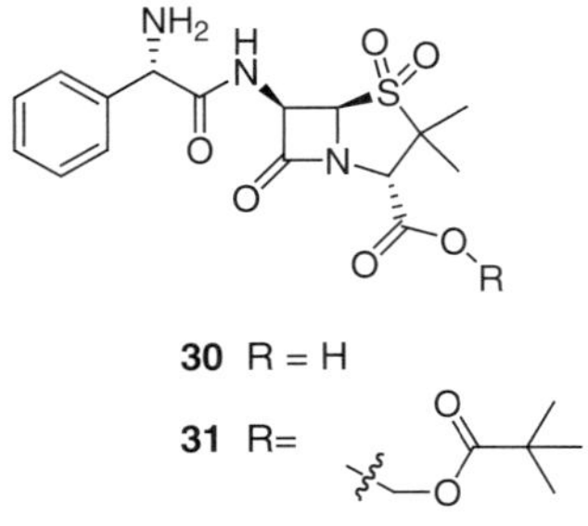

4. PRODRUGS TARGETING ACTIVE TRANSPORTERS

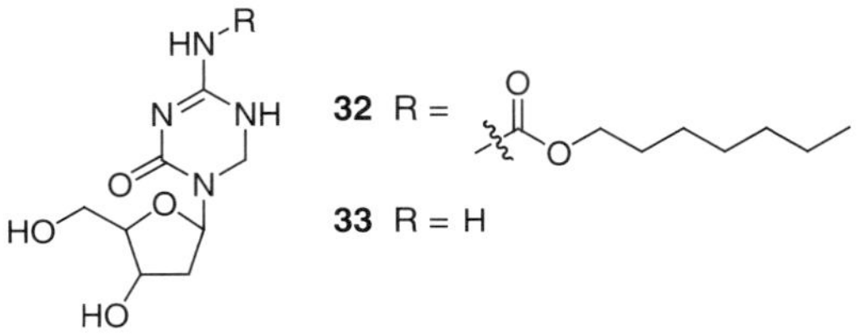

Mechanistic studies of intestinal transepithelial transport have revealed that the oral absorption of certain structural classes of drugs and prodrugs is enhanced by membrane-bound active transporters [7]. Such proteins include peptide transporters, amino acid transporters, nucleoside transporters, bile acid transporters, and monocarboxylic acid transporters. The clinically proven prodrug valacyclovir was originally discovered through random screening of various prodrug derivatives of acyclovir [26]. It was later found that valacyclovir is recognized and actively taken up by the peptide transporter PepT1, accounting for the improved oral bioavailability [27]. Since this discovery, targeting PepT1 or other transporters has been of continual interest in prodrug discovery research.

The unnatural amino acid **34** (LY354740), a group II metabotropic glutamate receptor agonist, has a poor oral bioavailability (10% in rats) due to low permeability across the intestinal epithelial membrane. With the goal of finding oligopeptides that are substrates of PepT1, a number of peptidic prodrug derivatives such as **35–37** were prepared by linking various amino acids and dipeptides to the C-2, C-6, and N-2 groups of the parent compound. Among them, the *N*-alanyl dipeptide **35** (LY544344) showed high affinity to human PepT1 (IC_{50} 0.12 mM vs. >19 mM for **34**) *in vitro* and complete conversion to parent when incubated with rat liver or human jejunum tissue preparations. These results correlated well with *in vivo* studies; the oral bioavailability of **34** from prodrug **35** in rats was 85%, with no prodrug circulating 30 min after oral administration [28,29].

Gabapentin (**38**), when administered orally, is absorbed by a low-capacity transporter localized in the upper small intestine. Saturation of the transporter at high therapeutic doses leads to non-linear plasma exposure and interpatient variability. To overcome this limitation, a prodrug approach was taken to target high-capacity transporters that are broadly distributed in the intestinal tract of humans. Prodrug derivatizations were conducted initially on both the amino and the carboxylic acid functional groups. The C-capped ester-type prodrug derivatives were disfavored because of their propensity to cyclization, forming the unwanted lactam **40**. Among the N-capped carbamate-type prodrug derivatives, compound **38** (XP13512) displayed the best profile for clinical development. *In vitro* studies demonstrated that **39** was efficiently absorbed through the actions of monocarboxylate transporter type 1 (MCT-1) and sodium-dependent multivitamin transporter (SMVT), and passive diffusion [30]. The oral bioavailability of gabapentin from **39** in monkeys at the highest anticipated clinical dose was significantly improved (84% vs. 25%), and the prodrug exposure was minimal [31].

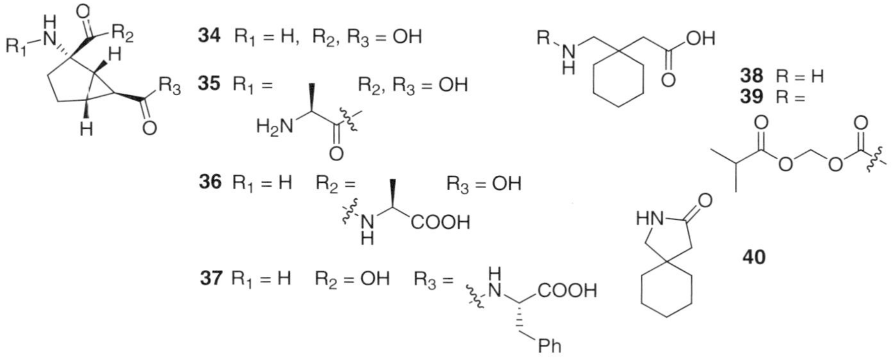

5. PRODRUGS FOR TARGETED TISSUE DELIVERY

One of the more recent endeavors in the oral prodrug field has been to selectively deliver active principles to target tissues. For this approach to be effective, the prodrugs must be absorbed intact through the intestinal mucosa, be resistant to metabolism by enterocytes and plasma, and be selectively taken up or reconverted to the parents by the target tissues. Results from recent studies on tissue distribution and metabolism of some prodrugs have strengthened the potential of this approach.

Capecitabine (**41**) is a recently developed oral prodrug of 5-fluorouracil (5-FU, **44**) with enhanced tumor selectivity. Initially, 5′-deoxy-5-fluorouridine (**43**) was developed as a prodrug of 5-FU. Compound **43** was hydrolyzed to 5-FU by thymidine phosphorylase, an enzyme that is found in higher concentration in tumors than in normal tissues, affording an improved therapeutic index. When administered orally, however, **43** caused dose-limiting GI side-effects such as diarrhea, due to the toxicity of 5-FU generated in the intestinal mucosa. Capecitabine was designed to address this issue. The carbamate prodrug is metabolically stable to enterocytes, readily absorbed intact by passive diffusion, and transformed to 5-FU by a cascade of reactions mediated by three enzymes: carboxylesterase in the liver (**41,42**), cytidine deaminase in the liver and preferentially in tumor tissues (**42,43**), and finally thymidine phosphorylase [32,33]. A similar design approach was applied for compound **46** (RO0094889), a tumor-activated prodrug of the dihydropyrimidine dehydrogenase inhibitor **45** that was intended to be used as a combination therapy with capecitabine [34].

GS7340 (**48**), a prodrug of the antiviral agent tenofovir (**47**), was designed to circulate systemically as the prodrug and undergo selective conversion to the parent inside target cells, such as peripheral blood mononuclear cells (PBMCs). GS7340 was much more stable (170-fold) in human plasma *in vitro* than the ester-type prodrug tenofovir disoproxil fumarate (**48**). The finding was corroborated by high plasma levels of **48** following oral dosing in dogs (>70% of the level achieved after an intravenous bolus administration of **48**). The oral AUC of the active parent **49** in PBMCs (as opposed to the plasma) from GS7340 dosed orally was >34-fold higher than that from **49**. A tissue distribution study with **48** in dogs showed preferential loading of the active parent into PBMCs and other lymphatic tissues [35]. Remofovir mesylate (**51**), a prodrug of the antiviral agent adefovir (**50**), was designed to deliver the active parent to the liver. This prodrug is activated by the cytochrome P450 isoenzyme CYP3A4 that is abundant in hepatocytes. Compared to the ester-type prodrug adefovir dipivoxil (**52**), an oral dose of radiolabeled **51** yielded 60-fold higher levels in the liver in a dog whole body autoradiography study [36]. Esterification of cidofovir with alkoxyalkanols diminished drug accumulation in the kidney, the site of dose-limiting toxicity, while providing increased oral bioavailability in mice [37].

The anticancer agents gemcitabine (**53**) and floxuridine (**55**) were derivatized to a series of amino acid ester prodrugs such as **54** and **56** not only to enhance oral absorption through affinity for the PepT1 transporter but also to selectively deliver the active parents to cancer cells or tissues overexpressing the same transporter. The optimal prodrugs showed improved permeability (>threefold) in the Caco-2 permeability assay and increased uptake by PepT1-overexpressing HeLa cells [38,39]. *In vivo* data to support this approach are not yet available.

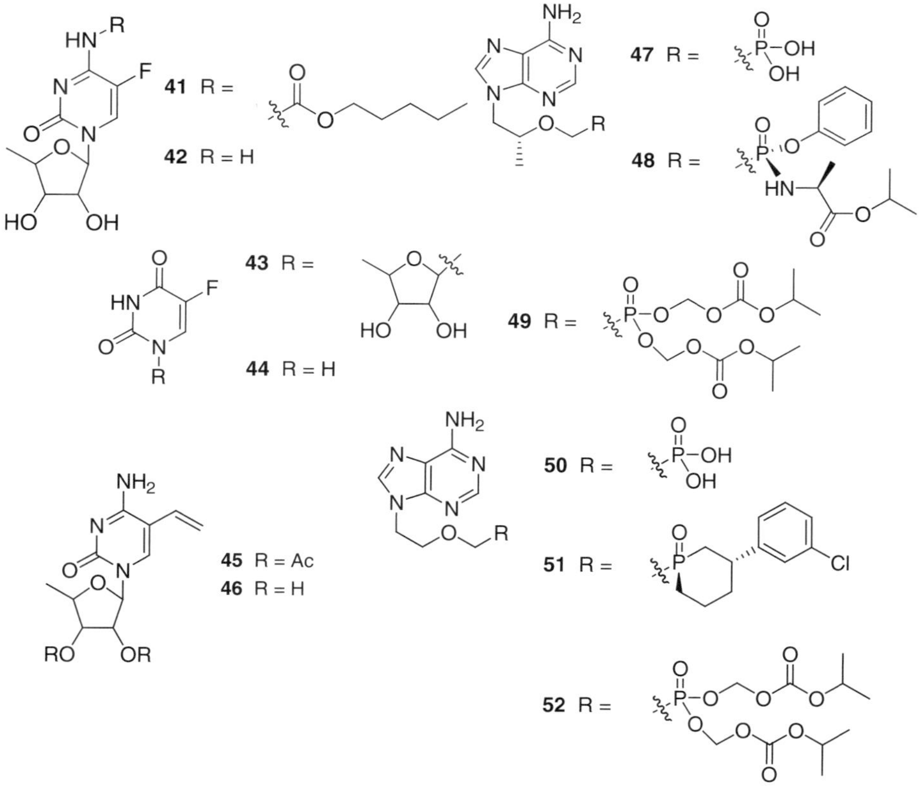

6. PRODRUGS WITH OTHER BENEFITS

Recent reports on inventive prodrug approaches targeting other benefits beyond improved oral bioavailability and selective tissue delivery are described in this section.

Although sustained release of an active agent has hitherto been solely the subject of drug formulation studies, efforts toward prolonging plasma half-life using prodrugs have recently been reported. A polyserine–naltrexone conjugate (**58**, in a 10:1 ratio of serine to naltrexone), when administered to rats orally, afforded a comparable total plasma AUC, yet with a threefold lower Cmax and sustained (>12 h vs. <8 h) plasma exposure of naltrexone (**57**) compared with the monomeric naltrexone control. This extended release effect is ascribed to slow hydrolysis of the polypeptide by enzymes associated with the intestinal mucosa. A polyglutamic acid–AZT conjugate (**60**) displayed a similar PK profile with a slight prolongation of plasma exposure and a sevenfold higher total plasma AUC of AZT (**59**) than the monomeric AZT control [40]. A polar, ionizable prodrug (**62**) of omeperazole was designed to achieve a sustained release of the parent (**61**), which has a short plasma half-life. Compound **62** had a 100-fold lower permeability than the parent in the Caco-2 permeability assay, resulting in slow oral absorption and thus longer plasma half-life (2.4 h vs. 0.7 h) of the parent in dogs [41]. A prodrug of gabapentin **63** is a substrate of bile transporters. Following oral administration in rats, plasma level of the intact prodrug reached a maximum at ~8 h and was sustained over 24 h due to efficient enterohepatic circulation mediated by both the intestine and the liver bile transporter systems. The plasma concentration of the released parent was similarly sustained through a 24-h period [42].

Mechanism-based GI side effects of non-steroidal antiinflammatory drugs have been previously addressed by a prodrug approach [43]. A similar strategy was applied to the antiviral castanospermine (**64**). This compound causes osmotic diarrhea due to inhibition of intestinal sucrases. The ester-type prodrug celgosivir (**65**) was relatively inactive at this enzyme and the GI side effects were minimized, allowing advanced clinical trials as an anti-HCV agent. Celgosivir was readily absorbed (>94% of the dose) orally, and converted rapidly to the parent (>92% of the dose) upon absorption [44].

Therapeutically beneficial effects of controlled substances have not been fully utilized because of the potential for drug abuse through intravenous and nasal administrations. Efforts to deal with these issues using prodrugs have recently been reported. Dextroamphetamine (**66**) was derivatized to a peptidic prodrug NPR-104 (**67**) that displayed excellent oral bioavailability of the parent in animals, combined with poor (<5%) nasal bioavailability and high chemical stability. Less than 10% of the parent was released following intravenous injection in rats. Compound **66** is currently being developed for the treatment of attention-deficit hyperactive disorder in pediatric populations [45]. Similarly, hydrocodone (**68**) was conjugated with ribose to form the prodrug **69**. The oral bioavailability of the active parent from the prodrug was 95% in rats, while it was less (<41%) after either nasal or intravenous administrations [46].

57 R = H

58 R =

59 R = H

60 R =

61 R = H

62 R =

63

64 R = H

65 R =

68 R = H (ketone)

66 R = H

67 R =

69 R =

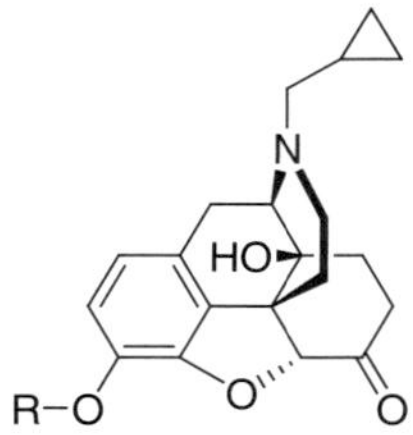

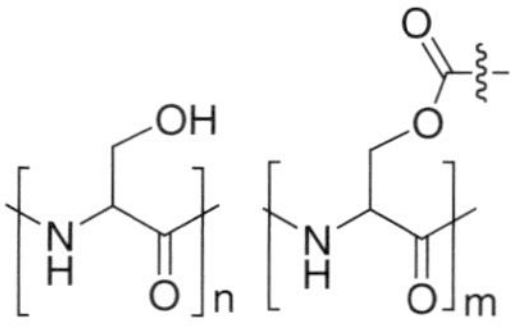

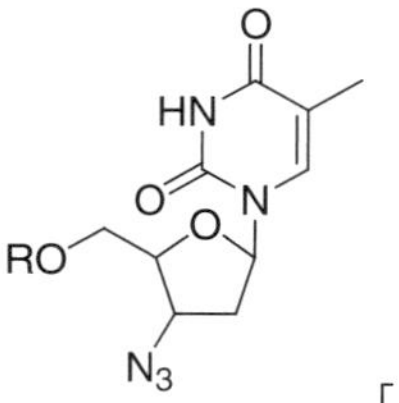

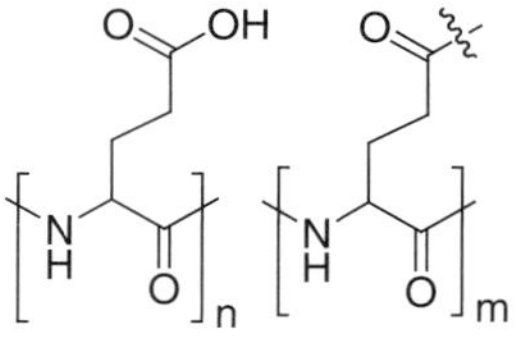

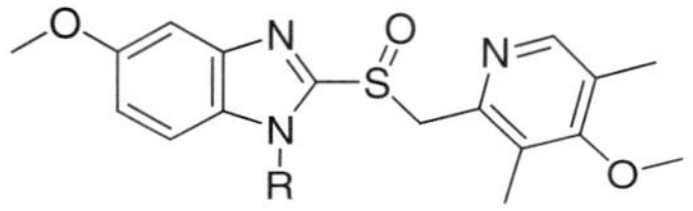

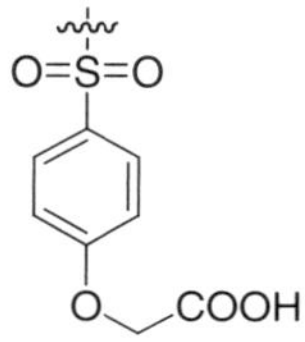

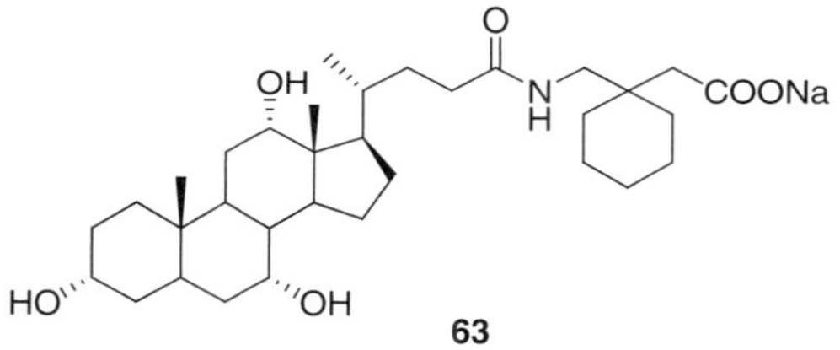

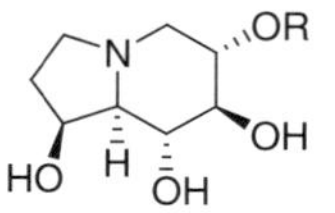

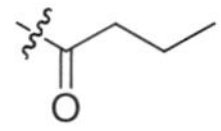

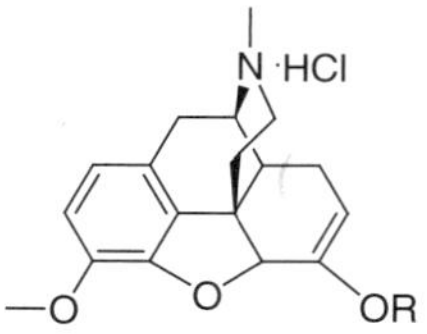

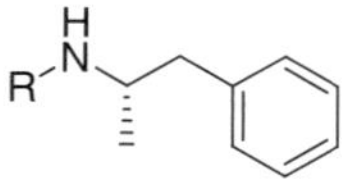

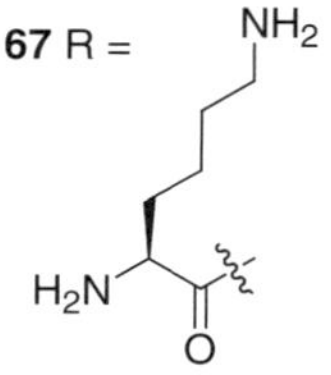

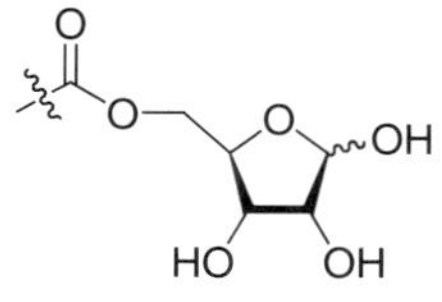

7. CONCLUSIONS

Oral prodrug discovery is a field of medicinal chemistry that has been continuously evolving. As discussed in this review, more detailed (sub)cellular functions of intestinal mucosa during prodrug absorption have been discovered and utilized in prodrug design. Additionally, a number of new prodrug tactics have been devised to address issues associated with parent drugs or lead compounds such as solubility, permeability, and tissue distribution. Despite the intrinsic challenges (e.g., added synthetic steps, interspecies variability in prodrug activation, added complexity of pharmacokinetic profiling, potential toxicity of the pro-moieties), oral prodrug research will continue to provide suitable avenues to expand the usefulness of therapeutic agents.

REFERENCES

[1] B. Testa and J. M. Mayer, *Hydrolysis in Drug and Prodrug Metabolism*, VHCA, Switzerland, 2003, p. 420.
[2] I. Gomez-Orellana, *Expert Opin. Drug Deliv.*, 2005, **2**, 419.
[3] K. Beaumont, R. Webster, I. Gardner and K. Dack, *Curr. Drug Metab.*, 2004, **4**, 416.
[4] P. Ettmayer, G. L. Amidon, B. Clement and B. Testa, *J. Med. Chem.*, 2004, **47**, 1.
[5] D. Fleisher, R. Bong and B. H. Stewart, *Adv. Drug Deliv. Rev.*, 1996, **19**, 115.
[6] M. D. Taylor, *Adv. Drug Deliv. Rev.*, 1996, **19**, 131.
[7] S. Majumdar, S. Duvvuri and A. K. Mitra, *Adv. Drug Deliv. Rev.*, 2004, **56**, 1437.
[8] O. H. Chan, H. L. Schmid, L. A. Stilgenbauer, W. Howson, D. C. Horwell and B. H. Stewart, *Pharmaceut. Res.*, 1998, **15**, 1012.
[9] E. S. Furfine, C. T. Baker, M. R. Hale, D. J. Reynolds, J. A. Salisbury, A. D. Searle, S. D. Studenberg, D. Todd, R. D. Tung and A. Spaltenstein, *Antimicrob. Agents Chemother.*, 2004, **48**, 791.
[10] Y. Ueda, T. P. Connolly, J. F. Kadow, N. A. Meanwell, T. Wang, C. H. Chen, K. Yeung, Z. Zhang, D. K. Leahy, S. K. Pack, N. Soundararajan, P. Sirard, K. Levesque and D. Thoraval, *US Patent* 2005/0209246 A1, 2005.
[11] J. Ohwada, M. Tsukazaki, T. Hayase, N. Oikawa, Y. Isshiki, H. Fukuda, E. Mizuguchi, M. Sakaitani, Y. Shiratori, T. Yamazaki, S. Ichihara, I. Umeda and N. Shimma, *Bioorg. Med. Chem. Lett.*, 2003, **13**, 191.
[12] A. Schmitt-Hoffmann, B. Roos, M. Heep, M. Schleimer, E. Weidekamm, T. Brown, M. Roehrle and C. Beglinger, *Antimicrob. Agents Chemother.*, 2006, **50**, 279.
[13] T. Hayase, S. Ichihara, Y. Isshiki, P. Liu, J. Ohwada, T. Sakai, N. Shimma, M. Tsukazaki, I. Umeda and T. Yamazaki, *US Patent* 6,300,353 B1, 2001.
[14] A. B. Nielsen, A. Buur and C. Larsen, *Eur. J. Pharm. Sci.*, 2005, **26**, 421.
[15] L. Schaller, T. Burnette, J. Cowan, P. Feldman, G. Freeman, H. Marr, B. Owens, K. Romines, J. Shepard, L. Boone, J. Chan, *Abstr. 43rd Intersci. Conf. Antimicrob. Agents Chemother.*, poster H-872, 2003.
[16] R. G. Ferris, R. J. Hazen, G. B. Roberts, M. H. Clair, J. H. Chan, K. R. Romines, G. A. Freeman, J. H. Tidwell, L. T. Schaller, J. R. Cowan, S. A. Short, K. L. Weaver, D. W. Selleseth, K. R. Moniri and L. R. Boone, *Antimicrob. Agents Chemother.*, 2005, **49**, 4046.
[17] D. E. Gingrich, D. R. Reddy, M. A. Igbal, J. Singh, L. D. Aimone, T. S. Angeles, M. Albom, S. Yang, M. A. Ator, S. L. Meyer, C. Robinson, B. A. Ruggeri, C. A. Dionne, J. L. Vaught, J. P. Mallamo and R. L. Hudkins, *J. Med. Chem.*, 2003, **46**, 5375.

[18] L. Bouerat, J. Fensholdt, X. Liang, S. Havez, S. F. Nielsen, J. R. Hansen, S. Bolvig and C. Andersson, *J. Med. Chem.*, 2005, **48**, 5412.
[19] V. Josyula, R. Gadwood, L. Thomasco, J. Kim and A. Choy, *WO Patent* 2005/028473 A1, 2005.
[20] A. Marfat, D. G. McLeod and J. P. O'Donnell, *US Patent* 2005/0004093 A1, 2005.
[21] S. A. Filla, M. A. Winter, K. W. Johnson, D. Bleakman, M. G. Bell, T. J. Bleisch, A. M. Castano, A. Clemens-Smith, M. del Prado, D. K. Dieckman, E. Dominguez, A. Escribano, K. H. Ho, K. J. Hudziak, M. A. Katofiasc, J. A. Martinez-Perez, A. Mateo, B. M. Mathes, E. L. Mattiuz, A. M. L. Ogden, L. A. Phebus, D. R. Stack, R. E. Stratford and P. L. Ornstein, *J. Med. Chem.*, 2002, **45**, 4383.
[22] C. Kondo, R. Onuki, H. Kusuhara, H. Suzuki, M. Suzuki, N. Okudaira, M. Kojima, K. Ishiwata, J. W. Jonker and Y. Sugiyama, *Pharmaceut. Res.*, 2005, **22**, 613.
[23] A. H. Dantzig, K. M. Hillgren and D. P. de Alwis, *Annu. Rep. Med. Chem.*, 2004, **39**, 279.
[24] H. Chanteux, F. V. Bambeke, M.-P. Mingeot-Leclercq and P. M. Tulkens, *Antimicrob. Agents Chemother*, 2005, **49**, 1279.
[25] R. Daifuku, A. Gall, D. Sergueev, D. Sologub, K. Harris, *US Patent* 2005/0014752 A1, 2005.
[26] K. R. Beutner, *Antivir. Res.*, 1995, **28**, 281.
[27] A. E. Thomsen, M. S. Christensen, M. A. Bagger and B. Steffansen, *Eur. J. Pharm. Sci.*, 2004, **23**, 319.
[28] A. B. Bueno, I. Collado, A. de Dios, C. Dominguez, J. A. Martin, L. M. Martin, M. A. Martinez-Grau, C. Montero, C. Pedregal, J. Catlow, D. S. Coffey, M. P. Clay, A. H. Dantzig, T. Lindstrom, J. A. Monn, H. Jiang, D. D. Schoepp, R. E. Stratford, L. B. Tobas, J. P. Tizzano, R. A. Wright and M. F. Herin, *J. Med. Chem.*, 2005, **48**, 5305.
[29] L. M. Rorick-Kehn, E. J. Perkins, K. M. Knitowski, J. C. Hart, B. G. Johnson, D. D. Schoepp and D. L. McKinzie, *J. Pharmacol. Exp. Ther.*, 2006, **316**, 905.
[30] K. C. Cundy, R. Branch, T. Chernov-Rogan, T. Dias, T. Estrada, K. Hold, K. Koller, X. Liu, A. Mann, M. Panuwat, S. P. Raillard, S. Upadhyay, Q. Q. Wu, J. Xiang, H. Yan, N. Zerangue, C. X. Zhou, R. W. Barrett and M. A. Gallop, *J. Pharmacol. Exp. Ther.*, 2004, **311**, 315.
[31] K. C. Cundy, T. Annamalai, L. Bu, J. De Vera, J. Estrela, W. Luo, P. Shirsat, A. Torneros, F. Yao, J. Zou, R. W. Barrett and M. A. Gallop, *J. Pharmacol. Exp. Ther.*, 2004, **311**, 324.
[32] M. Miwa, M. Ura, M. Nishida, N. Sawada, T. Ishikawa, K. Mori, N. Shimma, I. Umeda and H. Ishitsuka, *Eur. J. Cancer*, 1998, **34**, 1274.
[33] F. Desmoulin, V. Gilard, M. Malet-Martino and R. Martino, *Drug Metab. Dispos.*, 2002, **30**, 1221.
[34] K. Hattori, Y. Kohchi, N. Oikawa, H. Suda, M. Ura, T. Ishikawa, M. Miwa, M. Endoh, H. Eda, H. Tanimura, A. Kawashima, I. Horii, H. Ishitsuka and N. Shimma, *Bioorg. Med. Chem. Lett.*, 2003, **13**, 867.
[35] W. A. Lee, G. He, E. Eisenberg, T. Cihlar, S. Swaminathan, A. Mulato and K. C. Cundy, *Antimicrob. Agents Chemother.*, 2005, **49**, 1898.
[36] C. Lin, L. Yeh, D. Vitarella, Z. Hong and M. D. Erion, *Antivir. Chem. Chemother.*, 2004, **15**, 307.
[37] S. L. Ciesla, J. Trahan, W. B. Wan, J. R. Beadle, K. A. Aldern, G. R. Painter and K. Y. Hostetler, *Antivir. Res.*, 2003, **59**, 163.
[38] C. P. Landowski, X. Song, P. L. Lorenzi, J. M. Hilfinger and G. L. Amidon, *Pharmaceut. Res.*, 2005, **22**, 1510.
[39] X. Song, P. L. Lorenzi, C. P. Landowski, B. S. Vig, J. M. Hilfinger and G. L. Amidon, *Mol. Pharmaceutics*, 2005, **2**, 157.

[40] T. Piccariello, R. J. Kirk and L. P. Olon, *US Patent* 2004/0063628 A1, 2004.
[41] J. Chen, D. F. Welty and D. D. Tang-Liu, *US Patent* 2005/0075371 A1, 2005.
[42] K. C. Cundy, M. A. Gallop, C. X. Zhou, *US Patent* 2005/0272710 A1, 2005.
[43] M. Rodriguez-Tellez, F. Arguelles, J. M. Herrerias, Jr., D. Ledro, J. Esteban and J. M. Herrerias, *Curr. Pharm. Design*, 2001, **7**, 951.
[44] L. A. Sotbera, J. Castaner and L. Garcia-Capdevila, *Drugs Fut.*, 2005, **30**, 545.
[45] T. C. Mickle, S. Bera, S. Guenther, W. Hirschelman, S. Krishnan, C. Lauderback, S. Moncreif, J. Gill, D. Jones, D. Linn, A. Martin, C. Miller and D. Portlock, *Abstracts of Paper, MEDI-246, 230th ACS National Meeting,* Washington, DC, 2005.
[46] T. Mickle, T. Piccariello, J. S. Moncrief, N. J. Boerth and B. Bishop, *WO Patent* 2004/062614 A2, 2004.

Oxytocin Antagonists and Agonists

Alan D. Borthwick

GlaxoSmithKline Research and Development, Medicines Research Centre, Gunnels Wood Road, Stevenage, Herts, SG1 2NY, UK

Contents

1. Introduction	409
2. Oxytocin antagonists	410
2.1. Peptide antagonists	411
2.2. Small molecule antagonists	412
3. Oxytocin agonists	416
3.1. Peptide agonists	416
3.2. Small molecule agonists	417
4. Conclusions	419
References	419

1. INTRODUCTION

Oxytocin (OT) **1** is a cyclic nonapeptide hormone secreted by the posterior pituitary gland that acts at the OT receptor [1], a member of the super-family of seven-transmembrane (7TM) G-protein coupled receptors (GPCRs) that has no subtypes but is structurally related to the vasopressin receptors [2]. Oxytocin exhibits a range of physiological roles [1]. The OT receptor has been localised in central tissues, such as the paraventricular nucleus where it is involved in regulation of both male and female sexual response [1,3]. Within the periphery, the OT receptor is localised in a number of different organs including the uterus and mammary glands. In the uterus it is involved in the onset and progression of labour and has long been regarded as the pregnancy hormone as it stimulates labour and milk ejection. It is recognised as having a wide spectrum of functions outside pregnancy [4] including social and reproductive behaviour and emotions [5].

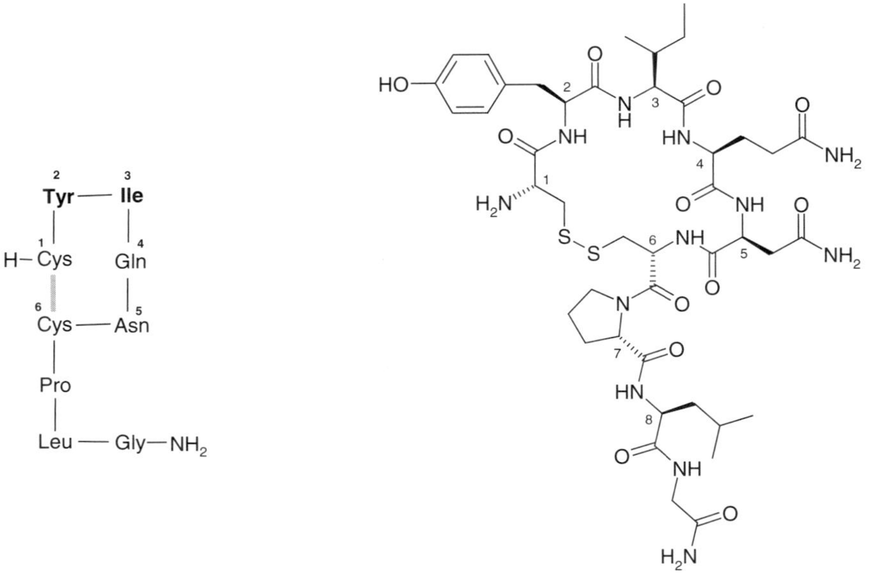

1

2. OXYTOCIN ANTAGONISTS

Oxytocin is a clinically proven inducer of labour in pregnant women. It works as a potent stimulant of uterine contractions via the interaction with OT receptors that are expressed in myometrial cells in the mammalian female uterus. These receptors in the uterus vastly increase in number during pregnancy. The agonist oxytocin, binds to the extracellular region and transmembrane domain of the receptor, which enables the intracellular part to couple to the G proteins and initiate a cascade of events liberating Ca^{2+} which causes contractions [2]. Oxytocin antagonists have been shown to inhibit uterine contractions and delay preterm delivery [6]. In the last decade, the intravenously administered peptidic oxytocin antagonist atosiban (Tractocile^{TM}) [7] has been established as an acute treatment of preterm labour and interest has increased in the search for orally bioavailable, selective, non-peptide antagonists. With the discovery that oxytocin has a wide spectrum of functions outside pregnancy [4,5], interest has also developed in oxytocin antagonists as a potential treatment of sexual dysfunction including premature ejaculation and the treatment or prevention of benign prostate hyperplasia. Oxytocin antagonists have been reviewed in 1997 [8], and recently covering the literature to the end of 2004 [9], while a general review of the condition of preterm labour and treatment options was published in 2003 [10]. This section will mainly focus on recent advances reported in late 2004 and 2005 towards the design and discovery of novel orally bioavailable non-peptide antagonists and the latest improvements in peptide antagonists.

2.1. Peptide antagonists

2.1.1. Atosiban

Atosiban **2** is the only oxytocin antagonist approved for the acute treatment of preterm labour. It is a cyclic peptide based on the natural hormone oxytocin. Early work established that modifications of the endogenous agonist oxytocin like capping of the 2-tyrosine hydroxyl group as a methyl or ethyl ether led to potent peptide antagonists [11]. The desamino ethoxy analogue was further modified at the 4th and 8th position by Ferring Laboratories [12] to give atosiban **2**, [1-deamino-D-2-Tyr(OEt)-4-Thr-8-Orn] -oxytocin. It is a mixed oxytocin/vasopressin V1a antagonist. In human recombinant receptors *in vitro*, atosiban had a K_i of 397 nM at the OT receptor and 4.7 nM at the V1a receptor, and low affinity at the V1b ($K_i = 256$ nM) and V2 ($K_i = 3195$ nM) receptors [13]. Atosiban is administered as an initial bolus followed by a continuous high-dose infusion over 3 h, then a lower dose continuous infusion for up to 45 hours.

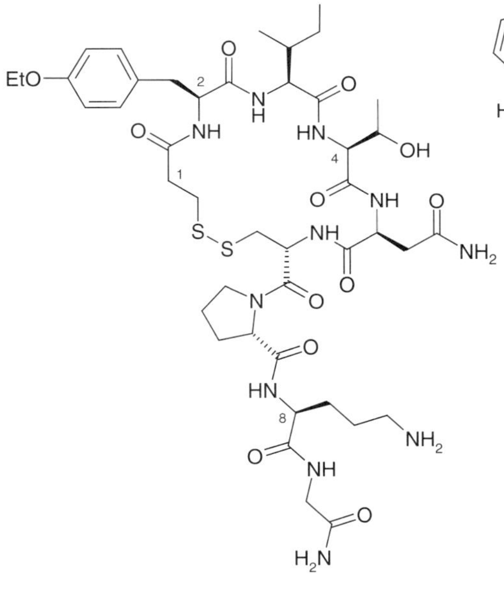

2 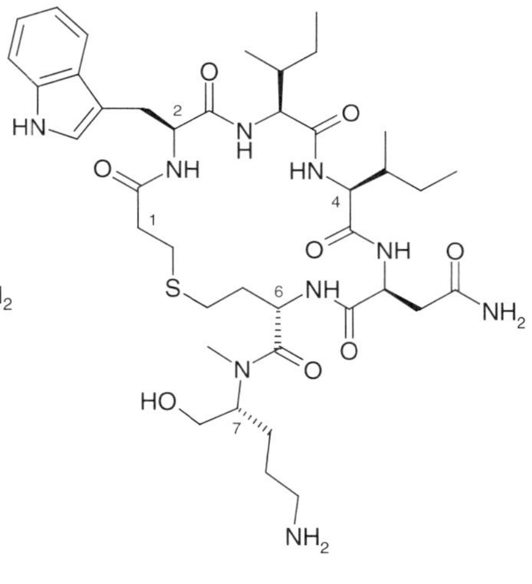   **3**

Clinical experience with this compound, has attested to its effectiveness in at-risk patients [14,15]. Atosiban was reviewed in 2004 [7]; since then several studies have been conducted comparing the efficacy and safety profile of atosiban as a tocolytic agent (a medication that arrests uterine contractions in preterm labour) to that of β-adrenergic agonists. All conclude that atosiban has a similar efficacy and a lower incidence of adverse effects, particularly those of a cardiovascular nature [16].

2.1.2. Barusiban

Although atosiban can be used to delay imminent preterm birth between 24 and 33 weeks of gestational age, it has a short duration of action and has to be dosed as a continuous infusion. In addition, atosiban is a mixed oxytocin/vasopressin V1a antagonist, being more potent at the vasopressin V1a receptor than at the OT receptor. In order to address the issues of short duration of action and lack of selectivity for the OT receptor over the vasopressin receptor seen with atosiban, further optimization of the OT nonapeptide template has been investigated. The disulphide bridge of desamino–oxytocin was replaced with a methylene sulphide unit in an effort to enhance the biological half-life. In addition 2nd, 4th and 6th positions on the six-membered cyclopeptide ring were modified as well as the three amino acids that extend from the 6th position which substantially shortened the exocyclic chain. The resulting barusiban **3** (FE 200440) showed improved oxytocin antagonist potency. It is 36-fold more potent than atosiban at the human cloned OT receptor with a $K_i = 0.31$ nM and 275-fold more selective for the human OT receptor (hOTR) versus the human vasopressin V1a receptor (hV1a) unlike atosiban which is 26-fold more selective for the hV1a receptor [17]. Barusiban was ~100 times more potent than atosiban at inhibiting oxytocin-induced contractions of the myometrium from preterm and term pregnant women [18]. In a preclinical study, barusiban **3** had a much longer duration of action (>13–15 h, compared with 1–3 h for atosiban) in the cynomolgus monkey model of preterm labour [19] where it was also four times more potent than atosiban.

2.2. Small molecule antagonists

In the last decade there has been considerable interest in overcoming the shortcomings of the first generation peptide antagonists, by identifying orally active non-peptide oxytocin antagonists with a higher degree of selectivity towards the vasopressin receptors (V1a, V1b, V2). As a result, several templates have been investigated as potential selective oxytocin antagonists [8–10]. Reflecting the close structural similarity between the oxytocin and vasopressin receptors especially V1a, most of these have been historically based on templates that are vasopressin receptor antagonists [20] and additional bulk and complexity have been added to achieve selectivity (10–100 fold) for the OT receptor [8,9]. More recently, by screening defined chemical classes for activity as oxytocin antagonists [22,27,35] newer templates have been discovered, some of which have been developed to achieve (from 500 to $>10,000$ fold) selectivity over the human vasopressin receptors [29].

2.2.1. Indolin-2-ones

Indolin-2-one derivatives are potent and selective oxytocin antagonists. The lead compound SSR126768A (**4**) has the same high affinity for both the rat and the hOTRs ($K_i = 0.44$ nM) and 100-fold lower affinity for the hV1a and hV1b receptors and even much lower (>500) for human V2 vasopressin receptors [21]. In rat-isolated myometrium, OT-induced uterine contractions were competitively antagonised by **4**

($pA_2 = 8.47$). In addition the indolin-2-one **4** has oral activity. Oral administration of a 3 mg/kg dose of SSR126768A to rats was found to be effective for up to 24 h in the competitive inhibition of uterine contractions. At higher doses **4** (30 mg/kg p.o.) was comparable in the delay of labour in pregnant rats to the β-adrenergic agonist tocolytic, ritodrine (10 mg/kg p.o.).

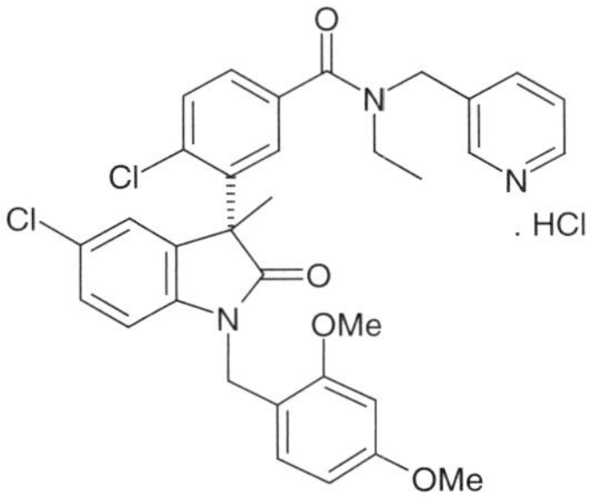

4

2.2.2. Hydrazone sulphanilides

Screening a library of compounds biased towards GPCRs, a series of isatin hydrazone sulphanilides that potently inhibited radioligand binding to the OT receptor was identified [22]. The most potent compound was **5** (hOTR: $K_i = 90$ nM). Structure-activity relationship (SAR) studies showed that the isatin hydrazone ring was required for good activity and para substitution in both the other aromatic rings was preferred for good potency. This gave the more potent oxytocin antagonist sulphanilide **6** (hOTR: $K_i = 14$ nM), which was 100-fold selective over the vasopressin receptor hV1a ($K_i = 1410$ nM). Oral bioavailability was achieved by incorporating water solubilising groups into the aryl ring of the sulphonamide by modifying the preferred 4-alkyl or 4-alkoxy groups. This furnished the lead compound **7**, (hOTR: $K_i = 0.65$ nM) which was 65-fold selective for the vasopressin receptor hV1a ($K_i = 42$ nM). It was also found to inhibit OT-induced uterine contractions in nonpregnant rats by 72% at 10 mg/kg (i.v.) and by 35% at 30 mg/kg (p.o.), and to reduce spontaneous uterine contractions in late-term pregnant rats by 30% at 30 mg/kg (p.o.).

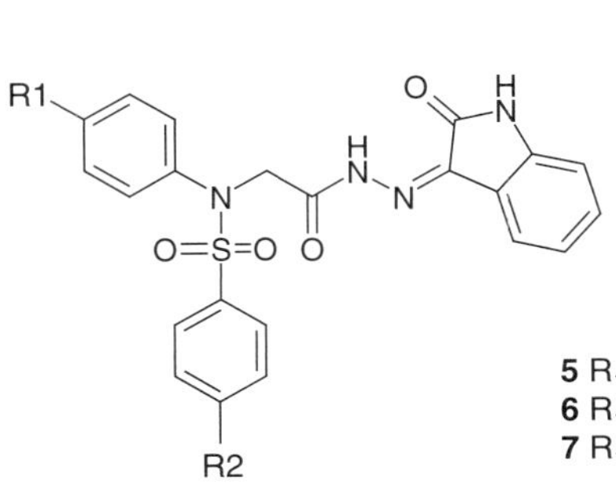

5 $R_1 = H$; $R_2 = H$
6 $R_1 = Me$; $R_2 = OEt$
7 $R_1 = C_1$; $R_2 = CH_2CH_2CONHCH_2CH_2CH_2NMe_2$

2.2.3. Aryl 1,3,4 triazoles

Swiss researchers originally claimed a series of *N*-aryl-1,3,4-triazoles as having oxytocin antagonism and hence of utility in the treatment of preterm labour [23]. The 2-benzylthio compound, **8**, had a K_i of 45 nM in a human oxytocin-binding assay. Recently, this template has been modified further to yield potent [24–26] oxytocin antagonists by focusing on their potential as treatment of sexual dysfunction as a therapeutic target. The possible application to the treatment of preterm labour is mentioned along with several other disease areas including prostate hyperplasia, hypertension, congestive heart failure (CHF) and neuropsychiatric disorders.

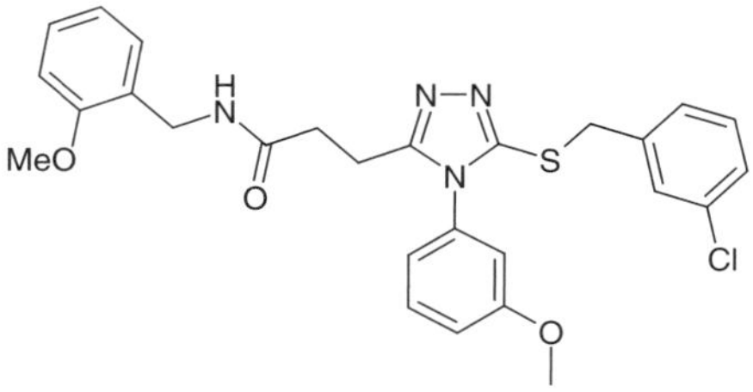

8

The initial reports in this series of compounds center on *N*4- [3-(2-methoxypyrid-5′-yl)] substituted 1,3,4-triazole in which the 3rd position in the triazole ring is a substituted methylene group and the 5th position is a para-substituted aryl ring **9** [24].

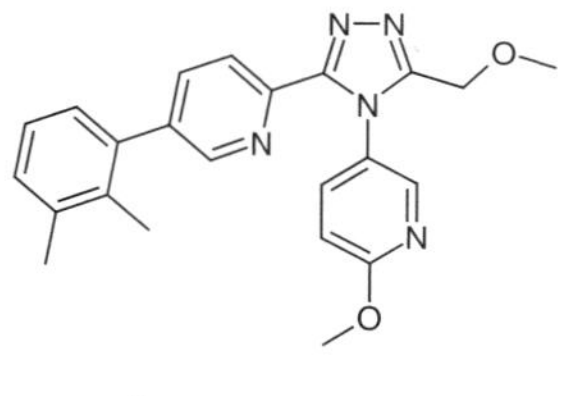

9 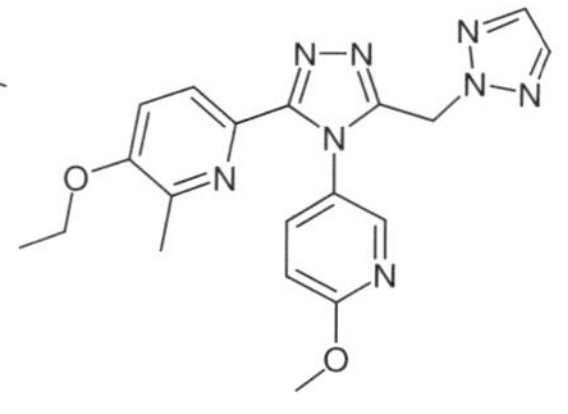   **10** 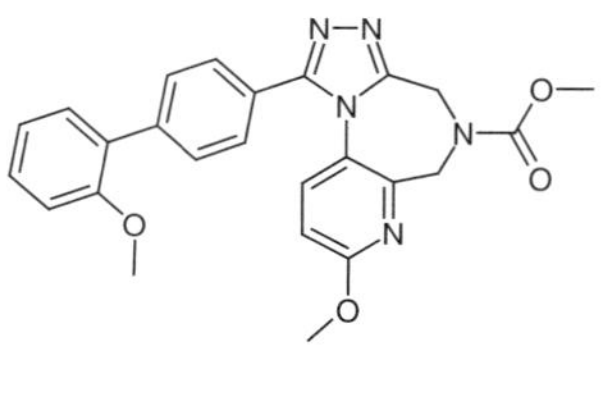   **11**

Further variations claimed on this template include, for example, triazoles at C3 and alkoxy pyridyl rings at C5 **10** [25]. The biological activity of oxytocin antagonists was determined in a colorimetric oxytocin receptor β-lactamase assay; typical compounds in the series, such as the analogue **9** had a K_i of 3 nM, while analog **10** had a K_i of 21 nM. In the most recent patent application covering this series, the early compounds have been further investigated and constraint of the methylene side chain by incorporation into a heterocyclic ring together with additional manipulation of the aryl groups is reported [26]. This has resulted in the identification of compounds of higher OT receptor affinity, such as the tricyclic analogue, **11** (K_i <3 nM).

2.2.4. 2,5-Diketopiperazines

A screening programme identified 2,5-diketopiperazines (DKPs) exemplified by **12** as novel templates for antagonists of human oxytocin (hOT). The lead, **12**, showed potency of pK_i 6.5 ($K_i = 300$ nM) as a mixture of isomers in the amide side-chain. Initial SAR studies led to the semi-rigid and chirally pure DKP **13** [pK_i 8.4 ($K_i = 3$ nM)], with *cis* disposed substituents at C3 and C6 and the R side-chain configuration. Optimal activity was shown to lie in the *RRR* series: the *RRS* isomer, where the stereochemistry in the amide side-chain is inverted, was 10-fold less potent. At C6, an indanyl group was preferred; its replacement by phenethyl and benzyl groups led to a progressive weakening of activity. At C3, a 4-carbon branched alkyl was shown to be optimal; smaller alkyl groups resulted in reduced antagonist activity [27].

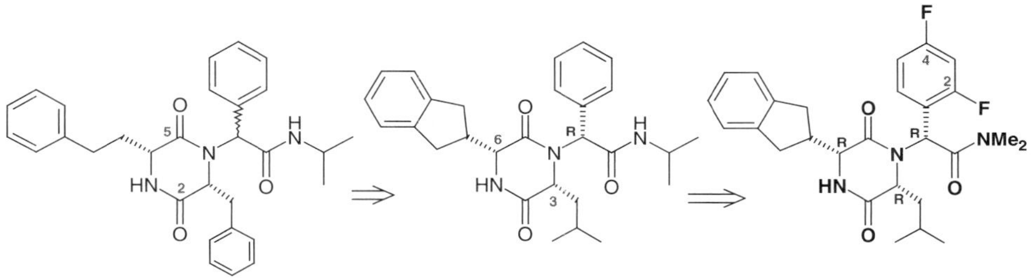

Further SAR studies on this chiral system revealed that, alkylation of the ring N-atom or removal of the C2 carbonyl group decreased potency. Potency was retained when the aryl group on the *N*4-glycinamide was replaced by five-membered heteroaryl or 5,6-fused heteroaryl systems; cycloalkyl or alkyl groups were not well tolerated. *Para* substitution was preferred on the arylglycinamide moiety and a broad range of groups at this position conferred good antagonist activity; *meta* substitution was less well tolerated [27,28]. Potency was maintained by a wide range of *N,N*-disubstituted glycinamides [28]. Optimisation of the pharmacokinetic profile of this template by analogy-and property-based design using an estimate of human oral absorption (EHOA) derived from measured lipophilicity (CHI log *D*) and calculated size (cMR) led to the 2′,4′-difluorophenyl-dimethylamide **14** with high levels of potency ($pK_i = 9.2$), good oral bioavailability in the rat (53%) and dog (51%) with low clearance. It was over 60-fold more potent than atosiban *in vitro* at the hOTR and had comparable potency to atosiban in the rat (IC_{50} 227 nM). In addition, **14** showed a high degree of selectivity towards the vasopressin receptors [>10,000 for hV1a/hV1b and ~500 for hV2] and had a satisfactory safety profile in the four-day oral toxicity test in rats [29]. In the most recent patent applications covering this series, compounds with heterocyclic variants of the arylglycinamide moiety, including the 1-methyl-1H-indazol-5-yl [30,31], 2-methyl-1,3-oxazol-4-yl [32] and the 2,6- dimethyl-3-pyridinyl [33] ring systems, are described.

2.2.5. Heteroaromatic glycinamides

The DKP class of chiral oxytocin antagonists was prepared in high diastereoselectivity using four-component Ugi reaction followed by ring closure [34]. Another group has used similar chemistry to produce large arrays of compounds as oxytocin antagonists, exemplified by **15**, which have partial similarity to the DKPs. Several potential applications including preterm labour were mentioned but the focus seemed to be on the treatment of sexual dysfunction. The patent [35] claims that all the compounds within its scope have a better than 70% inhibition of oxytocin at 10 μM in binding assays. Actual data is reported on only two compounds, **15** as a mixture of isomers and **16** as a single enatiomer, both compounds displayed a K_i value of 9.4 nM.

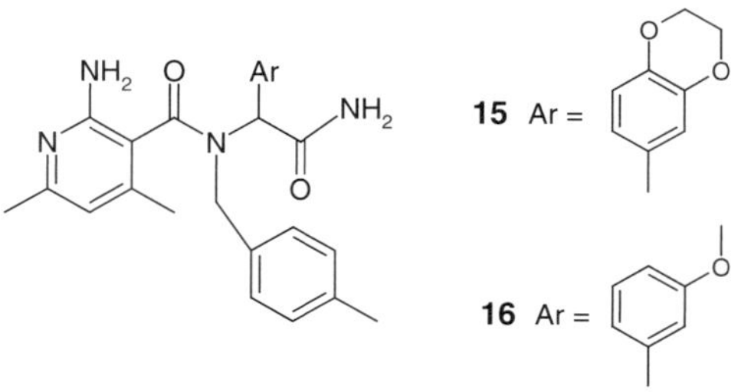

3. OXYTOCIN AGONISTS

The widespread distribution of OT receptors in the brain and the periphery has stimulated interest in the discovery of potent, selective and efficacious OT receptor agonists that may be used as pharmacological tools and as potential drugs. Recent efforts have increased our understanding of their physiological relevance [1–5], especially where they are involved in regulation of both male and female sexual response and in the onset and progress of labour.

3.1. Peptide agonists

3.1.1. Carbetocin

Carbetocin (**17** Duratocin) (1-deamino-1-carba-2-O-methyltyrosine-oxytocin), is an agonist with high stability and a long duration of action [36]. It has a half-life of 29–59 min compared to 4–10 min for oxytocin in non-pregnant women [37] and was prepared by modifications in the N-terminal and disulphide part of the oxytocin molecule which protects the molecule from aminopeptidase and disulphidase cleavage. Compared with oxytocin, carbetocin binds to the OT receptor with about 10-fold lower affinity [36], which is in good agreement with the recent results found from the binding profile of carbetocin and an analysis of its binding domains at the

OT receptor [38]. It was used therapeutically for the prevention of postpartum haemorrhage in the third stage of labour [39], and was shown to be better than oxytocin in prevention of uterine atony after Caesarean section [40].

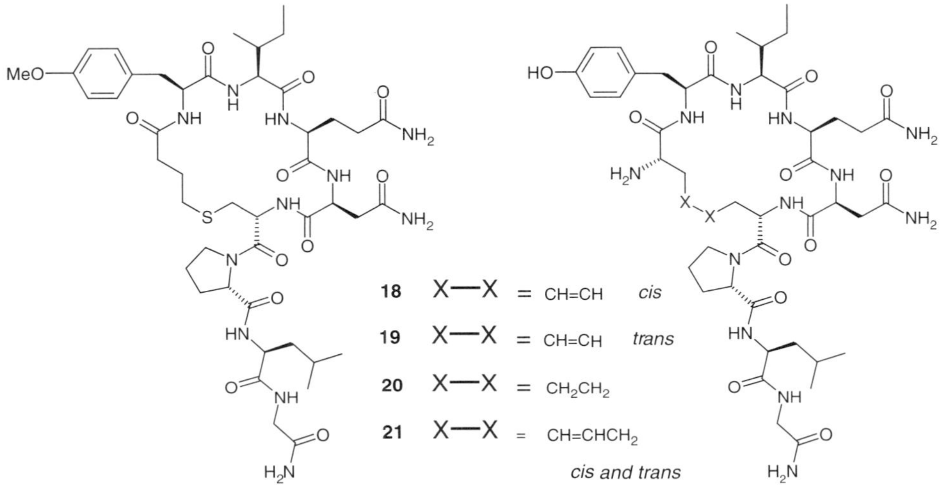

	X—X		
18	X—X	= CH=CH	cis
19	X—X	= CH=CH	trans
20	X—X	= CH$_2$CH$_2$	
21	X—X	= CH=CHCH$_2$	cis and trans

3.1.2. Carba analogs

Recently, the synthesis, activity and stability of a series of analogues of oxytocin have been investigated [41,42]. In each case, the disulphide bridge was replaced with methylene or methine (olefinic) units in an effort to enhance the biological half-life of the analogues without compromising inherent activity. The *cis*-alkene **18** is a potent agonist (EC$_{50}$ = 38 nM) that is ~10-fold less active than oxytocin, whereas the corresponding *trans*-alkene **19** and hydrogenated ethylene analogue **20** are both ~100-fold less active [41]. The analogue **21** as a 2:1 mixture of *cis/trans* isomers was 200-fold less active than oxytocin and has a greater half-life (8–11 min longer) than **1** when incubated in placental tissue from rats [42].

3.2. Small molecule agonists

Although the peptide oxytocin analogue carbetocin (**17**, Duratocin, K_d = 1.96 nM at OT receptor) is used in the clinic, it has affinity for the myometrial human vasopressin V1 receptor (K_d = 7.24 nM), and the renal vasopressin V2 receptor (K_d = 61.3 nM) [36]. In addition, carbetocin (**17**) has antagonistic properties against oxytocin *in vitro* (pA_2 = 8.21) and hence is considered a partial agonist/antagonist at the OT receptor [37]. Carbetocin (**17**) is given by intramuscular injection in contrast to oxytocin, which is given by intravenous infusion. Thus, there is a need to develop non-peptide compounds that are pure agonist and have the potential to be orally bioavailable and more selective against the vasopressin receptors. These have the potential to be drugs where the OT function is compromised, in particular in the treatment of various sexual disorders including the male erectile dysfunction. Additional indications for such compounds could be in promoting labour, controlling post-partum bleeding and increasing milk letdown.

3.2.1. Pyrazole-fused benzodiazepines

A series of pyrazole-fused benzodiazepine oxytocin agonists were developed from screening a vasopressin targeted library for oxytocin activity. This identified two hits **22** and **23,** structural elements of which were combined and optimised to give more potent compounds exemplified by **24** with an EC_{50} of 33 nM [43]. It had selectivity of > 25-fold over the human V2 receptor and it had no agonist activity at the human V1a or V1b receptors.

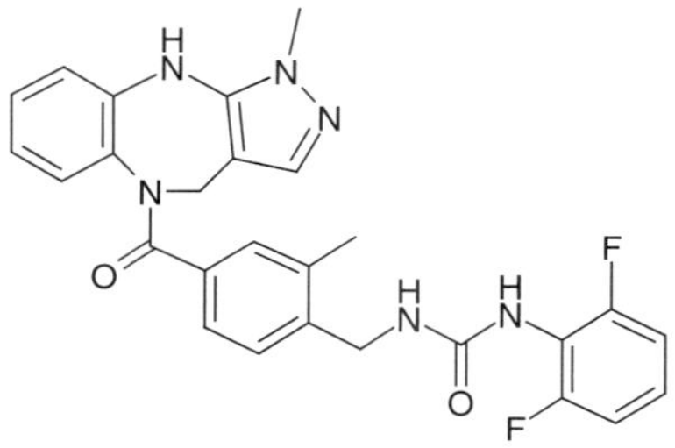

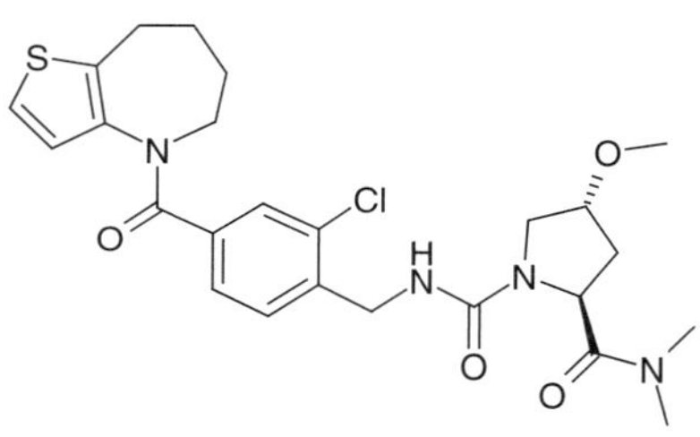

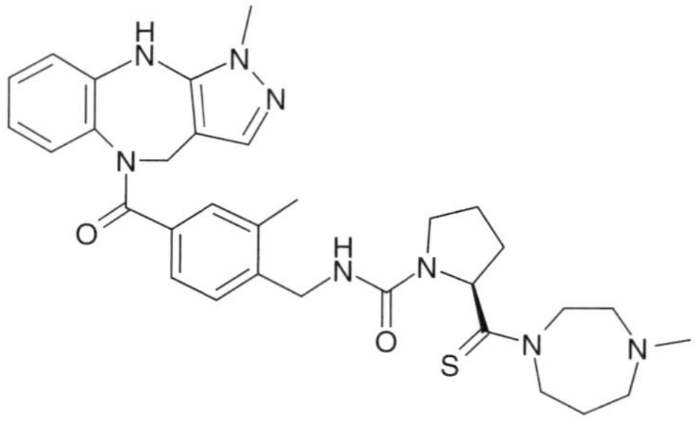

In a recently published patent application from the same group [44], pyrazole-fused benzodiazepines with a variety of substitution patterns, such as **25,** have been claimed for diverse therapeutic indications including sexual disorders, cancer of the prostate, breast, ovary and bones, and post-partum bleeding and depression. The biological activity of oxytocin agonists was assessed by measuring the extent of luciferase synthesis in CHO cells transfected with the hOTR and firefly luciferase gene. Selectivity for the OT receptor over the vasopessin V2 receptor is an issue with this series [45]. SAR studies showed that, that tricyclic azepines containing a pyrazole ring gave better OT vs V2 receptor selectivity. Compound **25** containing the pyrazole-fused benzodiazepine system was the most potent with an EC50 = 18 nM, its replacement with 5,6,7,8-tetrahydro-4H-thieno[3,2b]azepine led to a loss of~10-fold in potency, while replacement with acyclic/aliphatic groups lost all OT activity.

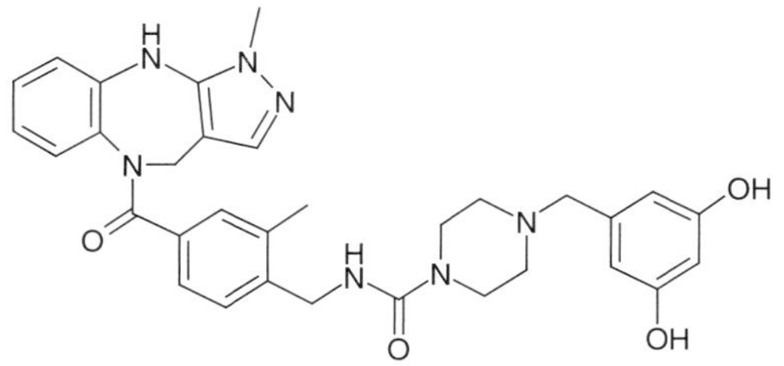

25

Replacement of the 3,5-dihydroxy benzyl group present in **25**, with hydroxylethyl or benzyl groups lost potency [both had an EC50 = 100 nM], while the 2,6-dihydroxy benzyl isomer was inactive. The 3-methyl group in the aryl ring of the central linker is optimal in this series of compounds. Replacement with either hydrogen, chloro, ethyl or a 3-methoxy groups abolished all OT agonist activity, while replacement with fluorine was fourfold less active and similar in potency to the 2-chloro and 2- fluoro isomers.

4. CONCLUSIONS

In the last few years, there has been an upsurge of interest in the natural neurohypophyseal hormone oxytocin, and new physiological and pathophysiological roles for this peptide have been indicated. During this period, interest has also increased in orally bioavailable oxytocin antagonist and agonists that could be used to treat a range of diseases in addition to the main therapeutic areas of preterm labour, benign prostatic hyperplasia and male sexual dysfunction. In addition, several oxytocin antagonists are being investigated in clinical trials for the prevention of preterm labor and treatment of dysmenorrhea; hence, we can expect more attention on oxytocin antagonists and agonists in the near future.

REFERENCES

[1] G. Gimpl and F. Fahrenholz, *Physiol. Rev.*, 2001, **81**, 629.
[2] H. H. Zingg and S. A. Laporte, *Trends Endocrinol. Metab.*, 2003, **14**, 222.
[3] K.-E. Andersson, *Phamacol. Rev.*, 2001, **53**, 417.
[4] T. H. Lippert, A. O. Mueck, H. Seeger and A. Pfaff, *Horm. Res.*, 2003, **60**, 262.

[5] K. Uvnas-Moberg, *The Oxytocin Factor*, Da Capo Press, MA, USA, 2003, p. 3.
[6] A. Coomarasamy, E. M. Knox, H. Gee and K. S. Khan, *Med. Sci. Mon.*, 2002, **8**, RA268.
[7] V. Tsatsaris, B. Carbonne and D. Cabrol, *Drugs*, 2004, **64**, 375.
[8] R. M. Freidinger and D. J. Pettibone, *Med. Res. Rev.*, 1997, **17**, 1.
[9] M. J. Allen, D. G. H. Livermore and J. E. Mordaunt, *Pro. Med. Chem.*, 2006, **44**, 335.
[10] M. K. Schwarz and P. Page, *Curr. Med. Chem.*, 2003, **10**, 1441.
[11] P. Melin, H. Vilhardt, G. Lindeberg, L. E. Larsson and M. Akerlund, *J. Endocrinol.*, 1981, **88**, 173.
[12] P. Melin, J. Trojnar, B. Johansson, H. Vilhardt and M. Akerlund, *J. Endocrinol.*, 1986, **111**, 125.
[13] M. Akerlund, T. Bossmar, R. Brouard, A. Kostrzewska, T. Laudanski, A. Lemancewicz, C. Serradeil-Le Gal and M. Steinwall, *Br. J. Obstet. Gynaecol.*, 1999, **106**, 1047.
[14] J. M. Moutquin, D. Sherman, H. Cohen, P. T. Mohide, D. Hochner-Celnikier, M. Fejgin, R. M. Liston, J. Dansereau, M. Mazor, E. Shalev, M. Boucher, M. Glezerman, E. Z. Zimmer and J. Rabinovici, *Am. J. Obstet. Gynecol.*, 2000, **182**, 1191.
[15] G. J. Valenzuela, L. Sanchez-Ramos, R. Romero, H. M. Silver, W. D. Koltun, L. Millar, J. Hobbins, W. Rayburn, G. Shangold, J. Wang, J. Smith and G. W. Creasy, *Am. J. Obstet. Gynecol.*, 2000, **182**, 1184.
[16] J. Chan, D. Cabrol, I. Ingemarsson, K. Marsal, J. M. Moutquin and N. Fisk, *7th World Congr. Controv. Obstet. Gynecol. Infertil.*, Athens, April 16, 2005, Abstract.
[17] L. Nilsson, T. Reinheimer, M. Steinwall and M. Akerlund, *Br. J. Obstet. Gynaecol.*, 2003, **110**, 1025.
[18] P. Pierzynski, A. Lemancewicz, T. Reinheimer, M. Akerlund and T. Laudanski, *J. Soc. Gynecol. Invest.*, 2004, **11**, 384.
[19] T. M. Reinheimer, W. H. Bee, J. C. Resendez, J. K. Meyer, G. J. Haluska and G. J. Chellman, *J. Clin. Endocrinol. Metab.*, 2005, **90**, 2275.
[20] E. J. Trybulski, *Ann. Rep. Med. Chem.*, 2001, **36**, 159.
[21] C. Serradeil-Le Gal, G. Valette, L. Foulon, G. Germain, C. Advenier, E. Naline, M. Bardou, J.-P. Martinolle, B. Pouzet, D. Raufaste, C. Garcia, E. Double-Cazanave, M. Pauly, M. Pascal, A. Barbier and B. Scatton, *J.-P. Maffrand and G. Le Fur J. Pharmacol. Exp. Ther.*, 2004, **309**, 414.
[22] A. Quattropani, J. Dorbais, D. Covini, P.-A. Pittet, V. Colovray, R. J. Thomas, R. Coxhead, S. Halazy, A. Scheer, M. Missotten, G. Ayala, C. Bradshaw, A.-M. De Raemy-Schenk, A. Nichols, R. Cirillo, E. G. Tos, C. Giachetti, L. Golzio, P. Marinelli, D. J. Church, C. Barberis, A. Chollet and M. K. Schwarz, *J. Med. Chem.*, 2005, **48**, 7882.
[23] A. Quattropani, M. Schwarz, R. J. Thomas and T. Coulter, *PCT Patent Appl.* WO 2003053437, 2003.
[24] A. D. Brown, D. Ellis and C. R. Smith, *PCT Patent Appl.* WO 2005028452 A1, 2005.
[25] A. D. Brown, A. A. Calabrese, D. Ellis and C. R. Smith, *PCT Patent Appl.* WO 2005082866 A2, 2005.
[26] A. D. Brown, A. A. Calabrese, D. Ellis and L. Watson, *PCT Patent Appl.* WO 2005121152 A1, 2005.
[27] P. G. Wyatt, M. J. Allen, A. D. Borthwick, D. E. Davies, A. M. Exall, R. D. Hatley, W. R. Irving, D. G. Livermore, N. D. Miller, F. Nerozzi, S. L. Sollis and A. K. Szardenings, *Bioorg. Med. Chem. Letts.*, 2005, **15**, 2579.
[28] A. D. Borthwick, D. E. Davies, A. M. Exall, D. G. Livermore, S. L. Sollis, F. Nerozzi, M. J. Allen, M. Perren, S. S. Shabbir, P. M. Woollard and P. G. Wyatt, *J. Med. Chem.*, 2005, **48**, 6956.

[29] A. D. Borthwick, D. E. Davies, A. M. Exall, R. D. Hatley, J. A. Hughes, W. R. Irving, D. G. Livermore, S. L. Sollis, F. Nerozzi, K. L. Valko, M. J. Allen, M. Perren, S. S. Shabbir, P. M. Woollard and M. A. Price, *J. Med. Chem.*, 2006, **49**, 4159–4170.

[30] A. D. Borthwick and S. L. Sollis, *PCT Patent Appl.* WO 2006000400 A1, 2006.

[31] A. D. Borthwick, D. M. B. Hickey, J. Liddle, A. M. Mason, D. R. Pollard and S. L. Sollis, *PCT Patent Appl.* WO 2006000759 A1, 2006.

[32] J. Liddle, *PCT Patent Appl.* WO 2005000840 A1, 2005.

[33] A. D. Borthwick, D. M. B. Hickey, J. Liddle and A. M. Mason, *PCT Patent Appl.* WO 2006000399 A1, 2006.

[34] D. E. Davies, R. J. D. Hatley, A. K. Richards, S. L. Sollis and C. M. Crawford, *Tet. Letts.*, 2006, submitted.

[35] D. R. Armour, A. S. Bell, P. J. Edwards, D. Ellis, D. Hepworth, M. L. Lewis and C. R. Smith, *PCT Patent Appl.* WO 2004020414 A1, 2004.

[36] T. Engstrom, T. Barth, P. Melin and H. Vilhardt, *Eur. J. Pharmacol.*, 1998, **355**, 203.

[37] G. Sweeney, A. M. Holbrook, M. Levine, M. Yip, K. Alfredsson, S. Cappi, M. Ohlin, P. Schulz and W. Wassenaar, *Curr. Ther. Res.*, 1990, **47**, 528.

[38] G. Gimpl, R. Postinaa, F. Fahrenholza and T. Reinheimerb, *Eur. J. Pharmacol.*, 2005, **510**, 9.

[39] Y.-S. Chong, L.-L. Su and S. Arulkumaran, *Curr. Opin. Obstet. Gynecol.*, 2004, **16**, 143.

[40] J. Dansereau, A. K. Joshi, M. E. Helewa, T. A. Doran, I. R. Lange, E. R. Luther, D. Farine, M. L. Schulz, G. L. Horbay, P. Griffin and W. Wassenaar, *Amer. J. Obstet. Gynecol.*, 1999, **180**, 670.

[41] J. L. Stymiest, B. F. Mitchell, S. Wong and J. C. Vederas, *Org. Lett.*, 2003, **5**, 47.

[42] J. L. Stymiest, B. F. Mitchell, S. Wong and J. C. Vederas, *J. Org. Chem.*, 2005, **70**, 7799.

[43] G. Pitt, A. Batt, R. Haigh, A. Penson, P. Robson, D. Rooker, A. Tartar, J. Trim, C. Yea and M. Roe, *Bioorg. Med. Chem. Lett.*, 2004, **14**, 4585.

[44] P. Hudson, G. Pitt, A. R. Batt and M. B. Roe, *PCT Patent Appl.* WO 2005023812 A2, 2005.

[45] A. Batt, D. Ashworth, A. Baxter, R. Haigh, P. Hudson, C. Heeney, A. Penson, G. Pitt, P. Robson, D. Rooker, A. Tartar, C. Yea and M. Roe, International Symposium on Advances in Synthetic, Combinatorial and Medicinal Chemistry, Moscow, May 5–8, 2004.

Section 7
Trends and Perspectives

Editor: Anthony Wood
Pfizer Global Research & Development
Sandwich Laboratories Sandwich
Kent UK

Knowledge and Intelligence in Drug Design

Andrew L. Hopkins[1] and Alex Polinsky[2]

[1]Pfizer Global Research & Development, Ramsgate Road, Sandwich, Kent CT13 9NJ, UK
[2]Pfizer Global Research & Development, 10770 Science Center Drive, San Diego, CA 92121, USA

Contents

1. Introduction	425
2. Possession of facts	426
2.1. Chemogenomics knowledge space	427
2.2. Open chemistry	428
3. Possession of skill	430
4. Artifical intelligence in drug design	432
5. Conclusion	434
References	435

1. INTRODUCTION

Peter Drucker, the preeminent management thinker of the twentieth century, described the pharmaceutical industry as an information industry [1]. Drucker's premise was that the value of the medicine lay not in the individual product, which may cost only pennies to manufacture, but in the knowledge, accrued through years of research and development, to create the medicine. The hierarchy from data through information to knowledge can be considered as the formulation of relationships, patterns and principles between each stage [2,3]. What then can be considered the "knowledge" in the field medicinal chemistry and drug design?

Medicinal chemists, by profession, are searching for new drugs – typically small organic molecules – that treat disease. The initial chemical starting point can come from (i) systematic screening of large numbers of compounds in biological assays, (ii) the selective optimization of off-target activities of known compounds on new pharmacological targets, (iii) modification of an existing lead or drug or (iv) the rational design of a drug from knowledge of the molecular mechanism. The vast majority of drug discovery projects today begin with a hypothesis targeting a specific molecular (usually protein) target.[1] A drug discovery project starts in which medicinal chemists attempt to achieve several goals

---

[1]Historically the exact molecular targets may have been unknown and only the specific effect of a compound on a biologically relevant assay was known. A return to approach of screening compounds blindly against phenotypic assays, irrespective of molecular assay is gaining popularity once more due to the rise of chemical biology [4].

- to find a drug molecule that affects the molecular target;
- to deliver this drug to the tissue/organ where the target is expressed;
- to maintain sufficient drug concentration *in vivo* for a desired time;
- to avoid affecting other molecular targets that could cause adverse effects.

In pursuit of those goals, medicinal chemists typically start with one or several lead compounds and then go through an iterative optimization process to turn lead compounds into clinical candidates. Medicinal chemists have to make a choice, at every iteration, about which compound(s) to make next. The sequence of these choices determines the path to an acceptable candidate, the quality of the candidate and the time and cost of finding it. The proposition by Ackoff and Emery that knowledge is essentially the efficiency of choice [2] is one that the medicinal chemistry can relate to. Ackoff and Emery propose that knowledge consists of at least two different senses: *possession of facts* (or awareness of a state of affairs) and *possession of skills*. Ackoff and Emery's dual nature of knowledge provides a useful framework for understanding of what medicinal chemistry knowledge is and how it enables chemists to be productive. The distinction between the possession of facts and the possession of skills is the distinction between ontology and epistemology. Ontology concerns what entities exist and what statements about them are true. Epistemology concerns how we can obtain knowledge about facts in the world and how can we ascertain their reliability.

2. POSSESSION OF FACTS

There are two major categories of facts that form a basis for medicinal chemist's choices (Table 1). One is related to our understanding of how the human body, an extremely complex system, operates at all levels – molecular, cellular, organ and whole organism. The other has to do with our ability to know precisely what chemical compounds actually do in the body. Concerning the former, in spite of accumulating a significant body of information about it, we are very far from the complete understanding. For each synthesized compound, a number of measurements are typically made (for example, *in vitro* and *in vivo* potency, selectivity against related targets, blood concentration, binding to plasma proteins, clearance,

Table 1. Examples of the types of information considered by the medicinal chemist in drug design

How the human body works	How drug molecules affect human body
Biological macromolecules	Pharmacological profiles
Cellular signaling pathways	Omics profiles
Cellular metabolic pathways	Pharmacokinetic/pharmacodynamic (PK/PD)
Protein interaction network	properties
Physiology	Toxicity profiles
Anatomy	Structure–activity/property relationships

presence or absence of some key adverse effects, etc.), but it is impossible to measure how the compound interacts with all components of the system, in the laboratory. Indeed, the most important properties of a drug are only discovered by randomized, double-blinded, controlled clinical trials in a large number of patients. Tragically, several idiosyncratic adverse drug reactions were only observed when a drug was tested across large patient populations with diverse genotypes.

Interestingly, in both categories, the number of available relevant facts, both known *a priori* and generated *in situ* during the course of drug discovery project, can be very large, and yet chemists never quite have complete information to make their choices with certainty. During the course of a drug discovery project the laboratory measurements for a series of compounds reveal the structure–activity relationships (SARs) and the structure–property relationships (SPRs). These relationships then provide *a priori* information for the next generation of drug discovery projects to learn from.

New knowledge is created within a project by exploration of the SARs and the testing of the compound through the hierarchy of biological and toxicity assay through to the clinic. However, the previous experience of medicinal chemistry also provides a rich source of information from which to derive new knowledge, useful to selecting and guiding drug-design programs. Prior knowledge provides an ontology that can be brought to bear on a problem from previous experience. *A priori* knowledge, expressed as ontology, is the patterns and principles derived from analysis of past experience. This knowledge is often probabilistic and conditional.

In many cases, the codification of knowledge takes form of empirical rules or filters that introduce awareness of potential downstream drug development problems into chemist's designs. Structural alerts have long been used to detect reactive molecules [5] or compounds with potential mutagenicity or carcinogenicity risks [6,7]. Predictive methods to detect potential labile sites of reactive metabolites are further developments on the initial observations [8]. Another example is Lipinski's famous "Rule-of-Five" filter [9] for identifying compounds with a lower likelihood of oral absorption. Further analysis has lead to filters based on polar surface area (PSA) that detect compounds with poor oral bioavailability and compounds that are unlikely to penetrate blood-brain barrier [10,11]. The recently introduced "ligand efficiency" parameter is used as a "ready reckoner" to evaluate the potential of low-molecular-weight compounds for further optimization [12]. Ligand efficiency is based on the observation that lower molecular weight (higher-ligand efficiency) compounds may suffer less attrition in clinical development than higher molecular weight compounds. Likewise the elucidation of the protein structure of a ligand-binding site [13,14] is a further example of how knowledge can reduce the "hypothesis space" to improve the efficiency of choice in design.

2.1. Chemogenomics knowledge space

The concept of deriving new medicinal chemistry knowledge from the global analysis of the relationship between all molecular targets and chemical structures has been termed the chemogenomics knowledge space [15–17]. A wide variety of external commercial and public source, of pharmacological data are now become available to the medicinal chemist, including SAR data extracted from the literature [18] (for reviews of available data sources, see Jacoby et al. [16] and Schuffenhauer and Jacoby [15]). Data availability is only part of the problem. In order to navigate and extract new knowledge, data need to be integrated into the chemogenomics knowledge space [15–17]. Paolini et al. have demonstrated how the principle of knowledge discovery in databases [19] can be used to integrate pharmacological data to explore the global relationships between chemical structure and biological targets, from diverse proprietary, public and commercial data sources [17]. To represent the chemogenomics knowledge space in a large-scale database, diverse data need to be integrated by semantic normalization of chemical structures, biological targets and disease names to common ontologies and taxonomies. Data integration is achieved through assigning unique canonical chemical structures, unique protein sequences and standardized disease indications to the construction of a comprehensive ligand-molecular target matrix [17].

The chemogenomics knowledge space, integrated by Paolini et al. identified chemical tools, leads and drugs for nearly 900 human proteins. Annotated chemical libraries of chemical tools that can be derived from such databases can be applied to target validation and chemical biology [20,21]. Importantly, Paolini et al. have demonstrated that new knowledge space can be created by analyzing SAR data in its entirety by mapping the interaction network between proteins in chemical space [17]. Thus, mapping the chemogenomics knowledge space enables the development of methods for the rational design of polypharmacology drugs [22,23]. The chemogenomics knowledge space also enables the ideas proposed by Frye [24] and Fliri et al. [25–27] of comparing the homology of SARs across a wide range of molecular targets to determine the relationship between structure, selectivity, efficacy and safety. The data in the chemogenomics knowledge space are not just a repository of previous experimental observation but can be used to derive probabilistic models for a large number of targets, enabling the biospectrum of a compound, off-target effects and polypharmacology to be predicted from *a priori* knowledge [17,27–30]. Using a holistic view of biologically active chemical space, new knowledge derived from the discovery of new relationships, new properties and new classifications can be formally encoded as ontologies [31–33].

2.2. Open chemistry

The evolution of drug discovery into a knowledge-based predictive science lies, in part, in the assembly and integration of all pharmacologically relevant information, at both the molecular and phenotype level. Since the possession of facts is essential to the medicinal chemist work, it is surprising how slow the progress of integrating medicinal chemistry information has been. Indeed, in some quarters, the open exchange an integration of chemical information has been actively discouraged. In contrast to medicinal chemistry, the molecular biology community has ensured public access to all gene and protein sequences and 3-dimensional protein structures. Public access and the free exchange of all sequence data culminated in the success of the "open genomics" model championed by Sultson and Waterston [34]. The open exchange of sequence data is encouraged by a culture determined by the behavior of editors and funding bodies. Submission of a new sequence to a sequence database is usually expected and often a prerequisite on publication in a research journal. For example, the US National Institutes of Health's GeneBank, founded in 1982, is the annotated repository for all publicly available gene sequences. At the time of writing over 50 million, freely available, gene sequences are stored in GenBank. Likewise, the protein crystallography and protein NMR communities use the Protein Data Bank (PDB) as the single international repository for macromolecular structure data. Founded in 1971 with the deposition of just seven structures, it now contains over 36,000 macromolecular structures at the beginning of 2005.

On the basis of wealth of publicly available sequence data, the field of bioinformatics has evolved to serve the molecular biology community. Bioinformatians have designed sophisticated database search and data-mining tools that are freely available and easy to use by laboratory-based molecular biologists. Indeed, bioinformatics groups in the pharmaceutical industry use open-source informatics tools to such an extent that their annual software budgets can be several orders of magnitude smaller than the cost of the licenses required to access chemical information and use computational chemistry software. The availability of comprehensive data in molecular biology has spurred computational biology and bioinformatics to discover new knowledge from the accumulated data. For example, our advanced understanding of evolutionary relationships between genes and genomes derived from the large-scale comparison of the publicly collated data. Entire new fields such as conceptual biology [35–37] and network biology [38–40] have emerged as a result of the developing new tools to find relationships in the wealth of public biological data.

While access to gene sequence data is widely available through the public global genome repositories, no such integrated public databanks exist for pharmaceutical structure–activity data, despite its direct relevance to improving human health. Most pharmacological data exist in proprietary screening databases, published documents such as journal articles and patents and a growing variety of commercial databases of varying quality. Murray-Rust *et al.* have demonstrated that much of the publicly disclosed data is not machine-readable or semantically indexed [41–45]. The lack of accepted data sharing standards and data integration across the industry is a significant barrier to learning and building predictive mathematical models from the output of the significant annual private and public investment in pharmacological research.

Recently, there has been a variety of public initiatives to openly provide access to chemical structure and screening data information, including PubChem [46], DrugBank [47], Harvard University's ChemBank Initiative [48], the US National Cancer Institute [NCI] Screening Database [49] and the US National Institute of Mental Health's Psychoactive Drug Screening Program (NIMH-PDSP) K_i Database [50], among others. New standards such as the International Chemical Identifier (InChI) [42,51] defined by the International Union of Pure and Applied Chemistry (IUPAC) and the US National Institutes of Standards and Technology (NIST) provide a unique machine-readable, text string for each compound that overcomes many of the issues of stereochemical and tautomeric representation.

The new public initiatives of disclosing screening results are to be welcomed. However, there still exists the fact there is no commonly adopted mechanism for collating new or previously published medicinal chemistry or chemical biology data for the benefit of the global community. To overcome this problem Murray-Rust et al. have proposed a manifesto for "open chemistry", which calls for chemists to "move towards publishing their collective knowledge in a systematic and easily accessible form for reuse and innovation" [41]. The open chemistry manifesto calls on authors, publishers and funding agencies to make chemical data accessible and machine-readable by adopting XML-based standards for data description, use of the InChI representation as unique compound identification and the explicit statement of the rights of published data reuse consistent with the Budapest Open Access Initiative [52]. The adoption of common standards and global repositories for published pharmacological data would be an essential step in building the Chemical Semantic Web: a vision where all relevant, disclosed, medicinal chemistry information is accessible by humans and semantically aware software [41,42,53,54]. How the medicinal chemistry community responds to the open chemistry manifesto will be a crucial factor in the future direction of knowledge-based drug discovery.

3. POSSESSION OF SKILL

The need to find an efficient path to a clinical candidate through iterative cycles of design (hypothesis generation), synthesis and testing makes the second component of medicinal chemistry knowledge – possession of skills – a critical one. Chemist's skills come from many disciplines that include, among others, modern synthetic chemistry and technology, biology, pharmacology and drug metabolism. Considering the vast size of chemical space to be explored, it is not surprising that experience and intuition are the characteristics that distinguish the most successful medicinal chemists.

Compound design where a hypothesis about properties of new chemicals is generated is the most important step in the process, and this is where a multitude of factors highly individual to each chemist come into play such as familiarity with certain synthetic methods, successes or failures in the past drug discovery projects, exposure to specific papers in the literature, ability to deal with high volumes of heterogeneous data, ability to innovate and invent. The tactics chemists employ in the design and the judgments they make evaluating different hypotheses can differ significantly between even experienced chemists [55] and constitute what is called *tacit* knowledge.

The concept of tacit knowledge as introduced by Polanyi [56,57] is a dimension of our knowledge that is derived from experience and practice. It is often encapsulated in a sentence "We can know more than we can tell". By definition tacit knowledge cannot be fully formalized or articulated and is often expressed as hunches or intuition. Polanyi, a chemist and a philosopher, argued that the act of discovery itself is tacit driven by personal commitment and passion. Drug hunters will recognize Polanyi emphasis on personal motivation. Critical importance of tacit knowledge in the practice of medicinal chemistry has driven both theoretical work and practical strategies for the conversion of tacit to explicit knowledge that can be recorded, or codified, and thus transferred from chemist to chemist [58].

Elements of tacit knowledge can also be recorded in the form of empirical generalizations or tactics that are instrumental for compound design. For example, the concept of privileged structures [59] guides molecular designers toward select structural types known to bind to multiple, unrelated classes of receptors or enzymes with high affinity. Topliss tree [60] is a substitution scheme that is useful in optimizing aromatic or aliphatic substituents using a fixed set of substituents. The bioisosterism rules (bioisosteres in most general sense are groups or molecules, which have chemical and physical similarities producing broadly similar biological effects [61]) assist medicinal chemists in the design of new analogues.

Over the last half century, the set of skills required for efficient prosecution of medicinal chemistry programs has changed significantly [62]. While chemists still face the same tasks and decisions such as selecting structural series to pursue or generating and analyzing SAR to optimize the properties of a clinical candidate, the way they go about doing that is quite different. Many more tools are available and need to be mastered, ranging from new catalysts and advanced analytical and purification technologies, to parallel chemistry, structure-based drug design and more advanced *in vitro* and *in vivo* assays. As a result, the amount of data informing chemist's decisions has drastically increased. Most importantly, linear optimization path where each compound was characterized in a sequence of tests, and was advanced to the next step only if it met certain criteria, has been replaced by the non-linear approach [63]. This approach is designed to shorten time to a drug candidate, and uses multiple information feedback loops such as primary and counter screen data, pharmacokinetic and drug metabolism information, functional cell-based or *in vivo* data, all arriving in time to inform the design of the next analogue.

Operating in such non-linear paradigm demands advanced data manipulation and analysis skills. Since chemists are trying to optimize multiple parameters at the same time, typically efficacy, dose, solubility, stability and permeability, even viewing multi-dimensional SAR datasets requires powerful visualization software. The introduction of high-speed parallel chemistry is changing the optimization philosophy of traditional medicinal chemistry by introducing a partly stochastic approach to find optimal molecules in a multi-dimensional space. Multi-parameter optimization techniques developed in engineering, such as desirability functions, could find a suitable application in medicinal chemistry [64–66]. There is an urgent need for tools assisting in hypothesis generation based on multi-dimensional data for large sets of analogues.

Emergence of chemical genomics [67,68] has brought additional dimensions to the information available to chemists, requiring further changes in the optimization philosophy. Traditionally, the structure-centric thinking has been natural for medicinal chemists trying to hit a specific molecular target. Most of the properties requiring optimization could be directly or indirectly related to the structure, and after all, it is the structure that chemists have control over and can change. In chemogenomics space, drug interactions with all components of the biological system are taken into account in order to understand the mechanism of drug's action and to detect possible adverse effects early. Biological system is represented as a network in which the nodes are genes, proteins, cofactors and metabolites [38]. Functional connections between nodes represent gene regulation, receptor inhibition or activation, physical interaction of proteins, etc. Chemical compounds perturb the state of the biological network, and new methods of measuring drug action by detecting patterns of changes in the network are emerging [17,39]. In this space, a stochastic approach to optimization using compound libraries should be more productive.

A problem-facing all-drug discovery organizations is how to capture the knowledge of experienced medicinal chemists and transfer it to the next generation of drug designers. Despite the emergence of new technologies that accelerate and rationalize drug discovery process, tacit knowledge of individual chemists will continue to define the unique path to the clinical candidate that they choose in the hypothesis space. Since tacit knowledge cannot be fully transferred from an experienced medicinal chemist to a younger one, it will continue to take years for young chemists to approach the level of design efficiency that experienced chemists have. The medicinal chemist learns through the experience of numerous discovery and optimization projects. If tacit knowledge is learnt through experience, then one solution could be the design of "drug design simulators" where one could test designs and decision-making in a compressed-time frame. The coupling of the chemogenomic knowledge space with interactive human–machine interfaces could be the basis of designing such drug designer simulators. Considering the average drug candidate can cost several million dollars, just in the pre-clinical stage, investment in drug design simulators may be an appropriate learning tool.

4. ARTIFICAL INTELLIGENCE IN DRUG DESIGN

Whilst there is a strong tacit dimension to the medicinal chemist's knowledge, many of the tactics and techniques used to navigate chemical space that derive from experience can be articulated and therefore potentially formalized. If we can articulate the tactics used to explore SARs efficiently then the next step is to formalize and codify such knowledge. Codifying medicinal chemistry knowledge enables the goal of evolving from computer-aided drug design (CADD) to knowledge-aided drug design (KADD).

Methods for developing KADD may be found in the field of artificial intelligence. Two features of medicinal chemistry hope that artificial intelligence methods could aide the drug designer. First, drug design is often a problem with a complete description and second the rigorous scientific method that medicinal chemists apply in during compound optimization. The computer scientist's definition of a problem with a complete description is one for which we can define the specification of the solution [69]. The medicinal chemistry solution – a desired drug candidate profile – can be accurately described as, for example, "an oral, rule-of-five compliant, inhibitor of phosphodiesterase 5 with 100-fold selectivity over phosphodiesterases and with no structural alerts present". Even though the problem can be completely described, the difficult challenge is to find an optimal solution. The difficulty of this class of problem, with complete descriptions, is one of effectively navigating the vast space of feasible solutions (referred to as a nondeterministic polynomial (NP) hard problem) [69].

Recent advances in artificial intelligence have demonstrated that parts of the iterative scientific discovery process can itself be the subject of automation. In particular, the "closed loop" of hypothesis generation and *in vitro* experimentation in drug design is potentially amenable to improvement through knowledge-based artificial intelligence. The automation of the iterative hypothetico-deductive process [70], using advances in inductive logic programming [71], has recently been demonstrated in the field of functional genomics [72]. A "robot scientist" was designed to elucidate cellular metabolic pathways. The system applied inductive logic programming to learn from a large knowledge base of metabolic pathway information in order to propose hypotheses of how the pathway could be constructed. The system tested the hypothesis by ordering reagent and running the experiments through an automated-laboratory robot. Results from the experiments were fed back into the hypothesis generation engine, to design the next round of experiments. Reagents costs could also be factored into the experimental design enabling the system to design a set of experiments that could either get to the answer by the quickest or cheapest route. There have also been theoretical developments in attempts to formalize the creative process in the field of computer-aided design in engineering [73]. These developments are intriguing as the hypothetico-deductive process of medicinal chemistry drug design is most analogous to the functional genomics problem solved by the robot scientist. Theoretically, what would be required to develop an intelligent interactive design tool for medicinal chemists?

In all creative fields the design process decreases the possible number of solutions between the current state of the product and its desired specifications. The medicinal chemist intuitively understands that identifying and selecting the "lead" is the crucial in step increasing the likelihood of the future success of a project. Selection of the right chemical lead reduces the search space enormously. The history of medicinal chemistry is replete with examples of one or two subsistent changes between the lead and the final drug product [74,75]. If we consider drug design as a multi-dimensional optimization problem then combinations of probabilistic predictions and heuristic search patterns (e.g. evolutionary and emergent algorithms) can be employed to explore pathways through "hypothesis space".

Inductive logic programming [71,72] is a suitable method of hypothesis generation that is capable of learning from large-scale knowledge bases. However, a key problem is what has been called the "knowledge-acquisition bottleneck". In order to reduce the search of "hypothesis space" medicinal chemists, utilize knowledge of "tactics", which can be applied in many situations, to make large conceptual jumps in the search space. Common tactics employed by the experienced medicinal chemist include "methylene shuffle", adding lipophilicity, adding chirality, searching for hydrogen-bond interactions, introducing or breaking conformational constraints, amongst many others [76]. The advent of large-scale semantically normalized integrated databases [17] provides a resource for such automatic knowledge acquisition and data mining for the discovery of new tactics and rules. In order for a design system to learn and apply the tactics of the medicinal chemist, a suitable "chemical algebra" to describe transformations in drug design would need to be derived [77]. Such a "chemical algebra" could be found in semiotics (the science of signs and processing of signs) which being explored, for example, in engineering to establishing a computational framework for design theory [73]. The advantage of applying semiotics is that algebraic semiotics [73,78] could be a suitable tool to manipulate such a "chemical algebra". In order to encode medicinal chemistry knowledge as semiotics a meta-level representation of chemical structure – what could be called "chemiotics" – is required. The costs of the iterative rounds of experiment can also be judged by connecting the hypothesis generation engine to various established retro-synthesis programs or virtual chemical libraries software that can estimate the synthetic accessibility.

Of the theoretical requirements for building an intelligent KADD system, as outlined above, the medicinal chemistry knowledge bases, the evolutionary search engines and the hypothesis generation methods are in existence today. What currently missing is the semiotic representation of medicinal chemistry knowledge in terms of tactics and rules. The application of knowledge-based artificial intelligence to medicinal chemistry design would enable the insights and knowledge of expert drug designers to be leveraged across a larger number of projects. Alternatively, less-experienced drug designers could be guided in their design by systems build on a wealth of historical knowledge, combined with predictive modeling [29,79]. By formalizing medicinal chemistry knowledge in an intelligent system, a radical improvement in productivity should be possible.

5. CONCLUSION

Ackoff described "intelligence" as the ability to increase efficiency and "wisdom" as the ability to increase effectiveness [3]. We have found the Ackoff and Emery description of knowledge [2] useful in describing the medicinal chemistry at the meta-science level. By considering drug design as a process, the core of medicinal chemistry knowledge can be abstracted as the possession of facts and the possession of skills. The possession of facts encompasses the knowledge gained by understanding the probability of predicted and observed properties of the chemical structure in relation to its desired properties. The possession of skills is the knowledge required to efficiently navigate the optimization path to achieve the desired balance of properties (optimize potency, selectivity, novelty, etc.). Each new piece of information, whether the crystal structure of a drug bound to a protein or the identification of a labile site of reactive metabolite or the general physicochemical properties of drug-space further restricts our hypothesis space in drug design. Practical skills such as mastery of design tactics or parallel chemistry methods help the medicinal chemists efficiently navigate a pathway through hypothesis space. Therefore by understanding and formalizing how knowledge is generated and used in medicinal chemistry, we can build a methodology to rationally improve the efficiency and effectiveness of drug design. Once we can articulate our knowledge, then the tools of ontology and semiotics enable us to formalize our methods of design. The first generation of automation in drug discovery has been capital intensive, leading to an increased throughput in compound screening and synthesis. The second generation of automation will be knowledge intensive, leading to automation of parts of the scientific process and cycle of iterative hypothesis generation [17,29,72].

REFERENCES

[1] P. F. Drucker, *Innovation and Entrepreneurship*, Harper & Row, New York, 1985.
[2] R. L. Ackoff and F. E. Emery, *On Purposeful Systems*, Travistock, London, 1972.
[3] R. L. Ackoff, *J. Appl. Sys. Anal.*, 1989, **16**, 3–9.
[4] B. Stockwell, *Nature*, 2004, **432**, 846–854.
[5] G. M. Rishton, *Drug Discov. Today*, 1997, **2**, 382–384.
[6] J. Ashby, *Environ. Mutagen.*, 1985, **7**, 919–921.
[7] D. P. Williams and D. J. Naisbitt, *Curr. Opin. Drug Discov. Devel.*, 2002, **5**, 104–115.
[8] G. Cruciani, E. Carosati, B. De Boeck, K. Ethirajulu, C. Mackie, T. Howe and R. Vianello, *J. Med. Chem.*, 2005, **48**, 6970–6979.
[9] C. A. Lipinski, F. Lombardo, B. W. Dominy and P. J. Feeney, *Adv. Drug Deliver. Revs.*, 1997, **23**, 3–25.
[10] K. Palm, P. Stenberg, K. Luthman and P. Artursson, *Pharmaceut. Res.*, 1997, **14**, 568–571.
[11] J. Kelder, P. D. Grootenhuis, D. M. Bayada, L. P. Delbressine and J. P. Ploemen, *Pharmaceut. Res.*, 1999, **16**, 1514–1519.
[12] A. L. Hopkins, C. R. Groom and A. Alex, *Drug Discov. Today*, 2004, **9**, 430–431.
[13] M. Congreve, C. W. Murray and T. L. Blundell, *Drug Discov. Today*, 2005, **10**, 895–907.
[14] S. P. Williams, L. F. Kuyper and K. H. Pearce, *Curr. Opin. Chem. Biol.*, 2005, **9**, 371–380.
[15] A. Schuffenhauer and E. Jacoby, *Drug Disc. Today: BIOSILICO*, 2004, **2**, 190–200.
[16] E. Jacoby, A. Schuffenhauer and P. Floersheim, *Drug News Perspect*, 2003, **16**, 93–102.
[17] G. V. Paolini, R. H. B. Shapland, W. P. van Hoorn, J. S. Mason and A. L. Hopkins, *Nat. Biotechnol*, 2006, **24** (7), 805–815.
[18] J. Overington and B. Al-Lazikani, *StARlite*, Inpharmatica Ltd, London, 2005.
[19] G. Piatetski-Shapiro and W. Frawley, *Knowledge Discovery in Databases*, MIT Press, Cambridge, 1992.
[20] D. E. Root, S. P. Flaherty, B. P. Kelley and B. Stockwell, *Chem. Biol.*, 2003, **10**, 881–892.
[21] N. P. Savchuk, K. V. Balakin and S. E. Tkachenko, *Curr. Opin. Chem. Biol.*, 2004, **8**, 412–417.
[22] S. Frantz, *Nature*, 2005, **437**, 942–943.
[23] A. L. Hopkins, J. S. Mason and J. P. Overington, *Curr. Opin. Struct. Biol.*, 2006, **16**, 127–136.
[24] S. V. Frye, *Chem. Biol.*, 1999, **6**, R3–R7.
[25] A. F. Fliri, W. T. Loging, P. F. Thadeio and R. A. Volkmann, *Proc. Natl. Acad. Sci. USA*, 2005, **102**, 261–266.
[26] A. F. Fliri, W. T. Loging, P. F. Thadeio and R. A. Volkmann, *J. Med. Chem.*, 2005, **48**, 6918–6925.
[27] A. F. Fliri, W. T. Loging, P. F. Thadeio and R. A. Volkmann, *Nat. Chem. Biol.*, 2005, **1**, 389–397.
[28] Nidhi, M. Glick, J. W. Davies and J. L. Jenkins, *J. Chem. Inf. Model*, 2006, **46** (3), 1124–1133.

[29] P. W. Swaan and S. Ekins, *Drug Discov. Today*, 2005, **10**, 1191–1200.
[30] D. Rogers, R. D. Brown and M. Hahn, *J. Biomol. Screen.*, 2005, **10**, 682–686.
[31] H. J. Feldman, M. Dumontier, S. Ling, N. Haider and C. W. Hogue, *FEBS Lett*, 2005, **579**, 4685–4691.
[32] A. Schuffenhauer, J. Zimmermann, R. Stroop, J.-J. van der Vyver, S. Lecchini and E. Jacoby, *J. Chem. Inf. Comp. Sci.*, 2002, **42**, 947–955.
[33] European Bioinformatics Institute. 2006. *Chemical Entities of Biological Interest (ChEBI)*. http://www.ebi.ac.uk/chebi/.
[34] J. Sulton and G. Ferry, *The Common Thread*, Bantam Press, London, 2002.
[35] T. Bekhuis, *Biomed. Digit. Libr.*, 2006, **3**, 2.
[36] M. Krallinger, R. A. Erhardt and A. Valencia, *Drug Discov. Today*, 2005, **10**, 439–445.
[37] J. Natarajan, D. Berrar, C. J. Hack and W. Dubitzky, *Crit. Rev. Biotechnol.*, 2005, **25**, 31–52.
[38] A. L Barabasi and Z. N. Oltvai, *Nat. Rev. Genet.*, 2004, **5**, 101–113.
[39] Y. Nikolsky, T. Nikolskaya and A. Bugrim, *Drug Discov. Today*, 2005, **10**, 653–662.
[40] R. Sharan and T. Ideker, *Nat. Biotechnol.*, 2006, **24**, 427–433.
[41] P. Murray-Rust, H. S. Rzepa, S. M. Tyrrell and Y. Zhang, *Org. Biomol. Chem.*, 2004, **2**, 3192–3203.
[42] S. J. Coles, N. E. Day, P. Murray-Rust, H. S. Rzepa and Y. Zhang, *Org. Biomol. Chem.*, 2005, **3**, 1832–1834.
[43] P. Murray-Rust, H. S. Rzepa, J. J. Stewart and Y. Zhang, *J. Mol. Model.*, 2005, **11**, 532–541.
[44] P. Murray-Rust, J. B. Mitchell and H. S. Rzepa, *BMC Bioinformatics*, 2005, **6**, 180.
[45] P. Murray-Rust, J. B. Mitchell and H. S. Rzepa, *BMC Bioinformatics*, 2005, **6** (1), 141.
[46] D. L. Wheeler, T. Barrett, D. A. Benson, S. H. Bryant, K. Canese, V. Chetvernin, D. M. Church, M. DiCuccio, R. Edgar, S. Federhen, L. Y. Geer, W. Helmberg, Y. Kapustin, D. L. Kenton, O. Khovayko, D. J. Lipman, T. L. Madden, D. R. Maglott, J. Ostell, K. D. Pruitt, G. D. Schuler, L. M. Schriml, E. Sequeira, S. T. Sherry, K. Sirotkin, A. Souvorov, G. Starchenko, T. O. Suzek, R. Tatusov, T. A. Tatusova, L. Wagner and E. Yaschenko, *Nucleic Acids Res.*, 2006, **34**, D173–D180.
[47] D. S. Wishart, C. Knox, A. C. Guo, S. Shrivastava, M. Hassanali, P. Stothard, Z. Chang and J. Woolsey, *Nucleic Acids Res.*, 2006, **34**, D668–D672.
[48] R. L. Strausberg and S. L. Schreiber, *Science*, 2003, **300**, 294–295.
[49] J. N. Weinstein, T. G. Myers, P. M. O'Connor, S. H. Friend, A. J. Fornace, Jr., K. W. Kohn, T. Fojo, S. E. Bates, L. V. Rubinstein, N. L. Anderson, J. K. Buolamwini, W. W. van Osdol, A. P. Monks, D. A. Scudiero, E. A. Sausville, D. W. Zaharevitz, B. Bunow, V. N. Viswanadhan, G. S. Johnson, R. E. Wittes and K. D. Paull, *Science*, 1997, **275** (5298), 343–349.
[50] B. L. Roth, W. K. Kroeze, S. Patel and E. Lopez, *Neuroscientist*, 2000, **6**, 252–262.
[51] S. E. Stein, S. R. Heller and D. Tchekhovoski, *An Open Standard for Chemical Structure Representation—The IUPAC Chemical Identifier*. Presented at Nimes International Chemical Information Conference Proceedings, 2003.
[52] Open Society Institute, 2001. Budapest Open Access Initiative. http://www.soros.org/openaccess/.
[53] T. Berners-Lee, J. Hendler and D. Lassila, *Scientific American*, 2001
[54] G. V. Gkoutos, P. Murray-Rust, H. S. Rzepa and M. Wright, *J. Chem. Inf. Comp. Sci.*, 2001, **41**, 1124–1130.
[55] M. S. Lajiness, G. M. Maggiora and V. Shanmugasundaram, *J. Med. Chem.*, 2004, **47**, 4891–4896.
[56] M. Polanyi, *The Tacit Dimension*, Routledge & Kegan Paul, London, 1966.

[57] M. Polanyi, *Personal Knowledge: Towards a Post-Critical Philosophy*, University of Chicago Press, Chicago, 1974.
[58] A. M. Sapienza and J. G. Lombardino, *Drug Des. Res.*, 2002, **57**, 51–57.
[59] B. E. Evans, K. E. Rittle, M. G. Bock, R. M. DiPardo, R. M. Freidinger, W. L. Whitter, G. F. Lundell, D. F. Veber, P. S. Anderson and K. D. Paull, *J. Med. Chem.*, 1988, **31**, 2235–2246.
[60] J. G. Topliss, *J. Med. Chem.*, 1972, **15**, 1006–1111.
[61] C. W. Thornber, *Chem. Soc. Rev.*, 1979, **8**, 563–580.
[62] J. G. Lombardino and J. A. I. Lowe, *Nat. Rev. Drug Discov.*, 2004, **3**, 853–862.
[63] M. MacCoss and T. A. Baillie, *Science*, 2004, **303**, 1810–1813.
[64] E. C. Harrington, *Ind. Qual. Control*, 1965, **21**, 494–498.
[65] G. C. Derringer and R. Suich, *J. Qual. Tech.*, 1980, 214–219.
[66] C. Le Bailly de Tilleghem, B. Beck, B. Boulanger and B. Govaerts, *J. Chem. Inf. Model.*, 2005, **45**, 758–767.
[67] F. R. Salemme, *Pharmacogenomics*, 2003, **4**, 257–267.
[68] F. Darvas, G. Dorman, P. Krajcsi, L. G. Puskas, Z. Kovari, Z. Lorincz and L. Urge, *Curr. Med. Chem.*, 2004, **11**, 3119–3145.
[69] K. Ueda, *Artif. Intell. Eng.*, 2001, **15**, 321–327.
[70] K. Popper, The Logic of Scientific Discovery, Hutchison & Co, London.
[71] S. Muggleton, *New Generat. Comput.*, 1991, **8**, 295–318.
[72] R. D. King, K. E. Whelan, F. M. Jones, P. G. Reiser, C. H. Bryant, S. H. Muggleton, D. B. Kell and S. G. Oliver, *Nature*, 2004, **247**, 247–252.
[73] V. V. Kryssanov, H. Tamaki and S. Kitamura, *Artif. Intell. Eng.*, 2001, **15**, 329–342.
[74] W. Sneader, *Drug Prototypes and their Exploitation*, Wiley, London.
[75] W. Sneader, *Drug Discovery—A History*, Wiley, Chichester.
[76] C. G. Wermuth, *Practice of Medicinal Chemistry*, Academic Press, London.
[77] S. Muggleton, *Nature*, 2006, **440**, 409–410.
[78] J. A. Goguen, in *Computation for Metaphor, Analogy and Agents Springer Lecture Notes in Artifical Intelligence* (ed. C Nehaniv), 1999, pp. 242–291.
[79] I. C. Parmee, *The Analyst*, 2005, **130**, 29–34.

To Market, To Market – 2005

Shridhar Hegde and Michelle Schmidt

Pfizer Global Research & Development, St. Louis, MO 63017, USA

Contents

1. Introduction	439
2. Ciclesonide (asthma, COPD) [6–9]	443
3. Clofarabine (anticancer) [10–14]	444
4. Darifenacin (urinary incontinence) [15–18]	445
5. Deferasirox (chronic iron overload) [19–21]	446
6. Doripenem (antibiotic) [22–25]	448
7. Eberconazole (antifungal) [26–28]	449
8. Entecavir (antiviral) [29–32]	450
9. Eszopiclone (hypnotic) [33–36]	451
10. Exenatide (anti-diabetic) [37–41]	452
11. Galsulfase (mucopolysaccharidosis VI) [42–44]	453
12. Luliconazole (antifungal) [45–47]	454
13. Lumiracoxib (anti-inflammatory) [48–51]	455
14. Nepafenac (anti-inflammatory) [52–55]	456
15. Pegaptanib (age-related macular degeneration) [56–64]	458
16. Pramlintide (anti-diabetic) [65–68]	460
17. Palifermin (mucositis) [69–72]	461
18. Ramelteon (insomnia) [73–78]	462
19. Rasagiline (Parkinson's disease) [79–87]	464
20. Sorafenib (anticancer) [88–92]	466
21. Tamibarotene (anticancer) [93–96]	467
22. Tigecycline (antibiotic) [97–103]	468
23. Tipranavir (HIV) [104–110]	470
24. Udenafil (erectile dysfunction) [111–115]	472
25. Ziconotide (severe chronic pain) [116–119]	473
References	474

1. INTRODUCTION

The year 2005 heralded the market entry of 22 new chemical entities (NCEs) and two new biological entities (NBEs), five more than the previous year [1–5]. The United States continued to be the most active market with the introduction of 15 new drug entities. The Japanese market saw the entry of three new drugs, with Brazil, Germany, Israel, South Korea, Spain, and the United Kingdom each supporting one apiece. Novartis was accredited with three introductions while Amgen, Aventis, Bayer, BMS, Eli Lilly, Pfizer, and Wyeth had a marketing or co-marketing role in one each. Amylin, Eli Lilly's collaborator on a novel anti-diabetic, solely launched a second peptide analog with an alternative mechanism of action for the treatment of diabetes mellitus. While new combinations of existing drugs continued to be popular in providing enhanced patient benefit, extensive coverage of these launches is beyond the scope of this article. Likewise, line extensions involving new formulations and new indications will not be fully addressed in this venue although this group comprised a significant percentage of new products in 2005.

The anti-infective therapeutic area enjoyed the introduction of five new drugs in 2005, including one first-in-class therapy. Tygacil™ (tigecycline) is the first broad-spectrum antibiotic from the glycylcycline class for the treatment of complicated intra-abdominal and skin infections. Having emerged from research to circumvent the efflux and ribosomal protection mechanisms of bacteria, it has proven to be effectual where traditional tetracyclines have encountered resistance. Aptivus® (tipranavir), an HIV protease inhibitor, received accelerated approval for combination retroviral treatment in HIV-1-infected adults with demonstrated resistance to multiple protease inhibitors. Its success, where other protease inhibitors have encountered resistance, is attributed to its design by structure-based analysis to possess increased flexibility making it amenable to conformational alterations at the binding site. In the azole class of antifungals, Ebernet® (eberconazole) and Lulicon® (luliconazole) were introduced as two new options for patients. The primary indication for both of these topical agents is athlete's foot (tinea pedis). The final anti-infective, Finibax (doripenem) was launched for the treatment of bacterial respiratory and urinary tract infections. While doripenem is the fourth carbapenem antibiotic on the market, it differentiates from its predecessors by possessing enhanced stability and improved antibacterial potency.

Another prolific therapeutic area in 2005 was CNS, also with five marketed drugs. In the field of pain management, Prialt® (ziconotide), delivered by intrathecal infusion, was launched as a non-opioid treatment for severe chronic pain, particularly in opioid refractory patients. The other pain reliever to debut last year was Prexige® (lumiracoxib), a COX-2 selective inhibitor for the treatment of osteoarthritis and acute pain. Two new psychopharmacological drugs entered the market for the treatment of insomnia. Lunesta™ (eszopiclone) is the optically active form of zopiclone, which had been previously marketed as a racemic mixture. It belongs to the class of non-benzodiazepine drugs that target the GABA-A receptor. Unlike other Schedule i.v. controlled hypnotics, however, eszopiclone has been approved for long-term use. The other new treatment for insomnia exploits a novel mechanism of action. Rozerem™ (ramelteon) targets the melatonin receptor (MT_1/MT_2). Since it has exhibited no evidence of abuse or dependence in clinical trials, ramelteon is not designated as a controlled substance. In the battle against Parkinson's disease, Azilect® (rasagiline) was introduced. It can be used as either an initial monotherapy in early Parkinson's patient or as an adjunct treatment in more advanced cases.

The oncolytic domain experienced the entry of drugs to not only treat the diseases but to also manage the severe side effects of chemotherapy and radiation. Clolar™ (clofarabine) is a second-generation purine nucleoside analog for the treatment of pediatric patients with relapsed or refractory acute lymphoblastic leukemia. Incorporation of a fluorine into the 2′-position of the sugar moiety confers increased stability compared to its predecessors. Amnolake® (tamibarotene), launched in Japan last year, has orphan drug status for the treatment of relapsed or refractory acute promyelocytic leukemia. Nexavar® (sorafenib), an inhibitor of multiple kinases involved in tumor angiogenesis and tumor cell proliferation, is currently indicated for renal cell carcinoma and has the potential to treat a wide variety of solid tumors. In addition to these primary cancer regimens, Kepivance™ (palifermin) promotes the healing of oral and gastrointestinal mucosa that is severely inflamed in patients receiving high dose chemotherapy and radiation for hematological malignancies.

In the field of endocrine and metabolic diseases, anti-diabetics led the way. In addition to the 2005 launch of Apidra® (insulin glulisine), a rapid-acting man-made insulin for managing mealtime blood sugar, two other synthetic peptides with different mechanisms of action were developed by Amylin. Symlin® (pramlintide) is an injectable human amylin analog that differs from its parent by the strategic substitution of Ala-25, Ser-28, and Ser-29 with prolines. These modifications serve to increase solubility and eliminate aggregation to provide a biologically active, pharmaceutically acceptable peptide to be used in conjunction with insulin. Byetta™ (exenatide) is an injectable incretin mimetic, the first in its class, approved as adjunctive therapy to improve glycemic control in patients where metformin and sulfonylureas have proven inadequate. Both of these peptides affect glucose homeostasis, albeit by different mechanisms, and both have the added benefit of potential weight loss. From the regimens described above, it comes as no surprise that the required multiple daily injections for effective disease management place a heavy burden of compliance on patients. Generex Biotechnology has developed a less invasive, pain-free alternative for insulin administration. Oral-Lyn™, a liquid aerosolized formulation of insulin, is sprayed directly into the patient's mouth via the RapidMist™ device. This product was launched initially in Ecuador as a treatment of both type 1 and type 2 diabetes, and it represents the first regulatory approval for a noninjectable insulin formulation. In the field of metabolic drugs, the recombinant protein Naglazyme™ (galsulfase) was launched as a treatment for mucopolysaccharidosis VI (also known as Maroteaux-Lamy syndrome), which is an inherited and life-threatening lysomal storage disorder caused by the deficiency of the enzyme required for the breakdown of glycosaminoglycans. Galsulfase is the first approved treatment for this rare disease.

The ophthalmic sector welcomed two NCEs: Macugen® (pegaptanib) and Nevanac™ (nepafenac). Pegaptanib has the unique distinction of being not only the first antiangeogenic agent for the treatment of neovascular age-related macular degeneration, but it is also the first RNA aptamer to be approved for a therapeutic application. Nepafenac is a pro-drug of the NSAID amfenac with the indication of pain and inflammation management following cataract surgery. In clinical studies, greater than 80% of patients treated with nepafenac were pain-free the day after surgery. In addition to these ophthalmic drugs, the world's first intravitreal drug implant for the treatment of uveitis, a leading cause of blindness, reached the market. Retisert™ (fluocinolone acetate intravitreal implant) functions by delivering sustained levels of the corticosteroid fluocinolone acetonide for 30 months directly to the back of the eye via a tiny drug reservoir to effectively treat the inflammation.

The other therapeutic area with two new introductions to the market was renal – urologic. While Emselex® (darifenacin) is another anti-muscarinic for the treatment of overactive bladder and urinary incontinence, it may differentiate from its predecessors by its higher selectivity for the M3 subtype. Joining the erectile dysfunction arsenal, Zydena® (udenafil) is a PDE5 inhibitor with a rapid onset and a long duration of action.

Of the remaining drugs, respiratory, gastrointestinal, and poison control all claimed one new entry. In the respiratory arena, Alvesco® (ciclesonide) was introduced as an inhaled corticosteroid for the prophylactic treatment of persistent asthma. Compared to other marketed corticosteroids, ciclesonide benefits from lower systemic exposure and longer duration of action. The gastrointestinal category was represented by Baraclude™ (entecavir), an orally active nucleoside analog for the treatment of chronic hepatitis B infection. For the treatment of chronic iron overload in blood transfusion patients over the age of two, Exjade® (deferasirox) was introduced. As an iron chelator that is administered as a dispersion in orange juice, apple juice, or water, deferasirox signifies a major advancement from the previous standard of care that requires subcutaneous infusion for 8–12 h per night for the duration of the transfusion or until excess iron is eliminated.

While the cardiovascular section did not establish an NCE in 2005, launching of Procoralan® (ivabradine), an oral treatment for angina, is expected early in 2006. Three existing drugs, however, found new cardiovascular indications. Revatio™ (sildenafil citrate) was approved as a treatment for pulmonary arterial hypertension; sildenafil has been on the market since 1998 under the name of Viagra® for the treatment of erectile dysfunction. Hyzaar®, a fixed-dose combination consisting of the angiotensin II receptor blocker losartan potassium and hydrochlorothiazide, is now also indicated for stroke risk reduction in patients with hypertension and left ventricular hypertrophy. Already approved for the treatment of hypertension, Aceon® (perindopril erbumine) use is now extended to patients with stable coronary artery disease to reduce the risk of cardiovascular mortality or nonfatal myocardial infarction.

In addition to the NCEs and NBEs described above, 2005 was a notably productive year for the introduction of new vaccines. A total of six new products were launched in this area: Bilive™, a combined hepatitis A and B vaccine; Fendrix®, an adjuvanted hepatitis B vaccine for patients with renal insufficiency; Rotarix®, a rotavirus vaccine for the prevention of gastroenteritis caused by rotavirus infection; Menactra™, a meningococcal vaccine; Proquad® for measles, mumps, rubella, and varicella; and Mearubik, a measles-rubella vaccine. Although not covered here in detail, these vaccines represent very significant advances in preventative care.

While all the drugs launched in 2005 anticipate providing long-term patient benefit, the class of 2004 suffered two market withdrawals. Exanta® (ximelagatran), developed for the prevention of venous thromboembolism in elective hip or knee replacement surgery, offered the promise of improved patient compliance over low-molecular weight heparin (subcutaneous injection) or warfarin (frequent laboratory monitoring and dose adjustment), but hepatotoxicity prompted AstraZeneca to voluntarily remove it from the market early in 2006. The withdrawal of Tysabri® (natalizumab), however, appears to be only temporary. Natalizumab is a humanized monoclonal antibody that binds to the α4 integrin on immune cell surfaces, thereby preventing the movement of immune cells across the blood–brain barrier. Its mechanism of action results in an effective treatment of relapsing forms of multiple sclerosis (MS). Due to reports of progressive multifocal leukoencephalopathy, a rare but frequently fatal demyelinating disease of the central nervous system, co-developers Elan and Biogen Idec voluntarily withdrew natalizumab from the market in February of 2005. Based on an extensive follow-up safety evaluation of the product, the companies submitted a supplemental BLA for the product to the FDA in September of 2005. In early 2006, the Peripheral and Central Nervous System Drugs Advisory Committee of the FDA voted unanimously to recommend reintroduction of this drug for treatment of relapsing forms of MS [1–5].

2. CICLESONIDE (ASTHMA, COPD) [6–9]

Country of origin	Spain
Originator	Recordati Espana
First introduction	UK
Introduced by	Altana/ Aventis/ Teijin
Trade name	Alvesco
CAS registry no	126544-47-6
Molecular weight	540.69

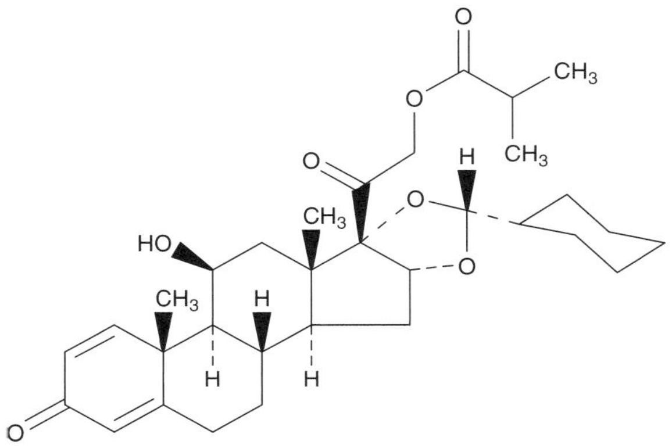

Ciclesonide, a new inhaled corticosteroid (ICS), is indicated for the prophylactic treatment of persistent asthma. ICS treatment is a widely accepted standard of care for maintenance therapy of chronic asthma, and the currently available agents include fluticasone propionate, budesonide, triamcinolone acetonide, flunisolide, and beclomethasone dipropionate. These agents exert their potent anti-inflammatory effects via modulation of the glucocorticoid receptor (GR). Although ICS drugs are generally safe and well tolerated compared with oral corticosteroids, many have measurable systemic exposures, and concerns over potential side effects resulting from it severely limit the dose at which they can be administered for long-term therapy. Systemic adverse effects associated with corticosteroids include HPA axis suppression, osteoporosis, abnormal glucose metabolism, cataracts, and glaucoma, some of which could potentially occur with the long-term use of high dose ICS. The key differentiators for ciclesonide relative to other ICS drugs are its longer duration of action and lower systemic exposure. Ciclesonide is an isobutyryl ester prodrug. It is cleaved by the endogenous esterases in the lung to des-isobutyryl ciclesonide (des-CIC), which is a potent GR agonist. The binding affinity of des-CIC for human GR ($K_i = 0.31$ nM) is similar to other ICS such as budesonide ($K_i = 0.44$ nM) and fluticasone propionate ($K_i = 0.24$ nM), while ciclesonide itself has about 100-fold lower affinity ($K_i = 37$ nM). In lung tissue, des-CIC undergoes reversible lipid conjugation to form oleate and palmitate ester conjugates, which act as a slow-release pool for the drug and increase the pulmonary residence time. This, in turn, contributes to the enhanced local effects and the long duration of action. Des-CIC has a pulmonary bioavailability of about 50%, whereas the oral bioavailability of both ciclesonide and des-CIC are <1%. Des-CIC has a high-clearance rate of 396 L/h, a volume of distribution of 1190 L, and a mean elimination half-life of 3.5 h. Any systemically available ciclesonide and Des-CIC undergo extensive first-pass liver metabolism, whereby they are rapidly hydoxylated by CYP3A4 to produce inactive metabolites. In addition, both ciclesonide and des-CIC have high protein binding (99%), which also contribute to low-systemic exposure of the free drug. The dosing regimen of ciclesonide is 80–320 μg once daily, and it is delivered in solution form via a hydrofluoroalkane metered-dose inhaler. In multiple 12-week clinical studies in asthma patients, 80–320 μg once daily ciclesonide was at least as effective as 400 μg once daily budesonide or 88 μg twice daily fluticasone at increasing forced expiratory volume in 1 s (FEV1) and forced vital capacity (FVC) from baseline. Inhaled ciclesonide was generally well tolerated in these clinical studies. Ciclesonide did not suppress biochemical markers of adrenal function in 52-week studies; however, the long-term (>52 weeks) systemic effects remain unknown. Ciclesonide is chemically produced via a semi-synthesis starting from 16-α-hydroxyprednisolone by first converting to a triisobutyryl ester intermediate with isobutyric anhydride, and subsequent reaction of the triester with cyclohexane carboxaldehyde and hydrochloric acid in dioxane. The latter step produces the cyclic ketal as a mixture of diastereomers, which is subjected to HPLC and fractional crystallization to produce ciclesonide.

3. CLOFARABINE (ANTICANCER) [10–14]

Country of origin	US
Originator	Southern Research Institute
First introduction	US
Introduced by	Genzyme/Bioenvision
Trade name	Clolar
CAS registry no	123318-82-1
Molecular weight:	303.68

Clofarabine is a new member of the purine nucleoside antimetabolite class of drugs, and it was launched last year as an intravenous infusion for treating pediatric patients (1–21 years old) with relapsed or refractory acute lymphoblastic leukemia (ALL) after at least two prior regimens. Adenosine-related antimetabolites, such as cladribine and fludarabine have proven successful in treating low-grade lymphomas, chronic lymphocytic leukemia, and hairy-cell leukemia. Although structurally similar to cladribine and fludarabine, a key differentiator for clofarabine is the presence of a fluorine in the C-2' position, which renders it less susceptible to phosphorolytic cleavage of the glycosydic bond and inactivation by purine nucleoside phosphorylases. In addition, the C-2' fluoro group improves the acid stability relative to its predecessors. As seen with other purine nucleoside analogs, the mechanism of action of clofarabine involves intracellular phosphorylation to active triphosphate by 2'-deoxycytidine kinase, and subsequent inhibition of RNA reductase and DNA polymerase α. Clofarabine is cytotoxic to rapidly proliferating and quiescent cancer cell types *in vitro*. It exhibits superior cytotoxicity ($IC_{50} = 5$ nM) than cladribine and fludarabine ($IC_{50} = 16$ and 460 nM, respectively) in K-562 cells.

The higher potency of clofarabine relative to other purine nucleoside analogs is attributed to the higher efficiency of its phosphorylation by deoxycytidine kinase, and the longer intracellular half-life of the triphosphate metabolite (>24 h). The chemical synthesis of clofarabine involves the conversion of 2-deoxy-2-β-fluoro-1,3,5-tri-*O*-benzoyl-1-α-D-arabinofuranose to the corresponding bromosugar with hydrogen bromide, subsequent coupling with 2-chloroadenine, and the removal of benzoyl protecting groups with catalytic sodium methoxide in methanol. The recommended pediatric dosage of clofarabine is 52 mg/m^2 daily, administered by i.v. infusion over 2 h, for 5 consecutive days. Treatment cycles are repeated following recovery or return to baseline organ function, approximately every 2–6 weeks. Clofarabine has a volume of distribution of 172 L/m^2, plasma protein binding of 47%, and a terminal half-life of about 5.2 h. It has a systemic clearance rate of 28.8 L/h/m^2. Elimination is primarily via renal excretion, with 49–60% of the dose excreted unchanged in the urine. *In vitro* studies using isolated human hepatocytes indicate very limited hepatic metabolism (0.2%); thus, non-renal pathways of elimination remain unknown. Clinical efficacy of clofarabine was evaluated in a single-arm study involving 49 pediatric patients, who had relapsed or failed two or more prior therapies. Fifteen patients (30.6%) demonstrated either a complete remission, a complete remission minus platelet recovery, or a partial response. For patients experiencing complete remission, the response lasted from 43 days to >160 days. Adverse events associated with clofarabine were similar to other chemotherapy agents, including vomiting, nausea, febrile neutropenia, and diarrhea.

4. DARIFENACIN (URINARY INCONTINENCE) [15–18]

Country of origin	US
Originator:	Pfizer
First introduction	Germany
Introduced by:	Novartis
Trade name	Enablex; Emselex
CAS registry no	133099-04-04
Molecular weight	426.55

Darifenacin is a novel muscarinic M3 selective antagonist for the once-daily oral treatment of urinary incontinence and overactive bladder. The majority of overactive bladder symptoms are thought to result from the overactivity of the detrusor muscle, which is primarily mediated by acetylcholine-induced stimulation of muscarinic M3 receptors in the bladder. Consequently, antimuscarinic agents have become the mainstay of overactive bladder treatment. Darifenacin has a higher level of M3 selectivity than the previously marketed antimuscarinic agents. It has K_i values of 16 nM for M1, 50 nM for M2, and 1.6 nM for M3 receptors. It is slightly more M3 selective than solifenacin (M1:K_i = 25 nM, M2:K_i = 126 nM, M3:K_i = 10 nM), which was launched in 2004. Darifenacin is significantly more selective than other muscarinics such as tolterodine, oxybutynin, and trospium, which are all essentially equipotent against M1, M2, and M3 receptors. In addition, darifenacin demonstrates greater effect on tissues in which the predominant receptor type is M3 rather than M1 or M2. *In vitro* darifenacin inhibits carbachol-induced contractions with greater potency in isolated guinea-pig bladder (M3) than in guinea-pig atria (M2) or dog saphenous vein (M1). In animal models, it shows greater selectivity for inhibition of detrusor contraction over salivation or tachycardia. The synthesis of darifenacin involves the coupling of 5-(2-bromoethyl)-2,3-dihydrobenzofuran with 3-(S)-(1-carbamoyl-1,1-diphenylmethyl)pyrrolidine as a key step. The latter intermediate is prepared from 3-(R)-hydroxypyrrolidine in a five-step sequence involving N-tosylation, Mitsunobu reaction to introduce a tosyloxy group in the 3-position with stereochemical inversion, anionic alkylation with diphenylacetonitrile, cleavage of the N-tosyl protecting group with HBr, and conversion of the cyano group to a carboxamide. Darifenacin is supplied as a controlled release formulation, and the recommended dosage is 7.5 mg once, daily. Darifenacin is rapidly and completely absorbed from the GI tract after oral administration, with maximum plasma levels achieved after about 7 h. The elimination half-life is approximately 3 h, but because of the controlled release characteristics of the formulation, the drug is suitable for once-daily dosing. Steady-state plasma levels are achieved within 6 days of commencing treatment. Darifenacin exhibits high-protein binding (98%), a volume of distribution of 163 L, and a clearance of 40 L/h. It has low oral bioavailability (15–19%) due to extensive first-pass metabolism by CYP3A4 and CYP2D6, but this can be saturated after multiple administrations. The major circulating metabolites are produced by monohydroxylation and N-dealkylation; however, none contribute significantly to the overall clinical effect of darifenacin. Approximately 58% of the dose is excreted in urine and 44% in feces; only a small percentage (3%) of the excreted dose is unchanged darifenacin. The clinical efficacy of darifenacin was established in three randomized, double-blind, placebo-controlled trials (n = 1059). For inclusion in these trials, patients had a six-month history of overactive bladder with at least eight micturitions per day, one episode of urinary urgency per day, and five episodes of urge urinary incontinence per week. In each trial, darifenacin reduced urinary incontinence by 8–11 episodes per week versus baseline (55–70%). However, due to a strong placebo response, this represented a decrease of only 2–4 episodes per week versus placebo (9–23%, $p<0.05$). The main adverse events associated with darifenacin included dry mouth, constipation, dyspepsia, and urinary tract infection.

5. DEFERASIROX (CHRONIC IRON OVERLOAD) [19–21]

Country of origin	Switzerland
Originator	Novartis
First introduction	US
Introduced by	Novartis
Trade name	Exjade
CAS registry no	201530-41-8
Molecular weight	373.4

Iron overload is a potentially life-threatening cumulative toxicity that occurs frequently in patients who receive multiple blood transfusions for the treatment of certain types of chronic anemias such as thalassemia and sickle-cell disease, as well as for the treatment of myelodysplastic syndromes. Progressive iron overload, if untreated, often causes injury to heart, liver, endocrine organs, joints, and other target cells and tissues. Iron overload is treated by administration of iron chelators, which mobilize the iron deposits into soluble complexes that can be excreted from the body. The current standard of care in iron chelation, deferoxamine (Desferal®), is effective, but typically requires subcutaneous infusion lasting eight to twelve hours per day, for five to seven days a week for as long as the patient continues to receive blood transfusions. In many patients, the need for transfusion and chelation therapy may be life-long. This has resulted in low patient acceptance of the product. Deferipone is an orally available iron chelator that has been marketed in certain parts of the world; however, it has a short duration of action and may be associated with serious side effects. Deferasirox, launched last year, is the latest entry in iron chelation therapy. It is an orally active tridentate ligand that binds iron with high affinity in a 2:1 ratio. The recommended dosage is 20 mg/kg of body weight once, daily. Deferasirox is synthesized in two steps starting from salicylamide, by first condensing with salicyloyl chloride to afford 2-(2-hydroxyphenyl)benz[1,3]oxazin-4-one, and subsequent reaction of this intermediate with 4-hydrazinobenzoic acid. Deferasirox is rapidly absorbed following oral administration, reaching peak plasma concentrations at 1.5–4 h. The C_{max} and AUC increase approximately linearly with dose after both single administration and under steady-state conditions. Deferasirox is highly protein bound (~99%), almost exclusively to serum albumin. It has a volume of distribution at steady state of 14.37 ± 2.69 L, mean elimination half-life ranging from 8 to 16 hours, and oral bioavailability of 70%. Glucuronidation is the main metabolic pathway for deferasirox, with minimal oxidative metabolism by CYP450 enzymes. Deferasirox and metabolites are primarily excreted in the feces via the bile (84% of the dose). Clinical trials involving more than 1000 adults and children demonstrated that deferasirox, at 20–30 mg/kg/day led to maintenance of or reductions in liver iron concentration, an indicator for body iron content in patients receiving blood transfusions. The most frequently reported adverse events were transient in nature and included mild to moderate nausea, vomiting, diarrhea, abdominal pain, and skin rash. In addition, patients treated with deferasirox experienced dose-related increase in serum creatinine at a greater frequency compared with deferoxamine-treated patients (38% vs. 15%, respectively). However, most of the creatinine elevations remained within the normal range. As with deferoxamine injection, deferasirox was associated with a low incidence of ocular and auditory disturbances.

6. DORIPENEM (ANTIBIOTIC) [22–25]

Country of origin	Japan
Originator	Shinogi
First introduction	Japan
Introduced by	Shinogi/ Peninsula
Trade name	Finibax
CAS registry no	148016-81-3
Molecular weight	420.5

Doripenem is a parenteral carbapenem antibiotic launched last year in Japan for the treatment of bacterial respiratory and urinary tract infections. It is a 1β-methyl carbapenem derivative, and it is the fourth analog to be marketed in this series following the launch of meropenem, biapenem, and ertapenem in previous years. The introduction of a 1β-methyl group to the carbapenem skeleton enhances metabolic stability to renal dehydropeptidase-1 (DHP-1) and leads to improved antibacterial potency. The mechanism of action is likely to involve covalent modification of peptidoglycan biosynthetic enzymes responsible for catalyzing the final transpeptidation step of cell wall biosynthesis. The chemical synthesis of doripenem involves the coupling of a commercially available 4-nitrobenzyl protected 1β-methylcarbapenem enolphosphate intermediate with a protected version of 2-(sulfamidomethyl)-4-mercaptopyrrolidine as the key step. The requisite pyrrolidine intermediate is prepared in six steps starting from *trans*-4-hydroxy-L-proline. *In vitro*, doripenem exhibits activity similar to that of imipenem against *Gram*-positive pathogens, and to that of meropenem against *Gram*-negative pathogens. The key differentiator for doripenem is its superior activity against *Pseudomonas aeruginosa* ($MIC_{90} = 3.13\,\mu g/mL$) as compared with meropenem and imipenem. Additionally, it possesses higher stability than imipenem or meropenem against mammalian dehydropeptidase I, and it is stable to most serine-based beta-lactamases. The route of administration of doripenem is either intravenous or intramuscular. It has low plasma protein binding (8.1%), and the primary route of elimination is renal. After single doses of doripenem 125–1000 mg in volunteers, C_{max} values were 8.1–63.0 µg/mL, and AUC was 8.7–75.6 µg h/mL. The mean urinary recovery of doripenem over 24 h was 75%. It has a half-life of approximately 1 h in normal individuals, and this value increases substantially in subjects with renal disease. In a phase II study in 55 patients with chronic respiratory tract infections (RTIs), parenteral doripenem (250–1000 mg/day twice daily or three times daily, for 15 days) resulted in the achievement of clinical efficacy in 40/42 evaluable patients (95.2%). Bacteriological eradication occurred in 21/24 (87.5%) isolates. A 95% overall clinical efficacy rate was achieved in 42 patients with chronic bronchitis, infected bronchiectasis and secondary infection of chronic respiratory disease treated with doripenem 125 mg, 250 mg, or 500 mg twice, daily.

Bacteriological eradication rate was 87.5%. The main adverse events associated with doripenem were drowsiness, headache, injection site reactions, and nausea. An inhaled formulation of doripenem is currently in development for the potential treatment of cystic fibrosis related infections.

7. EBERCONAZOLE (ANTIFUNGAL) [26–28]

Country of Origin	Spain
Originator	Chiesi Wassermann
First introduction	Spain
Introduced by	Laboratorios SALVAT SA
Trade name	Ebernet
CAS registry no	128326-82-9
Molecular Weight	329.22

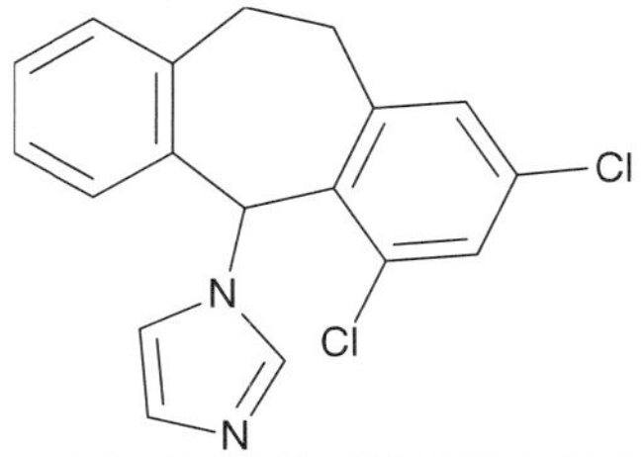

Eberconazole is a new member of the azole class of antifungal agents, and it is indicated for the topical treatment of cutaneous fungal infections, including tinea corporis (ringworm of the body), tinea cruris (ringworm of the groin) and tinea pedis (athlete's foot) infections. Its mode of action is similar to that of other azole antifungals, namely inhibition of fungal lanosterol 14α-demethylase. Eberconazole exhibits good *in vitro* activity against a wide range of *Candida* species, including *Candida. tropicalis*, dermatophytes and *Malassezia* spp. yeasts. It shows good activity against *Candida. Parapsilosis* ($MIC_{90} = 0.125\,\mu g/mL$), which is a relevant species in skin and nail disorders. In addition, eberconazole is effective against some of the highly triazole-resistant yeasts such as *Candida. glabrata* and *Candida. krusei*, as well as fluconazole-resistant *Candida. albicans*. However, eberconazole is less active than clotrimazole and ketoconazole against *Candida. neoformans* and a number of clinically relevant molds. Eberconazole is supplied as a 1% or 2% cream, and the topical application does not result in detectable serum, urine, or fecal levels. In a phase II study of 60 patients with tinea corporis and tinea cruris, treatment with topical eberconazole (1% or 2% cream), applied once or twice daily for 6 weeks, resulted in cure rates ranging from 73.3–93.3% at the end of therapy, and 66.7–100% six weeks post-therapy. In a phase III study of 157 patients with either cutaneous candidosis or dermatophyte skin infection, 1% eberconazole cream and 1% clotrimazole cream, both applied twice daily for 4 weeks, showed comparable overall effectiveness of 72% and 61%, respectively ($p = 0.15$). Eberconazole is synthesized in seven steps starting from 2-(carbomethoxy)benzyltriphenylphosphonium bromide, via a Wittig reaction with 3,5-dichlorobenzaldehyde, followed by ester hydrolysis, double-bond reduction, and cyclization to produce a dibenzocycloheptenone intermediate. Subsequent reduction of this intermediate to the corresponding alcohol, followed by conversion to the chloride, and alkylation with imidazole gives eberconazole.

8. ENTECAVIR (ANTIVIRAL) [29–32]

Country of origin	US	
Originator	BMS	
First introduction	US	
Introduced by	BMS	
Trade name	Baraclude	
CAS registry no	142217-69-4	
Molecular weight	295.3	

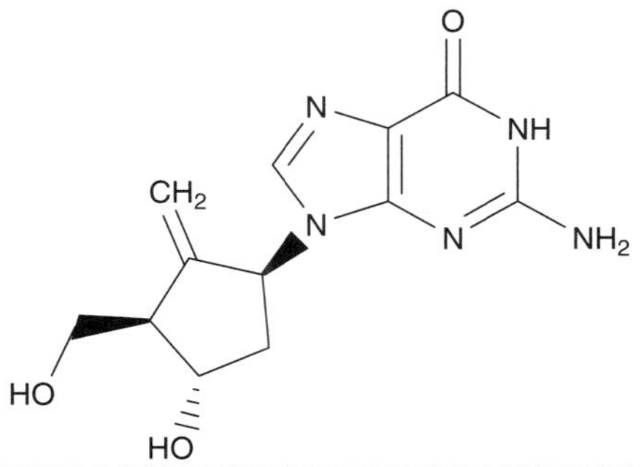

Entecavir is a cyclopentyl guanosine analog launched last year for the once-daily oral treatment of chronic hepatitis B virus (HBV) infection, and it is the third nucleoside or nucleotide analog to be marketed for this indication. Lamivudine, a deoxythiacytosine analog, and adefovir dipivoxil, a nucleotide analog, have been marketed since 1998 and 2002, respectively. Entecavir and adefovir are specifically indicated for HBV, whereas lamivudine is indicated for both HBV and HIV infections. In mammalian cells, entecavir is efficiently phosphorylated to the active triphosphate form, which competes with the natural substrate deoxyguanosine triphosphate and functionally inhibits all three activities of the HBV polymerase: (1) base priming, (2) reverse transcription of the negative strand from the pregenomic messenger RNA, and (3) synthesis of the positive strand of HBV DNA. Entecavir triphosphate is a potent inhibitor of HBV DNA polymerase ($K_i = 1.2$ nM), with little or no activity against cellular DNA polymerases α, β, and δ and mitochondrial DNA polymerase γ ($K_i = 18$ to $>160\,\mu$M). In human HepG2 cells transfected with wild-type HBV, entecavir inhibits HBV DNA synthesis with an EC_{50} of 4 nM. The median EC_{50} value against lamivudine-resistant HBV (rtL180 M, rtM204 V) is 26 nM (range 10–59 nM). In contrast to lamivudine, entecavir has no clinically relevant activity against HIV type 1 ($EC_{50} = >10\,\mu$M). Entecavir is produced in a 10-step asymmetric synthesis starting from cyclopentadiene. The key-step involves an asymmetric monohydroboration of 5-(benzyloxymethyl)cyclopentadiene with diisipinocamphenylborane reagent to the corresponding cyclopentenol (98% ee), which is subsequently converted to a cyclopentene oxide intermediate and coupled with 6-benzyloxy-2-aminopurine. The resultant cyclopentanol is oxidized to the ketone and methylenated to produce the entecavir scaffold. Following oral administration, entecavir peak plasma levels are achieved between 0.5 and 1.5 h, and the steady state is achieved after 6–10 days of once-daily dosing. Entecavir has low-protein binding (13%), essentially 100% oral bioavailability, an apparent volume of distribution in excess of total body water, and terminal elimination half-life of approximately 128–149 h. Entecavir is minimally metabolized, primarily to glucuronide and sulfate conjugates. It is predominantly eliminated by the kidney, with urinary recovery of unchanged drug at steady state ranging from 62% to 73% of the administered dose. In two randomized double-blind 1 year cinical studies involving nucleoside-naïve patients, 0.5 mg once-daily entecavir was superior to 100 mg once-daily lamivudine on the primary efficacy endpoint of histological improvement. Patients receiving entecavir achieved histologic improvement rates of 70–72%, compared with rates of 61–62% for patients receiving lamivudine ($p<0.05$). In another 1-year randomized, double-blind study involving lamivudine-refractory patients, 1 mg once-daily entecavir provided 55% histologic improvement as compared with a rate of 28% for patients receiving 100 mg once-daily lamivudine. The most common adverse events associated with the use of entecavir are similar to those typically seen with HBV therapy and include headache, abdominal pain, diarrhea, fatigue, and dizziness.

9. ESZOPICLONE (HYPNOTIC) [33–36]

Country of origin	France
Originator	Aventis
First introduction	US
Introduced by	Sepracor
Trade name	Lunesta
CAS registry no	138729-47-2
Molecular weight	388.81

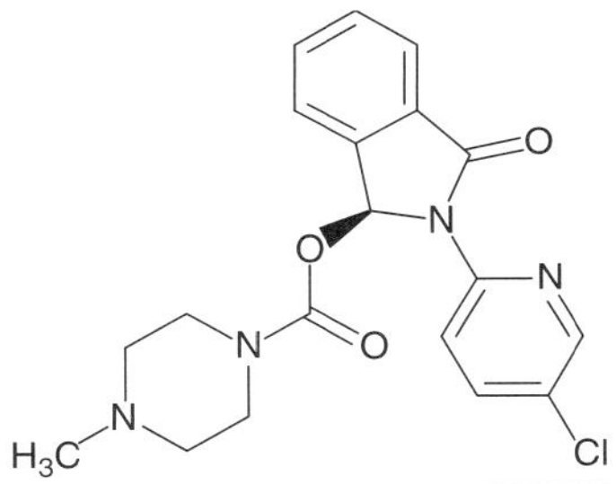

Eszopiclone is a non-benzodiazepine hypnotic agent indicated for the treatment of insomnia to induce sleep and for sleep maintenance. It has similar pharmacokinetic and pharmacodynamic parameters as the previously marketed non-benzodiazepine hypnotics zolpidem and zaleplon. However, unlike its predecessors, eszopiclone is not restricted to short-term treatment of insomnia. Clinical studies of up to 6 months of use show that patients do not develop tolerance to its effect. Eszopiclone is the (S)-enantiomer of zopiclone, which has been marketed as the racemic mixture in Europe for almost 20 years. These agents belong to the cyclopyrrolone class of drugs that act as agonists at the type A GABA receptor. Eszopiclone has approximately 50-fold higher binding affinity than its antipode (R)-zopiclone for GABA-A receptor ($IC_{50} = 21$ and 1130 nM, respectively). In addition, the two enantiomers exhibit significant differences in their pharmacokinetic parameters and *in vivo* efficacy. In healthy volunteers, eszopiclone has 2-fold higher C_{max} and 2-fold greater elimination half-life than the (R)-enantiomer. In assays for GABA-A mediated effects in monkeys, 0.1–10 mg/kg eszopiclone effectively substituted for the benzodiazepine midozalam, showing >80% midozalam-appropriate response, as compared with only a 45% efficacy for 100 mg/kg dose of the (R)-enantiomer. Eszopiclone is rapidly absorbed from the intestinal tract, with peak plasma concentrations occurring one hour after administration. Concomitant consumption of a high-fat meal may reduce the C_{max}, but does not alter the area under the curve.

Approximately, 55% of eszopiclone is bound to plasma proteins. Metabolism takes place in the liver through oxidation and demethylation by CYP3A4 and CYP2E1. The primary metabolites are (S)-zopiclone-N-oxide and N-desmethyl-(S)-zopiclone; the latter compound binds to GABA receptor with significantly lower potency than eszopiclone. (S)-zopiclone-N-oxide is inactive. The elimination half-life of eszopiclone is 6 h. Approximately 75% of an oral dose is excreted as a combination of unchanged eszopiclone and metabolites in urine. Less than 10% is excreted unchanged. The recommended daily dosage of eszopiclone is 2 mg for sleep induction and 3 mg for sleep maintenance in adults. The recommended dose for elderly patients is 1 mg for sleep induction and 2 mg for sleep maintenance. Eszopiclone's efficacy in reducing sleep latency and improving sleep maintenance was established in 6 placebo-controlled trials of patients ($n = 2100$, aged 18–86 years) with chronic and transient insomnia. Elderly patients ($n = 523$) were studied in 2 of these trials. The efficacy read-outs included objective measurement of wake time after sleep onset (WAS) and subjective measurement of total sleep time. Across all studies, eszopiclone significantly decreased sleep latency and improved measures of sleep maintenance. The two most frequent adverse events associated with eszopiclone treatment are unpleasant taste and headache. Other less frequent side effects include somnolence, dry mouth, and nausea.

10. EXENATIDE (ANTI-DIABETIC) [37–41]

Country of origin	US	H-His-Gly-Glu-Gly-Thr- Phe-Thr-
Originator	Amylin	Ser-Asp-Leu-Ser-Lys-Gln -Met-
First introduction	US	Glu-Glu-Glu-Ala-Val-Arg- Leu-
Introduced by	Amylin/Lilly	Phe-Ile-Glu-Trp-Leu-Lys -Asn-
Trade name:	Byetta	Gly-Gly-Pro-Ser-Ser-Gly -Ala-
CAS registry no	141758-74-9	Pro-Pro-Pro-Ser- NH_2
Molecular weight	4186.6	

Exenatide is the first drug in a new class of anti-diabetics known as the incretin mimetics, and it is indicated as adjunctive therapy to improve glycemic control in patients with type 2 diabetes who are taking metformin, a sulfonylurea, or both, but have not achieved adequate glycemic control. Exenatide is a functional analog of the human incretin Glucagon-Like Peptide-1 (GLP-1). GLP-1 is naturally released from cells in the GI tract in response to food intake and acts on its receptor on β-cells to potentiate glucose-stimulated insulin secretion. Exenatide is a long-acting agonist at the GLP-1 receptor. It is a synthetic version of a 39-amino acid peptide found in the salivary secretions of the Gila monster lizard. It has 53% amino acid homology with GLP-1, but unlike GLP-1, exenatide is less susceptible to degradation by neutral endopeptidase, has a longer half-life, binds with greater affinity to GLP-1 receptor, and is a more potent insulinotrope. In addition to stimulating glucose-dependent insulin secretion, exenatide reduces food intake and slows gastric emptying, thereby reducing the rate at which meal-derived glucose reaches the circulation. The drug also moderates peak serum glucagon levels during hyperglycemic periods following meals, but does not interfere with glucagon release in response to hypoglycemia. The dosing regimen for exenatide is 5 or 10 μg twice daily, administered as a subcutaneous injection within an hour before morning and evening meals. Following subcutaneous administration, peak plasma concentrations of exenatide are reached in 2.1 h, and the plasma pharmacokinetic profile is dose proportional. Exenatide has an apparent volume of distribution of 28.3 L, a clearance rate of 9.1 L/h, and an apparent *in vivo* half-life of 2.4 h. Exenatide levels are measurable for up to 10 h after administration. It is primarily excreted unchanged in the urine by glomerular filtration, after which it undergoes proteolytic degradation. The efficacy and safety of exenatide (5 or 10 μg as a twice-daily subcutaneous injection) has been evaluated in three 30-week double-blind, placebo-controlled clinical trials ($n = 1446$). The primary efficacy endpoint in each study was mean change from baseline in HbA1c (%) at 30 weeks. The exenatide 10 μg group demonstrated changes of -0.8, -0.9, and -0.8 ($p<0.0001$) in the three studies, compared with changes of -0.4, -0.5, and -0.6 ($p<0.05$, 0.05, and 0.0001), respectively, for the exenatide 5 μg group and changes of $+0.1$, $+0.1$, and $+0.2$ for the placebo group. The most common adverse events reported with exenatide include nausea, vomiting, diarrhea, feeling jittery, dizziness, headache, and dyspepsia.

11. GALSULFASE (MUCOPOLYSACCHARIDOSIS VI) [42–44]

Country of origin	US	Class:	Recombinant protein
Originator	BioMarin		
First introduction	US	Type:	Glycosidase
Introduced by	BioMarin	Molecular Weight:	56 kDa
Trade name	Naglazyme	Expression system:	CHO cell line
CAS registry No	55354-43-3	Manufacturer:	BioMarin

Mucopolysaccharide storage disorders are caused by the deficiency of specific lysosomal enzymes required for the catabolism of glycosaminoglycans (GAGs) such as chondroitin 4-sulfate and dermatan sulfate. Mucopolysaccharidosis VI (MPS VI, Maroteaux-Lamy syndrome) is characterized by the absence or marked reduction in *N*-acetylgalactosamine 4-sulfatase. The sulfatase activity deficiency results in the accumulation of the GAG substrate dermatan sulfate throughout the body. This accumulation leads to widespread cellular, tissue, and organ dysfunction. Galsulfase is a normal variant of the polymorphic human *N*-acetylgalactosamine 4-sulfatase and is intended to provide an exogenous source of the enzyme that will be taken up into lysosomes. Galsulfase is a glycoprotein that is produced by recombinant DNA technology in a CHO cell line. It is comprised of 495 amino acids and contains six asparagine-linked glycosylation sites, four of which carry a bis-mannose-6-phosphate mannose oligosaccharide for specific cellular recognition. Post-translational modification of Cys53 produces the catalytic amino acid residue Cα-formylglycine, which is required for sulfatase activity. Galsulfase uptake by cells into lysosomes is mediated by its binding to specific mannose-6-phosphate receptors. The dosing regimen of galsulfase is 1 mg/kg of body weight, administered once weekly as an intravenous infusion. After a single dose, galsulfase has a C_{max} of 0.8 µg/mL, clearance rate of 7.2 mL/kg/min, and a half-life of 9 min. A slight change in pharmacokinetic parameters is observed after 24 weeks of dosing ($C_{max} = 1.5$ µg/mL, clearance = 7.2 mL/kg/min, $t_{1/2} = 26$ min). In addition, changes in plasma AUCs are also noted from 1 week to 24 weeks. These changes are attributed to the fact that nearly all patients receiving galsulfase develop antibodies to galsulfase over the course of the treatment. In clinical studies, 1 mg/kg galsufase once-weekly for 24 weeks significantly increased baseline-adjusted mean 12 min walk distance compared to placebo, the study's primary endpoint (+92 m; $p = 0.025$). A positive, non-significant trend was observed in increasing performance in a 3 min stair climb test, vs. placebo (+5.7 stairs/min; $p = 0.053$). Additionally, patients receiving galsulfase showed significant reductions in urinary GAG secretion compared to placebo, although these reductions were not sufficient to reach normal, healthy-patient levels. Adverse events associated with the use of galsulfase included abdominal pain, ear pain, chest pain, conjunctivitis, dyspnea, and pharyngitis.

12. LULICONAZOLE (ANTIFUNGAL) [45–47]

Country of origin	Japan
Originator	Nihon Nohyaku
First introduction	Japan
Introduced by	Nihon Nohyaku/Pola
Trade name	Lulicon
CAS registry no	187164-19-8
Molecular weight	354.28

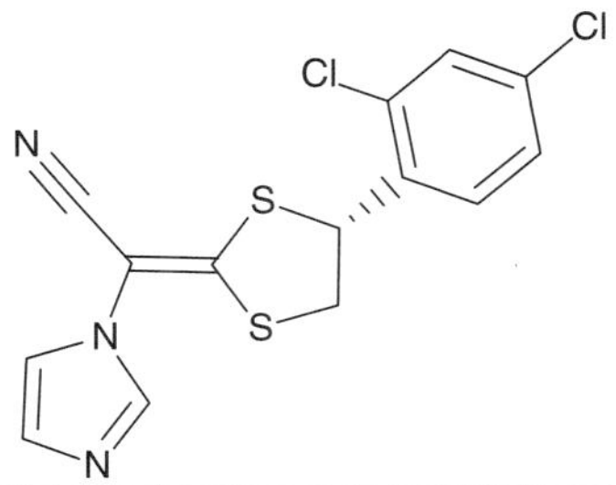

Luliconazole is a member of the imidazole class of antifungal agents, with specific utility as a dermatological antimycotic drug. It was launched last year in Japan as a topical agent for the treatment of athlete's foot. Luliconazole is an optically active drug with (R)-configuration at its chiral center. It is structurally related to lanoconazole, which has been marketed as a racemic mixture since 1994. As with other azole antifungal drugs, the mechanism of action of luliconazole is the inhibition of sterol 14-α-demethylase, and subsequently, inhibition of ergosterol biosynthesis. In *C. albicans*, luliconazole inhibits ergosterol biosynthesis with an IC_{50} of 14 nM, and it is about 2.5-fold more potent than lanoconazole ($IC_{50} = 36$ nM), and 28-fold more potent than bifonazole ($IC_{50} = 390$ nM). The corresponding (S)-enantiomer of luliconazole is virtually inactive. *In vitro*, luliconazole exhibits strong antifungal activity against *Trichophyton mentagrophytes* and *Trichophyton rubrum*, with mean minimum inhibitory concentration (MIC) values of 0.0093 and 0.0015 µg/mL, respectively. In general, it has high activity against filamentous fungi except zygomycetes, and good-to-moderate activity against yeast-like fungi. It also has activity against *C. albicans*, which is slightly lower than fluconazole, as well as activity against *Aspergillus fumigatus*, which is at least 60-fold lower than itraconazole. The *in vivo* activity of luliconazole against dermatophytes has been evaluated in the guinea pig model of tinea pedis. In this study, a 1% topical solution of luliconazole, administered once daily for seven days, achieved complete mycologic cure. Additionally, there were no occurrences of relapse for up to 16 weeks after the treatment. No data is currently available on the clinical efficacy of luliconazole. The chemical synthesis of luliconazole involves the condensation of 1-(cyanomethyl)imidazole with carbon disulfide to produce a dithioate intermediate, and subsequent alkylation with either the mesylate derivative of (*S*)-1-(2,4-dichlorophenyl)-2-bromoethanol or the bis-mesylate derivative of (*S*)-1-(2,4-dichlorophenyl)ethane-1,2-diol.

13. LUMIRACOXIB (ANTI-INFLAMMATORY) [48–51]

Country of origin	Switzerland
Originator	Novartis AG
First introduction	Brazil
Introduced by	Novartis AG
Trade name	Prexige
CAS registry no	220991-20-8
Molecular weight	293.72

As a second-generation, selective cyclooxygenase (COX-2) inhibitor, lumiracoxib is devoid of the gastrointestinal issues that plague other non-selective, non-steroidal, anti-inflammatory drugs (NSAIDs) that crossover to COX-1. As an inhibitor of the inducible COX-2 that is up-regulated in pathological processes of pain and inflammation, lumiracoxib blocks the conversion of arachidonic acid to prostaglandins, the mediators of the pathological effects. It's mode of binding to COX-2 has been found to differ from the other selective COX-2 inhibitors; the carboxylic acid forms hydrogen bonds with Tyr-385 and Ser-530 in the catalytic site rather than seeking interactions within the larger hydrophobic side pocket. In human whole blood, lumiracoxib displays an IC_{50} value of 0.13 µM for COX-2 versus 67 µM for COX-1, resulting in a selectivity ratio of 515. Regarding its synthesis, lumiracoxib may be prepared by a couple of routes. While the construction of the diarylamine core via a Buchwald–Hartwig coupling of a suitable aniline and an aryl halide is preferable on a small scale, the cost of palladium catalysts and ligands makes the procedure prohibitive on production scale. The Smiles rearrangement, utilizing a phenolic precursor instead of the aryl halide, is a viable alternative. The diarylamine, obtained from either route, is subsequently acylated with chloroacetyl chloride to provide the precursor for the intramolecular Friedel–Crafts alkylation. Hydrolysis of the resulting *N*-aryl oxindole affords lumiracoxib. The pharmacokinetic profile of lumiracoxib has been extensively documented. Lumiracoxib is absorbed rapidly with an oral bioavailability of approximately 74%. A median T_{max} is reached in 2 h, and C_{max} is proportional to dose (range of 25–800 mg). It has a relatively short plasma elimination half-life of 3–6 h and a mean plasma clearance of 8 L/h following i.v. administration. The volume of distribution is also moderate at 9 L. Lumiracoxib is metabolized predominantly by CYP2C9 with oxidation of the 5-methyl group and hydroxylation of the dihaloaromatic ring as the primary sites of biotransformation. In circulating plasma, the major metabolites 5-carboxy-lumiracoxib (10.8%) and 4'-hydroxy-5-carboxy-lumiracoxib (8.2%) are not pharmacologically active against either COX enzyme; however, 4'-hydroxy-lumiracoxib (5.8%) has similar potency and selectivity as the parent. Since 4'-hydroxy-lumiracoxib has comparable plasma protein binding (98–99%), it is unlikely that this metabolite is a major contributor to the drug's overall efficacy. Despite its short plasma half-life, clinical data support once-daily dosing with 100 mg/day emerging as the lowest effective dose for the treatment of osteoarthritis. In patients with rheumatoid arthritis, a dose of 200 mg/day appears to alleviate symptoms. These doses were superior to placebo in regards to clinical efficacy and comparable to the standard doses for the other coxibs. With regard to safety, most of this data was compiled from the Therapeutic Arthritis Research and Gastrointestinal Trial (TARGET), which enrolled 18,325 patients. A dose of 400 mg/day resulted in a cumulative 1-year incidence of ulcer complications that was approximately 3-fold less than naproxen (1000 mg/day) or ibuprofen (2400 mg/day). No significant difference in cardiovascular events, such as myocardial infarction, stroke, or cardiovascular death, were found between lumiracoxib and the combined comparator non-steroidal anti-inflammatory drug (NSAIDs); however, liver function test abnormalities were more frequent with lumiracoxib than with the comparator NSAIDs, raising the concern about a possible hepatotoxic potential of lumiracoxib. Since lumiracoxib is mainly metabolized by CYP2C9, a study evaluating the co-administration of lumiracoxib with fluconazole, a potent inhibitor of CYP2C9, was conducted, and it concluded that there was no need for lumiracoxib dose adjustment, since changes in the systemic exposure were not significant. No serious adverse effects were reported, but in the small number of cases where treatment was discontinued, Gastro intestinal (GI) and musculoskeletal complaints were common.

14. NEPAFENAC (ANTI-INFLAMMATORY) [52–55]

Country of origin	US
Originator	AH Robins
First introduction	US
Introduced by	Alcon
Trade name	Nevanac
CAS registry no	78281-72-8
Molecular weight	254.29

Nepafenac, launched last year by Alcon Laboratories, is a topical ophthalmic medication indicated for the treatment of ocular pain and inflammation associated with cataract surgery. Nepafenac is a prodrug of amfenac, which is an NSAID and a potent non-selective inhibitor of COX-1 ($IC_{50} = 0.25\,\mu M$)) and COX-2 ($IC_{50} = 0.15\,\mu M$). Nepefenac itself exhibits only weak activity against COX-1 ($IC_{50} = 64.3\,\mu M$). Amfenac (Fenazox) has been marketed in Japan since 1986 for the treatment of rheumatoid arthritis, post-surgical pain, and inflammation. With most NSAIDs that are currently being used as topical ophthalmic agents, the maximum drug concentration is achieved on the ocular surface, with progressively lower concentrations in the cornea, aqueous humor, vitreous, and retina. Nepafenac has been found to have a penetration coefficient that is 4 – 28 times greater than that achieved with conventional NSAIDs such as diclofenac, bromofenac, and ketorolac. In addition, the bioconversion of nepefenac to amfenac is primarily mediated by ocular tissue hydrolases, specifically in the iris, ciliary body, retina, and choroid. The enhanced permeability of nepefenac combined with rapid bioactivation in the ocular tissue translates into superior anti-inflammatory efficacy at the target sites. In preclinical models, a single topical ocular dose of nepefenac (0.1%) inhibits prostaglandin synthesis in the iris/ciliary body by 85–95% for more than 6 h, and in the retina/choroid by 55% for up to at least 4 h. By comparison, diclofenac (0.1%) shows 100% inhibition of prostaglandin synthesis in the iris/ciliary body for only 20 min, with 75% recovery observed within 6 h. Diclofenac's inhibition of prostaglandin synthesis in the retina/choroids is minimal. The recommended dose of nepafenac ophthalmic suspension is one drop in the affected eye(s) three times daily beginning one day prior to cataract surgery, continued on the day of surgery and through the first two weeks of the postoperative period. Although the drug is applied topically, low but quantifiable plasma concentrations of nepefenac and amfenac are observed in majority of the subjects following t.i.d. dosing of nepefenac ophthalmic solution. The clinical significance of the systemic absorption of nepefenac after ophthalmic administration is unknown. The efficacy of nepafenac was demonstrated in two placebo-controlled clinical studies involving over 680 patients. Nepafenac suspension was dosed three times daily, beginning one day prior to cataract surgery, continuing on the day of surgery, and for 14 days postoperatively. Approximately 80% of the patients receiving nepafenac were pain-free on the day following cataract surgery, compared to about 40–50% of patients in the placebo group. By day 14, 95% of drug-treated patients were pain-free, compared to about 45–60% of those on placebo. Additionally, 91% of patients treated with nepafenac had no clinically significant inflammation at day 14, compared to 47% of patients in the placebo group. The adverse events in these clinical studies, reported in 5–10% of the patients, were generally ocular, and included capsular opacity, decreased visual activity, foreign body sensation, increased intraocular pressure, and sticky sensation. Nepefenac is chemically produced in two synthetic steps starting from 2-aminobenzophenone. In the first step, 2-aminobenzophenone is ortho-alkylated with methylthioacetamide according to Gassman's method by using tert-butyl hypochlorite and triethylamine, and the resultant intermediate is subsequently desulfurized with Raney nickel.

15. PEGAPTANIB (AGE-RELATED MACULAR DEGENERATION) [56–64]

| Country of origin | US |

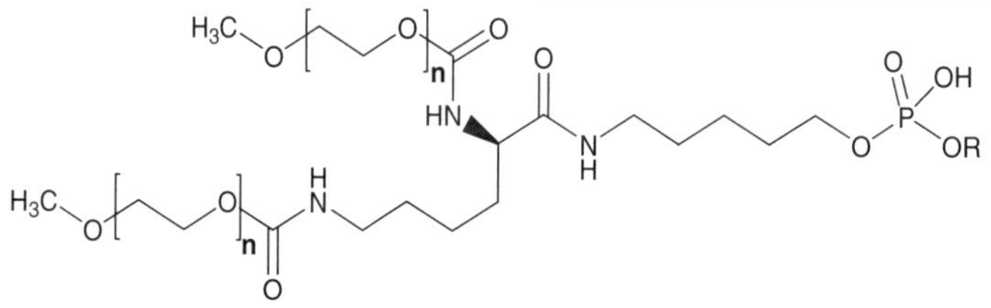

R = 28-mer RNA sequence =
$(2'\text{-deoxy-2'F})C\text{-}G_m\text{-}C_m\text{-}G_m\text{-}A\text{-}A\text{-}(2'\text{-deoxy-2'F})U\text{-}(2'\text{-deoxy-2'F})C\text{-}A_m\text{-}G_m\text{-}(2'\text{-deoxy-2'F})U\text{-}G_m\text{-}A_m\text{-}A_m\text{-}(2'\text{-deoxy-2'F})U\text{-}G_m\text{-}(2'\text{-deoxy-2'F})C\text{-}(2'\text{-deoxy-2'F})U\text{-}(2'\text{-deoxy-2'F})U\text{-}A_m\text{-}(2'\text{-deoxy-2'F})U\text{-}A_m\text{-}(2'\text{-deoxy-2'F})C\text{-}A_m\text{-}(2'\text{-deoxy-2'F})U\text{-}(2'\text{-deoxy-2'F})C\text{-}G_m\text{-}(3'\text{-}3')dT$

n = approx. 450

Originator	NeXstar
First introduction	US
Introduced by	Eyetech/ Pfizer
Trade name	Macugen
CAS Registry No	222716-86-1
Molecular weight	approx. 50 kD

Pegaptanib is the first anti-angiogenic agent launched for the treatment of age-related macular degeneration (AMD). Wet, or neovascular, AMD is the leading cause of vision loss in the elderly; vision loss is the consequence of choroidal neovascularization, leading to leakage of blood or serum, retinal detachment, and fibrovascular scarring. The underlying factor in the neovascularization is angiogenesis, the proliferation of new blood vessels from pre-existing vasculature. Vascular endothelial growth factor (VEGF) is one of the stimulators of this angiogenesis, and pegaptanib's mechanism of action involves the high-affinity binding ($K_d = 50$ pM) of extracellular VEGF, thereby, effectively sequestering the protein to ultimately prevent VEGF receptor activation. Pegaptanib is also classified as an RNA aptamer; and as such, it is the first aptamer to be approved for a therapeutic application. Aptamers are RNA or DNA oligonucleotides that are selected for their high-affinity binding to specific proteins. Pegaptanib was identified by the systematic evolution of ligands by the exponential enrichment (SELEX) technique. Since most aptamers are sensitive to nuclease degradation, both SELEX and post-SELEX modifications have been performed to confer nuclease resistance. Specifically, the sugar moieties of the ribonucleotides are modified at the 2'-positions with either a methoxy or a fluorine in most cases. Once the optimal sequence has been identified, the aptamer can be amplified (by reverse transcription, amplification using the polymerase chain reaction, and transcription) or synthesized by standard solid-phase oligonucleotide methods. While, modification of the sugar backbone protects against endonucleases, capping the termini further increases the stability. The addition of a deoxythymidine to the 3'-terminus via a 3'-3' linkage protects one end, and the di-PEGylation of the 5'-terminus via a lysine spacer imparts further exonuclease resistance. Pegaptanib is supplied in a sterile, single-dose, pre-filled syringe and is formulated as a 3.47 mg/mL solution (0.3 mg of active ingredient) for intravitreous injection. Following intravitreous injection of a single 3 mg dose (10 times the recommended dose) to AMD patients, a mean C_{max} of 80 ng/mL was achieved within 1–4 days, and the mean plasma half-life was 10 (± 4) days. At this dose, the mean AUC is approximately 25 µg·h/mL. The efficacy of pegaptanib was evaluated in two controlled, double-masked, randomized studies involving 1186 patients with neovascular AMD. Patients received either sham treatment or pegaptanib sodium (0.3, 1, or 3 mg) by intravitreous injection every 6 weeks for a 48-week duration. The proportion of patients losing less than 15 letters of visual acuity, from baseline up to 54-week assessment, was the primary efficacy endpoint. Statistically significant results were achieved in both trials; in one study, 73% of patients on pegaptanib met the endpoint compared to 60% on sham treatment, while 67% of pegaptanib patients in the second trial achieved the endpoint compared to 53% receiving sham treatment. While patients on the lowest dose of pegaptanib and sham-treated patients continued to experience vision loss, the rate of vision decline was slower in the pegaptanib population. After the first year, 1053 patients were re-randomized to either continue the same treatment or discontinue treatment for a second year. The results indicated that pegaptanib was less effective the second year than during the first year, and there is no data available regarding safety or efficacy beyond the second year. Serious adverse events, occurring in <1% of patients, were related to the injection procedure and included endophthalmitis, retinal detachment, and iatrogenic traumatic cataract. Other frequently reported adverse events, occurring in approximately 10–40% of patients, were anterior chamber inflammation, blurred vision, conjunctival hemorrhage, cataract, corneal edema, eye discharge, eye irritation, eye pain, hypertension, increased intraocular pressure, ocular discomfort, punctate keratitis, reduced visual acuity, visual disturbance, vitreous floaters, and vitreous opacities. Since the mode of drug delivery carries a risk of infection, pegaptanib is contraindicated in patients with ocular or periocular infections.

16. PRAMLINTIDE (ANTI-DIABETIC) [65–68]

Country of origin	US	┌─────────────┐
Originator:	Amylin	Lys-Cys-Asn-Thr-Ala-Thr-Cys-Ala-Thr-
First introduction:	US	Gln-Arg-Leu-Ala-Asn-Phe-Leu-Val-His-
Introduced by:	Amylin	Ser-Ser-Asn-Asn-Phe-Gly-Pro-Ile-Leu-
Trade name:	Symlin	Pro-Pro-Thr-Asn-Val-Gly-Ser-Asn-Thr-Tyr-NH_2
CAS registry no:	151126-32-8 (anhydrous free base)	
	196078-30-5 (acetate salt hydrate)	
	187887-46-3 (anhydrous acetate salt)	
Molecular weight:	3949.4 (acetate salt)	
	4027.49 (acetate salt hydrate)	

Pramlintide is an injectable human amylin analog that has been launched for the treatment of both type 1 and type 2 diabetes, in conjunction with insulin. While it is also a 37-amino acid peptide, it differs from its parent predecessor by the substitution of Ala-25, Ser-28, and Ser-29 with prolines. Not only do these modifications improve the solubility of the peptide, they also eliminate the aggregation observed with amylin, resulting in a stable synthetic analog with retention of biological activity that is suitable for pharmaceutical use. As an indication of potency, pramlintide inhibits the binding of radioiodinated rat amylin to rat nucleus accumbens membranes with a K_i value of 23 pM. Its mechanism of action mimics amylin; as a neurohormone that is co-secreted with insulin from the pancreatic β cells in response to meals, it is involved in glucose homeostasis. Both peptides lower postprandial glucose levels by inhibiting glucagon and by restraining the vagus-mediated rate of gastric emptying, thereby, slowing intestinal carbohydrate absorption. Furthermore, amylin, or pramlintide, has the added benefit of inducing postprandial satiety resulting in weight loss in the patients with type 2 diabetes. Linear pramlintide is prepared on methylbenzhydrylamine resin using standard Boc chemistry. The peptide is cyclized on the resin by treatment with thallium(III) trifluoroacetate to remove the acetamidomethyl (Acm) protecting groups of the cysteines with concomitant oxidation to the disulfide. Subsequent cleavage of the cyclic peptide from the resin with removal of side chain protecting groups is accomplished with liquid HF using dimethylsulfide and anisole scavengers. Reverse phase HPLC affords pure pramlintide. Following subcutaneous injection, the absolute bioavailability is approximately 30–40%. Dose-dependent plasma concentrations and clearance are observed. C_{max} is reached in approximately 20 min and decreases over a 3–h period. The primary route of excretion is via the kidneys with a mean elimination half-life of 40–50 min. Pramlintide doses between 60 and 120 μg correlate with plasma levels induced by the endogenous release of amylin following a meal. Because of its wide therapeutic window, pramlintide does not require constant dose adjustments like its partner insulin; patients remain on a constant dose regardless of meal size, carbohydrate content, or blood glucose concentrations. The efficacy of pramlintide was evaluated in several, long-term (26–52 weeks) clinical trials enrolling 2375 patients with type 1 diabetes and 1688 patients with type 2 diabetes. In patients with type 1 diabetes on fixed-dose insulin, 30 or 60 ug of pramlintide administered prior to meals resulted in a −0.43% change in baseline glycosylated hemoglobin (HbA_{1c}) compared to only a −0.10% change in patients receiving placebo with their fixed-dose insulin. In addition, patients receiving pramlintide experienced a −1.1 kg change in weight compared to a +0.6 kg weight change in the placebo group. Similar results were obtained in the clinical trials for type 2 diabetes; a 120 ug dose of pramlintide administered prior to meals effected a −0.57% change in baseline HbA_{1c} compared to a −0.17% change in patients receiving placebo with their fixed-dose insulin. Since pramlintide slows gastric emptying, it is contraindicated in patients with gastroparesis and in patients taking drugs that alter gastrointestinal motility (anticholinergic agents, such as, atropine) or slow down the intestinal absorption of nutrients (such as α-glucosidase inhibitors). Pramlintide is also contraindicated in patients with hypoglycemic tendencies; due to co-administration with insulin, severe insulin-induced hypoglycemia is a risk. The most commonly reported adverse events include nausea, vomiting, anorexia, headache, abdominal pain, fatigue, dizziness, coughing, and pharyngitis.

17. PALIFERMIN (MUCOSITIS) [69–72]

Country of origin	US	Class:	Recombinant protein
Originator	Amgen	Molecular Weight:	16.3 kDa
First introduction	US	Expression System:	*E. coli*
Introduced by	Amgen	Manufacturer:	Amgen
Trade name	Kepivance		
CAS registry no	162394-19-6		

Mucositis, an inflammation of the oral and gastrointestinal mucosa, is often induced in patients receiving high-dose chemotherapy and radiotherapy for hematological malignancies. Patients with head and neck cancer, as well as, patients receiving stem cell transplants are particularly at risk for developing this debilitating side effect that can lead to swallowing difficulty, thereby, necessitating total parenteral nutrition in severe cases. Other complications include potential infections, overuse of opioid analgesics to treat the pain, and extended hospitalization. Palifermin, a recombinant human keratinocyte growth factor (KGF), has been approved as a novel agent to treat this condition. As a member of the heparin-family of fibroblast growth factors, palifermin provides protection from the damaging effects of chemotherapy and radiation by selectively promoting epithelial cell proliferation, leading to an increased rate of healing. Palifermin is produced by recombinant DNA technology utilizing an expression vector encoding $KGF_{des1-23}$. As an N-terminal, truncated version of endogenous KGF having amino acids 1–23 deleted, palifermin has greater stability while retaining biological activity. In Phase I studies, i.v. administration of palifermin generated linear pharmacokinetic parameters with no accumulation over a dose range of 0.2–20 μg/kg/day for 3 days. The elimination half-life was 4 h. In another double-blind, placebo-controlled study, single escalating IV doses (60–250 μg/kg) resulted in dose-dependent exposure, as well as, a dose-dependent pharmacodynamic response as assessed by the proliferation of buccal mucosal cell epithelium. In a phase III study that enrolled 212 patients undergoing bone marrow transplantation treatment for hematological malignancies, the incidence of grade 3 or grade 4 oral mucositis (OM) was 63% with palifermin compared to 98% with placebo. In addition, the median duration of OM was 6 days in the palifermin-treated group and 9 days with placebo while the incidence of grade 4 OM was significantly less with palifermin, 20% compared to 62% with placebo. Other positive endpoints were the reductions in opioid analgesic use and in total parenteral nutrition in palifermin-treated patients. Serious adverse events were similar to those observed with placebo and were typically attributed to the underlying malignant disease. Minor side effects were skin rash ($<1\%$), dysesthesia, pruritus, tongue thickening, and transient taste changes. While keratinocyte growth-factor receptor is not expressed in hematological cancers, it is possible that secondary tumors that express this receptor could be advanced by palifermin treatment. Long-term evaluation of this risk is ongoing; however, at 12 months, the progression-free survival rates were similar for palifermin and placebo. The recommended dosage of palifermin is 60 Mg/kg/day, delivered by i.v. bolus injection, for 3 consecutive days before and three consecutive days following myelotoxic therapy, insuring that the third dose is delivered greater than 24 h before the commencement of the myelotoxic therapy. No formal drug–drug interaction studies have been conducted with palifermin.

18. RAMELTEON (INSOMNIA) [73–78]

Country of origin	Japan
Originator	Takeda
First introduction	US
Introduced by	Takeda
Trade name	Rozerem
CAS registry no	196597-26-9
Molecular weight	259.34

Unlike most treatments of insomnia that target the GABA (γ-aminobutyric acid) receptor complex, ramelteon is an agonist of the melatonin receptor. In particular, it has high selectivity for the MT_1 and MT_2 subtypes, which have been implicated in the maintenance of circadian rhythms, over the MT_3 receptor responsible for other melatonin functions. Its lack of affinity for not only the GABA receptor complex but also neurotransmitter, dopaminerigic, opiate, and benzodiazepine receptors suggests an improved safety profile devoid of the abuse potential of the hypnotic drugs that target these receptors. As such, ramelteon is not a scheduled drug. The development of ramelteon has come from a concerted effort to elucidate longer half-lived versions of the endogeneous ligand melatonin with improved oral bioavailability and selectivity. By incorporating the methoxy moiety of melatonin into a conformationally constrained indeno[5,4-*b*]furan, the oxygen is locked into an orientation that is favorable for optimal and selective binding to the MT_1 receptor. Furthermore, the *S*-configuration of the sidechain also contributes to the potency for MT_1 ($K_i = 13.8 \, pM$) and 190,000-fold selectivity over MT_3 ($K_i = 2.6 \, \mu M$). Starting from commercially available 6-methoxyindanone, a Horner–Emmons reaction employing diethyl cyanomethylphosphonate provides the unsaturated nitrile that is subsequently reduced to the exocyclic allyl amine. Following acylation with propionyl chloride, an asymmetric hydrogenation with a ruthenium-BINAP catalyst establishes the enantiomerically pure *S*-isomer. The fusion of the furan ring to the indan core is accomplished by a sequence of reactions involving the blocking of the C-5 position by bromination, demethylation, and allylation of the phenol for the subsequent Claisen rearrangement, ozonolysis of the vinyl moiety, removal of the bromine blocking group by hydrogenation, conversion of the primary alcohol to the mesylate, and finally ring closure by the phenolic displacement of the mesylate. Following a 16 mg oral dose, ramelteon undergoes rapid first-pass metabolism with a T_{max} of 0.3 h and a half-life of 1.2 h. The predominant mode of excretion is renal (84%). While the total absorption of ramelteon is at least 82% in humans, the absolute oral bioavailability is only 1.8% due to extensive Phase 1 metabolism. Primary metabolites include hydroxylation and oxidation to carbonyl species with secondary metabolites resulting from glucuronidation. Since CYP1A2 is the major isozyme involved in the hepatic metabolism of ramelteon, it should not be taken in combination with strong CYP1A2 inhibitors, such as fluvoxamine. Co-administration with either ketoconazole (a CYP3A4 inhibitor) or fluconazole (a potent CYP2C9 inhibitor) resulted in significant increases in AUC and C_{max}, but the extensive metabolism and highly variable plasma concentrations of ramelteon precluded the need for dose modification. The package insert, however, cautions patients about co-administration with potent CYP3A4 and CYP2C9 inhibitors. Utilizing polysomnography (PSG), the effects of ramelteon on subjects with chronic insomnia were studied in several randomized, double-blind trials. In one study, younger patients (18–64 years) were given either ramelteon (8 or 16 mg) or a matching placebo nightly for 35 days. PSG, performed on the first two nights of weeks one, three, and five, confirmed that ramelteon reduced the average latency to persistent sleep compared to placebo. In a three-period crossover trial in older adults (65 and older) with chronic insomnia, patients were administered either 4 or 8 mg of ramelteon or placebo. PSG on two consecutive nights of each study period proved that both doses reduced sleep latency compared to placebo. Evaluation in patients with transient insomnia also supported that 8 mg of ramelteon was sufficient in reducing sleep latency relative to placebo. Since many of the traditional treatments of insomnia have safety concerns, an abuse potential study was conducted in 14 patients with a history of sedative/hypnotic drug abuse. Even at doses up to 20 times the recommended therapeutic dose, there were no differences in the subjective responses of ramelteon-treated patients and placebo-treated patients in multiple tests of abuse potential. Based on the result of the clinical trials, the recommended dose of ramelteon is 8 mg taken within 30 min of going to bed. In addition to the precaution of co-administration with CYP inhibitors, it should not be used in patients with severe hepatic impairment. The adverse events, observed in 5% of patients in clinical studies, were somnolence, dizziness, nausea, fatigue, headache, and insomnia.

19. RASAGILINE (PARKINSON'S DISEASE) [79–87]

Country of origin	Israel
Originator	Teva/Eisai/Lundbeck
First introduction	Israel
Introduced by:	Teva/Eisai/Lundbeck
Trade name	Azilect
CAS registry no	136236-51-6 161735-79-1 (mesylate salt)
Molecular weight	171.24 267.33 (mesylate salt)

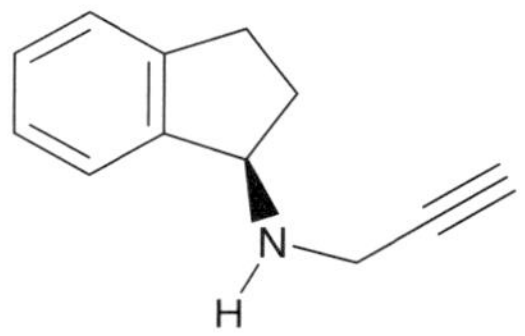

Rasagiline is a second-generation, irreversible monoamine oxidase type B (MAO-B) inhibitor that has been launched for the treatment of Parkinson's disease (PD). Unlike its predecessor selegiline, it is not metabolized to amphetamine derivatives and is, therefore, devoid of the sympathomimetic activity responsible for adverse side effects. Rasagiline is, however, similar to selegiline in the retention of the propargylamine moiety; this essential pharmacophore binds covalently to selectively form an irreversible bond with the flavin adenine dinucleotide portion of the MAO-B enzyme. SAR dictates that a distance of no more than two carbon units between the aromatic ring and the amine is essential for conferring MAO-B specificity (IC_{50} = 14 nM vs. 700 nM for MAO-A). The inhibition of MAO-B prevents the degradation of dopamine, thereby, prolonging the action of dopamine to reduce the effects of dopaminergic neuronal deficit. Rasagiline has been approved for both initial monotherapy in patients with early disease as well as an adjunct treatment in patients with advanced disease. While levodopa, a dopamine precursor, is a standard line of treatment for PD, many patients begin to experience motor complications after several years of artificial dopaminergic stimulation. As an adjunct therapy, rasagiline treats the fluctuations in motor symptoms. The R-enantiomer exhibits 4-times the potency of the S-enantiomer, so the synthetic method begins with the optical resolution of racemic N-benzyl-1-aminoindan using (R,R)-tartaric acid as the resolving agent. Once isolated, the enantiomerically-enriched salt is submitted to hydrogenolysis to afford 1(R)-aminoindane that is subsequently propargylated to provide rasagiline. It is formulated as its mesylate salt, and the recommended dosage of rasagiline is 1 mg/day, with or without levodopa. As an irreversible inhibitor, frequent dosing is not necessary since the duration of action is not driven by half-life; regeneration of MAO-B is the critical factor in the duration of action. Rasagiline is rapidly absorbed with a T_{max} of approximately 0.5 h and a C_{max} of approximately 10 ng/mL with an absolute oral bioavailability of 36%. Its major metabolite, 1(R)-aminoindane, is generated via CYP1A2-mediated deamination in the liver and accounts for 20% of the 63% of elimination via the urine with an elimination half-life of 0.6–2 h. While this metabolite does not possess MAO-B inhibitory activity, it has been implicated in the neuroprotective, anti-apoptotic propensity of rasagiline. Several clinical trials involving more than 1000 patients with PD assessed the efficacy of rasagiline as a monotherapy or as an adjunct therapy. In one study, drug-naïve patients were randomized to receive either placebo or rasagiline (1 or 2 mg/day) for 26 weeks. Both rasagiline-treated groups demonstrated statistically significant improvements in the mean change from baseline in the total score of the Unified Parkinson's Disease Rating Scale (UPDRS), the primary efficacy endpoint, compared to placebo. Another study involved the randomization of more advanced PD patients to receive placebo, rasagiline (1 mg/day), or entacapone (200 mg) concomitant with scheduled levodopa. For this study, the primary efficacy endpoint was the change from baseline in the mean number of hours spent in the "off" state, during the day. Entacapone, a catecholamine-O-methyltransferase inhibitor known as an effective add-on therapy for motor fluctuations, was used as a comparator. Rasagiline reduced the time spent in the "off" state while increasing the "on" time. Furthermore, the magnitude of these effects was comparable to that of entacapone. Overall, reported adverse events were comparable to placebo and included headache and arthalgia. Dyskinesia and accidental injury were also indicated as side effects when rasagiline was used as an adjunct to levodopa. As a substrate for CYP1A2, caution should be exercised when co-administering rasagiline with potent CYP1A2 inhibitors or inducers. The package insert advises against the concomitant use of rasagiline and antidepressants containing fluoxetine or fluvoxamine, suggesting a washout period of at least 5 weeks following the cessation of these CYP1A2 inhibitors. Also, rasagiline should not be used with other MAO inhibitors taken for the treatment of depression. Medical advice is required with over-the-counter drugs containing the cough suppressant dextromethorphan and nasal decongestants ephedrine and pseudoephedrine. Finally, the use of rasagiline is not recommended in patients under the age of 18.

20. SORAFENIB (ANTICANCER) [88–92]

Country of origin	Germany
Originator	Bayer/Onyx
First introduction	US
Introduced by	Bayer/Onyx
Trade name	Nexavar
CAS registry no	284461-73-0
Molecular weight	464.82

Sorafenib is a small molecular inhibitor of several kinases involved in tumor angiogenesis and proliferation, including, but not limited to, Raf ($IC_{50} = 12$ nM for Raf-1), VEGFR ($IC_{50} = 90$ nM for VEGFR-2 and $IC_{50} = 12$ nM for VEGFR-3), and platelet derived growth factor receptor ($IC_{50} = 57$ nM for PDGFR-β). Specifically, sorafenib blocks tumor progression by inhibiting cellular proliferation that is dependent on activation of the MAPK pathway (Raf) and/or inhibiting tumor angiogenesis through VEGFR and/or PDGFR. While it may be effective in the treatment of a variety of tumors, the first approvable indication is for renal cell carcinoma. Sorafenib was identified in a medicinal chemistry effort to target Raf-1 and was selected for further development based on Raf-1 potency and favorable kinase selectivity profile. It is prepared by a four-step synthesis in 63% overall yield. The final step involves the construction of the urea moiety by coupling 1-chloro-2-(trifluoromethyl)-4-isocyanotobenzene with 4-(4-aminophenoxy)-N-methylpyridine-2-carboxamide. It is formulated as its tosylate salt with each tablet containing 200 mg of sorafenib, and the recommended daily dose is 400 mg (2 × 200 mg). Following oral administration, peak plasma levels are achieved in approximately 3 h, but there is not a linear increase in C_{max} and AUC beyond doses of 400 mg twice, daily. The mean oral bioavailability is 38–49%, but taking sorafenib with a high-fat meal significantly reduces the bioavailability. It is, therefore, advised that sorafenib be administered without food. The mean elimination half-life is approximately 25–48 h. Sorafenib is metabolized primarily by CYP3A4 with the pyridine N-oxide being the major metabolite. Glucuronidation mediated by UGT1A9 accounts for 19% of the dose excreted in the urine. 77% of the dose is recovered in the feces with the largest percentage (51%) being unchanged sorafenib. The mean elimination half-life is approximately 25–48 h, and the plasma protein binding is 99.5%. A phase III, randomized, double-blind, placebo-controlled study in advanced renal cell carcinoma patients provided the best evaluation of the safety and efficacy of sorafenib. The primary study endpoints were overall survival and progression-free survival (PFS), with tumor response rate being a secondary endpoint. The PFS analysis included 384 sorafenib-treated patients (400 mg twice, daily) and 385 patients receiving placebo. The median PFS for sorafenib treatment was approximately twice that observed with placebo (167 days compared to 84 days). In evaluating the tumor response, 2% of sorafenib-treated patients had a confirmed partial response while no patients in the placebo group experienced a partial response, suggesting that the gain in PFS for the drug-treated group is primarily a reflection of stable disease. In this Phase III study, 30% of sorafenib-treated patients reported adverse events compared to 22% for placebo with the primary complaints in both groups including rash, diarrhea, hand-foot skin reaction, fatigue, and hypertension. Most of the observed laboratory abnormalities were comparable for sorafenib and placebo with the exception of hypophosphatemia (45% versus 11% with placebo). Regarding drug–drug interactions, caution should be exercised in co-administering drugs that inhibit CYP3A4 and UGT1A9 although limited data suggests that concomitant use of ketoconazole, an inhibitor of both CYP3A4 and UGT1A9, did not lead to an increase in the mean AUC of a 50 mg dose of sorafenib. Conversely, since sorafenib has been shown to inhibit CYP2B6 ($K_i = 6\,\mu M$) and CYP2C8 ($K_i = 1\text{-}2\,\mu M$), caution is recommended when co-administering substrates for these enzymes. The label does include a pregnancy category D warning since sorafenib has been shown to be teratogenic and to induce embryo-fetal toxicity in rats and rabbits. Overall, the drug appears to be well tolerated by the majority of patients at the 400 mg b.i.d. continuous dosing. As an inhibitor of multiple kinases vital for tumor progression, sorafenib may possess wide-spectrum anti-tumor properties and may emerge as an effective weapon against a variety of solid tumors.

21. TAMIBAROTENE (ANTICANCER) [93–96]

Country of origin	Japan
Originator	Toko Yakuhin Kogyo
First introduction	Japan
Introduced by	Nippon Shinyaku
Trade name	Amnolake
CAS registry no	094497-51-5
Molecular Weight	351.44

Tamibarotene, a selective agonist of the retinoic acid receptor, was launched in Japan last year as an oral treatment for relapsed or refractory acute promyelocytic leukemia (APL). APL is a form of acute myeloid leukemia (AML) characterized by a deficiency in mature blood cells and an excess of immature cells called promyelocytes in the bone marrow and peripheral blood. The current standard of care for APL includes treatment with all-*trans*-retinoic acid (ATRA), either alone or in combination with chemotherapy. ATRA is a high affinity ligand for two types of nuclear receptors, retinoic acid receptor (RAR) and retinoid X receptor (RXR), each of which has three subtypes (-α, -β, and -γ). Activation of RARα by ATRA causes promyelocytes to differentiate and mature, thereby inhibiting their proliferation and inducing disease remission. Although ATRA is one of the most clinically successful retinoids, its usage is hampered by the high rate of adverse effects, instability, and the appearance of ATRA-resistant leukemia cells. The rapid development of resistance to ATRA is partly attributed to a progressive decrease in plasma drug concentration because of the rapid upregulation of its catabolism, and partly to an increased expression of cytoplasmic retinoid-binding proteins (CRABP) which limit the amount of free drug. In an attempt to overcome these problems, tamibarotene was synthesized by the introduction of heteroatoms into ATRA-like structures. Tamibarotene exhibits an improved selectivity profile as compared with ATRA. It has slightly higher affinity for RARα ($EC_{50} = 45\,nM$) than RARβ and RARγ ($EC_{50} = 235$ and $591\,nM$, respectively), and it is inactive against RXR. In addition, tamibarotene has little affinity for CRABPs and is active against CRABP-rich ATRA-resistant cells. In a clinical study involving 24 APL patients who had relapsed from ATRA induced complete remission (CR), 6 mg/m2/day of oral tamibarotene resulted in 14 (58%) patients achieving a second CR. Adverse events included retinoic acid syndrome, hyperleukocytosis, xerosis, cheilitis, hypertriglyceridemia, and hypercholesterolemia; however, these side effects were generally milder than with ATRA, which all patients had received previously. Examination of human samples taken from Phase II and III clinical trials revealed that fecal excretion was the major elimination route, and the metabolism of tamibarotene occurred primarily through hydroxylation and taurine conjugation. In vitro, the plasma protein binding of tamibarotene is shown to be >98% in rats, dogs, and humans. Tamibarotene is synthesized from 5,5,8,8-tetramethyl-5,6,7,8-tetrahydronaphthalene in a four-step sequence consisting of regioselective nitration in the 2-position, reduction of the nitro group by hydrogenation to produce the corresponding aniline derivative, acylation of the aniline intermediate with 4-(carbomethoxy)benzoyl chloride, and hydrolysis of the methyl ester.

22. TIGECYCLINE (ANTIBIOTIC) [97–103]

Country of origin	US
Originator	Wyeth
First introduction	US
Introduced by	Wyeth
Trade name	Tygacil
CAS registry no	220620-09-7
Molecular weight	585.65

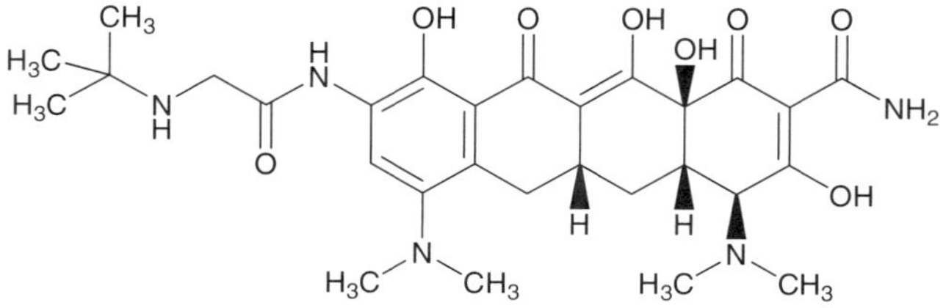

The emergence of drug-resistant bacteria has diminished the clinical utility of the tetracyclines. Research to circumvent the efflux and ribosomal protection mechanisms of bacteria has led to the development of the glycylcyclines. Tigecycline is the first glycylcycline antibiotic to launch for the parenteral treatment of baterial infection, including complicated intra-abdominal and skin infections. Its mechanism of action involves inhibiting protein translation in bacteria by binding to the 30S ribosomal subunit and blocking entry of amino-acyl tRNA molecules into the A site of the ribosome to effectively prevent incorporation of amino acid residues into elongating peptide chains. Presumably, ribosomal protection proteins are ineffective against tigecycline due to its higher affinity for ribosomal binding compared to tetracyclines (approximately 16-fold). In addition, tigecycline may be resistant to efflux mechanisms by either their inability to translocate it across the cytoplasmic membrane due to steric complications or simply by their failure to recognize the molecule. The essential glycylamido appendage, not found in any naturally occurring tetracycline and responsible for imparting certain microbiologic properties to tigecycline, is attached by reacting 9-amino-minocycline (derived from nitration of minocycline followed by reduction via catalytic hydrogenation) with t-butylaminoacetyl chloride. The purified product is isolated by gradient pH extraction between pH 4.5–6.5. Tigecyline is packaged in vials containing 50 mg of lyophilized powder for intravenous infusion. Following an intravenous infusion of a single, 100 mg dose over approximately 30–60 min, the following pharmacokinetic parameters have been measured: $C_{max} = 0.9$–$1.45\,\mu g/mL$, $t_{1/2} = 27\,h$, $CL = 22\,L/h$, and $AUC = 5.2\,\mu g.h/mL$. The serum protein binding of tigecycline is approximately 78%, and it is extensively distributed into the tissues with a steady state volume of distribution of 350–700 L. While tigecycline is not extensively metabolized, a glucuronide is found in urine. The primary route of elimination, however, is biliary excretion. Both *in vitro* and in clinical infections have demonstrated that tigecycline is active against a variety of commonly occurring aerobic and facultative Gram-positive, aerobic and facultative Gram-negative, and anaerobic microorganisms. In the Phase II trial of patients with complicated intra-abdominal infections, including perforated and gangrenous appendicitis, cholecystitis, perforated diverticulitis, and peritonitis, a loading dose of 100 mg i.v. was followed by 50 mg i.v. b.i.d. for 5–14 days resulting in clinical cure rates at the test of cure and end of treatment visits of 67% and 76%, respectively. The most common adverse events were nausea and vomiting. Tigecycline was also evaluated in patients with complicated skin and soft tissue infections in a separate phase II trial. Patients were randomized to receive either tigecycline 50 mg i.v. followed by 25 mg i.v. every 12 h or tigecycline 100 mg i.v. followed by 50 mg i.v. every 12 h. Slightly higher cure rates were observed in the 50 mg group (74%) compared to the 25 mg group (67%). Pathogens were also eradicated more frequently in the 50-mg group (69%) compared to the 25 mg group (56%). Both doses were well tolerated with nausea and vomiting again being the most common adverse effects. From the combined clinical data, the recommended dosage regimen for tigecycline is an initial dose of 100 mg, followed by 50 mg every 12 h by i.v. infusion over 30–60 min. Being structurally similar to the tetracycline class of antibiotics, similar adverse effects and warnings apply. As with tetracyclines, tigecycline should not be administered to pregnant women due to the potential for fetal harm. Furthermore, the use of tigecycline should be avoided during tooth development (last half of pregnancy, infancy, and childhood to the age of 8 years) because permanent discoloration of the teeth may occur. Since pseudomembranous colitis is a concern for nearly all antibacterial agents due to alteration of flora in the colon, it is important to consider this diagnosis in patients presenting with diarrhea. No potential drug–drug interactions are anticipated since tigecycline is not extensively metabolized nor does it inhibit any of the major P450 isoforms. It is recommended, however, that prothrombin time or other suitable anticoagulation tests should be monitored if tigecycline is co-administered with warfarin. Finally, as with all antibiotics, concurrent use with oral contraceptives may render the oral contraceptives less effective.

23. TIPRANAVIR (HIV) [104–110]

Country of origin	US
Originator	Pharmacia & Upjohn
First introduction	US
Introduced by	Boehringer Ingelheim
Trade name	Aptivus
CAS registry No	174484-41-4 191150-83-1
Molecular weight	602.66 (Disodium salt = 646.64)

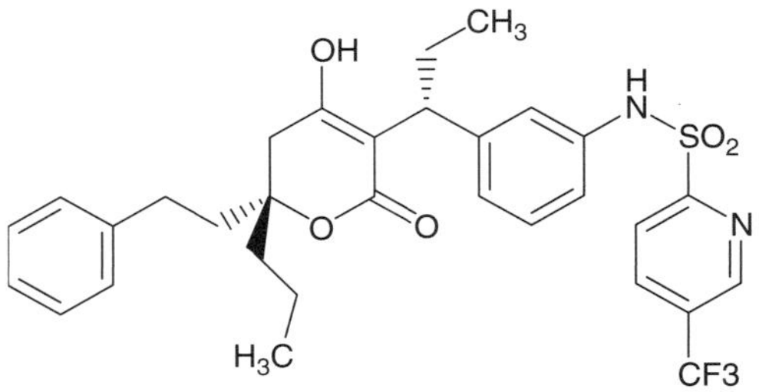

Tipranavir, an oral non-peptidic HIV protease inhibitor, was granted accelerated approval for use in combination with ritanovir and was subsequently launched the same month in 2005. The targeted patient population includes HIV-1 infected adults with evidence of viral replication and demonstrated resistance to multiple protease inhibitors. As with other protease inhibitors, binding at the HIV-1 protease's active site inhibits the virus-specific processing of the Gag and Gag-Pol polyproteins in HIV-1-infected cells resulting in the production of non-infectious virions. An effective regimen to reduce viral load and to preserve immune function typically consists of a cocktail of a protease inhibitor and at least one nucleoside reverse transcriptase inhibitor. It is believed that tipranavir has been effective where other protease inhibitors have encountered resistance because, as a non-peptidic inhibitor, it was designed by structure-based analysis to have increased flexibility making it acquiescent to conformational alterations at the binding site. Tipranavir is synthesized by a convergent route amenable to large-scale production. By splitting the molecule in half and constructing the C3–C4 bond of the dihydropyrone ring via a key aldol addition of a nitroaromatic ester to an aldehyde handle of the other half, the two stereocenters of the target can be set independently by prior enantiomeric resolution of each precursor. Furthermore, the shrewd selection of intermediates facilitates isolation and purification by crystallization. While, tipranavir has a K_i of <0.01 nM against HIV-1 and <1 nM against HIV-2, its poor solubility, impaired bioavailability, and high-protein binding ($>99\%$) initially resulted in a large pill burden, requiring between 4 and 8 capsules twice a day. Changing from the original 300 mg hard-filled capsule to a 250 mg self-emulsifying drug delivery system (SEDDS), significantly, improved the dissolution and bioavailability of tipranavir. A two-fold increase in systemic concentration was realized, thus, cutting the daily dose in half. Maximum plasma concentrations are, however, achieved with co-administration of ritonavir, particularly taken with a high-fat meal. Ritonavir inhibits CYP3A4, the predominant enzyme involved in tipranavir metabolism. The other factor responsible for the enhanced plasma levels is the fact that ritonavir also inhibits the P-glycoprotein efflux pump for which tipranavir is a substrate. Boosting with ritonavir (200 mg) increased tipranavir C_{min} 4-fold and C_{max} 20-fold. The primary route of excretion is through the feces (83%) with an effective mean elimination half-life of 4.8 h in healthy volunteers and 6.0 h in HIV-infected adults. The efficacy of tipranavir/ritonavir combination therapy has been evaluated in multiple clinical trials. The number of patients exhibiting <50 or $400\log_{10}$ copies/mL HIV RNA, change in viral load, and change in CD4 cell count were all considered primary endpoints. For example, the Phase III RESIST-2 trial enrolled more than 800 patients in Europe and South America. These patients had advanced HIV disease and were experiencing virologic failure. In addition, they had already endured a multiple drug regimen with documented protease inhibitor resistance. The results from the co-administration of tipranavir/ritonavir (tipranavir/r, 500 mg tipranavir (2×250 mg) + 200 mg ritonavir (2×100 mg)) were compared to a marketed protease inhibitor/ritonavir (CPI/r) combination. A decrease in viral load from baseline of 1 log10 or greater was achieved in 41% of patients who received tipranavir/r compared to 14.9% of patients in the CPI/r group. Furthermore, patients receiving tipranavir/r therapy experienced greater increases in CD4+ cell counts compared to the CPI/r group -31 cells/mm^3 and 1 cell/mm^3, respectively. While the adverse events were comparable between the two groups, the tipranavir/r patients displayed a higher rate of liver enzyme and lipid elevations; however, no observable symptoms resulted from these abnormal laboratory measurements. In general, the adverse events included nausea, vomiting, and diarrhea. While tipranavir is a CYP3A4 inducer and substrate, as previously stated, its co-administration with the CYP3A4 inhibitor ritonavir results in net inhibition; therefore, patients should avoid the concomitant use of drugs highly dependent on CYP3A4 for clearance. The complete list of contraindicated drugs can be found in the package insert, but the general classes include antiarrhythmics, antihistamines, antimycobacterials, neuroleptics, sedatives, ergot derivatives, GI motility agents, and the herbal supplement St. John's wort. Finally, as the elevated liver enzyme levels suggest, tipranavir should not be taken by patients with severe liver disease. Patients with clinical symptoms of hepatitis should immediately discontinue use of tipranavir. It is highly recommended that liver function tests be performed prior to treatment and throughout the course of therapy.

24. UDENAFIL (ERECTILE DYSFUNCTION) [111–115]

Country of origin	South Korea	
Originator	Dong-A	
First introduction	South Korea	
Introduced by	Dong-A	
Trade name	Zydena	
CAS registry no	268203-93-6	
Molecular Weight	516.66	

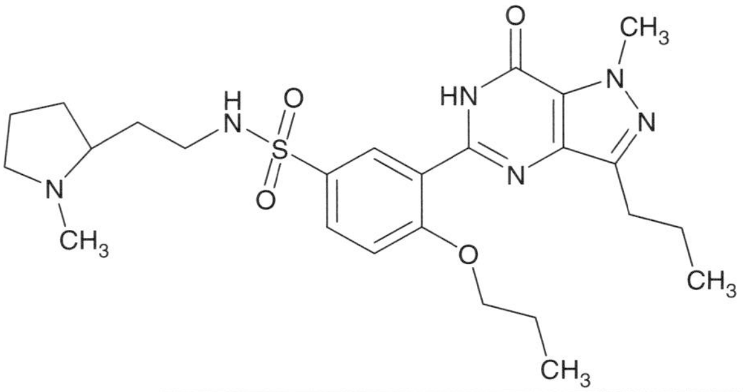

Udenafil is the fourth in a class of drugs targeting the inhibition of the enzyme phosphodiesterase 5 (PDE5) for the treatment of erectile dysfunction. Inhibition of PDE5 results in the increase in endogenous cyclic guanosine monophosphate (cGMP) concentrations in the penile corpus cavernosum. cGMP induces smooth muscle cell relaxation and subsequent increased blood flow leading to a sustainable erection. Udenafil is a potent antagonist of human PDE5 with an IC_{50} of 8.25 nM and a comparable selectivity profile as sildenafil for the other PDEs. Unlike tadalafil, it does not inhibit PDE11, which has been implicated in myalgia and testicular toxicity. The key steps in the synthesis of udenafil involve the coupling of 1-methyl-3-propyl-4-amino-5-carbamoyl pyrazole with 2-propoxybenzoyl chloride, chlorosulfonation of the resulting product, reaction of the sulfonyl chloride with 2-(2-aminoethyl)-1-methylpyrrolidine, and finally ring closure to build the pyrimidinone ring of the core. In Phase I studies to assess the safety, tolerability, and pharmacokinetics of udenafil, single doses of 12.5–400 mg and multiple doses of 100 and 200 mg/day were well-tolerated, and exposure increased in a dose-dependent manner. While the absolute oral bioavailability has not been determined in humans, it is absorbed rapidly; the oral bioavailability for multiple animal species averaged about 30%. With a T_{max} of 1 – 1.5 h and a terminal half-life of 11 – 13h, udenafil has a rapid onset with a long duration of action. While udenafil is biotransformed to three major metabolites by rat microsomes, the predominant active metabolite in humans is the N-dealkylated sulfonamide, generated via CYP3A4 metabolism. Co-administration with strong CYP3A4 inhibitors or inducers could, therefore, alter the metabolism of udenafil and significantly change the pharmacokinetics. In vitro studies in pooled human liver microsomes demonstrated that the selective CYP3A4 inhibitor ketoconazole effectively inhibited the N-dealkylation of udenafil. To assess efficacy, a randomized, double-blind, placebo-controlled, fixed-dose Phase II trial was conducted in 319 ED patients. The change in the International Index of Erectile Function (IIEF) score from baseline to the end of the 12-week study was used as the primary efficacy endpoint. After 12 weeks of therapy, the udenafil-treated group, receiving either 100 or 200 mg, had significantly higher IIEF scores than the placebo group. Furthermore, udenafil produced up to a 91% vaginal penetration success rate and up to a 67% intercourse completion rate compared to a 29% completion rate by placebo. Overall patient satisfaction, measured by a standard global assessment question, was 86% compared to only 26% in the placebo group. The most frequently recorded adverse events were mild-to-moderate facial flushing and headache.

25. ZICONOTIDE (SEVERE CHRONIC PAIN) [116–119]

Country of origin	US	Trade name	Prialt
Originator	Elan	CAS registry no	107452-89-1
First introduction	US	Molecular weight	2639.13
Introduced by	Elan		

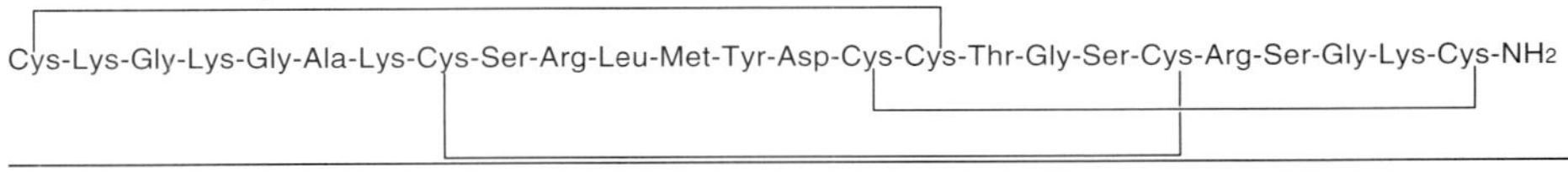

Cys-Lys-Gly-Lys-Gly-Ala-Lys-Cys-Ser-Arg-Leu-Met-Tyr-Asp-Cys-Cys-Thr-Gly-Ser-Cys-Arg-Ser-Gly-Lys-Cys-NH$_2$

Ziconotide, the synthetic equivalent of ω-conopeptide MVIIA isolated from marine snail venom, was launched as a novel non-opioid treatment for severe chronic pain, particularly in refractory patients. While early clinical studies also evaluated ziconotide's neuroprotective effects in preventing ischemic neurodegeneration in patients with stroke or brain trauma, pain was the only indication at launching. The therapeutic efficacy of ziconotide is derived from its potent and highly selective blockade of neural N-type voltage-sensitive calcium channels (IC$_{50}$ approximately 1.0 nM). Inhibition of pain-sensing nociceptors is the ultimate result of this blockade and is responsible for the drug's analgesic effect. Since ziconotide is a large, highly charged, hydrophilic peptide (overall +6 charge), membrane permeability is poor; intrathecal administration bypasses the blood-brain barrier to directly deliver the drug to the cerebrospinal fluid to efficiently block nerve transmission. While the presence of the C-terminal amide and the three disulfide bridges, establishing the three-dimensional structure of ziconotide responsible for its potency and selectivity, confer some resistance toward exopeptidases, bypassing systemic circulation by the intrathecal mode of delivery further lengthens the metabolic lifetime of ziconotide. In the most straightforward approach, the 25 amino acids of ziconotide are pieced together by standard solid-phase peptide synthesis to give the linear peptide which is ultimately cyclized and deprotected to afford crude ziconotide that is purified by RP-HPLC. Manipulation of the sequence ligation, folding, cyclizing, and deprotection strategy is possible and will influence the purity profile. In addition to high-binding affinity for N-type voltage-sensitive calcium channels, ziconotide has no appreciable affinity for other ion channels or opioid receptors. As a non-opioid therapeutic, it is devoid of tolerance issues associated with opioid therapy and has shown efficacy in opioid refractory patients. Following intrathecal administration, ziconotide has a half-life of 4.5 h in cerebrospinal fluid (CSF) with a clearance of 0.26 mL/min, comparable to the rate of turnover of human CSF. Likewise, the volume of distribution of 99 mL represents the total volume of human CSF. Rather than metabolism, the results suggest that bulk CSF flow is the primary driver of ziconotide clearance from the CSF. In the largest clinical study, 111 patients with cancer or AIDS who failed to obtain adequate relief from either oral or intrathecal opiates participated in this randomized, double-blind, and placebo-controlled study. At the start of the study, 92% of these patients were still on opioid therapy, despite the apparent lack of analgesic efficacy. The main measure of efficacy was the mean percentage change in visual analog scale of pain intensity (VASPI) scores. The VASPI scale ranges from 0 mm, corresponding to no pain, to 100 mm, defined as the worst imaginable pain. Intrathecal titration of ziconotide to the point of analgesia or to the maximum allowable dose over 5–6 days resulted in a 53% improvement in mean VASPI score (a reduction from 74 to 35 mm) for the treatment group. While pain relief was reported to be moderate or complete in 53% of ziconotide-treated patients, none of the placebo-treated patients experienced complete relief, but 17% reported modest effects. Regarding the concomitant use of opiates, the ziconotide-treated group experienced a 10% reduction while the placebo-treated group saw a 5% increase. Ziconotide doses initially started at 0.4 µg/h, but were subsequently reduced to 0.1 µg/h due to a high rate of adverse events. The most common adverse effects included dizziness, nystagmus, nausea, postural hypotension, somnolence, confusion, fever, headache, and urinary retention. In addition, due to drug or placebo delivery to immunocompromised patients via infusion though an external catheter, meningitis occurred in 5 drug-treated patients and 2 patients receiving placebo. Aseptic techniques are a must with this external pump system. Finally, since ziconotide is not circulated systemically, there is little risk of significant drug–drug interactions. Ziconotide is, however, contraindicated in patients with a history of psychosis, and the label features a prominent warning regarding the risk of severe psychiatric and neurological impairment.

REFERENCES

[1] The collection of new therapeutic entities first launched in 2005 originated from the following sources: (a) CIPSLINE, Prous database; (b) Iddb, Current Drugs database; (c) IMS R&D Focus; (d) Adis Business Intelligence R&D Insight; and (e) Pharmaprojects.
[2] A. I. Graul and J. R. Prous, *Drug News Perspect.*, 2006, **19**, 33.
[3] S. Hegde and M. Schmidt, *Ann. Rep. Med. Chem.*, 2005, **40**, 443.
[4] S. Hegde and J. Carter, *Ann. Rep. Med. Chem.*, 2004, **39**, 337.
[5] C. Boyer-Joubert, E. Lorthiois and F. Moreau, *Ann. Rep. Med. Chem.*, 2003, **38**, 347.
[6] M. Humbert, *Expert Opin. Invest. Drugs*, 2004, **13**, 1349.
[7] P. Christie, *Drugs Today*, 2004, **40**, 569.
[8] N. E. Mealy, M. Bayes and J. Castañer, *Drugs Future*, 2001, **26**, 1033.
[9] N. A. Reynolds and L. J. Scott, *Drugs*, 2004, **64**, 511.
[10] J. P. Kline and R. A. Larson, *Expert Opin. Pharmacother*, 2005, **6**, 2711.

[11] C.-H. Pui, S. Jeha and P. Kirkpatrick, *Nat. Rev. Drug Discov.*, 2005, **4**, 369.
[12] K. Chilman-Blair, N. E. Mealy and J. Castañer, *Drugs Future*, 2004, **29**, 112.
[13] Sternberg., *Curr. Opin. Invest. Drugs*, 2003, **4**, 1479.
[14] W. F. Bauta, B. E. Schulmeier, B. Burke, J. F. Puente, W. R. Cantrell, Jr., D. Lovett, J. Goebel, B. Anderson, D. Ionescu and R. Guo, *Org. Process Res. Dev.*, 2004, **8**, 889.
[15] C. R. Chapplet, *Expert Opin. Invest. Drugs*, 2004, **13**, 1493.
[16] J. L. Kirwin, *Formulary*, 2004, **39**, 291.
[17] P. E. Cross and A. R. MacKenzie, *US Patent* 5,096,890, 1992.
[18] F. Haab, *Drugs Today*, 2005, **41**, 441.
[19] J. A. McIntyre, J. Castañer, N. E. Mealy and M. Bayes, *Drugs Future*, 2004, **29**, 331.
[20] G. J. Kontoghiorghes, *Drugs Future*, 2005, **30**, 1241.
[21] J. C. Barton, *Curr. Opin. Invest. Drugs*, 2005, **6**, 327.
[22] M. N. Alekshun, *Expert Opin. Invest. Drugs*, 2005, **14**, 117.
[23] Y. Nishino, T. Komurasaki, T. Yuasa, M. Kakinuma, K. Izumi, M. Kobayashi, S. Fujiie, T. Gotoh, Y. Masui, M. Hajima, M. Takahira, A. Okuyama and T. Kataoka, *Org. Process Res. Dev.*, 2003, **7**, 649.
[24] Y. Nishino, M. Kobayashi, T. Shinno, K. Izumi, H. Yonezawa, Y. Masui and M. Takahira, *Org. Process Res. Dev.*, 2003, **7**, 846.
[25] R. N. Jones, H. K. Huynh, D. J. Biedenbach, T. R. Fritsche and H. S. Sader, *J. Antimicrob. Chemother.*, 2004, **54**, 144.
[26] X. Rabasseda, J. Castañer and E. Font, *Drugs Future*, 1996, **21**, 792.
[27] C. F. Gallemi, I. J. M. Bono and C. M. Vidal, *WO Patent* 9921838-A1, 1999.
[28] A. I. Rubin, B. Bagheri and R. K. Scher, *Am. J. Clin. Dermatol.*, 2002, **3**, 71.
[29] C. K. Opio, W. M. Lee and P. Kirkpatrick, *Nat. Rev. Drug Discov.*, 2005, **4**, 535.
[30] P. Honkoop and R. A. de Man, *Expert Opin. Invest. Drugs*, 2003, **12**, 683.
[31] G. S. Bisacchi, S. T. Chao, C. Bachard, J. P. Daris, S. Innaimo, G. A. Jacobs, O. Kocy, P. Lapointe, A. Martel, Z. Merchant, W. A. Slusarchyk, J. E. Sundeen, M. G. Young, R. Colonno and R. Zahler, *Bioorg. Med. Chem. Lett.*, 1997, **7**, 127.
[32] A. Rivkin, *Curr. Med. Res. Opin.*, 2005, **21**, 1845.
[33] M. Scharf, *Expert Opin. Pharmacother*, 2006, **7**, 345.
[34] C. Culy, J. Castañer and M. Bayes, *Drugs Future*, 2003, **28**, 640.
[35] A. Mack and J. Octavio Salazar, *Formulary*, 2003, **38**, 582.
[36] S. T. Melton, J. M. Wood and C. K. Kirkwood, *Ann. Pharmacother.*, 2005, **39**, 1659.
[37] K. A. Gryskiewicz and C. I. Coleman, *Formulary*, 2005, **40**, 86.
[38] M. B. Davidson, G. Bate and P. Kirkpatrick, *Nat. Rev. Drug Discov.*, 2005, **4**, 713.
[39] A. H. Barnett, *Drugs Today*, 2005, **41**, 563.
[40] G. M. Keating, *Drugs*, 2005, **65**, 1681.
[41] J. A. McIntyre and M. Bayes, *Drugs Future*, 2004, **29**, 23.
[42] J. J. Hopwood, G. Bate and P. Kirkpatrick, *Nat. Rev. Drug Discov.*, 2006, **5**, 101.
[43] M. Beck, *Therapy*, 2006, **3**, 9.
[44] Anonymous., *Drugs R. D.*, 2005, **6**, 312.
[45] K. Uchida, T. Tanaka and H. Yamaguchi, *Microbiol. Immunol.*, 2003, **47**, 143.
[46] Y. Niwano, N. Kuzuhara, H. Kodama, M. Yoshida, T. Miyazaki and H. Yamaguchi, *Antimicrob. Agents Chemother.*, 1998, **42**, 967.
[47] H. Kodama, Y. Niwano, K. Kanai and M. Yoshida, *WO Patent* 9702821, 1997.
[48] R. Lehmann, M. Brzosko, P. Kopsa, R. Nischik, A. Kreiss, H. Thurston, S. Litschig and V. S. Sloan, *Curr. Med. Res. Opin.*, 2005, **21**, 517.
[49] B. Bannwarth and F. Berenbaum, *Expert Opin. Invest. Drugs*, 2005, **14**, 521.
[50] C. M. Rordorf, L. Choi, P. Marshall and J. B. Mangold, *Clin. Pharmacokinet.*, 2005, **44**, 1247.

[51] M. Acemoglu, T. Allmendinger, J. Calienni, J. Cercus, O. Loiseleur, G. Sedelmeier and D. Xu, *WO Patent* 01/23346 A2, 2001.
[52] D. A. Gamache, G. Graff, M. T. Brady, J. M. Spellman and J. M. Yanni, *Inflammation*, 2000, **24**, 357.
[53] T.-L. Ke, G. Graff, J. M. Spellman and J. M. Yanni, *Inflammation*, 2000, **24**, 371.
[54] R. Lindstrom and T. Kim, *Curr. Med. Res. Opin.*, 2006, **22**, 397.
[55] D. A. Walsh, H. W. Moran, D. A. Shamblee, W. J. Welstead, Jr., J. C. Nolan, L. F. Sancilio and G. Graff, *J. Med. Chem.*, 1990, **33**, 2296.
[56] S. L. Fine, D. F. Martin and P. Kirkpatrick, *Nat. Rev. Drug Discov.*, 2005, **4**, 187.
[57] D. Jellinek, L. S. Green, C. Bell and N. Janjic, *Biochem.*, 1994, **33**, 10450.
[58] J. Ruckman, L. S. Green, J. Beeson, S. Waugh, W. Gillette, D. Henninger, L. Claesson-Welsh and N. Janjic, *J. Biol. Chem.*, 1998, **273**, 20556.
[59] S. A. Doggrell, *Expert Opin. Pharmacother*, 2005, **6**, 1421.
[60] F. W. Fraunfelder, *Drugs Today*, 2005, **41**, 703.
[61] S. A. Vinores, *Curr. Opin. Mol. Therapeut.*, 2003, **5**, 673.
[62] E. W. M. Ng, D. T. Shima, P. Calias, E. T. Cunningham, Jr., D. R. Guyer and A. P. Adamis, *Nat. Rev. Drug Discov.*, 2006, **5**, 123.
[63] H. Kourlas and D. S. Schiller, *Clin. Therapeut.*, 2006, **28**, 36.
[64] M. Friedlander, H. E. Agular and M. L. Dorrell, *WO Patent* 2005/117954 A2, 2005.
[65] G. J. Ryan, L. J. Jobe and R. Martin, *Clin. Therapeut.*, 2005, **27**, 1500.
[66] P. Norman, X. Rabasseda and P. A. Leeson, *Drugs Future*, 2001, **26**, 444.
[67] E. L. Kleppinger and E. M. Vivian, *Annals Pharmacotherapy*, 2003, **37**, 1082.
[68] L. Gaeta, H. Jones and E. Albrecht, *WO Patent* 9310146, 1993.
[69] D. A. Hussar, *J. Amer. Pharm. Assoc.*, 2005, **45**, 301.
[70] J. A. McIntyre, L. Martin and M. Bayes, *Drugs Future*, 2005, **30**, 117.
[71] R. Spielberger, P. Stiff, W. Bensinger, T. Gentile, D. Weisdorf, T. Kewalramani, T. Shea, S. Yanovich, K. Hansen, S. Noga, J. McCarty, C. F. LeMaistre, E. C. Sung, B. R. Blazar, D. Elhardt, M.-G. Chen and C. Emmanouilides, *N. Engl. J. Med.*, 2004, **351**, 2590.
[72] D. Gospodarowicz and F. Masiarz, *WO Patent* 9501434, 1995.
[73] Anonymous, *Drugs R. D.*, 2005, **6**, 186
[74] D. Buysse, G. Bate and P. Kirkpatrick, *Nature Rev. Drug Discov.*, 2005, **4**, 881.
[75] O. Uchikawa, K. Fukatsu, R. Tokunoh, M. Kawada, K. Matsumoto, Y. Imai, S. Hinuma, K. Kato, H. Nishikawa, K. Hirai, M. Miyamoto and S. Ohkawa, *J. Med. Chem.*, 2002, **45**, 4222.
[76] A. Karim, D. Tolbert and C. Cao, *J. Clin. Pharmacol.*, 2006, **46**, 140.
[77] K. Kato, K. Hirai, K. Nishiyama, O. Uchikawa, K. Fukatsu, S. Ohkawa, Y. Kawamata, S. Hinuma and M. Miyamoto, *Neuropharmacology*, 2005, **48**, 301.
[78] N. N. Nguyen, S. S. Yu and J. C. Song, *Formulary*, 2005, **40**, 146.
[79] A. Schapira, G. Bate and P. Kirkpatrick, *Nat. Rev. Drug Discov.*, 2005, **4**, 625.
[80] O. Rascol, *Expert Opin. Pharmacother*, 2005, **6**, 2061.
[81] F. Blandini, *CNS Drug Rev.*, 2005, **11**, 183.
[82] J. J. Chen and D. M. Swope, *J. Clin. Pharmacol.*, 2005, **45**, 878.
[83] O. Rascol, D. J. Brooks, E. Melamed, W. Oertel, W. Poewe, F. Stocchi and E. Tolosa, *Lancet*, 2005, **365**, 947.
[84] F. Hubalek, C. Binda, M. Li, Y. Herzig, J. Sterling, M. B. H. Youdim, A. Mattevi and D. E. Edmondson, *J. Med. Chem.*, 2004, **47**, 1760.
[85] C. Binda, F. Hubalek, M. Li, Y. Herzig, J. Sterling, M. B. H. Youdim, D. E. Edmondson and A. Mattevi, *J. Med. Chem.*, 2005, **48**, 8148.
[86] M. Toprakci and K. Yelekci, *Bio. Med. Chem. Lett.*, 2005, **15**, 4438.

[87] A. L. Gutman, I. Zaltzman, V. Ponomarev, M. Sotrihin and G. Nisnevich, *WO Patent* 02068376 A1, 2002.
[88] B. I. Rini, *Expert Opin. Pharmacother.*, 2006, **7**, 453.
[89] D. Strumberg, *Drugs Today*, 2005, **41**, 773.
[90] M. Beeram, A. Patnaik and E. K. Rowinsky, *J. Clin. Oncol.*, 2005, **23**, 6771.
[91] D. Strumberg, H. Richly, R. A. Hilger, N. Schleucher, S. Korfee, M. Tewes, M. Faghih, E. Brendel, D. Voliotis, C. G. Haase, B. Schwartz, A. Awada, R. Voigtmann, M. E. Scheulen and S. Seeber, *J. Clin. Oncol.*, 2005, **23**, 965.
[92] D. Bankston, J. Dumas, R. Natero, B. Riedl, M.-K. Monahan and R. Sibley, *Org. Proc. Res. Dev.*, 2002, **6**, 777.
[93] S. L. Davies, J. Castañer and L. Garcia-Capdevila, *Drugs Future*, 2005, **30**, 688.
[94] T. Sanda, T. Kuwano, S. Nakao, S. Lida, T. Ishida, H. Komatsu, K. Shudo, M. Kuwano, M. Ono and R. Ueda, *Leukemia*, 2005, **19**, 901.
[95] T. Tobita, A. Takeshita, K. Kitamura, K. Ohnishi, M. Yanagi, A. Hiraoka, T. Karasuno, M. Takeuchi, S. Miyawaki, R. Ueda, T. Naoe and R. Ohno, *Blood*, 1997, **90**, 967.
[96] K. Shudo, *Eur. Patent EP* 0170105, 1986.
[97] P.-E. Sum, V. J. Lee, R. T. Testa, J. J. Hlavka, G. A. Ellestad, J. D. Bloom, Y. Gluzman and F. P. Tally, *J. Med. Chem.*, 1994, **37**, 184.
[98] I. Chopra and M. Roberts, *Microbiol. Mol. Biol. Rev.*, 2001, **65**, 232.
[99] G. A. Noskin, *Clin. Infect. Dis.*, 2005, **41**, S303.
[100] J. Rello, *J. Chemotherapy*, 2005, **17**, 12.
[101] A. K. Meagher, P. G. Ambrose, T. H. Grasela and E. J. Ellis-Grosse, *Clin. Infect. Dis.*, 2005, **41**, S333.
[102] P.-E. Sum and P. Petersen, *Bio. Med. Chem. Lett.*, 1999, **9**, 1459.
[103] R. A. Squires and R. G. Postier, *Expert Opin. Invest. Drugs*, 2006, **15**, 155.
[104] B. Best and R. Haubrich, *Expert Opin. Invest. Drugs,*, 2006, **15**, 59.
[105] J. M. Ellis and J. Ross, *Formulary*, 2005, **40**, 104.
[106] M. Boffito, D. Maitland and A. Pozniak, *J. Clin. Pharmacol.*, 2006, **46**, 130.
[107] S. Thaisrivongs and J. Strohbach, *Biopolymers (Peptide Science)*, 1999, **51**, 51.
[108] S. R. Turner, J. W. Strohbach, R. A. Tommasi, P. A. Aristoff, P. D. Johnson, H. I. Skulnick, L. A. Dolak, E. P. Seest, P. K. Tomich, M. J. Bohanon, M.-M. Horng, J. C. Lynn, K.-T. Chong, R. R. Hinshaw, K. D. Watenpaugh, M. N. Janakiraman and S. Thaisrivongs, *J. Med. Chem.*, 1998, **41**, 3467.
[109] K. S. Fors, J. R. Gage, R. F. Heier, R. C. Kelly, W. R. Perrault and N. Wicnienski, *J. Org. Chem.*, 1998, **63**, 7348.
[110] J. J. Ferry, J. R. Baldwin and M. T. Borin, *WO Patent* 0025784, 2000.
[111] H. J. Shim, Y. C. Kim, J. H. Lee, K. J. Park, J. W. Kwon and W. B. Kim, *Biopharm. Drug Dispos.*, 2005, **26**, 161.
[112] H. Y. Ji, H. W. Lee, H. H. Kim, D. S. Kim, M. Yoo, W. B. Kim and H. S. Lee, *Xenobiotica*, 2004, **34**, 973.
[113] H. J. Shim, Y. C. Kim, J. H. Lee, J. W. Kwon, W. B. Kim, Y. G. Kim, S. H. Kim and M. G. Lee, *Biopharm. Drug Dispos.*, 2005, **26**, 269.
[114] Y. C. Kim, M. Yoo and M. G. Lee, *Drugs Future*, 2005, **30**, 678.
[115] M.-H. Yoo, W.-B. Kim, M.-S. Chang, S.-H. Kim, D.-S. Kim, C.-J. Bae, Y.-D. Kim, E.-H. Kim, *WO Patent* 0198304 A1, 2001.
[116] D. P. Wermeling, *Pharmacotherapy*, 2005, **25**, 1084.
[117] G. P. Miljanich, *Curr. Med. Chem.*, 2004, **11**, 3029.
[118] S. A. Doggrell, *Expert Opin. Invest. Drugs*, 2004, **13**, 875.
[119] D. J. Craik, N. L. Daly and J. J. Nielsen, *WO Patent* 0015654, 2000.

COMPOUND NAME, CODE NUMBER AND SUBJECT INDEX. VOL. 41

5HT2CR agonists, 88, 89
5HT6R antagonists, 90
△9-tetrahydrocannabinol, 78
A1 agonist, 175, 176
A-423579, 91
A-68930, 11
A-706149, 14
A-77636, 12
A-778193, 84
A-86929, 12
absorption kinetics, 385, 386
ACE inhibitors, 155, 157
ACE2, 192, 193
acetaminophen, 370, 371
acyl sulfonamide antiproliferatives (ASAP), 252, 253, 254, 260
adenosine, 175, 176
adenosine diphosphate (ADP), 142
adenosine triphosphate (ATP), 141
ADP, 143, 144
adrenaline, 245
adrogolide, 12
advair, 241, 245
ADX47273, 8
AG7088, 185, 186
AGI-1067, 200, 201, 206
alemcinal (ABT-229), 216
aliskiren, 158, 160, 161, 164
allergic conjunctivitis, 221, 222
allergic rhinitis, 221, 222, 224, 225, 227
allosteric modulators, 5, 8, 10
allylamines, 306
alosetron, 215
17-alpha-estradiol, 17-beta-estradiol, 52
altepase (Activase), 39
alvimopan, 217
ALX-5407, 7
amantadine, 288
ambroxol, 63
AMG-131, 108, 109
AMG-211, 135
amides, 131, 134
amino acids, 310
2-aminopyridine, 245
amiodarone, 177
amitriptyline, 59
AMPAkines, 5
angiotensin I, 156
angiotensin II, 155, 156
angiotensin receptor blockers, 155
angiotensin receptors, 156

angiotensinogen, 156, 157, 158
aniracetam, 5
5-HT_6 Antagonists, 13
antalarmin, 31
anticolinergics, 386
antifungal drug, 312
anti-mitotic, 263, 264
antiretroviral (ARV), 279
apoptosis, 40, 42, 48
aprepitant, 31
ARBs, 157, 161
AR-C68397AA, 245
ARV, 279, 280, 282
aryl sulfones, 135
AS-601245, 48
asthma, 221, 222, 223
ATC0065, 34
ATC0175, 34
ATI-7505, 213
atilmotin, 216
atopic dermatitis, 221
ATP, 142, 143, 144
avian influenza, 275, 276, 277, 278, 284
avian influenza vaccine, 278
azoles, 300, 303, 304, 306
BAL8557, 397
benzenesulfonamides, 135
benzimidazole, 244
benzodioxane, 242
beta-adrenergic agonists, 386
bezafibrate, 103, 104
biaryl blockers, 67
BILA 2157 BS, 158
BIO 1211, 206
bioactivation, 370, 372, 374, 375, 376, 377, 378, 379
biofilm, 312
BMS-601027, 266
BO-653, 201, 202
budesonide, 245
buspirone, 376
BVT-14225, 131
BVT-2733, 130, 131
BW A868C, 223
capecitabine, 401
carbamazepine, 59
carbenoxolone, 130, 131
carbetocin (duratocin), 416, 417
carmoterol, 239
caspase-3, 40, 41
caspases, 40

Ca$_V$2.2, 69
CB1R antagonists, 78, 81, 92
CCI (Bennett), 62, 63, 64
cDNAs, 331, 332, 333, 334
CDPPB, 8
celgosivir, 403
CEP-7055, 397
CFA, 62, 64
chalcones, 203, 206
chemical semantic web, 430
chemical tools, 428
chemoattractant receptor-homologous molecule expressed on Th2 cells, 222
chemogenomics, 427, 428, 431
CHF-4226, 239
chlolecystokinin CCKa receptor agonists, 215
chloroquine, 194
chlorpromazine, 177
cisapride (R 51619), 211, 212, 213
C-Jun-N-Terminal Kinase (JNK1, JNK2, JNK3), 47, 48
CK0106023, 263
classification of human cancers, 323
clozapine, 371, 372
CLX-0921, 110, 111
CMK inhibitor, 185
combivent, 245
COMPARE analysis, 253, 256
coronavirus, 183, 184, 185, 187, 189, 191, 192, 193, 194, 195
corticosteroids, 386
corticosterone, 128, 129, 130, 134
cortisol, 128, 133
cortisone, 128, 129, 133, 134
covalent binding, 370, 371, 372, 373, 379
CP-122721, 31
CP-154526, 31
CPPHA, 8
CRA0450, 32
cromakalim, 177
CRTH2, 222, 224, 225, 226, 227, 228, 229, 230, 231
CRTH2 Agonists, 225
CRTH2 Antagonists, 224, 226, 227, 229, 230, 231
CRX000143, 110, 111
CX-516, 5
CX-717, 5
cyclopenta-[e]azepine-4,10(1H,5H)-diones, 257
cyclothiazide, 6
cytochrome P450, 326, 369, 370
cytochrome P450, mechanism-based inactivation of, 378
cytokines, 46
cytotoxic, 387

D prostanoid, 222
DANA, 288, 289, 292
decursin, 257, 260
11-dehydrocorticosterone, 128, 129
dehydrocorticosterone, 129
deposition, 384, 386
desvenlafaxine, 24
dexloxiglumide, 215
DFB, 8
diamides blockers, 63
diarylquinolone, 281
diclofenac, 375, 376, 378
dihydrexidine, 11
dinapsoline, 12
dinoxyline, 12
domperidone, 212
dopamine D1/D5 receptor agonists, 11
dopamine D2 receptor antagonists, 212
dopamine D3 receptor agonists, 13
DOV-216303, 25
DOV-21947, 25
DP, 222, 223, 224, 225, 226, 227, 229, 231
DP Antagonists, 223, 224, 225, 226
DP knock-out mice, 222, 231
DP$_1$, 222
DP$_2$, 222
DPQ, 42
drug metabolizing enzymes (DMEs), 325
duloxetine, 24
echinocandins, 307
edema, 50, 51
emerging infections, 275, 284
EPC-2407 (MX-116407), 257
erythromycin, 212, 216
estrogen receptor (ER), 52
estrogens, 51, 52
ethacrynic acid, 191
excitotoxicity, 40
exubera (inhaled insulin), 383, 389
F-98214-TA, 24
farnesoid X receptor, 99
fenofibrate, 46, 103, 105, 106
fentanyl, 384
finasteride, 371
FK614, 108, 109
FLIPR, 69
fluticasone, 241, 245
flux assay, 70
FMS586, 86
formalin, 62, 63, 64, 67, 68
formamides, 240
formoterol, 237, 239, 240, 241, 245, 246
fosamprenavir, 396
FR247304, FR257516, FR197262, FR142057, 44

frequency-dependence, 61
fructose-1-phosphate (F-1-P), 144
fructose-6-phosphate (F-6-P), 144
FXR, 111, 117, 118
G-6-P, 142, 143, 144
gatifloxacin, 280, 281
geldanamycin, 50
gemfibrozil, 103, 104
gene set enrichment analysis (GSEA), 321, 322
genetic screen, 331, 332, 333
genomic data mining, 319
GHSR antagonists, 83, 84, 85
GKA, 142, 145, 146, 147, 148, 149, 150, 151, 152
GKRP, 144, 145, 146, 152
glucokinase activator (GKA), 146, 149
glucokinase regulatory protein (GKRP), 144
glucose-6-phosphate (G-6-P), 141
glucose-stimulated insulin release (GSIR), 142
glutathione adduct, 372, 373, 374, 375, 376, 377, 378, 379
glycine antagonists, 53
18β-glycyrrhetinic acid, 128
glycyrrhizin, 193
GlyT1 Inhibitors, 6
GP-6150, 43
G-protein, 222, 229
G-protein coupled receptors (GPCRs), 409, 413
GS7340, 401, 402
GSIR, 142, 144, 146, 147
guggulsterone, 117, 118
GW0742, 104, 105
GW3965, 113, 114, 115
GW4064, 117, 118
GW501516, 104, 105
GW-610, 256, 257
GW695634, 397
H3R antagonists, 91, 92
H5N1, 287, 294
halofenate, 108
HDAC inhibitors, 312
hemorrhagic stroke, 39, 40
heparin, 387, 390
hepatotoxicity, 370, 375, 376, 378
herbimycin A, 50
hexokinase, 141, 142
HR22C16, 268
human ether-a-go-go related gene (hERG), 147
human genome sequence, 338, 339
HUN-7293, 202
hydrofluoroalkane, 239
5-hydroxy indole, 10
8-hydroxycarbostyril, 238, 239, 244
7β-hydroxycholesterol, 129
hydroxypyridine-2-thione, 190

5-hydroxytryptamine, 239
hyperglycemia, 147
hypertension, 155, 156, 157, 161
hypoglycemia, 146, 152
IDRA-21, 6
IFN, 194
I_{Kr}, 169, 170, 173, 174
I_{Ks}, 169, 170, 173, 174
I_{Kur}, 169, 170, 173
imipramine, 177
InChI, 429, 430
indacaterol, 238, 239
inductive logic programming, 433
inflammation, 45, 46, 47
inflammatory cells migration, 222
influenza, 287, 288, 289, 290, 291, 292, 293, 294
INO-1001, 42
interferon, 194
interferon-alpha, 387, 389, 390
ipratropium, 245
irbesartan, 109, 110
isatin derivatives, 188
ischemic stroke, 39, 40, 43, 46, 48, 49, 51
isoprenaline, 243
ispinesib, SB-715992/CK0238273, 264
Itopride, 211, 212
kaletra, 190
KC11458, 216
KCNA5, 170
7-ketocholesterol, 129
ketones, 135
knowledge-aided drug design, 432
KR-31378, 205
KRP 297, 47
KRP-297, 105, 106
Kv1.3, 170
Kv1.5, 169, 170, 171, 172, 173
L-152804, 87
L759274, 31
L-888839, 225
lactam, 134
lamotrogine, 59
LC-MS/MS, 370, 373, 376, 377
lenalidomide, 251
leuprolide, 387, 388, 389, 390
LG100268, 111
licofelone, 205
lidocaine, 62
liver X receptor, 99
LL3858, 282
LL4858, 282
lopinavir, 190
loxiglumide, 215
Lu 36-274, 27

lung metabolism, 386
luteolin, 193
LXR, 111, 112, 113, 114, 118
LY181837, 9
LY-2121260, 145
LY2140023, 9
LY341495, 9
LY353381, 52
LY354740, 9
LY404039, 9
LY404187, 6
LY465608, 47
LY487379, 9
LY544344, 400
LY573636, 252, 255
M1/M4 agonist, 14
M-826, M-867, 40, 41
macromolecules, 386, 387, 390
maturity-onset diabetes of the young (MODY-2), 152
MB243, 82
MBX-102, 108, 109
MC4R agonists, 81, 82
MCH1R antagonists, 82
MCL0042, 33
MCL0129, 33
MDL28170, 193
ME3229, 399
metabolism, 372, 379
metabolite, reactive, 370, 371, 372, 373, 374, 375, 376, 377, 378, 379
metabotropic glutamate receptors, 7
metaglidasen, 108
methimazole, 205, 206
methoxyfluoroquinolone, 279, 280
N-methyl-D-aspartate (NMDA) receptor, 50
methylnaltrexone, 217
metoclopramide, 211, 212
mexiletine, 59
mGluR2, 9
mGluR5, 8
microRNA, 319, 323
mitotic kinesin, 263, 264, 271
MK-0767, 106
MK-499, 67
model organisms, 337, 339, 340, 347
monastrol, 267, 268, 270, 271
morphine, 384, 386
mosapride, 213
motilin receptor agonists, 216
moxifloxacin, 279, 280, 281, 282
Mtb72F/AS02A, 283
muraglitazar, 105, 106
MVA85A, 283
MX-116407 (EPC-2407), 257, 259

myopathy, 103, 105
Na/Ca exchanger (NCX), 175
naloxone, 87
naltrexone, 87
natalizumab, 206
$Na_v1.2$, 61, 63
$Na_v1.3$, 61, 69
$Na_v1.5$, 61, 63, 64
$Na_v1.7$, 61, 62, 63, 64, 66, 67
$Na_v1.8$, 61, 62, 63, 69, 70
$Na_v1.9$, 61
naveglitazar, 105, 106
NBI-30775, 32
NBI-35965, 32
NCX, 175
NCX1, 175
nefazodone, 375, 376, 377, 378
nelarabine, 251
nelfinavir, 194
neuraminidase, 288, 290, 291, 292, 293, 294
neuraminidase inhibitors, 277
SH-SY5Y neuroblastoma cells, 68
neuroprotection, 39, 40, 42, 48, 52
neuroprotective agent, 39
niclosamide, 190, 194
nitroimidazoles, 281, 282
nitroimidazo-oxazole, 282
nitroimidazopyran, 281, 282
nitrolinoleic acid, 101
nitrooleic acid, 101
S-nitroso-N-acetylpenicillamine, 193
NNC 38-1049, 91
NPR-104, 403
NPTS, 7
NPY, 86, 86
NSC-686288 (TK-2339), 257, 258
NXY-059, 40
olanzapine, 372
oncogenic addiction, 342
ONO-2231, ONO-1924H, 43
ONO-4127.Na, 224
OPC-67683, 282
open chemistry, 428, 430
opioid antagonists, 53
opioid receptor antagonists, 87
μ-opioid receptor antagonists, 217
oral glucose tolerance test (OGTT), 147, 148
orlistat, 78
oseltamivir, 277
oseltamivir, 288, 290, 293, 294
oxabispidine, 177
oxadiazole, 133
oxazoles, 135
oxazolones, 132
PA-824, 281, 282

PAT5a, 110
PD-173955, 50
PDE10A, 14
PDE4, 14
peptides, 383, 384, 385, 387
peramivir, 292, 294
permanent neonatal diabetes mellitus (PNDM), 152
peroxisome proliferator response elements, 100
peroxisome proliferator-activated receptors (PPARs), 45
peroxisome-proliferator-activated receptor, 99
persistent hyperinsulinemic hypoglycemia of infancy (PHHI), 152
PGD2, 221, 222, 223, 224, 225, 226, 227, 229, 230, 231, 232
PHA-543,613, 10
pharmacogenetics, 319, 325, 326, 328, 329
phasic block, 61
phenethyl, 242, 244
phenyloxyphenyl blockers, 65
phthalhydrazido, 187
pioglitazone, 45, 100, 101, 102, 109, 375
pivampicillin, 399
PJ34, 42
PNU, 190
PNU-120596, 10
PNU-282,987, 10
polanyi, 430
poly(ADP-ribose) polymerase-1 (PARP-1), 42, 43, 44
polyenes, 305
polypharmacology, 428
polypharmacy, 53
potassium channel activators, 53
PP, 78, 85, 86
PP1, PP2, 50
PPAR, 99, 100, 104, 107, 108, 111, 112, 118
preterm labor, 419
probucol, 199, 200, 201, 202
promazine, 190
prostaglandin D2, 221
protease, 184, 185, 186, 188, 190, 191, 194
prucalopride, 214
pumosetrag, 215
pyrazole, 133
pyrimidines, 309
pyrroles, 282
pyrrolidine, 135
pyrrolo [3,2-f]quinolin-9-one, 257
PYY, 87
PYY3-36, 85
QAB-149, 238
quetiapine, 372
quinolinones, 238

R121919, 31
R278995, 32
radicicol R3246, 50
ragaglitazar, 47, 105, 106
ralfinamide, 62
raloxifene, 374, 375, 378
ramatroban (BAY U3405), 222, 227
reactive metabolite, 355, 356, 360, 361
5-HT_{2C} receptor agonists, 14
recombinant tissue plasminogen activator (rTPA), 39
relenza, 288, 293, 294
remikiren, 158
remofovir, 402
renin receptor, 157
renzapride, 213
replicase, 184, 185
retinoid X receptor, 99
rhinovirus, 185, 188
ribavirin, 194
Riccardin C, 114
rimantadine, 288
rimonabant, 78, 79, 80
ritonavir, 190
RO 42-5892, 158
Ro 60-0175, 30
RO0094889, 401
RO0661132, 161
RO0661168, 162
RO-27-5145, 145
RO-28-0450, 146
RO-28-1674, 146
RO-28-1675, 146
rosiglitazone, 45, 46, 100, 101, 102, 105, 106, 109, 110, 111
RS 102221, 30
RXR, 99, 100, 111, 112, 118
RY764, 82
S-5751, 223
saligenin, 241, 243, 245
salmeterol, 237, 238, 239, 241, 242, 243, 245, 246
sarcosine, 7
saredutant, 31
SARS, 183, 184, 185, 186, 187, 188, 189, 190, 191, 192, 193, 194, 195
SARS-CoV 3CLpro, 185, 186, 187, 188, 189, 190, 191, 194
SARS-CoV PLpro, 192, 194
SB-242084, 30
SB-269970, 30
SB-271046, 13
SB-414796, 14
SB-649915, 28
SB-656104-A, 30

SB-743921, 264, 266, 271
SC-47921, 159
SC-51106, 158, 159
SCH23390, 11
schizophrenia, 3
selective PPARγ modulator, 108
serotonin 5HT$_3$receptor agonists, 214
serotonin 5HT$_4$ receptors agonists, 213
shRNA, 333
sibenadet, 245, 246
sibutramine, 78
signal transduction pathways, 340
siRNA, 323, 331, 333
SKF-38393, 11
SKF-812197, 11
SKI-606, 50, 51
SNAP-37889, 35
SNAP-398299, 35
SNAP-7941, 33
SNAP-94847, 34
SNL (Chung), 62
SNS-595, 251, 252, 255, 256
sodium channel blockers, 53
sorafenib, 251
sordarins, 310
SP-600125, 48
SPP100, 160
SPPARγM, 108, 109, 110
SQ 10,643, 191
src-family protein tyrosine kinases (SFKs), 50
SSR125543, 32
SSR126768A, 412, 413
SSR149415, 35
SSR180711A, 10
SSR-504734, 7
ST-280, 14
state dependence, 61
stroke, 39, 40, 42, 47, 49, 50, 52
sudoterb, 282, 283
sulfonamide, 244, 245
sulfonylureas, 147, 152
symbicort, 245
synthetic lethal interactions, 341
T0910792, 83
T0913659, 105, 104
T-131, 108, 109
T-1317, 113, 114, 116
TA-2005, 239
tacit knowledge, 430, 431, 432
tamiflu, 288, 293, 294
tamoxifen, 52
TD-5959, 246
tecadenoson, 176
tegaserod, 212, 214
telmisartan, 109, 110

tesaglitazar, 47, 106
tetrahydrocannabinoid, 384
tetra-O-galloyl-β-D-glugose, 193
tetrodotoxin, 60
TGEV 3CLpro, 185
thiadiazoles, 132
thiazoles, 131
thiazolidinediones (TZDs), 45
thiazolidinone, 100
thiazolones, 132
TIQ-A, 38, 43
TK-2339 (NSC-686288), 257
TMC207, 281
toxicity, idiosyncratic, 370, 372, 375, 379
toxicophore, 354, 355, 356, 357, 358, 360, 361, 362, 363, 365
transcriptional signature, 337
transporter, bile, 400, 403
transporter, Breast Cancer Resistant Protein, 399
transporter, monocarboxylate, 400
transporter, peptide, 400
transporter, sodium-dependent multivitamin, 401
trazodone, 376
triazole, 244
triazoles, 131, 133
N-(2-(trifluoro-methyl)-pyridin-4-yl) anthranilic ester, 257
troglitazone, 45, 100, 101
tuberculosis, 275, 278, 279, 280, 281, 282, 283, 284
tuberculosis vaccine, 283
TZD, 100, 101, 109, 110
TZD18, 106, 107
UC2, 190
V102862, 65
valacyclovir, 400
vascular endothelial growth factor (VEGF), 50
venlafaxine, 24
VER-2692, 14
veratridine, 68
VIPR, 67
WAY-161503, 30
WAY-163909, 14, 30
WIN-55,212-2, 204
XP13512, 400
Y2R agonists, 86
Y5R antagonists, 86
zafirlukast, 360
zanamivir, 277
zanamivir, 288, 289, 293, 294
zileuton, 360

CUMULATIVE CHAPTER TITLES KEYWORD INDEX, VOL. 1–41

acetylcholine receptors, 30, 41; 40, 3
acetylcholine transporter, 28, 247
acyl sulfonamide anti-proliferatives, 41, 251
adenylate cyclase, 6, 227, 233; 12, 172; 19, 293; 29, 287
adenosine, 33, 111
adenosine, neuromodulator, 18, 1; 23, 39
A_3 adenosine receptors, 38, 121
adjuvants, 9, 244
ADME by computer, 36, 257
ADME properties, 34, 307
adrenal steroidogenesis, 2, 263
adrenergic receptor antagonists, 35, 221
β-adrenergic blockers, 10, 51; 14, 81
β-adrenergic receptor agonists, 33, 193
$β_2$-adrenoceptor agonists, long acting, 41, 237
aerosol delivery, 37, 149
affinity labeling, 9, 222
$β_3$-agonists, 30, 189
AIDS, 23, 161, 253; 25, 149
AKT kinase inhibitors, 40, 263
alcohol consumption, drugs and deterrence, 4, 246
aldose reductase, 19, 169
alkaloids, 1, 311; 3, 358; 4, 322; 5, 323; 6, 274
allergic eosinophilia; 34, 61
allergy, 29, 73
alopecia, 24, 187
Alzheimer's Disease, 26, 229; 28, 49, 197, 247; 32, 11; 34, 21; 35, 31; 40, 35
Alzheimer's Disease Research, 37, 31
Alzheimer's Disease Therapies, 37, 197; 40, 35
aminocyclitol antibiotics, 12, 110
β-amyloid, 34, 21
amyloid, 28, 49; 32, 11
amyloidogenesis, 26, 229
analgesics (analgetic), 1, 40; 2, 33; 3, 36; 4, 37; 5, 31; 6, 34; 7, 31; 8, 20; 9, 11; 10, 12; 11, 23; 12, 20; 13, 41; 14, 31; 15, 32; 16, 41; 17, 21; 18, 51; 19, 1; 20, 21; 21, 21; 23, 11; 25, 11; 30, 11; 33, 11
androgen action, 21, 179; 29, 225
androgen receptor modulators, 36, 169
anesthetics, 1, 30; 2, 24; 3, 28; 4, 28; 7, 39; 8, 29; 10, 30; 31, 41
angiogenesis inhibitors, 27, 139; 32, 161
angiotensin/renin modulators, 26, 63; 27, 59
animal engineering, 29, 33
animal healthcare, 36, 319
animal models, anxiety, 15, 51
animal models, memory and learning, 12, 30
Annual Reports in Medicinal Chemistry, 25, 333

anorexigenic agents, 1, 51; 2, 44; 3, 47; 5, 40; 8, 42; 11, 200; 15, 172
antagonists, calcium, 16, 257; 17, 71; 18, 79
antagonists, GABA, 13, 31; 15, 41; 39, 11
antagonists, narcotic, 7, 31; 8, 20; 9, 11; 10, 12; 11, 23
antagonists, non-steroidal, 1, 213; 2, 208; 3, 207; 4, 199
Antagonists, PGD2, 41, 221
antagonists, steroidal, 1, 213; 2, 208; 3, 207; 4, 199
antagonists of VLA-4, 37, 65
anthracycline antibiotics, 14, 288
antiaging drugs, 9, 214
antiallergy agents, 1, 92; 2, 83; 3, 84; 7, 89; 9, 85; 10, 80; 11, 51; 12, 70; 13, 51; 14, 51; 15, 59; 17, 51; 18, 61; 19, 93; 20, 71; 21, 73; 22, 73; 23, 69; 24, 61; 25, 61; 26, 113; 27, 109
antianginals, 1, 78; 2, 69; 3, 71; 5, 63; 7, 69; 8, 63; 9, 67; 12, 39; 17, 71
anti-angiogenesis, 35, 123
antianxiety agents, 1, 1; 2, 1; 3, 1; 4, 1; 5, 1; 6, 1; 7, 6; 8, 1; 9, 1; 10, 2; 11, 13; 12, 10; 13, 21; 14, 22; 15, 22; 16, 31; 17, 11; 18, 11; 19, 11; 20, 1; 21, 11; 22, 11; 23, 19; 24, 11
antiapoptotic proteins, 40, 245
antiarrhythmic agents, 41, 169
antiarrhythmics, 1, 85; 6, 80; 8, 63; 9, 67; 12, 39; 18, 99; 21, 95; 25, 79; 27, 89
antibacterial resistance mechanisms, 28, 141
antibacterials, 1, 118; 2, 112; 3, 105; 4, 108; 5, 87; 6, 108; 17, 107; 18, .29, 113; 23, 141; 30, 101; 31, 121; 33, 141; 34, 169; 34, 227; 36, 89; 40, 301
antibacterial targets, 37, 95
antibiotic transport, 24, 139
antibiotics, 1, 109; 2, 102; 3, 93; 4, 88; 5, 75, 156; 6, 99; 7, 99, 217; 8, 104; 9, 95; 10, 109, 246; 11, 89; 11, 271; 12, 101, 110; 13, 103, 149; 14, 103; 15, 106; 17, 107; 18, 109; 21, 131; 23, 121; 24, 101; 25, 119; 37, 149
antibiotic producing organisms, 27, 129
antibodies, cancer therapy, 23, 151
antibodies, drug carriers and toxicity reversal, 15, 233
antibodies, monoclonal, 16, 243
antibody drug conjugates, 38, 229
anticancer agents, mechanical-based, 25, 129
anticancer drug resistance, 23, 265
anticoagulants, 34, 81; 36, 79; 37, 85
anticoagulant agents, 35, 83
anticoagulant/antithrombotic agents, 40, 85
anticonvulsants, 1, 30; 2, 24; 3, 28; 4, 28; 7, 39, 8, 29; 10, 30; 11, 13; 12, 10; 13, 21; 14, 22; 15, 22; 16, 31; 17, 11; 18, 11; 19, 11; 20, 11; 21, 11; 23, 19; 24, 11
antidepressants, 1, 12; 2, 11; 3, 14; 4, 13; 5, 13; 6, 15; 7, 18; 8, 11; 11, 3; 12, 1; 13, 1; 14, 1; 15, 1; 16, 1; 17, 41; 18, 41; 20, 31; 22, 21; 24, 21; 26, 23; 29, 1; 34, 1
antidepressant drugs, new, 41, 23
antidiabetics, 1, 164; 2, 176; 3, 156; 4, 164; 6, 192; 27, 219
antiepileptics, 33, 61
antifungal agents, 32, 151; 33, 173; 35, 157
antifungal drug discovery, 38, 163; 41, 299
antifungals, 2, 157; 3, 145; 4, 138; 5, 129; 6, 129; 7, 109; 8, 116; 9, 107; 10, 120; 11, 101; 13, 113; 15, 139; 17, 139; 19, 127; 22, 159; 24, 111; 25, 141; 27, 149
antiglaucoma agents, 20, 83
anti-HCV therapeutics, 34, 129; 39, 175

antihyperlipidemics, 15, 162; 18, 161; 24, 147
antihypertensives, 1, 59; 2, 48; 3, 53; 4, 47; 5, 49; 6, 52; 7, 59; 8, 52; 9, 57; 11, 61; 12, 60; 13, 71; 14, 61; 15, 79; 16, 73; 17, 61; 18, 69; 19, 61; 21, 63; 22, 63; 23, 59; 24, 51; 25, 51
antiinfective agents, 28, 119
antiinflammatory agents, 28, 109; 29, 103
anti-inflammatories, 37, 217
anti-inflammatories, non-steroidal, 1, 224; 2, 217; 3, 215; 4, 207; 5, 225; 6, 182; 7, 208; 8, 214; 9, 193; 10, 172; 13, 167; 16, 189; 23, 181
anti-ischemic agents, 17, 71
antimalarial inhibitors, 34, 159
antimetabolite cancer chemotherapies, 39, 125
antimetabolite concept, drug design, 11, 223
antimicrobial drugs—clinical problems and opportunities, 21, 119
antimicrobial potentiation, 33, 121
antimicrobial peptides, 27, 159
antimitotic agents, 34, 139
antimycobacterial agents, 31, 161
antineoplastics, 2, 166; 3, 150; 4, 154; 5, 144; 7, 129; 8, 128; 9, 139; 10, 131; 11, 110; 12, 120; 13, 120; 14, 132; 15, 130; 16, 137; 17, 163; 18, 129; 19, 137; 20, 163; 22, 137; 24, 121; 28, 167
anti-obesity agents, centrally acting, 41, 77
antiparasitics, 1, 136, 150; 2, 131, 147; 3, 126, 140; 4, 126; 5, 116; 7, 145; 8, 141; 9, 115; 10, 154; 11, 121; 12, 140; 13, 130; 14, 122; 15, 120; 16, 125; 17, 129; 19, 147; 26, 161
antiparkinsonism drugs, 6, 42; 9, 19
antiplatelet therapies, 35, 103
antipsychotics, 1, 1; 2, 1; 3, 1; 4, 1; 5, 1; 6, 1; 7, 6; 8, 1; 9, 1; 10, 2; 11, 3; 12, 1; 13, 11; 14, 12; 15, 12; 16, 11; 18, 21; 19, 21; 21, 1; 22, 1; 23, 1; 24, 1; 25, 1; 26, 53; 27, 49; 28, 39; 33, 1
antiradiation agents, 1, 324; 2, 330; 3, 327; 5, 346
anti-resorptive and anabolic bone agents, 39, 53
anti-retroviral chemotherapy, 25, 149
antiretroviral drug therapy, 32, 131
antiretroviral therapies, 35, 177; 36, 129
antirheumatic drugs, 18, 171
anti-SARS coronavirus chemistry, 41, 183
antisense oligonucleotides, 23, 295; 33, 313
antisense technology, 29, 297
antithrombotics, 7, 78; 8, 73; 9, 75; 10, 99; 12, 80; 14, 71; 17, 79; 27, 99; 32, 71
antithrombotic agents, 29, 103
antitumor agents, 24, 121
antitussive therapy, 36, 31
antiviral agents, 1, 129; 2, 122; 3, 116; 4, 117; 5, 101; 6, 118; 7, 119; 8, 150; 9, 128; 10, 161; 11, 128; 13, 139; 15, 149; 16, 149; 18, 139; 19, 117; 22, 147; 23, 161; 24, 129; 26, 133; 28, 131; 29, 145; 30, 139; 32, 141; 33, 163; 37, 133; 39, 241
antitussive therapy, 35, 53
anxiolytics, 26, 1
apoptosis, 31, 249
aporphine chemistry, 4, 331
arachidonate lipoxygenase, 16, 213
arachidonic acid cascade, 12, 182; 14, 178
arachidonic acid metabolites, 17, 203; 23, 181; 24, 71
arthritis, 13, 167; 16, 189; 17, 175; 18, 171; 21, 201; 23, 171, 181; 33, 203

arthritis, immunotherapy, 23, 171
aspartyl proteases, 36, 247
asthma, 29, 73; 32, 91
asymmetric synthesis, 13, 282
atherosclerosis, 1, 178; 2, 187; 3, 172; 4, 178; 5, 180; 6, 150; 7, 169; 8, 183; 15, 162; 18, 161; 21, 189; 24, 147; 25, 169; 28, 217; 32, 101; 34, 101; 36, 57; 40, 71
atherosclerosis HDL raising therapies, 40, 71
atherothrombogenesis, 31, 101
atrial natriuretic factor, 21, 273; 23, 101
attention deficit hyperactivity disorder, 37, 11; 39, 1
autoimmune diseases, 34, 257; 37, 217
autoreceptors, 19, 51
BACE inhibitors, 40, 35
bacterial adhesins, 26, 239
bacterial genomics, 32, 121
bacterial resistance, 13, 239; 17, 119; 32, 111
bacterial toxins, 12, 211
bacterial virulence, 30, 111
basophil degranulation, biochemistry, 18, 247
Bcl2 family, 31, 249; 33, 253
behavior, serotonin, 7, 47
benzodiazepine receptors, 16, 21
biofilm-associated infections, 39, 155
bioinformatics, 36, 201
bioisosteric groups, 38, 333
bioisosterism, 21, 283
biological factors, 10, 39; 11, 42
biological membranes, 11, 222
biological systems, 37, 279
biopharmaceutics, 1, 331; 2, 340; 3, 337; 4, 302; 5, 313; 6, 264; 7, 259; 8, 332
biosensor, 30, 275
biosimulation, 37, 279
biosynthesis, antibotics, 12, 130
biotechnology, drug discovery, 25, 289
biowarfare pathegens, 39, 165
blood-brain barrier, 20, 305; 40, 403
blood enzymes, 1, 233
bone, metabolic disease, 12, 223; 15, 228; 17, 261; 22, 169
bone metabolism, 26, 201
bradykinin-1 receptor antagonists, 38, 111
bradykinin B2 antagonists, 39, 89
brain, decade of, 27, 1
C5a antagonists, 39, 109
calcium antagonists/modulators, 16, 257; 17, 71; 18, 79; 21, 85
calcium channels, 30, 51
calmodulin antagonists, SAR, 18, 203
cancer, 27, 169; 31, 241; 34, 121; 35, 123; 35, 167
cancer chemosensitization, 37, 115
cancer chemotherapy, 29, 165; 37, 125
cancer cytotoxics, 33, 151

cancer, drug resistance, 23, 265
cancer therapy, 2, 166; 3, 150; 4, 154; 5, 144; 7, 129; 8, 128; 9, 139, 151; 10, 131; 11, 110; 12, 120; 13, 120; 14, 132; 15, 130; 16, 137; 17, 163; 18, 129; 21, 257; 23, 151; 37, 225; 39, 125
cannabinoid receptors, 9, 253; 34, 199
cannabinoid, receptors, CB1, 40, 103
carbohydrates, 27, 301
carboxylic acid, metalated, 12, 278
carcinogenicity, chemicals, 12, 234
cardiotonic agents, 13, 92; 16, 93; 19, 71
cardiovascular, 10, 61
caspases, 33, 273
catalysis, intramolecular, 7, 279
catalytic antibodies, 25, 299; 30, 255
Cathepsin K, 39, 63
CCR1 antagonists, 39, 117
CCR3 antagonists, 38, 131
cell adhesion, 29, 215
cell adhesion molecules, 25, 235
cell based mechanism screens, 28, 161
cell cycle, 31, 241; 34, 247
cell cycle kinases, 36, 139
cell invasion, 14, 229
cell metabolism, 1, 267
cell metabolism, cyclic AMP, 2, 286
cellular pathways, 37, 187
cellular responses, inflammatory, 12, 152
chemical tools, 40, 339
cheminformatics, 38, 285
chemogenomics, 38, 285
chemoinformatics, 33, 375
chemokines, 30, 209; 35, 191; 39, 117
chemotaxis, 15, 224; 17, 139, 253; 24, 233
chemotherapy of HIV, 38, 173
cholecystokinin, 18, 31
cholecystokinin agonists, 26, 191
cholecystokinin antagonists, 26, 191
cholesteryl ester transfer protein, 35, 251
chronic obstructive pulmonary disease, 37, 209
chronopharmacology, 11, 251
circadian processes, 27, 11
CNS medicines, 37, 21
CNS PET imaging agents, 40, 49
coagulation, 26, 93; 33, 81
cognition enhancers, 25, 21
cognitive disorders, 19, 31; 21, 31; 23, 29; 31, 11
collagenase, biochemistry, 25, 177
collagenases, 19, 231
colony stimulating factor, 21, 263
combinatorial chemistry, 34, 267; 34, 287
combinatorial libraries, 31, 309; 31, 319

combinatorial mixtures, 32, 261
complement cascade, 27, 199; 39, 109
complement inhibitors, 15, 193
complement system, 7, 228
conformation, nucleoside, biological activity, 5, 272
conformation, peptide, biological activity, 13, 227
conformational analysis, peptides, 23, 285
congestive heart failure, 22, 85; 35, 63
contrast media, NMR imaging, 24, 265
corticotropin-releasing factor, 25, 217; 30, 21; 34, 11
corticotropin-releasing hormone, 32, 41
cotransmitters, 20, 51
CXCR3 antagonists, 40, 215
cyclic AMP, 2, 286; 6, 215; 8, 224; 11, 291
cyclic GMP, 11, 291
cyclic nucleotides, 9, 203; 10, 192; 15, 182
cyclin-dependent kinases, 32, 171
cyclooxygenase, 30, 179
cyclooxygenase-2 inhibitors, 32, 211; 39, 99
cysteine proteases, 35, 309; 39, 63
cystic fibrosis, 27, 235; 36, 67
cytochrome P-450, 9, 290; 19, 201; 32, 295
cytokines, 27, 209; 31, 269; 34, 219
cytokine receptors, 26, 221
database searching, 3D, 28, 275
DDT-type insecticides, 9, 300
dermal wound healing, 24, 223
dermatology and dermatological agents, 12, 162; 18, 181; 22, 201; 24, 177
designer enzymes, 25, 299
diabetes, 9, 182; 11, 170; 13, 159; 19, 169; 22, 213; 25, 205; 30, 159; 33, 213; 39, 31; 40, 167
Diels-Alder reaction, intramolecular, 9, 270
dipeptidyl, peptidase 4, inhibitors, 40, 149
discovery indications, 40, 339
distance geometry, 26, 281
diuretic, 1, 67; 2, 59; 3, 62; 6, 88; 8, 83; 10, 71; 11, 71; 13, 61; 15, 100
DNA binding, sequence-specific, 27, 311; 22, 259
DNA vaccines, 34, 149
docking strategies, 28, 275
dopamine, 13, 11; 14, 12; 15, 12; 16, 11, 103; 18, 21; 20, 41; 22, 107
dopamine D_3, 29, 43
dopamine D_4, 29, 43
DPP-IV Inhibition, 36, 191
drug abuse, CNS agents, 9, 38
drug allergy, 3, 240
drug carriers, antibodies, 15, 233
drug carriers, liposomes, 14, 250
drug delivery systems, 15, 302; 18, 275; 20, 305
drug design, 34, 339
drug design, computational, 33, 397
drug design, knowledge and intelligence in, 41, 425

drug design, metabolic aspects, 23, 315
drug discovery, 17, 301; 34, ; 34, 307
drug discovery, bioactivation in, 41, 369
drug disposition, 15, 277
drug metabolism, 3, 227; 4, 259; 5, 246; 6, 205; 8, 234; 9, 290; 11, 190; 12, 201; 13, 196, 304; 14, 188; 16, 319; 17, 333; 23, 265, 315; 29, 307
drug receptors, 25, 281
drug resistance, 23, 265
drug safety, 40, 387
dynamic modeling, 37, 279
EDRF, 27, 69
elderly, drug action, 20, 295
electrospray mass spectrometry, 32, 269
electrosynthesis, 12, 309
enantioselectivity, drug metabolism, 13, 304
endorphins, 13, 41; 14, 31; 15, 32; 16, 41; 17, 21; 18, 51
endothelin, 31, 81; 32, 61
endothelin antagonism, 35, 73
endothelin antagonists, 29, 65, 30, 91
enzymatic monooxygenation reactions, 15, 207
enzyme induction, 38, 315
enzyme inhibitors, 7, 249; 9, 234; 13, 249
enzyme immunoassay, 18, 285
enzymes, anticancer drug resistance, 23, 265
enzymes, blood, 1, 233
enzymes, proteolytic inhibition, 13, 261
enzyme structure-function, 22, 293
enzymic synthesis, 19, 263; 23, 305
epitopes for antibodies, 27, 189
erectile dysfunction, 34, 71
estrogen receptor, 31, 181
ethnobotany, 29, 325
excitatory amino acids, 22, 31; 24, 41; 26, 11; 29, 53
ex-vivo approaches, 35, 299
factor VIIa, 37, 85
factor Xa, 31, 51; 34, 81
factor Xa inhibitors, 35, 83
Fc receptor structure, 37, 217
fertility control, 10, 240; 14, 168; 21, 169
filiarial nematodes, 35, 281
forskolin, 19, 293
free radical pathology, 10, 257; 22, 253
fungal nail infections, 40, 323
fungal resistance, 35, 157
G-proteins, 23, 235
G-proteins coupled receptor modulators, 37, 1
GABA, antagonists, 13, 31; 15, 41
galanin receptors, 33, 41
gamete biology, fertility control, 10, 240

gastrointestinal agents, 1, 99; 2, 91; 4, 56; 6, 68; 8, 93; 10, 90; 12, 91; 16, 83; 17, 89; 18, 89; 20, 117; 23, 201, 38, 89
gastrointestinal prokinetic agents, 41, 211
gender based medicine, 33, 355
gene expression, 32, 231
gene expression, inhibitors, 23, 295
gene targeting technology, 29, 265
gene therapy, 8, 245; 30, 219
genetically modified crops, 35, 357
gene transcription, regulation of, 27, 311
genomic data mining, 41, 319
genomics, 34, 227; 40, 349
ghrelin receptor modulators, 38, 81
glucagon, 34, 189
glucagon, mechanism, 18, 193
β-D-glucans, 30, 129
glucocorticoid receptor modulators, 37, 167
glucocorticosteroids, 13, 179
Glucokinase Activators, 41, 141
glutamate, 31, 31
glycoconjugate vaccines, 28, 257
glycogen synthase kinase-3 (GSK-3), 40, 135
glycopeptide antibiotics, 31, 131
glycoprotein IIb/IIIa antagonists, 28, 79
glycosylation, non-enzymatic, 14, 261
gonadal steroid receptors, 31, 11
gonadotropin releasing hormone, 30, 169; 39, 79
GPIIb/IIIa, 31, 91
G-Protein coupled receptor inverse agonists, 40, 373
G protein-coupled receptors, 35, 271
growth factor receptor kinases, 36, 109
growth factors, 21, 159; 24, 223; 28, 89
growth hormone, 20, 185
growth hormone secretagogues, 28, 177; 32, 221
guanylyl cyclase, 27, 245
hallucinogens, 1, 12; 2, 11; 3, 14; 4, 13; 5, 23; 6, 24
HDL cholesterol, 35, 251
health and climate change, 38, 375
heart disease, ischemic, 15, 89; 17, 71
heart failure, 13, 92; 16, 93; 22, 85
HCV antiviral agents, 39, 175
helicobacter pylori, 30, 151
hemoglobinases, 34, 159
hemorheologic agents, 17, 99
herbicides, 17, 311
heterocyclic chemistry, 14, 278
high throughput screening, 33, 293
histamine H_3 receptor agents, 33, 31; 39, 45
histone deacetylase inhibitors, 39, 145
hit-to-lead process, 39, 231

HIV co-receptors, 33, 263
HIV prevention strategies, 40, 277
HIV protease inhibitors, 26, 141; 29, 123
HIV reverse transcriptase inhibitors, 29, 123
HIV therapeutics, 40, 291
HIV vaccine, 27, 255
homeobox genes, 27, 227
hormones. glycoprotein, 12, 211
hormones. non-steroidal, 1, 191; 3, 184
hormones. peptide, 5, 210; 7, 194; 8, 204; 10, 202; 11, 158; 16, 199
hormones. steroid, 1, 213; 2, 208; 3, 207; 4, 199
host modulation, infection, 8, 160; 14, 146; 18, 149
Hsp90 inhibitors, 40, 263
5-HT2C receptor modulator, 37, 21
human gene therapy, 26, 315; 28, 267
human retrovirus regulatory proteins, 26, 171
11 β-hydroxysteroid dehydrogenase type 1 inhibitors, 41, 127
5-hydroxytryptamine, 2, 273; 7, 47; 21, 41
hypercholesterolemia, 24, 147
hypersensitivity, delayed, 8, 284
hypersensitivity, immediate, 7, 238; 8, 273
hypertension, 28, 69
hypertension, etiology, 9, 50
hypnotics, 1, 30; 2, 24; 3, 28; 4, 28; 7, 39; 8, 29; 10, 30; 11, 13; 12, 10; 13, 21; 14, 22; 15, 22, 16; 31; 17, 11; 18, 11; 19, 11; 22, 11
ICE gene family, 31, 249
IgE, 18, 247
Immune cell signaling, 38, 275
immune mediated idiosyncratic drug hypersensitivity, 26, 181
immune system, 35, 281
immunity, cellular mediated, 17, 191; 18, 265
immunoassay, enzyme, 18, 285
immunomodulatory proteins, 35, 281
immunophilins, 28, 207
immunostimulants, arthritis, 11, 138; 14, 146
immunosuppressants, 26, 211; 29, 175
immunosuppressive drug action, 28, 207
immunosuppressives, arthritis, 11, 138
immunotherapy, cancer, 9, 151; 23, 151
immunotherapy, infectious diseases, 18, 149; 22, 127
immunotherapy, inflammation, 23, 171
infections, sexually transmitted, 14, 114
infectious disease strategies, 41, 279
inflammation, 22, 245; 31, 279
inflammation, immunomodulatory approaches, 23, 171
inflammation, proteinases in, 28, 187
inflammatory bowel disease, 24, 167; 38, 141
inhibitors, anti-apoptotic proteins, 40, 245
inhibitors, complement, 15, 193
inhibitors, connective tissue, 17, 175

inhibitors, dipeptidyl peptidase 4, 40, 149
inhibitors, enzyme, 13, 249
inhibitors, influenza neuraminidase, 41, 287
inhibitors, irreversible, 9, 234; 16, 289
inhibitors, mitotic kinesin, 41, 263
inhibitors, platelet aggregation, 6, 60
inhibitors, proteolytic enzyme, 13, 261
inhibitors, renin, 41, 155
inhibitors, renin-angiotensin, 13, 82
inhibitors, reverse transcription, 8, 251
inhibitors, transition state analogs, 7, 249
inorganic chemistry, medicinal, 8, 294
inosine monophosphate dehydrogenase, 35, 201
inositol triphosphate receptors, 27, 261
insecticides, 9, 300; 17, 311
insulin, mechanism, 18, 193
integrins, 31, 191
β_2 –integrin Antagonist, 36, 181
integrin alpha 4 beta 1 (VLA-4), 34, 179
intellectual property, 36, 331
interferon, 8, 150; 12, 211; 16, 229; 17, 151
interleukin-1, 20, 172; 22, 235; 25, 185; 29, 205, 33, 183
interleukin-2, 19, 191
interoceptive discriminative stimuli, animal model of anxiety, 15, 51
intracellular signaling targets, 37, 115
intramolecular catalysis, 7, 279
ion channel modulators, 37, 237
ion channels, ligand gated, 25, 225
ion channels, voltage-gated, 25, 225
ionophores, monocarboxylic acid, 10, 246
ionotropic GABA receptors, 39, 11
iron chelation therapy, 13, 219
irreversible ligands, 25, 271
ischemia/reperfusion, CNS, 27, 31
ischemic injury, CNS, 25, 31
isotopes, stable, 12, 319; 19, 173
JAKs, 31, 269
β-lactam antibiotics, 11, 271; 12, 101; 13, 149; 20, 127, 137; 23, 121; 24, 101
β-lactamases, 13, 239; 17, 119
ketolide antibacterials, 35, 145
LDL cholesterol, 35, 251
learning, 3, 279; 16, 51
leptin, 32, 21
leukocyte elastase inhibitors, 29, 195
leukocyte motility, 17, 181
leukotriene biosynthesis inhibitors, 40, 199
leukotriene modulators, 32, 91
leukotrienes, 17, 291; 19, 241; 24, 71
LHRH, 20, 203; 23, 211
lipid metabolism, 9, 172; 10, 182; 11, 180; 12, 191; 13, 184; 14, 198; 15, 162

lipoproteins, 25, 169
liposomes. 14, 250
lipoxygenase, 16, 213; 17, 203
lymphocytes, delayed hypersensitivity, 8, 284
macrocyclic immunomodulators, 25, 195
macrolide antibacterials, 35, 145
macrolide antibiotics, 25, 119
macrophage migration inhibitor factor, 33, 243
magnetic resonance, drug binding, 11, 311
malaria, 31, 141; 34, 349, 38, 203
male contraception, 32, 191
managed care, 30, 339
MAP kinase, 31, 289
market introductions, 19, 313; 20, 315; 21, 323; 22, 315; 23, 325; 24, 295; 25, 309; 26, 297; 27, 321; 28, 325; 29, 331; 30, 295; 31, 337; 32, 305; 33, 327
mass spectrometry, 31, 319; 34, 307
mass spectrometry, of peptides, 24, 253
mass spectrometry, tandem, 21, 213; 21, 313
mast cell degranulation, biochemistry, 18, 247
matrix metalloproteinase, 37, 209
matrix metalloproteinase inhibitors, 35, 167
mechanism based, anticancer agents, 25, 129
mechanism, drug allergy, 3, 240
mechanisms of antibiotic resistance, 7, 217; 13, 239; 17, 119
medicinal chemistry, 28, 343; 30, 329; 33, 385; 34, 267
melanin-concentrating hormone, 40, 119
melanocortin-4 receptor, 38, 31
melatonin. 32, 31
melatonin agonists, 39, 21
membrane function, 10, 317
membrane regulators, 11, 210
membranes, active transport, 11, 222
memory, 3, 279; 12, 30; 16, 51
metabolism, cell, 1, 267; 2, 286
metabolism, drug, 3, 227; 4, 259; 5, 246; 6, 205; 8, 234; 9, 290; 11, 190; 12, 201; 13, 196, 304; 14, 188; 23, 265, 315
metabolism, lipid, 9, 172; 10, 182; 11, 180; 12, 191; 14, 198
metabolism, mineral, 12, 223
metabonomics, 40, 387
metabotropic glutamate receptor, 35, 1, 38, 21
metal carbonyls, 8, 322
metalloproteinases, 31, 231; 33, 131
metals, disease, 14, 321
metastasis, 28, 151
microbial genomics, 37, 95
microbial products screening, 21, 149
microtubule stabilizing agents, 37, 125
microwave-assisted chemistry, 37, 247
migraine, 22, 41; 32, 1
mitogenic factors, 21, 237

mitotic kinesin inhibitors, 39, 135
modified serum lipoproteins, 25, 169
molecular diversity, 26, 259, 271; 28, 315; 34, 287
molecular modeling, 22, 269; 23, 285
monoclonal antibodies, 16, 243; 27, 179; 29, 317
monoclonal antibody cancer therapies, 28, 237
monoxygenases, cytochrome P-450, 9, 290
multi-factorial diseases, basis of, 41, 337
multivalent ligand design, 35, 321
muscarinic agonists/antagonists, 23, 81; 24, 31; 29, 23
muscle relaxants, 1, 30; 2, 24; 3, 28; 4, 28; 8, 37
muscular disorders, 12, 260
mutagenicity, mutagens, 12, 234
mutagenesis, SAR of proteins, 18, 237
myocardial ischemia, acute, 25, 71
narcotic antagonists, 7, 31; 8, 20; 9, 11; 10, 12; 11, 23; 13, 41
natriuretic agents, 19, 253
natural products, 6, 274; 15, 255; 17, 301; 26, 259; 32, 285
natural killer cells, 18, 265
neoplasia, 8, 160; 10, 142
neurodegeneration, 30, 31
neurodegenerative disease, 28, 11
neurokinin antagonists, 26, 43; 31, 111; 32, 51; 33, 71; 34, 51
neurological disorders, 31, 11
neuronal calcium channels, 26, 33
neuronal cell death, 29, 13
neuropathic pain, 38, 1
neuropeptides, 21, 51; 22, 51
neuropeptide Y, 31, 1; 32, 21; 34, 31
neuropeptide Y receptor modulators, 38, 61
neuropeptide receptor antagonists, 38, 11
neuroprotection, 29, 13
neuroprotective agents, 41, 39
neurotensin, 17, 31
neurotransmitters, 3, 264; 4, 270; 12, 249; 14, 42; 19, 303
neutrophic factors, 25, 245; 28, 11
neutrophil chemotaxis, 24, 233
nicotinic acetylcholine receptor, 22, 281; 35, 41
nicotinic acetylcholine receptor modulators, 40, 3
nitric oxide synthase, 29, 83; 31, 221
NMR, 27, 271
NMR in biological systems, 20, 267
NMR imaging, 20, 277; 24, 265
NMR methods, 31, 299
NMR, protein structure determination, 23, 275
non-enzymatic glycosylation, 14, 261
non-HIV antiviral agents, 36, 119; 38, 213
non-nutritive, sweeteners, 17, 323
non-peptide agonists, 32, 277
non-peptidic δ-opinoid agonists, 37, 159

non-steroidal antiinflammatories, 1, 224; 2, 217; 3, 215; 4, 207; 5, 225; 6, 182; 7, 208; 8, 214; 9, 193; 10, 172; 13, 167; 16, 189
novel analgesics, 35, 21
NSAIDs, 37, 197
nuclear orphan receptors, 32, 251
nucleic acid-drug interactions, 13, 316
nucleic acid, sequencing, 16, 299
nucleic acid, synthesis, 16, 299
nucleoside conformation, 5, 272
nucleosides, 1, 299; 2, 304; 3, 297; 5, 333; 39, 241
nucleotide metabolism, 21, 247
nucleotides, 1, 299; 2, 304; 3, 297; 5, 333; 39, 241
nucleotides, cyclic, 9, 203; 10, 192; 15, 182
obesity, 1, 51; 2, 44; 3, 47; 5, 40; 8, 42; 11, 200; 15, 172; 19, 157; 23, 191; 31, 201; 32, 21
obesity therapeutics, 38, 239
obesity treatment, 37, 1
oligomerisation, 35, 271
oligonucleotides, inhibitors, 23, 295
oncogenes, 18, 225; 21, 159, 237
opioid receptor, 11, 33; 12, 20; 13, 41; 14, 31; 15, 32; 16, 41; 17, 21; 18, 51; 20, 21; 21, 21
opioids, 12, 20; 16, 41; 17, 21; 18, 51; 20, 21; 21, 21
opportunistic infections, 29, 155
oral pharmacokinetics, 35, 299
organocopper reagents, 10, 327
osteoarthritis, 22, 179
osteoporosis, 22, 169; 26, 201; 29, 275; 31, 211
oxazolidinone antibacterials, 35, 135
oxytocin antagonists and agonists, 41, 409
P38α MAP kinase, 37, 177
P-glycoprotein, multidrug transporter, 25, 253
parallel synthesis, 34, 267
parasite biochemistry, 16, 269
parasitic infection, 36, 99
patents in medicinal chemistry, 22, 331
pathophysiology, plasma membrane, 10, 213
PDE IV inhibitors, 31, 71
PDE7 inhibitors, 40, 227
penicillin binding proteins, 18, 119
peptic ulcer, 1, 99; 2, 91; 4, 56; 6, 68; 8, 93; 10, 90; 12, 91; 16, 83; 17, 89; 18, 89; 19, 81; 20, 93; 22, 191; 25, 159
peptide-1, 34, 189
peptide conformation, 13, 227; 23, 285
peptide hormones, 5, 210; 7, 194; 8, 204; 10, 202; 11, 158, 19, 303
peptide hypothalamus, 7, 194; 8, 204; 10, 202; 16, 199
peptide libraries, 26, 271
peptide receptors, 25, 281; 32, 277
peptide, SAR, 5, 266
peptide stability, 28, 285

peptide synthesis, 5, 307; 7, 289; 16, 309
peptide synthetic, 1, 289; 2, 296
peptide thyrotropin, 17, 31
peptidomimetics, 24, 243
periodontal disease, 10, 228
peroxisome proliferator – activated receptors, 38, 71
PET, 24, 277
PET imaging agents, 40, 49
PET ligands, 36, 267
pharmaceutics, 1, 331; 2, 340; 3, 337; 4, 302; 5, 313; 6, 254, 264; 7, 259; 8, 332
pharmaceutical innovation, 40, 431
pharmaceutical productivity, 38, 383
pharmaceutical proteins, 34, 237
pharmacogenetics, 35, 261; 40, 417
pharmacogenomics, 34, 339
pharmacokinetics, 3, 227, 337; 4, 259, 302; 5, 246, 313; 6, 205; 8, 234; 9, 290; 11, 190; 12, 201; 13, 196, 304; 14, 188, 309; 16, 319; 17, 333
pharmacophore identification, 15, 267
pharmacophoric pattern searching, 14, 299
phosphodiesterase, 31, 61
phosphodiesterase 4 inhibitors, 29, 185; 33, 91; 36, 41
phosphodiesterase 5 inhibitors, 37, 53
phospholipases, 19, 213; 22, 223; 24, 157
physicochemical parameters, drug design, 3, 348; 4, 314; 5, 285
pituitary hormones, 7, 194; 8, 204; 10, 202
plants, 34, 237
plasma membrane pathophysiology, 10, 213
plasma protein binding, 31, 327
plasminogen activator, 18, 257; 20, 107; 23, 111; 34, 121
plasmon resonance, 33, 301
platelet activating factor (PAF), 17, 243; 20, 193; 24, 81
platelet aggregation, 6, 60
polyether antibiotics, 10, 246
polyamine metabolism, 17, 253
polyamine spider toxins, 24, 287
polymeric reagents, 11, 281
positron emission tomography, 24, 277, 25, 261
potassium channel activators, 26, 73
potassium channel antagonists, 27, 89
potassium channel blockers, 32, 181
potassium channel openers, 24, 91, 30, 81
potassium channel modulators, 36, 11
potassium channels, 37, 237
privileged structures, 35, 289
prodrugs, 10, 306; 22, 303
prodrug discovery, oral, 41, 395
profiling of compound libraries, 36, 277
programmed cell death, 30, 239
prolactin secretion, 15, 202
prostacyclin, 14, 178

prostaglandins, 3, 290; 5, 170; 6, 137; 7, 157; 8, 172; 9, 162; 11, 80
prostanoid receptors, 33, 223
prostatic disease, 24, 197
proteases, 28, 151
proteasome, 31, 279
protein C, 29, 103
protein growth factors, 17, 219
proteinases, arthritis, 14, 219
protein kinases, 18, 213; 29, 255
protein kinase C, 20, 227; 23, 243
protein phosphatases, 29, 255
protein-protein interactions, 38, 295
protein structure determination, NMR, 23, 275
protein structure modeling, 39, 203
protein structure prediction, 36, 211
protein structure project, 31, 357
protein tyrosine kinases, 27, 169
protein tyrosine phosphatase, 35, 231
proteomics, 36, 227
psoriasis, 12, 162; 32, 201
psychiatric disorders, 11, 42
psychoses, biological factors, 10, 39
psychotomimetic agents, 9, 27
pulmonary agents, 1, 92; 2, 83; 3, 84; 4, 67; 5, 55; 7, 89; 9, 85; 10, 80; 11, 51; 12, 70; 13, 51; 14, 51; 15, 59; 17, 51; 18, 61; 20, 71; 21, 73; 22, 73; 23, 69; 24, 61; 25, 61; 26, 113; 27, 109
pulmonary disease, 34, 111
pulmonary hypertension, 37, 41
pulmonary inflammation, 31, 71
pulmonary inhalation technology, 41, 383
purine and pyrimide nucleotide (P2) receptors, 37, 75
purine-binding enzymes, 38, 193
purinoceptors, 31, 21
QT interval prolongation, 39, 255
quantitative SAR, 6, 245; 8, 313; 11, 301; 13, 292; 17, 281
quinolone antibacterials, 21, 139; 22, 117; 23, 133
radioimmunoassays, 10, 284
radioisotope labeled drugs, 7, 296
radioimaging agents, 18, 293
radioligand binding, 19, 283
radiosensitizers, 26, 151
ras farnesyltransferase, 31, 171
ras GTPase, 26, 249
ras oncogene, 29, 165
receptor binding, 12, 249
receptor mapping, 14, 299; 15, 267; 23, 285
receptor modeling, 26, 281
receptor modulators, nuclear hormone, 41, 99
receptor, concept and function, 21, 211
receptors, acetylcholine, 30, 41
receptors, adaptive changes, 19, 241

receptors, adenosine, 28, 295; 33, 111
receptors, adrenergic, 15, 217
receptors, β-adrenergic blockers, 14, 81
receptors, benzodiazepine, 16, 21
receptors, cell surface, 12, 211
receptors, drug, 1, 236; 2, 227; 8, 262
receptors, G-protein coupled, 23, 221, 27, 291,
receptors, G-protein coupled CNS, 28, 29
receptors, histamine, 14, 91
receptors, muscarinic, 24, 31
receptors, neuropeptide, 28, 59
receptors, neuronal BZD, 28, 19
receptors, neurotransmitters, 3, 264; 12, 249
receptors, neuroleptic, 12, 249
receptors, opioid, 11, 33; 12, 20; 13, 41; 14, 31; 15, 32; 16, 41; 17, 21
receptors, peptide, 25, 281
receptors, serotonin, 23, 49
receptors, sigma, 28, 1
recombinant DNA, 17, 229; 18, 307; 19, 223
recombinant therapeutic proteins, 24, 213
renal blood flow, 16, 103
renin, 13, 82; 20, 257
reperfusion injury, 22, 253
reproduction, 1, 205; 2, 199; 3, 200; 4, 189
resistant organisms, 34, 169
respiratory tract infections, 38, 183
retinoids, 30, 119
reverse transcription, 8, 251
RGD-containing proteins, 28, 227
rheumatoid arthritis, 11, 138; 14, 219; 18, 171; 21, 201; 23, 171, 181
ribozymes, 30, 285
RNAi, 38, 261
SAR, quantitative, 6, 245; 8, 313; 11, 301; 13, 292; 17, 291
same brain, new decade, 36, 1
schizophrenia, treatment of, 41, 3
secretase inhibitors, 35, 31; 38, 41
sedative-hypnotics, 7, 39; 8, 29; 11, 13; 12, 10; 13, 21; 14, 22; 15, 22; 16, 31; 17, 11; 18, 11; 19, 11; 22, 11
sedatives, 1, 30; 2, 24; 3, 28; 4, 28; 7, 39; 8, 29; 10, 30; 11, 13; 12, 10; 13, 21; 14, 22; 15; 22; 16, 31; 17, 11; 18, 11; 20, 1; 21, 11
sequence-defined oligonucleotides, 26, 287
serine proteases, 32, 71
SERMs, 36, 149
serotonergics, central, 25, 41; 27, 21
serotonergics, selective, 40, 17
serotonin, 2, 273; 7, 47; 26, 103; 30, 1; 33, 21
serotonin receptor, 35, 11
serum lipoproteins, regulation, 13, 184
sexually-transmitted infections, 14, 114
SH2 domains, 30, 227

SH3 domains, 30, 227
silicon, in biology and medicine, 10, 265
sickle cell anemia, 20, 247
signal transduction pathways, 33, 233
skeletal muscle relaxants, 8, 37
sleep, 27, 11; 34, 41
slow-reacting substances, 15, 69; 16, 213; 17, 203, 291
SNPs, 38, 249
sodium/calcium exchange, 20, 215
sodium channel blockers, 41, 59
sodium channels, 33, 51
solid-phase synthesis, 31, 309
solid state organic chemistry, 20, 287
solute active transport, 11, 222
somatostatin, 14, 209; 18, 199; 34, 209
spider toxins, 24, 287
SRS, 15, 69; 16, 213; 17, 203, 291
Statins, 37, 197; 39, 187
Statins, pleiotropic effects of, 39, 187
STATs, 31, 269
stereochemistry, 25, 323
steroid hormones, 1, 213; 2, 208; 3, 207; 4, 199
stroidogenesis, adrenal, 2, 263
steroids, 2, 312; 3, 307; 4, 281; 5, 192, 296; 6, 162; 7, 182; 8, 194; 11, 192
stimulants, 1, 12; 2, 11; 3, 14; 4, 13; 5, 13; 6, 15; 7, 18; 8, 11
stroke, pharmacological approaches, 21, 108
stromelysin, biochemistry, 25, 177
structural genomics, 40, 349
structure-based drug design, 27, 271; 30, 265; 34, 297
substance P, 17, 271; 18, 31
substituent constants, 2, 347
suicide enzyme inhibitors, 16, 289
superoxide dismutases, 10, 257
superoxide radical, 10, 257
sweeteners, non-nutritive, 17, 323
synthesis, asymmetric, 13, 282
synthesis, computer-assisted, 12, 288; 16, 281; 21, 203
synthesis, enzymic, 23, 305
T-cells, 27, 189; 30, 199; 34, 219
tachykinins, 28, 99
target identification, 41, 331
taxol, 28, 305
technology, providers and integrators, 33, 365
tetracyclines, 37, 105
thalidomide, 30, 319
therapeutic antibodies, 36, 237
thrombin, 30, 71; 31, 51; 34, 81
thrombolytic agents, 29, 93
thrombosis, 5, 237; 26, 93; 33, 81
thromboxane receptor antagonists, 25, 99

thromboxane synthase inhibitors, 25, 99
thromboxane synthetase, 22, 95
thromboxanes, 14, 178
thyrotropin releasing hormone, 17, 31
tissue factor pathway, 37, 85
TNF-α, 32, 241
TNF-α converting enzyme, 38, 153
topical microbicides, 40, 277
topoisomerase, 21, 247
toxicity, mathematical models, 18, 303
toxicity reversal, 15, 233
toxicity, structure activity relationships for, 41, 353
toxicology, comparative, 11, 242; 33, 283
toxins, bacterial, 12, 211
transcription factor NF-κB, 29, 235
transcription, reverse, 8, 251
transgenic animals, 24, 207
transgenic technology, 29, 265
translational control, 29, 245
transporters, drug, 39, 219
traumatic injury, CNS, 25, 31
trophic factors, CNS, 27, 41
TRPV1 vanilloid receptor, 40, 185
tumor classification, 37, 225
tumor necrosis factor, 22, 235
type 2 diabetes, 35, 211; 40, 167
tyrosine kinase, 30, 247; 31, 151
urinary incontinence, 38, 51
urokinase-type plasminogen activator, 34, 121
urotensin-II receptor modulators, 38, 99
vanilloid receptor, 40, 185
vascular cell adhesion molecule-1, 41, 197
vascular proliferative diseases, 30, 61
vasoactive peptides, 25, 89; 26, 83; 27, 79
vasoconstrictors, 4, 77
vasodilators, 4, 77; 12, 49
vasopressin antagonists, 23, 91
vasopressin receptor modulators, 36, 159
veterinary drugs, 16, 161
viruses, 14, 238
vitamin D, 10, 295; 15, 288; 17, 261; 19, 179
waking functions, 10, 21
water, structures, 5, 256
wound healing, 24, 223
xenobiotics, cyclic nucleotide metabolism, 15, 182
xenobiotic metabolism, 23, 315
X-ray crystallography, 21, 293; 27, 271

CUMULATIVE NCE INTRODUCTION INDEX, 1983–2005

GENERAL NAME	INDICATION	YEAR INTRO.	ARMC VOL., PAGE
abacavir sulfate	antiviral	1999	35, 333
abarelix	anticancer	2004	40, 446
acarbose	antidiabetic	1990	26, 297
aceclofenac	antiinflammatory	1992	28, 325
acemannan	wound healing agent	2001	37, 259
acetohydroxamic acid	hypoammonuric	1983	19, 313
acetorphan	antidiarrheal	1993	29, 332
acipimox	hypolipidemic	1985	21, 323
acitretin	antipsoriatic	1989	25, 309
acrivastine	antihistamine	1988	24, 295
actarit	antirheumatic	1994	30, 296
adalimumab	rheumatoid arthritis	2003	39, 267
adamantanium bromide	antiseptic	1984	20, 315
adefovir dipivoxil	antiviral	2002	38, 348
adrafinil	psychostimulant	1986	22, 315
AF-2259	antiinflammatory	1987	23, 325
afloqualone	muscle relaxant	1983	19, 313
agalsidase alfa	fabry's disease	2001	37, 259
alacepril	antihypertensive	1988	24, 296
alclometasone dipropionate	topical antiinflammatory	1985	21, 323
alefacept	plaque psoriasis	2003	39, 267
alemtuzumab	anticancer	2001	37, 260
alendronate sodium	osteoporosis	1993	29, 332
alfentanil HCl	analgesic	1983	19, 314
alfuzosin HCl	antihypertensive	1988	24, 296
alglucerase	enzyme	1991	27, 321
alitretinoin	anticancer	1999	35, 333
alminoprofen	analgesic	1983	19, 314
almotriptan	antimigraine	2000	36, 295
anakinra	antiarthritic	2001	37, 261
alosetron hydrochloride	irritable bowel syndrome	2000	36, 295
alpha-1 antitrypsin	protease inhibitor	1988	24, 297
alpidem	anxiolytic	1991	27, 322
alpiropride	antimigraine	1988	24, 296
alteplase	thrombolytic	1987	23, 326
amfenac sodium	antiinflammatory	1986	22, 315
amifostine	cytoprotective	1995	31, 338
aminoprofen	topical antiinflammatory	1990	26, 298
amisulpride	antipsychotic	1986	22, 316

GENERAL NAME	INDICATION	YEAR INTRO.	ARMC VOL., PAGE
amlexanox	antiasthmatic	1987	23, 327
amlodipine besylate	antihypertensive	1990	26, 298
amorolfine HCl	topical antifungal	1991	27, 322
amosulalol	antihypertensive	1988	24, 297
ampiroxicam	antiinflammatory	1994	30, 296
amprenavir	antiviral	1999	35, 334
amrinone	cardiotonic	1983	19, 314
amrubicin HCl	antineoplastic	2002	38, 349
amsacrine	antineoplastic	1987	23, 327
amtolmetin guacil	antiinflammatory	1993	29, 332
anagrelide HCl	hematological	1997	33, 328
anastrozole	antineoplastic	1995	31, 338
angiotensin II	anticancer adjuvant	1994	30, 296
aniracetam	cognition enhancer	1993	29, 333
anti-digoxin polyclonal antibody	antidote	2002	38, 350
APD	calcium regulator	1987	23, 326
apraclonidine HCl	antiglaucoma	1988	24, 297
aprepitant	antiemetic	2003	39, 268
APSAC	thrombolytic	1987	23, 326
aranidipine	antihypertensive	1996	32, 306
arbekacin	antibiotic	1990	26, 298
argatroban	antithrombotic	1990	26, 299
arglabin	anticancer	1999	35, 335
aripiprazole	neuroleptic	2002	38, 350
arotinolol HCl	antihypertensive	1986	22, 316
arteether	antimalarial	2000	36, 296
artemisinin	antimalarial	1987	23, 327
aspoxicillin	antibiotic	1987	23, 328
astemizole	antihistamine	1983	19, 314
astromycin sulfate	antibiotic	1985	21, 324
atazanavir	antiviral	2003	39, 269
atomoxetine	attention deficit hyperactivity disorder	2003	39, 270
atorvastatin calcium	dyslipidemia	1997	33, 328
atosiban	preterm labor	2000	36, 297
atovaquone	antiparasitic	1992	28, 326
auranofin	chrysotherapeutic	1983	19, 143
azacitidine	anticancer	2004	40, 447
azelnidipine	antihypertensive	2003	39, 270
azelaic acid	antiacne	1989	25, 310
azelastine HCl	antihistamine	1986	22, 316
azithromycin	antibiotic	1988	24, 298
azosemide	diuretic	1986	22, 316
aztreonam	antibiotic	1984	20, 315
balofloxacin	antibacterial	2002	38, 351

Cumulative NCE Introduction Index, 1983–2005 505

GENERAL NAME	INDICATION	YEAR INTRO.	ARMC VOL., PAGE
balsalazide disodium	ulcerative colitis	1997	33, 329
bambuterol	bronchodilator	1990	26, 299
barnidipine HCl	antihypertensive	1992	28, 326
beclobrate	hypolipidemic	1986	22, 317
befunolol HCl	antiglaucoma	1983	19, 315
belotecan	anticancer	2004	40, 449
benazepril HCl	antihypertensive	1990	26, 299
benexate HCl	antiulcer	1987	23, 328
benidipine HCl	antihypertensive	1991	27, 322
beraprost sodium	platelet aggreg. inhibitor	1992	28, 326
betamethasone butyrate prospinate	topical antiinflammatory	1994	30, 297
betaxolol HCl	antihypertensive	1983	19, 315
betotastine besilate	antiallergic	2000	36, 297
bevacizumab	anticancer	2004	40, 450
bevantolol HCl	antihypertensive	1987	23, 328
bexarotene	anticancer	2000	36, 298
biapenem	antibacterial	2002	38, 351
bicalutamide	antineoplastic	1995	31, 338
bifemelane HCl	nootropic	1987	23, 329
bimatoprost	antiglaucoma	2001	37, 261
binfonazole	hypnotic	1983	19, 315
binifibrate	hypolipidemic	1986	22, 317
bisantrene HCl	antineoplastic	1990	26, 300
bisoprolol fumarate	antihypertensive	1986	22, 317
bivalirudin	antithrombotic	2000	36, 298
bopindolol	antihypertensive	1985	21, 324
bortezomib	anticancer	2003	39, 271
bosentan	antihypertensive	2001	37, 262
brimonidine	antiglaucoma	1996	32, 306
brinzolamide	antiglaucoma	1998	34, 318
brodimoprin	antibiotic	1993	29, 333
bromfenac sodium	NSAID	1997	33, 329
brotizolam	hypnotic	1983	19, 315
brovincamine fumarate	cerebral vasodilator	1986	22, 317
bucillamine	immunomodulator	1987	23, 329
bucladesine sodium	cardiostimulant	1984	20, 316
budipine	antiParkinsonian	1997	33, 330
budralazine	antihypertensive	1983	19, 315
bulaquine	antimalarial	2000	36, 299
bunazosin HCl	antihypertensive	1985	21, 324
bupropion HCl	antidepressant	1989	25, 310
buserelin acetate	hormone	1984	20, 316
buspirone HCl	anxiolytic	1985	21, 324
butenafine HCl	topical antifungal	1992	28, 327
butibufen	antiinflammatory	1992	28, 327

GENERAL NAME	INDICATION	YEAR INTRO.	ARMC VOL., PAGE
butoconazole	topical antifungal	1986	22, 318
butoctamide	hypnotic	1984	20, 316
butyl flufenamate	topical antiinflammatory	1983	19, 316
cabergoline	antiprolactin	1993	29, 334
cadexomer iodine	wound healing agent	1983	19, 316
cadralazine	hypertensive	1988	24, 298
calcipotriol	antipsoriatic	1991	27, 323
camostat mesylate	antineoplastic	1985	21, 325
candesartan cilexetil	antihypertension	1997	33, 330
capecitabine	antineoplastic	1998	34, 319
captopril	antihypertensive agent	1982	13, 086
carboplatin	antibiotic	1986	22, 318
carperitide	congestive heart failure	1995	31, 339
carumonam	antibiotic	1988	24, 298
carvedilol	antihypertensive	1991	27, 323
caspofungin acetate	antifungal	2001	37, 263
cefbuperazone sodium	antibiotic	1985	21, 325
cefcapene pivoxil	antibiotic	1997	33, 330
cefdinir	antibiotic	1991	27, 323
cefditoren pivoxil	oral cephalosporin	1994	30, 297
cefepime	antibiotic	1993	29, 334
cefetamet pivoxil HCl	antibiotic	1992	28, 327
cefixime	antibiotic	1987	23, 329
cefmenoxime HCl	antibiotic	1983	19, 316
cefminox sodium	antibiotic	1987	23, 330
cefodizime sodium	antibiotic	1990	26, 300
cefonicid sodium	antibiotic	1984	20, 316
ceforanide	antibiotic	1984	20, 317
cefoselis	antibiotic	1998	34, 319
cefotetan disodium	antibiotic	1984	20, 317
cefotiam hexetil HCl	antibiotic	1991	27, 324
cefozopran HCl	injectable cephalosporin	1995	31, 339
cefpimizole	antibiotic	1987	23, 330
cefpiramide sodium	antibiotic	1985	21, 325
cefpirome sulfate	antibiotic	1992	28, 328
cefpodoxime proxetil	antibiotic	1989	25, 310
cefprozil	antibiotic	1992	28, 328
ceftazidime	antibiotic	1983	19, 316
cefteram pivoxil	antibiotic	1987	23, 330
ceftibuten	antibiotic	1992	28, 329
cefuroxime axetil	antibiotic	1987	23, 331
cefuzonam sodium	antibiotic	1987	23, 331
celecoxib	antiarthritic	1999	35, 335
celiprolol HCl	antihypertensive	1983	19, 317
centchroman	antiestrogen	1991	27, 324
centoxin	immunomodulator	1991	27, 325

GENERAL NAME	INDICATION	YEAR INTRO.	ARMC VOL., PAGE
cerivastatin	dyslipidemia	1997	33, 331
cetirizine HCl	antihistamine	1987	23, 331
cetrorelix	female infertility	1999	35, 336
cetuximab	anticancer	2003	39, 272
cevimeline hydrochloride	anti-xerostomia	2000	36, 299
chenodiol	anticholelithogenic	1983	19, 317
CHF-1301	antiparkinsonian	1999	35, 336
choline alfoscerate	nootropic	1990	26, 300
cibenzoline	antiarrhythmic	1985	21, 325
ciclesonide	asthma, COPD	2005	41, 443
cicletanine	antihypertensive	1988	24, 299
cidofovir	antiviral	1996	32, 306
cilazapril	antihypertensive	1990	26, 301
cilostazol	antithrombotic	1988	24, 299
cimetropium bromide	antispasmodic	1985	21, 326
cinacalcet	hyperparathyroidism	2004	40, 451
cinildipine	antihypertensive	1995	31, 339
cinitapride	gastroprokinetic	1990	26, 301
cinolazepam	hypnotic	1993	29, 334
ciprofibrate	hypolipidemic	1985	21, 326
ciprofloxacin	antibacterial	1986	22, 318
cisapride	gastroprokinetic	1988	24, 299
cisatracurium besilate	muscle relaxant	1995	31, 340
citalopram	antidepressant	1989	25, 311
cladribine	antineoplastic	1993	29, 335
clarithromycin	antibiotic	1990	26, 302
clobenoside	vasoprotective	1988	24, 300
cloconazole HCl	topical antifungal	1986	22, 318
clodronate disodium	calcium regulator	1986	22, 319
clofarabine	anticancer	2005	41, 444
clopidogrel hydrogensulfate	antithrombotic	1998	34, 320
cloricromen	antithrombotic	1991	27, 325
clospipramine HCl	neuroleptic	1991	27, 325
colesevelam hydrochloride	hypolipidemic	2000	36, 300
colestimide	hypolipidaemic	1999	35, 337
colforsin daropate HCl	cardiotonic	1999	35, 337
crotelidae polyvalent immune fab	antidote	2001	37, 263
cyclosporine	immunosuppressant	1983	19, 317
cytarabine ocfosfate	antineoplastic	1993	29, 335
dalfopristin	antibiotic	1999	35, 338
dapiprazole HCl	antiglaucoma	1987	23, 332
daptomycin	antibiotic	2003	39, 272
darifenacin	urinary incontinence	2005	41, 445

GENERAL NAME	INDICATION	YEAR INTRO.	ARMC VOL., PAGE
defeiprone	iron chelator	1995	31, 340
deferasirox	chronic iron overload	2005	41, 446
defibrotide	antithrombotic	1986	22, 319
deflazacort	antiinflammatory	1986	22, 319
delapril	antihypertensive	1989	25, 311
delavirdine mesylate	antiviral	1997	33, 331
denileukin diftitox	anticancer	1999	35, 338
denopamine	cardiostimulant	1988	24, 300
deprodone propionate	topical antiinflammatory	1992	28, 329
desflurane	anesthetic	1992	28, 329
desloratadine	antihistamine	2001	37, 264
dexfenfluramine	antiobesity	1997	33, 332
dexibuprofen	antiinflammatory	1994	30, 298
dexmedetomidine hydrochloride	sedative	2000	36, 301
dexmethylphenidate HCl	psychostimulant	2002	38, 352
dexrazoxane	cardioprotective	1992	28, 330
dezocine	analgesic	1991	27, 326
diacerein	antirheumatic	1985	21, 326
didanosine	antiviral	1991	27, 326
dilevalol	antihypertensive	1989	25, 311
dirithromycin	antibiotic	1993	29, 336
disodium pamidronate	calcium regulator	1989	25, 312
divistyramine	hypocholesterolemic	1984	20, 317
docarpamine	cardiostimulant	1994	30, 298
docetaxel	antineoplastic	1995	31, 341
dofetilide	antiarrhythmic	2000	36, 301
dolasetron mesylate	antiemetic	1998	34, 321
donepezil HCl	anti-Alzheimer	1997	33, 332
dopexamine	cardiostimulant	1989	25, 312
doripenem	antibiotic	2005	41, 448
dornase alfa	cystic fibrosis	1994	30, 298
dorzolamide HCL	antiglaucoma	1995	31, 341
dosmalfate	antiulcer	2000	36, 302
doxacurium chloride	muscle relaxant	1991	27, 326
doxazosin mesylate	antihypertensive	1988	24, 300
doxefazepam	hypnotic	1985	21, 326
doxercalciferol	vitamin D prohormone	1999	35, 339
doxifluridine	antineoplastic	1987	23, 332
doxofylline	bronchodilator	1985	21, 327
dronabinol	antinauseant	1986	22, 319
drospirenone	contraceptive	2000	36, 302
drotrecogin alfa	antisepsis	2001	37, 265
droxicam	antiinflammatory	1990	26, 302
droxidopa	antiparkinsonian	1989	25, 312
duloxetine	antidepressant	2004	40, 452

GENERAL NAME	INDICATION	YEAR INTRO.	ARMC VOL., PAGE
dutasteride	5α reductase inhibitor	2002	38, 353
duteplase	anticougulant	1995	31, 342
eberconazole	antifungal	2005	41, 449
ebastine	antihistamine	1990	26 302
ebrotidine	antiulcer	1997	33, 333
ecabet sodium	antiulcerative	1993	29, 336
edaravone	neuroprotective	2001	37, 265
efalizumab	psoriasis	2003	39, 274
efavirenz	antiviral	1998	34, 321
efonidipine	antihypertensive	1994	30, 299
egualen sodium	antiulcer	2000	36, 303
eletriptan	antimigraine	2001	37, 266
emedastine difumarate	antiallergic/antiasthmatic	1993	29, 336
emorfazone	analgesic	1984	20, 317
emtricitabine	antiviral	2003	39, 274
enalapril maleate	antihypertensive	1984	20, 317
enalaprilat	antihypertensive	1987	23, 332
encainide HCl	antiarrhythmic	1987	23, 333
enfuvirtide	antiviral	2003	39, 275
enocitabine	antineoplastic	1983	19, 318
enoxacin	antibacterial	1986	22, 320
enoxaparin	antithrombotic	1987	23, 333
enoximone	cardiostimulant	1988	24, 301
enprostil	antiulcer	1985	21, 327
entacapone	antiparkinsonian	1998	34, 322
entecavir	antiviral	2005	41, 450
epalrestat	antidiabetic	1992	28, 330
eperisone HCl	muscle relaxant	1983	19, 318
epidermal growth factor	wound healing agent	1987	23, 333
epinastine	antiallergic	1994	30, 299
epirubicin HCl	antineoplastic	1984	20, 318
eplerenone	antihypertensive	2003	39, 276
epoprostenol sodium	platelet aggreg. inhib.	1983	19, 318
eprosartan	antihypertensive	1997	33, 333
eptazocine HBr	analgesic	1987	23, 334
eptilfibatide	antithrombotic	1999	35, 340
erdosteine	expectorant	1995	31, 342
erlotinib	anticancer	2004	40, 454
ertapenem sodium	antibacterial	2002	38, 353
erythromycin acistrate	antibiotic	1988	24, 301
erythropoietin	hematopoetic	1988	24, 301
escitalopram oxolate	antidepressant	2002	38, 354
esmolol HCl	antiarrhythmic	1987	23, 334
esomeprazole magnesium	gastric antisecretory	2000	36, 303
eszopiclone	hypnotic	2005	41, 451
ethyl icosapentate	antithrombotic	1990	26, 303

GENERAL NAME	INDICATION	YEAR INTRO.	ARMC VOL., PAGE
etizolam	anxiolytic	1984	20, 318
etodolac	antiinflammatory	1985	21, 327
etoricoxibe	antiarthritic/analgesic	2002	38, 355
everolimus	immunosuppressant	2004	40, 455
exemestane	anticancer	2000	36, 304
exenatide	anti-diabetic	2005	41, 452
exifone	nootropic	1988	24, 302
ezetimibe	hypolipidemic	2002	38, 355
factor VIIa	haemophilia	1996	32, 307
factor VIII	hemostatic	1992	28, 330
fadrozole HCl	antineoplastic	1995	31, 342
falecalcitriol	vitamin D	2001	37, 266
famciclovir	antiviral	1994	30, 300
famotidine	antiulcer	1985	21, 327
fasudil HCl	neuroprotective	1995	31, 343
felbamate	antiepileptic	1993	29, 337
felbinac	topical antiinflammatory	1986	22, 320
felodipine	antihypertensive	1988	24, 302
fenbuprol	choleretic	1983	19, 318
fenoldopam mesylate	antihypertensive	1998	34, 322
fenticonazole nitrate	antifungal	1987	23, 334
fexofenadine	antiallergic	1996	32, 307
filgrastim	immunostimulant	1991	27, 327
finasteride	5α-reductase inhibitor	1992	28, 331
fisalamine	intestinal antiinflammatory	1984	20, 318
fleroxacin	antibacterial	1992	28, 331
flomoxef sodium	antibiotic	1988	24, 302
flosequinan	cardiostimulant	1992	28, 331
fluconazole	antifungal	1988	24, 303
fludarabine phosphate	antineoplastic	1991	27, 327
flumazenil	benzodiazepine antag.	1987	23, 335
flunoxaprofen	antiinflammatory	1987	23, 335
fluoxetine HCl	antidepressant	1986	22, 320
flupirtine maleate	analgesic	1985	21, 328
flurithromycin ethylsuccinate	antibiotic	1997	33, 333
flutamide	antineoplastic	1983	19, 318
flutazolam	anxiolytic	1984	20, 318
fluticasone propionate	antiinflammatory	1990	26, 303
flutoprazepam	anxiolytic	1986	22, 320
flutrimazole	topical antifungal	1995	31, 343
flutropium bromide	antitussive	1988	24, 303
fluvastatin	hypolipaemic	1994	30, 300
fluvoxamine maleate	antidepressant	1983	19, 319
follitropin alfa	fertility enhancer	1996	32, 307

Cumulative NCE Introduction Index, 1983–2005   511

GENERAL NAME	INDICATION	YEAR INTRO.	ARMC VOL., PAGE
follitropin beta	fertility enhancer	1996	32, 308
fomepizole	antidote	1998	34, 323
fomivirsen sodium	antiviral	1998	34, 323
fondaparinux sodium	antithrombotic	2002	38, 356
formestane	antineoplastic	1993	29, 337
formoterol fumarate	bronchodilator	1986	22, 321
fosamprenavir	antiviral	2003	39, 277
foscarnet sodium	antiviral	1989	25, 313
fosfosal	analgesic	1984	20, 319
fosfluconazole	antifungal	2004	40, 457
fosinopril sodium	antihypertensive	1991	27, 328
fosphenytoin sodium	antiepileptic	1996	32, 308
fotemustine	antineoplastic	1989	25, 313
fropenam	antibiotic	1997	33, 334
frovatriptan	antimigraine	2002	38, 357
fudosteine	expectorant	2001	37, 267
fulveristrant	anticancer	2002	38, 357
gabapentin	antiepileptic	1993	29, 338
gadoversetamide	MRI contrast agent	2000	36, 304
gallium nitrate	calcium regulator	1991	27, 328
gallopamil HCl	antianginal	1983	19, 319
galsulfase	mucopolysaccharidosis VI	2005	41, 453
ganciclovir	antiviral	1988	24, 303
ganirelix acetate	female infertility	2000	36, 305
gatilfloxacin	antibiotic	1999	35, 340
gefitinib	antineoplastic	2002	38, 358
gemcitabine HCl	antineoplastic	1995	31, 344
gemeprost	abortifacient	1983	19, 319
gemifloxacin	antibacterial	2004	40, 458
gemtuzumab ozogamicin	anticancer	2000	36, 306
gestodene	progestogen	1987	23, 335
gestrinone	antiprogestogen	1986	22, 321
glatiramer acetate	Multiple Sclerosis	1997	33, 334
glimepiride	antidiabetic	1995	31, 344
glucagon, rDNA	hypoglycemia	1993	29, 338
GMDP	immunostimulant	1996	32, 308
goserelin	hormone	1987	23, 336
granisetron HCl	antiemetic	1991	27, 329
guanadrel sulfate	antihypertensive	1983	19, 319
gusperimus	immunosuppressant	1994	30, 300
halobetasol propionate	topical antiinflammatory	1991	27, 329
halofantrine	antimalarial	1988	24, 304
halometasone	topical antiinflammatory	1983	19, 320
histrelin	precocious puberty	1993	29, 338
hydrocortisone aceponate	topical antiinflammatory	1988	24, 304

GENERAL NAME	INDICATION	YEAR INTRO.	ARMC VOL., PAGE
hydrocortisone butyrate	topical antiinflammatory	1983	19, 320
ibandronic acid	osteoporosis	1996	32, 309
ibopamine HCl	cardiostimulant	1984	20, 319
ibudilast	antiasthmatic	1989	25, 313
ibutilide fumarate	antiarrhythmic	1996	32, 309
ibritunomab tiuxetan	anticancer	2002	38, 359
idarubicin HCl	antineoplastic	1990	26, 303
idebenone	nootropic	1986	22, 321
iloprost	platelet aggreg. inhibitor	1992	28, 332
imatinib mesylate	antineoplastic	2001	37, 267
imidapril HCl	antihypertensive	1993	29, 339
imiglucerase	Gaucher's disease	1994	30, 301
imipenem/cilastatin	antibiotic	1985	21, 328
imiquimod	antiviral	1997	33, 335
incadronic acid	osteoporosis	1997	33, 335
indalpine	antidepressant	1983	19, 320
indeloxazine HCl	nootropic	1988	24, 304
indinavir sulfate	antiviral	1996	32, 310
indisetron	antiemetic	2004	40, 459
indobufen	antithrombotic	1984	20, 319
influenza virus (live)	antiviral vaccine	2003	39, 277
insulin lispro	antidiabetic	1996	32, 310
interferon alfacon-1	antiviral	1997	33, 336
interferon gamma-1b	immunostimulant	1991	27, 329
interferon, gamma	antiinflammatory	1989	25, 314
interferon, gamma-1α	antineoplastic	1992	28, 332
interferon, β-1a	multiple sclerosis	1996	32, 311
interferon, β-1b	multiple sclerosis	1993	29, 339
interleukin-2	antineoplastic	1989	25, 314
ioflupane	diagnosis CNS	2000	36, 306
ipriflavone	calcium regulator	1989	25, 314
irbesartan	antihypertensive	1997	33, 336
irinotecan	antineoplastic	1994	30, 301
irsogladine	antiulcer	1989	25, 315
isepamicin	antibiotic	1988	24, 305
isofezolac	antiinflammatory	1984	20, 319
isoxicam	antiinflammatory	1983	19, 320
isradipine	antihypertensive	1989	25, 315
itopride HCl	gastroprokinetic	1995	31, 344
itraconazole	antifungal	1988	24, 305
ivermectin	antiparasitic	1987	23, 336
ketanserin	antihypertensive	1985	21, 328
ketorolac tromethamine	analgesic	1990	26, 304
kinetin	skin photodamage/ dermatologic	1999	35, 341
lacidipine	antihypertensive	1991	27, 330

GENERAL NAME	INDICATION	YEAR INTRO.	ARMC VOL., PAGE
lafutidine	gastric antisecretory	2000	36, 307
lamivudine	antiviral	1995	31, 345
lamotrigine	anticonvulsant	1990	26, 304
landiolol	antiarrhythmic	2002	38, 360
lanoconazole	antifungal	1994	30, 302
lanreotide acetate	acromegaly	1995	31, 345
lansoprazole	antiulcer	1992	28, 332
laronidase	mucopolysaccaridosis I	2003	39, 278
latanoprost	antiglaucoma	1996	32, 311
lefunomide	antiarthritic	1998	34, 324
lenampicillin HCl	antibiotic	1987	23, 336
lentinan	immunostimulant	1986	22, 322
lepirudin	anticoagulant	1997	33, 336
lercanidipine	antihyperintensive	1997	33, 337
letrazole	anticancer	1996	32, 311
leuprolide acetate	hormone	1984	20, 319
levacecarnine HCl	nootropic	1986	22, 322
levalbuterol HCl	antiasthmatic	1999	35, 341
levetiracetam	antiepileptic	2000	36, 307
levobunolol HCl	antiglaucoma	1985	21, 328
levobupivacaine hydrochloride	local anesthetic	2000	36, 308
levocabastine HCl	antihistamine	1991	27, 330
levocetirizine	antihistamine	2001	37, 268
levodropropizine	antitussive	1988	24, 305
levofloxacin	antibiotic	1993	29, 340
levosimendan	heart failure	2000	36, 308
lidamidine HCl	antiperistaltic	1984	20, 320
limaprost	antithrombotic	1988	24, 306
linezolid	antibiotic	2000	36, 309
liranaftate	topical antifungal	2000	36, 309
lisinopril	antihypertensive	1987	23, 337
lobenzarit sodium	antiinflammatory	1986	22, 322
lodoxamide tromethamine	antiallergic ophthalmic	1992	28, 333
lomefloxacin	antibiotic	1989	25, 315
lomerizine HCl	antimigraine	1999	35, 342
lonidamine	antineoplastic	1987	23, 337
lopinavir	antiviral	2000	36, 310
loprazolam mesylate	hypnotic	1983	19, 321
loprinone HCl	cardiostimulant	1996	32, 312
loracarbef	antibiotic	1992	28, 333
loratadine	antihistamine	1988	24, 306
lornoxicam	NSAID	1997	33, 337
losartan	antihypertensive	1994	30, 302
loteprednol etabonate	antiallergic ophthalmic	1998	34, 324

GENERAL NAME	INDICATION	YEAR INTRO.	ARMC VOL., PAGE
lovastatin	hypocholesterolemic	1987	23, 337
loxoprofen sodium	antiinflammatory	1986	22, 322
luliconazole	antifungal	2005	41, 454
lumiracoxib	anti-inflammatory	2005	41, 455
Lyme disease	vaccine	1999	35, 342
mabuterol HCl	bronchodilator	1986	22, 323
malotilate	hepatoprotective	1985	21, 329
manidipine HCl	antihypertensive	1990	26, 304
masoprocol	topical antineoplastic	1992	28, 333
maxacalcitol	vitamin D	2000	36, 310
mebefradil HCl	antihypertensive	1997	33, 338
medifoxamine fumarate	antidepressant	1986	22, 323
mefloquine HCl	antimalarial	1985	21, 329
meglutol	hypolipidemic	1983	19, 321
melinamide	hypocholesterolemic	1984	20, 320
meloxicam	antiarthritic	1996	32, 312
mepixanox	analeptic	1984	20, 320
meptazinol HCl	analgesic	1983	19, 321
meropenem	carbapenem antibiotic	1994	30, 303
metaclazepam	anxiolytic	1987	23, 338
metapramine	antidepressant	1984	20, 320
mexazolam	anxiolytic	1984	20, 321
micafungin	antifungal	2002	38, 360
mifepristone	abortifacient	1988	24, 306
miglitol	antidiabetic	1998	34, 325
miglustat	gaucher's disease	2003	39, 279
milnacipran	antidepressant	1997	33, 338
milrinone	cardiostimulant	1989	25, 316
miltefosine	topical antineoplastic	1993	29, 340
miokamycin	antibiotic	1985	21, 329
mirtazapine	antidepressant	1994	30, 303
misoprostol	antiulcer	1985	21, 329
mitiglinide	antidiabetic	2004	40, 460
mitoxantrone HCl	antineoplastic	1984	20, 321
mivacurium chloride	muscle relaxant	1992	28, 334
mivotilate	hepatoprotectant	1999	35, 343
mizolastine	antihistamine	1998	34, 325
mizoribine	immunosuppressant	1984	20, 321
moclobemide	antidepressant	1990	26, 305
modafinil	idiopathic hypersomnia	1994	30, 303
moexipril HCl	antihypertensive	1995	31, 346
mofezolac	analgesic	1994	30, 304
mometasone furoate	topical antiinflammatory	1987	23, 338
montelukast sodium	antiasthma	1998	34, 326
moricizine HCl	antiarrhythmic	1990	26, 305
mosapride citrate	gastroprokinetic	1998	34, 326

GENERAL NAME	INDICATION	YEAR INTRO.	ARMC VOL., PAGE
moxifloxacin HCL	antibiotic	1999	35, 343
moxonidine	antihypertensive	1991	27, 330
mupirocin	topical antibiotic	1985	21, 330
muromonab-CD3	immunosuppressant	1986	22, 323
muzolimine	diuretic	1983	19, 321
mycophenolate mofetil	immunosuppressant	1995	31, 346
mycophenolate sodium	immunosuppressant	2003	39, 279
nabumetone	antiinflammatory	1985	21, 330
nadifloxacin	topical antibiotic	1993	29, 340
nafamostat mesylate	protease inhibitor	1986	22, 323
nafarelin acetate	hormone	1990	26, 306
naftifine HCl	antifungal	1984	20, 321
naftopidil	dysuria	1999	35, 344
nalmefene HCl	dependence treatment	1995	31, 347
naltrexone HCl	narcotic antagonist	1984	20, 322
naratriptan HCl	antimigraine	1997	33, 339
nartograstim	leukopenia	1994	30, 304
natalizumab	multiple sclerosis	2004	40, 462
nateglinide	antidiabetic	1999	35, 344
nazasetron	antiemetic	1994	30, 305
nebivolol	antihypertensive	1997	33, 339
nedaplatin	antineoplastic	1995	31, 347
nedocromil sodium	antiallergic	1986	22, 324
nefazodone	antidepressant	1994	30, 305
nelfinavir mesylate	antiviral	1997	33, 340
neltenexine	cystic fibrosis	1993	29, 341
nemonapride	neuroleptic	1991	27, 331
nepafenac	anti-inflammatory	2005	41, 456
neridronic acide	calcium regulator	2002	38, 361
nesiritide	congestive heart failure	2001	37, 269
neticonazole HCl	topical antifungal	1993	29, 341
nevirapine	antiviral	1996	32, 313
nicorandil	coronary vasodilator	1984	20, 322
nifekalant HCl	antiarrythmic	1999	35, 344
nilutamide	antineoplastic	1987	23, 338
nilvadipine	antihypertensive	1989	25, 316
nimesulide	antiinflammatory	1985	21, 330
nimodipine	cerebral vasodilator	1985	21, 330
nipradilol	antihypertensive	1988	24, 307
nisoldipine	antihypertensive	1990	26, 306
nitisinone	antityrosinaemia	2002	38, 361
nitrefazole	alcohol deterrent	1983	19, 322
nitrendipine	hypertensive	1985	21, 331
nizatidine	antiulcer	1987	23, 339
nizofenzone fumarate	nootropic	1988	24, 307
nomegestrol acetate	progestogen	1986	22, 324

GENERAL NAME	INDICATION	YEAR INTRO.	ARMC VOL., PAGE
norelgestromin	contraceptive	2002	38, 362
norfloxacin	antibacterial	1983	19, 322
norgestimate	progestogen	1986	22, 324
OCT-43	anticancer	1999	35, 345
octreotide	antisecretory	1988	24, 307
ofloxacin	antibacterial	1985	21, 331
olanzapine	neuroleptic	1996	32, 313
olimesartan Medoxomil	antihypertensive	2002	38, 363
olopatadine HCl	antiallergic	1997	33, 340
omalizumab	allergic asthma	2003	39, 280
omeprazole	antiulcer	1988	24, 308
ondansetron HCl	antiemetic	1990	26, 306
OP-1	osteoinductor	2001	37, 269
orlistat	antiobesity	1998	34, 327
ornoprostil	antiulcer	1987	23, 339
osalazine sodium	intestinal antinflamm.	1986	22, 324
oseltamivir phosphate	antiviral	1999	35, 346
oxaliplatin	anticancer	1996	32, 313
oxaprozin	antiinflammatory	1983	19, 322
oxcarbazepine	anticonvulsant	1990	26, 307
oxiconazole nitrate	antifungal	1983	19, 322
oxiracetam	nootropic	1987	23, 339
oxitropium bromide	bronchodilator	1983	19, 323
ozagrel sodium	antithrombotic	1988	24, 308
paclitaxal	antineoplastic	1993	29, 342
palifermin	mucositis	2005	41, 461
palonosetron	antiemetic	2003	39, 281
panipenem/betamipron	carbapenem antibiotic	1994	30, 305
pantoprazole sodium	antiulcer	1995	30, 306
parecoxib sodium	analgesic	2002	38, 364
paricalcitol	vitamin D	1998	34, 327
parnaparin sodium	anticoagulant	1993	29, 342
paroxetine	antidepressant	1991	27, 331
pazufloxacin	antibacterial	2002	38, 364
pefloxacin mesylate	antibacterial	1985	21, 331
pegademase bovine	immunostimulant	1990	26, 307
pegaptanib	age-related macular degeneration	2005	41, 458
pegaspargase	antineoplastic	1994	30, 306
pegvisomant	acromegaly	2003	39, 281
pemetrexed	anticancer	2004	40, 463
pemirolast potassium	antiasthmatic	1991	27, 331
penciclovir	antiviral	1996	32, 314
pentostatin	antineoplastic	1992	28, 334
pergolide mesylate	antiparkinsonian	1988	24, 308
perindopril	antihypertensive	1988	24, 309

GENERAL NAME	INDICATION	YEAR INTRO.	ARMC VOL., PAGE
perospirone HCL	neuroleptic	2001	37, 270
picotamide	antithrombotic	1987	23, 340
pidotimod	immunostimulant	1993	29, 343
piketoprofen	topical antiinflammatory	1984	20, 322
pilsicainide HCl	antiarrhythmic	1991	27, 332
pimaprofen	topical antiinflammatory	1984	20, 322
pimecrolimus	immunosuppressant	2002	38, 365
pimobendan	heart failure	1994	30, 307
pinacidil	antihypertensive	1987	23, 340
pioglitazone HCL	antidiabetic	1999	35, 346
pirarubicin	antineoplastic	1988	24, 309
pirmenol	antiarrhythmic	1994	30, 307
piroxicam cinnamate	antiinflammatory	1988	24, 309
pitavastatin	hypocholesterolemic	2003	39, 282
pivagabine	antidepressant	1997	33, 341
plaunotol	antiulcer	1987	23, 340
polaprezinc	antiulcer	1994	30, 307
porfimer sodium	antineoplastic adjuvant	1993	29, 343
pramipexole HCl	antiParkinsonian	1997	33, 341
pramiracetam H_2SO_4	cognition enhancer	1993	29, 343
pramlintide	anti-diabetic	2005	41, 460
pranlukast	antiasthmatic	1995	31, 347
pravastatin	antilipidemic	1989	25, 316
prednicarbate	topical antiinflammatory	1986	22, 325
pregabalin	antiepileptic	2004	40, 464
prezatide copper acetate	vulnery	1996	32, 314
progabide	anticonvulsant	1985	21, 331
promegestrone	progestogen	1983	19, 323
propacetamol HCl	analgesic	1986	22, 325
propagermanium	antiviral	1994	30, 308
propentofylline propionate	cerebral vasodilator	1988	24, 310
propiverine HCl	urologic	1992	28, 335
propofol	anesthetic	1986	22, 325
prulifloxacin	antibacterial	2002	38, 366
pumactant	lung surfactant	1994	30, 308
quazepam	hypnotic	1985	21, 332
quetiapine fumarate	neuroleptic	1997	33, 341
quinagolide	hyperprolactinemia	1994	30, 309
quinapril	antihypertensive	1989	25, 317
quinfamide	amebicide	1984	20, 322
quinupristin	antibiotic	1999	35, 338
rabeprazole sodium	gastric antisecretory	1998	34, 328
raloxifene HCl	osteoporosis	1998	34, 328
raltitrexed	anticancer	1996	32, 315
ramatroban	antiallergic	2000	36, 311

GENERAL NAME	INDICATION	YEAR INTRO.	ARMC VOL., PAGE
ramelteon	insomnia	2005	41, 462
ramipril	antihypertensive	1989	25, 317
ramosetron	antiemetic	1996	32, 315
ranimustine	antineoplastic	1987	23, 341
ranitidine bismuth citrate	antiulcer	1995	31, 348
rapacuronium bromide	muscle relaxant	1999	35, 347
rasagiline	parkinson's disease	2005	41, 464
rebamipide	antiulcer	1990	26, 308
reboxetine	antidepressant	1997	33, 342
remifentanil HCl	analgesic	1996	32, 316
remoxipride HCl	antipsychotic	1990	26, 308
repaglinide	antidiabetic	1998	34, 329
repirinast	antiallergic	1987	23, 341
reteplase	fibrinolytic	1996	32, 316
reviparin sodium	anticoagulant	1993	29, 344
rifabutin	antibacterial	1992	28, 335
rifapentine	antibacterial	1988	24, 310
rifaximin	antibiotic	1985	21, 332
rifaximin	antibiotic	1987	23, 341
rilmazafone	hypnotic	1989	25, 317
rilmenidine	antihypertensive	1988	24, 310
riluzole	neuroprotective	1996	32, 316
rimantadine HCl	antiviral	1987	23, 342
rimexolone	antiinflammatory	1995	31, 348
risedronate sodium	osteoporosis	1998	34, 330
risperidone	neuroleptic	1993	29, 344
ritonavir	antiviral	1996	32, 317
rivastigmin	anti-Alzheimer	1997	33, 342
rizatriptan benzoate	antimigraine	1998	34, 330
rocuronium bromide	neuromuscular blocker	1994	30, 309
rofecoxib	antiarthritic	1999	35, 347
rokitamycin	antibiotic	1986	22, 325
romurtide	immunostimulant	1991	27, 332
ronafibrate	hypolipidemic	1986	22, 326
ropinirole HCl	antiParkinsonian	1996	32, 317
ropivacaine	anesthetic	1996	32, 318
rosaprostol	antiulcer	1985	21, 332
rosiglitazone maleate	antidiabetic	1999	35, 348
rosuvastatin	hypocholesterolemic	2003	39, 283
roxatidine acetate HCl	antiulcer	1986	22, 326
roxithromycin	antiulcer	1987	23, 342
rufloxacin HCl	antibacterial	1992	28, 335
rupatadine fumarate	antiallergic	2003	39, 284
RV-11	antibiotic	1989	25, 318
salmeterol hydroxynaphthoate	bronchodilator	1990	26, 308

Cumulative NCE Introduction Index, 1983–2005

GENERAL NAME	INDICATION	YEAR INTRO.	ARMC VOL., PAGE
sapropterin HCl	hyperphenylalaninemia	1992	8, 336
saquinavir mesvlate	antiviral	1995	31 349
sargramostim	immunostimulant	1991	27, 332
sarpogrelate HCl	platelet antiaggregant	1993	29, 344
schizophyllan	immunostimulant	1985	22, 326
seratrodast	antiasthmatic	1995	31, 349
sertaconazole nitrate	topical antifungal	1992	28, 336
sertindole	neuroleptic	1996	32, 318
setastine HCl	antihistamine	1987	23, 342
setiptiline	antidepressant	1989	25, 318
setraline HCl	antidepressant	1990	26, 309
sevoflurane	anesthetic	1990	26, 309
sibutramine	antiobesity	1998	34, 331
sildenafil citrate	male sexual dysfunction	1998	34, 331
simvastatin	hypocholesterolemic	1988	24, 311
sivelestat	anti-inflammatory	2002	38, 366
SKI-2053R	anticancer	1999	35, 348
sobuzoxane	antineoplastic	1994	30, 310
sodium cellulose PO4	hypocalciuric	1983	19, 323
sofalcone	antiulcer	1984	20, 323
solifenacin	pollakiuria	2004	40, 466
somatomedin-1	growth hormone insensitivity	1994	30, 310
somatotropin	growth hormone	1994	30, 310
somatropin	hormone	1987	23, 343
sorafenib	anticancer	2005	41, 466
sorivudine	antiviral	1993	29, 345
sparfloxacin	antibiotic	1993	29, 345
spirapril HCl	antihypertensive	1995	31, 349
spizofurone	antiulcer	1987	23, 343
stavudine	antiviral	1994	30, 311
strontium ranelate	osteoporosis	2004	40, 466
succimer	chelator	1991	27, 333
sufentanil	analgesic	1983	19, 323
sulbactam sodium	β-lactamase inhibitor	1986	22, 326
sulconizole nitrate	topical antifungal	1985	21, 332
sultamycillin tosylate	antibiotic	1987	23, 343
sumatriptan succinate	antimigraine	1991	27, 333
suplatast tosilate	antiallergic	1995	31, 350
suprofen	analgesic	1983	19, 324
surfactant TA	respiratory surfactant	1987	23, 344
tacalcitol	topical antipsoriatic	1993	29, 346
tacrine HCl	Alzheimer's disease	1993	29, 346
tacrolimus	immunosuppressant	1993	29, 347
tadalafil	male sexual dysfunction	2003	39, 284
talaporfin sodium	anticancer	2004	40, 469

GENERAL NAME	INDICATION	YEAR INTRO.	ARMC VOL., PAGE
talipexole	antiParkinsonian	1996	32, 318
taltirelin	CNS stimulant	2000	36, 311
tamibarotene	anticancer	2005	41, 467
tamsulosin HCl	antiprostatic hypertrophy	1993	29, 347
tandospirone	anxiolytic	1996	32, 319
tasonermin	anticancer	1999	35, 349
tazanolast	antiallergic	1990	26, 309
tazarotene	antipsoriasis	1997	33, 343
tazobactam sodium	β-lactamase inhibitor	1992	28, 336
tegaserod maleate	irritable bowel syndrome	2001	37, 270
teicoplanin	antibacterial	1988	24, 311
telithromycin	antibiotic	2001	37, 271
telmesteine	mucolytic	1992	28, 337
telmisartan	antihypertensive	1999	35, 349
temafloxacin HCl	antibacterial	1991	27, 334
temocapril	antihypertensive	1994	30, 311
temocillin disodium	antibiotic	1984	20, 323
temoporphin	antineoplastic/ photosensitizer	2002	38, 367
temozolomide	anticancer	1999	35, 349
tenofovir disoproxil fumarate	antiviral	2001	37, 271
tenoxicam	antiinflammatory	1987	23, 344
teprenone	antiulcer	1984	20, 323
terazosin HCl	antihypertensive	1984	20, 323
terbinafine HCl	antifungal	1991	27, 334
terconazole	antifungal	1983	19, 324
tertatolol HCl	antihypertensive	1987	23, 344
thymopentin	immunomodulator	1985	21, 333
tiagabine	antiepileptic	1996	32, 319
tiamenidine HCl	antihypertensive	1988	24, 311
tianeptine sodium	antidepressant	1983	19, 324
tibolone	anabolic	1988	24, 312
tigecycline	antibiotic	2005	41, 468
tilisolol HCl	antihypertensive	1992	28, 337
tiludronate disodium	Paget's disease	1995	31, 350
timiperone	neuroleptic	1984	20, 323
tinazoline	nasal decongestant	1988	24, 312
tioconazole	antifungal	1983	19, 324
tiopronin	urolithiasis	1989	25, 318
tiotropium bromide	bronchodilator	2002	38, 368
tipranavir	HIV	2005	41, 470
tiquizium bromide	antispasmodic	1984	20, 324
tiracizine HCl	antiarrhythmic	1990	26, 310
tirilazad mesylate	subarachnoid hemorrhage	1995	31, 351

GENERAL NAME	INDICATION	YEAR INTRO.	ARMC VOL., PAGE
tirofiban HCl	antithrombotic	1998	34, 332
tiropramide HCl	antispasmodic	1983	19, 324
tizanidine	muscle relaxant	1984	20, 324
tolcapone	antiParkinsonian	1997	33, 343
toloxatone	antidepressant	1984	20, 324
tolrestat	antidiabetic	1989	25, 319
topiramate	antiepileptic	1995	31, 351
topotecan HCl	anticancer	1996	32, 320
torasemide	diuretic	1993	29, 348
toremifene	antineoplastic	1989	25, 319
tositumomab	anticancer	2003	39, 285
tosufloxacin tosylate	antibacterial	1990	26, 310
trandolapril	antihypertensive	1993	29, 348
travoprost	antiglaucoma	2001	37, 272
treprostinil sodium	antihypertensive	2002	38, 368
tretinoin tocoferil	antiulcer	1993	29, 348
trientine HCl	chelator	1986	22, 327
trimazosin HCl	antihypertensive	1985	21, 333
trimegestone	progestogen	2001	37, 273
trimetrexate glucuronate	*Pneumocystis carinii* pneumonia	1994	30, 312
troglitazone	antidiabetic	1997	33, 344
tropisetron	antiemetic	1992	28, 337
trovafloxacin mesylate	antibiotic	1998	34, 332
troxipide	antiulcer	1986	22, 327
ubenimex	immunostimulant	1987	23, 345
udenafil	erectile dysfunction	2005	41, 472
unoprostone isopropyl ester	antiglaucoma	1994	30, 312
valaciclovir HCl	antiviral	1995	31, 352
vadecoxib	antiarthritic	2002	38, 369
vaglancirclovir HCL	antiviral	2001	37, 273
valrubicin	anticancer	1999	35, 350
valsartan	antihypertensive	1996	32, 320
vardenafil	male sexual dysfunction	2003	39, 286
venlafaxine	antidepressant	1994	30, 312
verteporfin	photosensitizer	2000	36, 312
vesnarinone	cardiostimulant	1990	26, 310
vigabatrin	anticonvulsant	1989	25, 319
vinorelbine	antineoplastic	1989	25, 320
voglibose	antidiabetic	1994	30, 313
voriconazole	antifungal	2002	38, 370
xamoterol fumarate	cardiotonic	1988	24, 312
ximelagatran	anticoagulant	2004	40, 470
zafirlukast	antiasthma	1996	32, 321
zalcitabine	antiviral	1992	28, 338

GENERAL NAME	INDICATION	YEAR INTRO.	ARMC VOL., PAGE
zaleplon	hypnotic	1999	35, 351
zaltoprofen	antiinflammatory	1993	29, 349
zanamivir	antiviral	1999	35, 352
ziconotide	severe chronic pain	2005	41, 473
zidovudine	antiviral	1987	23, 345
zileuton	antiasthma	1997	33, 344
zinostatin stimalamer	antineoplastic	1994	30, 313
ziprasidone hydrochloride	neuroleptic	2000	36, 312
zofenopril calcium	antihypertensive	2000	36, 313
zoledronate disodium	hypercalcemia	2000	36, 314
zolpidem hemitartrate	hypnotic	1988	24, 313
zomitriptan	antimigraine	1997	33, 345
zonisamide	anticonvulsant	1989	25, 320
zopiclone	hypnotic	1986	22, 327
zuclopenthixol acetate	antipsychotic	1987	23, 345

CUMULATIVE NCE INTRODUCTION INDEX, 1983–2005 (BY INDICATION)

GENERIC NAME	INDICATION	YEAR INTRO.	ARMC VOL., (PAGE)
gemeprost	ABORTIFACIENT	1983	19 (319)
mifepristone		1988	24 (306)
lanreotide acetate	ACROMEGALY	1995	31 (345)
pegvisomant		2003	39 (281)
pegaptanib	AGE-RELATED MACULAR DEGENERATION	2005	41 (458)
nitrefazole	ALCOHOL DETERRENT	1983	19 (322)
omalizumab	ALLERGIC ASTHMA	2003	39 (280)
tacrine HCl	ALZHEIMER'S DISEASE	1993	29 (346)
quinfamide	AMEBICIDE	1984	20 (322)
tibolone	ANABOLIC	1988	24 (312)
mepixanox	ANALEPTIC	1984	20 (320)
alfentanil HCl	ANALGESIC	1983	19 (314)
alminoprofen		1983	19 (314)
dezocine		1991	27 (326)
emorfazone		1984	20 (317)
eptazocine HBr		1987	23 (334)
etoricoxib		2002	38 (355)
flupirtine maleate		1985	21 (328)
fosfosal		1984	20 (319)
ketorolac tromethamine		1990	26 (304)
meptazinol HCl		1983	19 (321)
mofezolac		1994	30 (304)
parecoxib sodium		2002	38 (364)
propacetamol HCl		1986	22 (325)
remifentanil HCl		1996	32 (316)
sufentanil		1983	19 (323)
suprofen		1983	19 (324)
desflurane	ANESTHETIC	1992	28 (329)
propofol		1986	22 (325)
ropivacaine		1996	32 (318)
sevoflurane		1990	26 (309)
levobupivacaine hydrochloride	ANESTHETIC, LOCAL	2000	36 (308)
azelaic acid	ANTIACNE	1989	25 (310)
betotastine besilate	ANTIALLERGIC	2000	36 (297)
emedastine difumarate		1993	29 (336)
epinastine		1994	30 (299)
fexofenadine		1996	32 (307)
nedocromil sodium		1986	22 (324)
olopatadine hydrochloride		1997	33 (340)
ramatroban		2000	36 (311)
repirinast		1987	23 (341)
suplatast tosilate		1995	31 (350)

GENERIC NAME	INDICATION	YEAR INTRO.	ARMC VOL., (PAGE)
tazanolast		1990	26 (309)
lodoxamide tromethamine	ANTIALLERGIC	1992	28 (333)
rupatadine fumarate		2003	39 (284)
loteprednol etabonate	OPHTHALMIC	1998	34 (324)
donepezil hydrochloride	ANTI-ALZHEIMERS	1997	33 (332)
rivastigmin		1997	33 (342)
gallopamil HCl	ANTIANGINAL	1983	19 (319)
cibenzoline	ANTIARRHYTHMIC	1985	21 (325)
dofetilide		2000	36 (301)
encainide HCl		1987	23 (333)
esmolol HCl		1987	23 (334)
ibutilide fumarate		1996	32 (309)
landiolol		2002	38 (360)
moricizine hydrochloride		1990	26 (305)
nifekalant HCl		1999	35 (344)
pilsicainide hydrochloride		1991	27 (332)
pirmenol		1994	30 (307)
tiracizine hydrochloride		1990	26 (310)
anakinra	ANTIARTHRITIC	2001	37 (261)
celecoxib		1999	35 (335)
etoricoxib		2002	38 (355)
meloxicam		1996	32 (312)
leflunomide		1998	34 (324)
rofecoxib		1999	35 (347)
valdecoxib		2002	38 (369)
amlexanox	ANTIASTHMATIC	1987	23 (327)
emedastine difumarate		1993	29 (336)
ibudilast		1989	25 (313)
levalbuterol HCl		1999	35 (341)
montelukast sodium		1998	34 (326)
pemirolast potassium		1991	27 (331)
seratrodast		1995	31 (349)
zafirlukast		1996	32 (321)
zileuton		1997	33 (344)
balofloxacin	ANTIBACTERIAL	2002	38 (351)
biapenem		2002	38 (351)
ciprofloxacin		1986	22 (318)
enoxacin		1986	22 (320)
ertapenem sodium		2002	38 (353)
fleroxacin		1992	28 (331)
gemifloxacin		2004	40 (458)
norfloxacin		1983	19 (322)
ofloxacin		1985	21 (331)
pazufloxacin		2002	38 (364)
pefloxacin mesylate		1985	21 (331)
pranlukast		1995	31 (347)
prulifloxacin		2002	38 (366)
rifabutin		1992	28 (335)
rifapentine		1988	24 (310)
rufloxacin hydrochloride		1992	28 (335)
teicoplanin		1988	24 (311)

Cumulative NCE Introduction Index, 1983–2005 (by Indication) 525

GENERIC NAME	INDICATION	YEAR INTRO.	ARMC VOL., (PAGE)
temafloxacin hydrochloride		1991	27 (334)
tosufloxacin tosylate		1990	26 (310)
arbekacin	ANTIBIOTIC	1990	26 (298)
aspoxicillin		1987	23 (328)
astromycin sulfate		1985	21 (324)
azithromycin		1988	24 (298)
aztreonam		1984	20 (315)
brodimoprin		1993	29 (333)
carboplatin		1986	22 (318)
carumonam		1988	24 (298)
cefbuperazone sodium		1985	21 (325)
cefcapene pivoxil		1997	33 (330)
cefdinir		1991	27 (323)
cefepime		1993	29 (334)
cefetamet pivoxil hydrochloride		1992	28 (327)
cefixime		1987	23 (329)
cefmenoxime HCl		1983	19 (316)
cefminox sodium		1987	23 (330)
cefodizime sodium		1990	26 (300)
cefonicid sodium		1984	20 (316)
ceforanide		1984	20 (317)
cefoselis		1998	34 (319)
cefotetan disodium		1984	20 (317)
cefotiam hexetil hydrochloride		1991	27 (324)
cefpimizole		1987	23 (330)
cefpiramide sodium		1985	21 (325)
cefpirome sulfate		1992	28 (328)
cefpodoxime proxetil		1989	25 (310)
cefprozil		1992	28 (328)
ceftazidime		1983	19 (316)
cefteram pivoxil		1987	23 (330)
ceftibuten		1992	28 (329)
cefuroxime axetil		1987	23 (331)
cefuzonam sodium		1987	23 (331)
clarithromycin		1990	26 (302)
dalfopristin		1999	35 (338)
dirithromycin		1993	29 (336)
doripenem		2005	41 (448)
erythromycin acistrate		1988	24 (301)
flomoxef sodium		1988	24 (302)
flurithromycin ethylsuccinate		1997	33 (333)
fropenam		1997	33 (334)
gatifloxacin		1999	35 (340)
imipenem/cilastatin		1985	21 (328)
isepamicin		1988	24 (305)
lenampicillin HCl		1987	23 (336)
levofloxacin		1993	29 (340)
linezolid		2000	36 (309)

GENERIC NAME	INDICATION	YEAR INTRO.	ARMC VOL., (PAGE)
lomefloxacin		1989	25 (315)
loracarbef		1992	28 (333)
miokamycin		1985	21 (329)
moxifloxacin HCl		1999	35 (343)
quinupristin		1999	35 (338)
rifaximin		1985	21 (332)
rifaximin		1987	23 (341)
rokitamycin		1986	22 (325)
RV-11		1989	25 (318)
sparfloxacin		1993	29 (345)
sultamycillin tosylate		1987	23 (343)
telithromycin		2001	37 (271)
temocillin disodium		1984	20 (323)
tigecycline		2005	41 (468)
trovafloxacin mesylate		1998	34 (332)
meropenem	ANTIBIOTIC,	1994	30 (303)
panipenem/betamipron	CARBAPENEM	1994	30 (305)
mupirocin	ANTIBIOTIC, TOPICAL	1985	21 (330)
nadifloxacin		1993	29 (340)
abarelix	ANTICANCER	2004	40 (446)
alemtuzumab		2001	37 (260)
alitretinoin		1999	35 (333)
arglabin		1999	35 (335)
azacitidine		2004	40 (447)
belotecan		2004	40 (449)
bevacizumab		2004	40 (450)
bexarotene		2000	36 (298)
bortezomib		2003	39 (271)
cetuximab		2003	39 (272)
clofarabine		2005	41 (444)
denileukin diftitox		1999	35 (338)
erlotinib		2004	40 (454)
exemestane		2000	36 (304)
fulvestrant		2002	38 (357)
gemtuzumab ozogamicin		2000	36 (306)
ibritumomab tiuxetan		2002	38 (359)
letrazole		1996	32 (311)
OCT-43		1999	35 (345)
oxaliplatin		1996	32 (313)
pemetrexed		2004	40 (463)
raltitrexed		1996	32 (315)
SKI-2053R		1999	35 (348)
sorafenib		2005	41 (466)
talaporfin sodium		2004	40 (469)
tamibarotene		2005	41 (467)
tasonermin		1999	35 (349)
temozolomide		1999	35 (350)
topotecan HCl		1996	32 (320)
tositumomab		2003	39 (285)
valrubicin		1999	35 (350)
angiotensin II	ANTICANCER ADJUVANT	1994	30 (296)

GENERIC NAME	INDICATION	YEAR INTRO.	ARMC VOL., (PAGE)
chenodiol	ANTICHOLELITHOGENIC	1983	19 (317)
duteplase	ANTICOAGULANT	1995	31 (342)
lepirudin		1997	33 (336)
parnaparin sodium		1993	29 (342)
reviparin sodium		1993	29 (344)
ximelagatran		2004	40 (470)
lamotrigine	ANTICONVULSANT	1990	26 (304)
oxcarbazepine		1990	26 (307)
progabide		1985	21 (331)
vigabatrin		1989	25 (319)
zonisamide		1989	25 (320)
bupropion HCl	ANTIDEPRESSANT	1989	25 (310)
citalopram		1989	25 (311)
duloxetine		2004	40 (452)
escitalopram oxalate		2002	38 (354)
fluoxetine HCl		1986	22 (320)
fluvoxamine maleate		1983	19 (319)
indalpine		1983	19 (320)
medifoxamine fumarate		1986	22 (323)
metapramine		1984	20 (320)
milnacipran		1997	33 (338)
mirtazapine		1994	30 (303)
moclobemide		1990	26 (305)
nefazodone		1994	30 (305)
paroxetine		1991	27 (331)
pivagabine		1997	33 (341)
reboxetine		1997	33 (342)
setiptiline		1989	25 (318)
sertraline hydrochloride		1990	26 (309)
tianeptine sodium		1983	19 (324)
toloxatone		1984	20 (324)
venlafaxine		1994	30 (312)
acarbose	ANTIDIABETIC	1990	26 (297)
epalrestat		1992	28 (330)
exenatide		2005	41 (452)
glimepiride		1995	31 (344)
insulin lispro		1996	32 (310)
miglitol		1998	34 (325)
mitiglinide		2004	40 (460)
nateglinide		1999	35 (344)
pioglitazone HCl		1999	35 (346)
pramlintide		2005	41 (460)
repaglinide		1998	34 (329)
rosiglitazone maleate		1999	35 (347)
tolrestat		1989	25 (319)
troglitazone		1997	33 (344)
voglibose		1994	30 (313)
acetorphan	ANTIDIARRHEAL	1993	29 (332)
anti-digoxin polyclonal antibody	ANTIDOTE	2002	38 (350)
crotelidae polyvalent		2001	37 (263)

GENERIC NAME	INDICATION	YEAR INTRO.	ARMC VOL., (PAGE)
immune fab			
fomepizole		1998	34 (323)
aprepitant	ANTIEMETIC	2003	39 (268)
dolasetron mesylate		1998	34 (321)
granisetron hydrochloride		1991	27 (329)
indisetron		2004	40 (459)
ondansetron hydrochloride		1990	26 (306)
nazasetron		1994	30 (305)
palonosetron		2003	39 (281)
ramosetron		1996	32 (315)
tropisetron		1992	28 (337)
felbamate	ANTIEPILEPTIC	1993	29 (337)
fosphenytoin sodium		1996	32 (308)
gabapentin		1993	29 (338)
levetiracetam		2000	36 (307)
pregabalin		2004	40 (464)
tiagabine		1996	32 (320)
topiramate		1995	31 (351)
centchroman	ANTIESTROGEN	1991	27 (324)
caspofungin acetate	ANTIFUNGAL	2001	37 (263)
eberconazole		2005	41 (449)
fenticonazole nitrate		1987	23 (334)
fluconazole		1988	24 (303)
fosfluconazole		2004	40 (457)
itraconazole		1988	24 (305)
lanoconazole		1994	30 (302)
luliconazole		2005	41 (454)
micafungin		2002	38 (360)
naftifine HCl		1984	20 (321)
oxiconazole nitrate		1983	19 (322)
terbinafine hydrochloride		1991	27 (334)
terconazole		1983	19 (324)
tioconazole		1983	19 (324)
voriconazole		2002	38 (370)
amorolfine hydrochloride	ANTIFUNGAL, TOPICAL	1991	27 (322)
butenafine hydrochloride		1992	28 (327)
butoconazole		1986	22 (318)
cloconazole HCl		1986	22 (318)
liranaftate		2000	36 (309)
flutrimazole		1995	31 (343)
neticonazole HCl		1993	29 (341)
sertaconazole nitrate		1992	28 (336)
sulconizole nitrate		1985	21 (332)
apraclonidine HCl	ANTIGLAUCOMA	1988	24 (297)
befunolol HCl		1983	19 (315)
bimatroprost		2001	37 (261)
brimonidine		1996	32 (306)
brinzolamide		1998	34 (318)
dapiprazole HCl		1987	23 (332)
dorzolamide HCl		1995	31 (341)
latanoprost		1996	32 (311)

GENERIC NAME	INDICATION	YEAR INTRO.	ARMC VOL., (PAGE)
levobunolol HCl		1985	21 (328)
travoprost		2001	37 (272)
unoprostone isopropyl ester		1994	30 (312)
acrivastine	ANTIHISTAMINE	1988	24 (295)
astemizole		1983	19 (314)
azelastine HCl		1986	22 (316)
cetirizine HCl		1987	23 (331)
desloratadine		2001	37 (264)
ebastine		1990	26 (302)
levocabastine hydrochloride		1991	27 (330)
levocetirizine		2001	37 (268)
loratadine		1988	24 (306)
mizolastine		1998	34 (325)
setastine HCl		1987	23 (342)
alacepril	ANTIHYPERTENSIVE	1988	24 (296)
alfuzosin HCl		1988	24 (296)
amlodipine besylate		1990	26 (298)
amosulalol		1988	24 (297)
aranidipine		1996	32 (306)
arotinolol HCl		1986	22 (316)
azelnidipine		2003	39 (270)
barnidipine hydrochloride		1992	28 (326)
benazepril hydrochloride		1990	26 (299)
benidipine hydrochloride		1991	27 (322)
betaxolol HCl		1983	19 (315)
bevantolol HCl		1987	23 (328)
bisoprolol fumarate		1986	22 (317)
bopindolol		1985	21 (324)
bosentan		2001	37 (262)
budralazine		1983	19 (315)
bunazosin HCl		1985	21 (324)
candesartan cilexetil		1997	33 (330)
carvedilol		1991	27 (323)
celiprolol HCl		1983	19 (317)
cicletanine		1988	24 (299)
cilazapril		1990	26 (301)
cinildipine		1995	31 (339)
delapril		1989	25 (311)
dilevalol		1989	25 (311)
doxazosin mesylate		1988	24 (300)
efonidipine		1994	30 (299)
enalapril maleate		1984	20 (317)
enalaprilat		1987	23 (332)
eplerenone		2003	39 (276)
eprosartan		1997	33 (333)
felodipine		1988	24 (302)
fenoldopam mesylate		1998	34 (322)
fosinopril sodium		1991	27 (328)
guanadrel sulfate		1983	19 (319)
imidapril HCl		1993	29 (339)

GENERIC NAME	INDICATION	YEAR INTRO.	ARMC VOL., (PAGE)
irbesartan		1997	33 (336)
isradipine		1989	25 (315)
ketanserin		1985	21 (328)
lacidipine		1991	27 (330)
lercanidipine		1997	33 (337)
lisinopril		1987	23 (337)
losartan		1994	30 (302)
manidipine hydrochloride		1990	26 (304)
mebefradil hydrochloride		1997	33 (338)
moexipril HCl		1995	31 (346)
moxonidine		1991	27 (330)
nebivolol		1997	33 (339)
nilvadipine		1989	25 (316)
nipradilol		1988	24 (307)
nisoldipine		1990	26 (306)
olmesartan medoxomil		2002	38 (363)
perindopril		1988	24 (309)
pinacidil		1987	23 (340)
quinapril		1989	25 (317)
ramipril		1989	25 (317)
rilmenidine		1988	24 (310)
spirapril HCl		1995	31 (349)
telmisartan		1999	35 (349)
temocapril		1994	30 (311)
terazosin HCl		1984	20 (323)
tertatolol HCl		1987	23 (344)
tiamenidine HCl		1988	24 (311)
tilisolol hydrochloride		1992	28 (337)
trandolapril		1993	29 (348)
treprostinil sodium		2002	38 (368)
trimazosin HCl		1985	21 (333)
valsartan		1996	32 (320)
zofenopril calcium		2000	36 (313)
captopril	ANTIHYPERTENSIVE AGENT	1982	13 (086)
daptomycin	ANTI INFECTIVE	2003	39 (272)
aceclofenac	ANTIINFLAMMATORY	1992	28 (325)
AF-2259		1987	23 (325)
amfenac sodium		1986	22 (315)
ampiroxicam		1994	30 (296)
amtolmetin guacil		1993	29 (332)
butibufen		1992	28 (327)
deflazacort		1986	22 (319)
dexibuprofen		1994	30 (298)
droxicam		1990	26 (302)
etodolac		1985	21 (327)
flunoxaprofen		1987	23 (335)
fluticasone propionate		1990	26 (303)
interferon, gamma		1989	25 (314)
isofezolac		1984	20 (319)
isoxicam		1983	19 (320)
lobenzarit sodium		1986	22 (322)

Cumulative NCE Introduction Index, 1983–2005 (by Indication)

GENERIC NAME	INDICATION	YEAR INTRO.	ARMC VOL., (PAGE)
loxoprofen sodium		1986	22 (322)
lumiracoxib		2005	41 (455)
nabumetone		1985	21 (330)
nepafenac		2005	41 (456)
nimesulide		1985	21 (330)
oxaprozin		1983	19 (322)
piroxicam cinnamate		1988	24 (309)
rimexolone		1995	31 (348)
sivelestat		2002	38 (366)
tenoxicam		1987	23 (344)
zaltoprofen		1993	29 (349)
fisalamine	ANTIINFLAMMATORY, INTESTINAL	1984	20 (318)
osalazine sodium		1986	22 (324)
alclometasone dipropionate	ANTIINFLAMMATORY, TOPICAL	1985	21 (323)
aminoprofen		1990	26 (298)
betamethasone butyrate propionate		1994	30 (297)
butyl flufenamate		1983	19 (316)
deprodone propionate		1992	28 (329)
felbinac		1986	22 (320)
halobetasol propionate		1991	27 (329)
halometasone		1983	19 (320)
hydrocortisone aceponate		1988	24 (304)
hydrocortisone butyrate propionate		1983	19 (320)
mometasone furoate		1987	23 (338)
piketoprofen		1984	20 (322)
pimaprofen		1984	20 (322)
prednicarbate		1986	22 (325)
pravastatin	ANTILIPIDEMIC	1989	25 (316)
arteether	ANTIMALARIAL	2000	36 (296)
artemisinin		1987	23 (327)
bulaquine		2000	36 (299)
halofantrine		1988	24 (304)
mefloquine HCl		1985	21 (329)
almotriptan	ANTIMIGRAINE	2000	36 (295)
alpiropride		1988	24 (296)
eletriptan		2001	37 (266)
frovatriptan		2002	38 (357)
lomerizine HCl		1999	35 (342)
naratriptan hydrochloride		1997	33 (339)
rizatriptan benzoate		1998	34 (330)
sumatriptan succinate		1991	27 (333)
zolmitriptan		1997	33 (345)
dronabinol	ANTINAUSEANT	1986	22 (319)
amrubicin HCl	ANTINEOPLASTIC	2002	38 (349)
amsacrine		1987	23 (327)
anastrozole		1995	31 (338)
bicalutamide		1995	31 (338)
bisantrene hydrochloride		1990	26 (300)
camostat mesylate		1985	21 (325)

GENERIC NAME	INDICATION	YEAR INTRO.	ARMC VOL., (PAGE)
capecitabine		1998	34 (319)
cladribine		1993	29 (335)
cytarabine ocfosfate		1993	29 (335)
docetaxel		1995	31 (341)
doxifluridine		1987	23 (332)
enocitabine		1983	19 (318)
epirubicin HCl		1984	20 (318)
fadrozole HCl		1995	31 (342)
fludarabine phosphate		1991	27 (327)
flutamide		1983	19 (318)
formestane		1993	29 (337)
fotemustine		1989	25 (313)
geftimib		2002	38 (358)
gemcitabine HCl		1995	31 (344)
idarubicin hydrochloride		1990	26 (303)
imatinib mesylate		2001	37 (267)
interferon gamma-1α		1992	28 (332)
interleukin-2		1989	25 (314)
irinotecan		1994	30 (301)
lonidamine		1987	23 (337)
mitoxantrone HCl		1984	20 (321)
nedaplatin		1995	31 (347)
nilutamide		1987	23 (338)
paclitaxal		1993	29 (342)
pegaspargase		1994	30 (306)
pentostatin		1992	28 (334)
pirarubicin		1988	24 (309)
ranimustine		1987	23 (341)
sobuzoxane		1994	30 (310)
temoporphin		2002	38 (367)
toremifene		1989	25 (319)
vinorelbine		1989	25 (320)
zinostatin stimalamer		1994	30 (313)
porfimer sodium	ANTINEOPLASTIC ADJUVANT	1993	29 (343)
masoprocol	ANTINEOPLASTIC, TOPICAL	1992	28 (333)
miltefosine		1993	29 (340)
dexfenfluramine	ANTIOBESITY	1997	33 (332)
orlistat		1998	34 (327)
sibutramine		1998	34 (331)
atovaquone	ANTIPARASITIC	1992	28 (326)
ivermectin		1987	23 (336)
budipine	ANTIPARKINSONIAN	1997	33 (330)
CHF-1301		1999	35 (336)
droxidopa		1989	25 (312)
entacapone		1998	34 (322)
pergolide mesylate		1988	24 (308)
pramipexole hydrochloride		1997	33 (341)
ropinirole HCl		1996	32 (317)
talipexole		1996	32 (318)
tolcapone		1997	33 (343)

Cumulative NCE Introduction Index, 1983–2005 (by Indication)

GENERIC NAME	INDICATION	YEAR INTRO.	ARMC VOL., (PAGE)
lidamidine HCl	ANTIPERISTALTIC	1984	20 (320)
gestrinone	ANTIPROGESTOGEN	1986	22 (321)
cabergoline	ANTIPROLACTIN	1993	29 (334)
tamsulosin HCl	ANTIPROSTATIC HYPERTROPHY	1993	29 (347)
acitretin	ANTIPSORIATIC	1989	25 (309)
calcipotriol		1991	27 (323)
tazarotene		1997	33 (343)
tacalcitol	ANTIPSORIATIC, TOPICAL	1993	29 (346)
amisulpride	ANTIPSYCHOTIC	1986	22 (316)
remoxipride hydrochloride		1990	26 (308)
zuclopenthixol acetate		1987	23 (345)
actarit	ANTIRHEUMATIC	1994	30 (296)
diacerein		1985	21 (326)
octreotide	ANTISECRETORY	1988	24 (307)
adamantanium bromide	ANTISEPTIC	1984	20 (315)
drotecogin alfa	ANTISEPSIS	2001	37 (265)
cimetropium bromide	ANTISPASMODIC	1985	21 (326)
tiquizium bromide		1984	20 (324)
tiropramide HCl		1983	19 (324)
argatroban	ANTITHROMBOTIC	1990	26 (299)
bivalirudin		2000	36 (298)
defibrotide		1986	22 (319)
cilostazol		1988	24 (299)
clopidogrel hydrogensulfate		1998	34 (320)
cloricromen		1991	27 (325)
enoxaparin		1987	23 (333)
eptifibatide		1999	35 (340)
ethyl icosapentate		1990	26 (303)
fondaparinux sodium		2002	38 (356)
indobufen		1984	20 (319)
limaprost		1988	24 (306)
ozagrel sodium		1988	24 (308)
picotamide		1987	23 (340)
tirofiban hydrochloride		1998	34 (332)
flutropium bromide	ANTITUSSIVE	1988	24 (303)
levodropropizine		1988	24 (305)
nitisinone	ANTITYROSINAEMIA	2002	38 (361)
benexate HCl	ANTIULCER	1987	23 (328)
dosmalfate		2000	36 (302)
ebrotidine		1997	33 (333)
ecabet sodium		1993	29 (336)
egualen sodium		2000	36 (303)
enprostil		1985	21 (327)
famotidine		1985	21 (327)
irsogladine		1989	25 (315)
lansoprazole		1992	28 (332)
misoprostol		1985	21 (329)
nizatidine		1987	23 (339)
omeprazole		1988	24 (308)
ornoprostil		1987	23 (339)

GENERIC NAME	INDICATION	YEAR INTRO.	ARMC VOL., (PAGE)
pantoprazole sodium		1994	30 (306)
plaunotol		1987	23 (340)
polaprezinc		1994	30 (307)
ranitidine bismuth citrate		1995	31 (348)
rebamipide		1990	26 (308)
rosaprostol		1985	21 (332)
roxatidine acetate HCl		1986	22 (326)
roxithromycin		1987	23 (342)
sofalcone		1984	20 (323)
spizofurone		1987	23 (343)
teprenone		1984	20 (323)
tretinoin tocoferil		1993	29 (348)
troxipide		1986	22 (327)
abacavir sulfate	ANTIVIRAL	1999	35 (333)
adefovir dipivoxil		2002	38 (348)
amprenavir		1999	35 (334)
atazanavir		2003	39 (269)
cidofovir		1996	32 (306)
delavirdine mesylate		1997	33 (331)
didanosine		1991	27 (326)
efavirenz		1998	34 (321)
emtricitabine		2003	39 (274)
enfuvirtide		2003	39 (275)
entecavir		2005	41 (450)
famciclovir		1994	30 (300)
fomivirsen sodium		1998	34 (323)
fosamprenavir		2003	39 (277)
foscarnet sodium		1989	25 (313)
ganciclovir		1988	24 (303)
imiquimod		1997	33 (335)
indinavir sulfate		1996	32 (310)
interferon alfacon-1		1997	33 (336)
lamivudine		1995	31 (345)
lopinavir		2000	36 (310)
nelfinavir mesylate		1997	33 (340)
nevirapine		1996	32 (313)
oseltamivir phosphate		1999	35 (346)
penciclovir		1996	32 (314)
propagermanium		1994	30 (308)
rimantadine HCl		1987	23 (342)
ritonavir		1996	32 (317)
saquinavir mesylate		1995	31 (349)
sorivudine		1993	29 (345)
stavudine		1994	30 (311)
tenofovir disoproxil fumarate		2001	37 (271)
valaciclovir HCl		1995	31 (352)
zalcitabine		1992	28 (338)
zanamivir		1999	35 (352)
zidovudine		1987	23 (345)
influenza virus live	ANTIVIRAL VACCINE	2003	39 (277)

Cumulative NCE Introduction Index, 1983–2005 (by Indication)

GENERIC NAME	INDICATION	YEAR INTRO.	ARMC VOL., (PAGE)
cevimeline hydrochloride	ANTI-XEROSTOMIA	2000	36 (299)
alpidem	ANXIOLYTIC	1991	27 (322)
buspirone HCl		1985	21 (324)
etizolam		1984	20 (318)
flutazolam		1984	20 (318)
flutoprazepam		1986	22 (320)
metaclazepam		1987	23 (338)
mexazolam		1984	20 (321)
tandospirone		1996	32 (319)
ciclesonide	ASTHMA, COPD	2005	41 (443)
atomoxetine	ATTENTION DEFICIT HYPERACTIVITY DISORDER	2003	39 (270)
flumazenil	BENZODIAZEPINE ANTAG.	1987	23 (335)
bambuterol	BRONCHODILATOR	1990	26 (299)
doxofylline		1985	21 (327)
formoterol fumarate		1986	22 (321)
mabuterol HCl		1986	22 (323)
oxitropium bromide		1983	19 (323)
salmeterol hydroxynaphthoate		1990	26 (308)
tiotropium bromide		2002	38 (368)
APD	CALCIUM REGULATOR	1987	23 (326)
clodronate disodium		1986	22 (319)
disodium pamidronate		1989	25 (312)
gallium nitrate		1991	27 (328)
ipriflavone		1989	25 (314)
neridronic acid		2002	38 (361)
dexrazoxane	CARDIOPROTECTIVE	1992	28 (330)
bucladesine sodium	CARDIOSTIMULANT	1984	20 (316)
denopamine		1988	24 (300)
docarpamine		1994	30 (298)
dopexamine		1989	25 (312)
enoximone		1988	24 (301)
flosequinan		1992	28 (331)
ibopamine HCl		1984	20 (319)
loprinone hydrochloride		1996	32 (312)
milrinone		1989	25 (316)
vesnarinone		1990	26 (310)
amrinone	CARDIOTONIC	1983	19 (314)
colforsin daropate HCL		1999	35 (337)
xamoterol fumarate		1988	24 (312)
cefozopran HCL	CEPHALOSPORIN, INJECTABLE	1995	31 (339)
cefditoren pivoxil	CEPHALOSPORIN, ORAL	1994	30 (297)
brovincamine fumarate	CEREBRAL VASODILATOR	1986	22 (317)
nimodipine		1985	21 (330)
propentofylline		1988	24 (310)
succimer	CHELATOR	1991	27 (333)
trientine HCl		1986	22 (327)
fenbuprol	CHOLERETIC	1983	19 (318)
deferasirox	CHRONIC IRON OVERLOAD	2005	41 (446)

Cumulative NCE Introduction Index, 1983–2005 (by Indication)

GENERIC NAME	INDICATION	YEAR INTRO.	ARMC VOL., (PAGE)
auranofin	CHRYSOTHERAPEUTIC	1983	19 (314)
taltirelin	CNS STIMULANT	2000	36 (311)
aniracetam	COGNITION ENHANCER	1993	29 (333)
pramiracetam H_2SO_4		1993	29 (343)
carperitide	CONGESTIVE HEART FAILURE	1995	31 (339)
nesiritide		2001	37 (269)
drospirenone	CONTRACEPTIVE	2000	36 (302)
norelgestromin		2002	38 (362)
nicorandil	CORONARY VASODILATOR	1984	20 (322)
dornase alfa	CYSTIC FIBROSIS	1994	30 (298)
neltenexine		1993	29 (341)
amifostine	CYTOPROTECTIVE	1995	31 (338)
nalmefene HCL	DEPENDENCE TREATMENT	1995	31 (347)
ioflupane	DIAGNOSIS CNS	2000	36 (306)
azosemide	DIURETIC	1986	22 (316)
muzolimine		1983	19 (321)
torasemide		1993	29 (348)
atorvastatin calcium	DYSLIPIDEMIA	1997	33 (328)
cerivastatin		1997	33 (331)
naftopidil	DYSURIA	1999	35 (343)
alglucerase	ENZYME	1991	27 (321)
udenafil	ERECTILE DYSFUNCTION	2005	41 (472)
erdosteine	EXPECTORANT	1995	31 (342)
fudosteine		2001	37 (267)
agalsidase alfa	FABRY'S DISEASE	2001	37 (259)
cetrorelix	FEMALE INFERTILITY	1999	35 (336)
ganirelix acetate		2000	36 (305)
follitropin alfa	FERTILITY ENHANCER	1996	32 (307)
follitropin beta		1996	32 (308)
reteplase	FIBRINOLYTIC	1996	32 (316)
esomeprazole magnesium	GASTRIC ANTISECRETORY	2000	36 (303)
lafutidine		2000	36 (307)
rabeprazole sodium		1998	34 (328)
cinitapride	GASTROPROKINETIC	1990	26 (301)
cisapride		1988	24 (299)
itopride HCL		1995	31 (344)
mosapride citrate		1998	34 (326)
imiglucerase	GAUCHER'S DISEASE	1994	30 (301)
miglustat		2003	39 (279)
somatotropin	GROWTH HORMONE	1994	30 (310)
somatomedin-1	GROWTH HORMONE INSENSITIVITY	1994	30 (310)
factor VIIa	HAEMOPHILIA	1996	32 (307)
levosimendan	HEART FAILURE	2000	36 (308)
pimobendan		1994	30 (307)
anagrelide hydrochloride	HEMATOLOGIC	1997	33 (328)
erythropoietin	HEMATOPOETIC	1988	24 (301)
factor VIII	HEMOSTATIC	1992	28 (330)
malotilate	HEPATOPROTECTIVE	1985	21 (329)
mivotilate		1999	35 (343)

Cumulative NCE Introduction Index, 1983–2005 (by Indication) 537

GENERIC NAME	INDICATION	YEAR INTRO.	ARMC VOL., (PAGE)
tipranavir	HIV	2005	41 (470)
buserelin acetate	HORMONE	1984	20 (316)
goserelin		1987	23 (336)
leuprolide acetate		1984	20 (319)
nafarelin acetate		1990	26 (306)
somatropin		1987	23 (343)
zoledronate disodium	HYPERCALCEMIA	2000	36 (314)
cinacalcet	HYPERPARATHYROIDISM	2004	40 (451)
sapropterin hydrochloride	HYPERPHENYL-ALANINEMIA	1992	28 (336)
quinagolide	HYPERPROLACTINEMIA	1994	30 (309)
cadralazine	HYPERTENSIVE	1988	24 (298)
nitrendipine		1985	21 (331)
binfonazole	HYPNOTIC	1983	19 (315)
brotizolam		1983	19 (315)
butoctamide		1984	20 (316)
cinolazepam		1993	29 (334)
doxefazepam		1985	21 (326)
eszopiclone		2005	41 (451)
loprazolam mesylate		1983	19 (321)
quazepam		1985	21 (332)
rilmazafone		1989	25 (317)
zaleplon		1999	35 (351)
zolpidem hemitartrate		1988	24 (313)
zopiclone		1986	22 (327)
acetohydroxamic acid	HYPOAMMONURIC	1983	19 (313)
sodium cellulose PO4	HYPOCALCIURIC	1983	19 (323)
divistyramine	HYPOCHOLESTEROLEMIC	1984	20 (317)
lovastatin		1987	23 (337)
melinamide		1984	20 (320)
pitavastatin		2003	39 (282)
rosuvastatin		2003	39 (283)
simvastatin		1988	24 (311)
glucagon, rDNA	HYPOGLYCEMIA	1993	29 (338)
acipimox	HYPOLIPIDEMIC	1985	21 (323)
beclobrate		1986	22 (317)
binifibrate		1986	22 (317)
ciprofibrate		1985	21 (326)
colesevelam hydrochloride		2000	36 (300)
colestimide		1999	35 (337)
ezetimibe		2002	38 (355)
fluvastatin		1994	30 (300)
meglutol		1983	19 (321)
ronafibrate		1986	22 (326)
modafinil	IDIOPATHIC HYPERSOMNIA	1994	30 (303)
bucillamine	IMMUNOMODULATOR	1987	23 (329)
centoxin		1991	27 (325)
thymopentin		1985	21 (333)
filgrastim	IMMUNOSTIMULANT	1991	27 (327)
GMDP		1996	32 (308)
interferon gamma-1b		1991	27 (329)

GENERIC NAME	INDICATION	YEAR INTRO.	ARMC VOL., (PAGE)
lentinan		1986	22 (322)
pegademase bovine		1990	26 (307)
pidotimod		1993	29 (343)
romurtide		1991	27 (332)
sargramostim		1991	27 (332)
schizophyllan		1985	22 (326)
ubenimex		1987	23 (345)
cyclosporine	IMMUNOSUPPRESSANT	1983	19 (317)
everolimus		2004	40 (455)
gusperimus		1994	30 (300)
mizoribine		1984	20 (321)
muromonab-CD3		1986	22 (323)
mycophenolate sodium		2003	39 (279)
mycophenolate mofetil		1995	31 (346)
pimecrolimus		2002	38 (365)
tacrolimus		1993	29 (347)
ramelteon	INSOMNIA	2005	41 (462)
defeiprone	IRON CHELATOR	1995	31 (340)
alosetron hydrochloride	IRRITABLE BOWEL SYNDROME	2000	36 (295)
tegasedor maleate		2001	37 (270)
sulbactam sodium	β-LACTAMASE INHIBITOR	1986	22 (326)
tazobactam sodium		1992	28 (336)
nartograstim	LEUKOPENIA	1994	30 (304)
pumactant	LUNG SURFACTANT	1994	30 (308)
sildenafil citrate	MALE SEXUAL DYSFUNCTION	1998	34 (331)
gadoversetamide	MRI CONTRAST AGENT	2000	36 (304)
telmesteine	MUCOLYTIC	1992	28 (337)
laronidase	MUCOPOLYSACCARIDOSIS	2003	39 (278)
galsulfase	MUCOPOLYSACCHARIDOSIS VI	2005	41 (453)
palifermin	MUCOSITIS	2005	41 (461)
interferon ß-1a	MULTIPLE SCLEROSIS	1996	32 (311)
interferon ß-1b		1993	29 (339)
glatiramer acetate		1997	33 (334)
natalizumab		2004	40 (462)
afloqualone	MUSCLE RELAXANT	1983	19 (313)
cisatracurium besilate		1995	31 (340)
doxacurium chloride		1991	27 (326)
eperisone HCl		1983	19 (318)
mivacurium chloride		1992	28 (334)
rapacuronium bromide		1999	35 (347)
tizanidine		1984	20 (324)
naltrexone HCl	NARCOTIC ANTAGONIST	1984	20 (322)
tinazoline	NASAL DECONGESTANT	1988	24 (312)
aripiprazole	NEUROLEPTIC	2002	38 (350)
clospipramine hydrochloride		1991	27 (325)
nemonapride		1991	27 (331)
olanzapine		1996	32 (313)
perospirone hydrochloride		2001	37 (270)

GENERIC NAME	INDICATION	YEAR INTRO.	ARMC VOL., (PAGE)
quetiapine fumarate		1997	33 (341)
risperidone		1993	29 (344)
sertindole		1996	32 (318)
timiperone		1984	20 (323)
ziprasidone hydrochloride		2000	36 (312)
rocuronium bromide	NEUROMUSCULAR BLOCKER	1994	30 (309)
edaravone	NEUROPROTECTIVE	1995	37 (265)
fasudil HCL		1995	31 (343)
riluzole		1996	32 (317)
bifemelane HCl	NOOTROPIC	1987	23 (329)
choline alfoscerate		1990	26 (300)
exifone		1988	24 (302)
idebenone		1986	22 (321)
indeloxazine HCl		1988	24 (304)
levacecarnine HCl		1986	22 (322)
nizofenzone fumarate		1988	24 (307)
oxiracetam		1987	23 (339)
bromfenac sodium	NSAID	1997	33 (329)
lornoxicam		1997	33 (337)
OP-1	OSTEOINDUCTOR	2001	37 (269)
alendronate sodium	OSTEOPOROSIS	1993	29 (332)
ibandronic acid		1996	32 (309)
incadronic acid		1997	33 (335)
raloxifene hydrochloride		1998	34 (328)
risedronate sodium		1998	34 (330)
strontium ranelate		2004	40 (467)
tiludronate disodium	PAGET'S DISEASE	1995	31 (350)
rasagiline	PARKINSON'S DISEASE	2005	41 (464)
tadalafil	PDE5 INHIBITOR	2003	39 (284)
vardenafil		2003	39 (286)
temoporphin	PHOTOSENSITIZER	2002	38 (367)
verteporfin		2000	36 (312)
alefacept	PLAQUE PSORIASIS	2003	39 (267)
beraprost sodium	PLATELET AGGREG. INHIBITOR	1992	28 (326)
epoprostenol sodium		1983	19 (318)
iloprost		1992	28 (332)
sarpogrelate HCl	PLATELET ANTIAGGREGANT	1993	29 (344)
trimetrexate glucuronate	*PNEUMOCYSTIS CARINII* PNEUMONIA	1994	30 (312)
solifenacin	POLLAKIURIA	2004	40 (466)
histrelin	PRECOCIOUS PUBERTY	1993	29 (338)
atosiban	PRETERM LABOR	2000	36 (297)
gestodene	PROGESTOGEN	1987	23 (335)
nomegestrol acetate		1986	22 (324)
norgestimate		1986	22 (324)
promegestrone		1983	19 (323)
trimegestone		2001	37 (273)
alpha-1 antitrypsin	PROTEASE INHIBITOR	1988	24 (297)

GENERIC NAME	INDICATION	YEAR INTRO.	ARMC VOL., (PAGE)
nafamostat mesylate		1986	22 (323)
adrafinil	PSYCHOSTIMULANT	1986	22 (315)
dexmethylphenidate HCl		2002	38 (352)
dutasteride		2002	38 (353)
efalizumab	PSORIASIS	2003	39 (274)
finasteride	5α-REDUCTASE INHIBITOR	1992	28 (331)
surfactant TA	RESPIRATORY SURFACTANT	1987	23 (344)
Adalimumab	RHEUMATOID ARTHRITIS	2003	39 (267)
dexmedetomidine hydrochloride	SEDATIVE	2000	36 (301)
ziconotide	SEVERE CHRONIC PAIN	2005	41 (473)
kinetin	SKIN PHOTODAMAGE/ DERMATOLOGIC	1999	35 (341)
tirilazad mesylate	SUBARACHNOID HEMORRHAGE	1995	31 (351)
APSAC	THROMBOLYTIC	1987	23 (326)
alteplase		1987	23 (326)
balsalazide disodium	ULCERATIVE COLITIS	1997	33 (329)
darifenacin	URINARY INCONTINENCE	2005	41 (445)
tiopronin	UROLITHIASIS	1989	25 (318)
propiverine hydrochloride	UROLOGIC	1992	28 (335)
Lyme disease	VACCINE	1999	35 (342)
clobenoside	VASOPROTECTIVE	1988	24 (300)
falecalcitriol	VITAMIN D	2001	37 (266)
maxacalcitol		2000	36 (310)
paricalcitol		1998	34 (327)
doxercalciferol	VITAMIN D PROHORMONE	1999	35 (339)
prezatide copper acetate	VULNERARY	1996	32 (314)
acemannan	WOUND HEALING AGENT	2001	37 (257)
cadexomer iodine		1983	19 (316)
epidermal growth factor		1987	23 (333)

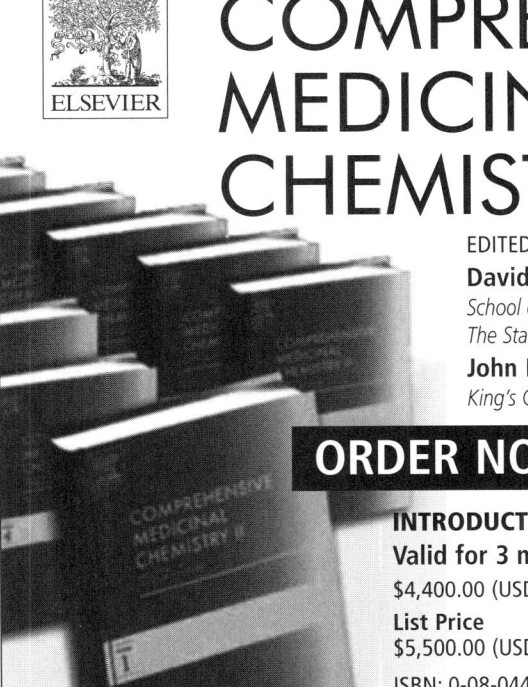

COMPREHENSIVE MEDICINAL CHEMISTRY II

8-Volume Set

EDITED BY
David J. Triggle
School of Pharmacy and Pharmaceutical Sciences, The State University of New York, Buffalo, USA
John B. Taylor
King's College London, UK

ORDER NOW AND SAVE 20%

INTRODUCTORY PRICE
Valid for 3 months after publication
$4,400.00 (USD) / €3,650.00 (EUR) / £2,515.00 (GBP)
List Price
$5,500.00 (USD) / €4,560.00 (EUR) / £3,145.00 (GBP)
ISBN: 0-08-044513-6   ISBN-13: 978-0-08-044513-7

Publishing November 2006

Reviews for the First Edition

'Beautifully produced six-part magnum opus of medicinal chemistry ... The preface boasts "a milestone in the literature of the subject in terms of coverage, clarity and presentation" with which I agree.'
THE LANCET

'A most impressive collection of up to date, excellently referenced reviews covering virtually every aspect of biological science of relevance to the production of medicines ...'
CHEMISTRY IN BRITAIN

Providing a global and current perspective of today's drug discovery processes, these volumes cover the strategy, technologies, principles and applications of medicinal chemistry in a single work.

The new edition has been refocused, revised and expanded to reflect the significant developments and changes over the past decade in genomics, proteomics, bioinformatics, combinatorial chemistry, high-throughput screening and pharmacology.

Online Version
Forthcoming on ScienceDirect